Universitext

T0178461

Springer
New York
Berlin
Heidelberg
Barcelona
Hong Kong
London
Milan
Paris
Singapore
Tokyo

Universitext

Editors (North America): S. Axler, F.W. Gehring, and K.A. Ribet

(continued after index)

Paulo Ribenboim

Classical Theory of Algebraic Numbers

 Springer

Paulo Ribenboim
Department of Mathematics
Queen's University
Kingston, Ontario K7L 3N6
Canada

Mathematics Subject Classification (2000): 11-01, 11Sxx

Library of Congress Cataloging-in-Publication Data
Ribenboim, Paulo.
   Classical theory of algebraic numbers / Paulo Ribenboim. — 1st ed.
      p.   cm. — (Universitext)
   Rev. ed. of: Algebraic numbers. 1972.
   Includes bibliographical references and indexes.
   ISBN 978-1-4419-2870-2   e-ISBN 978-0-387-21690-4
   1. Algebraic number theory.   I. Ribenboim, Paulo. Algebraic Numbers.   II. Title.
QA247.R465   2001
512′.74—dc21   00-040044

Printed on acid-free paper.

Printed in the United States of America.

9 8 7 6 5 4 3 2 1

Springer-Verlag   New York Berlin Heidelberg
A member of BertelsmannSpringer Science+Business Media GmbH

Preface

The theory of algebraic numbers is one of the monuments of nineteenth century mathematics. The work of Gauss on quadratic forms led to the study of quadratic fields. The deep investigations of Fermat's last theorem by Kummer led to cyclotomic fields. Then came Dedekind, Dirichlet, Hermite, Kronecker, Hurwitz, Weber, Hilbert, and other eminent mathematicians who developed a beautiful theory.

The aim of this book is to present a detailed self-contained exposition of the classical theory of algebraic numbers. I use today's language including groups, modules, rings, but I shy away from more sophisticated methods unnecessary at this level, preferring to keep closer to the classical origins. Many suggestions for further reading and study directions are proposed at the end of the book. My point of view is that these modern developments are rooted in the classical ideas, so it is advisable to acquire a solid background. Of course, time has not stopped and the student is encouraged to progress forward and to study the modern techniques. In music, this is comparable to a solid study of Bach before Bartók.

Thinking about graduate students, I began the book with two introductory algebraic chapters, followed by two chapters on elementary number theory. These two chapters contain material usable at undergraduate level.

The subject proper begins in Part Two and all the basic aspects of the theory are carefully derived—algebraic integers, ideals, units, class groups, class numbers and the Hilbert's theory of decomposition, inertia and ramification in Galois extensions. This study culminates in the deep theorem of Kronecker and Weber, which is also the cornerstone of class field theory. This book contains only a summary of the results in class field theory but I make suggestions for the study of this more advanced theory. Chapter 16, which ends Part Two, is very special—it is entirely devoted to specific numerical examples and is highly recommended to the reader, not only for its content, but also as a means of testing the reader's understanding before continuing to the final part of the book.

In Part Three, I examine the theory of cyclotomic fields, and how it was developed by Kummer to lead to the proof of Fermat's last theorem for regular prime exponents. This study includes the λ-adic local methods, Lagrange resolvents, and the Jacobi cyclotomic function.

Bernoulli numbers are the subject of a whole chapter which is followed by the derivation of Kummer's regularity criterion. These matters are of great importance in modern developments and are treated in detail in this book.

Part Three also contains the beautiful theory of characters by Dirichlet in a sleek presentation; introductory results about the Riemann and Dedekind zeta-functions, as well as Dirichlet and Hecke L-series associated to characters. Nevertheless, I restricted my attention only to real arguments $s > 0$. For my purpose, I do not require complex arguments (as Riemann did), nor do I consider the question of analytical continuation or functional equations. These are, of course, of the utmost importance and may be found in books dealing with analytical number theory and the prime number theorems. In no way should the reader be unaware of such developments—but this book is already too voluminous as it is!

I include the magnificent achievement which is the theorem of Dirichlet on primes in arithmetic progressions. In his proof, Dirichlet introduced many major new ideas which have influenced dramatically the development of both algebraic and analytic number theory. The reader is encouraged to carefully study this proof.

The Frobenius automorphism is studied and a proof of the density theorem of Chebotarev is given, without appealing to class field theory. As in all the preceding theorems, all the details of proofs—even when intricate—are given to the reader who should be able to fully understand the proofs.

The last four chapters concern the class numbers. First, I deduce—all steps included—the famous formulas for the class number of quadratic fields (result of Dirichlet) and of cyclotomic fields (as developed by Kummer). The approach is different in the last two chapters, where I present, sometimes without proof, results about divisibility and other arithmetical properties of the class number of quadratic and cyclotomic fields. These chapters are not meant to be updated surveys of the questions, but rather introductions to problems and research. Two especially rich lists of references are found in the Bibliography at the end of the book.

For the readers who have come to this point, there is a description of various avenues of study with the indication of excellent books.

The exercises have been included so that the reader may develop a certain familiarity with the concepts; they are therefore recommended as a useful complement to the text.

This book has evolved over a long period, and many of its parts have been taught in various courses, mostly for master and doctoral students. The first sixteen chapters, now with slight improvements, constituted the book *Algebraic Numbers*, published in 1972, which enjoyed considerable esteem, until it went out of print. The present book has double the size of the former one. More important, it contains results obtained with analytical means.

As the development of the theory has uncovered, many of the deeper results are now obtainable only with analytical methods—therefore the reader may learn these theorems in this book.

A reader may take this book to a desert island. Free of distractions and with enough time available, he will be able, all by himself, to understand each proof and to master the classical theory of algebraic numbers. And if—as I hope—he succeeds in quitting his island he will have a solid background on which to learn any one of the modern developments.

Paulo Ribenboim
October, 1999

Acknowledgments

I am greatly indebted to Jerzy Browkin for his invaluable help, suggesting numerous improvements in the text and supervising with great care and patience the material preparation of this book. This task was expertly done by Sławomir Browkin.

It is also my delight to acknowledge the help of several colleagues who suggested improvements on preliminary drafts of this book. In alphabetical order: Karl Dilcher, Wulf-Dieter Geyer, the late Kustaa Inkeri, Claude Levesque, Tauno Metsänkylä, and Dimitrij Ugrin-Sparač.

Contents

Index of Notations

The following notations are used in the text without explanation:

Notation	Explanation
$\mathbb{Z}$	ring of integers
$\mathbb{Q}$	field of rational numbers
$\mathbb{R}$	field of real numbers
$\mathbb{C}$	field of complex numbers
$\mathbb{A}$	field of all algebraic numbers

The following notations are listed in the order that they appear in the book:

Page	Notation	Explanation
5	$a\|b$	a divides b
5	$a \sim b$	a, b are associated
5	$K[X]$	ring of polynomials with coefficients in K
5	K	multiplicative group of the field K
6	$\gcd(a, b)$	greatest common divisor of a and b
6	$\operatorname{lcm}(a, b)$	least common multiple of a and b
7	$J + J'$	sum of ideals J and J'
7	$J \cdot J'$, JJ'	product of ideals J and J'
7	(a)	principal ideal generated by a
7	$(a_1, \ldots, a_n)$	ideal generated by $a_1, \ldots, a_n$
7	$a \equiv b \pmod{J}$	congruence relation modulo an ideal
14	$K(S)$	smallest field containing K and S
14	$[L : K]$	degree of L over K

Introduction

Introduction

The purpose of the Introduction is to gather, for the convenience of the reader, results about commutative integral domains and fields, which will be required at various places in this book. There is no attempt to provide proofs. The reader may wish to go to Part One after a superficial acquaintance with Chapters 1 and 2 and return to these chapters when required.

1

Unique Factorization Domains, Ideals, and Principal Ideal Domains

1.1 Unique Factorization Domains

Let A be a domain, that is, a commutative ring with unit element (different from 0), having no zero-divisors (except 0). Let K be its field of quotients.

If $a, b \in K$ we say that a *divides* b (with respect to the domain A) when there exists an element $c \in A$ such that $a \cdot c = b$. We write $a|b$ to express this fact. Thus, every element of K divides 0 and $1|a$ for every $a \in A$.

We have $a|a$, for every $a \in K$ and if $a, b, c \in K$, from $a|b$, $b|c$ we deduce that $a|c$. However, we may well have $a \neq b$, but $a|b$ and $b|a$. For example, a divides $-a$ and $-a$ divides a. We say that a, b are *associated* whenever it is true that $a|b$ and $b|a$; we write $a \sim b$ to express this fact.

The set $U = \{a \in K \mid a \sim 1\}$ is a subgroup of $K^{\cdot}$ (multiplicative group of nonzero elements of K), and $U \subseteq A$. The elements of U are therefore *invertible in* A, and they are called the *units* of the ring A. For example, if $A = \mathbb{Z}$ then the units are precisely 1, -1. If $A = K[X]$ (ring of polynomials with coefficients in a field K) then the units are the nonzero elements of K.

A nonzero element p of a domain A is said to be an *irreducible element* of A if p is not a unit of A and it satisfies the following condition: if $a \in A$, $a|p$, then either $a \sim 1$ or $a \sim p$.

For example, if $A = \mathbb{Z}$ then the irreducible elements are p, $-p$, where p is any prime number. If $A = L[X]$ or more generally $A = L[X_1, \ldots, X_n]$, where L is a field, the irreducible elements are the irreducible polynomials.

The domain A is said to be a *unique factorization domain* when the following statement is true in A:

Let S be a set of irreducible elements of A such that every irreducible element of A is associated with one and only one element of S. Then every element $x \in K$, $x \neq 0$, may be written uniquely in the form

$$x = u \prod_{p \in S} p^{v_p(x)},$$

where u is a unit of A, $v_p(x) \in \mathbb{Z}$ for each p and $v_p(x) \neq 0$ for at most a finite subset of irreducible elements in S (this finite subset depending on the element x). It follows that $x \in A$ if and only if $v_p(x) \geq 0$ for every $p \in S$.

The domains $\mathbb{Z}$, $L[X]$, and more generally $L[X_1, \ldots, X_n]$ (where L is a field) are well-known examples of unique factorization domains.

If A is a unique factorization domain, if an irreducible element p divides a product ab, with $a, b \in A$, then either $p|a$ or $p|b$.

If A is a unique factorization domain, any two elements $a, b \in A$ have a *greatest common divisor* d (which is unique up to unit elements); by definition d satisfies the following properties: $d|a$, $d|b$; if $d' \in A$ and $d'|a$, $d'|b$ then $d'|d$. We write $d = \gcd(a, b)$.

Similarly, $a, b \in A$ have a *least common multiple* m (unique up to a unit element): $a|m$, $b|m$; if $m' \in A$ and $a|m'$, $b|m'$ then $m|m'$. We write $m = \operatorname{lcm}(a, b)$.

The elements $a, b \in A$ are said to be *relatively prime* or *coprime* when 1 is a greatest common divisor of a, b.

1.2 Ideals

In our study we shall encounter more general types of commutative rings than the unique factorization domains.

Let A be a commutative ring with unit element 1. A subset J of A is called an *ideal* of A when it satisfies the following properties:

(a) If $a, b \in J$, then $a + b \in J$.

(b) If $b \in J$, $a \in A$, then $ab \in J$.

In particular, J is also an additive subgroup of A.

Among the ideals of A we have the *zero ideal* 0 (consisting only of the element 0) and the *unit ideal* A.

The ring A is a field if and only if its only ideals are 0 and A.

Every ideal J of A gives rise to the ring A/J, whose elements are the cosets $a + J = \{a + b \mid b \in J\}$ for every $a \in A$. We have $a + J = a' + J$ if and only if $a - a' \in J$. The operations of A/J are defined by

$$(a + J) + (a' + J) = (a + a') + J, \qquad (a + J) \cdot (a' + J) = aa' + J.$$

The mapping $\varphi : A \to A/J$, $\varphi(a) = a + J$ for every $a \in A$, is the canonical ring-homomorphism from A onto A/J.

An ideal P of A is said to be a *prime ideal* when it satisfies the following conditions:

(a) $P \neq A$.

(b) If $a, b \in A$, $a \cdot b \in P$, then either $a \in P$ or $b \in P$.

Thus P is a prime ideal of A if and only if A/P is a domain.

An ideal P of A is said to be a *maximal ideal* when:

(a) $P \neq A$.

(b) There exists no ideal J of A such that $P \subset J \subset A$.

It is easily seen that P is a maximal ideal of A if and only if A/P is a field. Hence every maximal ideal is a prime ideal.

We define the operations of *addition* and *multiplication* between ideals in the following way:

$$J + J' = \{a + a' \mid a \in J, a' \in J'\},$$

$$J \cdot J' = \left\{ \sum_{i=1}^{n} a_i a_i' \,\middle|\, n > 0, a_i \in J, a_i' \in J' \right\},$$

(we also write JJ').

Then $J + J'$ and $J \cdot J'$ are ideals of A and we have the following properties: $J + J'$ contains J and J' and if any ideal of A contains J and J' then it contains $J + J'$; $J \cdot J' \subseteq J \cap J'$; if J'' is another ideal of A then $(J + J') + J'' = J + (J' + J'')$ and $(J \cdot J') \cdot J'' = J \cdot (J' \cdot J'')$; $A \cdot J = J$ for every ideal J.

We note also that if P is a prime ideal, if J, J' are ideals of A such that $J \cdot J' \subseteq P$, then $J \subseteq P$ or $J' \subseteq P$. In fact, otherwise there would exist elements $a \in J$, $a \notin P$, $a' \in J'$, $a' \notin P$, such that $aa' \in J \cdot J' \subseteq P$, which is impossible since P is a prime ideal.

Any intersection of ideals of the ring A is still an ideal of A. Thus, given any subset S of A the intersection of all ideals of A containing S is the smallest ideal of A containing S; it is called the *ideal generated by S*.

If $S = \{a\}$, the ideal generated by $\{a\}$ is $Aa = \{xa \mid x \in A\}$, it is called the *principal ideal generated by a* and also denoted by (a).

We note that $Aa \cdot Ab = Aab$ for any $a, b \in A$.

If $S = \{a_1, \ldots, a_n\}$ the ideal generated by S is $Aa_1 + \cdots + Aa_n = \{\sum_{i=1}^{n} x_i a_i \mid x_i \in A \text{ for } i = 1, 2, \ldots, n\}$; it is also denoted by $(a_1, \ldots, a_n)$.

If J is any ideal of the ring A, then we define the *congruence relation modulo J* as follows. For $a, b \in A$, let

$$a \equiv b \pmod{J}$$

when $a - b \in J$.

For $J = 0$ this is the equality relation; for $J = A$, it is the trivial relation. It is straightforward to verify that the congruence relation modulo J is an equivalence relation and, also that from $a \equiv b \pmod{J}$, $a' \equiv b' \pmod{J}$, it follows that $a + a' \equiv b + b' \pmod{J}$, $-a \equiv -b \pmod{J}$, and $aa' \equiv bb'$

(mod J). Moreover, if $J \neq A$ the congruence classes modulo J are identified with the elements of the quotient ring A/J.

1.3 Principal Ideal Domains

Let A be a domain. If every ideal of A is a principal ideal, A is called a *principal ideal domain*.

The domains $\mathbb{Z}, K[X]$ (where K is a field) are principal ideal domains. However, $K[X, Y]$ is not a principal ideal domain.

Let A be a principal ideal domain. We have $Aa \subseteq Bb$ if and only if $b|a$; thus $Aa = Ab$ if and only if $a \sim b$. In particular, $Aa = A$ if and only if a is a unit element.

Moreover, Ap is a prime ideal if and only if p is an irreducible element of A or $p = 0$. It follows that every nonzero prime ideal is a maximal ideal.

Every principal ideal domain is a unique factorization domain. But the converse is not true: $K[X, Y]$ (where K is a field) is a unique factorization domain, which is not a principal ideal domain.

If $a, b \in A$ and $d = \gcd(a, b)$, then $Aa + Ab = Ad$, hence there exist elements $x, y \in A$ such that $d = xa + yb$. In particular, if a, b are relatively prime, there exist $x, y \in A$ such that $1 = xa + yb$.

If A is a principal ideal domain, if $a, b, c \in A$, if $\gcd(a, b) = 1$ and $a|bc$, then $a|c$.

E X E R C I S E S

1. Show that every integer $n \geq 1$ has only finitely many divisors.

2. Show that there exist infinitely many prime numbers.

3. Prove: If $a, m, n \in \mathbb{Z}$, if a divides mn, and a is relatively prime to m, then a divides n. In particular, if p is a prime number dividing mn and not dividing m, then p divides n.

4. Prove (*Euclidean algorithm*): If a, b are positive integers, there exist integers q, r, with $0 \leq q$, $0 \leq r < b$ such that $a = qb + r$. Moreover, if q', r' are also such that $0 \leq q'$, $0 \leq r' < b$, and $a = q'b + r'$, then $q' = q$, $r' = r$. Here q is called the *quotient of a by b* and r is the *remainder* of the division of a by b.

5. Show that every natural number $m > 0$ may be written in a unique way (except for the order of the factors) as a product of powers of prime numbers.

6. If m, n are nonzero integers, show that there exists a greatest common divisor d of m and n. Show that if d, d' are greatest common divisors of m, n, then $d = d'$ or $d = -d'$.

7. Determine a computational procedure to find the greatest common divisor d of integers $m, n \geq 1$, by means of the Euclidean algorithm.

Hint: Divide m by n (when $m \geq n$) and n by the remainder (if not zero) and so on.

8. Show that if m, n are nonzero integers, d a greatest common divisor of m, n, then there exist integers a_0, b_0 such that $d = a_0 m + b_0 n$. Conversely, every number of the form $am + bn$ (with $a, b \in \mathbb{Z}$) is a multiple of d (this is known as the *Bézout property* of the integers).

9. Let d be a greatest common divisor of the integers $m, n \geq 1$. Show that m/d, n/d are relatively prime.

10. If m, n are nonzero integers, show that there exists a least common multiple l of m, n; if l' is any other one, then $l = l'$ or $l = -l'$.

11. Generalize Exercises 6 to 10 for the case of nonzero integers $m_1, \ldots, m_r$.

12. Let d be a greatest common divisor and l a least common multiple of m, n. Show that $mn = \pm ld$.

13. Let d, l be positive integers. Show that there exist integers a, b such that $\gcd(a, b) = d$ and $\operatorname{lcm}(a, b) = l$ if and only if d divides l. In this case, show that the number of possible pairs of positive integers a, b with the above properties is 2^r, where r is the number of distinct prime factors of l/d.

14. Determine the highest power of the prime number p which divides $n!$.

15. Show that the product of n successive integers is a multiple of $n!$.

16. If p is a prime number, prove that the binomial coefficients $\binom{p}{k}$, with $1 \leq k < p$, are multiples of p. More generally, if $1 \leq k < p^m$, $m \geq 1$, if p^r divides k but p^{r+1} does not divide k, then p^{m-r} is the exact power of p dividing $\binom{p^m}{k}$.

17. The *Fibonacci numbers* $1, 1, 2, 3, 5, 8, 13, 21, 34, \ldots$ are defined inductively by the relation

$$a_n = a_{n-1} + a_{n-2} \quad \text{for} \quad n > 2 \quad (\text{with } a_1 = 1, \ a_2 = 1).$$

Prove:

(a) Two consecutive Fibonacci numbers are relatively prime.

(b) $a_n^2 - a_{n-1}a_{n+1} = (-1)^{n-1}$ for $n > 1$.

18. The *Lucas numbers* 1, 3, 4, 7, 11, 18, 29, 47, ... are defined inductively by the relation

$$b_n = b_{n-1} + b_{n-2} \qquad \text{for} \quad n > 2 \qquad (\text{with } b_1 = 1, \ b_2 = 3).$$

Prove:

(a) Two consecutive Lucas numbers are relatively prime.

(b) $b_n^2 - b_{n-1}b_{n+1} = 5(-1)^n$ for every $n \geq 1$.

19. Let a_n denote the nth Fibonacci number and b_n the nth Lucas number. Prove:

(a) $2a_{m+n} = a_m b_n + a_n b_m$.

(b) $b_n^2 - 5a_n^2 = 4(-1)^n$ and $b_{n+1}b_n - 5a_{n+1}a_n = 2(-1)^n$.

(c) a_n and b_n have the same parity.

(d) If they are both odd then $\gcd(a_n, b_n) = 1$; otherwise, $\gcd(a_n, b_n) = 2$.

(e) If m divides n then a_m divides a_n.

(f) If $\gcd(m, n) = d$ then $\gcd(a_m, a_n) = a_d$.

(g) If $\gcd(m, n) = 1$ then $a_m a_n$ divides a_{mn}.

20. Let K be a field. Prove (*Euclidean algorithm*):

(a) If f, g are nonzero elements of $K[x]$, there exist $q, r \in K[x]$, such that $f = qg + r$, where $r = 0$ or $\deg(r) < \deg(g)$. If $q', r' \in K[x]$ are such that $f = q'g + r'$ with $r' = 0$ or $\deg(r') < \deg(g)$, then $q = q'$ and $r = r'$.

(b) $K[x]$ is a principal ideal domain.

21. Show that every principal ideal domain is a unique factorization domain.

22. If $f = a_n X^n + a_{n-1} X^{n-1} + \cdots + a_0 \in \mathbb{Z}[X]$, let $\gamma(f) = \gcd(a_0, a_1, \ldots, a_n)$ be the *content of* f. We say that f is *primitive* when $\gamma(f) = 1$. Prove *Gauss' lemma*: If $f, g \in \mathbb{Z}[X]$ are primitive polynomials then fg is also primitive.

23. Show that if $f, g \in \mathbb{Z}[X]$ then their contents satisfy $\gamma(fg) = \gamma(f) \cdot \gamma(g)$.

24. Show that if $f \in \mathbb{Z}[X]$ is a primitive polynomial then it is the product of primitive polynomials which are irreducible; this decomposition is unique (up to the order of the factors and their signs).

25. Show that if $f \in \mathbb{Z}[X]$ then it may be written in a unique way (up to the order of the factors and up to signs of h_j) in the form

$$f = \pm p_1 p_2 \cdots p_r h_1 h_2 \cdots h_s,$$

where each $p_i \in \mathbb{Z}$ is a prime number and each $h_j \in \mathbb{Z}[X]$ is a primitive irreducible polynomial.

26. Show that if $f \in \mathbb{Z}[X]$ is irreducible in $\mathbb{Z}[X]$ then it is also irreducible in $\mathbb{Q}[X]$.

27. Generalize Exercises 22 through 26, replacing $\mathbb{Z}$ by any unique factorization domain R. Conclude that if R is a unique factorization domain then $R[X]$ is also a unique factorization domain.

28. Prove that if R is a unique factorization domain and if $(X_i)_{i \in I}$ is any family of indeterminates, then $R[X_i]_{i \in I}$ is a unique factorization domain.

2

Commutative Fields

For the convenience of the reader we recall some definitions and facts about commutative fields.

2.1 Algebraic Elements

Let L be a field, K a subfield of L.

The element $x \in L$ is said to be *algebraic over* K when there exists a nonzero polynomial f, with coefficients in K such that $f(x) = 0$; dividing by the leading coefficient, we may assume that f is monic. In other words, there exist elements $a_1, \ldots, a_n \in K$ (with $n > 0$) such that $x^n + a_1 x^{n-1} + \cdots + a_n = 0$.

If $x \in L$ but x is not algebraic over K, we say that it is *transcendental over* K.

We shall be mainly concerned with algebraic elements.

If $x \in L$, $K[x]$ shall denote the subring of L of all elements of the form $\sum_{i=0}^{m} c_i x^i$ where $m \geq 0$, $c_i \in K$ for $i = 0, 1, \ldots, m$.

If $x \in L$ is algebraic over K let $J = \{f \in K[X] \mid f(x) = 0\}$. J contains a unique monic polynomial $f_0 = X^n + a_1 X^{n-1} + \cdots + a_n$ of smallest degree; f_0 is irreducible over K and if $x \neq 0$ then $a_n \neq 0$. A polynomial f belongs to J if and only if f is a multiple of f_0. Thus J is the principal ideal of $K[X]$ generated by f_0 and since f_0 is irreducible, J is a maximal ideal. The polynomial f_0 defined above is called the *minimal polynomial of x over K*. Its degree is called the *degree of x over* K.

The set of elements $\{1, x, \ldots, x^{n-1}\}$ is a basis of the K-vector space $K[x]$. Indeed, x^n is expressible in terms of the lower powers of x because $x^n + a_1 x^{n-1} + \cdots + a_n = 0$. On the other hand, no relation with coefficients in K may exist between 1, x, $\ldots$, x^{n-1}, because f_0 is a polynomial of minimal degree in the ideal J.

The mapping $\varphi : K[X] \to K[x]$, defined by $\varphi(f) = f(x)$ for every polynomial $f \in K[X]$, is a ring-homomorphism with kernel J, and $K[X]/J \cong$

$K[x]$. Since J is a maximal ideal then $K[x]$ is a field. Explicitly, if $x \neq 0$, then x is invertible in $K[x]$, namely

$$x \cdot [-a_n^{-1}(x^{n-1} + a_1 x^{n-2} + \cdots + a_{n-1})] = 1.$$

2.2 Algebraic Extensions, Algebraically Closed Fields

Let L be a field, K a subfield of L. If every element of L is algebraic over K, then L is said to be an *algebraic extension* of K. Otherwise, L is called a *transcendental extension* of K.

If the K-vector space L is of finite dimension $n = [L : K]$, we say that n is the *degree of* L *over* K.

If K is a subfield of L, if S is any subset of L, we denote by $K(S)$ the smallest subfield of L which contains K and S. If $S = \{x_1, \ldots, x_n\}$, we write $K(x_1, \ldots, x_n)$. L is said to be *finitely generated over* K if there exist elements $x_1, \ldots, x_n \in L$ such that $L = K(x_1, \ldots, x_n)$. We say that L is obtained from K by adjoining $x_1, \ldots, x_n$.

Every extension of finite degree must be algebraic and finitely generated; the converse is also true.

If K is a subfield of L, and S a set of elements of L which are algebraic over K, then $K(S)$ is an algebraic extension of K. In particular, the set of all elements of L which are algebraic over K is itself a field, called the *algebraic closure of* K *in* L.

More generally, Steinitz' theorem states: given any field K there exists a field $\overline{K}$ with the following properties:

(a) $\overline{K}$ is an algebraic extension of K.

(b) If L is any algebraic extension of $\overline{K}$ then $L = \overline{K}$.

(c) If $\widetilde{K}$ is any field satisfying properties (a), (b) above, then there exists a K-isomorphism between $\widetilde{K}$, $\overline{K}$ (that is, an isomorphism leaving invariant all the elements of the subfield K).

$\overline{K}$ is called the *algebraic closure* of K.

Any field satisfying property (b) only of $\overline{K}$ is called an *algebraically closed field*. This implies that if f is a polynomial with coefficients in $K \subseteq \overline{K}$, then all its roots belong to $\overline{K}$; also, if L is any algebraic extension of K, then there exists a K-isomorphism from L onto a subfield of $\overline{K}$. Thus, for the purpose of studying algebraic extensions of K we may restrict our attention to the subfields of an algebraic closure $\overline{K}$. We note also that if L is an algebraically closed field containing K, if $\overline{K}$ is the subfield of all elements of L which are algebraic over K, then $\overline{K}$ is an algebraic closure of K.

2.3 Algebraic Number Fields

A very important theorem (sometimes called the *Fundamental Theorem of Algebra*) states that the field $\mathbb{C}$ of complex numbers is algebraically closed.

We say that a complex number x is an *algebraic number* if it is algebraic over the field $\mathbb{Q}$ of rational numbers. The set $\mathbb{A}$ of all algebraic numbers is a field, which is an algebraic closure of $\mathbb{Q}$.

Let us note that $[\mathbb{A} : \mathbb{Q}] = \infty$, since by Eisenstein's irreducibility criterion there exist irreducible polynomials of arbitrary degree over $\mathbb{Q}$ (for example, $X^n - p$, where p is any prime number). Every algebraic extension of finite degree over $\mathbb{Q}$ is called an *algebraic number field*, or simply a *number field*.

It is an easy matter to show that the field $\mathbb{A}$ must be countable. Since $\mathbb{C}$ is not countable, there exist uncountably many transcendental complex numbers (for example, π, e). Hermite showed in 1873 that e is transcendental, while Lindemann proved the transcendence of π in 1882. Hilbert, Gelfond, and Schneider gave more recent proofs (see Lang [16, Appendix]).

2.4 Characteristic and Prime Fields

Let K be a field, let 1 denote its unit element; for every integer $n > 0$, let $n \cdot 1 = 1 + \cdots + 1$ (n times). If $n \cdot 1 \neq 0$ for every $n > 0$, we say that K has *characteristic* 0. If $n > 0$ is the smallest integer such that $n \cdot 1 = 0$, then we say that K has *characteristic* n; it is easily seen that n must be a prime number $n = p$.

The field $\mathbb{Q}$ of rational numbers has characteristic 0. For every prime p, the field $\mathbb{F}_p$ of residue classes modulo p, namely $\mathbb{F}_p = \{\overline{1}, \overline{2}, \ldots, \overline{p-1}\}$ has characteristic p (see Chapter 3).

If K is any field, the intersection of all its subfields is again a subfield, called the *prime field of* K. It is isomorphic to $\mathbb{Q}$ when K has characteristic 0, or to $\mathbb{F}_p$ when K has characteristic p.

2.5 Normal Extensions, Splitting Fields

Let K be a field and $\overline{K}$ an algebraic closure of K. The elements $x, x' \in \overline{K}$ are *conjugate over* K whenever they have the same minimal polynomial over K. This happens if and only if there exists a K-isomorphism σ from $K(x)$ onto $K(x')$ such that $\sigma(x) = x'$.

Let L, L' be extensions of K contained in $\overline{K}$. We say that L, L' are *conjugate over* K if there exists a K-isomorphism σ from L onto L'.

If $[L : K]$ is finite the number of K-isomorphisms from L into an algebraic closure $\overline{K}$ of K is at most equal to the degree $[L : K]$.

If L is equal to all its conjugates over K, then L is said to be a *normal extension* of K. For example, $\overline{K}$ is a normal extension of K. Every extension of degree 2 is also normal.

A typical normal extension of K is obtained by considering an arbitrary polynomial $f \in K[X]$ and adjoining to K the roots of f. The resulting field L is equal to all its conjugates over K, hence it is a normal extension of K. It is called the *splitting field* of f over K.

Conversely, if $L|K$ is a normal extension of finite degree, then $L = K(x_1, \ldots, x_n)$, where x_1, $\ldots$, x_n are algebraic elements. If f_i is the minimal polynomial of x_i over K and $f = f_1 \cdot f_2 \cdots f_n$ then all the roots of f are still in L, so L is the splitting field of f.

Given any field $L, K \subseteq L \subseteq \overline{K}$, the intersection L' of all normal extensions of K between L and $\overline{K}$ is the smallest normal extension of K containing L and contained in $\overline{K}$. If L has finite degree over K then L' has also finite degree over K.

If L is a normal extension of K, the set of K-automorphisms of L forms a group (under composition), which we denote by $G(L|K)$.

2.6 Separable Extensions

Let $K \subseteq L \subseteq \overline{K}$ where $[L : K] < \infty$; we say that L is *separable over K* if the number of distinct K-isomorphisms from L into $\overline{K}$ is exactly equal to the degree $[L : K]$. More generally, if L is any field, $K \subseteq L \subseteq \overline{K}$, we say that L is separable over K if every subfield F of L, such that F contains K and $[F : K] < \infty$, is separable over K (in the sense just defined).

In particular, the algebraic extension $L = K(x)$ is separable over K whenever x has $n = [L : K]$ distinct conjugates over K. In this case, we say that x is a *separable element over K*.

It follows that an algebraic extension L of K is separable if and only if every element x of L is separable over K.

If K is a field of characteristic 0, all the roots of any irreducible polynomial f over K are necessarily distinct. Hence every algebraic extension of a field of characteristic 0 is separable.

If L is a separable extension of finite degree of a field K, then there exists an element t in L such that $L = K(t)$; t is called a *primitive element* of L over K. In particular, this theorem holds when $K = \mathbb{Q}$.

If L is an algebraic number field of degree n over $\mathbb{Q}$, we have $L = \mathbb{Q}(t)$, where t has a minimal polynomial f with coefficients in $\mathbb{Q}$ and degree n. Let $t_1 = t$, t_2, $\ldots$, t_n be the roots of f, which belong to $\mathbb{C}$. If t_i is not a real number, then its complex conjugate $\bar{t}_i$ is also a root of f; thus the nonreal roots of f occur in pairs.

Let r_1 denote the number of real roots of f, let $2r_2$ denote the number of nonreal roots of f; then

$$n = r_1 + 2r_2.$$

An element $x \in K$ is called *totally positive* if all its real conjugates are positive. Thus, if an element has no real conjugates, then it is totally positive.

2.7 Galois Extensions

An algebraic extension L over K is said to be a *Galois extension* whenever L is normal and separable over K. If K has characteristic 0, this means that L is a normal extension over K. The group $G(L|K)$ of all K-automorphisms of a Galois extension L of K is called the *Galois group of L over K*. It is denoted also by $\mathrm{Gal}(L|K)$. If $[L : K] = n$ then $G(L|K)$ has precisely n elements. L is called an *Abelian extension* (respectively, a *cyclic extension*) of K when its Galois group is Abelian (respectively, cyclic).

If $f \in K[X]$ is a polynomial having distinct roots $x_1, \ldots, x_n$, then $K(x_1, \ldots, x_n)|K$ is a Galois extension and its Galois group is called *the Galois group of f over K*.

Let L be a Galois extension of finite degree of K, let $G = G(L|K)$. If G' is any subgroup of G, let us consider the subfield $fi(G')$ of all elements $x \in L$ which are invariant under G' (that is $\sigma(x) = x$ for every $\sigma \in G'$). Similarly, if K' is any field, $K \subseteq K' \subseteq L$, then L is a Galois extension of K' and we may consider the Galois group $gr(K') = G(L|K')$ of all K'-automorphisms of L.

The *fundamental theorem of Galois theory* states:

(a) G' is equal to the Galois group $G(L'|K')$ where $K' = fi(G')$,

(b) K' is the field of invariants of the group $G' = G(L|K')$.

Moreover, G' is a normal subgroup of G if and only if the corresponding field K' is a Galois extension of K. Then $G(K'|K) \cong G/G'$.

Let $\widetilde{K}$ be an extension of K, let L, L' be subfields of $\widetilde{K}$ containing K. The *compositum* of L, L' in $\widetilde{K}$ is the smallest subfield of $\widetilde{K}$ containing L and L'. We denote it by LL' or $L'L$.

If $L|K$ is a Galois extension, then the compositum LL' is a Galois extension of L' and $G(LL'|L') \cong G(L|L \cap L')$. Explicitly, every $(L \cap L')$-automorphism of L may be uniquely extended to an L'-automorphism of LL'.

If $L|K$ and $L'|K$ are Galois extensions, then the compositum LL' is also a Galois extension of K. The mapping $\sigma \in G(LL'|K) \to (\sigma_L, \sigma_{L'}) \in G(L|K) \times G(L'|K)$ (where σ_L, $\sigma_{L'}$ are the restrictions of σ to L and L', respectively) is a group-isomorphism onto the subset of all pairs $(\tau, \tau') \in$

$G(L|K) \times G(L'|K)$, with $\tau_{L\cap L'} = \tau'_{L\cap L'}$. In particular, $G(LL'|L \cap L') \cong$
$G(L|L \cap L') \times G(L'|L \cap L')$. So, if $L \cap L' = K$, we have

$$G(LL'|K) \cong G(L|K) \times G(L'|K).$$

2.8 Roots of Unity

Let K be a field and n an integer, $n > 0$. An element $x \in K$ such that
$x^n = 1$ is called an nth *root of unity*. The set $W_{n,K}$ of all nth roots of
unity in K is a multiplicative group. If K is a subfield of L then $W_{n,K}$ is a
subgroup of $W_{n,L}$. If m divides n then $W_{m,K} \subseteq W_{n,K}$.

We denote by W_K the set of all roots of unity in K; that is,

$$W_K = \bigcup_{n \geq 1} W_{n,K}.$$

Let $x \in W_K$ and let n be its order in the multiplicative group W_K,
that is, the smallest integer such that $x^n = 1$. Then we say that x is a
primitive nth root of unity. It is easily seen that if K has characteristic
p and $x \in W_K$ then its order n is not a multiple of p.

To deal simultaneously with fields of any characteristic, it is customary
to say that K has *characteristic exponent* 1 when its characteristic is 0,
and *characteristic exponent* p when its characteristic is p.

Let K be an algebraically closed field and let n be a natural number
relatively prime to the characteristic exponent of K. Thus $X^n - 1$ has
exactly n distinct roots in K, which are the elements of the group $W_{n,K}$.
This group is cyclic, each generator being a primitive nth root of unity.

If ζ is such a generator then $W_{n,K} = \{\zeta, \zeta^2, \zeta^3, \ldots, \zeta^{n-1}, 1\}$. ζ^a is a
primitive nth root of unity if and only if a, n are relatively prime. Thus the
number of primitive nth roots of unity is equal to $\varphi(n)$, where φ denotes
Euler's function (see Chapter 3, Definition 2). It follows that if L is a
subfield of K then $W_{n,L}$ is also a cyclic group and its order divides n.

If ζ is a primitive nth root of unity in K, then its conjugates over the
prime subfield K_0 of K are also primitive nth roots of unity. The polynomial

$$\Phi_{n,K_0} = \prod_{\substack{\gcd(a,n)=1 \\ 1 \leq a \leq n}} (X - \zeta^a)$$

is called the nth *cyclotomic polynomial* (over K_0). It has degree $\varphi(n)$. The
coefficients of Φ_{n,K_0} belong to K_0, since they are invariant by conjugation.

If K has characteristic 0 it is known that $\Phi_{n,\mathbb{Q}} = \Phi_n$ is irreducible over
$\mathbb{Q}$; that is, $\mathbb{Q}(\zeta)$ has degree $\varphi(n)$ over $\mathbb{Q}$ (see Chapter 16, (**B**)). For example,

$$\Phi_1 = X - 1,$$
$$\Phi_p = X^{p-1} + X^{p-2} + \cdots + X + 1,$$
$$\Phi_{p^r} = X^{p^{r-1}(p-1)} + X^{p^{r-1}(p-2)} + \cdots + X^{p^{r-1}} + 1.$$

Moreover,

$$X^n - 1 = \prod_{d|n} \Phi_d,$$

since both sides are equal to the product of all linear factors $X - \zeta^a$ for $a = 1, 2, \ldots, n$. Thus $X^{p^k} - 1 = (X^{p^{k-1}} - 1)\Phi_{p^k}$ where $k \geq 1$. From this relation it is possible to compute Φ_n by recurrence.

$\mathbb{Q}(\zeta)$ is a Galois extension of $\mathbb{Q}$, having a Galois group isomorphic to the multiplicative group $P(n)$ of prime residue classes modulo n (see Chapter 3). $\mathbb{Q}(\zeta)$ is therefore an Abelian extension of $\mathbb{Q}$.

It is worth mentioning that if p does not divide n it may occur that $\Phi_{n,\mathbb{F}_p}$ is reducible over $\mathbb{F}_p$.

2.9 Finite Fields

We describe now the finite fields. If K is a finite field, it has characteristic p (for some prime p), and the number of its elements is a power p^n, $n > 0$. Precisely, n is the degree of K over the prime field $\mathbb{F}_p$. For every n there exists exactly one field with p^n elements, which we denote by $\mathbb{F}_{p^n}$. Its nonzero elements are the roots of $X^{p^n - 1} - 1$ and $\mathbb{F}_{p^n} = \mathbb{F}_p(\zeta)$, where ζ is a primitive $(p^n - 1)$th root of unity.

$\mathbb{F}_{p^n}$ is a Galois extension of $\mathbb{F}_p$ with a cyclic Galois group. A canonical generator of $G(\mathbb{F}_{p^n}|\mathbb{F}_p)$ is the *Frobenius automorphism* σ, defined by $\sigma(x) = x^p$ for every $x \in \mathbb{F}_{p^n}$.

In particular, if m divides n, then $\mathbb{F}_{p^n}$ is a Galois extension of $\mathbb{F}_{p^m}$ with a cyclic Galois group generated by σ^m, where $\sigma^m(x) = x^{p^m}$ for every $x \in \mathbb{F}_{p^n}$.

2.10 Trace and Norm of Elements

Let K be a field, and let L be a separable extension of degree n. Thus there exist n distinct K-isomorphisms $\sigma_1 = \varepsilon$ (identity), σ_2, ..., σ_n of L into an algebraic closure $\overline{K}$ of K (which may be assumed to contain L).

If $x \in L$ the *trace of x* (relative to K) is defined as

$$\mathrm{Tr}_{L|K}(x) = \sum_{i=1}^{n} \sigma_i(x)$$

and the *norm of x* (relative to K) is

$$N_{L|K}(x) = \prod_{i=1}^{n} \sigma_i(x).$$

If x is a primitive element of $L|K$, if $f = X^n + a_1 X^{n-1} + \cdots + a_n$ is the minimal polynomial of x over K, then $\mathrm{Tr}_{L|K}(x) = -a_1$, $N_{L|K}(x) = (-1)^n a_n$.

Among the properties of the trace and norm, we note

$$\mathrm{Tr}_{L|K}(x + y) = \mathrm{Tr}_{L|K}(x) + \mathrm{Tr}_{L|K}(y);$$
$$N_{L|K}(xy) = N_{L|K}(x) \cdot N_{L|K}(y).$$

If $x \in K$ then $\mathrm{Tr}_{L|K}(x) = nx$, $N_{L|K}(x) = x^n$. If $K \subseteq L \subseteq L'$ are fields and L' is separable over K, then we have the transitivity of the trace and the norm

$$\mathrm{Tr}_{L|K}(\mathrm{Tr}_{L'|L}(x)) = \mathrm{Tr}_{L'|K}(x)$$

and

$$N_{L|K}(N_{L'|L}(x)) = N_{L'|K}(x).$$

If $L|K$ is a separable extension, there always exists an element $x \in L$ such that $\mathrm{Tr}_{L|K}(x) \neq 0$.

Indeed, since $\mathrm{Tr}_{L|K}(1) = [L : K] \cdot 1$, if K has characteristic 0 then $\mathrm{Tr}_{L|K}(1) \neq 0$. Similarly, if K has characteristic p and p does not divide $[L : K]$ then $\mathrm{Tr}_{L|K}(1) \neq 0$. The remaining case, where K has characteristic p and $[L : K]$ is a multiple of p, requires a nontrivial proof based on Dedekind's theorem on independence of the K-isomorphisms of L.

If σ_1, $\ldots$, σ_n are the K-isomorphisms from L into an algebraic closure $\overline{K}$ of K, if $x_1, \ldots, x_n \in \overline{K}$ and $\sum_{i=1}^{n} x_i \sigma_i = 0$ (null mapping from L to $\overline{K}$) then necessarily $x_1 = \cdots = x_n = 0$.

It is also useful to know that if $L|K$ is a separable extension of finite degree n and the characteristic of K does not divide n, then there exists a primitive element t such that $\mathrm{Tr}_{L|K}(t) = 0$.

2.11 The Discriminant

We shall now consider the notion of the discriminant.

Let $L|K$ be a separable extension of degree n and let $(x_1, \ldots, x_n)$ be an n-tuple of elements in L. We define its *discriminant* (in $L|K$) to be

$$\mathrm{discr}_{L|K}(x_1, \ldots, x_n) = \det(\mathrm{Tr}_{L|K}(x_i x_j)),$$

that is, the determinant of the matrix whose (i, j)-entry is equal to $\mathrm{Tr}_{L|K}(x_i x_j)$ for $i, j = 1, \ldots, n$. Thus, the discriminant belongs to K.

If $(x_1', \ldots, x_n')$ is another n-tuple of elements in L, and $x_j' = \sum_{i=1}^{n} a_{ij} x_i$ for all $j = 1, \ldots, n$, where $a_{ij} \in K$, then

$$\mathrm{discr}_{L|K}(x_1', \ldots, x_n') = [\det(a_{ij})]^2 \cdot \mathrm{discr}_{L|K}(x_1, \ldots, x_n).$$

Another expression for the discriminant may be obtained in terms of the K-isomorphisms $\sigma_1, \ldots, \sigma_n$ of L:

$$\mathrm{discr}_{L|K}(x_1, \ldots, x_n) = [\det(\sigma_i(x_j))]^2.$$

In order that $x_1, \ldots, x_n \in L$ be linearly independent over K it is necessary and sufficient that $\mathrm{discr}_{L|K}(x_1, \ldots, x_n) \neq 0$.

Consider the special basis $\{1, t, \ldots, t^{n-1}\}$ where t is a primitive element, $L = K(t)$; we obtain

$$\mathrm{discr}_{L|K}(1, t, \ldots, t^{n-1}) = \prod_{i<j}(t_i - t_j)^2,$$

where $t_1 = t$, $t_2, \ldots, t_n$ are the conjugates of t.

In this situation, it is customary to call the above expression the *discriminant of t* (over K). Thus, all the conjugates of t have the same discriminant.

2.12 Discriminant and Resultant of Polynomials

If $f \in K[X]$ is a polynomial of degree n, with roots $x_1, \ldots, x_n$ and leading coefficient a_0, we define the *discriminant of f* by

$$\mathrm{discr}(f) = a_0^{2n-2}\prod_{i<j}(x_i - x_j)^2 = (-1)^{n(n-1)/2}a_0^{2n-2}\prod_{i\neq j}(x_i - x_j).$$

Thus $\mathrm{discr}(f) \in K$ and $\mathrm{discr}(f) \neq 0$ if and only if f has distinct roots. If t is a separable element over K and if f is its minimal polynomial, then $\mathrm{discr}_{L|K}(t) = \mathrm{discr}(f)$.

It is important to compute the discriminant of an irreducible polynomial $f = a_0 X^n + a_1 X^{n-1} + \cdots + a_n$ without knowing a priori its roots. If $p_k = x_1^k + \cdots + x_n^k$ (where $x_1, \ldots, x_n$ are the roots of f) for $k = 0, 1, 2, \ldots$, then $p_0 = n$, $p_1 = -a_1/a_0$, and $p_2, p_3, \ldots$ may be computed recursively (without computing the roots) by the well-known Newton formulas (see Exercise 17). Then

$$\mathrm{discr}(f) = a_0^{2n-2}\begin{vmatrix} p_0 & p_1 & \cdots & p_{n-1} \\ p_1 & p_2 & \cdots & p_n \\ \vdots & \vdots & & \vdots \\ p_{n-1} & p_n & \cdots & p_{2n-2} \end{vmatrix}.$$

In some cases, the computation of the above determinant is rather awkward. However, it is possible to compute the discriminant of f by the more direct formula

$$\mathrm{discr}(f) = (-1)^{n(n-1)/2}a_0^{n-2}N_{K(x)|K}(f'(x)),$$

where x is any root of f and f' is the derivative of the polynomial f.

Just like the discriminant of f detects if f has a multiple root, the resultant of two polynomials pinpoints when the polynomials are relatively prime.

Let

$$f = a_0 X^m + a_1 X^{m-1} + \cdots + a_m,$$
$$g = b_0 X^n + b_1 X^{n-1} + \cdots + b_n,$$

with $a_i, b_j \in K$, $a_0 b_0 \neq 0$, and $m, n > 0$.

The eliminant of f, g is the matrix with $m + n$ rows and columns:

$$E\ell(f,g) = \begin{pmatrix} a_0 & a_1 & \cdots & a_m & 0 & 0 & \cdots & 0 & 0 & \cdots \\ 0 & a_0 & \cdots & a_{m-1} & a_m & 0 & \cdots & 0 & 0 & \cdots \\ \cdots & \cdots & \cdots & \cdots & \cdots & \cdots & \cdots & \cdots & \cdots & \cdots \\ b_0 & b_1 & \cdots & & \cdots & \cdots & \cdots & b_n & 0 & \cdots \\ 0 & b_0 & \cdots & & \cdots & \cdots & \cdots & b_{n-1} & b_n & \cdots \\ \cdots & \cdots & \cdots & & \cdots & \cdots & \cdots & \cdots & \cdots & \cdots \end{pmatrix},$$

with n rows with the coefficients a_i and m rows with the coefficients b_j.

The determinant of this matrix is called the *resultant* of f, g and denoted by $R(f, g)$; it is an element of K.

Clearly, $R(f, f) = 0$. We extend this definition, letting $R(f, b_0) = b_0^m$ (where $b_0 \in K$, $m = \deg(f) > 0$) and $R(a_0, g) = a_0^n$ (where $a_0 \in K$, $n = \deg(g) > 0$).

If $x_1, \ldots, x_m$ are the roots of f and $y_1, \ldots, y_n$ are the roots of g, then

$$R(f,g) = a_0^n b_0^m \prod_{i=1}^{m} \prod_{j=1}^{n} (x_i - y_j)$$

$$= a_0^n \prod_{i=1}^{m} g(x_i) = (-1)^{mn} b_0^m \prod_{j=1}^{n} f(y_j).$$

As said, the vanishing of the discriminant of a polynomial f indicates if f has a multiple root, that is, f and f' have a common root. More generally, we introduce the resultant of two polynomials to indicate if f, g have a common factor.

Let K be any field. The polynomials $f, g \in K[X]$ are relatively prime if and only if $R(f, g) \neq 0$.

The discriminant of a polynomial f may be expressed in terms of the resultant of f and its derivative f':

$$R(f, f') = (-1)^{m(m-1)/2} a_0 \, \mathrm{discr}(f).$$

We quote the following properties of the resultant. Let $f, g, h, k \in K[X]$, $\deg(f) = m$, and $\deg(g) = n$. Then:

(1) $R(g, f) = (-1)^{mn} R(f, g)$.

(2) If $\deg(f) + \deg(h) \leq \deg(g)$ then $R(f,g) = R(f, g + fh)$.

(3) $R(hk, g) = R(h, g) \cdot R(k, g)$, $R(f, hk) = R(f, h) \cdot R(f, k)$.

(4) $\mathrm{discr}(fg) = (-1)^{mn} \mathrm{discr}(f) \cdot \mathrm{discr}(g) \cdot [R(f,g)]^2$.

2.13 Inseparable Extensions

Even though almost all extensions considered in this book are separable we shall have occasion to deal with extensions, which are not separable. At that point the reader may wish to return to this section.

Let K be a field and let x be algebraic over K. If the minimal polynomial $f \in K[X]$ of x over K has multiple roots, then x is said to be *inseparable over* K. If an inseparable element exists, then the characteristic of K is not 0.

An algebraic extension $L|K$ is said to be *purely inseparable* if every $x \in L$, $x \notin K$ is inseparable over K. If $L|K$ is purely inseparable of finite degree, then $[L : K]$ is a power of the characteristic p of K. Let $L|K$ be any algebraic extension and let S be the set of all elements $x \in L$ which are separable over K—clearly S is a field and $S|K$ is a separable extension. Then $L|S$ is a purely inseparable extension.

If $F|K$ is a separable extension and $L|K$ is a purely inseparable extension then $LF|L$ is separable, while $LF|F$ is purely inseparable. Moreover if $L|K$ and $F|K$ have finite degree then $[LF : F] = [L : K]$ and $[LF : L] = [F : K]$. More precisely, a basis of the K-vector space F is also a basis of the L-vector space LF.

2.14 Perfect Fields

A field K is said to be *perfect* if K has characteristic 0, or if K has characteristic $p \neq 0$, for every element $x \in K$ there exists $y \in K$ such that $y^p = x$. In this situation y is unique with the above property. We write $K = K^p$ to express the above condition.

All finite fields and algebraically closed fields (of any characteristic) are perfect fields. A field K is perfect if and only if every algebraic extension $L|K$ is separable. Hence every algebraic extension of a perfect field is again a perfect field.

Every field K of characteristic p is contained in a unique minimal perfect field, which is a purely inseparable extension of K.

2.15 The Theorem of Steinitz

If K is any field, if K_0 is any subfield of K, then there exists a family $(x_i)_{i \in I}$ of elements in K, which are algebraically independent over K_0, such that $K|K_0(x_i)_{i \in I}$ is an algebraic extension. Thus $K_0(x_i)_{i \in I} \cong K_0(X_i)_{i \in I}$ where $(X_i)_{i \in I}$ is a family of indeterminates, because if $i_1, \ldots, i_m \in I$, if f is a polynomial in $K_0[T_1, \ldots, T_m]$ such that $f(x_{i_1}, \ldots, x_{i_m}) = 0$, then $f = 0$ (the zero polynomial). The family $(x_i)_{i \in I}$ is called a *transcendence basis* of $K|K_0$. Any two transcendence bases of $K|K_0$ have the same cardinality, called the *transcendence degree* of $K|K_0$.

If K, K' are two algebraically closed fields, such that $K|K_0$ and $K'|K_0$ have the same transcendence degree, if $(x_i)_{i \in I}$ and $(x_i')_{i \in I'}$ are transcendence bases of $K|K_0$, respectively, of $K'|K_0$, if $\varphi_0 : K_0(x_i)_{i \in I} \rightarrow K_0(x_i')_{i \in I'}$ is any isomorphism of fields, then it may be extended to an isomorphism $\varphi : K \rightarrow K'$.

2.16 Orderable Fields

Let K be a field and let $\leq$ be a total order on K, compatible with the operations. Then $(K, \leq)$ is said to be an *ordered field*. Then K has characteristic 0. If $0 \leq x$ we say that x is *positive* and if $x \leq 0$, x is said to be *negative*. 0 is the only element which is both positive and negative. Every element of K is either positive or negative. Every square is positive; but the converse need not be true. If K' is a subfield of K and $\leq$ is an order as above, its restriction to K' is also a compatible order; we say that $(K', \leq)$ is an ordered subfield of $(K, \leq)$.

The field $\mathbb{R}$ of real numbers is ordered by letting $x \leq y$ when $y - x$ is a square in $\mathbb{R}$. Thus, with the induced order, every subfield of $\mathbb{R}$ is also ordered. The above order on $\mathbb{R}$ is the only possible. Similarly, the ordinary order on $\mathbb{Q}$ is the unique compatible order. On the other hand, $\mathbb{Q}(\sqrt{2})$ has the orders $\leq$, $\leq'$ where $\leq$ is the restriction of the order of $\mathbb{R}$ and $\leq'$ is the order such that $0 \leq' a + b\sqrt{2}$ whenever $0 \leq a - b\sqrt{2}$.

A field K is *orderable* if there exists a compatible order $\leq$ on K. Clearly, if K is orderable then the following equivalent properties hold:

(i) If $x_1^2 + \cdots + x_n^2 = 0$ (each $x_i \in K$) then $x_1 = x_2 = \cdots = x_n = 0$.

(ii) -1 is not a sum of squares in K.

Artin and Schreier showed that if (i) or (ii) hold then K is orderable.

Thus, the field $\mathbb{C}$ of complex numbers is not orderable, since -1 is a square.

2.17 The Theorem of Artin.

Let L be an algebraically closed field and assume that it has a proper subfield K such that $[L : K] < \infty$. Artin showed L has characteristic 0, $[L : K] = 2$, K is a real closed field and $L = K(\sqrt{-1})$.

E X E R C I S E S

1. Show that $\sqrt{2}$ is not a rational number.

2. Let $m, n \geq 1$ be integers. Show that $\sqrt[n]{m}$ is a rational number if and only if m is an nth power of a natural number.

3. Prove *Eisenstein's irreducibility criterion*: The polynomial

$$f = X^n + a_{n-1}X^{n-1} + \cdots + a_1 X + a_0,$$

with coefficients in $\mathbb{Z}$, is irreducible over $\mathbb{Q}$ provided there exists a prime p dividing $a_0, \ldots, a_{n-1}$ but such that p^2 does not divide a_0.

4. Show that $\Phi_p = X^{p-1} + X^{p-2} + \cdots + X + 1$ (where p is a prime number) is irreducible over $\mathbb{Q}$.

Hint: Apply Eisenstein's irreducibility criterion to the polynomial $\Phi_p(X + 1)$.

5. Show that if p is a prime number then

$$\Phi_{p^r} = X^{p^{r-1}(p-1)} + X^{p^{r-1}(p-2)} + \cdots + X^{p^{r-1}} + 1$$

is irreducible over $\mathbb{Q}$.

6. Discuss whether the following polynomials are irreducible over $\mathbb{Q}$:
$$X^3 + X + 1; \quad X^4 + X^2 + 1; \quad X^3 + X^2 + X + 1; \quad X^6 + X^4 + 1;$$
$$X^4 - 4X^2 + 8X - 4.$$

7. Let $K \subseteq K' \subseteq K''$ be fields. Show that $[K'' : K] = [K'' : K'] \cdot [K' : K]$.

8. Let $K \subseteq K' \subseteq K''$ be fields. Show that $K''|K$ is an algebraic extension if and only if $K'|K$ and $K''|K'$ are algebraic extensions.

9. Prove that $L|K$ has finite degree if and only if L is a finitely generated algebraic extension of K.

10. Find the degree over $\mathbb{Q}$ of the following fields: $\mathbb{Q}(i, \sqrt{3})$, where
$$i = \sqrt{-1}; \quad \mathbb{Q}((1+i)/2); \quad \mathbb{Q}(\sqrt{2}, \sqrt{5}, \sqrt{10}); \quad \mathbb{Q}(\sqrt{2 + \sqrt{2 + \sqrt{2}}}, i).$$

11. Show that the following numbers are algebraic over $\mathbb{Q}$, determine the minimal polynomial and the conjugates over $\mathbb{Q}$:

$$1 + \sqrt{2}; \quad \sqrt{2 + \sqrt{2}}; \quad \sqrt{2 + \sqrt{2 + \sqrt{2}}}; \quad 1 - 2\sqrt{-1}; \quad \sqrt{2} + \sqrt{3} + \sqrt{5}.$$

12. Let $L|K$ be an algebraic extension. Prove:

    (a) if K is finite then L is countable;

    (b) if K is infinite then $\#(L) = \#(K)$.

13. Show that there exist uncountably many transcendental complex numbers.

    Hint: Use the previous exercise.

14. Show that an algebraically closed field cannot be finite.

15. Let X_1, X_2, $\ldots$, X_n be indeterminates and consider the symmetric polynomials

$$s_1 = X_1 + X_2 + \cdots + X_n,$$
$$s_2 = X_1X_2 + X_1X_3 + \cdots + X_2X_3 + \cdots + X_{n-1}X_n,$$
$$s_3 = X_1X_2X_3 + X_1X_2X_4 + \cdots + X_{n-2}X_{n-1}X_n,$$
$$\cdots$$
$$s_k = \sum_{i_1 < i_2 < \cdots < i_k} X_{i_1} X_{i_2} \cdots X_{i_k} \quad \text{(where } k \leq n\text{)},$$
$$\cdots$$
$$s_n = X_1 X_2 \cdots X_n.$$

Show that if Y is any indeterminate then

$$Y^n - s_1 Y^{n-1} + s_2 Y^{n-2} - \cdots + (-1)^k s_k Y^{n-k} + \cdots + (-1)^n s_n = \prod_{k=1}^{n} (Y - X_k).$$

16. Let R be a domain. A polynomial $f \in R[X_1, \ldots, X_n]$ is said to be *symmetric* if for every permutation σ of $\{1, 2, \ldots, n\}$ we have

$$f(X_{\sigma(1)}, X_{\sigma(2)}, \ldots, X_{\sigma(n)}) = f(X_1, X_2, \ldots, X_n).$$

Show that if f is a symmetric polynomial, there exists a polynomial $g \in R[X_1, \ldots, X_n]$ such that $f(X_1, \ldots, X_n) = g(s_1, s_2, \ldots, s_n)$.

    Hint: Define the weight of a monomial $aX_1^{e_1} X_2^{e_2} \cdots X_n^{e_n}$ as being $e_1 + 2e_2 + \cdots + ne_n$; define the weight of a polynomial as being the maximum of the weights of its monomials; the proof is done by double induction on n and on the degree d of f; consider $f(X_1, \ldots, X_{n-1}, 0)$, express it as

$g_1(s_1^0, s_2^0, \ldots, s_{n-1}^0)$ where $s_1^0, s_2^0, \ldots, s_{n-1}^0$ are the elementary symmetric polynomials on the indeterminates $X_1, \ldots, X_{n-1}$; observe that

$$g_1(s_1, s_2, \ldots, s_{n-1})$$

has degree at most d in $X_1, \ldots, X_n$; then $f_1(X_1, \ldots, X_n) = f(X_1, \ldots, X_n) - g_1(s_1, \ldots, s_{n-1})$ has degree at most d and is symmetric; also $f_1(X_1, \ldots, X_{n-1}, 0) = 0$ so f_1 is a multiple of X_n and by symmetry also a multiple of $X_1, X_2 \cdots X_{n-1}$; define f_2 by $f_1 = s_n f_2$ hence its degree is less than d; continue by induction.

17. Let $p_0 = n$ and $p_k = X_1^k + X_2^k + \cdots + X_n^k$ where $k \geq 1$ and $X_1, X_2, \ldots, X_n$ are indeterminates. Prove the *Newton formulas*:

(a) If $k \leq n$ then

$$p_k - p_{k-1}s_1 + p_{k-2}s_2 - \cdots + (-1)^{k-1}p_1 s_{k-1} + (-1)^k k s_k = 0.$$

(b) If $k > n$ then

$$p_k - p_{k-1}s_1 + \cdots + (-1)^n p_{k-n}s_n = 0.$$

Hint: Let $f(T) = \prod_{i=1}^n (T - X_i)$, where T is a new indeterminate; write the quotient $f'(T)/f(T)$ as a rational fraction in $T, X_1, \ldots, X_n$ (where f' denotes the derivative of f with respect to T); then develop in formal power series and after multiplying both sides by $f(T)/T^n$, equate the coefficients of equal powers of T.

18. Let $f = X^3 + X^2 - X + 1$ and let x_1, x_2, x_3 be the roots of f in $\mathbb{C}$. Determine $x_1^3 + x_2^3 + x_3^3$.

19. Let K be a field of characteristic p. Show that the mapping $\theta : K \to K$, defined by $\theta(x) = x^p$ for every $x \in K$, is an isomorphism from K into K. Moreover, if K is finite it is an automorphism.

20. Show that if K is a field of characteristic p then $K(X)$ is an inseparable extension of $K(X^p)$.

21. Show that if K is a field of characteristic 0 and $f \in K[X]$ is an irreducible polynomial then its roots are all distinct.

22. Let K be a field of characteristic p, $f \in K[X]$ a polynomial such that there exists an integer $e \geq 1$, and a polynomial $g \in K[X]$ for which $f(X) = g(X^{p^e})$. Then every root of f has multiplicity at least p^e.

23. Show that if K is any infinite field and $f \in K[X_1, \ldots, X_n]$, there exist infinitely many n-tuples $x = (x_1, \ldots, x_n) \in K^n$ such that $f(x_1, \ldots, x_n) \neq 0$.

Hint: Proceed by recurrence on n.

24. Let V be a vector space of dimension n over K; let W_1, ..., W_m be subspaces of V, distinct from V. Show that if K is an infinite field then $W_1 \cup \cdots \cup W_m \neq V$.

Hint: Use the previous exercise.

25. Prove the theorem of the primitive element: If L is a separable extension of finite degree over a field K, then there exists an element $t \in L$ such that $L = K(t)$.

Hint: Consider first the case when K is finite; then, letting K be infinite, consider the sets $\{x \in L \mid \sigma_i(x) = \sigma_j(x)\}$ where σ_i, σ_j are distinct K-isomorphisms from L into an algebraic closure of K; conclude using the previous exercise.

26. Find a primitive element over $\mathbb{Q}$ for each of the following fields:

$$\mathbb{Q}(\sqrt{2}, i); \quad \mathbb{Q}(\sqrt{2}, \sqrt{3}); \quad \mathbb{Q}(\sqrt{2} + \sqrt{3}, \sqrt{2} - \sqrt{5}).$$

27. Determine the smallest normal extension K of $\mathbb{Q}$ containing $\sqrt[3]{2}$; what is the degree of K over $\mathbb{Q}$? Find a primitive element of K over $\mathbb{Q}$.

28. Give an example of an extension of degree 4 of $\mathbb{Q}$ which is not normal.

29. Determine the $\mathbb{Q}$-isomorphisms of the following fields:

$$\mathbb{Q}(\sqrt[3]{2}); \quad \mathbb{Q}((1+i)/2); \quad \mathbb{Q}(\sqrt{2}, \sqrt{3}); \quad \mathbb{Q}(\sqrt{2}+\sqrt{3}); \quad \mathbb{Q}(\sqrt{2 + \sqrt{2}}).$$

30. Determine the Galois groups over $\mathbb{Q}$ of the following polynomials:

$$X^3 + X + 1; \quad X^4 - X^2 + 1; \quad X^3 - 2; \quad X^3 - 1; \quad X^3 + 1; \quad X^4 - 5.$$

31. For each of the above polynomials, determine the subgroups of the Galois group and the corresponding fields of invariants.

32. Determine the Galois group of $(X^2 - p_1)(X^2 - p_2) \cdots (X^2 - p_n)$ over $\mathbb{Q}$, where p_1, p_2, ..., p_n are distinct primes.

33. If L is a finite separable extension of K, prove that there exist only finitely many fields K' such that $K \subseteq K' \subseteq L$.

34. Determine which roots of unity belong to the following fields:

$$\mathbb{R}; \quad \mathbb{Q}(i); \quad \mathbb{Q}(\sqrt{2}); \quad \mathbb{Q}(\sqrt{-3}); \quad \mathbb{Q}(\sqrt{-5}).$$

35. Let K be a finite extension of $\mathbb{Q}$. Show that K contains only finitely many roots of unity.

36. Prove that if K is an algebraically closed field of characteristic 0 or p, then the group of nth roots of unity in K is cyclic.

Hint: Let $n = p_1^{e_1} p_2^{e_2} \cdots p_r^{e_r}$ be the decomposition of n into prime factors; let $n_i = n/p_i$; show that there exists $x_i \in K$ such that $x_i^n = 1$ but $x_i^{n_i} \neq 1$; next note that if $m_i = n/p_i^{e_i}$ then $x_i^{m_i}$ has order $p_i^{e_i}$ for $i = 1, \ldots, r$; conclude with Lemma 1 of Chapter 3; pay special attention to the case when K has characteristic p dividing n.

37. Compute Φ_6, Φ_{12}, and Φ_{18}.

38. Prove that $\Phi_{n,\mathbb{Q}}$ is a monic polynomial with coefficients in $\mathbb{Z}$.

Hint: Use the expression for $X^n - 1$ as a product of cyclotomic polynomials.

39. Prove that $\Phi_{n,\mathbb{Q}}$ is irreducible over $\mathbb{Q}$.

Hint: Let f be the minimal polynomial of ζ over $\mathbb{Q}$; by Gauss' lemma $f \in \mathbb{Z}[X]$ and $\Phi_n = f \cdot g$, with $g \in \mathbb{Z}[X]$ monic. Show that for every prime p, not dividing n, it must be $f(\zeta^p) = 0$; if this is not true, consider $h = g(X^p)$, deduce that f divides h, and conclude that Φ_n, reduced modulo p, would have a double root. Show that this is not the case.

40. Let ζ be a primitive nth root of unity. Show that $\mathbb{Q}(\zeta)$ is a Galois extension of degree $\varphi(n)$ over $\mathbb{Q}$ and that $\mathrm{Gal}(\mathbb{Q}(\zeta)|\mathbb{Q})$ is isomorphic to the group $P(n)$ of residue classes $\bar{a}$ modulo n, where a is an integer relatively prime to n (see Chapter 3).

Hint: Use the previous exercise; then consider the mapping $\theta : \mathrm{Gal}(\mathbb{Q}(\zeta)|\mathbb{Q}) \to P(n)$, defined as follows: if $\sigma(\zeta) = \zeta^a$ then $\theta(\sigma) = \bar{a}$.

41. Let p be a prime not dividing n. Show that $\Phi_{n,\mathbb{F}_p}$ is irreducible over $\mathbb{F}_p$ if and only if the class of p modulo n has order $\varphi(n)$ in the multiplicative group $P(n)$ of prime residue classes modulo n.

Hint: $\Phi_{n,\mathbb{F}_p}$ is irreducible if and only if $[\mathbb{F}_p(\zeta) : \mathbb{F}_p] = \varphi(n)$, and using a property of finite fields, this means that $\varphi(n)$ is the smallest integer such that n divides $p^{\varphi(n)} - 1$.

42. Show that $\Phi_{8,\mathbb{F}_7}$ is reducible over $\mathbb{F}_7$ and find its decomposition.

43. Show that Φ_{12} is reducible over the fields $\mathbb{F}_5$, $\mathbb{F}_7$, $\mathbb{F}_{11}$, $\mathbb{F}_{13}$ and find in each case its decomposition into irreducible polynomials.

44. Determine the minimal polynomial over $\mathbb{F}_3$ for a primitive eighth root of unity.

45. Determine the minimal polynomial over $\mathbb{F}_2$ for a primitive seventh root of unity.

46. Let K be a finite field. Show that every element of K is the sum of two squares of elements of K.

47. Let $L|K$ be a separable extension of degree n, and assume that the characteristic of K does not divide n. Show that there exists $t \in L$ such that $L = K(t)$ and $\mathrm{Tr}_{L|K}(t) = 0$.

Hint: If $L = K(t')$ and $X^n + a_1 X^{n-1} + \cdots + a_n$ is the minimal polynomial of t' over K, show that $t = t' + a_1/n$ satisfies the required property.

48. Prove the following theorem of Warning and Chevalley:

Let K be a finite field of characteristic p, let $f \in K[X_1, \ldots, X_n]$ be a polynomial without constant term, of degree $d < n$. Then the number of zeros of f in K is a multiple of p; in particular, there exist $x_1, \ldots, x_n \in K$, not all x_i being equal to zero, and such that $f(x_1, \ldots, x_n) = 0$.

Hint:

(a) Let $g = X_1^{a_1} \cdots X_n^{a_n}$ (with $0 < \sum_{i=1}^{n} a_i < (q-1)n$), where $q = p^r = \#(K)$; show that $\sum_{x \in K^n} g(x) = 0$ by computing $\sum_{y \in K} y^a$ for different values of the integer $a \geq 0$.

(b) Note that the number N of zeros of f in K is given by

$$N = \sum_{x \in K^n} [1 - f(x)^{q-1}] \equiv - \sum_{x \in K^n} f(x)^{q-1} \pmod{p}$$

and then apply (a) for the monomials of f^{q-1}.

49. Determine the discriminant of the following polynomials: $aX^2 + bX + c$; $aX^3 + bX^2 + cX + d$; $aX^4 + bX^3 + cX^2 + dX + e$.

Hint: Use the Newton formulas.

50. Prove that if $f \in K[X]$ is a monic irreducible polynomial of degree n, if t is any root of f and f' is the derivative of f, then

$$\mathrm{discr}(f) = (-1)^{n(n-1)/2} \cdot N_{K(t)|K}(f'(t)).$$

Hint: Let $t = t_1, \ldots, t_n$ be the conjugates of t over K, compute explicitly $f'(t_i)$ and compare its values with $\mathrm{discr}(f)$.

51. Determine the discriminant of $\Phi_{n,\mathbb{Q}}$.

Hint: Use the previous exercise.

52. Let K be a field and $\overline{K}$ an algebraic closure of K.

Prove that the subfield generated by the union of all Abelian extensions of K is an Abelian extension K' of K; it is called the *Abelian closure* of K.

53. Let A be an integral domain and let $R(f, g)$ be the resultant of polynomials $f, g \in A[X]$ as defined in Chapter 2, Section 11. Denote $n = \deg f$, $m = \deg g$. Show:

(a) $R(g, f) = (-1)^{mn} R(f, g)$.

(b) If $h \in A[X]$ then $R(f, gh) = R(f, g) \cdot R(f, h)$.

(c) If $f(X) = a_0 \prod_{i=1}^{n} (X - \alpha_i)$ and $g(X) = b_0 \prod_{j=1}^{m} (X - \beta_j)$ then

$$R(f, g) = a_0^m \prod_{i=1}^{n} g(\alpha_i) = (-1)^{mn} b_0^n \prod_{j=1}^{m} f(\beta_j)$$

$$= a_0^m b_0^n \prod_{i=1}^{n} \prod_{j=1}^{m} (\alpha_i - \beta_j).$$

54. Let $f(X) = X^n + a_1 X^{n-1} + \cdots + a_n$ with $n \geq 1$ ($a_1, \ldots, a_n$ are in an integral domain). Let $f'(X)$ denote the derivative of $f(X)$. Show that the discriminant of $f(X)$ (as defined in Chapter 2, Section 11) is equal to $(-1)^{n(n-1)/2} R(f, f')$.

Part One

In Part One we discuss facts from the Arithmetic of rational integers of an elementary nature, mainly residue classes and quadratic residues. It is appropriate to establish the main results about finite Abelian groups, which will be used throughout this book.

If x, y are rational numbers, we say that x *divides* y if there exists an integer $n \in \mathbb{Z}$ such that $nx = y$. We write $x|y$ to express that x divides y and $x \nmid y$ if x does not divide y.

If p is a prime number, x is a rational number, $m \in \mathbb{Z}$ and if $p^m|x$ but $p^{m+1} \nmid x$ (so $x \neq 0$), we write $p^m \| x$ and also $v_p(x) = m$. We call $v_p(x)$ the p-adic *value of x*. By convention, $v_p(0) = +\infty$. It is clear that $v_p(xy) = v_p(x) + v_p(y)$, $v_p(x + y) \geq \min\{v_p(x), v_p(y)\}$, and if $v_p(x) < v_p(y)$ then $v_p(x + y) = v_p(x)$ (we also used the convention that $+\infty + m = +\infty$ for every $m \in \mathbb{Z}$).

We have devoted our book (see Ribenboim [26]) to the study of p-adic valuations and more general types of valuations.

3

Residue Classes

3.1 Congruences

In this chapter, we study residue classes modulo a natural number. This leads to the consideration of groups. Therefore it is convenient to recall that if G is a finite group, the number of elements of G is called the *order* of G, denoted by $\#(G)$.

If G is an additive group, if $g \in G$ and there exists $m \geq 1$ such that $mg = 0$, then the *order* of g is the smallest $m \geq 1$ with the above property. For a multiplicative group, the order of g is the smallest $m \geq 1$ such that $g^m = e$, where e denotes the unit element of G. For a finite group G every element g has an order.

Lagrange's theorem asserts:

If G is a finite group, then the order of each element of G divides the order of G.

Definition 1. Let m be a positive integer. We say that the integers a, b are *congruent modulo m* if they leave the same remainder when divided by m. We write $a \equiv b \pmod{m}$ to express this fact. It is equivalent to saying that m divides $a - b$. The relation of "congruence modulo m" is reflexive, symmetric, and transitive; that is, it is an equivalence relation.

The equivalence class of a *modulo m* is the set $\bar{a} = \{a + km \mid k \in \mathbb{Z}\}$. $\bar{a}$ is called the *residue class of a modulo m*. Let $\mathbb{Z}/m$ denote the set of equivalence classes modulo m.

If $a \equiv a' \pmod{m}$, $b \equiv b' \pmod{m}$, then $a + b \equiv a' + b' \pmod{m}$, $a \cdot b \equiv a' \cdot b' \pmod{m}$. Hence we may define the addition and multiplication of equivalence classes: $\bar{a} + \bar{b} = \overline{a + b}$, $\bar{a} \cdot \bar{b} = \overline{a \cdot b}$. With these operations, if $m > 1$ then $\mathbb{Z}/m$ is a ring. Its unit element is $\bar{1} = \{a \in \mathbb{Z} \mid a \equiv 1 \pmod{m}\}$.

It follows that the mapping $a \to \bar{a}$ is a ring-homomorphism from $\mathbb{Z}$ onto $\mathbb{Z}/m$. The kernel of this homomorphism is $(m) = \mathbb{Z}m$ (the ideal of multiples of m). So $\mathbb{Z}/m$ is the quotient ring of $\mathbb{Z}$ by the ideal (m), and we may also use the notation $\mathbb{Z}/(m)$.

The additive group of $\mathbb{Z}/m$ is cyclic with m elements, generated by $\overline{1}$.
In fact $m \cdot \overline{1} = \overline{0}$ ($m \cdot \overline{1}$ denotes $\overline{1} + \overline{1} + \cdots + \overline{1}$, m times) and if $n \cdot \overline{1} = \overline{0}$ then n is a multiple of m.

Conversely, we have the following result:

A.   *Every cyclic group with m elements is isomorphic to $\mathbb{Z}/m$.*

Proof: Let G be a cyclic group (written additively) with m elements. Let g be a generator of G and consider the mapping $\theta : \mathbb{Z} \rightarrow G$ defined as follows: $\theta(n) = ng$ for every $n \in \mathbb{Z}$. Then θ is a group-homomorphism onto G; we have $n\overset{\circ}{g} = 0$ (in G) if and only if n is a multiple of m, because g has order m. Thus, the kernel of θ is (m) and so $G \cong \mathbb{Z}/(m)$. ∎

Now, we shall determine all the generators of the cyclic group $\mathbb{Z}/m$.

B.   *Let a be a positive integer; then $\overline{a}$ is a generator of $\mathbb{Z}/m$ if and only if a, m are relatively prime.*

Proof: Let $\overline{a}$ be a generator of $\mathbb{Z}/m$ and let $d = \gcd(a, m)$ (the greatest common divisor of a, m). Then

$$\frac{m}{d} \cdot a = m \cdot \frac{a}{d} \equiv 0 \pmod{m}.$$

Since $\overline{a}$ has order m, then m divides m/d, that is, $d = 1$.

Conversely, if $\gcd(a, m) = 1$ and m' is the order of $\overline{a}$, then m divides $m'a$ (because $m'a \equiv 0 \pmod{m}$) and therefore m divides m'. On the other hand, $m\overline{a} = \overline{0}$, so the order m' of $\overline{a}$ divides m. Thus $m' = m$. ∎

We are interested in counting the generators of the additive group $\mathbb{Z}/m$; this number depends on m, and we may introduce the following definition:

Definition 2.   For every $m \geq 1$, let $\varphi(m)$ denote the number of integers a, $1 \leq a \leq m$, such that $\gcd(a, m) = 1$.

φ is called the *Euler function* or *totient function*. Thus, $\varphi(m)$ is the number of generators of the additive group $\mathbb{Z}/m$. For example, $\varphi(1) = 1$, $\varphi(2) = 1$, $\varphi(3) = 2$ and, more generally, for every prime p, $\varphi(p) = p-1$. In order to determine $\varphi(m)$, for an arbitrary $m > 1$, we shall prove some interesting results about residue classes.

Theorem 1.   *Let $m = \prod_{i=1}^{r} p_i^{e_i}$ be the decomposition of $m > 1$ into prime powers (with $e_i > 0$ for $i = 1, \ldots, r$). Then there exists a ring-isomorphism*

$$\overline{\theta} : \mathbb{Z}/m \ \xrightarrow{\sim} \ \prod_{i=1}^{r} \mathbb{Z}/p_i^{e_i}$$

(Cartesian product of the rings $\mathbb{Z}/p_i^{e_i}$ for $i = 1, \ldots, r$).

Proof: Let $\theta : \mathbb{Z} \rightarrow \prod_{i=1}^{r} \mathbb{Z}/p_i^{e_i}$ be the mapping defined as follows:
$\theta(n) = (\nu_1, \ldots, \nu_r)$ where ν_i denotes the residue class of n modulo $p_i^{e_i}$.

θ is a ring-homomorphism whose kernel is equal to the ideal (m) of multiples of m, because n is a multiple of m if and only if n is a multiple of each $p_i^{e_i}$ $(i = 1, \ldots, r)$. Thus θ induces a one-to-one homomorphism $\overline{\theta}$ from $\mathbb{Z}/m$ to $\prod_{i=1}^{r} \mathbb{Z}/p_i^{e_i}$.

Finally, the number of elements in $\prod_{i=1}^{r} \mathbb{Z}/p_i^{e_i}$ is $\prod_{i=1}^{r} p_i^{e_i} = m$. Thus $\overline{\theta}$ maps $\mathbb{Z}/m$ onto $\prod_{i=1}^{r} \mathbb{Z}/p_i^{e_i}$. ■

As a corollary we have the so-called *"Chinese remainder theorem"* below.

C.   *Given $r \geq 1$ distinct prime numbers p_1, ..., p_r, given integers $e_1 \geq 1$, ..., $e_r \geq 1$ and $a_1, \ldots, a_r \in \mathbb{Z}$, there exists $n \in \mathbb{Z}$ satisfying all the congruences*

$$n \equiv a_i \pmod{p_i^{e_i}} \qquad for \quad i = 1, \ldots, r.$$

Moreover, $n' \in \mathbb{Z}$ satisfies the above congruences if and only if

$$n' \equiv n \left(\mathrm{mod} \ \prod_{i=1}^{r} p_i^{e_i} \right).$$

Proof: Let ν_i be the residue class of a_i modulo $p_i^{e_i}$, let $\nu \in \mathbb{Z}/m$ be the unique residue class modulo m such that $\overline{\theta}(\nu) = (\nu_1, \ldots, \nu_r)$. Then n satisfies the congruences

$$n \equiv a_i \pmod{p_i^{e_i}} \qquad \text{for} \quad i = 1, \ldots, r$$

if and only if ν is the class of n modulo m. The last assertion is obvious. ■

In order to determine $\varphi(m)$, we consider the set $P(m)$ of all nonzero residue classes $\overline{a}$ modulo m, where $\gcd(a, m) = 1$. These are called the *prime* or *irreducible classes modulo m*, and $P(m)$ has therefore $\varphi(m)$ elements. We also use the notation $(\mathbb{Z}/m)^{.}$ for $P(m)$. We remark that if $\overline{a} \in P(m)$ then every element $a + km$ in the class $\overline{a}$ is relatively prime to m.

D.   *$P(m)$ is the multiplicative group of all invertible elements of the ring $\mathbb{Z}/m$.*

Proof: Let $a \in \mathbb{Z}$ be such that $\gcd(a, m) = 1$. Then there exist integers s, t such that $s \cdot a + t \cdot m = 1$. Hence $\overline{s} \cdot \overline{a} = \overline{1}$ and $\overline{a}$ is invertible in the ring $\mathbb{Z}/m$.

Conversely, if $\overline{a} \in \mathbb{Z}/m$ is invertible in the ring $\mathbb{Z}/m$, then $\overline{s} \cdot \overline{a} = \overline{1}$ (for some $s \in \mathbb{Z}$), so $sa \equiv 1 \pmod{m}$ and $\gcd(a, m) = 1$, that is, $\overline{a} \in P(m)$.

It is easy to see that in any ring the set of all invertible elements constitutes a multiplicative group. Thus, $P(m)$ is a multiplicative group. ■

Let us note in particular that $\mathbb{Z}/p$ is a field (when p is prime), since every nonzero residue class modulo p is invertible. Actually $\mathbb{Z}/p$ is the prime field of characteristic p, and as such we denote it also by $\mathbb{F}_p$.

As a corollary, we obtain the following congruence (from Euler):

E.   *If* $\gcd(a, m) = 1$ *then*

$$a^{\varphi(m)} \equiv 1 \pmod{m}.$$

Proof: Since $\bar{a} \in P(m)$, a multiplicative group of order $\varphi(m)$, then by Lagrange's theorem on finite groups $\bar{a}^{\varphi(m)} = \bar{1}$, that is, $a^{\varphi(m)} \equiv 1 \pmod{m}$.   ∎

In particular, for any prime number p we have (Fermat)

$$a^{p-1} \equiv 1 \pmod{p}.$$

F.   *If* $m = p_1^{e_1} \cdots p_r^{e_r}$ *is the prime-power decomposition of* m, *then*

$$P(m) \cong \prod_{i=1}^{r} P(p_i^{e_i}),$$

and

$$\varphi(m) = \prod_{i=1}^{r} \varphi(p_i^{e_i}) = m \prod_{i=1}^{r} \left(1 - \frac{1}{p_i}\right).$$

Proof: We consider again the ring-isomorphism $\bar{\theta} : \mathbb{Z}/m \xrightarrow{\sim} \prod_{i=1}^{r} \mathbb{Z}/p_i^{e_i}$ of Theorem 1. A residue class $\bar{a}$ modulo m is invertible if and only if its image $\bar{\theta}(\bar{a})$ is invertible in the Cartesian product of rings $\mathbb{Z}/p_i^{e_i}$; in other words, each component of $\bar{\theta}(\bar{a})$ is invertible in the corresponding component $\mathbb{Z}/p_i^{e_i}$. Thus $\bar{\theta}(P(m)) = \prod_{i=1}^{r} P(p_i^{e_i})$, proving the isomorphism of multiplicative groups $P(m) \cong \prod_{i=1}^{r} P(p_i^{e_i})$.

By counting the number of elements in these groups it follows that $\varphi(m) = \prod_{i=1}^{r} \varphi(p_i^{e_i})$.

To evaluate $\varphi(p^e)$ (for a prime p and $e \geq 1$), we just note that if $1 \leq a \leq p^e$, $\gcd(a, p^e) = 1$ then a is not a multiple of p, and conversely. Thus $P(p^e)$ has $p^e - p^{e-1}$ elements; that is, $\varphi(p^e) = p^e[1 - (1/p)]$. Hence $\varphi(m) = \prod_{i=1}^{r} p_i^{e_i} \cdot \prod_{i=1}^{r}[1 - (1/p_i)] = m \prod_{i=1}^{r}[1 - (1/p_i)]$.   ∎

Incidentally, if $\gcd(m, n) = 1$ then $\varphi(mn) = \varphi(m) \cdot \varphi(n)$, as follows immediately from the above. We express this fact by saying that φ is a *multiplicative function*.

The Euler function also possesses the following interesting property:

G.   *If* $n \geq 1$ *then* $n = \sum_{d|n} \varphi(d)$ (*the sum being taken over all divisors* $d \geq 1$ *of* n).

Proof: Let $S = \{1, 2, \ldots, n\}$ and, for each divisor $d \geq 1$ of n, let $C(d) = \{s \in S \mid \gcd(n, s) = d\}$. This gives rise to a partition of S into pairwise disjoint subsets. Therefore $n = \#S = \sum_{d|n} \#C(d)$. Now, if $n = dd'$ then $s \in C(d')$ if and only if $s/d' \leq n/d' = d$ and $\gcd(s/d', d) = 1$. Thus, the number of elements in $C(d')$ is equal to $\varphi(d)$; therefore $n = \sum_{d|n} \varphi(d)$.   ∎

Up to now, we have studied the additive group of residue classes modulo m, which is cyclic; its generators are the elements of the multiplicative group $P(m)$.

3.2 The Group of Invertible Residue Classes

In the sequel, we shall study the structure of the group $P(m)$. First, we point out another way of describing $P(m)$:

H.   *The group* $\text{Aut}(\mathbb{Z}/m)$ *of automorphisms of the additive group* $\mathbb{Z}/m$ *is isomorphic to* $P(m)$.

Proof: We will show that the ring of endomorphisms $\text{End}(\mathbb{Z}/m)$ of the additive group $\mathbb{Z}/m$ is isomorphic to the ring $\mathbb{Z}/m$. Since $\text{Aut}(\mathbb{Z}/m)$ is just the group of invertible elements of $\text{End}(\mathbb{Z}/m)$, then the conclusion follows from (**D**).

Let $\theta : \text{End}(\mathbb{Z}/m) \to \mathbb{Z}/m$ be defined by $\theta(f) = f(\bar{1})$. Evidently $\theta(0) = \bar{0}$ and $\theta(f + g) = (f + g)(\bar{1}) = f(\bar{1}) + g(\bar{1}) = \theta(f) + \theta(g)$ so θ is a homomorphism of additive groups. The kernel of θ is 0, for if $f(\bar{1}) = \bar{0}$ and $n > 0$ then $f(\bar{n}) = f(n \cdot \bar{1}) = n \cdot f(\bar{1}) = 0$, so $f = 0$. Thus θ is a one-to-one mapping. Moreover, given $\bar{a} \in \mathbb{Z}/m$ if we define $h : \mathbb{Z}/m \to \mathbb{Z}/m$ by $h(\bar{b}) = \bar{a} \cdot \bar{b}$ then $h \in \text{End}(\mathbb{Z}/m)$ and $\theta(h) = h(\bar{1}) = \bar{a}$. Thus θ is an isomorphism of additive groups. Clearly the image of the identity automorphism of $\mathbb{Z}/m$ is the unit element $\bar{1}$.

It only remains to show that θ preserves multiplication. Accordingly, suppose that $\theta(f) = \bar{a}$, $\theta(g) = \bar{b}$ with $a, b > 0$; then $\theta(f \cdot g) = f(g(\bar{1})) = f(\bar{b}) = f(b \cdot \bar{1}) = b \cdot f(\bar{1}) = b \cdot \bar{a} = \bar{b} \cdot \bar{a}$ as desired.   ∎

In view of (**F**), it suffices to study the groups $P(p^e)$, where p is a prime number, $e \geq 1$. For this purpose, we require two lemmas from the theory of groups.

Lemma 1.   *Let G be a multiplicative group, let $x, y \in G$ be elements such that $x \cdot y = y \cdot x$, and the orders h of x and k of y are relatively prime. Then the order of $x \cdot y$ is kh.*

Proof: From $x \cdot y = y \cdot x$ we deduce $(x \cdot y)^{hk} = (x^h)^k \cdot (y^k)^h = e$ (identity element of G). Thus $x \cdot y$ has finite order l dividing hk.

Since $(x \cdot y)^l = x^l \cdot y^l = e$ then $x^l = y^{-l}$. The order of this last element divides both h and k, hence it must be 1 (because h, k are relatively prime). Thus $x^l = y^{-l} = e$, so h divides l, k divides l, and therefore hk divides l. This shows that $x \cdot y$ has order $l = hk$.   ∎

The least upper bound of the orders of the elements of group G is called the *exponent* of G. If the exponent is finite, it is a positive integer, equal to the maximum of the orders of the elements of G. By Lagrange's theorem, if G is finite the exponent of G divides the order of G.

Lemma 2.   *Let G be a multiplicative Abelian group with finite exponent g. Then the order of every element of G divides g.*

Proof: Let x be an element of order g and let y be an element of order k, not dividing g. Then there exists a prime p and integers $n > m \geq 0$, such that $g = p^m g'$ (with p not dividing g') and $k = p^n k'$. Let $x' = x^{p^m}$, $y' = y^{k'}$, so x' has order g' and y' has order p^n.

By Lemma 1, $x'y'$ has order $p^n g' > p^m g' = g$, which is contrary to the assumption, thus establishing the lemma.   ∎

I.   *$P(p)$ is a cyclic group.*

Proof: We know that $P(p)$ has $\varphi(p) = p - 1$ elements. It is enough to show that the exponent h of $P(p)$ is equal to $p - 1$, so there will exist an element of order $p - 1$.

As we said before, h divides $p-1$. By Lemma 2, the order of every element of $P(p) = \mathbb{F}_p$ divides h. That is, $\overline{x}^h = \overline{1}$ for every $\overline{x} \in P(p)$. Therefore, every element of $P(p)$ is a root of the polynomial $X^h - 1 \in \mathbb{F}_p[X]$. This polynomial has at most h roots; thus $p - 1 \leq h$. This shows that $h = p - 1$.   ∎

From the proof, it follows that every element of $P(p)$ may be viewed as a $(p - 1)$th root of unity.

Since $P(p)$ is cyclic with $p - 1$ elements, it is isomorphic to the additive group $\mathbb{Z}/(p - 1)$ (by (**A**)), hence it has $\varphi(p - 1)$ generators (by (**B**)). Each generator of $P(p)$ is called a *primitive root modulo p*. It is also customary to say that the integer x, $1 \leq x \leq p - 1$, is a primitive root modulo p, when its class modulo p generates $P(p)$. If $\overline{x}$ is a primitive root modulo p then $\overline{x}^h$ is a primitive root modulo p if and only if $\gcd(h, p - 1) = 1$.

Let us remark that no quick procedure is known for the determination of the smallest integer a, $1 \leq a \leq p - 1$, such that $\overline{a} \in P(p)$ is a primitive root modulo p. See Ribenboim [25, Chapter 2, Section II A].

Next, we consider the groups $P(p^e)$, where $e \geq 2$. First, we treat the case where $p \neq 2$.

J.   *If $p \neq 2$, $e \geq 1$, then $P(p^e)$ is a cyclic group and*

$$P(p^e) \cong \mathbb{Z}/(p - 1) \times \mathbb{Z}/p^{e-1}.$$

Proof: We may assume that $e \geq 2$.

Let $\overline{a}$ denote the residue class of a modulo p^e, and $\overline{\overline{a}}$ the residue class of a modulo p. Let $f : P(p^e) \to P(p)$ be defined by $f(\overline{a}) = \overline{\overline{a}}$. It is obviously a well-defined group-homomorphism onto $P(p)$. Its kernel is

$$C = \{\overline{a} \in P(p^e) \mid a \equiv 1 \pmod{p}\}.$$

C is a subgroup of $P(p^e)$ having order $\varphi(p^e)/\varphi(p) = p^{e-1}$ because

$$P(p^e)/C \cong P(p).$$

C is a cyclic group with generator $\overline{1+p}$. It is enough to show that $(\overline{1+p})^{p^{e-2}} \neq \overline{1}$, that is, $(1+p)^{p^{e-2}} \not\equiv 1 \pmod{p^e}$. This is true for $e = 2$ and let us assume it is true for $e-1$, that is, $(1+p)^{p^{e-3}} \not\equiv 1 \pmod{p^{e-1}}$ and $(1+p)^{p^{e-3}} \equiv 1 \pmod{p^{e-2}}$. Therefore $(1+p)^{p^{e-3}} = 1+rp^{e-2}$ where $r \not\equiv 0 \pmod{p}$. Raising to the pth power, we have

$$(1+p)^{p^{e-2}} = 1 + \binom{p}{1}rp^{e-2} + \binom{p}{2}r^2 p^{2(e-2)} + \cdots + r^p p^{p(e-2)}$$

$$= 1 + rp^{e-1} + sp^e. \text{ *}$$

Hence $(1+p)^{p^{e-2}} \not\equiv 1 \pmod{p^e}$ and $(1+p)^{p^{e-2}} \equiv 1 \pmod{p^{e-1}}$.

Now, let $B = \{\overline{a} \in P(p^e) \mid \overline{a}^{p-1} = \overline{1}\}$. B is obviously a subgroup of $P(p^e)$ and $B \cap C = \{\overline{1}\}$ since B has no element (except $\overline{1}$) of order a power of p. Thus $P(p^e)$ contains the subgroup $BC \cong B \times C$. Then B has order at most $\varphi(p^e)/p^{e-1} = p - 1$.

We have $\overline{a}^{p^{e-1}} \in B$ for every $\overline{a} \in P(p)$. Since $f(\overline{a}^{p^{e-1}}) = \overline{\overline{a}}^{p^{e-1}} = \overline{\overline{a}}$ then B contains the distinct elements $\overline{1}^{p^{e-1}}, \overline{2}^{p^{e-1}}, \ldots, \overline{(p-1)}^{p^{e-1}}$ (since they have distinct images by f). Thus B has $p-1$ elements and $P(p^e) = BC \cong B \times C$.

Finally, B is a cyclic group. Indeed, let b be a primitive root modulo p, then $b^{p^{e-1}}$ has order d dividing $p-1$. From $b^{p^{e-1}} \equiv b \pmod{p}$ it follows that $b^d \equiv (b^d)^{p^{e-1}} \equiv 1 \pmod{p}$ so $p-1$ divides d, thus $d = p - 1$.

By Lemma 1, $P(p^e)$ has an element of order $(p-1)p^{e-1} = \varphi(p^e)$, so it is a cyclic group and is isomorphic to $\mathbb{Z}/(p-1) \times \mathbb{Z}/p^{e-1}$. ∎

We need a special treatment for the case $p = 2$.

K. $P(4) = \{\overline{1}, \overline{3}\}$ *is a cyclic group generated by* $\overline{3}$. *If* $e \geq 3$ *then* $P(2^e) \cong \mathbb{Z}/2 \times \mathbb{Z}/2^{e-2}$ *and is not a cyclic group.*

Proof: The first assertion is obvious. Let us assume $e \geq 3$.

Let $\overline{a}$ denote the residue class of a modulo 2^e and $\overline{\overline{a}}$ the residue class of a modulo 4. Let $f : P(2^e) \to P(4)$ be defined by $f(\overline{a}) = \overline{\overline{a}}$. It is obviously a well-defined group-homomorphism onto $P(4)$. Its kernel is

$$C = \{\overline{a} \in P(2^e) \mid a \equiv 1 \pmod 4\}.$$

C is a subgroup of $P(2^e)$ having order $\varphi(2^e)/\varphi(4) = 2^{e-2}$ because $P(2^e)/C \cong P(4)$.

C is a cyclic group generated by $\overline{5}$. It is enough to show that $\overline{5}^{2^{e-3}} \neq \overline{1}$. Indeed $5^{2^{e-3}} \equiv (1+2^2)^{2^{e-3}} \equiv 1 + 2^{e-3} \cdot 2^2 \not\equiv 1 \pmod{2^e}$.

---

* This is false if and only if $p = 2$ and $e = 3$.

We show now that $P(2^e)$ is isomorphic to the Cartesian product

$$\{1, -1\} \times C.$$

Let $\theta : P(2^e) \to \{1, -1\} \times C$ be the mapping so defined: $\theta(\bar{a}) = ((-1)^r, \bar{a}^*)$ where

$$\bar{a}^* = \begin{cases} \bar{a} & \text{when} \quad a \equiv 1 \pmod 4, \\ -\bar{a} & \text{when} \quad a \equiv -1 \pmod 4, \end{cases}$$

(we remark that if $\bar{a} \in P(2^e)$ then a is odd) and

$$r = \begin{cases} 0 & \text{when} \quad a \equiv 1 \pmod 4, \\ 1 & \text{when} \quad a \equiv -1 \pmod 4. \end{cases}$$

It is obvious that θ is a group-homomorphism and from $\bar{a} = (-1)^r \bar{a}^*$ we conclude that θ is one-to-one. Since $P(2^e)$ and $\{1, -1\} \times C$ have 2^{e-1} elements, it follows that θ is an isomorphism.

Since $\{1, -1\}$ and C are isomorphic to the additive groups $\mathbb{Z}/2$ and $\mathbb{Z}/2^{e-2}$, respectively, then $P(2^e) \cong \mathbb{Z}/2 \times \mathbb{Z}/2^{e-2}$.

To see that $P(2^e)$ is not cyclic, we just observe that the order of every element of $P(2^e)$ divides 2^{e-2}. ∎

As a consequence, we indicate the values of m for which $P(m)$ is a cyclic group. In such a case, each generator of $P(m)$, or each integer a of this residue class, $1 \leq a \leq m - 1$, is called a *primitive root modulo m*.

L.   *$P(m)$ is a cyclic group if and only if $m = 2, 4, p^e, 2p^e$, where $e \geq 1$ and p is an odd prime.*

Proof: By **(I)**, **(J)**, and **(K)**, $P(m)$ is cyclic for each of the given values of m, noting that $P(2p^e) \cong P(2) \times P(p^e) = P(p^e)$.

To prove the converse, we note that if p is any prime then $P(p^e)$ has even order, except when $p = 2$, $e = 1$. By **(F)** it suffices to show that if G has order $2r$ and H has order $2s$ then $G \times H$ is not cyclic. Indeed, for every $x \in G$, $y \in H$ we have $(x, y)^{2rs} = (1, 1)$. Therefore, no element of $G \times H$ has order $(2r) \cdot (2s)$. ∎

The following related lemma will be useful:

Lemma 3.   *If g is a primitive root modulo p, there exist g_1 and g_2 such that $g_1 \equiv g \pmod p$, $g_2 \equiv g \pmod p$ and $g_1^{p-1} \equiv 1 \pmod{p^2}$, $g_2^{p-1} \not\equiv 1 \pmod{p^2}$.*

Proof: If $g^{p-1} = 1 + bp$ with $b \in \mathbb{Z}$, let $a \in \mathbb{Z}$ and consider the congruence $(g + ap)^{p-1} \equiv g^{p-1} + (p-1)g^{p-2}ap \equiv 1 + bp - g^{p-2}ap \pmod{p^2}$. Choosing a_1 such that $g^{p-2}a_1 \equiv b \pmod p$, then $g_1 = g + a_1 p$ has the required property. Choosing a_2 such that $g^{p-2}a_2 \not\equiv b \pmod p$, then $g_2 = g + a_2 p$ is such that $g_2^{p-1} \not\equiv 1 \pmod{p^2}$. Moreover, $g_1 \equiv g_2 \equiv g \pmod p$. ∎

3.3 Finite Abelian Groups

We conclude this chapter by proving the structure theorem of finite Abelian groups. It is a theoretical result analogous to the theorems of structure of $P(m)$. We shall not require this fact until Chapter 8.

Theorem 2.   *Let G be a finite Abelian group (written multiplicatively). Then G is isomorphic to a Cartesian product of cyclic groups.*

Proof (Artin):   Let k be an integer with the following properties:

  (1) there exist elements $x_1, \ldots, x_k$ in G such that every element of G is a product of powers of $x_1, \ldots, x_k$;

  (2) k is the smallest integer satisfying (1).

If $k = 1$ then G consists of all the powers of x_1, so G is a cyclic group.

Let us assume the theorem true for all groups having a system of generators with less than k elements.

Since G is a finite group, there exist integers $e_1, \ldots, e_k$, not all equal to zero, such that

$$\prod_{i=1}^{k} x_i^{e_i} = 1. \tag{3.1}$$

Let b be the minimum of the absolute values of all nonzero exponents, which appear in all possible relations of type (3.1); thus $b > 0$ and by renumbering and taking inverses, if necessary, we may assume that $b = e_1$, for some relation (3.1). Let

$$\prod_{i=1}^{k} x_i^{f_i} = 1 \tag{3.2}$$

be any other relation, with $f_i \in \mathbb{Z}$, not all equal to 0. Then $b = e_1$ divides f_1. In fact, if $f_1 = qe_1 + r$ with $0 < r < e_1 = b$, dividing (3.2) by the qth power of (3.1) we obtain

$$x_1^{r} \prod_{i=2}^{k} x_i^{f_i - qe_i} = 1,$$

which is contrary to the definition of b.

Similarly, $b = e_1$ divides all exponents e_i. In fact, if $e_i = q_i b + r_i$ with $0 < r_i < b$, we consider the system of generators $\{x_1 x_i^{q_i}, x_2, \ldots, x_k\}$ of G; it satisfies the relation

$$(x_1 x_i^{q_i})^{e_1} x_2^{e_2} \cdots x_i^{r_i} \cdots x_k^{e_k} = 1 \tag{3.3}$$

with $0 < r_i < b$, which is a contradiction.

Now, let $e_i = q_i b$ and $y_1 = x_1 x_2^{q_2} \cdots x_k^{q_k}$, so $\{y_1, x_2, \ldots, x_k\}$ is also a system of generators of G and the element y_1 has order $b = e_1$. In fact, if

$y_1^f = 1$ then $x_1^f x_2^{fq_2} \cdots x_k^{fq_k} = 1$ and, therefore, by what we proved above, b divides f; on the other hand, $y_1^b = x_1^{e_1} x_2^{e_1 q_2} \cdots x_k^{e_1 q_k} = 1$.

The group G', generated by $\{x_2, \ldots, x_k\}$ is isomorphic to a Cartesian product of cyclic groups (by induction on k).

Next, we show that $G \cong G_1 \times G'$, where G_1 is the group generated by y_1. Indeed, if $x \in G$, we may write $x = y_1^{c_1} y'$ with $c_1 \in \mathbb{Z}$, $y' \in G'$. The elements $y_1^{c_1}$, y' are uniquely determined by x, for if $y_1^{c_1} y' = y_1^{d_1} z'$ then $y_1^{c_1 - d_1} \cdot (y' z'^{-1}) = 1$; this is a relation of type $x_1^{c_1 - d_1} \cdot x_2^{f_2} \cdots x_k^{f_k} = 1$, thus by the above proof b divides $c_1 - d_1$, so $y_1^{c_1 - d_1} = 1$, and $y_1^{c_1} = y_1^{d_1}$ so also $y' = z'$.

Therefore, the mapping $x \mapsto (y_1^{c_1}, y')$ is an isomorphism between G and $G_1 \times G'$, so G is a Cartesian product of cyclic groups.   ∎

The preceding theorem contains no uniqueness assertion. For example, Theorem 1 states that every cyclic group of order m is isomorphic to the Cartesian product of cyclic groups of prime-power order. So, any uniqueness statement can at most hold for decompositions into cyclic groups of prime-power order.

We shall prove that this is indeed true, and for this purpose we shall first consider the Abelian groups of prime-power order.

The first basic fact to note also holds for non-Abelian groups, being a consequence of the first Sylow theorem. As we only require this fact for Abelian groups, we shall indicate here a simpler direct proof:

M.   *If G is a finite Abelian group of order m and p is a prime dividing m, then G has an element of order p.*

Proof: By Theorem 2, $G \cong G_1 \times \cdots \times G_k$ where G_i is a cyclic group (for $i = 1, \ldots, k$), with order m_i. Since p divides $m = m_1 m_2 \cdots m_k$ then p divides m_i (for some index i, $1 \leq i \leq k$). If x_i is a generator of G_i and $m_i = p m_i'$, if $y_i \in G$ corresponds to

$$(1, \ldots, x_i^{m_i'}, \ldots, 1) \in G_1 \times \cdots \times G_i \times \cdots \times G_k,$$

then y_i has order p.   ∎

Another proof independent of Theorem 2 is the following:

Let n be the exponent of the group G. We shall prove by induction that the order of G, $\#(G)$ divides a power of n. We may assume that $\#(G) > 1$.

Let $x \in G$, $x \neq e$ (identity element of G), then $x^n = e$. If H is the subgroup of G generated by x, then $\#(H)$ divides n. Since the order of every element of the quotient group G/H divides n, the exponent n' of G/H divides n. By induction, $\#(G/H)$ divides a power of n', hence of n. Thus

$$\#(G) = \#(G/H) \cdot \#(H)$$

divides a power of n.

Now, if p is a prime dividing $m = \#(G)$ then p divides a power of the exponent n, so p divides n; that is, $n = pn'$. If $x \in G$ has order n then $x^{n'}$ has order p. ∎

If p is a prime number, a finite group G is said to be a *p-group* when its order is a power of p. In view of (**M**) we deduce:

N. *A finite Abelian group G is a p-group if and only if the order of every element of G is a power of p.* *

Proof: If G is a p-group, by Lagrange's theorem the order of every element of G must be a power of p. Conversely, if q is a prime different from p and dividing the order of G, by (**M**) there exists an element in G having order q. ∎

In order to derive the uniqueness theorem, we first prove the uniqueness of decomposition into the Cartesian product of p-groups (for different primes p):

O. *Let G be a finite Abelian group of order $m = \prod_{i=1}^{r} p_i^{e_i}$ (with $r > 0$, $e_i \geq 1$, and p_1, ..., p_r distinct primes). Then G is isomorphic to the Cartesian product of the p_i-groups $G_{(p_i)} = \{x \in G \mid \text{order of } x \text{ is a power of } p_i\}$:*

$$G \cong G_{(p_1)} \times \cdots \times G_{(p_r)}.$$

This decomposition is unique in the following strong sense: if θ is an isomorphism $\theta : G \to G_1 \times \cdots \times G_s$, where each G_i is a q_i-group (q_1, ..., q_s being distinct primes), then $s = r$, there exists a permutation π of $\{1, \ldots, r\}$ such that $q_{\pi(i)} = p_i$, and if $F_i = \theta^{-1}(\{1\} \times \cdots \times G_i \times \cdots \times \{1\})$ then $F_{\pi(i)} = G_{(p_i)}$ for $i = 1, \ldots, r$.

Proof: If $r = 1$ then $G = G_{(p_1)}$, as follows from Lagrange's theorem. So, we may assume that $r \geq 2$.

For every prime p let $G_{(p)} = \{x \in G \mid \text{order of } x \text{ is a power of } p\}$. By (**M**) and Lagrange's theorem $G_{(p)} \neq \{1\}$ if and only if p divides m; that is, $p = p_i$ for some $i, 1 \leq i \leq r$. By (**N**), each $G_{(p_i)}$ is a p_i-group.

We shall show that G is the internal direct product of the subgroups $G_{(p_1)}$, ..., $G_{(p_r)}$.

For every i let $m_i = m/p_i^{e_i}$, hence the integers m_1, ..., m_r are relatively prime and there exist integers h_1, ..., h_r such that $\sum_{i=1}^{r} h_i m_i = 1$.

If $x \in G$ we may write

$$x = x^{\sum_{i=1}^{r} h_i m_i} = \prod_{i=1}^{r} x^{h_i m_i},$$

---

* This result also holds for non-Abelian groups.

and $(x^{h_i m_i})^{p^{c_i}} = x^{h_i m} = 1$ (by Lagrange's theorem). Thus, $x^{h_i m_i}$ has
an order power of p_i, so $x^{h_i m_i} \in G_{(p_i)}$. From $x = \prod_{i=1}^{r} x^{h_i m_i}$ it follows
that the subgroup of G generated by $G_{(p_1)}, \ldots, G_{(p_r)}$ is equal to G itself.
Moreover, for every i we have $G_{(p_i)} \cap H_i = \{1\}$, where H_i denotes the
subgroup generated by all $G_{(p_j)}$ with $j \neq i$. In fact, if $y \in G_{(p_i)} \cap H_i$ then
the order of y is $p_i^{f_i}$ for some $f_i \geq 0$. Since $y \in H_i$ then $y = \prod_{j \neq i} y_j$
where $y_j \in G_{(p_j)}$ for every $j \neq i$. Let $p_j^{f_j}$ be the order of y_j, therefore by
Lemma 1, $k = \prod_{j \neq i} p_j^{f_j}$ is the order of y. Therefore $k = p_i^{f_i}$. We conclude
that the order of y is necessarily 1, so $y = 1$. Thus, we have shown that
G is the internal direct product of the subgroups $G_{(p_1)}, \ldots, G_{(p_r)} : G \cong$
$G_{(p_1)} \times \cdots \times G_{(p_r)}$. In particular, since the order of $G_{(p_i)}$ is a power of
p_i and $m = \prod_{i=1}^{r} p_i^{e_i}$, by the uniqueness of the decomposition of m into
prime factors, it follows that $\#(G_{(p_i)}) = p_i^{e_i}$ for every $i = 1, \ldots, r$.

Let us assume now that $\theta : G \to G_1 \times \cdots \times G_s$ is an isomorphism,
where G_i is a q_i-group, and $q_1, \ldots, q_s$ are distinct primes. Since $m =$
$\prod_{i=1}^{r} p_i^{e_i}$, and also $m = \#(G) = \prod_{i=1}^{s} q_i^{f_j}$ then necessarily $s = r$, there
is a permutation π of $\{1, \ldots, s\}$ such that $q_{\pi(i)} = p_i$, $G_{\pi(i)}$ is a p_i-group,
and $\#(G_{\pi(i)}) = p_i^{e_i}$. Let $F_{\pi(i)}$ be the subgroup of G which corresponds
by the isomorphism θ to the subgroup $\{1\} \times \cdots \times G_{\pi(i)} \times \cdots \times \{1\}$;
then $F_{\pi(i)}$ is a p_i-group contained in G, hence $F_{\pi(i)} \subseteq G_{(p_i)}$ and since
$\#(F_{\pi(i)}) = \#(G_{(p_i)}) = p_i^{e_i}$, we have $F_{\pi(i)} = G_{(p_i)}$.   ∎

Now we are ready to prove the main uniqueness theorem:

Theorem 3.   *Every finite Abelian group is isomorphic to a Cartesian
product of cyclic groups with prime-power orders. Moreover, if $G \cong G_1 \times$
$\cdots \times G_r \cong G_1' \times \cdots \times G_s'$ where each G_i, G_i' is a cyclic group of prime-
power order, then $r = s$ and there exists a permutation π of $\{1, \ldots, r\}$
such that $G'_{\pi(i)} \cong G_i$.*

Proof: By Theorem 2, G is isomorphic to a Cartesian product of cyclic
groups; by Theorem 1, G is isomorphic to a Cartesian product of cyclic
groups of prime-power order.

In view of (**O**), in order to prove the uniqueness of the decomposition,
there is no loss of generality in assuming that the group itself is a p-group.
In fact, the p-groups appearing in any decomposition of G correspond to
the primes dividing $\#(G) = m$, and for every such prime p the product of
p-groups in the decomposition must be isomorphic to the subgroup $G_{(p)}$.

Thus, let G be a p-group

$$G \cong G_1 \times \cdots \times G_r \cong G_1' \times \cdots \times G_s',$$

where G_i, G_i' are cyclic groups, with generators x_i, x_i' having orders
$\#(G_i) = p^{e_i}$, $\#(G_i') = p^{f_i}$, respectively. There is no loss of generality
if we assume that

$$e_1 \geq e_2 \geq \cdots \geq e_r, \qquad f_1 \geq f_2 \geq \cdots \geq f_s.$$

Let $G^* = \{x \in G \mid x^p = 1\}$, so G^* is a subgroup of G. From

$$G \cong G_1 \times \cdots \times G_r$$

it follows that $G^* \cong G_1^* \times \cdots \times G_r^*$ where G_i^* is the subgroup of G_i generated by $x_i^{p^{e_i-1}}$. Thus $\#(G^*) = \prod_{i=1}^r \#(G_i^*) = p^r$ since each G_i^* has order p. Similarly, from $G \cong G_1' \times \cdots \times G_s'$ we deduce that $\#(G^*) = p^s$; therefore $r = s$.

Now, let us prove that $e_1 = f_1$, ..., $e_r = f_r$. It is enough to show that $e_i \geq f_i$ for every $i = 1, \ldots, r$, since $\prod_{i=1}^r p^{e_i} = \#(G) = \prod_{i=1}^r p^{f_i}$.

If j is the smallest index such that $e_j < f_j$, let $G^{**} = \{x \in G \mid$ there exists $y \in G$ such that $y^{p^{e_j}} = x\}$, then G^{**} is a subgroup of G. From $G \cong G_1 \times \cdots \times G_r$, it follows that $G^{**} \cong G_1^{**} \times \cdots \times G_{j-1}^{**}$ where each G_i^{**} is the cyclic group generated by $x_i^{p^{e_j}} (i = 1, \ldots, j-1)$. On the other hand, from $G \cong G_1' \times \cdots \times G_r'$ it follows that

$$G^{**} \cong G_1'^{**} \times \cdots \times G_{j-1}'^{**} \times G_j'^{**} \times \cdots ,$$

where each $G_i'^{**}$ is the cyclic group generated by

$$x_i'^{p^{e_j}} (i = 1, \ldots, j-1, j, \ldots)$$

and certainly $G_j'^{**}$ is not trivial. By what we have just proved, the number of cyclic p-groups in any decomposition of G^{**} must be invariant, and so we have arrived at a contradiction. Thus $e_i \geq f_i$ for every $i = 1, \ldots, r$ showing the actual equality.   ∎

Exercises

1. Show that for every natural number m there exist only finitely many numbers n such that $\varphi(n) = m$. In particular, find all integers n such that $\varphi(n) = 2$, $\varphi(n) = 3$, $\varphi(n) = 4$, $\varphi(n) = 6$.

2. Determine the positive integers such that:

(a) $\varphi(n) = n/2$.

(b) $\varphi(n) = n - 1$.

(c) $\varphi(n) = \varphi(2n)$.

(d) $\varphi(n) = \varphi(4n) = \varphi(6n)$.

(e) $\varphi(n) = 12$.

(f) $\varphi(n)$ divides n.

(g) $2\varphi(n)$ divides n.

(h) $\varphi(n) \equiv 2 \pmod{4}$.

3. Show that if $d = \gcd(m, n)$ then $\varphi(mn) = (d\varphi(m)\varphi(n))/\varphi(d)$.

4. Determine the sum of all integers m such that $1 \leq m \leq n$ and
$$\gcd(m, n) = 1.$$

5. Prove that if d divides n then $\varphi(d)$ divides $\varphi(n)$.

6. Let p, q be distinct prime numbers. Prove that
$$p^{q-1} + q^{p-1} \equiv 1 \pmod{pq}.$$

7. Let m, n be relatively prime positive integers. Prove that
$$m^{\varphi(n)} + n^{\varphi(m)} \equiv 1 \pmod{mn}.$$

8. Determine explicitly the multiplication table of the integers modulo 12; verify which residue classes are invertible and find their order in the multiplicative group $P(12)$.

9. Solve the following numerical congruences:

    (a) $3x \equiv 12 \pmod{17}$.

    (b) $4x \equiv 16 \pmod{57}$.

    (c) $5x \equiv 12 \pmod{18}$.

    (d) $20x \equiv 60 \pmod{80}$.

10. Solve the following system of congruences:
$$\begin{cases} x \equiv 1 \pmod{2} \\ x \equiv 2 \pmod{3} \\ x \equiv 3 \pmod{5}. \end{cases}$$

11. Solve the following system of congruences:
$$\begin{cases} 2x \equiv 5 \pmod{7} \\ 4x \equiv 4 \pmod{9} \\ 2x \equiv 6 \pmod{25}. \end{cases}$$

12. Solve the following system of congruences:
$$\begin{cases} x \equiv 1 \pmod{2} \\ x \equiv 2 \pmod{3} \\ x \equiv 3 \pmod{4} \\ x \equiv 4 \pmod{5}. \end{cases}$$

13. Let $\{x_0, x_1, \ldots, x_{n-1}\}$ be a complete set of residues modulo $n > 1$ and let a be an integer relatively prime to n and b any integer. Show that
$$\{ax_0 + b, ax_1 + b, \ldots, ax_{n-1} + b\}$$

is a complete set of residues modulo n.

14. Let m, n be relatively prime positive integers. Let $\{x_0, x_1, \ldots, x_{m-1}\}$, $\{y_0, y_1, \ldots, y_{n-1}\}$ be complete sets of residues modulo m and n, respectively. Show that $\{nx_i + my_j \mid i = 0, 1, \ldots, m - 1; \; j = 0, 1, \ldots, n - 1\}$ is a complete set of residues modulo mn.

15. Let p be a prime number. Show that

$$1^n + 2^n + \cdots + (p - 1)^n \equiv \begin{cases} -1 \pmod{p} & \text{when } p - 1 \text{ divides } n, \\ 0 \pmod{p} & \text{otherwise.} \end{cases}$$

16. Let p be a prime number and let n be a natural number dividing $p - 1$. Show that the congruence $x^n \equiv 1 \pmod{p}$ has exactly n roots.

17. Prove that any integer

$$a = a_0 + a_1 \cdot 10 + a_2 \cdot 10^2 + \cdots + a_n \cdot 10^n$$

is congruent modulo 9 to the sum of its digits: $a \equiv a_0 + a_1 + \cdots + a_n \pmod{9}$. Establish this as a particular case of a more general fact.

18. Prove *Wilson's theorem*: $(m - 1)! \equiv -1 \pmod{m}$ if and only if m is a prime number.

19. Show that if p is an odd prime number, then

$$\left\{ \left(\frac{p-1}{2} \right)! \right\}^2 \equiv (-1)^{(p+1)/2} \pmod{p}.$$

20. Let $f \in \mathbb{Z}[X]$ be a polynomial of degree greater than 0. Show that there exist infinitely many integers n such that $f(n)$ is not a prime number.

21. Let $f \in \mathbb{Z}[X]$ have degree n. Assume that there exists an integer m such that the prime number p divides $f(m)$, $f(m + 1)$, $f(m + 2)$, $\ldots$, $f(m + n)$. Show that p divides $f(x)$ for every integer x.

22. Determine all the generators of the cyclic groups $P(17)$, $P(31)$, and $P(27)$.

23. Let q be equal to 2, 4, p^e, or $2p^e$ (where p is an odd prime), and let a be an integer relatively prime to q. Show that the congruence $x^n \equiv a \pmod{q}$ has a solution if and only if $a^{\varphi(q)/d} \equiv 1 \pmod{q}$, where $d = \gcd(n, \varphi(q))$.

24. Let p be a prime number and let a be a positive integer such that p does not divide a. Show that if $a^p \equiv a \pmod{p^2}$, then $(a + p)^p \not\equiv a + p \pmod{p^2}$.

25. Show that there exist primitive roots r modulo p such that

$$r^{p-1} \not\equiv 1 \pmod{p^2}.$$

26. Let $p \neq 2$ be a prime number and let r be a primitive root modulo p such that $r^{p-1} \not\equiv 1 \pmod{p^2}$. Show that $r \bmod p^m$ generates $P(p^m)$ for every $m > 1$.

27. Let $a, b \geq 3$ be relatively prime integers and let b be odd. Show that $P(ab)$ is not a cyclic group.

28. Let p be a prime number and let e be a positive integer dividing $p-1$. Show that there are exactly $\varphi(e)$ residue classes modulo p having order e in the multiplicative group $P(p)$.

29. Let p be a prime number and let n, r be any natural numbers. Show that

$$\binom{p^r n}{p^r} \equiv n \pmod{pn}$$

($\binom{a}{b}$ denotes the number of combinations of a letters in groups of b letters).

Hint: Show that

$$\binom{p^r n - 1}{p^r - 1} = \prod_{k=1}^{p^r-1} \left(\frac{p^r n}{k} - 1\right) \equiv (-1)^{p^r-1} \pmod{p}.$$

30. Let r be a given primitive root modulo p. If m is an integer, $1 \leq m \leq p-1$, and $m \equiv r^a \pmod{p}$, where $0 \leq a < p-1$, we say that a is the *index* of m with respect to r (modulo p), and we write $a = \mathrm{ind}_r(m)$. Show that $\mathrm{ind}_r(mn) \equiv \mathrm{ind}_r(m) + \mathrm{ind}_r(n) \pmod{p-1}$.

31. Let r be the smallest primitive root modulo 29; determine the indices of the prime residue classes modulo 29, with respect to r.

32. Using the table of indices, compute the least positive residue modulo 17 of the following integers:

(a) $a = 432 \times 8328$.
(b) $b = 38^{919}$.
(c) $c = ((3^3)^3)^3$.

33. Using the table of indices, compute the least positive residue modulo 29 of the following integers:

(a) $a = 583 \times 1875$.
(b) $b = 1051^{875}$.
(c) $c = (5^5)^5$.

34. Using the table of indices, solve the congruence

$$25x \equiv 15 \pmod{29}.$$

35. Let $a_1, \ldots, a_r$ be pairwise relatively prime positive integers, $a = \prod_{i=1}^{r} a_i$, $t = \operatorname{lcm}\{\varphi(a_1), \ldots, \varphi(a_r)\}$. Show that if $b \in P(a)$ then $b^t \equiv 1 \pmod{a}$.

36. Compute the four last digits of:

(a) $((9^9)^9)^9$.

(b) $9^{(9^{(9^9)})}$.

Hint: For (b) use the previous exercise.

37. Let p be a prime number, $r \geq 2$ an integer, $f \in \mathbb{Z}[X]$ and assume that $f(a) \equiv 0 \pmod{p^{r-1}}$, where $0 \leq a \leq p^{r-1} - 1$. Let $f' \in \mathbb{Z}[X]$ be the derivative of f. Show that:

(a) If $f'(a) \not\equiv 0 \pmod{p}$ then there exists a unique integer b such that $b \equiv a \pmod{p^{r-1}}$, $f(b) \equiv 0 \pmod{p^r}$, and $0 \leq b \leq p^r - 1$.

(b) If $f'(a) \equiv 0 \pmod{p}$ then there are p integers $b_0, \ldots, b_{p-1}$ such that $f(b_i) \equiv 0 \pmod{p^r}$, $0 \leq b_i \leq p^r - 1$, when $f(a) \equiv 0 \pmod{p^r}$ and these integers satisfy $b_i \equiv a \pmod{p^{r-1}}$; on the other hand, if $f(a) \not\equiv 0 \pmod{p^r}$ there exists no integer b, $b \equiv a \pmod{p^{r-1}}$ such that $f(b) \equiv 0 \pmod{p^r}$.

38. Solve the congruences:

(a) $x^3 + 5x - 8 \equiv 0 \pmod{5^2}$.

(b) $x^3 - 3x + 2 \equiv 0 \pmod{245}$.

39. Find the decomposition into irreducible factors of the following polynomials:

(a) $X^4 + 2X - 3 \in \mathbb{F}_5[X]$.

(b) $X^4 + 3X^3 - 2X^2 - 9X - 3 \in \mathbb{F}_{11}[X]$.

(c) $X^4 - 3X^3 - 4X^2 + X - 4 \in \mathbb{F}_7[X]$.

40. Any mapping f from the set of positive integers with values in a domain R is called an *arithmetical function*. Moreover, if $f(mn) = f(m) \cdot f(n)$ when m, n are relatively prime, then f is said to be a *multiplicative function*. Show that if f is multiplicative, if g is defined by $g(n) = \sum_{d|n} f(d)$, then g is also multiplicative.

41. Let $\sigma(n)$ denote the sum of positive divisors of $n \geq 1$, and let $\tau(n)$ denote the number of positive divisors of n. Show that σ, τ are multiplicative functions.

    Hint: Use the preceding exercise.

42. If $n = \prod_{i=1}^{r} p_i^{e_i}$ is the prime decomposition of n, show that

$$\sigma(n) = \prod_{i=1}^{r} \frac{p_i^{e_i+1} - 1}{p_i - 1},$$

$$\tau(n) = \prod_{i=1}^{r}(e_i + 1).$$

43. Show that $\sum_{d|n}[\tau(d)]^3 = [\sum_{d|n} \tau(d)]^2$.

44. Let R be a domain and let A be the set of all arithmetical functions with values in R; that is, the set of sequences $s = (s_1, s_2, \ldots, s_n, \ldots)$ of elements of R. On the set A we define the following operations:

$$(s_n) + (s'_n) = (s_n + s'_n),$$

and

$$(s_n) * (s'_n) = (t_n) \qquad \text{where} \quad t_n = \sum_{dd'=n} s_d s'_{d'}.$$

Besides, if $r \in R$, $(s_n) \in A$, we define a scalar multiplication as

$$r(s_n) = (rs_n).$$

    Show:

    (a) A is a commutative ring and also an R-module; the zero element is $0 = (0, 0, \ldots)$ and the unit element is $e = (1, 0, 0, \ldots)$.

    (b) For every $s = (s_n)$, $s \neq 0$, let $\pi(s)$ be the smallest integer n such that $s_n \neq 0$; let $\pi(0) = \infty$, where $\infty \cdot n = \infty \cdot \infty = \infty$ for every integer $n > 0$. Show that $\pi(s * s') = \pi(s) \cdot \pi(s')$, hence A is a domain.

    (c) $s \in A$ is invertible in A if and only if $\pi(s) = 1$ and s_1 is invertible in the ring R.

45. Let $u = (1, 1, 1, \ldots) \in A$ and let $\mu \in A$ be its inverse; that is, $u * \mu = e$. The arithmetical function μ is called the *Möbius function*. Prove that $\mu(1) = 1$, $\mu(n) = 0$ when some square greater than 1 divides n, and $\mu(n) = (-1)^r$ when

$$n = p_1 p_2 \ldots p_r$$

(product of r distinct primes); deduce that μ is a multiplicative function.

46. Show the Möbius inversion formula: If $s, s' \in A$ then $u * s = s'$ if and only if $s = \mu * s'$. As an application, prove the following relations:

(a) $\sum_{d|n} \mu(d) = \begin{cases} 1 & \text{when } n = 1, \\ 0 & \text{when } n \neq 1. \end{cases}$

(b) $\varphi(n) = \sum_{d|n} \mu\left(\dfrac{n}{d}\right) \cdot d.$

(c) $n = \sum_{d|n} \sigma(d)\mu\left(\dfrac{n}{d}\right).$

(d) $1 = \sum_{d|n} \tau(d)\mu\left(\dfrac{n}{d}\right).$

47. Prove:

(a) $\tau(n)$ is odd if and only if n is a square.

(b) $\tau(2^n - 1) \geq \tau(n)$ when $n \geq 1$.

(c) $\tau(2^n + 1)$ is at least equal to the number of odd positive divisors of n.

48. Show that if $d = \gcd(m, n)$ then

$$\sigma(m)\sigma(n) = \sum_{t|d} t\sigma\left(\frac{mn}{t^2}\right).$$

Hint: First consider the case where m, n are powers of the same prime number.

49. Prove that if $m > 0$, $n > 1$, then

$$\frac{\sigma(m)}{m} < \frac{\sigma(mn)}{mn} \leq \frac{\sigma(m)\sigma(n)}{mn}.$$

50. Prove that n is equal to the product of its proper positive divisors if and only if $n = p^3$ or $n = p_1 p_2$, $p_1 \neq p_2$ (where p, p_1, p_2 are prime numbers).

51. Let Λ be the *von Mangoldt arithmetical function* defined as follows:

$\begin{cases} \Lambda(n) = \log p & \text{when } n \text{ is a power of some prime number } p, \\ \Lambda(n) = 0 & \text{otherwise.} \end{cases}$

Prove:

(a) $\log n = \sum_{d|n} \Lambda(d).$

(b) $\Lambda(n) = \sum_{d|n} \mu(d) \log\left(\frac{n}{d}\right) = -\sum_{d|n} \mu(d) \log d.$

52. Let λ be the *Liouville arithmetical function*, defined as follows:
$\lambda(1) = 1$, and if $n = \prod_{i=1}^{r} p_i^{e_i}$, each p_i prime number, $e_i \geq 1$, set

$$\lambda(n) = \begin{cases} 1 & \text{when } \sum_{i=1}^{r} e_i \text{ is even,} \\ -1 & \text{when } \sum_{i=1}^{r} e_i \text{ is odd.} \end{cases}$$

Prove:

(a) $\lambda(mn) = \lambda(m)\lambda(n)$ for any positive integers m, n.

(b) $\sum_{d|n} \lambda(d) = \begin{cases} 1 & \text{when } n \text{ is a square,} \\ 0 & \text{otherwise.} \end{cases}$

53. For every real number x let $[x]$ denote the unique integer such that

$$[x] \leq x < [x] + 1.$$

Prove:

(a) $[x] + [y] \leq [x + y]$.

(b) $[x/n] = [[x]/n]$ for every positive integer n.

(c) The number of multiples of n which do not exceed x is $[x/n]$.

(d) $[x] + [y] + [x + y] \leq [2x] + [2y]$.

54. Let f, g be arithmetical functions such that $f(n) = \sum_{d|n} g(d)$. Show that

$$\sum_{m=1}^{n} f(m) = \sum_{m=1}^{n} [n/m]g(m).$$

55. Prove that

$$\sum_{m=1}^{n} \left[\frac{n}{m}\right] \varphi(m) = n(n+1)/2.$$

56. Prove that if $n \geq 1$ then $\sum_{m=1}^{n} \tau(m) = \sum_{m=1}^{n} [n/m]$.

57. Let $n \geq 1$ and $k = [\sqrt{n}\,]$. Show that

$$\sum_{m=1}^{n} \tau(m) = 2\left(\sum_{m=1}^{k} [n/m]\right) - k^2.$$

58. The numbers $F_n = 2^{2^n} + 1$ ($n \geq 0$ integer) are called *Fermat numbers*.

Prove:

(a) If $r > 0$ then F_n divides $F_{n+r} - 2$.

(b) Any two Fermat numbers are relatively prime.

(c) If $a \geq 2$ and $a^r + 1$ is prime then a is even and r is a power of 2.

(d) 641 divides F_5 (Euler).

We note in this respect that the only Fermat numbers which are known to be prime are F_0, F_1, F_2, F_3, F_4. On the other hand, it has been shown that F_n is not prime when $5 \leq n \leq 23$ and for many other values of n. The largest known composite Fermat number is F_n with $n = 23471$ (see Ribenboim [25, Chapter 2, Section VI]).

It is an open question whether the number of Fermat primes is finite.

59. A natural number n is said to be *perfect* when n is equal to the sum of its proper divisors. Prove: n is an even perfect number if and only if

$$n = 2^{p-1}(2^p - 1),$$

where p, $2^p - 1$ are primes. Give examples of even perfect numbers.

Note: It is not known whether there exists any odd perfect number. Any such number must have at least eight distinct prime factors and be greater than 10^{300}. Any prime q of the type $q = 2^p - 1$ (where p is a prime) is called a *Mersenne prime*. The known Mersenne primes correspond to $p = 2, 3, 5, \ldots, 19\,937, 21\,701, 23\,209, 44\,497, 86\,243, 110\,503, 132\,049, 216\,091, 756\,839, 859\,433, 1\,257\,787, 1\,398\,269, 2\,976\,221, 3\,021\,377, 6\,972\,593$. It is not yet known whether there are infinitely many Mersenne primes; equivalently, it is not known whether there are infinitely many even perfect numbers (see Ribenboim [25, Chapter 2, Section VII]).

60. Show that if n is odd and has at most two distinct prime factors, then $\sigma(n) < 2n$, hence n is not a perfect number.

61. Let n be an odd perfect number. Prove that $n = p^r m^2$, where p is a prime number not dividing m and $p \equiv 1 \pmod 4$. On the other hand, given m, there is at most one odd perfect number of the type $p^r m^2$, with p prime not dividing m.

62. Prove that the nth cyclotomic polynomial is expressible directly by means of the Möbius function as follows:

$$\Phi_n = \prod_{d|n} (X^d - 1)^{\mu(n/d)}.$$

63. Let f, g, h be arithmetical functions, $f * g = h$, and assume that h is multiplicative. As an application deduce anew that if $g(n) = \sum_{d|n} f(d)$ and g is multiplicative then so is f.

64. If R is a unique factorization domain, show that every arithmetical function with values in R is the product of a finite number of "prime" arithmetical functions.

65. Show that the ring of arithmetical functions with values in the domain R is isomorphic to the ring of *unrestricted formal power series* $S = R[[[X_1, \ldots, X_n, \ldots]]]$. Explicitly, S consists of all countable infinite formal sums of monomials in the variables X_i with coefficients in R. The addition in S is componentwise, while the multiplication follows the same pattern as for polynomials (note that for every monomial $m \in S$ there exist only finitely many monomials $m', m'' \in S$ such that $m'm'' = m$).

   Note: Cashwell and Everett* established that if R is a field then the ring of arithmetical functions with values in this field is a unique factorization domain.

66. Let G be an Abelian group of order m and assume that for every prime p dividing m, G has exactly $p - 1$ elements of order p. Show that G is a cyclic group.

67. Show that a finite Abelian group G is cyclic if and only if

$$G \cong \mathbb{Z}/p_1^{e_1} \times \cdots \times \mathbb{Z}/p_r^{e_r},$$

where $p_1, \ldots, p_r$ are distinct prime numbers.

68. We say that a finite Abelian group G is *indecomposable* if it is not possible to write $G \cong G_1 \times G_2$, where G_1, G_2 are Abelian groups of strictly smaller order. Show that G is indecomposable if and only if it is cyclic of prime-power order.

69. Show that the number of pairwise nonisomorphic Abelian groups of order $m = \prod_{i=1}^r p_i^{e_i}$ is $\prod_{i=1}^r \pi(e_i)$, where we define $\pi(e)$ as follows: it is the number of nonincreasing sequences of integers $n_1 \geq n_2 \geq \cdots \geq n_j > 0$ such that $\sum_{i=1}^j n_i = e$.

70. Determine the number of pairwise nonisomorphic Abelian groups of order 8, 16 200.

71. Let $G \cong \mathbb{Z}/p^{e_1} \times \mathbb{Z}/p^{e_2} \times \cdots \times \mathbb{Z}/p^{e_k}$ where $e_1 \geq e_2 \geq \cdots \geq e_k > 0$. Let $G_r = \{x \in G \mid p^r x = 0\}$. Show that G_r is a subgroup of G having order

$$p^{ri + e_{i+1} + \cdots + e_k}$$

where i is the unique index such that $e_i \geq r > e_{i+1}$ (with the convention that $e_0 = \sum_{j=1}^k e_j$, $e_{k+1} = 0$).

---

* The Ring of Number Theoretic Functions, *Pacific J. Math.*, **9**, 1959, 975–985.

72. Let G be as in the previous exercise. Show that the number of elements of order p^r in G is equal to

$$p^{ri+e_{i+1}+\cdots+e_k} - p^{(r-1)j+e_{j+1}+\cdots+e_k},$$

where i, j are such that $e_i \geq r > e_{i+1}$, $e_j \geq r - 1 > e_{j+1}$ (with the same convention as in the previous exercise). As a corollary, show that G has $p^k - 1$ elements of order p; thus G is cyclic if and only if it has $p - 1$ elements of order p.

73. Let G be an Abelian group such that every nonzero element has order p. Show that G is a vector space over the field $\mathbb{F}_p$. If G is finite, show that $G \cong \mathbb{Z}/p \times \cdots \times \mathbb{Z}/p$ (for a finite number of copies of $\mathbb{Z}/p$). Such Abelian groups are called *elementary Abelian p-groups*.

74. Show that if G is an elementary Abelian p-group of order p^n and $1 \leq r \leq n$ then the number of distinct subgroups of order p^r of G is

$$\frac{(p^n - 1)(p^n - p)(p^n - p^2), \dots, (p^n - p^{r-1})}{(p^r - 1)(p^r - p)(p^r - p^2), \dots, (p^r - p^{r-1})} .$$

In particular, there are $1 + p + p^2 + \cdots + p^{n-1}$ subgroups of order p^{n-1} or also of order p. Moreover, the number of subgroups of order p^r is the same as the number of subgroups of order p^{n-r}.

4

Quadratic Residues

4.1 The Legendre Symbol and Gauss' Reciprocity Law

In this chapter we investigate the following question. Let $m > 1$ and let a be an integer relatively prime to m. When is the residue class $\bar{a}$ a square in the multiplicative group $P(m)$? In other words, when does there exist an integer x such that $x^2 \equiv a \pmod{m}$?

Definition 1. If $m > 1$ and a are integers, and $\gcd(a, m) = 1$, we say that a is a *quadratic residue modulo* m when $\bar{a}$ is a square in $P(m)$. Otherwise, we say that a is a *quadratic nonresidue modulo* m.

The first results will reduce the problem to that of finding the quadratic residues modulo an odd prime p or 4 or 8.

A. *If $m = p_1^{e_1} \ldots p_r^{e_r}$ is the prime-power decomposition of m, if a is an integer relatively prime to m, then a is a quadratic residue modulo m if and only if it is a quadratic residue modulo $p_i^{e_i}$ for all $i = 1, \ldots, r$.*

Proof: By Chapter 3, **(F)**, $P(m) \cong \prod_{i=1}^{r} P(p_i^{e_i})$.

An element of a Cartesian product of groups is a square if and only if its components are squares. If $\bar{a} \in P(m)$, its component in the group $P(p_i^{e_i})$, by the above isomorphism, is the residue class of a modulo $p_i^{e_i}$ for all $i = 1, \ldots, r$. ∎

The study of squares in $P(p^e)$ will now be reduced to that of squares in $P(p)$, when $p \neq 2$:

B. *Let p be an odd prime, $e > 1$, and let w_0 be a primitive root modulo p. If a is an integer prime to p, the following conditions are equivalent:*

(1) *a is a quadratic residue modulo p^e.*

(2) *a is a quadratic residue modulo p.*

(3) *$a \equiv w_0^t \pmod{p}$ where t is even.*

Proof: $(1) \rightarrow (2)$ Let x be an integer such that $a \equiv x^2 \pmod{p^e}$. Then $a \equiv x^2 \pmod{p}$.

(2) $\to$ (3) From $a \equiv x^2$ (mod p) and $a \equiv w_0^t$ (mod p), $x \equiv w_0^u$ (mod p) it follows that $t \equiv 2u$ (mod $p - 1$) because the group $P(p)$ has order $p - 1$. Thus t is even.

(3) $\to$ (1) By Chapter 3, (**J**), $P(p^e)$ is the cyclic group generated by $\overline{w(1 + p)}$ where $w = w_0^{p^{e-1}}$. Hence $a \equiv w^s(1 + p)^s$ (mod p^e) for some s, $1 \le s < p^e$. Since $w_0^{p^{e-1}} \equiv w_0$ (mod p) and $(1 + p)^s \equiv 1$ (mod p) then $a \equiv w_0^s$ (mod p). By hypothesis $a \equiv w_0^t$ (mod p), hence $s \equiv t$ (mod $p - 1$) and from t we even deduce that s is even, say, $s = 2u$. We conclude that $a \equiv [w^u(1 + p)^u]^2$ (mod p^e).   ∎

For $p = 2$, we have:

C.   *Let a be an odd integer. Then:*

(1) *a is a quadratic residue modulo 4 if and only if $a \equiv 1$ (mod 4).*

(2) *a is a quadratic residue modulo 8 if and only if $a \equiv 1$ (mod 8).*

(3) *a is a quadratic residue modulo 2^e (where $e > 3$) if and only if a is a quadratic residue modulo 8.*

Proof: Since $P(4) = \{\overline{1}, \overline{3}\}$, $P(8) = \{\overline{1}, \overline{3}, \overline{5}, \overline{7}\}$, then the only square in $P(4)$, $P(8)$ is the residue class of 1.

Let $e > 3$. If a is an integer, by Chapter 3, (**K**), we may write

$$a \equiv (-1)^{e'} 5^{e''} \pmod{2^e},$$

where $e' \in \{0, 1\}$, $0 \le e'' < 2^{e-2}$.

If x is an integer such that $x^2 \equiv a$ (mod 2^e), letting $x \equiv (-1)^{f'} 5^{f''}$ (mod 2^e), where $f' \in \{0, 1\}$, $0 \le f'' < 2^{e-2}$, it follows that $2f' \equiv e'$ (mod 2) and $2f'' \equiv e''$ (mod 2^{e-2}). These congruences have a solution if and only if e', e'' are even, that is, $a \equiv 5^{e''}$ (mod 2^e), where e'' is even. This is equivalent to $a \equiv 1$ (mod 8).   ∎

Let us note that, for $e > 3$, in $P(2^e)$ there are exactly 2^{e-3} squares, hence $2^{e-1} - 2^{e-3} = 3 \cdot 2^{e-3}$ nonsquares.

Putting together these results, we have:

D.   *Let $m > 1$ and a be relatively prime integers. Then a is a quadratic residue modulo m if and only if:*

(1) *a is a quadratic residue modulo p, for every odd prime p dividing m;*

(2) *$a \equiv 1$ (mod 4) if 4 divides m but 8 does not divide m;*

(3) *$a \equiv 1$ (mod 8) if 8 divides m.*

Proof: This results immediately from (**A**), (**B**), and (**C**).   ∎

In order to determine the quadratic residues modulo p, we introduce the following terminology:

Definition 2.   Let p be an odd prime and let a be a nonzero integer not a multiple of p. We define the *Legendre symbol* $\left(\dfrac{a}{p}\right)$ of a, relative to p, as follows:

$$\left(\frac{a}{p}\right) = \begin{cases} 1 & \text{when } a \text{ is a quadratic residue modulo } p, \\ -1 & \text{when } a \text{ is a quadratic nonresidue modulo } p. \end{cases}$$

For typographical reasons, we also use the notation $\left(\dfrac{a}{p}\right) = (a/p)$.

E.   *The Legendre symbol has the following properties:*

   (1) *if $a \equiv b \pmod{p}$ then $(a/p) = (b/p)$.*
   (2) $(ab/p) = (a/p)(b/p)$.

Proof: The first assertion is immediate.

Let w be a primitive root modulo p. If a, b are integers, not multiples of p, we may write $a \equiv w^r \pmod{p}$, $b \equiv w^s \pmod{p}$ where $0 \leq r,\ s < p-1$. By **(B)**, we have $(a/p) = 1$ if and only if r is even, and similarly $(b/p) = 1$ when s is even. Since $ab \equiv w^t \pmod{p}$, with $t \equiv r + s \pmod{p-1}$, then t is even if and only if r, s have the same parity. This proves the second assertion.   ∎

F.   *For every odd prime p, there are as many quadratic residue classes as there are quadratic nonresidue classes modulo p^e where $e \geq 1$.*

Proof: First we assume $e = 1$. Consider the mapping $\sigma : P(p) \to P(p)^2$, defined by $\sigma(\overline{x}) = \overline{x}^2$. Then $\sigma(\overline{x}) = \sigma(\overline{y})$ if and only if $\overline{x} = \overline{y}$ or $\overline{x} = -\overline{y}$, because $P(p) = \mathbb{F}_p^{\cdot}$. Since $p \neq 2$ this shows that $P(p)$ has twice as many elements as $P(p)^2$, so there are as many square residues as nonresidues modulo p.

If $e > 1$ we consider the group homomorphism $f : P(p^e) \to P(p)$, defined by $f(\overline{a}) = \overline{\overline{a}}$, where $\overline{a}$ is the residue class of a modulo p^e and $\overline{\overline{a}}$ is the residue class of a modulo p. By **(B)**, $\overline{a}$ is a square if and only if $\overline{\overline{a}}$ is a square. Therefore $P(p^e)$ has as many squares as nonsquares.   ∎

Let us note that the above result does not hold for $p = 2$ (as we have already remarked) as well as for products of different primes (for example, when $m = 15$ there are only two quadratic residue classes and six quadratic nonresidue classes modulo 15).

We may find whether an integer is a quadratic residue modulo p by explicit determination of the multiplication in the group $P(p)$ or by first determining a primitive root modulo p. For large primes this may be rather involved. We shall be interested in simpler methods.

G.   (Euler's Criterion). *Let p be an odd prime and let a be an integer not a multiple of p. Then*

$$(a/p) \equiv a^{(p-1)/2} \pmod{p}.$$

Proof: Let $a \equiv w^t \pmod{p}$, where w is a primitive root modulo p and $0 \leq t < p - 1$. Since $\bar{w}$ is not a square in $P(p)$ (as follows from **(B)**) and $w^{(p-1)/2} \equiv -1 \pmod{p}$, we have

$$\left(\frac{a}{p}\right) = \left(\frac{w^t}{p}\right) = \left(\frac{w}{p}\right)^t = (-1)^t \equiv (w^{(p-1)/2})^t = (w^t)^{(p-1)/2}$$

$$\equiv a^{(p-1)/2} \pmod{p}. \qquad \blacksquare$$

H. -1 *is a square modulo p if and only if $p \equiv 1 \pmod 4$.*

Proof: $(-1/p) \equiv (-1)^{(p-1)/2} \pmod{p}$ implies the equality $(-1/p) = (-1)^{(p-1)/2}$ (since these integers are either 1 or -1). So $(-1/p) = 1$ exactly when $p \equiv 1 \pmod 4$. $\qquad \blacksquare$

If p is large Euler's criterion is not convenient, since it gives rise to lengthy computations.

A better criterion is due to Gauss. If p does not divide a there exists a unique integer s, $1 \leq s \leq (p-1)/2$, such that $a \equiv s \pmod{p}$ or $a \equiv -s \pmod{p}$.

I.   (Gauss' Criterion). *Let p be an odd prime, let a be an integer, not a multiple of p, and let ν be the number of elements ka in the set*

$$\left\{ a, 2a, \ldots, \frac{p-1}{2}\, a \right\}$$

such that $ka \equiv -s \pmod{p}$ where $0 < s \leq (p-1)/2$. Then $(a/p) = (-1)^\nu$.

Proof: If $1 \leq k < k' \leq (p-1)/2$ then $ka \not\equiv k'a \pmod{p}$, otherwise p would divide $k - k'$. Also, $ka \not\equiv -k'a \pmod{p}$, because p does not divide $k + k'$.

Thus, all integers $s = 1, \ldots, (p-1)/2$ are such that s or $-s$ is congruent modulo p to some multiple ka $(1 \leq k \leq (p-1)/2)$. Taking into account the definition of ν we have

$$a \cdot 2a \cdots \cdots \frac{p-1}{2}\, a \equiv (-1)^\nu \cdot 1 \cdot 2 \cdots \cdots \frac{p-1}{2} \pmod{p}.$$

We deduce that $a^{(p-1)/2} \equiv (-1)^\nu \pmod{p}$. By **(G)**, we conclude that $(a/p) = (-1)^\nu$. $\qquad \blacksquare$

We can use this criterion to determine when 2 is a quadratic residue:

J.   *2 is a quadratic residue modulo p if and only if $p \equiv \pm 1$ (mod 8); explicitly $(2/p) = (-1)^{(p^2-1)/8}$.*

Proof: We apply Gauss' criterion. Among the integers 2, 4, ..., $p-1$, those satisfying $p/2 < 2k \leq p-1$ are the ones such that $2k \equiv -s$ (mod p), with $0 < s \leq (p-1)/2$. Their number ν is equal to the number of integers k, such that $p/4 < k \leq (p-1)/2$. If $p \equiv 1$ (mod 4) there are $(p-1)/2 - (p+3)/4 + 1 = (p-1)/4$ such integers. If $p \equiv -1$ (mod 4), there are $(p-1)/2 - (p+1)/4 + 1 = (p+1)/4$ such integers.

Now, if $p \equiv 1$ (mod 8) then $(p-1)/4$ is even, if $p \equiv 5$ (mod 8) then $(p-1)/4$ is odd, if $p \equiv 3$ (mod 8) then $(p+1)/4$ is odd, and if $p \equiv 7$ (mod 8) then $(p+1)/4$ is even. In conclusion, $(2/p) = 1$ if and only if $p \equiv \pm 1$ (mod 8). The last expression is obvious, if we note that $(p^2-1)/8$ is even exactly when $p \equiv \pm 1$ (mod 8).   ∎

We shall indicate now a relationship between Legendre symbols relative to different primes; this will be used later as the basis for a very satisfactory method of computation of the Legendre symbol.

Gauss gave several proofs of this theorem which is a special case of the profound reciprocity law in class field theory.

Theorem 1 (Gauss' Quadratic Reciprocity Law).   *If p, q are distinct odd primes, then*

$$(p/q)(q/p) = (-1)^{\frac{p-1}{2} \cdot \frac{q-1}{2}}.$$

Proof: By (**I**), we have $(q/p) = (-1)^{\nu}$ where ν is the number of integers x, $1 \leq x \leq (p-1)/2$, such that $qx = py + r$ where $-p/2 < r < 0$, and y is an integer.

We must have $1 \leq y \leq (q-1)/2$, because y is neither 0 nor negative and

$$py = xq - r < \frac{p-1}{2}q + \frac{p}{2} < \frac{p}{2}(q+1),$$

hence $y < (q+1)/2$, so $y \leq (q-1)/2$.

Similarly $(p/q) = (-1)^{\mu}$, where μ is the number of integers y, $1 \leq y \leq (q-1)/2$, such that $py = qx + s$ where $-q/2 < s < 0$ and x is an integer; again, we have $1 \leq x \leq (p-1)/2$.

Therefore $(q/p)(p/q) = (-1)^{\nu+\mu}$.

We observe that $\nu + \mu$ is the number of pairs of integers (x, y) such that $1 \leq x \leq (p-1)/2$, $1 \leq y \leq (q-1)/2$, and $-p/2 < qx - py < q/2$.

Let us consider the following sets of pairs of integers:

$$S = \{(x,y) \mid 1 \leq x \leq (p-1)/2, \ 1 \leq y \leq (q-1)/2\},$$

$$S_1 = \{(x,y) \in S \mid qx - py \leq -p/2\},$$

$$S_0 = \{(x,y) \in S \mid -p/2 < qx - py < q/2\},$$

$$S_1' = \{(x,y) \in S \mid q/2 \leq qx - py\}.$$

The mapping $\theta : S \to S$, defined by $\theta(x, y) = (x', y')$ where $x' = (p+1)/2 - x$, $y' = (q+1)/2 - y$ has the following properties (which are easy to verify): θ is a one-to-one mapping from S onto S, θ^2 is the identity mapping, $\theta(S_1) = S_1'$, $\theta(S_1') = S_1$, so that $\theta(S_0) = S_0$.

Therefore, $\#(S) = \#(S_1) + \#(S_0) + \#(S_1') \equiv \#(S_0) \pmod{2}$ so

$$\frac{p-1}{2} \cdot \frac{q-1}{2} \equiv \nu + \mu \pmod{2}.$$

Thus,

$$(p/q) \cdot (q/p) = (-1)^{\frac{p-1}{2} \cdot \frac{q-1}{2}}. \qquad \blacksquare$$

We may also rewrite the above relation as follows:

$$(p/q) = (q/p)(-1)^{\frac{p-1}{2} \cdot \frac{q-1}{2}}.$$

This form is obtained by noting that $(q/p)^2 = 1$.

The following corollary is immediate: if p, q are distinct odd primes and p or q is congruent to 1 modulo 4 then $(p/q) = (q/p)$. Otherwise, $(p/q) = -(q/p)$.

We have now a very effective method of computation of the Legendre symbol (a/p), where $p \neq 2$. Indeed, if a is a nonzero integer not a multiple of p, $a = (-1)^d 2^e \prod_{i=1}^{r} q_i^{e_i}$ where $d \in \{0, 1\}$, $e \geq 0$, $e_i \geq 1$, and each q_i is an odd prime distinct from p, then by (E) we have

$$(a/p) = (-1/p)^d (2/p)^e \prod_{i=1}^{r} (q_i/p)^{e_i}.$$

Using (H), (J), we have only to compute (q_i/p); this may be done by application of (E) and successive reductions using Gauss' reciprocity law.

We illustrate the method by means of a numerical example. Let $p = 2311$, $a = 1965 = 3 \times 5 \times 131$. Then

$$\left(\frac{1965}{2311}\right) = \left(\frac{3}{2311}\right)\left(\frac{5}{2311}\right)\left(\frac{131}{2311}\right),$$

$$\left(\frac{3}{2311}\right) = \left(\frac{2311}{3}\right)(-1)^{1 \times 1155} = -\left(\frac{2311}{3}\right) = -\left(\frac{1}{3}\right) = -1,$$

where we used Gauss' reciprocity law and the congruence $2311 \equiv 1 \pmod 3$:

$$\left(\frac{5}{2311}\right) = \left(\frac{2311}{5}\right)(-1)^{2 \times 1155} = \left(\frac{2311}{5}\right) = \left(\frac{1}{5}\right) = 1.$$

by using the reciprocity law and the congruence $2311 \equiv 1 \pmod 5$:

$$\left(\frac{131}{2311}\right) = \left(\frac{2311}{131}\right)(-1)^{65 \times 1155} = -\left(\frac{2311}{131}\right)$$

$$= -\left(\frac{84}{131}\right) = -\left(\frac{2}{131}\right)^2 \cdot \left(\frac{3}{131}\right) \cdot \left(\frac{7}{131}\right)$$

$$= -\left(\frac{3}{131}\right) \cdot \left(\frac{7}{131}\right) = -\left(\frac{131}{3}\right)\left(\frac{131}{7}\right)(-1)^{65 \times 1}(-1)^{65 \times 3}$$

$$= -\left(\frac{131}{3}\right)\left(\frac{131}{7}\right) = -\left(\frac{2}{3}\right)\left(\frac{5}{7}\right)$$

$$= -(-1)\left(\frac{7}{5}\right)(-1)^{2 \times 3} = \left(\frac{7}{5}\right) = \left(\frac{2}{5}\right) = -1$$

using several times the reciprocity law, the congruences $2311 \equiv 84$ (mod 131), $131 \equiv 2 \pmod 3$, $131 \equiv 5 \pmod 7$, $7 \equiv 2 \pmod 5$, and the fact that 3 and 5 are not congruent to ± 1 modulo 8. Therefore $(1965/2311) = 1$, that is, 1965 is a square modulo 2311.

Recapitulating our results, we arrive at the following interesting observation. The fact that -1 is a quadratic residue modulo p depends only on the residue class of p modulo 4. For 2, it depends on the residue class of p modulo 8. Finally, for an odd prime q, $q \neq p$, it depends only on the residue class of p modulo $4q$.

This is easily seen, for if p' is a prime of the form $p' = p + 4kq$ (k integer) then

$$\left(\frac{q}{p'}\right) = \left(\frac{p'}{q}\right)(-1)^{\frac{q-1}{2} \cdot \frac{p+4kq-1}{2}}$$

$$= \left(\frac{p}{q}\right)(-1)^{\frac{p-1}{2} \cdot \frac{q-1}{2}}(-1)^{\frac{q-1}{2} \cdot 2kq}$$

$$= \left(\frac{p}{q}\right)(-1)^{\frac{p-1}{2} \cdot \frac{q-1}{2}} = \left(\frac{q}{p}\right).$$

We may now consider the following inverse problem. Given -1, or 2, or an odd prime q, how many odd primes p does there exist such that -1, 2, or q is a quadratic residue modulo p?

We have:

$(-1/p) = 1$ if and only if p is a prime in the arithmetic progression $\{1, 5, 9, \ldots, 4n + 1, \ldots\}$.

$(2/p) = 1$ if and only if p is a prime in one of the arithmetic progressions $\{1, 9, 17, \ldots, 8n + 1, \ldots\}$ or $\{7, 15, \ldots, 8n - 1, \ldots\}$.

Similarly, let p_0 be an odd prime dividing $q - 1$ or $q - 4$ (if $q - 1$ is a power of 2) then $(q/p_0) = (1/p_0) = 1$, or $(q/p_0) = (4/p_0) = 1$; therefore there exists at least one prime p_0 such that q is a quadratic residue modulo p_0; then for every prime p in the arithmetic progression $\{p_0, p_0 + 4q, p_0 + 8q, \ldots, p_0 + 4nq, \ldots\}$ we have $(q/p) = 1$.

Dirichlet's theorem on arithmetic progressions states:

In any arithmetical progression $\{k, k+m, k+2m, \ldots, k+nm, \ldots\}$ where $0 < k \le m$, and k, m are relatively prime, there exist infinitely many prime numbers.

We shall give a proof of this remarkable theorem in Chapter 20.

Applying this theorem, we obtain at once the answer to the problem considered above:

K.   *For each of the numbers -1, 2, or any prime $q \ne 2$, there exist infinitely many primes p such that -1, 2, or q is a quadratic residue modulo p.*

For the case of an odd prime q, we shall compute explicitly (q/p), where p is any odd prime.

L.   *Let q be an odd prime; q is a quadratic residue modulo the odd prime $p \ne q$ if and only if p is congruent modulo $4q$ to one of the following integers: $\pm 1^2$, $\pm 3^2$, $\pm 5^2$, $\ldots$, $\pm(q-2)^2$.*

Proof: If $p \equiv (2a+1)^2 \pmod{4q}$ then $p \equiv 1 \pmod 4$ and by Gauss' reciprocity law

$$\left(\frac{q}{p}\right) = \left(\frac{p}{q}\right)(-1)^{\frac{p-1}{2} \cdot \frac{q-1}{2}} = (-1)^{\frac{p-1}{2} \cdot \frac{q-1}{2}} = 1.$$

If $p \equiv -(2a+1)^2 \pmod{4q}$ then $p \equiv -1 \pmod 4$, and again

$$\left(\frac{q}{p}\right) = \left(\frac{p}{q}\right)(-1)^{\frac{p-1}{2} \cdot \frac{q-1}{2}} = \left(\frac{-1}{q}\right)(-1)^{\frac{p-1}{2} \cdot \frac{q-1}{2}}$$

$$= (-1)^{\frac{q-1}{2} \cdot \left(1 + \frac{p-1}{2}\right)} = 1$$

(by Euler's criterion).

Conversely, let us assume that $(q/p) = 1$. By Gauss' reciprocity law and Euler's criterion we deduce that

$$\left(\frac{p}{q}\right) = (-1)^{\frac{p-1}{2} \cdot \frac{q-1}{2}} = \left(\frac{(-1)^{(p-1)/2}}{q}\right)$$

hence $\left(p(-1)^{(p-1)/2} / q\right) = 1$. Thus there exists x such that

$$p(-1)^{(p-1)/2} \equiv x^2 \pmod q.$$

Since $x^2 \equiv (q-x)^2 \pmod q$ and x or $q-x$ is odd, we may assume, for example, that x is odd, hence $x^2 \equiv 1 \pmod 4$. If $p \equiv 1 \pmod 4$ then $p \equiv x^2 \pmod q$; from $x^2 \equiv 1 \pmod 4$ we deduce that $p \equiv x^2 \pmod{4q}$.

If $p \equiv -1 \pmod 4$ then $p \equiv -x^2 \pmod q$; from $p \equiv -x^2 \pmod 4$ we conclude that $p \equiv -x^2 \pmod {4q}$.   ■

We may illustrate this result with some numerical examples. If $q = 3$ then

$$\left(\frac{3}{p}\right) = \begin{cases} 1 & \text{when } p \equiv \pm 1 \pmod{12}, \\ -1 & \text{when } p \equiv \pm 5 \pmod{12}. \end{cases}$$

Indeed, $a = 1$ is the only number such that $0 < a < 12$, $a \equiv 1 \pmod 4$ and $(a/3) = 1$. Thus, $(3/p) = 1$ exactly when $p \equiv \pm 1 \pmod{12}$.

Similarly, if $q = 11$, we consider the squares 1^2, 3^2, 5^2, 7^2, 9^2, whose residue classes modulo 44 are 1, 9, 25, 5, 37. By our result, $(11/p) = 1$ when p is congruent modulo 44 to any of the integers ± 1, ± 5, ± 9, ± 25, ± 37.

The result which follows may be called the *"global property of quadratic residues."* Its interest lies in the fact that a property is deduced for an integer whenever a similar property relative to residue classes modulo p, holds for all primes p.

Theorem 2.   *A nonzero integer is a square if and only if it is positive and also a quadratic residue modulo p, for every prime p.*

Proof: If $a = b^2$ then $a > 0$ and $a \equiv b^2 \pmod p$ for every prime p.

Conversely, let us assume that a is not a square, so it is of form $a < 0$ or $a = m^2 p_1 \ldots p_r$ where $p_1, \ldots, p_r$ are distinct primes, $r \geq 1$, and if $i < r$ then $p_i \neq 2$.

Case 1: $a > 0$.

We shall show that $(a/p) = -1$ for some prime p.

First we prove that if q is an odd prime there exists an integer u such that $u \equiv 1 \pmod 4$, q does not divide u, and $(u/q) = -1$. Indeed, we exclude from the set of q integers $\{1, 5, 9, \ldots, 4q - 3\}$ those which are the least positive residues modulo q. We also exclude q when $q \equiv 1 \pmod 4$ or $3q$ when $q \equiv -1 \pmod 4$. There remains a set with $q - (q - 1)/2 - 1 = (q - 1)/2 \geq 1$ elements. If u belongs to this set we have $(u/q) = -1$. We apply this fact for $q = p_r$ when p_r is odd. If $p_r = 2$ we take $u = 5$.

By the Chinese remainder theorem there exists an integer x satisfying the following congruences:

$$\begin{cases} x \equiv 1 \pmod{p_1}, \\ \cdots \\ x \equiv 1 \pmod{p_{r-1}}, \\ x \equiv u \pmod{4p_r}. \end{cases}$$

By Dirichlet's theorem there exists a prime p such that

$$p \equiv x \pmod{4p_1 \cdots p_{r-1}p_r}.$$

Then $a = m^2 p_1 p_2 \cdots p_r$ satisfies

$$\left(\frac{a}{p}\right) = \left(\frac{p_1}{p}\right) \cdots \left(\frac{p_{r-1}}{p}\right)\left(\frac{p_r}{p}\right)$$

$$= \left(\frac{p}{p_1}\right)(-1)^{\frac{p_1-1}{2} \cdot \frac{p-1}{2}} \cdots \left(\frac{p}{p_{r-1}}\right)(-1)^{\frac{p_{r-1}-1}{2} \cdot \frac{p-1}{2}} \cdot \left(\frac{p_r}{p}\right) = -1,$$

since $p \equiv x \equiv 1 \pmod{p_i}$ for all $i = 1, \ldots, r - 1$; $p \equiv x \equiv 1 \pmod 4$; $(2/p) = -1$, since $p \equiv 5 \pmod 8$ when $p_r = 2$;

$$\left(\frac{p_r}{p}\right) = \left(\frac{p}{p_r}\right)(-1)^{\frac{p_r-1}{2} \cdot \frac{p-1}{2}}\left(\frac{u}{p_r}\right) = -1$$

because $p \equiv u \pmod{p_r}$.

Case 2: $a < 0$.
    If $a = -m^2$, let p be a prime such that $p \equiv -1 \pmod 4$ (for example, $p = 3$); then $(a/p) = (-1/p) = -1$.
    If $a = -m^2 p_1 \cdots p_r$ where $p_1, \ldots, p_r$ are distinct primes and $r \geq 1$, we consider a prime p such that $p \equiv 1 \pmod 4$ and $(-a/p) = -1$, which exists by the first case. Then $(a/p) = (-1/p)(-a/p) = -1$.   ∎

4.2 Gaussian Sums

We now want to present another more penetrating proof of Gauss' reciprocity law. This proof will illustrate the possibility of deriving properties of integers by considerations in algebraic extensions of the field $\mathbb{Q}$ of rational numbers. This is just one instance of a very fruitful method, and we shall later encounter more applications of this idea.

    We assume that the reader has a familiarity with the basic concepts of the theory of commutative fields as found in the Introduction and in several textbooks.

    Let p be an odd prime and let $K_0 = \mathbb{Q}$ or $K_0 = \mathbb{F}_q$ (where q is prime distinct from p). Let ζ be a primitive pth root of unity (in an algebraic closure of K_0). Thus $1, \zeta, \zeta^2, \ldots, \zeta^{p-1}$ are all the pth roots of 1 in the algebraic closure, and $\zeta^p = 1$. We agree to write $\zeta^{\bar{a}} = \zeta^a$, where $\bar{a}$ denotes the residue class of a modulo p. From $0 = \zeta^p - 1 = (\zeta - 1)(\zeta^{p-1} + \zeta^{p-2} + \cdots + \zeta + 1)$, $\zeta \neq 1$, it follows that $\zeta^{p-1} + \zeta^{p-2} + \cdots + \zeta + 1 = 0$.

    For every $\bar{a} \in P(p)$ we shall consider the sum

$$\tau(\bar{a}) = \sum_{\bar{x} \in P(p)} \left(\frac{x}{p}\right)\zeta^{ax},$$

which is an element of the field $K = K_0(\zeta)$. It is called the *Gaussian sum over K_0 belonging to $\bar{a}$*. The *principal Gaussian sum* is

$$\tau(\bar{1}) = \sum_{\bar{x} \in P(p)} \left(\frac{x}{p}\right) \zeta^x.$$

M.   *For every $\bar{a} \in P(p)$ we have $\tau(\bar{a}) = (a/p)\tau(\bar{1})$.*

Proof: Let $\bar{a} \cdot \bar{x} = \bar{y}$ for every $\bar{x} \in P(p)$. Since $P(p)$ is a finite multiplicative group then

$$\left(\frac{a}{p}\right)\tau(\bar{a}) = \left(\frac{a}{p}\right) \sum_{\bar{x} \in P(p)} \left(\frac{x}{p}\right) \zeta^{ax}$$

$$= \sum_{\bar{x} \in P(p)} \left(\frac{ax}{p}\right) \zeta^{ax} = \sum_{\bar{y} \in P(p)} \left(\frac{y}{p}\right) \zeta^y = \tau(\bar{1}).$$

We deduce, by multiplication with (a/p), that $\tau(\bar{a}) = (a/p) \cdot \tau(\bar{1})$.   ∎

Now, we compute the square of the principal Gaussian sum over the field K_0. It is convenient to denote by $\tilde{1}$ the unit element of K_0.

N. $\tau(\bar{1})^2 = (-1)^{(p-1)/2} p \cdot \tilde{1}$, *or explicitly*

$$\tau(\bar{1})^2 = \begin{cases} p \cdot \tilde{1} & when \ p \equiv 1 \pmod 4, \\ -p \cdot \tilde{1} & when \ p \equiv 3 \pmod 4. \end{cases}$$

In particular, since K_0 has characteristic different from p, then $\tau(\bar{1}) \neq 0$.

Proof: The statement is proved by a straightforward computation

$$\tau(\bar{1})^2 = \left[\sum_{\bar{x} \in P(p)} \left(\frac{x}{p}\right) \zeta^x\right] \cdot \left[\sum_{\bar{y} \in P(p)} \left(\frac{y}{p}\right) \zeta^y\right]$$

$$= \sum_{\bar{x},\bar{y} \in P(p)} \left(\frac{xy}{p}\right) \zeta^{x+y}.$$

Let us write $\bar{y} = \bar{x} \cdot \bar{t}$ (which is possible, since $P(p)$ is a multiplicative group); hence

$$\left(\frac{xy}{p}\right) = \left(\frac{x^2 t}{p}\right) = \left(\frac{t}{p}\right),$$

so

$$\tau(\bar{1})^2 = \sum_{\bar{x},\bar{t} \in P(p)} \left(\frac{t}{p}\right) \zeta^{x(1+t)}$$

$$= \sum_{\bar{t} \in P(p)} \left(\frac{t}{p}\right) \left[\sum_{\bar{x} \in P(p)} \zeta^{x(1+t)}\right].$$

If $\bar{1} + \bar{t} \neq \bar{0}$ then $\{\bar{x}(\bar{1} + \bar{t}) \mid \bar{x} \in P(p)\} = P(p)$, thus,

$$\sum_{\bar{x} \in P(p)} \zeta^{x(1+t)} = \zeta + \zeta^2 + \cdots + \zeta^{p-1} = -\tilde{1}.$$

If $\bar{1} + \bar{t} = \bar{0}$ then $\sum_{\bar{x} \in P(p)} \zeta^{x(1+t)} = (p-1) \cdot \tilde{1}$. Therefore,

$$\tau(\bar{1})^2 = \left(\frac{-1}{p}\right)(p-1) \cdot \tilde{1} + \sum_{-\bar{1} \neq \bar{t} \in P(p)} \left(\frac{t}{p}\right)(-\tilde{1})$$

$$= \left(\frac{-1}{p}\right) p \cdot \tilde{1} - \sum_{\bar{t} \in P(p)} \left(\frac{t}{p}\right) \cdot \tilde{1} = \left(\frac{-1}{p}\right) p \cdot \tilde{1} = (-1)^{(p-1)/2} p \cdot \tilde{1}$$

because there are as many quadratic residues as nonresidues modulo p (by (F), (H)). In particular, since K_0 has characteristic different from p, then $\tau(\bar{1}) \neq 0$. ∎

The above expression of $\tau(\bar{1})^2$ will soon be used in an important instance.

O.   *Let q be an odd prime different from p. For the principal Gaussian sum (for the prime p) over $\mathbb{F}_q$ we have*

$$\tau(\bar{1})^{q-1} = \left(\frac{q}{p}\right) \cdot \tilde{1}.$$

Proof:

$$\tau(\bar{1})^q = \left[\sum_{\bar{x} \in P(p)} \left(\frac{x}{p}\right)\zeta^x\right]^q = \sum_{\bar{x} \in P(p)} \left(\frac{x}{p}\right)\zeta^{qx}$$

$$= \left(\frac{q}{p}\right)\left[\sum_{\bar{x} \in P(p)} \left(\frac{qx}{p}\right)\zeta^{qx}\right] = \left(\frac{q}{p}\right)\tau(\bar{1}),$$

since $\mathbb{F}_p(\zeta)$ has characteristic q. We conclude that $\tau(\bar{1})^{q-1} = (q/p) \cdot \tilde{1}$ because $\tau(\bar{1}) \neq 0$. ∎

We are now ready to indicate a *new proof of the quadratic reciprocity law*, which is also due to Gauss.

Let p, q be distinct odd primes. We compute in the algebraic closure of $\mathbb{F}_q$ the value of the Legendre symbol (q/p). Let $p^* = (-1)^{(p-1)/2}p$ and let $\tilde{1}$ be the unit element of $\mathbb{F}_q$. By (**O**):

$$\left(\frac{q}{p}\right) \cdot \tilde{1} = \tau(\bar{1})^{q-1} = [\tau(\bar{1})^2]^{(q-1)/2} = (p^*)^{(q-1)/2} \cdot \tilde{1}.$$

By Euler's criterion we have the equality in the field $\mathbb{F}_q$:

$$\left(\frac{q}{p}\right) = \left(\frac{(-1)^{(p-1)/2}p}{q}\right) = \left(\frac{-1}{q}\right)^{(p-1)/2}\left(\frac{p}{q}\right) = (-1)^{\frac{p-1}{2} \cdot \frac{q-1}{2}}\left(\frac{p}{q}\right),$$

so that $(p/q)(q/p) = (-1)^{\frac{p-1}{2} \cdot \frac{q-1}{2}}$.   ∎

This method of proof implies the following interesting result:

P.   *If $K|\mathbb{Q}$ is an algebraic extension of degree 2, then there exists a root of unity ζ such that $K \subseteq \mathbb{Q}(\zeta)$.*

Proof: We may assume that $K = \mathbb{Q}(\sqrt{d})$ where d is an integer with no square factors. Indeed, by the theorem of the primitive element (see Chapter 2, Section 6) there exists an element t such that $K = \mathbb{Q}(t)$. The minimal polynomial of t is of degree 2, $X^2 + aX + b$. Replacing t by $t' = t + (a/2)$, it follows that $K = \mathbb{Q}(t')$ and t' is a root of $X^2 - d_1/d_2$, where $d_1/d_2 = a^2/4 - b$, with d_1, $d_2 \in \mathbb{Z}$, $d_2 \neq 0$. So $K = \mathbb{Q}(\sqrt{d_1/d_2}) = \mathbb{Q}(\sqrt{d_1 d_2})$, hence $K = \mathbb{Q}(\sqrt{d})$, where d is an integer with no square factors. Thus $d = \pm 2^e p_1 \cdots \cdot p_r$, where $e = 0$ or 1, $r \geq 0$, and each p_i is an odd prime. It follows that $K = \mathbb{Q}(\sqrt{d}) \subseteq \mathbb{Q}(\sqrt{-1}, \sqrt{2}, \sqrt{p_1}, \ldots, \sqrt{p_r})$. Here $\sqrt{-1}$ is a primitive root of unity ζ_4 and $\sqrt{2}$ is expressible in terms of a primitive eighth root of unity ζ_8 because

$$(\zeta_8 + \zeta_8^{-1})^2 = \zeta_8^2 + \zeta_8^{-2} + 2 = \zeta_4 + \zeta_4^{-1} + 2 = 2,$$

thus $\mathbb{Q}(\sqrt{2}) \subseteq \mathbb{Q}(\zeta_8)$.

Finally, if p is any prime, $p \neq 2$, then $\pm p = [\tau(\bar{1})]^2$ (by (**N**)), therefore, $\sqrt{p} = \tau(\bar{1}) \in \mathbb{Q}(\zeta_p)$ or $\sqrt{p} = \sqrt{-1}\tau(\bar{1}) \in \mathbb{Q}(\zeta_4, \zeta_p)$, where ζ_p is a primitive pth root of unity.

Combining these facts, $K = \mathbb{Q}(\sqrt{d}) \subseteq \mathbb{Q}(\zeta_8, \zeta_{p_1}, \ldots, \zeta_{p_r}) \subseteq \mathbb{Q}(\zeta_m)$, where $m = 8p_1 \cdots \cdot p_r$ and ζ_m is a primitive mth root of unity.   ∎

The preceding result has a far-reaching generalization, which is the classical Kronecker and Weber theorem:

If $K|\mathbb{Q}$ is an Abelian extension (that is, a Galois extension with Abelian Galois group) of finite degree, then there exists a root of unity ζ, such that $K \subseteq \mathbb{Q}(\zeta)$.

We postpone the proof of this theorem until Chapter 15, since it requires deep considerations of an arithmetical nature.

4.3 The Jacobi Symbol

We conclude this chapter by indicating a generalization of Legendre's symbol, which is useful in the study of quadratic number fields.

Let a be a nonzero integer, and let b be an odd integer relatively prime to a, $|b| = \prod_{p|b} p^{\beta_p}$ (p odd prime, $\beta_p \geq 1$).

We define the *Jacobi symbol* $\left[\dfrac{a}{b}\right]$ (also denoted $[a/b]$) by

$$\left[\frac{a}{b}\right] = \left[\frac{a}{-b}\right] = \prod_{p|b}\left(\frac{a}{p}\right)^{\beta_p}.$$

In particular $[a/1] = [a/-1] = 1$. Let us note that since $\gcd(a, b) = 1$, if $\beta_p \geq 1$, then p does not divide a, so that the Legendre symbol (a/p) has a meaning.

The Jacobi symbol has value 1 or -1. If $b = p$ is an odd prime number then $[a/p] = (a/p)$. We write $\left(\dfrac{a}{b}\right)$ or (a/b), instead of $\left[\dfrac{a}{b}\right]$ or $[a/b]$, without ambiguity.

Below we list some of the properties of the Jacobi symbol (under the above assumptions about the numerator and denominator):

Q.   (1) *If $a \equiv a' \pmod b$ then $(a/b) = (a'/b)$.*

(2) $\left(\dfrac{aa'}{b}\right) = \left(\dfrac{a}{b}\right)\left(\dfrac{a'}{b}\right)$.

(3) $\left(\dfrac{a}{bb'}\right) = \left(\dfrac{a}{b}\right)\left(\dfrac{a}{b'}\right)$.

(4) *If the class of a modulo $b > 1$ is a square in $P(b)$ then $(a/b) = 1$.*

Proof: These properties follow easily from the corresponding properties for the Legendre symbol.

To show (4) we note that if $a \equiv x^2 \pmod b$ and if p is a prime dividing b then $a \equiv x^2 \pmod p$, so $(a/p) = 1$. Thus $(a/b) = 1$.   ∎

Let us observe, however, that it may happen that a modulo b is not a square in $P(b)$ and yet $(a/b) = 1$. For example, $(2/9) = (2/3)^2 = 1$, but 2 is not a square modulo 9. Similarly $(2/15) = 1$, but 2 is not a square modulo 15.

To deduce other properties we first observe the following simple facts. If $b = \prod_{p|b} p^{\beta_p}$ is an odd integer then $b \equiv 1 \pmod 4$ if and only if there is an even number of primes p such that β_p is odd and $p \equiv 3 \pmod 4$. Hence,

$$\sum_{p|b} \beta_p \cdot (p-1)/2 \equiv (b-1)/2 \pmod 2.$$

Similarly, $b \equiv \pm1 \pmod 8$ if and only if there is an even number of primes p such that β_p is odd and $p \equiv \pm3 \pmod 8$. Thus

$$\sum_{p|b} \beta_p(p^2-1)/8 \equiv (b^2-1)/8 \pmod 2.$$

R.   *For odd integers b we have*

$$\left(\frac{-1}{b}\right) = (-1)^{(|b|-1)/2} = \begin{cases} 1 & when \ |b| \equiv 1 \pmod 4, \\ -1 & when \ |b| \equiv 3 \pmod 4, \end{cases}$$

$$\left(\frac{2}{b}\right) = (-1)^{(b^2-1)/8} = \begin{cases} 1 & when \ b \equiv \pm 1 \pmod 8, \\ -1 & when \ b \equiv \pm 3 \pmod 8. \end{cases}$$

Proof: By definition and (**H**) we have

$$\left(\frac{-1}{b}\right) = \left(\frac{-1}{|b|}\right) = \prod_{p|b}\left(\frac{-1}{p}\right)^{\beta_p} = (-1)^{\sum_{p|b}\beta_p(p-1)/2} = (-1)^{(|b|-1)/2}.$$

In the same way

$$\left(\frac{2}{b}\right) = \prod_{p|b}\left(\frac{2}{p}\right)^{\beta_p} = (-1)^{\sum_{p|b}\beta_p(p^2-1)/8} = (-1)^{(b^2-1)/8}. \qquad \blacksquare$$

S.   *The reciprocity law for the Jacobi symbol is the following:*

$$\left(\frac{a}{b}\right) = \varepsilon(-1)^{\frac{a-1}{2}\cdot\frac{b-1}{2}}\left(\frac{b}{a}\right),$$

where a, b are relatively prime odd integers and

$$\varepsilon = \begin{cases} 1, & when \ a \ or \ b \ is \ positive, \\ -1, & when \ a < 0, \ b < 0. \end{cases}$$

Proof: First we assume that $a > 0$ and $b > 0$, and let $a = \prod_{q|a} q^{\alpha_q}$, $b = \prod_{p|b} p^{\beta_p}$ be the prime decompositions of a, b. The primes dividing b are different from those dividing a. Thus by the quadratic reciprocity law for the Legendre symbol

$$\left(\frac{a}{b}\right) = \prod_{p|b}\left(\frac{a}{p}\right)^{\beta_p} = \prod_{p|b}\prod_{q|a}\left(\frac{q}{p}\right)^{\alpha_q\beta_p}$$

$$= \prod_{p|b}\prod_{q|a}\left(\frac{p}{q}\right)^{\alpha_q\beta_p}(-1)^{\sum_{p,q}\alpha_q\cdot\frac{q-1}{2}\cdot\beta_p\cdot\frac{p-1}{2}}$$

$$= \prod_{q|a}\left(\frac{b}{q}\right)^{\alpha_q}(-1)^{\sum_{p,q}\alpha_q\cdot\frac{q-1}{2}\cdot\beta_p\cdot\frac{p-1}{2}}.$$

Since

$$\sum_{q|a}\alpha_q\cdot\frac{q-1}{2} \equiv \frac{a-1}{2} \pmod 2,$$

and

$$\sum_{p|b}\beta_p\cdot\frac{p-1}{2} \equiv \frac{b-1}{2} \pmod 2,$$

we have

$$\left(\frac{a}{b}\right) = \prod_{q|a} \left(\frac{b}{q}\right)^{\alpha_q} (-1)^{\frac{a-1}{2}\cdot\frac{b-1}{2}} = \left(\frac{b}{a}\right)(-1)^{\frac{a-1}{2}\cdot\frac{b-1}{2}}.$$

Now, if $a < 0$ and $b > 0$ then

$$\left(\frac{a}{b}\right) = \left(\frac{-1}{b}\right)\left(\frac{|a|}{b}\right) = (-1)^{(b-1)/2}(-1)^{\frac{b-1}{2}\cdot\frac{|a|-1}{2}} \cdot \left(\frac{b}{|a|}\right)$$

$$= (-1)^{\frac{b-1}{2}\left\{1+\frac{|a|-1}{2}\right\}}\left(\frac{b}{a}\right).$$

Since $1 + (|a| - 1)/2 \equiv (a - 1)/2 \pmod{2}$ when $a < 0$ and odd, then

$$\left(\frac{a}{b}\right) = (-1)^{\frac{b-1}{2}\cdot\frac{a-1}{2}}\left(\frac{b}{a}\right).$$

In the final case $a < 0$, $b < 0$ we have

$$\left(\frac{a}{b}\right) = \left(\frac{a}{-b}\right) = (-1)^{\frac{a-1}{2}\cdot\frac{-b-1}{2}}\left(\frac{-b}{a}\right)$$

$$= (-1)^{\frac{a-1}{2}\cdot\frac{-b-1}{2}} \cdot (-1)^{\frac{-a-1}{2}}\left(\frac{b}{a}\right).$$

From

$$\frac{a-1}{2}\cdot\frac{-b-1}{2} + \frac{-a-1}{2} \equiv \frac{a-1}{2}\cdot\frac{b+1}{2} + \frac{a+1}{2}$$

$$\equiv \frac{a-1}{2}\cdot\frac{b-1}{2} + 1 \pmod{2},$$

for a, b odd, it follows that

$$\left(\frac{a}{b}\right) = \varepsilon(-1)^{\frac{a-1}{2}\cdot\frac{b-1}{2}}\left(\frac{b}{a}\right),$$

since $\varepsilon = -1$ in this case.   ∎

EXERCISES

1. Determine the squares in the following groups: $P(7)$, $P(11)$, $P(12)$, $P(16)$, $P(49)$.

2. Compute the following Legendre symbols:

$$\left(\frac{6}{11}\right), \quad \left(\frac{8}{17}\right), \quad \left(\frac{18}{23}\right).$$

3. Compute the following Legendre symbols:

$$\left(\frac{205}{307}\right), \quad \left(\frac{18}{461}\right), \quad \left(\frac{753}{811}\right), \quad \left(\frac{48}{1117}\right).$$

4. Determine the primes p modulo 68 such that 17 is a quadratic residue modulo p.

5. Determine the primes p modulo 20 for which 5 is a quadratic residue modulo p.

6. Determine the odd primes p for which 7 is a quadratic residue modulo p.

7. Determine the primes p modulo 12 such that -3 is a quadratic residue modulo p.

8. Determine the odd primes p for which 10 is a quadratic residue modulo p.

9. Determine the odd primes p for which -2 is a quadratic residue modulo p.

10. Prove that 7 is a primitive root modulo every prime $p = 2^{2^k} + 1$ ($k \geq 1$).

Hint: First show that $2^k \equiv 2$ or 4 (mod 6), then $2^{2^k} \equiv 2$ or 4 (mod 7); next compute $(7/p)$.

11. If p is a prime number, $p \equiv 1$ (mod 4), prove that there exists an integer x such that $1 + x^2 = mp$, with $0 < m < p$.

12. Let p be an odd prime. Show that there exist integers x, y, and m, $0 < m < p$, such that $1 + x^2 + y^2 = mp$.

13. Prove that for every integer $a > 1$ there exist infinitely many integers n which are not prime and such that $a^{n-1} \equiv 1$ (mod n).

Hint: For every odd prime p not dividing $a(a^2 - 1)$ consider $n = (a^{2p} - 1)/(a^2 - 1)$.

14. Prove that $2^{1092} \equiv 1$ (mod 1093^2). (The congruence was discovered by actual calculation by Meissner; the proof below is by Landau.) In this respect, let us mention that it seems to be only rarely that a prime number p satisfies $2^{p-1} \equiv 1$ (mod p^2); namely, $p = 1093$, $p = 5311$ are the only primes less than 4×10^{12} for which this is true.

Hint: In order to show that $2^{182} \equiv -1$ (mod p^2) establish that $3^{14} \equiv 4p+1$ (mod p^2), $3^2 \cdot 2^{26} \equiv -469p-1$ (mod p^2), $3^{14} \cdot 2^{182} = -4p-1$ (mod p^2).

15. Let p be an odd prime. Show that

$$\sum_{n=1}^{p-2} \left(\frac{n(n+1)}{p} \right) = -1.$$

Hint: Make use of the fact that n has an inverse modulo p.

16. Let p be an odd prime. Prove that the number of pairs of consecutive integers n, $n+1$, with $1 \leq n \leq p-2$, which are quadratic residues modulo p is equal to $\frac{1}{4}(p - 4 - (-1/p))$.

Hint: Use the previous exercise after evaluating $\sum_{n=1}^{p-2}(1 + (n/p)) \cdot (1 + ((n+1)/p))$.

17. Let p be an odd prime. Show that the product of the quadratic residues a modulo p, where $1 \leq a \leq p - 1$, is congruent to $-(-1/p)$ modulo p.

Hint: Group the quadratic residues by pairs.

18. Let $a, b, c \in \mathbb{Z}$, and let p be an odd prime not dividing a. Show that there exists $x \in \mathbb{Z}$, such that $ax^2 + bx + c \equiv 0 \pmod{p}$ if and only if $b^2 - 4ac$ is a quadratic residue modulo p.

19. Prove that there are $\frac{1}{2}p(p-1)$ quadratic residues modulo p^2 and these are the solutions of the congruence

$$x^{p(p-1)/2} \equiv 1 \pmod{p^2}.$$

20. Let p be a prime number, $p \equiv 1 \pmod 4$, and let a_1, a_2, $\ldots$, $a_{(p-1)/2}$ be all the quadratic residues modulo p such that $1 < a_i \leq p - 1$. Prove that

$$\sum_{i=1}^{(p-1)/2} a_i = p(p-1)/4.$$

21. Show that if the prime number p divides $839 = 38^2 - 5 \cdot 11^2$, then $(5/p) = 1$. From this deduce that 839 is a prime number.

Hint: Determine the primes for which 5 is a quadratic residue.

22. Show with the same method that 757 is a prime number.

Hint: Write $757 = 55^2 - 7 \cdot 18^2$.

23. Let p be a prime number, a an integer not multiple of p, and assume that there exist integers x, y such that $p = x^2 + ay^2$. Show that $-a$ is a quadratic residue modulo p.

24. Assume that the congruence $x^n \equiv a \pmod{m}$, where $a > 0$, is solvable for every integer $m > 1$. Prove that a is the nth power of a natural number.

25. Let p be an odd prime, $1 \leq a < p$ and $n = ap + 1$ or $n = ap^2 + 1$. Prove that if $2^a \not\equiv 1 \pmod{n}$ and $2^{n-1} \equiv 1 \pmod{n}$ then n is a prime number.

   Hint: Show that the order d of 2 modulo n is a multiple of p, hence p divides $\varphi(n)$; noting that p does not divide n, deduce that $n = qm$, where q is a prime number, $q \equiv 1 \pmod{p}$; conclude by proving that $m > 1$ leads to a contradiction in both cases $n \equiv 1 \pmod{p}$, $n \equiv 1 \pmod{p^2}$.

26. Let p be an odd prime, $m \geq 2$, $1 \leq a < 2^m$, $n = 2^m a + 1$, and assume that $(n/p) = -1$. Prove that n is a prime number if and only if

$$p^{(n-1)/2} \equiv -1 \pmod{n}.$$

   Hint: If n is prime, use Gauss' reciprocity law and Euler's criterion. Conversely, consider the order d of p modulo q (a prime factor of n); show that 2^m divides d, hence $q \equiv 1 \pmod{2^m}$ and from this conclude that $n = q$.

27. Apply the previous exercise to show that the Fermat number F_n (see Exercise 58, Chapter 3) is prime if and only if F_n divides

$$3^{(F_n-1)/2} + 1.$$

28. Find the smallest prime p which may be written simultaneously in the forms $p = x_1^2 + y_1^2 = x_2^2 + 2y_2^2 = x_3^2 + 3y_3^2$, where $x_i, y_i \in \mathbb{Z}$.

29. Let p be a prime number, $p > 7$, $p \equiv 3 \pmod{4}$. Prove (Euler):
   (a) $2p + 1$ is a prime if and only if $2^p \equiv 1 \pmod{2p + 1}$.
   (b) If $2p + 1$ is prime then the Mersenne number $M_p = 2^p - 1$ is not a prime number.
   (c) Show successively that $23 | M_{11}$, $47 | M_{23}$, $167 | M_{83}$, $263 | M_{131}$, $359 | M_{179}$, $383 | M_{191}$, $479 | M_{239}$, and $503 | M_{251}$.

   Hint: Use Exercise 25, this chapter.

30. Show that if n is not a square, there exist infinitely many prime numbers p such that n is not a quadratic residue modulo p.

31. Prove the following particular case of Dirichlet's theorem: There exist infinitely many prime numbers in the arithmetic progression

$$\{12k + 7 \mid k = 0, 1, 2, \ldots\}.$$

Hint: After computing the Legendre symbol $(-3/p)$, show that $4a^2 + 3$ is divisible by a prime in the given arithmetic progression; conclude considering integers of the form $4(p_1 p_2 \ldots p_n)^2 + 3$.

32. Prove the following particular case of Dirichlet's theorem: In the arithmetic progression $\{1 + 2^n k \mid k = 0, 1, 2, \ldots\}$, $n \geq 1$, there exist infinitely many primes.

Hint: First establish that if $p \neq 2$ is prime and p divides $2^{2^{n-1}} + 1$ then $p \equiv 1 \pmod{2^n}$.

33. Prove the following particular case of Dirichlet's theorem: In the arithmetic progression $\{1 + q^n k \mid k = 0, 1, 2, \ldots\}$, $n \geq 1$, q a prime number, there exist infinitely many prime numbers.

Hint: First establish that if p is a prime number, $p \neq q$, dividing $1 + x^{q^{n-1}} + x^{2q^{n-1}} + \cdots + x^{(q-1)q^{n-1}}$ then $p \equiv 1 \pmod{q^n}$. For this purpose, write $y = x^{q^{n-1}}$ and note that

$$1 + y + y^2 + \cdots + y^{q-1} = (y-1)^{q-1} + \binom{q}{1}(y-1)^{q-2}$$

$$+ \binom{q}{2}(y-1)^{q-3} + \cdots + \binom{q}{q-2}(y-1) + q.$$

34. *The Kronecker symbol.* Let $a \neq 0$ and let p be any prime number. We define the Kronecker symbol $\left\{\dfrac{a}{p}\right\} = \{a/p\}$ as follows: if $p|a$ then $\{a/p\} = 0$; if p is odd and $p \nmid a$ then $\{a/p\} = (a/p)$ (the Legendre symbol); next,

$$\left\{\frac{a}{2}\right\} = \begin{cases} +1, & \text{if } a \equiv 1 \pmod 8, \\ -1, & \text{if } a \equiv 5 \pmod 8, \\ \text{undefined}, & \text{if } a \equiv 3 \pmod 4. \end{cases}$$

If $b = p_1 \cdots p_r$ ($p_1, \ldots, p_r$ odd primes, not necessarily distinct), we define

$$\left\{\frac{a}{b}\right\} = \left\{\frac{a}{-b}\right\} = \prod_{i=1}^{r} \left\{\frac{a}{p_i}\right\}.$$

If $b = 2^e b'$, where $e \geq 1$, b' is odd, we define

$$\left\{\frac{a}{b}\right\} = \begin{cases} \left\{\dfrac{a}{b'}\right\}, & \text{if } e \text{ is even,} \\[2ex] \left\{\dfrac{a}{2}\right\}\left\{\dfrac{a}{b'}\right\}, & \text{if } e \text{ is odd.} \end{cases}$$

(note that $\{a/b\} = 0$ if e is odd and a is even, and it is undefined when e is odd and $a \equiv 3 \pmod 4$).

Assuming that the Kronecker symbols below are defined, prove:

(a) If $a \equiv a' \pmod{4b}$ then $\left\{\dfrac{a}{b}\right\} = \left\{\dfrac{a'}{b}\right\}$.

(b) $\left\{\dfrac{a}{bb'}\right\} = \left\{\dfrac{a}{b}\right\} \cdot \left\{\dfrac{a}{b'}\right\}$.

(c) $\left\{\dfrac{aa'}{b}\right\} = \left\{\dfrac{a}{b}\right\} \cdot \left\{\dfrac{a'}{b}\right\}$.

(d) Let $b = 2^e b'$ with $e \geq 0$, b' odd. Then

$$\left\{\dfrac{-1}{b}\right\} = \begin{cases} (-1)^{(|b'|-1)/2} & \text{if } e \text{ is even,} \\ \text{undefined} & \text{if } e \text{ is odd.} \end{cases}$$

(e) $\left\{\dfrac{2}{b}\right\} = \begin{cases} (-1)^{(b'^2-1)/8} & \text{if } e \text{ is even,} \\ 0 & \text{if } e \text{ is odd.} \end{cases}$

(f) $\left\{\dfrac{a}{2}\right\} = \left\{\dfrac{2}{a}\right\}$ when $a \equiv 1 \pmod 4$.

(g) If $\gcd(a, b) = 1$, b is odd, and $a \equiv 1 \pmod 4$, then

$$\left\{\dfrac{a}{b}\right\} = \varepsilon\left\{\dfrac{b}{a}\right\},$$

where $\varepsilon = -1$ if $a < 0$, $b < 0$, and $\varepsilon = 1$ otherwise.

(h) If $\gcd(a, b) = 1$, $b > 0$, $a = 2^e a'$ with $e \geq 0$, a' odd, then

$$\left\{\dfrac{a}{b}\right\} = \left\{\dfrac{2}{b}\right\}^e (-1)^{\frac{a'-1}{2} \cdot \frac{b-1}{2}} \left\{\dfrac{b}{a'}\right\}.$$

(i) If b_1, b_2 are odd, $b_1 \equiv b_2 \pmod{4a}$, and $b_1 b_2 > 0$, then

$$\left\{\dfrac{a}{b_1}\right\} = \left\{\dfrac{a}{b_2}\right\}.$$

(j) For every a not a square there exists b such that $\{a/b\} = -1$.

35. Prove the following properties of the Kronecker symbol where $a \equiv 0$ or $1 \pmod 4$:

(a) $\left\{\dfrac{a}{|a|-1}\right\} = \begin{cases} 1 & \text{when } a > 0, \\ -1 & \text{when } a < 0. \end{cases}$

(b) If $b \equiv -b' \pmod{|a|}$ then

$$\left\{\dfrac{a}{b}\right\} = \begin{cases} \left\{\dfrac{a}{b'}\right\} & \text{when } a > 0, \\ -\left\{\dfrac{a}{b'}\right\} & \text{when } a < 0. \end{cases}$$

Part Two

Part Two

5

Algebraic Integers

5.1 Integral Elements, Integrally Closed Domains

The arithmetic of the field of rational numbers is mainly the study of divisibility properties with respect to the ring of integers.

Similarly, the arithmetic of an algebraic number field K is concerned with divisibility properties of algebraic numbers relative to some subring of K, which plays the role of the integers. Accordingly, we shall define the concept of an algebraic integer.

More generally, we introduce the following definition:

Definition 1. Let R be a ring,* and A a subring of R. We say that the element $x \in R$ is an *integer over* A when there exist elements $a_1, \ldots, a_n \in A$ such that $x^n + a_1 x^{n-1} + \cdots + a_n = 0$.

For example, if $A = K$ and $R = L$ are fields, then $x \in L$ is integral over K if and only if it is algebraic over K.

The first basic result about integral elements is the following:

A. *Let R be a ring, A a subring of R, and $x \in R$. Then the following properties are equivalent:*

(1) *x is integral over A.*

(2) *The ring $A[x]$ is a finitely generated A-module.*

(3) *There exists a subring B of R such that $A[x] \subseteq B$ and B is a finitely generated A-module.*

Proof: (1) $\rightarrow$ (2) Let us assume that $x^n + a_1 x^{n-1} + \cdots + a_n = 0$ with $a_1, \ldots, a_n \in A$. We shall show that $\{1, x, \ldots, x^{n-1}\}$ is a system of generators of the A-module $A[x]$. Indeed, from $x^n = -(a_1 x^{n-1} + \cdots + a_n)$ it follows that x^{n+1}, x^{n+2}, ... are expressible as linear combinations of 1, x, ..., x^{n-1} with coefficients in A.

---

* We shall consider commutative rings with unit element, and the image of the unit element by all ring-homomorphisms is again the unit element.

(2) $\rightarrow$ (3)   It is enough to take $B = A[x]$.

(3) $\rightarrow$ (1)   Let $B = Ay_1 + \cdots + Ay_n$.

Since $x, y_i \in B$ then $xy_i \in B$; thus there exist elements a_{ij} $(j = 1, \ldots, n)$ such that $xy_i = \sum_{j=1}^{n} a_{ij}y_j$ for all $i = 1, \ldots, n$. Therefore, letting $\delta_{ij} = 1$ when $i = j$, $\delta_{ij} = 0$ when $i \neq j$, we may write $\sum_{j=1}^{n}(\delta_{ij}x - a_{ij})y_j = 0$ for all $i = 1, \ldots, n$.

In other words, the system of linear equations $\sum_{j=1}^{n}(\delta_{ij}x - a_{ij})Y_j = 0$ for all $i = 1, \ldots, n$ has the solution $(y_1, \ldots, y_n)$.

Let d be the determinant of the matrix $(\delta_{ij}x - a_{ij})_{i,j}$. By Cramer's rule, we must have $dy_j = 0$ for all $j = 1, \ldots, n$.

Since $1 \in B$, it may be written in the form $1 = \sum_{j=1}^{n} c_j y_j$ with $c_j \in A$, hence $d = d \cdot 1 = \sum_{j=1}^{n} c_j dy_j = 0$.

Computing d explicitly:

$$
d = \det \begin{pmatrix} x - a_{11} & -a_{12} & \cdots & -a_{1n} \\ -a_{21} & x - a_{22} & \cdots & -a_{2n} \\ \vdots & \vdots & & \vdots \\ -a_{n1} & -a_{n2} & \cdots & x - a_{nn} \end{pmatrix},
$$

we deduce that d is of the form $0 = d = x^n + b_1 x^{n-1} + \cdots + b_n$ where each $b_i \in A$. This shows that x is integral over A.   ∎

With this result we are able to deduce readily several properties of integral elements.

Definition 2.   Let R be a ring and A a subring. We say that R is *integral over* A when every element of R is integral over A.

The following fact is evident:

Let R be a ring integral over the subring A, let $\theta : R \rightarrow R'$ be a homomorphism from R onto the ring R' and $\theta(A) = A'$. Then R' is integral over the subring A'.

Definition 3.   Let R be a ring and A a subring of R. If every element of R which is integral over A belongs to A, then A is said to be *integrally closed* in R.

If A is a domain, $R = K$ (the field of quotients of A) and A is integrally closed in K, we say that A is an *integrally closed domain*.

The following properties are easy to establish:

B.   *Let R be a ring, A a subring, and let $x_1, \ldots, x_n \in R$. If x_1 is integral over A, if x_2 is integral over $A[x_1]$, $\ldots$, if x_n is integral over $A[x_1, \ldots, x_{n-1}]$, then $A[x_1, \ldots, x_n]$ is a finitely generated A-module.*

Proof: By (A), $A[x_1]$ is a finitely generated A-module and $A[x_1, x_2]$ is a finitely generated $A[x_1]$-module. Hence, $A[x_1, x_2]$ is a finitely generated A-module.

The remainder of the proof is done similarly.   ∎

C.   *Let $A \subseteq B \subseteq C$ be rings. If C is integral over B and B is integral over A, then C is also integral over A.*

Proof: Let $x \in C$, so there exist elements $b_1, \ldots, b_n \in B$ such that $x^n + b_1 x^{n-1} + \cdots + b_n = 0$.

But this means that x is integral over the subring $A[b_1, \ldots, b_n]$. By (**A**), the ring $A[b_1, \ldots, b_n, x]$ is a finitely generated module over $A[b_1, \ldots, b_n]$. By (**B**), $A[b_1, \ldots, b_n]$ is a finitely generated A-module, hence

$$A[x] \subseteq A[b_1, \ldots, b_n, x],$$

which is a finitely generated A-module. Thus, by (**A**) again, we deduce that x is integral over A, proving our statement. ∎

D.   *Let R be a ring, A a subring, and let A' be the set of all elements $x \in R$ which are integral over A. Then A' is a subring of R, which is integrally closed in R, and integral over A.*

Proof: Clearly $A \subseteq A'$, because every element $a \in A$ is a root of the polynomial $X - a$.

If $x, y \in A'$, then $x + y$, $x - y$, xy belong to the ring $A[x, y]$. By (**B**), $A[x, y]$ is a finitely generated A-module, hence by (**A**), $x + y$, $x - y$, xy are integral over A, so belong to A'.

By (**C**), A' is integrally closed in R. ∎

This result justifies the following definition:

Definition 4.   Let R be a ring, and A a subring. The ring A' of all elements of R which are integral over A is called the *integral closure of A in R*.

We examine these notions in the special case of domains and fields.

E.   *If R is a domain, which is integral over the subring A, if J is a nonzero ideal of R, then $J \cap A \neq 0$.*

Proof: Let $x \in J$, $x \neq 0$; by hypothesis, there exists a monic polynomial $f = X^n + a_1 X^{n-1} + \cdots + a_n \in A[X]$ such that $f(x) = 0$. We may take one such polynomial of minimal degree. Then $a_n \neq 0$, otherwise

$$x^{n-1} + a_1 x^{n-2} + \cdots + a_{n-1} = 0$$

(since $x \neq 0$ and R is a domain). Hence

$$a_n = -(x^{n-1} + a_1 x^{n-2} + \cdots + a_{n-1})x \in J \cap A.$$ ∎

F.   *Let R be a domain which is integral over the subring A. Then, R is a field if and only if A is a field.*

Proof: If R is a field, if $x \in A$, $x \neq 0$, consider its inverse x^{-1} in R. It is integral over A, hence there exist elements $a_i \in A$ such that

$$(x^{-1})^n + a_1 (x^{-1})^{n-1} + \cdots + a_n = 0.$$

Multiplying by x^{n-1} we obtain

$$x^{-1} + a_1 + a_2 x + \cdots + a_n x^{n-1} = 0,$$

hence $x^{-1} \in A$.

Conversely, let A be a field, let $x \in R$, and $x \neq 0$. By (**E**) there exists $a \in Rx \cap A$, $a \neq 0$; so $a = bx$, with $b \in R$.

Let $a' \in A$ be the inverse of a, so $1 = a'a = (a'b)x$; hence x is invertible in R and R is a field.   ∎

As a corollary:

G.   *Let R be a ring integral over the subring A and let Q be a prime ideal of R. Then Q is a maximal ideal of R if and only if $Q \cap A$ is a maximal ideal of A.*

Proof: If Q is a maximal ideal of R then R/Q is a field, which is integral over the subring $A/(Q \cap A)$; thus $A/(Q \cap A)$ is a field and $Q \cap A$ is a maximal ideal of A.

Conversely, if $Q \cap A$ is a maximal ideal of A then the domain R/Q is integral over the field $A/(Q \cap A)$. So R/Q is a field and Q is a maximal ideal of R.   ∎

The important situation for algebraic numbers is the case where A is an integrally closed domain, with field of quotients K, and L is an algebraic extension of K.

H.   *Let A be an integrally closed domain with field of quotients K and let L be an algebraic extension of K. If $x \in L$ is integral over A, then its minimal polynomial over K has all its coefficients in A; all conjugates of x over K are also integral over A. If B is the integral closure of A in L then $B \cap K = A$.*

Proof: Let $f \in K[X]$ be the minimal polynomial of x over K. Since x is integral over A, there exists a monic polynomial g with coefficients in A, such that $g(x) = 0$. Hence f divides g.

Let L' be the splitting field of f over K, that is, the field generated over K by the roots of f. Let A' be the integral closure of A in L'. Then $A' \cap K$ is integral over A, hence must be equal to A (which is assumed integrally closed).

The conjugates of x are also roots of g, hence integers over A, so they belong to A'.

The coefficients of f are, up to sign, equal to the elementary symmetric polynomials in the conjugates of x, hence f has coefficients in $A' \cap K = A$.

The last assertion follows from the hypothesis that A is integrally closed.   ∎

Another useful property follows:

I.   *Let A be an integrally closed domain with field of quotients K and let L be an algebraic extension of K. If B denotes the integral closure of A in L, then every element of L is of the form b/d, where $b \in B$, $d \in A$ $(d \neq 0)$.*

Proof: Let $x \in L$, $x \neq 0$, so x is algebraic over K, hence there exist elements $c_i/d_i \in K$ with $c_i, d_i \in A$, $d_i \neq 0$ for $i = 1, \ldots, n$, such that $x^n + (c_1/d_1)x^{n-1} + \cdots + c_n/d_n = 0$.

Let $d = d_1 \cdots d_n \in A$, then

$$d^n x^n + (d^{n-1}d_1' c_1)x^{n-1} + \cdots + d^{n-1}d_n' c_n = 0$$

where $d_i' = d/d_i \in A$ for $i = 1, \ldots, n$. It follows that

$$(dx)^n + (d_1' c_1)(dx)^{n-1} + \cdots + d^{n-1}d_n' c_n = 0,$$

with $d^{i-1}d_i' c_i \in A$ for $i = 1, \ldots, n$. Thus dx is integral over A, so $dx = b \in B$, $b \neq 0$, and $x = b/d$.   ∎

Let us now note some important types of integrally closed domains.

J.   *Every unique factorization domain is an integrally closed domain.*

Proof: Let K be the field of quotients of the unique factorization domain A. Let $x \in K$, $x \neq 0$, so $x = a/b$ with $a, b \in A$, $a, b \neq 0$, and we may assume that $\gcd(a, b) = 1$.

If x is integral over A, there exist elements $c_1, \ldots, c_n \in A$ such that $(a/b)^n + c_1(a/b)^{n-1} + \cdots + c_n = 0$, thus

$$a^n + c_1 b a^{n-1} + \cdots + c_n b^n = 0.$$

It follows that $a^n = -b(c_1 a^{n-1} + \cdots + c_n b^{n-1})$, so b divides $a^n = a \cdot a^{n-1}$. Hence, repeating the argument, b divides a, that is, $x = a/b \in A$, proving that A is integrally closed.   ∎

In particular, since every principal ideal domain is a unique factorization domain, then every principal ideal domain is also integrally closed.

By imitating the procedure in the case of the rational integers, we may prove the following result:

K.   *Let A be a domain, satisfying the following property: if*

$$Aa_1 \subseteq Aa_2 \subseteq \cdots \subseteq Aa_n \subseteq \cdots$$

is an increasing chain of principal ideals of A, then there exists an integer n such that

$$Aa_n = Aa_{n+1} = \cdots.$$

Then every nonzero element of A which is not a unit may be written as a product of indecomposable elements.

Proof: Let $a \in A$, $a \neq 0$. If a is not a unit element of A, then either a is indecomposable or there exists an element $a_1 \in A$, $a_1 \neq 0$, a_1 not a unit and not associated with a, such that $a_1 | a$; hence $Aa \subset Aa_1$. This argument may be repeated with a_1: either a_1 is indecomposable, or there exists $a_2 \in A$, $a_2 \neq 0$, a_2 not a unit and not associated with a_1, such that $a_2 | a_1$; hence $Aa \subset Aa_1 \subset Aa_2$. In virtue of the hypothesis, repeating the argument, there exists n, such that a_n is indecomposable. Thus, we have shown that if $a \in A$, $a \neq 0$, there exists an indecomposable element p_1 such that $p_1 | a$; hence there exists $b_1 \in A$, $b_1 \neq 0$, such that $a = p_1 b_1$, so $Aa \subset Ab_1$. But again if b_1 is not a unit there exists an indecomposable element p_2 such that $b_1 = p_2 b_2$, so $a = p_1 p_2 b_2$, $Aa \subset Ab_1 \subset Ab_2$. By the hypothesis, there exists m such that b_m is a unit, hence $a = p_1 p_2 \cdots (p_m b_m)$, that is a product of indecomposable elements.   ∎

Let us note this reformulation of the unique factorization:

L.   *Let A be a domain such that every element is a product of indecomposable elements. Then the following statements are equivalent:*

(1) *If $p_1 \cdots p_r = p_1' \cdots p_s'$ (where p_i, p_j' are indecomposable elements of A) then $s = r$ and there is a permutation σ of $\{1, \ldots, r\}$ such that $p_i \sim p_{\sigma(i)}'$.*

(2) *If p is indecomposable and p divides xy (where $x, y \in A$) then either $p | x$ or $p | y$.*

Proof: The proof of this statement is very similar to the case where A is the ring of integers, therefore we leave the details to the reader.   ∎

The Euclidean division algorithm is an important and useful tool in the study of rings of algebraic integers.

Definition 5.   Let A be a domain. It is said to be a *Euclidean domain*, when for every element $a \in A$, $a \neq 0$, a positive integer $\delta(a)$ is associated, and the following properties are satisfied:

(1) If $a, b \in A$ are nonzero elements, then $\delta(ab) \geq \delta(a)$.

(2) If $a, b \in A$, $b \neq 0$, there exist elements $q, r \in A$ such that $a = bq + r$ and $r = 0$ or $\delta(r) < \delta(b)$.

$\mathbb{Z}$ is a Euclidean domain, by choosing $\delta(a) = |a|$ for every $a \in \mathbb{Z}$, $a \neq 0$. This statement is a rephrasement of the possibility of performing Euclidean division in $\mathbb{Z}$ with respect to the absolute value (that is, $\delta(x) = |x|$ for $x \neq 0$).

By imitating the usual proof in $\mathbb{Z}$, we obtain:

M.   *Every Euclidean domain is a principal ideal domain.*

Proof: Let J be a nonzero ideal of the Euclidean domain A. Consider the set of positive integers $\{\delta(a) \mid a \in J, a \neq 0\}$. Let m be the minimum of the integers in this set and let $b \in J$, $b \neq 0$, such that $\delta(b) = m$.

Now if $a \in J$, there exist $q, r \in A$ such that $a = bq + r$, with $r = 0$ or $\delta(r) < \delta(b) = m$. But $r = a - bq \in J$, so by the minimality of m, $r = 0$, that is, $a \in Ab$, showing that $J = Ab$. ∎

5.2 Rings of Algebraic Integers

We shall now consider explicitly the above definitions and results for the case of algebraic number fields.

Definition 6.   An algebraic number, which is a root of a monic polynomial with coefficients in the ring $\mathbb{Z}$ of integers, is called an *algebraic integer.*

It is also customary to say that the elements of $\mathbb{Z}$ are *rational integers.* We have seen that $\mathbb{Z}$ is integrally closed since it is a principal ideal domain. We may also give an independent direct proof that:

If $x \in \mathbb{Q}$ is an algebraic integer over $\mathbb{Z}$, then $x \in \mathbb{Z}$.

Proof: We may assume that $x > 0$ and that

$$x^n + a_1 x^{n-1} + \cdots + a_n = 0,$$

with $a_1, \ldots, a_n \in \mathbb{Z}$. There exists an integer $k > 0$ such that $kx, kx^2, \ldots, kx^{n-1} \in \mathbb{Z}$; we choose the smallest k with this property.

Let $[x]$ denote the unique integer such that $[x] \leq x < [x] + 1$. Let

$$k' = k(x - [x]) = kx - k[x] \in \mathbb{Z},$$

so $0 \leq k' < k$. Also

$$k'x = k(x - [x])x = kx^2 - k[x]x \in \mathbb{Z}$$

and similarly

$$k'x^2, \ldots, k'x^{n-2} \in \mathbb{Z}.$$

Also

$$k'x^{n-1} = k(x - [x])x^{n-1} = kx^n - [x]kx^{n-1}$$
$$= -k(a_1 x^{n-1} + \cdots + a_{n-1}x + a_n) - [x]kx^{n-1} \in \mathbb{Z}.$$

By the minimality of k, $k' = 0$, that is, $x = [x]$ is in $\mathbb{Z}$. ∎

From our results, we see that the conjugates of an algebraic integer are algebraic integers.

If K is any field of algebraic numbers (of arbitrary degree over $\mathbb{Q}$), if $A_K = A$ denotes the ring of all algebraic integers in K, then A is an integrally closed domain, so $A \cap \mathbb{Q} = \mathbb{Z}$. We deduce that the trace (equal to the sum of conjugates) and the norm (equal to the product of conjugates) of an algebraic integer are rational integers (since they belong to $A \cap \mathbb{Q} = \mathbb{Z}$).

We shall be concerned with the arithmetic in an algebraic number field, relative to the subring of its algebraic integers. The natural question to ask is whether such rings must necessarily be unique factorization domains, and if this is not the case, how is it possible to describe their arithmetic.

Concerning the units of the ring of algebraic integers, we have the following easy fact:

N.   *An algebraic integer x is a unit if and only if its norm $N(x) = \pm 1$.*

Proof: If x is a unit, there exists an algebraic integer x' such that $xx' = 1$; taking the norms, we obtain $N(x) \cdot N(x') = 1$, so $N(x)$ is a unit in the ring $\mathbb{Z}$, that is, $N(x) = \pm 1$.

Conversely, if $N(x) = \pm 1$, then letting x' be the product of all conjugates of x, distinct from x, we have $x \cdot x' = \pm 1$; but x' is an algebraic integer, so x divides 1 in the ring A of algebraic integers.   ∎

We deduce:

O.   *Let A be the ring of algebraic integers of an algebraic number field K. Then every $a \in A$, $a \neq 0$, which is not a unit, is the product of indecomposable elements.*

Proof: We apply (**K**). Let $Aa_1 \subseteq Aa_2 \subseteq \cdots$ be the chain of principal ideals of A. Since $a_1 = ba_2$, taking norms in the extension $K|\mathbb{Q}$, then $N(a_1) = N(ba_2) = N(b) \cdot N(a_2)$. Since the norms are rational integers, then $N(a_2)$ divides $N(a_1)$. This argument may be repeated, so $N(a_n)$ divides $N(a_1)$ for every $n \geq 1$.

Since each rational integer has only finitely many divisors, there exists n such that $|N(a_{n+1})| = |N(a_n)|$. Thus $a_n = ca_{n+1}$ with $c \in A$ and $N(c) = \pm 1$. So by (**N**), c is a unit and $Aa_n = Aa_{n+1}$, which was required to be proved.   ∎

However, we have not established the *uniqueness* of decomposition into indecomposable elements. As we shall see this is actually not true.

We add the following definition:

Definition 7.   The algebraic number field K is said to be *Euclidean* when the ring A of algebraic integers is a Euclidean domain.

Often, we take the function δ to be equal to the norm, but this need not be so.

5.3 Arithmetic in the Field of Gaussian Numbers

An important example of a number field is $\mathbb{Q}(\sqrt{-1})$, which is called the field of *Gaussian numbers*. Its elements are of the form $a + bi$, where $a, b \in \mathbb{Q}$ and $i = \sqrt{-1}$, that is, $i^2 = -1$. Then $[\mathbb{Q}(i) : \mathbb{Q}] = 2$.

We shall describe the integers, units, and indecomposable elements in $\mathbb{Q}(i)$.

P.   *The ring of Gaussian integers is* $\mathbb{Z}[i] = \{a + bi \mid a, b \in \mathbb{Z}\}$.

Proof: Since $i = \sqrt{-1}$ is a root of $x^2 + 1$, then it is a Gaussian integer. Thus $\mathbb{Z}[i]$ is contained in the ring A of Gaussian integers.

Conversely, let $a, b \in \mathbb{Q}$ and $x = a + bi \in A$. Then $\text{Tr}(x) = 2a$ and $N(x) = a^2 + b^2$ are in $\mathbb{Z}$. Thus $(2a)^2 + (2b)^2 \in \mathbb{Z}$, so $(2b)^2 \in \mathbb{Z}$ and therefore $2b \in \mathbb{Z}$. Since $(2a)^2 + (2b)^2 \equiv 0 \pmod 4$, then necessarily $2a$, $2b$ are even. This shows that $a, b \in \mathbb{Z}$. ∎

Now we determine the units of the ring $\mathbb{Z}[i]$ of Gaussian integers.

Q.   *The units of* $\mathbb{Z}[i]$ *are* 1, -1, i, $-i$. *They are roots of unity.*

Proof: If $a + bi \in \mathbb{Z}[i]$ is a unit, by **(N)** $N(a + bi) = a^2 + b^2$ must be equal to ± 1, hence to 1. The only possibilities are $a = \pm 1$, $b = 0$, or $a = 0$, $b = \pm 1$. These give 1, -1, i, $-i$, which are indeed roots of unity. ∎

Now we shall prove that $\mathbb{Z}[i]$ is a Euclidean domain with respect to the norm.

R.   *If* $x, y \in \mathbb{Z}[i]$, $y \neq 0$, *there exist* $q, r \in \mathbb{Z}[i]$ *such that* $x = qy + r$, $|N(r)| < |N(y)|$. *In other words,* $\mathbb{Q}(i)$ *is a Euclidean field with respect to the norm.*

Proof: Consider the Gaussian number $x/y \in \mathbb{Q}(i)$, so we may write $x/y = a' + b'i$ with $a', b' \in \mathbb{Q}$.

Let $a, b \in \mathbb{Z}$ be the best approximations to a', b', that is,

$$|a' - a| \leq \tfrac{1}{2}, \qquad |b' - b| \leq \tfrac{1}{2}.$$

Let $q = a + bi \in \mathbb{Z}[i]$, so

$$x = qy + \left[(a' - a) + (b' - b)i\right]y$$

and

$$N(\left[(a' - a) + (b' - b)i\right]y) = \left[(a' - a)^2 + (b' - b)^2\right]N(y)$$
$$\leq \tfrac{1}{2}N(y) < N(y)$$

because $y \neq 0$ implies $N(y) \neq 0$. ∎

It follows from **(M)**, **(R)** that the ring $\mathbb{Z}[i]$ is a principal ideal domain, hence also a unique factorization domain. By **(L)**, it is true that if p is an indecomposable element in $\mathbb{Z}[i]$ and $p|xy$, then either $p|x$ or $p|y$.

To complete our description of the arithmetic in $\mathbb{Q}(i)$ we still need to determine the indecomposable elements in $\mathbb{Z}[i]$.

S.   *A Gaussian integer is indecomposable if and only if it is associated with one and only one of the following Gaussian integers:*

(1) *Any rational prime* p, *such that* $p \equiv 3 \pmod 4$.

(2) $1 + i$.

(3) $a + bi$ where $a, b \in \mathbb{Z}$, satisfy $a > 0$, $b \neq 0$, a is even, $a^2 + b^2 = p$ where p is a rational prime, $p \equiv 1 \pmod 4$.

Proof: First we show that every indecomposable Gaussian integer x divides one and only one rational prime.

In fact, $N(x) \in \mathbb{Z}$, so

$$N(x) = \pm p_1 \cdots p_r,$$

where each p_i is a rational prime; since x divides $N(x)$ (relatively to $\mathbb{Z}[i]$) then x divides some p_i.

Next, if x divides the distinct rational primes p, p', since there exist rational integers m, m' such that $mp + m'p' = 1$, it follows that x divides 1, that is, x is a unit, which is contrary to our hypothesis.

Now, we shall determine all indecomposable Gaussian integers $x = a + bi$ dividing a rational prime p.

If $p = 2$, from $x|p$ it follows that $a^2 + b^2 = N(x)$ divides $N(p) = p^2 = 4$. The only possibilities are: $a = \pm 1$, $b = \pm 1$, or $a = \pm 2$, $b = 0$, or $a = 0$, $b = \pm 2$. Since ± 1, $\pm i$ are units in $\mathbb{Z}[i]$, then all integers $1 + i$, $1 - i$, $-1 + i$, $-1 - i$ are associated. Finally, $1 + i$ is indecomposable, since its norm is the rational prime 2; so, if $1 + i = yz$, with Gaussian integers y, z, then $N(y)$ or $N(z)$ is ± 1, so y or z is a unit. In the other cases, we get ± 2, or $\pm 2i$, which are associated, and from $2 = (1 + i)(1 - i) = -i(1 + i)^2$, it follows that 2 is a decomposable Gaussian integer.

Let $p \equiv 1 \pmod 4$. By Chapter 4, (**H**), -1 is a square modulo p, that is, $p|n^2 + 1$ for some $n \in \mathbb{Z}$. But $n^2 + 1 = (n + i)(n - i)$. If x is an indecomposable Gaussian integer dividing p, then by (**L**), $x|n + i$ or $x|n - i$.

x and p are not associated in $\mathbb{Z}[i]$, because this would imply that p divides either $n + i$ or $n - i$, hence

$$\frac{n}{p} + \frac{1}{p}i \quad \text{or} \quad \frac{n}{p} - \frac{1}{p}i$$

would be a Gaussian integer, which is not the case.

It follows that $N(x) \neq N(p)$ and since x divides p then $N(x) = p$. Thus, if $x = a + bi$ then $p = N(x) = a^2 + b^2$. From $p \equiv 1 \pmod 4$ it follows that exactly one of a or b is even.

Among such numbers we have $x = a + bi$ with $a^2 + b^2 = p$, $a > 0$, a even, $b \neq 0$.

If b is even, then $\mp i(a + bi) = \pm b \mp ai$, so $a + bi$ is associated with $b - ai$ (or $-b + ai$) with b even, and $b > 0$ (or $-b > 0$).

Moreover, if $x = a + bi$, $y = c + di$ are such that $a^2 + b^2 = p$, $c^2 + d^2 = p$, a, c are even, $a > 0$, $c > 0$, $b \neq 0$, $d \neq 0$, and if $x \sim y$ then $x = y$. Indeed $x = uy$ with $u = 1$ or -1 or i or $-i$. But $u = -1$ would imply $a = -c < 0$, $u = i$ would imply $a = -d$ is odd, and $u = -i$ would imply that $a = d$ is odd. This establishes which are the indecomposable Gaussian integers dividing p, when $p \equiv 1 \pmod 4$.

Finally, let $p \equiv 3 \pmod 4$. If $x = a + bi$ is an indecomposable Gaussian integer dividing p, then $1 \neq N(x) = a^2 + b^2$ divides p^2, so either $a^2 + b^2 = p$ or $a^2 + b^2 = p^2$. But, from $p \equiv 3 \pmod 4$, it is not possible that $a^2 + b^2 = p$ (since $n^2 \equiv 0$ or $n^2 \equiv 1$ modulo 4, for every $n \in \mathbb{Z}$). Thus $N(x) = N(p)$ hence x/p is a unit, so x is associated with p. ∎

Part (3) of (S) may be phrased more explicitly, and constitutes an interesting theorem about rational integers, discovered by Fermat:

T.   *A positive integer $n = p_1^{k_1} \cdots p_s^{k_s}$ is the sum of squares of two integers if and only if k_j is even when $p_j \equiv 3 \pmod 4$.*

Proof: First we note that $2 = 1^2 + 1^2$. If p is a prime number congruent to 1 modulo 4, let $x = a + bi$ be an indecomposable Gaussian integer dividing p. Then $p = N(x) = a^2 + b^2$. Now we observe that if

$$n_1 = a_1^2 + b_1^2, \qquad n_2 = a_2^2 + b_2^2,$$

with $a_j, b_j \in \mathbb{Z}$, then also

$$n_1 n_2 = (a_1^2 + b_1^2)(a_2^2 + b_2^2) = (a_1 a_2 - b_1 b_2)^2 + (a_1 b_2 + a_2 b_1)^2.$$

Altogether, we have shown that if $n = p_1^{k_1} \cdots p_s^{k_s}$, if k_j is even when $p_j \equiv 3 \pmod 4$, then n is the sum of two squares.

Conversely, if $p \equiv 3 \pmod 4$ and $n = p^{2k+1}m$, with $k \geq 0$ and m not divisible by p, then n cannot be the sum of two squares. Indeed, let $n = a^2 + b^2$, let $d = \gcd(a,b)$, so $a = da_1$, $b = db_1$, and $\gcd(a_1, b_1) = 1$. Writing $a_1^2 + b_1^2 = n_1$ then $d^2 n_1 = n$. The exact power of p dividing d^2 has even exponent, hence p divides n_1. Then a_1 and b_1 are not multiples of p, since they are relatively prime.

Let $c \in \mathbb{Z}$ be such that $a_1 c \equiv b_1 \pmod p$. Then

$$n_1 = a_1^2 + b_1^2 \equiv a_1^2(1 + c^2) \equiv 0 \pmod p,$$

and therefore $c^2 \equiv -1 \pmod p$, that is, -1 is a quadratic residue modulo p. Therefore p would be either 2 or $p \equiv 1 \pmod 4$, which is contrary to the hypothesis. ∎

Due to its historical importance, it is now worthwhile to give Fermat's own proof of this theorem.

First we prove:

T'.   *A prime number p is a sum of two squares if and only if $p = 2$ or $p \equiv 1 \pmod 4$.*

Proof: If $p \neq 2$ and $p = a^2 + b^2$, then a, b cannot be both even—otherwise 4 divides p. If a, b are both odd, then $p \equiv 1 + 1 = 2 \pmod 4$, since every odd square is congruent to 1 modulo 4. Thus $p = 2$. If, say, a is odd and b is even, then $p \equiv 1 + 0 = 1 \pmod 4$.

Conversely, $2 = 1^2 + 1^2$, so let $p \equiv 1 \pmod 4$. By Chapter 4, (**H**), -1 is a square modulo p, so there exists x, $1 \leq x \leq p-1$, such that $x^2+1 = mp$, with $1 \leq m \leq p-1$.

Hence the set $\{m \mid 1 \leq m \leq p-1$, such that $mp = x^2 + y^2$ for some integers x, $y\}$ is not empty.

Let m_0 be the smallest integer in this set, so $1 \leq m_0 \leq p-1$. We show that $m_0 = 1$, hence p is a sum of two squares. Assume, on the contrary, that $1 < m_0$. We write

$$\begin{cases} x = cm_0 + x_1, \\ y = dm_0 + y_1, \end{cases}$$

with $-m_0/2 < x_1$, $y_1 \leq m_0/2$ and integers c, d. We observe that x_1 or y_1 is not 0. Otherwise m_0^2 divides $x^2 + y^2 = m_0 p$, hence m_0 divides p, thus $m_0 = p$, which is absurd.

We have

$$0 < x_1^2 + y_1^2 \leq \frac{m_0^2}{4} + \frac{m_0^2}{4} = \frac{m_0^2}{2} < m_0^2$$

and

$$x_1^2 + y_1^2 = m_0 m',$$

with $1 \leq m' < m_0$. But

$$m_0 p = x^2 + y^2, \qquad m_0 m' = x_1^2 + y_1^2,$$

hence

$$m_0^2 m' p = (x^2 + y^2)(x_1^2 + y_1^2) = (xx_1 + yy_1)^2 + (xy_1 - yx_1)^2.$$

We also have

$$xx_1 + yy_1 = x(x - cm_0) + y(y - dm_0)$$
$$= (x^2 + y^2) - m_0(xc + yd) = m_0 t,$$
$$xy_1 - yx_1 = x(y - dm_0) - y(x - cm_0)$$
$$= -m_0(xd - yc) = m_0 u$$

for some integers t, u. Hence $m' p = t^2 + u^2$, with $1 \leq m' < m_0$. This is a contradiction and concludes the proof. ∎

Now we may complete Fermat's proof of (**T**).

Let $n = p_1^{k_1} \cdots p_s^{k_s}$ and assume that k_j is even if $p_j = 3 \pmod 4$. Then $n = n_0^2 n_1$ where $n_0 \geq 1$, $n_1 \geq 1$, and n_1 is the product of distinct primes, which are either equal to 2, or congruent to 1 modulo 4. By (**T'**), each factor of n_1 is a sum of two squares; by the two-squares identity, n_1, and therefore also n, is a sum of two squares.

Conversely, let $n = x^2 + y^2$; the statement is trivial if $x = 0$ or $y = 0$. Let x, y be nonzero, let $d = \gcd(x, y)$, so d^2 divides n.

Let $n = d^2 n'$, $x = dx'$, $y = dy'$, hence $\gcd(x', y') = 1$ and $n' = x'^2 + y'^2$. If p divides n', then p does not divide x'—otherwise p would also divide y'.

Let k be such that $kx' \equiv y' \pmod{p}$. Then $x'^2 + y'^2 \equiv x'^2(1 + k^2) \equiv 0$ $\pmod{p}$. Thus p divides $1 + k^2$, that is, -1 is a square modulo p, so $p = 2$ or $p \equiv 1 \pmod 4$, by Chapter 4, **(H)**.

It follows that if $p_j \equiv 3 \pmod 4$ then p_j does not divide n', hence p_j divides d, so the exponent k_j must be even. ∎

5.4 Integers of Quadratic Number Fields

Let K be a quadratic extension of $\mathbb{Q}$, that is, $[K : \mathbb{Q}] = 2$.

As we already indicated in Chapter 4, **(P)**, we have $K = \mathbb{Q}(\sqrt{d})$, where d is a square-free integer.

Every element of K is of type $a + b\sqrt{d}$, where $a, b \in \mathbb{Q}$. The conjugate of $a + b\sqrt{d}$ is $a - b\sqrt{d}$. Let A denote the ring of all algebraic integers of $\mathbb{Q}(\sqrt{d})$.

U. $a + b\sqrt{d} \in A$ *if and only if* $2a = u \in \mathbb{Z}$, $2b = v \in \mathbb{Z}$, *and* $u^2 - dv^2 \equiv 0 \pmod 4$.

Proof: If $x = a + b\sqrt{d} \in A$ then its conjugate $x' = a - b\sqrt{d}$ is also an algebraic integer. So $x + x' = 2a \in A \cap \mathbb{Q} = \mathbb{Z}$, $x \cdot x' = a^2 - b^2 d \in A \cap \mathbb{Q} = \mathbb{Z}$.

It follows that $(2a)^2 - (2b)^2 d \in 4\mathbb{Z}$ and since $(2a)^2 \in \mathbb{Z}$ then $(2b)^2 d \in \mathbb{Z}$; but d is square-free, thus $2b$ has denominator equal to 1, that is, $v = 2b \in \mathbb{Z}$.

Conversely, these conditions imply $a^2 - b^2 d \in \mathbb{Z}$, and since x is a root of $X^2 - 2aX + (a^2 - b^2 d)$, then x is an algebraic integer. ∎

The previous result may be reformulated as follows:

V.   *Let* $K = \mathbb{Q}(\sqrt{d})$ *where* d *is a square-free integer; let* A *be the ring of all algebraic integers of* K. *If* $d \equiv 2 \pmod 4$ *or* $d \equiv 3 \pmod 4$ *then* $A = \{a + b\sqrt{d} \mid a, b \in \mathbb{Z}\}$. *If* $d \equiv 1 \pmod 4$ *then*

$$A = \left\{ \frac{u + v\sqrt{d}}{2} \;\middle|\; u, v \in \mathbb{Z}, u \text{ and } v \text{ have the same parity} \right\}.$$

Proof: We examine all the possible cases in succession.

If $d \equiv 2 \pmod 4$:

u	even	even	odd	odd	
v	even	odd	even	odd	
$u^2 - dv^2 \equiv$	0	2	1	3	(mod 4)

If $d \equiv 3 \pmod 4$:

u	even	even	odd	odd	
v	even	odd	even	odd	
$u^2 - dv^2 \equiv$	0	1	1	2	$\pmod 4$

If $d \equiv 1 \pmod 4$:

u	even	even	odd	odd	
v	even	odd	even	odd	
$u^2 - dv^2 \equiv$	0	3	1	0	$\pmod 4$

Therefore, by means of (**U**), we deduce statement (**V**).   ∎

Let us note incidentally the following fact, which will be generalized later:

W.   *A is a free Abelian group. If $d \equiv 2 \pmod 4$ or $d \equiv 3 \pmod 4$ then $\{1, \sqrt{d}\}$ is a basis of A. If $d \equiv 1 \pmod 4$ then $\{1, (1 + \sqrt{d})/2\}$ is a basis of A.*

Proof: The statement is obvious when $d \equiv 2 \pmod 4$ or $d \equiv 3 \pmod 4$. Let us assume now that $d \equiv 1 \pmod 4$ and let us show that every algebraic integer $(u + v\sqrt{d})/2$ (with u, v integers of the same parity) is a linear combination of 1 and $(1 + \sqrt{d})/2$, with coefficients in $\mathbb{Z}$.

If u, v are even, $u = 2a$, $v = 2b$ with $a, b \in \mathbb{Z}$, so

$$\frac{u + v\sqrt{d}}{2} = a + b\sqrt{d} = (a - b)1 + 2b\left(\frac{1 + \sqrt{d}}{2}\right).$$

If u, v are odd, then $u - 1$, $v - 1$ are even, so

$$\frac{u + v\sqrt{d}}{2} = \frac{1 + \sqrt{d}}{2} + \left[\frac{u - 1}{2} + \frac{v - 1}{2}\sqrt{d}\right]$$

and this last summand is a linear combination of 1, $(1 + \sqrt{d})/2$ with coefficients in $\mathbb{Z}$.   ∎

It has been shown that if d is square-free and $\mathbb{Q}(\sqrt{d})$ is a Euclidean field (with respect to the norm), then

$$d = -11, \ -7, \ -3, \ -2, \ -1, \ 2, \ 3, \ 5, \ 6, \ 7, \ 11,$$
$$13, \ 17, \ 19, \ 21, \ 29, \ 33, \ 37, \ 41, \ 57, \ 73.$$

In these cases, it follows from (**N**) that the ring of algebraic integers is a principal ideal domain.

This rather negative result does not yet exclude the possibility that for all quadratic fields, the ring of algebraic integers is a principal ideal domain. The following classical example shows that this is not the case:

Example: Let us consider the field $K = \mathbb{Q}(\sqrt{-5})$, so the ring of algebraic integers A consists of the numbers of the type $a + b\sqrt{-5}$, where $a, b \in \mathbb{Z}$ (by (**V**)).

We may write

$$21 = 3 \cdot 7 = (1 + 2\sqrt{-5})(1 - 2\sqrt{-5}).$$

Each of the numbers 3, 7, $1 + 2\sqrt{-5}$, $1 - 2\sqrt{-5}$ is indecomposable in $\mathbb{Z}[\sqrt{-5}]$. For example, if $3 = xy$, with x, y not units, then taking norms, $9 = N(x) \cdot N(y)$, therefore $N(x) = N(y) = 3$. If

$$x = a + b\sqrt{-5}$$

then

$$N(x) = a^2 + 5b^2 = 3,$$

but this is impossible with $a, b \in \mathbb{Z}$.

Also the numbers 3, 7, $1 + 2\sqrt{-5}$, $1 - 2\sqrt{-5}$ are pairwise nonassociated, since $N(3) = 9$, $N(7) = 49$, $N(1 + 2\sqrt{-5}) = N(1 - 2\sqrt{-5}) = 21$, and

$$\frac{1 + 2\sqrt{-5}}{1 - 2\sqrt{-5}} = \frac{-19 + 4\sqrt{-5}}{21} \notin \mathbb{Z}[\sqrt{-5}].$$

This example shows what seems to be even stronger, namely, $\mathbb{Z}[\sqrt{-5}]$ is not a unique factorization domain. (Later, we shall prove that the ring A of all algebraic integers of a field of algebraic numbers is a principal ideal domain if and only if it is a unique factorization domain.)

These hopes being dashed, we may still ask which are the quadratic fields $\mathbb{Q}(\sqrt{d})$ having a principal ideal domain as the ring of integers?

Besides the domains with Euclidean algorithms, if $d < 0$ then it must be equal to $d = -19$, -43, -67, or -163 (see Chapter 13, Section 7).

The story of this proof is very interesting. Heilbronn and Linfoot proved in 1934 that there could exist at most one more value of $d < 0$ for which the ring of integers of $\mathbb{Q}(\sqrt{d})$ would be a principal ideal domain. Lehmer used analytical methods to show that any other possible d could not be small in absolute value, say $|d| > 5 \times 10^9$. Heegner proved in 1952 that no other d could exist, but his proof, using modular forms, was flawed. In 1966, Baker used his results on bounds of linear forms in logarithms, to prove that no other $d < 0$ could exist. In 1967 Stark gave another proof and Deuring corrected the errors in Heegner's proof.

We have mentioned these facts, since they serve to illustrate the need of appealing to delicate analytical methods. This is a recurring characteristic in the theory of algebraic numbers.

On the other hand, Gauss conjectured that there exist infinitely many fields $\mathbb{Q}(\sqrt{d})$, with $d > 0$, whose ring of algebraic integers is a principal ideal domain. No proof has yet been found for this statement.

5.5 Integers of Cyclotomic Fields

Let $K = \mathbb{Q}(\zeta)$, where ζ is a primitive pth root of unity and p is an odd prime number (for $p = 2$ the results are trivial).

The minimal polynomial of ζ over $\mathbb{Q}$ is

$$\Phi_p = X^{p-1} + X^{p-2} + \cdots + X + 1,$$

hence ζ belongs to the ring A of integers of $\mathbb{Q}(\zeta)$. The roots of Φ_p are $\zeta, \zeta^2, \ldots, \zeta^{p-1}$, thus

$$\Phi_p = \prod_{i=1}^{p-1} (X - \zeta^i);$$

in particular, $p = \Phi_p(1) = \prod_{i=1}^{p-1}(1 - \zeta^i)$. Let us note that the elements $1 - \zeta$, $1 - \zeta^2$, $\ldots$, $1 - \zeta^{p-1}$ are associated. In fact, if $1 \le i$, $j \le p - 1$ then there exists integer k such that $j \equiv ik \pmod{p}$; thus

$$\frac{1 - \zeta^j}{1 - \zeta^i} = \frac{1 - \zeta^{ik}}{1 - \zeta^i} = 1 + \zeta^i + \zeta^{2i} + \cdots + \zeta^{(k-1)i} \in A;$$

similarly, $(1 - \zeta^i)/(1 - \zeta^j) \in A$, so $1 - \zeta^i = u_i(1 - \zeta)$, where u_i is a unit of A. We conclude that $p = u(1 - \zeta)^{p-1}$ where $u = u_1 \cdots u_{p-1}$ is a unit of A.

The element $1 - \zeta$ is not invertible in A, otherwise p would have an inverse, which belongs to $A \cap \mathbb{Q} = \mathbb{Z}$. Hence $A(1 - \zeta) \cap \mathbb{Z} = \mathbb{Z}p$, since the ideal $A(1 - \zeta) \cap \mathbb{Z}$ contains p and is not equal to the unit ideal.

X.   *A is a free Abelian group with basis* $\{1, \zeta, \ldots, \zeta^{p-2}\}$, *so* $A = \mathbb{Z}[\zeta]$.

Proof: Obviously, 1, ζ, $\ldots$, ζ^{p-2} are linearly independent over $\mathbb{Q}$, otherwise ζ would be a root of a polynomial of degree at most $p-2$, contradicting the fact that Φ_p is its minimal polynomial.

If $x \in A$ there exist uniquely defined rational numbers a_0, a_1, $\ldots, a_{p-2}$ such that $x = a_0 + a_1\zeta + \cdots + a_{p-2}\zeta^{p-2}$. We shall prove that each $a_i \in \mathbb{Z}$.

We have $x\zeta = a_0\zeta + a_1\zeta^2 + \cdots + a_{p-2}\zeta^{p-1}$ and subtracting:

$$x(1 - \zeta) = a_0(1 - \zeta) + a_1(\zeta - \zeta^2) + \cdots + a_{p-2}(\zeta^{p-2} - \zeta^{p-1}).$$

We note that the traces (in $\mathbb{Q}(\zeta)|\mathbb{Q}$) of ζ, ζ^2, $\ldots$, ζ^{p-1} are all equal (since these elements are conjugate). Hence

$$\mathrm{Tr}(x(1 - \zeta)) = \mathrm{Tr}(a_0(1 - \zeta)) = a_0 \cdot \mathrm{Tr}(1 - \zeta)$$
$$= a_0[(p - 1) + 1] = a_0 p.$$

To show that $a_0 \in \mathbb{Z}$, we compute $\text{Tr}(x(1 - \zeta))$.

Let $x_1 = x, x_2, \ldots, x_{p-1} \in A$ be the conjugates of x; so

$$\text{Tr}(x(1 - \zeta)) = x_1(1 - \zeta) + x_2(1 - \zeta^2) + \cdots + x_{p-1}(1 - \zeta^{p-1})$$
$$= (1 - \zeta)x' \in A(1 - \zeta),$$

since $(1 - \zeta^{i+1})/(1-\zeta) = 1 + \zeta + \cdots + \zeta^i \in A$. But $\text{Tr}(x(1-\zeta)) \in A \cap \mathbb{Q} = \mathbb{Z}$, hence $\text{Tr}(x(1 - \zeta)) \in A(1 - \zeta) \cap \mathbb{Z} = \mathbb{Z}p$, that is, $a_0 \in \mathbb{Z}$.

Now, we show by induction that also $a_1, \ldots, a_{p-2} \in \mathbb{Z}$. To prove that $a_j \in \mathbb{Z}$, we multiply by ζ^{p-j}, obtaining $x\zeta^{p-j} = a_0\zeta^{p-j} + a_1\zeta^{p-j+1} + \cdots + a_{j-1}\zeta^{p-1} + a_j + a_{j+1}\zeta + \cdots + a_{p-2}\zeta^{p-j-2}$, and expressing ζ^{p-1} in terms of the lower powers of ζ, we may write $x\zeta^{p-j}$ in the form

$$x\zeta^{p-j} = (a_j - a_{j-1}) + a_1'\zeta + a_2'\zeta^2 + \cdots + a_{p-2}'\zeta^{p-2}.$$

By induction $a_{j-1} \in \mathbb{Z}$, so that by the same argument, $a_j - a_{j-1} \in \mathbb{Z}$, thus $a_j \in \mathbb{Z}$.   ∎

EXERCISES

1. Let x be a root of $X^3 - 2X + 5$. Compute the norm and trace of $2x - 1$ in the extension $\mathbb{Q}(x)|\mathbb{Q}$.

2. Let x be an algebraic integer, $x \neq 0$. Let $f \in \mathbb{Q}[X]$ be the minimal polynomial of x. Show that x^{-1} is an algebraic integer if and only if $f(0) = \pm 1$.

3. Let $f \in \mathbb{Z}[X]$ be a monic polynomial and let x be an algebraic number. Show that if $f(x)$ is an algebraic integer then x is an algebraic integer.

4. Let J be a nonzero ideal of the ring A of integers of an algebraic number field K. Show that there exists a positive integer m belonging to J.

5. Give an example of a Gaussian number $x = a + bi$ such that $N(x) = 1$ but x is not an algebraic integer.

6. Find the quotient and the remainder of the following divisions:
   (a) $2 + 3i$ by $1 + i$.
   (b) $3 - 2i$ by $1 + 2i$.
   (c) $4 + 5i$ by $2 - i$.

7. Find the greatest common divisor of the following pairs of Gaussian integers:

(a) $15 + 12i, \ 3 - 9i$.

(b) $6 + 7i, \ 12 - 3i$.

(c) $3 + 8i, \ 12 + i$.

8. Find the decomposition into prime factors of the following Gaussian integers: $12 + i, \ 6 + 2i, \ 35 - 12i, \ 3 + 5i$.

9. Show that if $x = a + bi$ is an indecomposable Gaussian integer then either $ab = 0$ or $\gcd(a, b) = 1$.

10. Determine the indecomposable elements of the ring of algebraic integers of $\mathbb{Q}(\sqrt{2})$.

11. Determine the indecomposable elements of the ring of algebraic integers of $\mathbb{Q}(\sqrt{5})$.

12. Let $K = \mathbb{Q}(\sqrt{d})$ and let A be the ring of integers of K. Show that A is a Euclidean domain if and only if for every $x + y\sqrt{d} \in K$ there exists $a + b\sqrt{d} \in A$ such that $|N((x + y\sqrt{d}) - (a + b\sqrt{d}))| < 1$. Consider the cases where $d \equiv 1 \pmod 4$ and $d \not\equiv 1 \pmod 4$ and derive explicit relations.

13. Prove that $\mathbb{Q}(\sqrt{-1}), \ \mathbb{Q}(\sqrt{-2}), \ \mathbb{Q}(\sqrt{-3}), \ \mathbb{Q}(\sqrt{-7}), \ \mathbb{Q}(\sqrt{-11})$ are the only Euclidean fields $\mathbb{Q}(\sqrt{d})$, with $d < 0$.

Hint: Use the previous exercise.

14. Prove that if $d = 2, \ 3, \ 5, \ 6, \ 7, \ 13, \ 17, \ 21,$ and 29 then $\mathbb{Q}(\sqrt{d})$ is a Euclidean field.

Hint: Use Exercise 12.

15. Prove that there exist only finitely many integers d such that $d \equiv 2 \pmod 4$ or $d \equiv 3 \pmod 4$ and $\mathbb{Q}(\sqrt{d})$ is a Euclidean field.

16. Prove that $\mathbb{Q}(\sqrt{-19})$ and $\mathbb{Q}(\sqrt{23})$ are not Euclidean fields.

17. Determine the ring of integers of the field $\mathbb{Q}(\sqrt{2}, i)$. Prove that this is a Euclidean domain.

18. Determine the ring of integers of the field $\mathbb{Q}(\sqrt{2}, \sqrt{3})$ and prove that this is a Euclidean domain.

19. Let ζ be a primitive fifth root of unity. Prove that $\mathbb{Q}(\zeta)$ is a Euclidean field.

20. Determine the ring of integers of the field $\mathbb{Q}(\sqrt[4]{2})$.

21. Let ω be a primitive cubic root of unity. Determine the norm of an arbitrary element $a + b\omega\,(a, b \in \mathbb{Q})$. Determine the ring of algebraic integers and the units of $\mathbb{Q}(\omega)$.

22. Let ω be a primitive cubic root of unity. Prove that $\mathbb{Q}(\omega)$ is a Euclidean field.

23. Determine the indecomposable elements of the ring of algebraic integers of $\mathbb{Q}(\omega)$, where ω is a primitive cubic root of unity.

24. In the ring $\mathbb{Z}\big[\sqrt{-5}\big]$ consider the following ideals: $I = (3, 4 + \sqrt{-5})$, $I' = (3, 4 - \sqrt{-5})$, $J = (7, 4 + \sqrt{-5})$, $J' = (7, 4 - \sqrt{-5})$. Prove that $I \cdot I' = (3)$, $J \cdot J' = (7)$, $I \cdot J = (4 + \sqrt{-5})$, $I' \cdot J' = (4 - \sqrt{-5})$, and show that I, I', J, J' are prime ideals.

25. Find algebraic integers $x, y \in \mathbb{Q}(\sqrt{-5})$ such that x, y have no common factor different from a unit, but $xr + ys \neq 1$ for all algebraic integers $r, s \in \mathbb{Q}(\sqrt{-5})$.

26. Show that the ring of algebraic integers of $\mathbb{Q}(\sqrt{10})$ is not a unique factorization domain.

27. Let K be a Galois extension of degree n over $\mathbb{Q}$. Show that for every nonzero algebraic integer x of K we have

$$\sum_{i=1}^{n} \sigma_i(x) \cdot \overline{\sigma_i(x)} \geq n,$$

(where $\sigma_1, \ldots, \sigma_n$ are the automorphisms of K and $\bar{a}$ denotes the complex conjugate of a).

   Hint: Consider the norm of x and use the fact that the geometric mean of positive numbers does not exceed their arithmetic mean.

28. Let $\zeta_1, \ldots, \zeta_n$ be roots of unity over $\mathbb{Q}$. Show that $|\zeta_1 + \cdots + \zeta_n| \leq n$ and the equality holds if and only if $\zeta_1 = \cdots = \zeta_n$. Conclude that if $x = \zeta_1 + \cdots + \zeta_n$ and $|x| < n$ then for every conjugate $\sigma(x)$ of x we have $|\sigma(x)| < n$.

29. Let R be a domain, and let $\mathbf{K}_R$ be the free R-module with basis $\{1, i, j, k\}$.

   On $\mathbf{K}_R$ we define an operation of multiplication which is bilinear and has the following multiplication table: 1 is the unit element; $i^2 = j^2 = k^2 = -1$; $ij = -ji = k$; $jk = -kj = i$; $ki = -ik = j$. With this operation $\mathbf{K}_R$ is an R-algebra, called the *algebra of quaternions over R*. We identify R with the subring $R \cdot 1$ of multiples of 1. In $\mathbf{K}_R$ we have the conjugation, defined as follows: if $\alpha = a_0 + a_1 i + a_2 j + a_3 k$ then $\bar{\alpha} = a_0 - a_1 i - a_2 j - a_3 k$. Finally, let $N : \mathbf{K}_R \to R$ be the norm mapping, defined by

$$N(\alpha) = \alpha\bar{\alpha} = a_0^2 + a_1^2 + a_2^2 + a_3^2 \in R.$$

Prove:
  (a) $\overline{\alpha + \beta} = \overline{\alpha} + \overline{\beta}$, $\overline{\alpha\beta} = \overline{\alpha}\,\overline{\beta}$, $\overline{\overline{\alpha}} = \alpha$.
  (b) $N(\alpha\beta) = N(\alpha) \cdot N(\beta)$.
  (c) α is invertible in $\mathbf{K}_R$ if and only if $N(\alpha)$ is invertible in R.
  (d) If R is a field in which 0 is not a sum of four squares, of which some can be equal to 0, then $N(\alpha) = 0$ if and only if $\alpha = 0$; hence $\mathbf{K}_R$ is a *skew-field*, that is, a ring which is not commutative and such that every nonzero element is invertible.
  (e) The product of two sums of four squares in R is a sum of four squares in R.

30. Prove *Euler's identity*: In any commutative ring we have

$$(x_1^2 + x_2^2 + x_3^2 + x_4^2)(y_1^2 + y_2^2 + y_3^2 + y_4^2)$$
$$= (x_1 y_1 - x_2 y_2 - x_3 y_3 - x_4 y_4)^2 + (x_1 y_2 + x_2 y_1 + x_3 y_4 - x_4 y_3)^2$$
$$+ (x_1 y_3 + x_3 y_1 + x_4 y_2 - x_2 y_4)^2 + (x_1 y_4 + x_4 y_1 + x_2 y_3 - x_3 y_2)^2.$$

31. A quaternion

$$\alpha = \tfrac{1}{2}(a_0 + a_1 i + a_2 j + a_3 k) \in \mathbf{K}_\mathbb{Q}$$

is said to be *integral* when all coefficients a_i are integers with the same parity. Let A be the set of all integral quaternions. Show:
  (a) A is a subring of $\mathbf{K}_\mathbb{Q}$ containing $\mathbf{K}_\mathbb{Z}$.
  (b) A is a free $\mathbb{Z}$-module with basis $\{\tfrac{1}{2}(1 + i + j + k), i, j, k\}$.
  (c) If $\alpha \in A$ then $N(\alpha) \in \mathbb{Z}$.
  (d) The only units in the ring A (that is, quaternions α such that α and α^{-1} belong to A) are the following 24 quaternions: ± 1, $\pm i$, $\pm j$, $\pm k$, $\tfrac{1}{2}(\pm 1 \pm i \pm j \pm k)$.

32. Let $\alpha, \beta \in \mathbf{K}_\mathbb{Q}$. We say that β is a *right-hand factor* of α when there exists $\gamma \in A$ such that $\alpha = \gamma\beta$. Similarly, we define a left-hand factor. We say that α, β are *associates* when there exists a unit ε of A such that $\alpha = \beta\varepsilon$ or $\alpha = \varepsilon\beta$. Prove:
  (a) If $\alpha, \beta \in A$, $\beta \neq 0$, there exists $\gamma, \rho \in A$ such that $\alpha = \beta\gamma + \rho$ and $N(\rho) < N(\beta)$.
  (b) Let J be a right ideal of A, that is, $J \pm J \subseteq J$, $J \cdot A \subseteq J$. Show that J is principal; that is, there exists $\alpha \in A$ such that $J = \alpha A$.
  (c) If $\alpha, \beta \in A$, not both equal to 0, there exists a greatest common right-hand factor δ; δ is unique up to a left-hand unit factor and may be expressed in the form $\delta = \mu\alpha + \vartheta\beta$ with $\mu, \vartheta \in A$ (Bézout property).

(d) Let $\alpha \in A$, let $b \in \mathbb{Z}$, $b > 0$; show that the greatest common right-hand factor of α, b is equal to 1 if and only if $\gcd(N(\alpha), b) = 1$.

(e) If $\alpha \in A$ show that there exists one associate α' of α which belongs to $\mathbf{K}_\mathbb{Z}$.

33. An integral quaternion π is said to be *indecomposable* whenever the only factors of π are units or associates of π. Prove:

(a) If $p \in \mathbb{Z}$ is a prime number then p is not an indecomposable quaternion.

(b) An integral quaternion $\pi \in A$ is indecomposable if and only if $N(\pi)$ is a prime number.

34. Prove the following theorem of Lagrange:

Every natural number is the sum of squares of four nonnegative integers.

Hint: Use Exercise 30 to reduce to the case of a prime number; then use Exercises 32 and 33 to express p as the norm of an indecomposable quaternion.

6

Integral Basis, Discriminant

We have seen in the numerical examples of the preceding chapter that the ring of algebraic integers of a quadratic number field, and also of the cyclotomic field $\mathbb{Q}(\zeta)$ (where ζ is a primitive pth root of unity), are free finitely generated Abelian groups.

In this chapter, we shall prove this fact in general and establish other interesting properties of the ring of algebraic integers. For this purpose, it will be necessary to develop theories which belong properly to algebra. However, we think that their inclusion in the text will be convenient to certain readers. Others, already well versed in these facts, may just take note of our notation and terminology.

To conclude the chapter, we shall introduce the discriminant, which is a rational integer associated with every algebraic number field. It will be a recurring procedure in the theory to attach numerical invariants to algebraic number fields, and in each case they will serve to measure a certain phenomenon.

6.1 Finitely Generated Modules

To begin, let us recall the notion of rank of a module. Let R be a domain, M an R-module. If n is the maximum number of linearly independent elements of M, then n is called *the rank of M*.

If F is the field of quotients of R, if it is known that the R-module M is contained in a vector space V over F, let FM denote the subspace of V generated by M (it consists of all elements of the form $\sum_{i=1}^{r} a_i x_i$, with $a_i \in F$, $x_i \in M \subseteq V$ for $i = 1, \ldots, r$).

Thus for every element $y \in FM$ there exists $a \in R$, $a \neq 0$, such that $ay \in M$. We note that the elements $x_1, \ldots, x_n$ of M are linearly independent over R if and only if they are also linearly independent over F. Hence, M has rank n exactly when the F-vector space FM has dimension n.

Thus, everything is as natural as possible, when M is contained in a vector space V over F. There are several instances in which this occurs, but we just want to mention the very simplest of these cases:

A.   *Let R be a domain, F its field of quotients. If M is a free R-module having a basis with n elements, then there exists an F-vector space V containing M; any two bases of M have the same number of elements, equal to the rank of M.*

Proof: We take $V = F\xi_1 \oplus \cdots \oplus F\xi_n$ to be the set of all "formal" linear combinations of symbols $\xi_1, \ldots, \xi_n$, with coefficients in F; thus, elements of V may be written uniquely in the form $\sum_{i=1}^n a_i\xi_i$, with $a_i \in F$.

Let $\{x_1, \ldots, x_n\}$ be any basis of the R-module M, so

$$M = Rx_1 \oplus \cdots \oplus Rx_n.$$

The mapping $\theta : M \to V$, defined by $\theta(\sum_{i=1}^n a_i x_i) = \sum_{i=1}^n a_i\xi_i$, is an isomorphism from the R-module M into V. Thus, replacing M by its image, we may consider M as contained in V.

If $y_1, \ldots, y_r$ are elements of $M \subseteq V$, linearly independent over R then by our previous considerations $r \le n$, the dimension of the vector space V over F. In particular, any other basis of the R-module M has at most n elements. By symmetry, any two bases of the R-module M have n elements and n is the rank of M.   ∎

Let us prove a weak result, which already hints of the main theorem:

B.   *Let R be an integrally closed domain, F its field of quotients, let K be a separable extension of degree n of F, and let A be the integral closure of R in K. Then there exist free R-modules M and M' of rank n, such that $M' \subseteq A \subseteq M$. Explicitly, if $K = F(t)$ with $t \in A$, if d is the discriminant of t in $K|F$, then*

$$M' = R \oplus Rt \oplus \cdots \oplus Rt^{n-1} = R[t], \qquad M = (1/d)R[t].$$

Proof: Let t' be a primitive element of K over F, so $K = F(t')$. By Chapter 5, (**I**), we may write $t' = t/b$, with $t \in A$, $b \in R$. Thus $K = F(t)$ and therefore $\{1, t, \ldots, t^{n-1}\}$ is an F-basis of K. Since $t \in A$, then A contains the free R-module M' generated by this basis: $M' = R \oplus Rt \oplus \cdots \oplus Rt^{n-1} \subseteq A$; evidently M' is free of rank n.

To prove that A is contained in a free R-module M, we proceed as follows:

Let d be the discriminant of t in $K|F$, that is, $d = \prod_{i<j}(t_i - t_j)^2$, where $t_1 = t, t_2, \ldots, t_n$ are the conjugates of t over F; so $d \in F$, $d \ne 0$, since the extension is separable.

We shall prove that $A \subseteq M \subseteq R(1/d) \oplus R(t/d) \oplus \cdots \oplus R(t^{n-1}/d)$, and M is a free module of rank n. Let $y \in A$, so we may write $y = \sum_{j=0}^{n-1} c_j t^j$, where $c_j \in F$ for all $j = 0, 1, \ldots, n-1$. Then $y = \sum_{j=0}^{n-1} dc_j(t^j/d)$ and our task is now to show that each element dc_j belongs to R. We have $dc_j \in F$, and since R is integrally closed, it is enough to show that dc_j is integral over R.

Let K' be the smallest normal extension of F containing K, let $y = y_1, \ldots, y_n$ be the conjugates of y over F, so $y_i, t_i \in K'$. From $y = \sum_{j=0}^{n-1} c_j t^j$ we deduce $y_i = \sum_{j=0}^{n-1} c_j t_i^j$ for all $i = 1, \ldots, n$.

This set of relations indicates that c_0, ..., c_{n-1} is the solution of the system of linear equations $\sum_{j=0}^{n-1} t_i^j X_j = y_i$ $(i = 1, \ldots, n)$ with coefficients $t_i^j \in K'$. To apply Cramer's rule, we note that

$$\delta = \det(t_i^j) = \prod_{i<j} (t_i - t_j)$$

(as a Vandermonde determinant) so $\delta^2 = d$ and $\delta c_j = e_j$ where e_j is the determinant of the matrix obtained from (t_i^j) replacing the column of jth powers by the elements y_1, ..., y_n. Since each y_i, t_i^j is integral over R, then δ and e_j are integral over R. Therefore, $dc_j = \delta e_j$ is integral over R, proving the proposition. ∎

The preceding result leads to the following question: if R is a ring, if M is an R-module, when are all submodules of M finitely generated? We treat this question in a roundabout way, giving a name to the modules with the above property, and finding sufficient conditions for modules to belong to the class in question.

Definition 1.   Let R be a ring; an R-module M is said to be *Noetherian* whenever every submodule of M is finitely generated.

In particular, M itself is finitely generated.

C.   *Let R be a ring; M an R-module. Then the following properties are equivalent:*

(1) *M is a Noetherian R-module.*

(2) *Every strictly increasing chain $N_1 \subset N_2 \subset N_3 \subset \cdots$ of submodules of M is finite.*

(3) *Every nonempty family of submodules of M has a maximal element (with respect to the inclusion relation).*

Proof: (1) → (2) Let $N_1 \subset N_2 \subset N_3 \subset \cdots$ be an increasing sequence of submodules of M, and let N be the union of all these submodules. By hypothesis, N is finitely generated, say by the elements x_1, ..., x_n. For every index $i = 1, \ldots, n$, there exists an index j_i such that $x_i \in N_{j_i}$. If $m \geq j_i$ for all $i = 1, \ldots, n$, then each $x_i \in N_m$ $(i = 1, \ldots, n)$ hence $N = N_m$, so $N_m = N_{m+1} = \cdots$.

(2) → (3) Let $\mathcal{M}$ be a nonempty family of submodules of M. Let $N_1 \in \mathcal{M}$; if N_1 is not a maximal element of $\mathcal{M}$, there exists $N_2 \in \mathcal{M}$ such that $N_1 \subset N_2$; if N_2 is not a maximal element of $\mathcal{M}$, there exists $N_3 \in \mathcal{M}$, such that $N_1 \subset N_2 \subset N_3$. This procedure must lead to a maximal element

of $\mathcal{M}$, otherwise there would exist an infinite strictly increasing chain of submodules of $\mathcal{M}$, which is contrary to the hypothesis.

(3) $\rightarrow$ (1) Let us assume that there exists a submodule N of M which is not finitely generated. Let $\mathcal{M}$ be the family of all finitely generated submodules of M which are contained in N (for example, $0 \in \mathcal{M}$); let N' be a maximal element of $\mathcal{M}$, so $N' \neq N$. If $x \in N$, $x \notin N'$, the module $N' + Rx$ is still finitely generated, and contained in N, hence $N' + Rx \in \mathcal{M}$, with $N' \subset N' + Rx$; this contradicts the maximality of N'.   ∎

An important particular case is obtained by considering the R-module $M = R$:

Definition 2.   A ring R is said to be *Noetherian* when every ideal of R is finitely generated.

Thus we may rephrase (**C**):

C′.   *Let R be a ring. The following properties are equivalent:*

(1) *R is a Noetherian ring.*

(2) *Every strictly increasing chain $J_1 \subset J_2 \subset J_3 \subset \cdots$ of ideals of R is finite.*

(3) *Every nonempty family of ideals of R has a maximal element (with respect to the inclusion relation).*

Thus, in particular, every principal ideal domain is a Noetherian ring.

It follows at once that if R is a Noetherian ring and J is an ideal, $J \neq R$, then J is contained in a maximal ideal.

Indeed, it suffices to consider the nonempty family of all ideals I of R, such that $J \subseteq I \neq R$ and then to apply the above property (3).

We shall now develop properties of Noetherian modules:

D.   *Every submodule and every quotient module of a Noetherian module are Noetherian modules.*

Proof: Let M be a Noetherian R-module and N a submodule. Since every submodule of N is also a submodule of M, the first assertion follows from (**C**).

Similarly, there is a one-to-one correspondence, preserving inclusion, between the submodules of the quotient module M/N and the submodules of M containing N. Then the second assertion also follows from (**C**), part (2).   ∎

In order to be able to use inductive arguments, we now establish the following result:

E.   *Let M be an R-module having a submodule N such that N and M/N are Noetherian modules. Then M itself is a Noetherian module.*

Proof: Let $\varphi : M \rightarrow M/N$ be the canonical homomorphism from M onto M/N. Let M' be any submodule of M. Since M/N is a Noetherian module, there exist finitely many elements $x_1, \ldots, x_n \in M'$ such that their classes modulo N generate the submodule $(M' + N)/N$ of M/N.

If $y \in M'$ then $\varphi(y) \in (M' + N)/N$ so there exist elements $a_1, \ldots, a_n \in R$ such that $\varphi(y) = \sum_{i=1}^{n} a_i \varphi(x_i)$, so $y - \sum_{i=1}^{n} a_i x_i \in \varphi^{-1}(0) = N$; but, on the other hand, $y - \sum a_i x_i \in M'$. Since N is a Noetherian module, the submodule $M' \cap N$ is finitely generated, say by the elements $y_1, \ldots, y_m \in M' \cap N$; hence there exist elements $b_1, \ldots, b_m \in R$ such that

$$y = \sum_{i=1}^{n} a_i x_i + \sum_{j=1}^{m} b_j y_j.$$

This shows that $\{x_1, \ldots, x_n, y_1, \ldots, y_m\}$ is a system of generators of M', and so M is a Noetherian module. ∎

As a corollary, we have:

F.   *If $M_1, M_2, \ldots, M_n$ are Noetherian R-module s, then the Cartesian product $M_1 \times M_2 \times \cdots \times M_n$ is also Noetherian.*

Proof: It is enough to show the statement for two modules M_1, M_2.

$M_1 \times M_2$ has the Noetherian submodule M_1 such that the quotient module $(M_1 \times M_2)/M_1 \cong M_2$ is also Noetherian. Hence, by (**E**), $M_1 \times M_2$ is a Noetherian module. ∎

G.   *If R is a Noetherian ring, if M is a finitely generated R-module, then M is a Noetherian module.*

Proof: Let $x_1, \ldots, x_n$ be generators of the R-module M. Let $R^n = R \times \cdots \times R$ be the Cartesian product of n copies of the R-module R, let $\varphi : R^n \to M$ be the homomorphism from R^n onto M such that

$$\varphi(a_1, \ldots, a_n) = \sum_{i=1}^{n} a_i x_i.$$

Then $M \cong R^n/\mathrm{Ker}(\varphi)$, where

$$\mathrm{Ker}(\varphi) = \left\{ (a_1, \ldots, a_n) \,\middle|\, \sum_{i=0}^{n} a_i x_i = 0 \right\}$$

(kernel of the mapping φ).

By (**F**), R^n is a Noetherian R-module, and by (**D**), M is a Noetherian module. ∎

Let R be a domain, K its field of quotients. An R-module M is said to be *torsion-free* when the following property holds: if $a \in R$, $x \in M$ and $ax = 0$, $x \neq 0$ then $a = 0$.

We have:

H.   *If R is a domain and M is a free R-module, then M is torsion-free.*

Proof: Let $(x_i)_{i \in I}$ be a basis of the R-module M. Thus if $x \in M$ it may be written in a unique way in the form

$$x = \sum_{i \in I} a_i x_i,$$

with $a_i \in R$ and $a_i = 0$ except for a finite set of indices $\{i_1, \ldots, i_n\} \subsetneq I$. Moreover, if $x \neq 0$, then $n \geq 1$.

If $a \in R$, $x \in M$, $x \neq 0$, and $ax = 0$, then by the uniqueness of the representation of 0 we have $aa_i = 0$ for every $i \in I$. Since $x \neq 0$ then $a_{i_1} \neq 0$. From the hypothesis that R is a domain, we conclude that $a = 0$, showing that M is a torsion-free module. ∎

The converse holds, when R is a principal ideal domain:

Theorem 1.   *Every finitely generated torsion-free module M over a principal ideal domain R is a free module.*

Proof: If $M = 0$ then it is a free module, with an empty basis.

Let $\{x_1, \ldots, x_n\}$ be a set of nonzero generators of M, with $n \geq 1$. If $n = 1$ then $M = Rx_1$ is a free R-module, because if $a_1 x_1 = 0$, then $a_1 = 0$.

We assume now that the theorem holds for modules with less than n generators, where $n > 1$. Let $S = \{y \in M \mid \text{ there exists } a \in R,\ a \neq 0,$ with $ay \in Rx_n\}$. Since R is a domain, S is a submodule of M, containing x_n. The quotient module M/S is torsion-free; indeed, if $a \in R$, $a \neq 0$, and $a\overline{y} = \overline{0}$ in M/S (where $\overline{y}$ denotes the image of y in M/S), then $ay \in S$, so there exists $a' \in R$, $a' \neq 0$, with $a'ay \in Rx_n$; since $a'a \neq 0$ then $y \in S$, that is, $\overline{y} = \overline{0}$.

The module M/S is finitely generated by the images $\overline{x}_1, \ldots, \overline{x}_{n-1}$. By induction, M/S is a free R-module with a finite basis. In the following lemma, we shall prove that $M \cong S \oplus (M/S)$.

It will be enough now to show that S is itself a free R-module; then a basis of S, together with a basis of M/S, will constitute a basis of M.

We show that S is isomorphic to a submodule of the field of quotients K of R. Namely, if $y \in S$, $y \neq 0$, let $a, b \in R$, $a \neq 0$, be such that $ay = bx_n$; if $a', b' \in R$, $a' \neq 0$ are also such that $a'y = b'x_n$, then $ba'y = bb'x_n = ab'y$, so $(ba' - ab')y = 0$, thus $ba' = ab'$, that is, $b/a = b'/a'$ (in K). This allows us to define a mapping $\theta : S \to K$ by putting $\theta(y) = b/a$ (where $y \neq 0$, $ay = bx_n$ with $a, b \in R$, $a \neq 0$) and $\theta(0) = 0$. It is easy to check that θ is an isomorphism of the R-module S into K. But M is a finitely generated R-module and R is a principal ideal domain, hence a Noetherian ring; by (**G**), M is a Noetherian R-module, so S and $\theta(S)$ are finitely generated R-modules. If $b_1/a_1, \ldots, b_m/a_m$ are generators of $\theta(S)$, then $S \cong \theta(S) \subseteq Rz$, where $z = 1/a_1 \ldots a_m$.

We shall soon prove in a lemma that every submodule of a cyclic R-module is again cyclic (when R is a principal ideal domain). This implies therefore that S is a cyclic module, $S = Ry \supseteq Rx_n \neq 0$, and since S is torsion-free, it must be free.

The proof of the theorem is completed, except for two lemmas.   ∎

Now, we fill the gaps in the above proof establishing the necessary lemmas.

Lemma 1.   *Let R be any domain, and let M be an R-module, M' a free R-module, and assume that there exists a homomorphism φ from M onto M'. Then there exists an isomorphism ψ from M' into M, such that*

$$M = \mathrm{Ker}(\varphi) \oplus \psi(M').$$

Proof: Let $\{x_i'\}_{i \in I}$ be a basis of M' and for every $i \in I$ let us choose arbitrarily an element $x_i \in M$ such that $\varphi(x_i) = x_i'$.

We define $\psi : M' \to M$ as follows: if $x' \in M'$, let $x' = \sum_{i \in I} a_i x_i'$ be the unique expression of x' as a linear combination of the basis; we put $\psi(x') = \sum_{i \in I} a_i x_i$. Then ψ is a homomorphism from M' into M, and $\varphi \circ \psi$ is the identity mapping of M', hence ψ is one-to-one.

Now, we may prove that $M = \mathrm{Ker}(\varphi) \oplus \psi(M')$. If $x \in M$, let $x' = \varphi(x) = \sum_{i \in I} a_i x_i'$, then

$$x = \left(x - \sum_{i \in I} a_i x_i \right) + \sum_{i \in I} a_i x_i = (x - \psi(x')) + \psi(x'),$$

with $x - \sum_{i \in I} a_i x_i \in \mathrm{Ker}(\varphi)$, $\psi(x') \in \psi(M')$. Also $\mathrm{Ker}(\varphi) \cap \psi(M') = 0$, showing the lemma.   ∎

Lemma 2.   *The ring R is a principal ideal domain, if and only if every submodule of a cyclic R-module is again cyclic.*

Proof: Assume that R is a principal ideal domain and let N be a cyclic R-module, so $N = Rx$; if $\theta : R \to N$ is defined by $\theta(a) = ax$, then θ is a homomorphism of R-modules, having a kernel which is an ideal J of R. Since $\theta(R) = N$ then $N \cong R/J$.

If N' is a submodule of N, then its inverse image by θ is an ideal J' of R, $J \subseteq J'$. But R is a principal ideal domain, so $J' = Rb$ and $N' = \theta(J') = R \cdot \theta(b)$, so it is a cyclic R-module.

The converse holds, because R is a cyclic R-module and by assumption every submodule of R, that is, every ideal of R, has to be principal.   ∎

We apply this theorem, obtaining the following corollary:

I.   *If R is a principal ideal domain then every submodule of a free R-module of rank n is again free, of rank at most n.*

Proof: Let M be free of rank n, so M is torsion-free (by **(H)**). If N is a submodule of M, it is also torsion-free, and finitely generated (by **(G)**), hence by Theorem 1, N is a free R-module.

The assertion concerning the rank of the submodule is trivial.   ∎

6.2 Integral Basis

We apply the preceding results to the ring of algebraic integers of an algebraic number field K of degree n over $\mathbb{Q}$.

To begin, we note:

J.   *The ring of algebraic integers of an algebraic number field is a Noetherian ring.*

Proof: Let A be the ring of algebraic integers, so by (**B**), $A \subseteq G$, where G is a free Abelian group (that is, a $\mathbb{Z}$-module) of finite rank. Since $\mathbb{Z}$ is a Noetherian ring, by (**G**), G is a Noetherian $\mathbb{Z}$-module, and by (**D**) A is a Noetherian $\mathbb{Z}$-module. Since $\mathbb{Z} \subseteq A$, every A-submodule of A (that is, ideal) is also a $\mathbb{Z}$-submodule of A (that is, a subgroup). Using (**C**), part (2), A is a Noetherian A-module, that is, a Noetherian ring.   ∎

Thus, every ideal of the ring of algebraic integers is finitely generated. We shall show later that, even when an ideal is not principal, it may still be generated by two elements.

From the preceding results we also obtained the following important theorem:

Theorem 2.   *If K is an algebraic number field of degree n, then the ring A of algebraic integers is a free Abelian group of rank n.*

Proof: By (**B**), there exist free Abelian groups M', M of rank n, such that $M' \subseteq A \subseteq M$. By (**I**), A is a free Abelian group of rank necessarily equal to n.   ∎

Definition 3.   Any basis of the free Abelian group A (ring of algebraic integers) is called an *integral basis* of K.

An integral basis is therefore also a basis of the vector space K over $\mathbb{Q}$, since it has $n = [K : \mathbb{Q}]$ elements.

We may apply (**I**), to obtain:

K.   *Let A be the ring of integers of an algebraic number field of degree n. Then every nonzero ideal J of A is a free Abelian group of rank n.*

Proof: By (**I**), J is a free Abelian group of rank at most equal to n. However, if $\{x_1, \ldots, x_n\}$ is an integral basis of A, if $a \in J$, $a \neq 0$, then $\{ax_1, \ldots, ax_n\}$ is a linearly independent set of elements of J. Thus J has rank equal to n.   ∎

Actually (**K**) may be considerably improved by a statement which relates two bases of the Abelian groups A and J.

L.   *Let R be a principal ideal domain, let M be a free R-module of rank n, and let M' be a submodule of M of rank m. Then there exists a basis $\{x_1, \ldots, x_n\}$ of M and nonzero elements $f_1, \ldots, f_m \in R$ such that $\{f_1x_1, \ldots, f_mx_m\}$ is a basis of M'. Moreover, we may choose the elements $f_1, \ldots, f_m$ so that f_i divides f_{i+1} for $i = 1, \ldots, m - 1$.*

Proof: The proof will be done by induction on m. The statement is trivial for $m = 0$.

Let $\{y_1, \ldots, y_n\}$ be any basis of the free R-module M. Every element $y \in M$ may be written in a unique way in the form

$$y = \sum_{i=1}^{n} a_iy_i, \qquad \text{with} \quad a_i \in R.$$

Let $p_i : M \to R$ be a mapping defined by $p_i(y) = a_i$, so p_i is a linear transformation, namely, the ith projection. Since $M' \neq 0$ there exists an index i such that $p_i(M') \neq 0$.

So, we may consider the nonempty set of linear transformations $u : M \to R$ such that $u(M') \neq 0$. Since R is a principal ideal domain, hence a Noetherian ring, there exists a maximal element in the family of principal ideals $u(M') \neq 0$ so obtained, say $u_1(M') = Rf_1 \neq 0$.

Let $z_1 \in M'$ be such that $u_1(z_1) = f_1$. We shall show that if u is any other linear transformation from M to R we have $f = u(z_1) \in Rf_1$. Indeed, R is a principal ideal domain, therefore the ideal generated by f_1, f is principal, say equal to Rf'; thus there exist elements $r_1, r \in R$ such that $f' = r_1f_1 + rf = (r_1u_1 + ru)(z_1)$. For the linear transformation $r_1u_1 + ru : M \to R$ we have $Rf_1 \subseteq Rf' \subseteq (r_1u_1 + ru)(M')$. By the maximality of Rf_1 we conclude that $(r_1u_1 + ru)(M') = Rf_1$ hence $f' \in Rf_1$ and $Rf \subseteq Rf_1$.

In particular, considering the projections p_i associated with the basis $\{y_1, \ldots, y_n\}$ of M, we have $p_i(z_1) \in Rf_1$ for every $i = 1, \ldots, n$. Therefore $z_1 = \sum_{i=1}^{n} p_i(z_1) \cdot y_i = f_1x_1$ for some element $x_1 \in M$; we conclude that $f_1 = u_1(z_1) = f_1 \cdot u_1(x_1)$, thus $u_1(x_1) = 1$.

Let N be the kernel of u_1; then $M = Rx_1 \oplus N$ and $M' = Rf_1x_1 \oplus (N \cap M')$. Indeed, $Rx_1 \cap N = 0$ since $rx_1 \in N$ implies $r = u_1(rx_1) = 0$; in particular, $Rf_1x_1 \cap (N \cap M') = 0$. On the other hand, if $y \in M$ then we may write $y = u_1(y)x_1 + (y - u_1(y)x_1)$ with $u_1(y)x_1 \in Rx_1$ and $u_1(y - u_1(y)x_1) = 0$; this shows that $M = Rx_1 \oplus N$. Also, if $y' \in M'$ then $u_1(y') \in u_1(M') = Rf_1$, hence $u_1(y')x_1 \in Rf_1x_1 = Rz_1 \subseteq M'$ and $y' - u_1(y')x_1 \in M' \cap N$.

Now $M' \cap N$ is a module of rank $m - 1$ contained in the free module N of rank $n - 1$. By induction, there exists a basis $\{x_2, \ldots, x_n\}$ of N and nonzero elements $f_2, \ldots, f_m \in R$ such that $\{f_2x_2, \ldots, f_mx_m\}$ is a basis of

$M' \cap N$ and f_i divides f_{i+1} for $i = 2, \ldots, m - 1$. So, $\{x_1, x_2, \ldots, x_n\}$ is a basis of M and $\{f_1 x_1, f_2 x_2, \ldots, f_m x_m\}$ is a basis of M'.

We still have to show that f_1 divides f_2. We prove first that if $u : M \to R$ is any linear transformation, then $u(M' \cap N) \subseteq Rf_1$. Assuming the contrary, we consider the linear transformation $v : M \to R$, which coincides with u on N and coincides with u_1 on Rx_1; explicitly, if $r_1 \in R$, $y \in N$, then $v(r_1 x_1 + y) = r_1 + u(y)$. Then $v(M') = Rf_1 + u(N \cap M')$, and because $u(N \cap M')$ is not contained in Rf_1 we conclude that $v(M')$ contains properly Rf_1, contrary to the maximality of the ideal Rf_1; this is a contradiction.

Now let p_2' be the linear transformation from M to R defined by $p_2'(x_2) = 1$, $p_2'(x_i) = 0$ when $i \neq 2$. Then $Rf_2 = p_2'(M' \cap N) \subseteq Rf_1$ by the preceding considerations, and f_1 divides f_2. ■

It is possible to prove that if $\{x_1', \ldots, x_n'\}$ is another basis of M and $f_1', \ldots, f_m' \in R$ are nonzero elements such that f_i' divides f_{i+1}' for $i = 1, \ldots, m - 1$ and $\{f_1' x_1', \ldots, f_m' x_m'\}$ is a basis of M', then f_i, f_i' are associated for $i = 1, \ldots, m$.

6.3 The Discriminant

Our purpose now is to introduce and to give elementary properties of the discriminant, which is a numerical invariant attached to the algebraic number field.

M.   *Let K be a field of algebraic numbers, A the ring of algebraic integers, and $\{x_1, \ldots, x_n\}$ an integral basis. If $x_1', \ldots, x_n' \in A$ then*

$$\mathrm{discr}_{K|\mathbb{Q}}(x_1, \ldots, x_n) = \mathrm{discr}_{K|\mathbb{Q}}(x_1', \ldots, x_n')$$

if and only if $\{x_1', \ldots, x_n'\}$ is an integral basis.

Proof: By definition (see Chapter 2, Section 11):

$$\mathrm{discr}_{K|\mathbb{Q}}(x_1, \ldots, x_n) = \det(\mathrm{Tr}_{K|\mathbb{Q}}(x_i x_j)) \neq 0.$$

We have $x_j' = \sum_{i=1}^{n} a_{ij} x_i$ (with $a_{ij} \in \mathbb{Z}$) for $j = 1, \ldots, n$, hence

$$\mathrm{discr}_{K|\mathbb{Q}}(x_1', \ldots, x_n') = \left[\det(a_{ij})\right]^2 \mathrm{discr}_{K|\mathbb{Q}}(x_1, \ldots, x_n).$$

The matrix $(a_{ij})_{i,j}$ has determinant ± 1 if and only if it is invertible, or equivalently $\{x_1', \ldots, x_n'\}$ is an integral basis. ■

Definition 4.   The discriminant (in $K|\mathbb{Q}$) of any integral basis is called the *discriminant of the field K*, and denoted by $\delta_{K|\mathbb{Q}} = \delta_K$. Thus, $\delta_K \in \mathbb{Z}$, $\delta_K \neq 0$.

We shall see later the significance of the discriminant. For the moment, we are interested in computing the discriminant in some special cases.

For this purpose the following remarks will be useful.

N.   *Let $K = \mathbb{Q}(t)$ be an extension of degree n, t an algebraic integer. Then δ_K divides $\mathrm{discr}_{K|\mathbb{Q}}(t)$ and the quotient is the square of an integer.*

Proof: Let $\{x_1, \ldots, x_n\}$ be an integral basis of $K|\mathbb{Q}$. We may write

$$t^j = \sum_{i=1}^{n} a_{ij} x_i \quad (j = 0, 1, \ldots, n-1) \text{ with } a_{ij} \in \mathbb{Z}.$$

Hence, $\mathrm{discr}_{K|\mathbb{Q}}(t) = \mathrm{discr}_{K|\mathbb{Q}}(1, t, \ldots, t^{n-1}) = (\det(a_{ij}))^2 \cdot \delta_K$ and therefore δ_K divides $\mathrm{discr}_{K|\mathbb{Q}}(t) \in \mathbb{Z}$, and the quotient is the square of an integer.   ∎

This limits the search of the discriminant, once a primitive integral element has been found. A further limitation comes from the following result due to Stickelberger:

O.   *Let $K|\mathbb{Q}$ be an extension of degree n, and let $\{y_1, \ldots, y_n\}$ be any $\mathbb{Q}$-basis of K, where each y_i is an algebraic integer. Then*

$$\mathrm{discr}_{K|\mathbb{Q}}(y_1, \ldots, y_n) \equiv 0 \text{ or } 1 \pmod 4.$$

Proof: By definition, if $\sigma_1, \ldots, \sigma_n$ are the $\mathbb{Q}$-isomorphisms of K then

$$\mathrm{discr}_{K|\mathbb{Q}}(y_1, \ldots, y_n) = \left[\det(\sigma_i(y_j))\right]^2 = (P - N)^2 = (P + N)^2 - 4PN,$$

where P (respectively, N) denotes the sum of the terms with positive sign (respectively, negative sign) in the expression of the determinant. Since $P + N$ and PN remain unchanged by the action of each σ_i, then by Galois theory $P + N, PN \in \mathbb{Q}$. On the other hand, $P + N, PN$ are algebraic integers, hence $P + N, PN \in \mathbb{Z}$. Therefore, $\mathrm{discr}_{K|\mathbb{Q}}(y_1, \ldots, y_n)$ is congruent to 0 or to 1 modulo 4.   ∎

6.4 Discriminant of Quadratic Fields

P.   *Let $K = \mathbb{Q}(\sqrt{d})$, where d is a square-free integer. If $d \equiv 2 \pmod 4$ or $d \equiv 3 \pmod 4$ then the discriminant of K is $\delta_K = 4d$. If $d \equiv 1 \pmod 4$ then $\delta_K = d$.*

Proof: By Chapter 5, (**W**), if $d \equiv 2 \pmod 4$ or $d \equiv 3 \pmod 4$ then $\{1, \sqrt{d}\}$ is an integral basis. Therefore

$$\delta_K = \begin{vmatrix} \mathrm{Tr}_{K|\mathbb{Q}}(1) & \mathrm{Tr}_{K|\mathbb{Q}}(\sqrt{d}) \\ \mathrm{Tr}_{K|\mathbb{Q}}(\sqrt{d}) & \mathrm{Tr}_{K|\mathbb{Q}}(d) \end{vmatrix} = \begin{vmatrix} 2 & 0 \\ 0 & 2d \end{vmatrix} = 4d.$$

If $d \equiv 1 \pmod 4$, then $\{1, (1+\sqrt{d})/2\}$ is an integral basis, so

$$\delta_K = \begin{vmatrix} \mathrm{Tr}_{K|\mathbb{Q}}(1) & \mathrm{Tr}_{K|\mathbb{Q}}\left(\dfrac{1+\sqrt{d}}{2}\right) \\[3mm] \mathrm{Tr}_{K|\mathbb{Q}}\left(\dfrac{1+\sqrt{d}}{2}\right) & \mathrm{Tr}_{K|\mathbb{Q}}\left(\dfrac{1+\sqrt{d}}{2}\right)^2 \end{vmatrix}$$

$$= \begin{vmatrix} 2 & 1 \\[2mm] 1 & \dfrac{1+d}{2} \end{vmatrix} = (1+d) - 1 = d. \qquad\blacksquare$$

Thus, the possible discriminants of quadratic fields are integers δ such that $\delta \equiv 1 \pmod 4$, or $\delta \equiv 8 \pmod{16}$ or $\delta \equiv 12 \pmod{16}$.

Let us also note in all cases:

Q.   *The integers of $\mathbb{Q}(\sqrt{d})$ may be written in the form $(a+b\sqrt{\delta})/2$ where $a, b \in \mathbb{Z}$, $a^2 \equiv \delta b^2 \pmod 4$, and conversely.*

Proof: Indeed, by Chapter 5, (U), this is true when $d \equiv 1 \pmod 4$, since $\delta = d$. If $d \equiv 2 \pmod 4$ or $d \equiv 3 \pmod 4$, then $\delta = 4d$ and if $a + b\sqrt{d}$ is an algebraic integer, $a, b \in \mathbb{Z}$, so

$$a + b\sqrt{d} = (2a + 2b\sqrt{d})/2 = (2a + b\sqrt{\delta})/2$$

with $(2a)^2 \equiv \delta b^2 \pmod 4$; and conversely. $\blacksquare$

6.5 Discriminant of Cyclotomic Fields

R.   *Let $K = \mathbb{Q}(\zeta)$, where ζ is a primitive pth root of 1, and p is an odd prime number. Then the discriminant of $K|\mathbb{Q}$ is*

$$\delta = (-1)^{(p-1)/2} p^{p-2}.$$

Proof: By Chapter 5, (X), $\{1, \zeta, \zeta^2, \ldots, \zeta^{p-2}\}$ is an integral basis of $\mathbb{Q}(\zeta)$. The minimal polynomial of ζ over $\mathbb{Q}$ is the pth cyclotomic polynomial $\Phi_p = X^{p-1} + X^{p-2} + \cdots + X + 1$. By Chapter 2, Section 11:

$$\delta = \mathrm{discr}(\Phi_p) = (-1)^{[(p-1)(p-2)]/2} N_{\mathbb{Q}(\zeta)|\mathbb{Q}}(\Phi_Q'(\zeta))$$

where Φ_Q' is the derivative of Φ_p. But $X^p - 1 = (X-1)\Phi_p$, hence taking derivatives, $pX^{p-1} = \Phi_p + (X-1)\Phi_p'$; thus for every root ζ^j of Φ_p ($j = 1, 2, \ldots, p-1$), we have $p(\zeta^j)^{p-1} = (\zeta^j - 1)\Phi_p'(\zeta^j)$.

Now, $N_{\mathbb{Q}(\zeta)|\mathbb{Q}}(\Phi'_p(\zeta)) = \prod_{j=1}^{p-1} \Phi'_p(\zeta^j)$, so we compute

$$\prod_{j=1}^{p-1} \zeta^j = N_{\mathbb{Q}(\zeta)|\mathbb{Q}}(\zeta) = (-1)^{p-1} \cdot 1 = 1,$$

$$\prod_{j=1}^{p-1}(\zeta^j - 1) = \prod_{j=1}^{p-1}(1 - \zeta^j) = \Phi_p(1) = p;$$

hence,

$$N_{\mathbb{Q}(\zeta)|\mathbb{Q}}(\Phi'_p(\zeta)) = \frac{p^{p-1} \cdot 1}{p} = p^{p-2}.$$

and therefore

$$\delta = (-1)^{(p-1)(p-2)/2}p^{p-2} = (-1)^{(p-1)/2}p^{p-2}$$

since $[(p-1)(p-2)]/2 \equiv (p-1)/2 \pmod 2$. ∎

EXERCISES

1. Let t be an algebraic integer of degree n over $\mathbb{Q}$, and let $x_1, \ldots, x_n \in \mathbb{Z}[t]$ be such that $\operatorname{discr}(x_1, \ldots, x_n)$ is a square-free rational integer. Prove that $\{x_1, \ldots, x_n\}$ is a basis of the Abelian group $\mathbb{Z}[t]$.

2. Let d be a square-free rational integer, and let δ be the discriminant of the field $\mathbb{Q}(\sqrt{d})$. Show that $\{1, (\delta + \sqrt{\delta})/2\}$ is a basis of the Abelian group of all algebraic integers of $\mathbb{Q}(\sqrt{d})$.

3. Let J be a nonzero ideal of the ring A of integers of an algebraic number field K. Let $\{x_1, \ldots, x_n\}$ be an integral basis of K. Show that J has a basis $\{y_1, \ldots, y_n\}$ over $\mathbb{Z}$ of the following type:

$$\begin{cases} y_1 = a_{11}x_1, \\ y_2 = a_{21}x_1 + a_{22}x_2, \\ \quad \cdots \\ y_n = a_{n1}x_1 + a_{n2}x_2 + \cdots + a_{nn}x_n, \end{cases}$$

with $a_{ij} \in \mathbb{Z}$ satisfying $0 \le a_{ij} < a_{jj} \le a_{11}$ for every $i, j = 1, \ldots, n$.

Hint: **1.** Construct inductively a basis $\{y'_1, \ldots, y'_n\}$ for the $\mathbb{Z}$-module J in the following way: using Exercise 4 of Chapter 5, show that there exists the smallest positive integer a'_{11} such that $y'_1 = a'_{11}x_1 \in J$; then consider the smallest positive integer a'_{22} for which a linear combination $y'_2 = a'_{21}x_1 + a'_{22}x_2$ belongs to J; define $y'_3, \ldots, y'_n$ in a similar manner.

2. Note that if $\{y_1', \ldots, y_n'\}$ is a basis of the $\mathbb{Z}$-module J and $i \neq j$, $m \in \mathbb{Z}$, then

$$\left\{ y_1', \ldots, y_i' - my_j', y_{i+1}', \ldots, y_n' \right\}$$

is also a basis of J; using this fact, subtract a suitable multiple of y_{n-1}' from y_n', so that the new coefficient of x_{n-1} is not negative and smaller than $a_{n-1,n-1}'$; repeat this procedure.

4. Let $K = \mathbb{Q}(\sqrt{-5})$, and A the ring of integers of K. Find a basis over $\mathbb{Z}$ for each of the following ideals of A:

   (a) The ideal generated by 3 and $4 + \sqrt{-5}$.

   (b) The ideal generated by 7 and $4 + \sqrt{-5}$.

5. Let $\{x_1, \ldots, x_n\}$ be a linearly independent set of algebraic integers of the field K of degree n over $\mathbb{Q}$. Prove:

   (a) $\operatorname{discr}_{K|\mathbb{Q}}(x_1, \ldots, x_n)$ is a multiple of the discriminant δ of the field K.

   (b) $\{x_1, \ldots, x_n\}$ is an integral basis if and only if $|\operatorname{discr}_{K|\mathbb{Q}}(x_1, \ldots, x_n)| = |\delta|$.

6. Let $K = \mathbb{Q}(t)$ be an extension of degree n, $t \in A$ (the ring of algebraic integers of K), and let d be the discriminant of t in $K|\mathbb{Q}$. For each $i = 1, \ldots, n$, among all integral elements of the form

$$\frac{1}{d}(a_0 + a_1 t + \cdots + a_i t^i) \qquad \text{(with any } a_j \in \mathbb{Z} \text{ and } a_i \neq 0)$$

let x_i be one such that $|a_i|$ is the least possible. Show that $\{x_1, \ldots, x_n\}$ is an integral basis of K.

7. Let $K|\mathbb{Q}$ be an extension of degree n, let $\{x_1, \ldots, x_n\}$ be a $\mathbb{Q}$-basis of K composed of algebraic integers. Show that $\{x_1, \ldots, x_n\}$ is an integral basis of $K|\mathbb{Q}$ if and only if $|\operatorname{discr}_{K|\mathbb{Q}}(x_1, \ldots, x_n)| \leq |\operatorname{discr}_{K|\mathbb{Q}}(y_1, \ldots, y_n)|$ for every $\mathbb{Q}$-basis $\{y_1, \ldots, y_n\}$ of K composed of algebraic integers.

8. Compute the discriminant of the field $\mathbb{Q}(\sqrt{2}, \sqrt{3})$.

9. Compute the discriminant of the field $\mathbb{Q}(\sqrt{2}, i)$.

10. Compute the discriminant of the field $\mathbb{Q}(\sqrt[4]{2})$.

11. Compute the discriminant of the field $\mathbb{Q}(\sqrt[5]{2})$.

12. Let R be a ring, and M an R-module. If $x \in M$, let

$$\operatorname{Ann}(x) = \{a \in R \mid ax = 0\}$$

(the *annihilator* of x). We call x a *torsion element* when $\text{Ann}(x) \neq 0$ and a *torsion-free element* when $\text{Ann}(x) = 0$. M is called a *torsion module* if all its elements are torsion elements. Show:

(a) For every $x \in M$, $\text{Ann}(x)$ is an ideal of R.

(b) If R is a domain, the set T of all torsion elements of M forms a torsion submodule of M, and M/T is a torsion-free R-module.

13. Let R be a commutative ring, let M be an R-module and let N be a submodule of M. We say that N is a *pure submodule* of M when $N \cap aM = aN$ for every $a \in R$. Prove:

(a) If $M = N \oplus N'$ then N, N' are pure submodules of M.

(b) If M is a torsion-free module then the submodule N is pure if and only if $ax \in N$, $a \in R$, $a \neq 0$, $x \in M$ imply $x \in N$.

(c) Let M be a torsion-free module; N is a pure submodule of M if and only if M/N is a torsion-free module.

14. Let R be a principal ideal domain, let M be a free R-module of rank n, and let N be a submodule of M. Show that N is a pure submodule of M if and only if there exists a submodule N' of M such that $M \cong N \oplus N'$.

15. Let R be a principal ideal domain, let M be a free R-module of rank n, and let N be a submodule of M (hence N is also a free R-module). Show that N is a pure submodule of M if and only if every basis of N may be extended to a basis of M.

16. Let $K \subseteq L$ be algebraic number fields, and let $A \subseteq B$ be their rings of algebraic integers. Show that the Abelian group A is a pure subgroup of B. Conclude that every integral basis of A is a part of an integral basis of B.

7

The Decomposition of Ideals

7.1 Dedekind's Theorem

We have shown that the ring A of algebraic integers of an algebraic number field is Noetherian and integrally closed. However, it is not true in general that A is a principal ideal domain. What can be said about ideals which are not principal?

We shall imitate the theory of divisibility, replacing elements of A by ideals of A, and more generally, elements of K (not in A) by a more general type of ideals.

Definition 1.   Let A be any domain and K its field of quotients. An A-module M, contained in K, is said to be a *fractional ideal of A* when there exists an element $a \in A$, $a \neq 0$, such that $a \cdot M \subseteq A$.

Thus, every ideal of A is also a fractional ideal (taking $a = 1$), and if necessary, we shall call it an *integral ideal*.

However, K itself is not a fractional ideal of A (unless $A = K$), otherwise there exists $a \in A$, $a \neq 0$, such that $K = A(1/a)$. Then $1/a^2 = b/a$, with $b \in A$, and so $1/a = b \in A$, showing that $K = A(1/a) = A$.

The set $\mathcal{F}$ of nonzero fractional ideals of A is endowed with an operation of multiplication: $M \cdot M' = \{\sum_{i=1}^{n} x_i x_i' \mid n \geq 1, x_i \in M, x_i' \in M'\}$.

It is easy to verify that $M \cdot M'$ is again a fractional ideal. If M, M' are integral ideals, so is $M \cdot M'$.

This operation is commutative, associative, and has a unit element, namely the ideal A itself: $M \cdot A = M$. It generalizes the operation defined for integral ideals in Chapter 1, Section 2.

We say that a nonzero fractional ideal M is *invertible* when there exists a fractional ideal M' such that $M \cdot M' = A$. We shall return later to this concept. It is enough to observe now the following general fact: M_1, M_2 are invertible fractional ideals if and only if $M_1 \cdot M_2$ is invertible.

Among the fractional ideals, we consider the *principal fractional ideals*, namely those of type Ax, where $x \in K$. Every nonzero principal fractional ideal is invertible. We denote by $\mathcal{Pr}$ the set of nonzero principal fractional ideals of A.

If M, N are nonzero fractional ideals of A, we say that M *divides* N, and write $M|N$, when there exists an integral ideal $J \subseteq A$ such that $M \cdot J = N$.

The following properties are immediate to verify: $M|M$ for every nonzero fractional ideal M; if M, M', M'' are fractional ideals, and $M|M'$, $M'|M''$ then $M|M''$; if $M|M'$ then $M' \subseteq M$ (because $M' = M \cdot J \subseteq M \cdot A = M$); if $M|M'$ and $M'|M$ then $M = M'$.

Thus, the relation of divisibility between nonzero fractional ideals is an order relation, which implies the reverse inclusion. Actually, we shall see soon that for rings of algebraic integers the divisibility relation is equivalent to the reverse inclusion.

Let us also note that if $a, b \in K^{\cdot}$, then a divides b if and only if the principal fractional ideal Aa divides Ab.

In the theory of divisibility of $\mathbb{Z}$ the prime numbers play a basic role. As we know, $n \in \mathbb{Z}$ is a prime number if and only if $\mathbb{Z}/n$ is a domain different from $\mathbb{Z}$ (and then also a field).

Thus the nonzero prime ideals of $\mathbb{Z}$ are $\mathbb{Z}p$, for all prime numbers. So, the nonzero prime ideals of the ring A of algebraic integers of an algebraic number field K are candidates to play a similar role as prime elements do in the case of the ring $\mathbb{Z}$.

A.   *Let A be the ring of integers of an algebraic number field K. Then every nonzero prime ideal P of A is a maximal ideal, and it contains exactly one prime number p. Moreover A/P is a finite field containing $\mathbb{F}_p = \mathbb{Z}/p$.*

Proof: Let P be a nonzero prime ideal of A. Since A is integral over $\mathbb{Z}$, by Chapter 5, (**E**), $P \cap \mathbb{Z}$ is a nonzero ideal which is obviously prime. Thus $P \cap \mathbb{Z} = \mathbb{Z}p$ for some prime number p. No other prime $p' \neq p$ is contained in P, otherwise taking m, n integers such that $1 = mp + np'$ we would have $1 \in P$, which is not the case.

But A/P is integral over the field $\mathbb{F}_p = \mathbb{Z}/p$. Hence, by Chapter 5, (**F**), A/P is a field and P is a maximal ideal of A.   ∎

Summarizing, we have shown that A is a Noetherian, integrally closed domain in which every nonzero prime ideal is maximal. These properties are already very strong and have significant implications on the arithmetic of the ring A.

We shall soon establish a fundamental theorem, which encompasses the results of Dedekind, Noether, Krull, and Matusita. We begin with some easy generalities.

We recall first (see the remark after Chapter 6, (**C′**)): If A is any Noetherian ring, every ideal $J \neq A$ is contained in a prime ideal, because every maximal ideal is a prime ideal.

Let A be any domain and K its field of quotients. If J is any nonzero ideal of A, let $J^{-1} = \{x \in K \mid xJ \subseteq A\}$.

Then J^{-1} is a fractional ideal, $A \subseteq J^{-1}$ and $J \subseteq JJ^{-1} \subseteq A$.

Indeed, J^{-1} is obviously an A-module contained in K and if $a \in J$, $a \neq 0$, then $aJ^{-1} \subseteq JJ^{-1} \subseteq A$, so J^{-1} is a fractional ideal. From $1 \in J^{-1}$, we deduce that $A \subseteq J^{-1}$.

If J is a nonzero ideal then J is an invertible ideal if and only if $JJ^{-1} = A$.

B.   *Let R be an arbitrary domain, let P_1, ..., P_r be invertible prime ideals, and let $J = P_1 \cdots P_r$. Then if $J = P_1' \cdots P_r'$, with P_j' prime ideals, we have $s = r$ and after some permutation of indices, $P_j' = P_j$, for all $j = 1, \ldots, r$.*

Proof: Let P_1 be a minimal ideal among P_1, P_2, ..., P_r (relative to inclusion). From $P_1 \supseteq J = P_1' \cdots P_s'$, it follows that there exists an index j, say $j = 1$, such that $P_1 \supseteq P_1'$. Similarly $P_1' \supseteq J = P_1 \cdots P_r$, hence $P_1' \supseteq P_i$ for some i; thus $P_1 \supseteq P_i$ and by the minimality $i = 1$ and $P_1 = P_1'$.

Now, since P_1 is invertible, multiplying the relation $P_1 \cdots P_r = J = P_1' \cdots P_s'$ by P_1^{-1}, we have $P_2 \cdots P_r = P_1^{-1}J = P_2' \cdots P_s'$ where $P_1^{-1}J$ is still an integral ideal. We proceed inductively and arrive at the required conclusion.   ∎

C.   *If A is a Noetherian ring and if J is an ideal of A, $J \neq A$, there exist prime ideals P_1, ..., P_r of A such that $P_i \supseteq J$ for all $i = 1, \ldots, r$, but $J \supseteq P_1 P_2 \cdots P_r$.*

Proof: The assertion holds if J is a prime ideal. Let $\mathcal{S}$ be the set of all ideals $J \neq A$ for which the statement does not hold. If $\mathcal{S} \neq \varnothing$, since A is a Noetherian ring, by (**C'**) of Chapter 6 there exists an ideal $J \in \mathcal{S}$ which is maximal in $\mathcal{S}$ (with respect to inclusion). Then J is not a prime ideal. Thus, there exist elements $a, a' \in A$, $a, a' \notin J$, but $aa' \in J$. Let $I = J + Aa$, $I' = J + Aa'$, so $I, I' \supset J$, hence $I, I' \notin \mathcal{S}$. Also $II' \subseteq J$. Thus, there exist prime ideals P_1, ..., P_s, P_1', ..., P_r' of A, such that $P_i \supseteq I \supseteq P_1 \cdots P_s$ and $P_j' \supseteq I' \supseteq P_1' \cdots P_r'$ for each $i = 1, \ldots, s$ and $j = 1, \ldots, r$. Hence $P_i, P_j' \supseteq J \supseteq II' \supseteq P_1 \cdots P_s P_1' \cdots P_r'$ for all i, j. Then $J \notin \mathcal{S}$, which is a contradiction and concludes the proof.   ∎

Now we are ready to prove the following basic theorem:

Theorem 1.   *Let A be a domain. Then the following properties are equivalent:*

(1) *A is Noetherian, integrally closed and every nonzero prime ideal of A is maximal.*

(2) *Every nonzero (integral) ideal of A is expressible in a unique way as the product of prime ideals.*

(3) *Every nonzero (integral) ideal of A is the product of prime ideals.*

(4) *The set of nonzero fractional ideals of A is a multiplicative group.*

Proof: (1) → (2) First we show:

(a) Every nonzero prime ideal P of A is invertible.

We have seen that P^{-1} is a fractional ideal of A, $PP^{-1} \subseteq A \subseteq P^{-1}$. We show now that $P^{-1} \neq A$. Let $c \in P$, $c \neq 0$, hence by (**C**) there exist prime ideals $P_1, \ldots, P_r$ such that $P \supseteq Ac \supseteq P_1 \cdots P_r$, with $P_i \supseteq Ac$ for every $i = 1, \ldots, r$. We may choose r to be the minimum possible, for which this property holds. But P must contain one of these prime ideals, say $P \supseteq P_1$, hence $P = P_1$, since by the hypothesis every nonzero prime ideal is maximal. By the minimality of r, we have $Ac \not\supseteq P_2 \cdots P_r$, so there exists $a \in P_2 \cdots P_r$, $a \notin Ac$, that is, $a/c \notin A$. However $a/c \in P^{-1}$, because $(a/c)P \subseteq (1/c)P_1 P_2 \cdots P_r \subseteq A$, so $A \subset P^{-1}$.

Next, we prove that $PP^{-1} = A$. At any rate $P = P \cdot A \subseteq PP^{-1} \subseteq A$, and PP^{-1} is an ideal of A, so either $PP^{-1} = A$ or $PP^{-1} = P$. If this second case takes place, then $PP^{-2} = (PP^{-1})P^{-1} = PP^{-1} = P$ and similarly $PP^{-n} = P$ for every $n \geq 1$. Thus, if $a \in P$, $b \in P^{-1}$ then $ab^n \in PP^{-n} = P$ for every $n \geq 1$. The ideal $J = \sum_{n=0}^{\infty} Aab^n$ is finitely generated, because A is a Noetherian ring, so there exists $n > 1$ and elements $c_0, \ldots, c_{n-1} \in A$ such that $ab^n = \sum_{i=0}^{n-1} c_i ab^i$, hence $b^n - \sum_{i=0}^{n-1} c_i b^i = 0$; this shows that b is integral over A, hence $b \in A$ because A is integrally closed. Thus $P^{-1} \subseteq A$, which is contrary to the fact that $A \subset P^{-1}$. Therefore, $PP^{-1} = A$, proving (a).

Now we may prove the implication (1) → (2)

If $J = A$ the assertion is trivially verified.

If J is a nontrivial ideal of A, let $P_1, \ldots, P_r$ be prime ideals of A such that $P_i \supseteq J$ $(i = 1, \ldots, r)$ and $J \supseteq P_1 \cdots P_r$ (by (**C**)); we may assume that r is the minimum possible, $r \geq 1$.

We shall prove by induction on r that J is expressible as a product of prime ideals.

If $r = 1$ then $P_1 \supseteq J \supseteq P_1$, hence $J = P_1$.

If the statement holds for ideals containing a product of at most $r - 1$ prime ideals, let P be a prime ideal of A, such that $P \supseteq J$; from $P \supseteq J \supseteq P_1 \cdots P_r$, it follows that P contains some of the ideals P_i, say $P \supseteq P_r$ hence $P = P_r$; from (a), we have $A = PP^{-1} \supseteq JP_r^{-1} \supseteq P_1 \cdots P_r P_r^{-1} = P_1 \cdots P_{r-1}$. By induction, $JP_{r-1}^{-1} = P_1' \cdots P_s'$ where each P_i' is a prime ideal, and so $J = JP_r^{-1}P_r = P_1' \cdots P_s' P_r$.

The uniqueness of the decomposition into a product of prime ideals was established in (**B**).

(2) → (3) This statement is trivial.

(3) → (4) It will be enough to show that every nonzero prime ideal of A is invertible. In fact, if J is any nonzero fractional ideal of A, let $d \in A$, $d \neq 0$ be such that $d \cdot J \subseteq A$. Then $J = (Ad)^{-1} \cdot (Ad \cdot J)$ where $(Ad)^{-1} = Ad^{-1}$ is the inverse of the principal ideal Ad. By the hypothesis, $Ad \cdot J$ and Ad are products of nonzero prime ideals; if each nonzero prime

ideal of A is invertible, then the same holds for J. Thus the set of all nonzero fractional ideals of A is a multiplicative group.

Now, let P be a nonzero prime ideal, and let $a \in P$, $a \neq 0$; by hypothesis Aa is a product of prime ideals, so $P \supseteq Aa = P_1 \cdots P_r$. Since Aa is invertible then each ideal P_i is invertible. But P is prime, hence it contains some ideal P_i. The proof will be finished if we show that P_i is a maximal ideal, thus $P = P_i$ is invertible.

So, we are led to prove:

(b) If P is an invertible prime ideal of A, then P is maximal.

Let $a \in A$, $a \notin P$, and consider the ideals $J = P + Aa$, $J' = P + Aa^2$. By hypothesis (3), $J = P_1 \cdots P_m$, $J' = P_1' \cdots P_n'$, where P_i, P_j' are prime ideals, which must contain J, J', respectively, and so contain P. In the domain $\overline{A} = A/P$, we consider the image $\overline{a}$ of a. Then $\overline{A} \cdot \overline{a} = \overline{P_1} \cdots \overline{P_m}$, $\overline{A} \cdot \overline{a}^2 = \overline{P_1'} \cdots \overline{P_n'}$, where $\overline{P_i} = P_i/P$, $\overline{P_j'} = P_j'/P$ are invertible prime ideals of $\overline{A}$ (because $\overline{a} \neq 0$ so $\overline{A}\overline{a}$, $\overline{A}\overline{a}^2$ are invertible ideals). Since $\overline{A} \cdot \overline{a}^2 = (\overline{A}\,\overline{a})^2$, by (B), we must have $2m = n$ and, after renumbering, we have $\overline{P_{2i-1}'} = \overline{P_{2i}'} = \overline{P_i}$ for every $i = 1, \dots, m$; therefore $P_{2i-1}' = P_{2i}' = P_i$ and so $(P + Aa)^2 = P + Aa^2$, thus $P \subseteq P + Aa^2 = (P + Aa)(P + Aa) \subseteq P^2 + Aa$. This implies that every element $x \in P$ may be written in the form $x = y + za$ with $y \in P^2$, $z \in A$. So $za = x - y \in P$ and since $a \notin P$ then $z \in P$. So, actually $P \subseteq P^2 + P \cdot Aa \subseteq P$, therefore $P = P(P + Aa)$ and since P is invertible then $A = P + Aa$. This is true for every $a \in A$, $a \notin P$, so P is a maximal ideal.

(4) $\rightarrow$ (1)   We show first that A is a Noetherian ring. Let J be any (integral) ideal of A, $J \neq 0$; let J^{-1} be its inverse, so $A = JJ^{-1}$, therefore there exist elements $x_1, \dots, x_n \in J$ and $y_1, \dots, y_n \in J^{-1}$ such that $1 = \sum_{i=1}^{n} x_i y_i$. If $a \in J$ we have $a = \sum_{i=1}^{n} x_i(y_i a)$ with $y_i a \in J^{-1}J = A$, so the elements $x_1, \dots, x_n$ generate the ideal J.

Next, we prove that A is integrally closed. Let $x \in K$ be a root of a monic polynomial, with coefficients in A : $X^m + a_1 X^{m-1} + \cdots + a_m$. Then $x^m = -(a_1 x^{m-1} + \cdots + a_{m-1}x + a_m)$ belongs to the fractional ideal J generated by 1, x, $\dots$, x^{m-1}. From this fact, we have $J^2 \subseteq J$, and since $J \neq 0$ is invertible, then $J \subseteq A$; in particular, $x \in A$ and so A is integrally closed.

Finally, let P be a nonzero prime ideal of A, let $a \in A$, $a \notin P$ and consider the ideal $J = P + Aa$. Since J is invertible, then $J' = J^{-1}P$ is such that $JJ' = P$, so $J' \supseteq P$.

We note now that $J' = J^{-1}P \subseteq J^{-1}J = A$.

On the other hand, if $y \in J'$, then $ay \in P$; from $a \notin P$ it follows that $y \in P$, showing the other inclusion $J' \subseteq P$. Thus $J' = P$ and $JP = P$ implies that $J = A$. This is true for every $a \in A$, $a \notin P$, so P is a maximal ideal. ∎

According to the previous results, the ring A of algebraic integers of an algebraic number field K of finite degree satisfies condition (1) in Theorem 1. Thus we may state:

Theorem 2.   *Let A be the ring of integers of an algebraic number field K of finite degree. Then every nonzero fractional ideal J of K is, in a unique way, of the form $J = \prod_{i=1}^{r} P_i^{e_i}$, where $P_1, \ldots, P_r$ are distinct prime ideals and $e_1, \ldots, e_r$ are nonzero integers. Moreover, J is an integral ideal if and only if $e_i > 0$ for each $i = 1, \ldots, r$.*

Proof: By definition, there exists $a \in A$, $a \neq 0$, such that $Aa \cdot J$ is an integral ideal. By Theorem 1, Aa and $Aa \cdot J$ may be written, in a unique way, as products of prime ideals. By part (4) of Theorem 1, J may be written, in a unique way, in the form indicated. The last assertion is obvious.   ∎

The unique factorization of ideals of the ring of algebraic integers was discovered by Dedekind, so this fact is usually known as Dedekind's theorem. Hurwitz gave a direct algebraic proof, and later Noether found which purely algebraic properties of a domain imply the unique factorization of ideals into prime ideals; Krull improved the form of Noether's properties, showing that (1) implies (2). Matusita proved the equivalence of conditions (2) and (3).

Since the domains with the properties of Theorem 1 are so important, we introduce a definition:

Definition 2.   A domain A is said to be a *Dedekind domain* whenever it satisfies the equivalent properties (1), (2), (3), and (4) of Theorem 1.

Thus the domain A is a Dedekind domain if and only if the group $\mathcal{F}$ of nonzero fractional ideals is the free Abelian group generated by the set of nonzero prime ideals of A.

With this terminology, we may say that the ring of algebraic integers of an algebraic number field of finite degree is a Dedekind domain.

7.2 Dedekind Domains

Many of the results, which we may prove for the ring of algebraic integers, are actually valid for arbitrary Dedekind domains; thus, we shall state these results in full generality. The reader may consider Dedekind domains as a natural generalization of the principal ideal domains and as a counterpart of the unique factorization domains.

As a bonus of Theorem 1, we arrive at several interesting consequences.

Let us look at the prime and indecomposable elements of A. The element p ($p \neq 0$ and is not invertible) is said to be *indecomposable* whenever if $a, b \in A$ and $p = ab$ then p is associated to a or to b. The element $p \neq 0$ is said to be *prime* if the principal ideal Ap is prime.

If p is a prime element then it is indecomposable. If p is indecomposable then the principal ideal Ap is not the product of (more than one) principal ideals (distinct from A). Thus, a reasonable generalization would be the following: an integral ideal P of the domain A is said to be *indecomposable* when it is not equal to a product $P = J \cdot J'$, where J, J' are integral ideals different from P. But, actually, these are no new kinds of ideals:

D.   *Let A be a Dedekind domain. An ideal of A is indecomposable if and only if it is prime.*

Proof: The following implication holds for an arbitrary domain. Let P be an ideal, which is not indecomposable. Then we may write $P = JJ'$, where J, J' are integral ideals different from P. From $J|P$, $J'|P$, it follows that $P \subset J$, $P \subset J'$, so there exist elements $a \in J$, $a' \in J'$, such that $a \notin P$, $a' \notin P$, however, $aa' \in JJ' = P$. Thus, P is not a prime ideal.

Conversely, if A is a Dedekind domain, and if J is an indecomposable ideal, either $J = 0$ (hence it is prime) or $J \neq 0$, and then J is a product of nonzero prime ideals; but since J is indecomposable, it must be a prime ideal.   ∎

Next:

E.   *If P is a prime ideal in a Dedekind domain A, if J, J' are ideals of A, and $P|JJ'$, then $P|J$ or $P|J'$.*

Proof: By the uniqueness of the prime ideal decomposition of JJ', from $JJ' = PI$ (where I is an integral ideal), it follows that P is present among the prime ideals in the decomposition of JJ'. Hence, either $P|J$ or $P|J'$.   ∎

The divisibility between ideals means the reverse inclusion:

F.   *Let M, M' be nonzero fractional ideals of a Dedekind domain. Then M divides M' if and only if $M \supseteq M'$.*

Proof: We have already shown that if $M|M'$ then $M \supseteq M'$. Conversely, if $M \supseteq M'$, since the nonzero fractional ideals form a group, there exists the fractional ideal M^{-1} and $M'M^{-1} \subseteq MM^{-1} = A$, so $M'M^{-1}$ is an integral ideal such that $(M'M^{-1})M = M'$; that is, M divides M'.   ∎

G.   *If J is any nonzero integral ideal of a Dedekind domain A, then there exist only finitely many integral ideals of A which divide J.*

Proof: Let $J = \prod_{i=1}^{r} P_i^{e_i} \subseteq A$. Then the integral ideals dividing J are precisely those of the form $\prod_{i=1}^{r} P_i^{f_i}$, where $0 \leq f_i \leq e_i$ for every $i = 1, \ldots, r$. Hence, there are only finitely many integral ideals dividing J.   ∎

We may now consider the greatest common divisor and least common multiple of fractional ideals of a Dedekind domain.

H.   *If M, M' are nonzero fractional ideals of the Dedekind domain A, then $M + M'$ is the greatest common divisor of M and M' and $M \cap M'$ is the least common multiple of M, M'.*

Proof: We begin by observing that if M, M' are nonzero fractional ideals then $M + M'$ and $M \cap M'$ are also nonzero fractional ideals (for example, if $b, b' \in A$ are nonzero elements such that $bM \subseteq A$ and $b'M' \subseteq A$ then $bb'(M + M') \subseteq A$; also, if $a \in M$, $a' \in M'$ are nonzero elements, then $0 \neq (ba)(b'a') \in AM \cap AM' \subseteq M \cap M'$).

By (**F**), $M + M'$ divides both M and M' and if $M''|M$, $M''|M'$ then $M'' \supseteq M$, $M'' \supseteq M'$, so $M'' \supseteq M + M'$, that is, M'' divides $M + M'$; thus $M + M'$ is the greatest common divisor of M and M'.

The proof is analogous for the least common multiple.   ∎

I.   *A domain, which is both a Dedekind and a unique factorization domain, must be a principal ideal domain, and conversely.*

Proof: The converse has already been mentioned, so let us assume that the Dedekind domain A is also a unique factorization domain.

It is enough to prove that every nonzero prime ideal P is principal. Let $a \in P$, $a \neq 0$. Since a is not a unit, it is a product of prime elements, $a = \prod_{i=1}^{r} p_i^{e_i} \geq 1$, $r > 0$.

From $\prod_{i=1}^{r} p_i^{e_i} = a \in P$ it follows that $p_i \in P$ (for some index i), hence $Ap_i \subseteq P$. But Ap_i is a prime ideal, because if $a, b \in A$, $ab \in Ap_i$ then $p_i|ab$, thus $p_i|a$ or $p_i|b$, noting that A is a unique factorization domain. By Theorem 1, Ap_i is a maximal ideal, so $P = Ap_i$ is a principal ideal.
    ∎

Thus, we have justified our previous remark, namely, the ring of integers of an algebraic number field is a unique factorization domain if and only if it is a principal ideal domain.

Another useful fact follows:

J.   *If A is a Dedekind domain, for every nonzero fractional ideal M there exists a fractional ideal M', such that MM' is any given nonzero principal ideal.*

Proof: Given $Aa \neq 0$, since the nonzero fractional ideals form a group, there exists M' such that $MM' = Aa$.   ∎

One of the basic theorems in Dedekind domains is the generalization of the Chinese remainder theorem (see Chapter 3, Theorem 1 and (**C**)). We may prove an even more general result:

K.   *Let A be a ring, let J be any ideal of A, and assume that $J = \bigcap_{i=1}^{r} Q_i$, where the ideals Q_i are distinct from A and satisfy the relation $Q_j + \bigcap_{i \neq j} Q_i = A$ for every $j = 1, \ldots, r$. Then $A/J \cong \prod_{i=1}^{r} A/Q_i$.*

Proof: Let $\theta : A \to \prod_{i=1}^{r} A/Q_i$ be the mapping defined as follows: $\theta(a) = (\theta_1(a), \ldots, \theta_r(a))$ where $\theta_i(a)$ denotes the image of a by the natural mapping $A \to A/Q_i$.

θ is a ring-homomorphism, whose kernel is equal to $\bigcap_{i=1}^{r} Q_i = J$. It remains to prove that θ maps A onto $\prod_{i=1}^{r} A/Q_i$; in other words, given elements $x_i \in A$ $(i = 1, \ldots, r)$, we shall show the existence of an element $x \in A$ such that $x - x_i \in Q_i$ for all $i = 1, \ldots, r$.

Special Case: There exists an index j such that $x_i = 0$ for all $i \neq j$, $1 \leq i \leq r$.

Since $Q_j + \bigcap_{i \neq j} Q_i = A$, we may write $x_j = y + z$ with $y \in Q_j$, $z \in \bigcap_{i \neq j} Q_i$, hence $z - x_j = -y \in Q_j$, and $z - x_i = z \in Q_i$ for $i \neq j$. We may therefore choose $x = z$.

General Case: For every $j = 1, \ldots, r$, we have shown the existence of an element $z_j \in A$ such that $z_j - x_j \in Q_j$, $z_j \in Q_i$ for all $i \neq j$. Let $x = \sum_{j=1}^{r} z_j$, then $x - x_i = \sum_{j \neq i} z_j + (z_i - x_i) \in Q_i$ for all $i = 1, \ldots, r$. ∎

Theorem 3. *Let A be a Dedekind domain, let J be an integral ideal, and $J = \prod_{i=1}^{r} P_i^{e_i}$ where $r \geq 1$, $e_i \geq 1$, P_i are distinct nonzero prime ideals of A. Then*

$$A/J \cong \prod_{i=1}^{r} A/P_i^{e_i}.$$

Proof: It is enough to note that $\prod_{i=1}^{r} P_i^{e_i} = \bigcap_{i=1}^{r} P_i^{e_i}$ and $P_j^{e_j} + \prod_{i \neq j} P_i^{e_i} = A$, when $P_1, \ldots, P_r$ are distinct prime ideals of the Dedekind domain A. This follows at once from (**H**). ∎

We may sharpen slightly the main assertion of the preceding theorem:

L. *Let A be a Dedekind domain and let $P_1, \ldots, P_r$ be distinct nonzero prime ideals of A; let $x_1, \ldots, x_r \in A$ and let $e_1, \ldots, e_r$ be integers, $e_i \geq 0$ for every index i. Then there exists an element $x \in A$ such that*

$$x - x_i \in P_i^{e_i}, \qquad x - x_i \notin P_i^{e_i+1} \qquad \text{for all} \quad i = 1, \ldots, r.$$

Proof: For every $i = 1, \ldots, r$, we have $P_i^{e_i} \subset P_i^{e_i+1}$, so there exists an element $a_i \in P_i^{e_i}$, $a_i \notin P_i^{e_i+1}$. By Theorem 2, there exists $x \in A$ such that $x - (x_i + a_i) \in P_i^{e_i+1}$ for every $i = 1, \ldots, r$. Therefore, $x - x_i = [x - (x_i + a_i)] + a_i \in P_i^{e_i}$ but $x - x_i \notin P_i^{e_i+1}$. ∎

We state a special case: with the same hypothesis and notations, there exists $x \in A$ such that $x \in P_1$, but $x \notin P_i$ for $i = 2, \ldots, r$.

Here is a useful variant:

M. *Let A be a Dedekind domain and let $P_1, \ldots, P_r$ be distinct nonzero prime ideals of A; let $e_1, \ldots, e_r$ be integers, $e_i \geq 0$ for every*

$i = 1, \ldots, r$. Then there exists $x \in A$, such that $Ax = J \cdot \prod_{i=1}^{r} P_i^{e_i}$, where $\gcd(J, \prod_{i=1}^{r} P_i) = A$.

Proof: We apply (**L**) with $x_1 = \cdots = x_r = 0$. Since $Ax \subseteq \prod_{i=1}^{r} P_i^{e_i}$ but $x \notin P_i^{e_i+1}$ for every $i = 1, \ldots, r$, then $Ax = J \cdot \prod_{i=1}^{r} P_i^{e_i}$ by (**F**). Moreover, P_i does not divide J, for each $i = 1, \ldots, r$, so $\gcd(J, \prod_{i=1}^{r} P_i) = A$. ∎

An interesting corollary concerns Dedekind domains, which have only finitely many prime ideals:

N.   *If a Dedekind domain has only finitely many prime ideals then it is a principal ideal domain.*

Proof: Since every nonzero ideal of the Dedekind domain A is a product of nonzero prime ideals, it is enough to show that these are principal ideals.

Thus, let $P_1, \ldots, P_r$ be the nonzero prime ideals of A. By (**L**), for every h, $1 \le h \le r$, there exists an element $y_h \in A$, such that $y_h \in P_h$, $y_h \notin P_h^2$, and $y_h \notin P_j$ for all $j \ne h$. This is just the statement of (**L**), with $x_1 = \cdots = x_r = 0$, $e_h = 1$, $e_j = 0$ for $j \ne h$.

Thus $P_h | Ay_h$, P_h^2 does not divide Ay_h and P_j does not divide Ay_h (for $j \ne h$). Hence the decomposition of Ay_h into prime ideals is $Ay_h = P_h$, proving that each prime ideal P_h is principal. ∎

Since a Dedekind domain is Noetherian, all ideals are finitely generated. Actually, we may prove a much better result:

O.   *If A is a Dedekind domain, then every fractional ideal of A may be generated by at most two elements: one of these elements may be arbitrarily chosen.*

Proof: It is clearly enough to prove this statement for nonzero integral ideals J. Let $a \in J$, $a \ne 0$ so $Aa \subseteq J$ hence J divides Aa, by (**F**). Let $Aa = \prod_{i=1}^{r} P_i^{e_i}$ where $P_1, \ldots, P_r$ are distinct prime ideals and $e_i > 0$ for each $i = 1, \ldots, r$. Thus $J = \prod_{i=1}^{r} P_i^{f_i}$ with $0 \le f_i \le e_i$ for each $i = 1, \ldots, r$. By (**M**), there exists $b \in A$ such that $Ab = \prod_{i=1}^{r} P_i^{f_i} \cdot J'$, where J' is an ideal such that $\gcd(\prod_{i=1}^{r} P_i, J') = A$ and $Aa + Ab = \gcd(Aa, Ab) = \prod_{i=1}^{r} P_i^{f_i} = J$. ∎

We conclude by studying the integral closure of a Dedekind domain in a separable extension of its field of quotients:

P.   *Let A be a Dedekind domain, and K its field of quotients. Let $L|K$ be a separable extension of degree n and let B be the integral closure of A in L. Then B is a Dedekind domain.*

Proof: B is integrally closed, by definition.

From Chapter 6, (**B**), we know that B is a submodule of a free A-module M of rank n; since A is a Noetherian ring then M is a Noetherian A-module (Chapter 6, (**G**)), hence B is a Noetherian A-module; therefore, B is a Noetherian B-module, that is, a Noetherian ring.

Finally, if Q is a nonzero prime ideal of B, then $Q \cap A = P$ is a prime ideal of A and $Q \cap A \neq 0$ (Chapter 5, (E)), so P is a maximal ideal of A and by Chapter 5, (G), Q is a maximal ideal of B.

By Theorem 1, B is a Dedekind domain.   ∎

To conclude, we extend the relation of congruence. Let J be a nonzero integral ideal of the Dedekind domain A. Let $a/a', b/b' \in K$ be such that $\gcd(Aa, Aa') = A$, $\gcd(Ab, Ab') = A$, $\gcd(Aa', J) = A$, and $\gcd(Ab', J) = A$. We define

$$\frac{a}{a'} \equiv \frac{b}{b'} \pmod{J}$$

if $ab' - a'b \in J$. It is straightforward to verify that the relation of congruence modulo J is an equivalence relation and if

$$\frac{a}{a'} \equiv \frac{b}{b'} \pmod{J}, \qquad \frac{a_1}{a_1'} \equiv \frac{b_1}{b_1'} \pmod{J}$$

(with a_1, a_1', b_1, b_1' subjected to the same hypothesis), then

$$\frac{a}{a'} + \frac{a_1}{a_1'} \equiv \frac{b}{b'} + \frac{b_1}{b_1'} \pmod{J}$$

and

$$\frac{a}{a'} \cdot \frac{a_1}{a_1'} \equiv \frac{b}{b'} \cdot \frac{b_1}{b_1'} \pmod{J}.$$

Exercises

1. Let A be the ring of algebraic integers of an algebraic number field. Prove that A has infinitely many prime ideals.

2. Let A be the ring of algebraic integers of an algebraic number field K. Prove successively the following facts:

(a) If x is a root of $h \in A[X]$ then $h/(X - x)$ has coefficients in A.

Hint: Let d be the leading coefficient of h and m the degree of h; first note that dx is an algebraic integer; proceed by induction on m, considering the polynomial $h - dX^{m-1}(X - x)$.

(b) If $h = d(X - x_1)(X - x_2) \cdots (X - x_m) \in A[X]$ then $dx_1 x_2 \cdots x_k$ is an algebraic integer (for every k, $1 \leq k \leq m$).

(c) If $f, g \in A[X]$, if $c \in A$, $c \neq 0$, divides every coefficient of fg then, for every coefficient a of f and b of g, c divides ab.

Hint: Express f, g as products of the leading coefficient and linear factors, consider $fg/c \in A[X]$; apply (b) to this polynomial and conclude noting that the coefficients of any polynomial are expressible in terms of elementary symmetric functions of its roots.

3. Apply the previous exercise to obtain another proof, due to Hurwitz, that the ring A of algebraic integers of K is a Dedekind domain. Prove successively the following facts:

(a) For every nonzero ideal I of A there exists a nonzero ideal J of A such that IJ is a principal ideal.

(b) If I, I', I'' are nonzero ideals of A and $I \cdot I' = I \cdot I''$ then $I' = I''$.

(c) If I, J are ideals of A such that $I \cdot J$ is a multiple of the prime ideal P of A then either P divides I or P divides J.

(d) Every nonzero ideal I of A is equal to a product of prime ideals of A; this representation is unique, except for the order of the factors.

4. Prove, using transfinite induction (or Zorn's lemma) that every proper ideal I of a ring A (with unit element) is contained in a maximal ideal. Assuming that A is Noetherian, indicate a simpler proof for the same fact.

5. Prove the *Hilbert basis theorem*: If A is a Noetherian ring then the ring $A[X_1, \ldots, X_n]$ of polynomials in n indeterminates is also Noetherian.

6. Prove that if the ring A' is a homomorphic image of a Noetherian ring A then A' is also a Noetherian ring.

7. Let B be a ring, A a Noetherian subring of B, and let $x_1, \ldots, x_n \in B$. Prove that the subring $A[x_1, \ldots, x_n]$ of B generated by A and $x_1, \ldots, x_n$ is also a Noetherian ring.

8. Let A be a domain, and let I, J be ideals of A. The *conductor of J into I* is the set of all elements $x \in A$ such that $x \cdot J \subseteq I$. It is denoted by $I : J$.

Prove:

(a) $I : J$ is an ideal of A containing I.

(b) If J' is an ideal of A, then $J \cdot J' \subseteq I$ if and only if $J' \subseteq I : J$.

For any ideals of A, show:

(c) $\left(\bigcap_{i=1}^{n} I_i \right) : J = \bigcap_{i=1}^{n} (I_i : J).$

(d) $I : \left(\sum_{j=1}^{n} J_j \right) = \bigcap_{j=1}^{n} (I : J_j).$

(e) $I : JJ' = (I : J) : J'.$

9. Let A be a domain and let I be an ideal of A. The *root* $\sqrt{I}$ of the ideal I is the set of all elements $x \in A$ having some power x^n $(n \geq 1)$ in I.
Prove:

(a) $\sqrt{I}$ is an ideal containing I.

(b) $\sqrt{\sqrt{I}} = \sqrt{I}.$

(c) If I, J are ideals of A and there exists $n \geq 1$ such that $I^n \subseteq J$, then $\sqrt{I} \subseteq \sqrt{J}.$

(d) $\sqrt{I \cdot J} = \sqrt{I \cap J} = \sqrt{I} \cap \sqrt{J}.$
$\sqrt{I + J} = \sqrt{\sqrt{I} + \sqrt{J}}.$

10. An ideal I of a ring A is said to be *primary* if the following condition is satisfied: If $a, b \in A$, $ab \in I$, $a \notin I$, then there exists an integer $m \geq 1$ such that $b^m \in I$.
Prove:

(a) Every prime ideal is a primary ideal.

(b) The root $\sqrt{I}$ of a primary ideal I is a prime ideal P; in this situation, if $a, b \in A$, $ab \in I$, $a \notin I$, then $b \in P$.

(c) If I is a primary ideal of A, if J, J' are ideals of A such that $J \cdot J' \subseteq I$, $J \nsubseteq I$, then $J' \subseteq \sqrt{I}.$

(d) If A is a Noetherian ring and I is a primary ideal of A, show that there exists an integer $m \geq 1$ such that $(\sqrt{I})^m \subseteq I.$

11. Let I, P be ideals of a ring A and suppose that:

(a) $I \subseteq P \subseteq \sqrt{I}.$

(b) If $ab \in I$, $a \notin I$, then $b \in P$.

Show that I is a primary ideal and $P = \sqrt{I}.$

12. Let B be a ring and A a subring of B. Let I, I' be ideals of A and J, J' ideals of B. For every ideal I of A let $B \cdot I$ denote the ideal of B generated by I. Prove the following facts:

(a) $B \cdot I$ is the set of sums $\sum_{i=1}^{n} b_i x_i$, with $n \geq 1$, $x_i \in I$, $b_i \in B$.

(b) $B(J \cap A) \subseteq J$ and $B(J \cap A) \cap A = J \cap A$.

(c) $B(B \cdot I \cap A) = B \cdot I$.

(d) $(J + J') \cap A \supset J \cap A + J' \cap A$ and $B(I + I') = B \cdot I + B \cdot I'$.

(e) $(J \cap J') \cap A = (J \cap A) \cap (J' \cap A)$ and $B(I \cap I') \subseteq B \cdot I \cap B \cdot I'$.

(f) $J \cdot J' \cap A \supseteq (J \cap A)(J' \cap A)$ and $B \cdot (I \cdot I') = (B \cdot I)(B \cdot I')$.

(g) $(J : J') \cap A \subseteq (J \cap A : J' \cap A)$ and $B(I : I') \subseteq (B \cdot I) : (B \cdot I')$.

(h) $\sqrt{J} \cap A = \sqrt{J \cap A}$ and $B\sqrt{I} \subseteq \sqrt{B \cdot I}$.

(i) If J is a primary ideal of B, then $J \cap A$ is a primary ideal of A and $\sqrt{J \cap A} = \sqrt{J} \cap A$.

13. Prove that the powers of any maximal ideal of a ring A are primary ideals.

14. Show that the primary ideals of any Dedekind domain A are precisely the powers of the prime ideals of A.

15. Let I, J be integral ideals of the Dedekind domain A. Prove that there exists an integral ideal I' such that $I \cdot I'$ is a principal ideal and $J + I' = A$.

16. Let A be a Dedekind domain, and let I be a nonzero integral ideal of A. Show that the ring A/I satisfies the *descending chain condition for ideals* (every strictly descending chain $J_1 \supset J_2 \supset \cdots \supset J_n \supset \cdots$ of ideals is finite).

17. Let A be a domain satisfying the following properties:

(a) A is a Noetherian ring.

(b) A is integrally closed.

(c) For every nonzero ideal I of A the ring A/I satisfies the descending chain condition for ideals.

Prove that A is a Dedekind domain.

18. Let A be a Dedekind domain. Show that the ideals of A satisfy the following distributive laws:

$$I \cap (J + J') = (I \cap J) + (I \cap J')$$

and

$$I + (J \cap J') = (I + J) \cap (I + J').$$

19. Prove the following *general form of the Chinese remainder theorem*: Let A be a Dedekind domain, and let $I_1, \ldots, I_n$ be ideals of A. The system of congruences $x \equiv a_i \pmod{I_i}$ (for $i = 1, \ldots, n$) has a solution in A if and only if $a_i \equiv a_j \pmod{I_i + I_j}$ for any indices i, j.

20. Let M be an additive subgroup of $\mathbb{Q}$. Prove that the following statements are equivalent:

(a) M is a finitely generated additive group.

(b) M is a fractional ideal of $\mathbb{Z}$.

(c) M is a principal fractional ideal of $\mathbb{Z}$.

21. Let p be a prime number. For each $a = p^r m \in \mathbb{Z}$, with p not dividing m, $r \geq 0$, let $v_p(a) = r$. For $x = a/b$ with $a, b \in \mathbb{Z}$, $a \neq 0$, $b \neq 0$, $\gcd(a, b) = 1$, let $v_p(x) = v_p(a) - v_p(b)$. If I is any nonzero fractional ideal of $\mathbb{Z}$ then, by the previous exercise, $I = \mathbb{Z}x$, $x \in \mathbb{Q}$, $x \neq 0$. Define $v_p(I) = v_p(x)$.

Prove:

(a) For every $I \neq 0$ we have $v_p(I) = 0$ except at most for a finite number of primes p.

If I, J are nonzero fractional ideals of $\mathbb{Z}$ then:

(b) $I \subseteq J$ if and only if $v_p(I) \geq v_p(J)$ for every prime p.

(c) $v_p(I + J) = \inf\{v_p(I), v_p(J)\}$.

(d) $v_p(I \cap J) = \sup\{v_p(I), v_p(J)\}$.

(e) $v_p(I \cdot J) = v_p(I) + v_p(J)$.

22. If M is a nonzero additive subgroup of $\mathbb{Q}$ and p is a prime number, define $v_p(M) = \inf\{v_p(x) \mid x \in M\}$. Show:

(a) $v_p(M) \in \mathbb{Z} \cup \{-\infty\}$; for every M we have $v_p(M) \leq 0$ except at most for a finite number of primes p.

(b) $M = \{x \in \mathbb{Q} \mid v_p(x) \geq v_p(M) \text{ for every prime } p\}$.

23. Let S be the set of sequences $\nu = (n_p)_{p \text{ prime}}$, $n_p \in \mathbb{Z} \cup \{-\infty\}$, such that $n_p \leq 0$ except at most for a finite number of primes p. Let $\mathcal{M}$ denote the set of all nonzero additive subgroups of $\mathbb{Q}$, and let $\theta : \mathcal{M} \to S$ be the mapping defined by $\theta(M) = (v_p(M))_{p \text{ prime}}$.

If $M, M' \in \mathcal{M}$ show:

(a) $M \subseteq M'$ if and only if $v_p(M) \geq v_p(M')$ for every prime p.

(b) θ is a one-to-one mapping.

(c) θ maps $\mathcal{M}$ onto S.

(d) M is a nonzero fractional ideal of $\mathbb{Z}$ if and only if $\theta(M) = (n_p)_p$ is such that each n_p is an integer and $n_p = 0$ except at most for a finite number of primes p.

(e) M is a nonzero integral ideal of $\mathbb{Z}$ if and only if $\theta(M) = (n_p)_p$ satisfies the above condition and, moreover, $n_p \geq 0$ for every prime p.

(f) M is a subring of $\mathbb{Q}$ if and only if $v_p(M) = 0$ or $-\infty$.

(g) $\mathbb{Q}$ has $2^{\aleph_0}$ distinct subrings.

(h) For every nonzero additive subgroup M of $\mathbb{Q}$ there exists a largest subring R of $\mathbb{Q}$ for which M is an R-module.

24. Let A be a Dedekind domain, P any nonzero prime ideal of A. If I is a nonzero fractional ideal of A, let $I = \prod_P P^{v_P(I)}$ be its decomposition into prime ideals (where $v_P(I) \in \mathbb{Z}$ and $v_P(I) = 0$ except at most for finitely many prime ideals, depending on I). Define also $v_P(x) = v_P(Ax)$ for every $x \in A$, $x \neq 0$; finally, let $v_P(0) = \infty$. If I, J are nonzero fractional ideals of A, show:

(a) $v_P(xy) = v_P(x) + v_P(y)$.

(b) $v_P(x + y) \geq \min\{v_P(x), v_P(y)\}$.

(c) If $v_P(x) \neq v_P(y)$ then $v_P(x + y) = \min\{v_P(x), v_P(y)\}$.

(d) I divides J if and only if $v_P(I) \leq v_P(J)$ for every prime ideal $P \neq 0$ of A.

(e) $v_P(I) = \min\{v_P(x) \mid x \in I\}$.

(f) $v_P(I \cdot J) = v_P(I) + v_P(J)$.

(g) $v_P(I + J) = \min\{v_P(I), v_P(J)\}$.

(h) $v_P(I \cap J) = \max\{v_P(I), v_P(J)\}$.

25. Let A be a Dedekind domain, K its field of quotients, and M any A-submodule of K; let P be any nonzero prime ideal of A. Define $v_P(M) = \inf\{v_P(I) \mid$ for every fractional ideal $I \subseteq M\}$. Show:

(a) $v_P(M) \in \mathbb{Z} \cup \{-\infty\}$; for every M we have $v_P(M) \leq 0$ except at most for a finite number of prime ideals P.

(b) $M = \{x \in K \mid v_P(x) \geq v_P(M)$ for every nonzero prime ideal $P\}$.

26. Let $\mathcal{P}$ be the set of nonzero prime ideals of the Dedekind domain A, and let $\mathcal{M}$ denote the set of all nonzero A-submodules of the quotient field K of A. Let $\mathcal{S}$ be the set of all sequences $(n_P)_{P \in \mathcal{P}}$, where $n_P \in \mathbb{Z} \cup \{-\infty\}$ and $n_P \leq 0$ except at most for a finite number of prime ideals P. Finally, let $\theta : \mathcal{M} \to \mathcal{S}$ be the mapping defined by $\theta(M) = (v_P(M))_{P \in \mathcal{P}}$.

If M, $M' \in \mathcal{M}$ show:

(a) $M \subseteq M'$ if and only if $v_P(M) \geq v_P(M')$ for every $P \in \mathcal{P}$.

(b) θ is a one-to-one mapping.

(c) θ maps $\mathcal{M}$ onto $\mathcal{S}$.

(d) M is a nonzero fractional ideal of A if and only if $\theta(M) = (n_P)_P$ where each $n_P \in \mathbb{Z}$ and $n_P = 0$ except at most for a finite number of prime ideals $P \in \mathcal{P}$.

(e) M is a nonzero integral ideal of A if and only if the preceding condition is satisfied and moreover $n_P \geq 0$ for every $P \in \mathcal{P}$.

(f) M is a subring of K if and only if $v_P(M) = 0$ or $-\infty$ for every $P \in \mathcal{P}$.

(g) If A is a Dedekind domain with countably many prime ideals then K has $2^{\aleph_0}$ distinct subrings.

(h) For every nonzero additive subgroup M of K there exists a largest subring R of K for which M is an R-module.

8

The Norm and Classes of Ideals

We know already that the ring A of integers of an algebraic number field K need not be a principal ideal domain. In this chapter, we associate with every field K a numerical invariant h, which measures the extent to which A deviates from being a principal ideal domain. h will be equal to 1 if and only if A is a principal ideal domain.

To begin, we introduce the important concept of the norm of an ideal.

8.1 The Norm of an Ideal

In Chapter 7, (**A**), we have seen that if P is any nonzero prime ideal of the ring A of integers of an algebraic number field K, then the quotient domain A/P (which is a field) is necessarily finite. If $P \cap \mathbb{Z} = \mathbb{Z}p$, then $\#(A/P)$ is a power of p.

More generally:

A. *If P is a nonzero prime ideal of A, $e \geq 1$, then $\#(A/P^e) = \#(A/P)^e$.*

Proof: The result is true for $e = 1$. We shall proceed by induction on e, assuming it true for $e - 1$. P^{e-1}/P^e is an ideal of the ring A/P^e, hence by the isomorphism theorem for rings, we have

$$(A/P^e)/(P^{e-1}/P^e) \cong A/P^{e-1}.$$

It follows that $\#(A/P^e) = \#(A/P^{e-1}) \cdot \#(P^{e-1}/P^e)$. By induction, it suffices to show that $\#(P^{e-1}/P^e) = \#(A/P)$. We observe that P^{e-1}/P^e is a vector space over the field A/P, with a scalar multiplication so defined: if $\tilde{a} \in A/P$, $\overline{x} \in P^{e-1}/P^e$, then $\tilde{a}\,\overline{x} = \overline{ax}$ (the reader may easily check the details). Thus it is enough to show that P^{e-1}/P^e has dimension at most 1, since we know that $P^{e-1} \supset P^e$, the dimension of P^{e-1}/P^e is at least 1; this will imply that $P^{e-1}/P^e \cong A/P$, from which we deduce the required relation. Now, let $x \in P^{e-1}$, $x \notin P^e$. Then $P^e \subset P^e + Ax \subseteq P^{e-1}$, so there exist (integral) ideals J, J' such that $(P^e + Ax)J = P^e$ and

$P^{e-1}J' = P^e + Ax$; moreover, $J \neq A$. Then $P^{e-1}JJ' = P^e$ and therefore $JJ' = P$, hence $J = P$, $J' = A$. So $P^{e-1} = P^e + Ax$. This implies that if $\bar{y} \in P^{e-1}/P^e$ there exists $a \in A$ such that $\bar{y} = \tilde{a}\,\bar{x}$, which was required to be proved. ∎

Let us note incidentally the following useful result:

B.   *Let P be a nonzero prime ideal of A, $e \geq 1$. Let S be a system of representatives of A modulo P, such that $0 \in S$; let $t \in P$, $t \notin P^2$. Then*

$$R = \{s_0 + s_1 t + \cdots + s_{e-1}t^{e-1} \mid s_i \in S \text{ for } i = 0, 1, \ldots, e-1\}$$

is a system of representatives of A modulo P^e.

Proof: The result is proved by induction on e; it is trivial for $e = 1$, and we now assume it true for $e - 1$. We show that if $s_0, s_0' \in S$, if

$$r_1 = s_1 + s_2 t + \cdots + s_{e-1}t^{e-2}, \qquad r_1' = s_1' + s_2' t + \cdots + s_{e-1}' t^{e-2}$$

(with $s_i, s_i' \in S$ for every $i = 1, \ldots, e-1$), if $r = s_0 + r_1 t$, $r' = s_0' + r_1' t$, and $r \neq r'$ then $r - r' \notin P^e$. Indeed, assume that $r - r' \in P^e$, then

$$(s_0 - s_0') + (r_1 - r_1')t \in P^e,$$

so

$$s_0 - s_0' \in At + P^e \subseteq P,$$

hence $s_0 = s_0'$ and therefore $(r_1 - r_1')t \in P^e$, so P^e divides $A(r_1 - r_1') \cdot At$. Since $t \in P$, $t \notin P^2$, then $P|At$, $P^2 \nmid At$, thus $P^{e-1}|A(r_1 - r_1')$ hence $r_1 - r_1' \in P^{e-1}$. By induction, $r_1 = r_1'$ and we conclude that $r = r'$.

Thus R already contains $\#(A/P)^e$ different representatives of A modulo P^e. From (**A**) if follows that R is a system of representatives of A modulo P^e. ∎

By means of Theorem 2 in Chapter 7, we deduce:

C.   *If J is any nonzero integral ideal of A, $J = \prod_{i=1}^{r} P_i^{e_i}$ then*

$$\#(A/J) = \prod_{i=1}^{r} \#(A/P_i)^{e_i}.$$

Proof: We just combine (**A**) with the above-mentioned theorem. ∎

Thus, we have associated with every nonzero integral ideal a positive integer:

Definition 1.   The *norm* of an ideal J of the ring A of integers of an algebraic number field is defined to be the positive integer $N(J) = \#(A/J)$.

Thus, if P is a nonzero prime ideal of A, $e \geq 1$, then $N(P^e) = N(P)^e$, as shown in (**A**). More generally, by means of (**C**), we obtain:

D.   *If J, J' are ideals of A, then $N(JJ') = N(J) \cdot N(J')$.*

Proof: If $J = \prod_{i=1}^{r} P_i^{e_i}$, $J' = \prod_{i=1}^{r} P_i^{e_i'}$, where $e_i \geq 0$, $e_i' \geq 0$, then $JJ' = \prod_{i=1}^{r} P_i^{e_i+e_i'}$, so by means of **(C)** we deduce that $N(JJ') = N(J) \cdot N(J')$. ∎

The above property is usually called the *multiplicativity of the norm*.

Previously, we have defined the norm of an element $x \in K$ (relative to the extension $K|\mathbb{Q}$); in the case where $x \in A$, $x \neq 0$, we shall compare $N_{K|\mathbb{Q}}(x)$ and $N(Ax)$.

For this purpose we need an important relation between the norm of an ideal and the discriminant of any basis of this ideal.

E.   *Let J be a nonzero integral ideal of A. Let $\{y_1, \ldots, y_n\}$ be any basis of the free Abelian group J, and δ the discriminant of the field K. Then:*

$$N(J)^2 = \frac{\mathrm{discr}_{K|\mathbb{Q}}(y_1, \ldots, y_n)}{\delta}.$$

Proof: By Chapter 6, **(L)**, there exists an integral basis $\{x_1, \ldots, x_n\}$ and integers $f_1, \ldots, f_n$ such that $\{f_1 x_1, \ldots, f_n x_n\}$ is a basis of the free Abelian group J. Thus, $A = \mathbb{Z}x_1 \oplus \cdots \oplus \mathbb{Z}x_n$, while $J = \mathbb{Z}f_1 x_1 \oplus \cdots \oplus \mathbb{Z}f_n x_n$. Then the quotient Abelian group is $A/J \cong \prod_{i=1}^{n} \mathbb{Z}/\mathbb{Z}f_i$, the isomorphism being obviously the one which comes from the mapping

$$A \to \prod_{i=1}^{n} \mathbb{Z}/\mathbb{Z}f_i, \qquad \sum_{i=1}^{n} a_i x_i \mapsto (a_i \bmod f_i)_{1 \leq i \leq n},$$

by noting that the kernel is J. Therefore $N(J) = \#(A/J) = \prod_{i=1}^{n} |f_i|$.

Since $\mathrm{discr}(f_1 x_1, \ldots, f_n x_n) = (\prod_{i=1}^{n} f_i)^2 \cdot \mathrm{discr}(x_1, \ldots, x_n)$, noting that $\mathrm{discr}(x_1, \ldots, x_n) = \delta$ (discriminant of the field K), and

$$\mathrm{discr}_{K|\mathbb{Q}}(f_1 x_1, \ldots, f_n x_n) = \mathrm{discr}_{K|\mathbb{Q}}(y_1, \ldots, y_n)$$

for any basis $\{y_1, \ldots, y_n\}$ of the Abelian group J, we deduce that

$$N(J)^2 = \frac{\mathrm{discr}_{K|\mathbb{Q}}(y_1, \ldots, y_n)}{\delta}. \qquad ∎$$

Now, we have:

F.   *For every element $y \in A$, $y \neq 0$:*

$$N(Ay) = |N_{K|\mathbb{Q}}(y)|.$$

Proof: Let $\{x_1, \ldots, x_n\}$ be an integral basis, so $\{y x_1, \ldots, y x_n\}$ is a basis of the Abelian group Ay. Hence, by **(E)**:

$$N(Ay)^2 = \frac{\mathrm{discr}_{K|\mathbb{Q}}(y x_1, \ldots, y x_n)}{\delta}.$$

However, by Chapter 2, Section 11:

$$\text{discr}_{K|\mathbb{Q}}(yx_1, \ldots, yx_n) = \det(\sigma_i(yx_j))^2$$
$$= \det(\sigma_i(y\delta_{ij}))^2 \cdot \det(\sigma_i(x_j))^2$$
$$= (N_{K|\mathbb{Q}}(y))^2 \cdot \delta$$

(where $\delta_{ij} = 1$ when $i = j$, $\delta_{ij} = 0$ when $i \neq j$). Therefore, $N(Ay)^2 = (N_{K|\mathbb{Q}}(y))^2$, and since $N(Ay) > 0$ then $N(Ay) = |N_{K|\mathbb{Q}}(y)|$.   ∎

One of the main facts about norms is the following:

G.   *If J is a nonzero integral ideal of A, then J divides the principal ideal $A \cdot N(J)$. For every integer $m > 0$ there are only finitely many ideals with norm equal to m.*

Proof: Let $N(J) = m = \#(A/J)$; thus in the quotient group A/J the order of every element divides m; therefore, if $x \in A$ then $mx \in J$. In particular $m = m \cdot 1 \in J$, so J divides Am.

Since the ideal Am has only finitely many divisors (by Chapter 7, **(E)**) then there exist only finitely many ideals J with norm equal to m.   ∎

The following property will be used in a crucial way in the next section. Let $\{x_1, \ldots, x_n\}$ be an integral basis of the field K. For each $i = 1, \ldots, n$, let $x_i^{(1)} = x_i$, $x_i^{(2)}$, $\ldots$, $x_i^{(n)}$ be the images of x_i by the n isomorphisms from K (into $\mathbb{C}$). Let

$$\mu = \prod_{j=1}^{n} \sum_{i=1}^{n} |x_i^{(j)}|;$$

so μ is a positive real number depending on K (and the given integral basis).

Now we prove:

H.   *Let K be any algebraic number field, and let μ be defined as above. Then for every nonzero integral ideal J of A, there exists an element $a \in J$, $a \neq 0$, such that*

$$|N_{K|\mathbb{Q}}(a)| \leq N(J) \cdot \mu.$$

Proof: Let $\{x_1, \ldots, x_n\}$ be an integral basis of K and let $\mu > 0$ be defined as above.

If J is any nonzero integral ideal of A, let k be an integer such that $k^n \leq N(J) = \#(A/J) < (k+1)^n$. Consider the set S of all elements $\sum_{i=1}^{n} d_i x_i$ where $0 \leq d_i \leq k$. Since $\#(S) = (k+1)^n > \#(A/J)$, there must exist $b, c \in S$, $b \neq c$, such that $a = b - c = \sum_{i=1}^{n} a_i x_i \in J$. We note that $|a_i| \leq k$ for each $i = 1, \ldots, n$. It follows that $|N_{K|\mathbb{Q}}(a)| = \prod_{j=1}^{n} |\sum_{i=1}^{n} a_i x_i^{(j)}| \leq \prod_{j=1}^{n} k(\sum_{i=1}^{n} |x_i^{(j)}|) = k^n \mu \leq N(J)\mu.$ ∎

8.2 Classes of Ideals

As before let $\mathcal{F}$ denote the multiplicative Abelian group of nonzero fractional ideals of A, let $\mathcal{Pr}$ be the subgroup of nonzero principal fractional ideals. We may therefore consider the quotient group $\mathcal{F}/\mathcal{Pr}$.

Explicitly, two nonzero fractional ideals M, M' are said to be *equivalent* when there exists $x \in K$, $x \neq 0$, such that $M' = Ax \cdot M$. We write $M \sim M'$.

This is clearly an equivalence relation, and if $M_1 \sim M_2$, $M_1' \sim M_2'$, then $M_1 \cdot M_1' \sim M_2 \cdot M_2'$. $\mathcal{Pr}$ is precisely the subgroup of those ideals equivalent to the unit ideal A. Each element of $\mathcal{F}/\mathcal{Pr}$ is called an *ideal class* of K, and $\mathcal{F}/\mathcal{Pr}$ is the *ideal class group* of K. It is denoted by $C(K)$ or also by Cl_K, or some similar notation. Roughly speaking, the larger the group $\mathcal{F}/\mathcal{Pr}$ is, the more the ring A fails to be a principal ideal domain. So, a natural question is the following: is it possible that $\mathcal{F}/\mathcal{Pr}$ is an infinite group?

We now prove a fundamental theorem due to Minkowski. The special case of quadratic fields is due to Gauss, while Kummer proved the theorem in the case of cyclotomic fields.

Theorem 1.   *The number of classes of ideals of an algebraic number field is finite.*

Proof: The norm of every nonzero integral ideal is a positive integer. Given the real number μ, defined in **(H)**, by **(G)** there exist only finitely many nonzero ideals $J_1, \ldots, J_k$ such that $N(J_i) \leq \mu$.

We shall prove that if I is any nonzero ideal of A, then I is equivalent to some ideal J_i; therefore the number of classes of ideals is at most k, hence it is finite.

Now, let I^{-1} denote the fractional ideal inverse of I, so there exists an element $c \in A$, $c \neq 0$, such that cI^{-1} is an integral ideal.

By **(H)**, there exists an element $b \in cI^{-1}$, $b \neq 0$, such that $N(Ab) \leq N(cI^{-1}) \cdot \mu$. Multiplying by $N(I)$ and observing that $Ibc^{-1} \subseteq A$, we obtain

$$N(Ibc^{-1}) \cdot N(Ac) = N(Ibc^{-1} \cdot Ac) = N(Ib)$$
$$= N(Ab) \cdot N(I) \leq N(cI^{-1}) \cdot N(I)\mu = N(Ac)\mu,$$

thus $N(Ibc^{-1}) \leq \mu$, so $Ibc^{-1} = J_i$ for some index i.   ∎

We introduce the following numerical invariant:

Definition 2.   The number of classes of ideals of an algebraic number field K is called the *class number* of K and denoted by $h = h_K$.

An easy corollary follows:

I.   *If J is any nonzero fractional ideal of the ring A then J^h is a principal fractional ideal of A.*

Proof: Just note that h is the order of the multiplicative group $\mathcal{F}/\mathcal{P}r$ of classes of ideals; thus the hth power of every fractional ideal is principal. ∎

The following result gives the justification of the word "ideal":

J. *Let K be an algebraic number field with class number h. There exists an extension K' of degree at most h over K with the following property: for every nonzero fractional ideal J of K there exists an element $x \in K'$ such that $J = K \cap A'x$ (where A' denotes the ring of algebraic integers of K').*

Proof: The Abelian multiplicative group $\mathcal{F}/\mathcal{P}r$ of classes of ideals has order h and by the structure theorem for finite Abelian groups (Chapter 3, Theorem 3), $\mathcal{F}/\mathcal{P}r$ is the product of cyclic groups $\mathcal{C}_1, \ldots, \mathcal{C}_m$ having orders $h_1, \ldots, h_m$, respectively; moreover, $h = h_1 h_2 \ldots h_m$. Let $J_1, \ldots, J_m$ be nonzero fractional ideals whose ideal classes are the generators of $\mathcal{C}_1, \ldots, \mathcal{C}_m$. Then $J_i^{h_i} = Aa_i$ where $a_i \in A$ for every $i = 1, \ldots, m$. Let x_i be a root of the polynomial $X^{h_i} - a_i$ (in $\mathbb{C}$). Let $K' = K(x_1, \ldots, x_n)$ and let A' denote the ring of algebraic integers of K'. We note that the field K' depends on the choice of a_i, x_i and has degree $[K' : K] \leq h_1 h_2 \cdots h_m = h$.

Let J be any nonzero fractional ideal of K. Then there exist $c \in K$, $c \neq 0$, and integers $e_1, \ldots, e_m$ with $0 \leq e_i < h_i$ for every $i = 1, \ldots, m$ such that $J = Ac \cdot J_1^{e_1} \cdots J_m^{e_m}$. Writing $h = h_i h_i'$ $(i = 1, \ldots, m)$, we have $J^h = Ac^h \cdot J_1^{he_1} \cdots J_m^{he_m} = A(c^h a_1^{h_1'e_1} \cdots a_m^{h_m'e_m})$, thus $J^h = Aa$ with $a = c^h a_1^{h_1'e_1} \cdots a_m^{h_m'e_m}$.

Then the element $x = cx_1^{e_1} \cdots x_m^{e_m} \in K'$ is such that $J = K \cap A'x$. In fact, we begin by noting that $x^h = c^h x_1^{he_1} \cdots x_m^{he_m} = a$. Now, if $y \in J$ then $y^h \in J^h = Aa$ so $(y/x)^h = y^h/a \in A$, thus $y/x \in A'$, since it is integral over A and belongs to K'. Thus $y \in K \cap A'x$.

Conversely, if $y \in K \cap A'x$ then $y/x \in A'$, and $y^h/a = (y/x)^h \in A' \cap K = A$, so $(Ay \cdot J^{-1})^h$ is an integral ideal, hence $Ay \cdot J^{-1}$ is also an integral ideal of the Dedekind ring A, showing that $y \in J$. ∎

The preceding result tells us that for every ideal J, not necessarily principal in K, there exists some element x in a field K' of degree at most h over K, such that J is the set of multiples in K of this element x; when J is not principal then x does not belong to K, so if we restrict our attention only to K, the element x is an "ideal element" and its multiples form the "ideal" J in K. This is the origin of this terminology introduced by Kummer, now so widespread.

Since not every ring of algebraic integers is a principal ideal domain, the possibility still remains that in any particular case we may enlarge the field K to another field K' whose ring A' of integers is a principal ideal domain.

In other words, is it always possible to find a finite extension K' of K such that the class number of K' is equal to 1?

In *class field theory*, which is an advanced branch of the theory of algebraic numbers, with every algebraic number field K is associated an extension K' of degree h (the class number of K) over K. Among many important properties, for every fractional ideal J of K, the ideal $A'J$ of K' is principal. However, there may exist fractional ideals in K', not generated by ideals in K, which are not principal. The next natural thing to do is to repeat the procedure just indicated, obtaining the tower of class fields: $K \subset K' \subset K'' \subset \cdots$ and the question arises whether, after a finite number of steps one reaches a field with class number 1; that is, whether the above chain is finite. This has been referred to as the "class field tower problem."

In 1966, Shafarevich and Golod showed that there exist infinite class field towers.

In this respect, we make the following remark. Let $\mathcal{T}$ be the set of algebraic number fields having finite class field towers. Let $\mathcal{K}_1$ denote the set of algebraic number fields having class number equal to 1. Then $\mathcal{T}$ is infinite if and only if $\mathcal{K}_1$ is infinite.

Indeed, $\mathcal{K}_1 \subseteq \mathcal{T}$, so if $\mathcal{T}$ is finite, then $\mathcal{K}_1$ is finite. Conversely, if $\mathcal{K}_1 = \{K_1, \ldots, K_m\}$, if $L \in \mathcal{T}$, there exists $K_1 \in \mathcal{K}_1$ such that $L \subseteq K_1$. Thus L belongs to the finite set of fields contained in some K_i for $1 \le i \le m$. Gauss conjectured that there exist infinitely many real quadratic fields with class number 1 (see Chapter 28).

To conclude this section, we introduce more general class groups.

Let J be a nonzero integral ideal, let $\mathcal{F}_J = \mathcal{F}_J(K)$ denote the set of all nonzero fractional ideals, which may be written as quotient I/I' of integral ideals $I, I' \ne 0$, with $\gcd(I, J) = \gcd(I', J) = A$. Clearly, $\mathcal{F}_J$ is a subgroup of $\mathcal{F}$.

Let $\mathcal{P}r_J = \mathcal{P}r_J(K)$ be the set of nonzero principal fractional ideals of the form Aa/Aa' where $a, a' \ne 0$, $\gcd(Aa, J) = \gcd(Aa', J) = A$, and $a \equiv a'$ (mod J). Equivalently,

$$\mathcal{P}r_J = \{Ax \mid x \in K,\ x \ne 0,\ x \equiv 1 \ (\mathrm{mod}\ J)\}.$$

Thus $\mathcal{P}r_J$ is a subgroup of $\mathcal{F}_J$. For $J = A$ we have $\mathcal{F}_A = \mathcal{F}$, $\mathcal{P}r_A = \mathcal{P}r$. It is clear that $\mathcal{P}r_J \subseteq \mathcal{F}_J \cap \mathcal{P}r$.

The quotient group $\mathcal{C}_J = \mathcal{C}_J(K) = \mathcal{F}_J/\mathcal{P}r_J$ is called the *group of classes of fractional ideals associated to J*. It follows from Theorem 1, that $\mathcal{C}_J$ is finite (see (**K**) below). We denote by $h_J = h_J(K)$ the *number of classes of ideals associated to J*.

We now introduce still another type of classes of ideals.

Let J be a nonzero integral ideal. Let $\mathcal{P}r_{J,+} = \mathcal{P}r_{J,+}(K)$ denote the subset of $\mathcal{P}r_J$, consisting of all $Ax \in \mathcal{P}r_J$ such that x is totally positive (that is, all real conjugates of x are positive). So $\mathcal{P}r_{J,+}$ is a subgroup of $\mathcal{P}r_J$. If $J = A$, then $\mathcal{P}r_+ = \mathcal{P}r_{A,+}$ consists of all the nonzero principal fractional ideals Ax with x totally positive. Thus $\mathcal{P}r_{J,+} \subseteq \mathcal{F}_J \cap \mathcal{P}r_+$.

Let $\mathcal{C}_{J,+} = \mathcal{C}_{J,+}(K)$ be the quotient group $\mathcal{C}_{J,+} = \mathcal{F}_J/\mathcal{P}r_{J,+}$.

We note the canonical isomorphisms: $\mathcal{C}_J = (\mathcal{F}_J/\mathcal{Pr}_{J,+})/(\mathcal{Pr}_J/\mathcal{Pr}_{J,+})$ and $\mathcal{C} = (\mathcal{F}/\mathcal{Pr}_+)/(\mathcal{Pr}/\mathcal{Pr}_+)$. Thus, $\mathcal{C}_J$ is a quotient of $\mathcal{C}_{J,+}$, respectively, $\mathcal{C}$ is a quotient of $\mathcal{C}_+ = \mathcal{F}/\mathcal{Pr}_+$. We show:

K.   *For every integral ideal $J \neq 0$, the groups $\mathcal{C}_{J,+}$, $\mathcal{C}_+$ are finite.*

Proof: By Theorem 1, the group $\mathcal{C}$ is finite. The group $\mathcal{C}_J$ is also finite since there is only a finite number of residues modulo J. Now we show that $\mathcal{Pr}/\mathcal{Pr}_+$ is a finite group. Let $S = \{1, -1\}$ be the multiplicative group with two elements. Let $\sigma_1, \ldots, \sigma_{r_1}$ be the $r_1 \geq 0$ isomorphisms from K into $\mathbb{R}$. For each $x \in K$, $x \neq 0$ let the signature of x be

$$\mathrm{sgn}(x) = (\varepsilon_1, \varepsilon_2, \ldots, \varepsilon_{r_1}) \in S^{r_1},$$

where

$$\varepsilon_i = \begin{cases} +1 & \text{if } \sigma_i(x) > 0, \\ -1 & \text{if } \sigma_i(x) < 0. \end{cases}$$

For each $C \in \mathcal{Pr}/\mathcal{Pr}_+$, choose $x \in K$, $x \neq 0$, such that $Ax\mathcal{Pr}_+ = C$.

We define $\Sigma : \mathcal{Pr}/\mathcal{Pr}_+ \to S^{r_1}$ by letting $\Sigma(C) = \mathrm{sgn}(x)$. Then if $C \neq C'$ we must have $\mathrm{sgn}(x) \neq \mathrm{sgn}(x')$. Otherwise x/x' is totally positive and $Ax = A(x/x') \cdot Ax' \in \mathcal{Pr}_+ \cdot Ax$, contrary to the hypothesis. So Σ is an injective mapping. Since S^{r_1} is finite, then so is $\mathcal{Pr}/\mathcal{Pr}_+$ and so $\mathcal{C}_+$ is a finite group.

The proof for $\mathcal{C}_{J,+}$ is similar, we should only consider the homomorphism

$$\Sigma : \mathcal{Pr}_J/\mathcal{Pr}_{J,+} \to (A/J)^* \times S^{r_1}$$

defined by $\Sigma(C) = (x \pmod J, \mathrm{sgn}(x))$, where $C \in \mathcal{Pr}_J/\mathcal{Pr}_{J,+}$ is given by $Ax\mathcal{Pr}_{J,+} = C$, for some $x \in K$, $x \neq 0$.   ∎

We use the notations $h_+ = \#(\mathcal{C}_+)$, $h_{J,+} = \#(\mathcal{C}_{J,+})$. h_+ is called the *number of restricted classes of ideals of K* and we have $h \leq h_+$. h_J (respectively, $h_{J,+}$) is called the *number of classes of ideals* (respectively, *restricted classes of ideals*) *associated* to J.

Obviously, if K has no conjugate contained in $\mathbb{R}$, then $h_{J,+} = h_J$, $h_+ = h$.

EXERCISES

1. Let A be the ring of algebraic integers of the field K, let I be a nonzero integral ideal of A, and let $a, b \in A$. We say that a, b are congruent modulo I, and we write $a \equiv b \pmod I$ when $a - b \in I$. Prove:

  (a) If $Aa + I = A$ there exists $x \in A$ such that $ax \equiv b \pmod I$; moreover, if $y \in A$ satisfies $ay \equiv b \pmod I$ then $x \equiv y \pmod I$.

(b) If $I_1, \ldots, I_n$ are integral ideals such that $I_i + I_j = A$ for $i \neq j$, if $a_1, \ldots, a_n \in A$, then there exists $x \in A$ such that $x \equiv a_i$ (mod I_i) for $i = 1, \ldots, n$.

2. Let I be an integral ideal of the ring A of algebraic integers of a field K. Let $\varphi(I)$ denote the number of congruence classes modulo I of elements $a \in A$ such that $Aa + I = A$. Prove:

(a) If the ideals I, J are such that $I + J = A$, then $\varphi(I \cdot J) = \varphi(I) \cdot \varphi(J)$.

(b) If P is a prime ideal, then $\varphi(P) = N(P) - 1$.

(c) If $I = P^e$ ($e \geq 1$), then

$$\varphi(I) = N(I) \left[1 - \frac{1}{N(P)} \right].$$

(d) If $I = P_1^{e_1} \cdots P_r^{e_r}$ (where $P_1, \ldots, P_r$ are distinct prime ideals), then

$$\varphi(I) = N(I) \cdot \prod_{i=1}^{r} \left[1 - \frac{1}{N(P_i)} \right].$$

(Note that in the particular case where $K = \mathbb{Q}$ then φ is identical with the Euler function.)

3. Let I be a nonzero ideal of the ring A of algebraic integers of K. Show that if $a \in A$ and $Aa + I = A$ then

$$a^{\varphi(I)} \equiv 1 \pmod{I}.$$

Deduce *Fermat's little theorem* for ideals: If P is a nonzero prime ideal of A, if $a \in A$, $a \notin P$, then

$$a^{N(P)-1} \equiv 1 \pmod{P}.$$

4. Let P be a nonzero prime ideal of the ring A of algebraic integers of the field K. Show that the set of residue classes $\bar{a} = a + P$ of elements $a \in A$, $a \notin P$ forms a cyclic multiplicative group (of order $N(P) - 1$).

5. Let P be a nonzero prime ideal of the ring A of algebraic integers of K, let $a \in A$. Show that there exists an integer $m \in \mathbb{Z}$ such that $a \equiv m$ (mod P) if and only if $a^p \equiv a$ (mod P) where $P \cap \mathbb{Z} = \mathbb{Z}p$.

6. Let P be a nonzero prime ideal of the ring A of algebraic integers of a field K. Let $a_1, \ldots, a_m \in A$. Show that there exist at most m pairwise noncongruent integers $x \in A$ such that

$$x^m + a_1 x^{m-1} + \cdots + a_{m-1} x + a_m \equiv 0 \pmod{P}.$$

7. Let I be an ideal of the ring A of algebraic integers of an algebraic number field K. Show that there exist $a, b \in I$ such that

$$N(I) = \gcd(N_{K|\mathbb{Q}}(a), N_{K|\mathbb{Q}}(b)).$$

8. Let P be a prime ideal of A. Show that there exists $a \in A$ with the following property: For every $r \geq 1$ and $b \in A$ there exists $g \in \mathbb{Z}[X]$ such that $b \equiv g(a) \pmod{P^r}$.

9. Show that if P is a prime ideal, $N(P) = p^f$, if $a \in A$ is the element with the property indicated in the preceding exercise, then there exists $g \in \mathbb{Z}[X]$, $\deg(g) = f$ such that $P = (p, g(a))$.

10. Let I be an ideal of the ring A of algebraic integers of the field K. Show that in every class of ideals of K there exists an integral ideal J such that $I + J = A$.

11. Let J be an integral ideal of the algebraic number field K. Show that if there exists $a \in J$ such that $N(J) = |N_{K|\mathbb{Q}}(a)|$ then $J = Aa$.

12. Let J be a nonzero ideal of the ring of integers A of the real quadratic field $K = \mathbb{Q}(\sqrt{d})$. We assume that $J^2 = Aa$, $N_{K|\mathbb{Q}}(a) < 0$, and $N_{K|\mathbb{Q}}(u) > 0$ for every unit u of A. Show that J is not a principal ideal of A.

13. Let h be the class number of the field K. Let I, J be nonzero fractional ideals of K, p a prime number not dividing h. Prove that if $I^p \sim J^p$ then $I \sim J$.

14. Let A be the ring of algebraic integers of the algebraic number field K, and let $x, y, x', y' \in K$. Show that the ideals $J = (x, y)$ and $J' = (x', y')$ coincide if and only if there exist elements $a, b, c, d \in A$, such that $ad - bc$ is a unit of A and

$$\begin{cases} x' = ax + by \\ y' = cx + dy. \end{cases}$$

15. Let p be an odd prime, ζ a primitive pth root of unity, $K = \mathbb{Q}(\zeta)$, and A the ring of integers of K. Show that if $x \in A$ there exists $m \in \mathbb{Z}$ such that $x^p \equiv m \pmod{A(1 - \zeta)^p}$.

16. With the same notations as in the previous exercise, let $x \in A$, $x \notin A(1 - \zeta)$. Show that there exists a positive integer e such that $\zeta^e x \equiv m \pmod{A(1 - \zeta)^2}$ for some $m \in \mathbb{Z}$.

17. Let K be an algebraic number field, and let $\mathcal{C}(K)$ be the ideal class group of K. Show that there exists an integer $q \geq 0$ and ideal classes $\widetilde{I}_1, \ldots, \widetilde{I}_q$ such that every ideal class $\widetilde{I}$ of K may be written in a unique way in the form $\widetilde{I} = \widetilde{I}_1^{k_1} \cdots \widetilde{I}_q^{k_q}$, with $0 \leq k_i \leq h_i - 1$ where $h_1 h_2 \cdots h_q =$

h (class number of K), each h_i is the power of a prime number, and $\widetilde{I}_i^{h_i}$ is the unit element of $\mathcal{C}(K)$. Moreover, q is uniquely defined; it is called the *rank* of the class group $\mathcal{C}(K)$.

9

Estimates for the Discriminant

In this chapter we study the discriminant. A method of "Geometry of Numbers" is used to provide sharper estimates for the discriminant.

9.1 The Theorem of Minkowski

In the n-dimensional vector space $\mathbb{R}^n$ of all n-tuples of real numbers we shall consider certain additive subgroups.

Definition 1. A *lattice* A in $\mathbb{R}^n$ $(n \geq 1)$ is the set of all linear combinations, with coefficients in $\mathbb{Z}$, of n $\mathbb{R}$-linearly independent elements $a^{(1)}, \ldots, a^{(n)} \in \mathbb{R}^n$.

In other words, A is a free Abelian group of rank n, contained in $\mathbb{R}^n$, having a basis which is also an $\mathbb{R}$–basis of $\mathbb{R}^n$.

The *fundamental parallelotope* of A is the set

$$\Pi = \left\{ \sum_{k=1}^{n} x_k a^{(k)} \;\middle|\; x_k \in \mathbb{R},\; 0 \leq x_k \leq 1 \text{ for } k = 1, \ldots, n \right\}.$$

The volume of Π shall be denoted by $\mu = \mu(\Pi)$.

Definition 2. A subset S of $\mathbb{R}^n$ is said to be *convex* when it satisfies the following property: If $y, y' \in S$ then the line segment joining $y,\ y'$ is contained in S.

We can also phrase this property as follows: If $y, y' \in S$, if $\lambda, \lambda' \in \mathbb{R}$, $0 \leq \lambda, \lambda' \leq 1$, and $\lambda + \lambda' = 1$, then $\lambda y + \lambda' y' \in S$.

We recall now some well-known notions:

If S is any set, and $\alpha \in \mathbb{R}$, $\alpha > 0$, we define the *homothetic image* of S (with ratio α) to be the set $\alpha S = \{\alpha x = (\alpha x_1, \ldots, \alpha x_n) \mid x = (x_1, \ldots, x_n) \in S\}$. Then any homothetic image of a convex set is convex.

A subset S of $\mathbb{R}^n$ is said to be a *bounded set* when there exists a suffi-
ciently large real number γ such that $|x| \leq \gamma$ for every $x \in S$.* In other
words, S is contained in the n-dimensional sphere of center 0, and radius
$\gamma > 0$.

If S is bounded, and $\alpha > 0$ then αS is also a bounded set.

A subset S of $\mathbb{R}^n$ is said to be *closed* when it satisfies the following
property: If $y^{(1)}$, ..., $y^{(k)}$, ... is any sequence of elements in S then every
accumulation point of this sequence still belongs to S. This means that S is
a closed set in the topological space $\mathbb{R}^n$ (with its natural topology). Again,
any homothetic image of a closed set is closed.

Definition 3.   A nonempty convex, bounded, and closed subset of $\mathbb{R}^n$
is called a *convex body*.

If S is a subset of $\mathbb{R}^n$ such that when $y \in S$ then $-y \in S$, we say that S
is a *symmetric subset*.

It is very easy to indicate some symmetric convex bodies. For example,
the closed n-dimensional cube $S = \{x = (x_1, \ldots, x_n) \in \mathbb{R}^n \mid -1 \leq x_i \leq 1$
for $i = 1, \ldots, n\}$ and the closed n-dimensional sphere $S = \{x \in \mathbb{R}^n \mid |x| \leq
1\}$ are symmetric convex bodies. Similarly, if $n = 2$, if C is an ellipse of
center at the origin, the closed bounded region S of $\mathbb{R}^2$ determined by C
(and including the points of C) is a symmetric convex body. This is true,
whatever the radii of the ellipse are and for any slopes of the axes; thus,
we may also consider very elongated ellipses, with center at the origin and
axes with irrational slopes.

The "form" of a convex body may be rather difficult to describe. However, our considerations will depend on the volume, rather than on the form
of the convex body.

Here is not the place to enter into lengthy discussions on the concept of
volume; let us only say that it will be used in an intuitive way.

For the convenience of the reader, we state the main pertinent facts:

(1) Let $e_1 = (1, 0, \ldots, 0)$, $e_2 = (0, 1, \ldots, 0)$, ..., $e_n = (0, 0, \ldots, 1)$ be the standard basis of $\mathbb{R}^n$; then the parallelotope determined by the origin and e_1, ..., e_n has volume equal to 1.

(2) If φ is an invertible linear transformation of $\mathbb{R}^n$, then the volume of the parallelotope determined by 0, $\varphi(e_1)$, ..., $\varphi(e_n)$ is equal to $|d|$, where $d = \det(a_{ji})$ and $\varphi(e_i) = \sum_{j=1}^{n} a_{ji} e_j$ $(i = 1, \ldots, n)$.

(3) If S_1, S_2 are subsets of $\mathbb{R}^n$ such that $S_1 \subseteq S_2$ and if their volumes are defined, then $\text{vol}(S_1) \leq \text{vol}(S_2)$.

(4) If S is a subset of $\mathbb{R}^n$, $\alpha \in \mathbb{R}$, $\alpha > 0$, let $\alpha S = \{\alpha x \mid x \in S\}$; if the volume of S is defined so is $\text{vol}(\alpha S)$ and $\text{vol}(\alpha S) = \alpha^n \cdot \text{vol}(S)$.

---

* If $x = \{x_1, \ldots, x_n\}$ then $|x|$ denotes the positive square root $\sqrt{x_1^2 + \cdots + x_n^2}$, which
is the distance from x to the origin.

(5) If S is a subset of $\mathbb{R}^n$ whose volume is defined, if $x \in \mathbb{R}^n$, and $x + S = \{x + y \mid y \in S\}$ then the volume of $x + S$ is defined and equal to the volume of S.

(6) If S is a bounded subset of $\mathbb{R}^n$ and its volume is defined, then $0 \leq \text{vol}(S) < \infty$.

(7) If S_1, S_2 are disjoint subsets of $\mathbb{R}^n$ and their volumes are defined, so is the volume of $S_1 \cup S_2$, and $\text{vol}(S_1 \cup S_2) = \text{vol}(S_1) + \text{vol}(S_2)$.

It will be quite clear that for all the subsets S which we shall encounter, it is possible to define the volume in an unambiguous way. More generally, a theorem due to Blaschke states that *it is possible to define the volume of every convex body.*

Before proceeding, let us note this easy fact:

Lemma 1.   *If y, y' are points in the interior of a convex set S, then every point of the line segment joining y, y' also lies in the interior of S.*

Proof: We recall the concept of an interior point. We say that $y \in S$ is an *interior point* if there exists a real number $c > 0$ such that, for every element $x \in \mathbb{R}^n$ with $|x - y| < c$, we have $x \in S$.

If y, y' are interior points of the convex set S, we may choose $c \in \mathbb{R}$, $c > 0$, so small that if $x \in \mathbb{R}^n$, $|x - y| < c$, then $x \in S$ and also if $|x - y'| < c$, then $x \in S$.

Consider now any point $\lambda y + \lambda' y'$ of the segment joining y, y', so $0 \leq \lambda$, $\lambda' \leq 1$, $\lambda + \lambda' = 1$; since S is convex then $\lambda y + \lambda' y' \in S$. Let $x \in \mathbb{R}^n$ be such that if $z = x - (\lambda y + \lambda' y')$ then $|z| < c$. We have $x = z + \lambda y + \lambda' y' = \lambda(y + z) + \lambda'(y' + z)$ with $|(y + z) - y| < c$, $|(y' + z) - y'| < c$; since y, y' are in the interior of S, then $y + z$, $y' + z$ are in S and therefore $x \in S$, showing that $\lambda y + \lambda' y'$ is an interior point of S.   ∎

Theorem 1 (Minkowski).   *Let Λ be a lattice in $\mathbb{R}^n$ ($n \geq 1$), and let μ be the volume of the fundamental parallelotope of Λ.*

If S is a symmetric convex body having volume $\text{vol}(S) > 2^n \mu$, then there exists a point of Λ, distinct from the origin, which belongs to the interior of S.

However if $\text{vol}(S) = 2^n \mu$, it can only be said that there is a point of Λ, distinct from the origin, which is in S (but not necessarily in the interior of S).

Proof: We show that if there exists no point $y \in \Lambda$, $y \neq 0$ in the interior of S then $\text{vol}(S) \leq 2^n \mu$; that is,

$$\text{vol}(\tfrac{1}{2} S) = \frac{1}{2^n} \text{vol}(S) \leq \mu.$$

For every $y \in \Lambda$, let us consider the convex body $y + \tfrac{1}{2} S$ with center y, obtained by translation from $\tfrac{1}{2} S$.

If $y_1, y_2 \in \Lambda$, $y_1 \neq y_2$, we show that the interiors of $y_1 + \frac{1}{2}S$, $y_2 + \frac{1}{2}S$ are disjoint, or equivalently, if $y \in \Lambda$, $y \neq 0$, then the interiors of $\frac{1}{2}S$, $y + \frac{1}{2}S$ are disjoint. In fact, if x should be in the interior of $\frac{1}{2}S$ and of $y + \frac{1}{2}S$, then $y - x$ would be in the interior of $y - (y + \frac{1}{2}S) = -\frac{1}{2}S = \frac{1}{2}S$ (since S is symmetric), hence $\frac{1}{2}y = \frac{1}{2}(y - x) + \frac{1}{2}x$ would be in the interior of $\frac{1}{2}S$ (by Lemma 1), so $y \in \Lambda$, $y \neq 0$, would be in the interior of S contrary to the hypothesis.

We denote by $a^{(1)}, \ldots, a^{(n)} \in \mathbb{R}^n$ the generators of the lattice Λ.

Let $m > 0$ be a sufficiently large integer, let Λ_m be the set of all lattice points $y = \sum_{i=1}^{n} y_i a^{(i)}$ such that $y_i \in \mathbb{Z}$, $-m \leq y_i \leq m$ for all $i = 1, \ldots, n$; thus, Λ_m consists of $(2m + 1)^n$ elements.

Since S is bounded, there exists a sufficiently large real number $\gamma > 0$, such that if $x = \sum_{i=1}^{n} x_i a^{(i)} \in S$, $x_i \in \mathbb{R}$, then $|x_i| \leq \gamma$ for all $i = 1, \ldots, n$. Then for every point $y = (y_1, \ldots, y_n) \in \Lambda_m$ we have $|y_i + \frac{1}{2}x_i| \leq m + \frac{1}{2}\gamma$ for every point $x \in S$. Thus $\Lambda_m + \frac{1}{2}S$ is contained in the parallelotope

$$\Pi' = \left\{ \sum_{k=1}^{n} x_k a^{(k)} \;\middle|\; |x_k| \leq m + \frac{1}{2}\gamma \right\}.$$

But, $\Lambda_m + \frac{1}{2}S = \bigcup_{y \in \Lambda_m} [y + \frac{1}{2}S]$. Since the sets $y + \frac{1}{2}S$ have disjoint interiors, by the invariance of the volume by translation, we have

$$\mathrm{vol}[\Lambda_m + \tfrac{1}{2}S] = (2m + 1)^n \cdot \mathrm{vol}[\tfrac{1}{2}S] \leq (2m + \gamma)^n \cdot \mu,$$

this last quantity being the volume of the parallelotope Π'.

The above relation is true for every sufficiently large integer m, therefore,

$$\mathrm{vol}(\tfrac{1}{2}S) \leq \lim_{m \to \infty} \left(\frac{2m + \gamma}{2m + 1} \right)^n \cdot \mu = \mu.$$

Now we prove the second assertion. Let $\mathrm{vol}(S) = 2^n \mu$. For every $n \geq 1$, $\mathrm{vol}((1 + 1/n)S) > \mathrm{vol}(S) = 2^n \mu$, hence by the first assertion, there exists $x_n \in A$, $x_n \neq 0$, such that $x_n \in (1 + 1/n)S \subseteq 2S$. Since $2S$ is bounded, then $2S \cap \Lambda$ is finite. Hence there exist $n_1 < n_2 < \cdots$ such that $x_{n_1} = x_{n_2} = \cdots$; call this element x. Thus for infinitely many indices n, $x = (1 + 1/n)s_n$. If m is sufficiently large, namely $|s| \leq m$ for all $s \in S$, it follows that $|x - s_n| \leq (1/n)|s_n| \leq m/n$. So x is a limit point of S, hence $x \in S \cap \Lambda$, $x \neq 0$. ∎

We apply this theorem to show the existence of solutions for a system of linear inequalities; for $n = 1$ the following statement is trivial:

A.   *Let $n > 1$, let $L_i = \sum_{j=1}^{n} a_{ij} X_j$ $(i = 1, \ldots, n)$ be n linear forms with real coefficients, such that $d = \det(a_{ij}) \neq 0$.*

If $\tau_1, \ldots, \tau_n$ are positive real numbers such that $\tau_1 \cdots \tau_n \geq |d|$, given any index i_0, $1 \leq i_0 \leq n$, there exist integers $x_1, \ldots, x_n \in \mathbb{Z}$, not all equal to 0, such that $|L_i(x_1, \ldots, x_n)| < \tau_i$ for $i \neq i_0$, and $|L_{i_0}(x_1, \ldots, x_n)| \leq \tau_{i_0}$.

Proof: For simplicity of notation we may assume that $i_0 = n$. If the assertion is not true, for all n-tuples $(x_1, \ldots, x_n)$ of integers, not all equal to 0, we have either:

(1) there exists i, $1 \le i \le n-1$ such that $|L_i(x_1, \ldots, x_n)| \ge \tau_i$; or

(2) if $1 \le i \le n-1$ then $|L_i(x_1, \ldots, x_n)| < \tau_i$, but $|L_n(x_i, \ldots, x_n)| > \tau_n$.

Let T be the set of all n-tuples $(x_1, \ldots, x_n)$ of integers satisfying condition (2). If $T = \varnothing$ we have case (1). If $T \ne \varnothing$, let $(x_1, \ldots, x_n) \in T$ and τ_n' be such that $|L_n(x_x, \ldots, x_n)| < \tau_n'$.

Thus the subset T' of T consisting of all n-tuples $(x_1, \ldots, x_n)$ such that $|L_n(x_1, \ldots, x_n)| < \tau_n'$ is not empty; actually T' is a finite set, because the coordinates of its points are integers and the determinant of the coefficients of the linear forms is not zero. Therefore the minimum of the quantities $|L_n(x_1, \ldots, x_n)|$ when $(x_1, \ldots, x_n) \in T'$ is $\tau_n + \delta$ for some $\delta > 0$, and the same holds for all points in T. Thus, we have shown the existence of $\delta > 0$ such that if $x_1, \ldots, x_n \in \mathbb{Z}$, not all equal to 0, then either $|L_i(x_1, \ldots, x_n)| \ge \tau_i$ for some $i = 1, \ldots, n-1$, or $|L_n(x_1, \ldots, x_n)| \ge \tau_n + \delta$.

Let $\theta \colon \mathbb{R}^n \to \mathbb{R}^n$ be the linear transformation defined as follows: If $x = (x_1, \ldots, x_n)$ then $\theta(x) = (L_1(x), \ldots, L_n(x))$. Since $\det(a_{ij}) = d \ne 0$, then θ transforms the linearly independent vectors $e^{(1)} = (1, 0, \ldots, 0)$, $e^{(2)} = (0, 1, \ldots, 0)$, $\ldots$, $e^{(n)} = (0, 0, \ldots, 1)$ into linearly independent vectors $\theta(e^{(1)})$, $\theta(e^{(2)})$, $\ldots$, $\theta(e^{(n)})$. Let Λ be the lattice defined by these vectors, and let the volume of its fundamental parallelotope be $\mu = |d|$.

Let S be the set of all elements $x = (x_1, \ldots, x_n) \in \mathbb{R}^n$ such that $|x_i| \le \tau_i$ for $i = 1, \ldots, n-1$ and $|x_n| \le \tau_n + \delta$. Thus, S is a symmetric convex body, having volume $\mathrm{vol}(S) = 2\tau_1 \cdot 2\tau_2 \cdots 2\tau_{n-1} \cdot 2(\tau_n + \delta) > 2^n \tau_1 \cdots \tau_n \ge 2^n |d| = 2^n \cdot \mu$.

By Minkowski's theorem, there is a point $\theta(y) \in \Lambda$, $\theta(y) \ne 0$, belonging to the interior of S. In other words, there exist integers y_1, $\ldots$, y_n, not all zero, such that $|L_i(y)| < \tau_i$ for $i = 1, \ldots, n-1$ and $|L_n(y)| < \tau_n + \delta$, which is a contradiction. ∎

A similar result holds for linear forms with complex coefficients:

B.   *Let $n > 1$ and let $L_i = \sum_{i=1}^{n} a_{ij} X_i$ ($i = 1, \ldots, n$) be n linear forms with complex coefficients such that $d = \det(a_{ij}) \ne 0$. Let us assume that for every i there exists an index i' such that $\overline{L}_i = \sum_{j=1}^{n} \overline{a}_{ij} X_j$ (the complex conjugate form of L_i) is equal to $L_{i'}$; moreover, if L_i has real coefficients then $i' = i$.*

Let τ_1, $\ldots$, τ_n be real positive numbers such that if $\overline{L}_i = L_{i'}$ then $\tau_i = \tau_{i'}$. If $\tau_1 \cdots \tau_n \ge |d|$, given any index i_0 such that $\overline{L}_{i_0} = L_{i_0}$, there exist integers $x_1, \ldots, x_n \in \mathbb{Z}$, not all equal to 0, such that $|L_i(x_1, \ldots, x_n)| < \tau_i$ for $i \ne i_0$, and $|L_{i_0}(x_1, \ldots, x_n)| \le \tau_{i_0}$.

Proof: We begin by associating real forms L'_i and constants τ'_i ($i = 1, \ldots, n$) to the given forms L_i and constants τ_i in the following way: If L_i is a real form let $L'_i = L_i$ and $\tau'_i = \tau_i$; if L_i is not a real form and $L_{i'} = \bar{L}_i$ we put $L'_i = (L_i + L_{i'})/2$, $L'_{i'} = (L_i - L_{i'})/2\sqrt{-1}$ and $\tau'_i = \tau'_{i'} = \tau_i/\sqrt{2}$. Thus L'_i, $L'_{i'}$ are real forms and $L_i = L'_i + \sqrt{-1}\, L'_{i'}$, $L_{i'} = L'_i - \sqrt{-1}\, L'_{i'}$ and $\tau'_1 \cdots \tau'_n = 2^{-r_2}\tau_1 \cdots \tau_n$, where r_2 denotes the number of pairs of nonreal complex conjugate forms L_i, $L_{i'}$. If d' is the determinant of the coefficients of the forms L'_i ($i = 1, \ldots, n$) then $|d'| = 2^{-r_2}|d|$. The reader may easily do this verification. From $\tau_1\tau_2 \cdots \tau_n \geq |d|$ it follows that

$$\tau'_1 \cdot \tau'_2 \cdots \tau'_n \geq 2^{-r_2}|d| = |d'|.$$

We are therefore in a position to apply (**A**), deducing the existence of integers $x_1, \ldots, x_n$, not all equal to 0, such that $|L'_i(x_1, \ldots, x_n)| < \tau'_i$ for $i \neq i_0$ and $|L'_{i_0}(x_1, \ldots, x_n)| \leq \tau'_{i_0}$.

It follows that if $i \neq i_0$ and $\bar{L}_i = L_{i'}$ with $i \neq i'$ then

$$|L_i(x)|^2 = |L_{i'}(x)|^2 = (L'_i(x))^2 + (L'_{i'}(x))^2 < \frac{\tau_i^2}{2} + \frac{\tau_i^2}{2} = \tau_i^2;$$

if $\bar{L}_i = L_i$, then obviously $|L_i(x_i, \ldots, x_n)| < \tau_i$ for $i \neq i_0$, and $|L_{i_0}(x_1, \ldots, x_n)| \leq \tau_{i_0}$.   ∎

9.2 Estimates of the Discriminant

As a first application of these methods, we may refine the result of Chapter 8, (**H**):

C.   *For every algebraic number field K, distinct from $\mathbb{Q}$, and for every nonzero integral ideal J of A there exists an element $a \in J$, $a \neq 0$, such that*

$$|N_{K|\mathbb{Q}}(a)| < N(J) \cdot \sqrt{|\delta|},$$

where δ is the discriminant of K.

Proof: Let $\{a_1, \ldots, a_n\}$ be a basis of the free Abelian group J (Chapter 6, (**K**)), and let us consider the n linear forms $L_i = \sum_{j=1}^{n} a_j^{(i)} X_j$ ($i = 1, \ldots, n$) where $a_j^{(1)} = a_j, a_j^{(2)}, \ldots, a_j^{(n)}$ are the conjugates of $a_j \in K$ over $\mathbb{Q}$. The determinant of the coefficients is equal, in absolute value, to $|\det(a_j^{(i)})| = N(J) \cdot \sqrt{|\delta|} \neq 0$ (by Chapter 8, (**E**)).

Let $\tau_1 = \tau_2 = \cdots = \tau_n = |\det(a_j^{(i)})|^{1/n}$. Since $n > 1$, by (**B**) there exist integers $x_1, \ldots, x_n$, not all equal to 0, such that

$$\left| \sum_{j=1}^{n} a_j^{(i)} x_j \right| \leq |\det(a_j^{(i)})|^{1/n} \qquad \text{for all} \quad i = 1, \ldots, n,$$

with at most one equality, hence with at least one strict inequality.

Hence, letting $a = \sum_{j=1}^{n} x_j a_j \in J$ we have $a \neq 0$ (since not all x_j are equal to 0), and

$$|N_{K|\mathbb{Q}}(a)| = \prod_{i=1}^{n} \left| \sum_{j=1}^{n} a_j^{(i)} x_j \right| < |\det(a_j^{(i)})| = N(J) \cdot \sqrt{|\delta|}. \qquad \blacksquare$$

Let us observe at this point, that (**C**) *is equivalent to the fact that in every class of ideals there exists a nonzero integral ideal I such that $N(I) < \sqrt{|\delta|}$*.

Indeed, let $J \neq 0$ be an ideal in the given class, and let $b \in A$, $b \neq 0$ be such that $bJ^{-1} \subseteq A$; applying (**C**), there exists $a \in bJ^{-1}$, $a \neq 0$, for which

$$|N_{K|\mathbb{Q}}(a)| < N(Ab \cdot J^{-1})\sqrt{|\delta|};$$

then letting $I = Aab^{-1} \cdot J \subseteq A$ we have $N(I) < \sqrt{|\delta|}$.

Conversely, let J be a nonzero integral ideal and let I be an integral ideal in the class of J^{-1} such that $N(I) < \sqrt{|\delta|}$; then there exists $a \in K$, $a \neq 0$ such that $aJ^{-1} = I \subseteq A$; hence $a \in J$ and $|N_{K|\mathbb{Q}}(a)| < N(J) \cdot \sqrt{|\delta|}$.

By Chapter 8, (**G**), there are only finitely many integral ideals I_i ($i = 1, \ldots, t$) such that $N(I_i) < \sqrt{|\delta|}$; therefore, every ideal is in the same class as some of the ideals $I_1, \ldots, I_t$; this provides a new proof of Theorem 1 of Chapter 8.

A very important corollary follows now:

D.   *If K is an algebraic number field, distinct from $\mathbb{Q}$, then $|\delta| \geq 2$.*

Proof: We take J equal to the unit ideal, $J = A$. Then $N(J) = 1$ and therefore $1 \leq |N_{K|\mathbb{Q}}(a)| < N(J) \cdot \sqrt{|\delta|} = \sqrt{|\delta|}$ for some element $a \in A$, $a \neq 0$. Since δ is an integer, then $|\delta| \geq 2$. $\qquad \blacksquare$

We apply this method to obtain sharper estimates for the discriminant.

Let $K^{(1)}$, $K^{(2)}$, $\ldots$, $K^{(n)}$ be all the conjugates of the field K (over $\mathbb{Q}$). As we know, the complex conjugate of any field is a conjugate (in the algebraic sense), because if x is a root of a polynomial f, then its complex conjugate $\overline{x}$ is also a root of f. Of course, if a field is contained in the field of real numbers, then it coincides with its complex conjugate. Thus, the conjugates of K may be grouped and numbered as follows: let $r_1 \geq 0$ be such that $K^{(1)}$, $K^{(2)}$, $\ldots$, $K^{(r_1)}$ are contained in $\mathbb{R}$; let $r_2 \geq 0$ be such that

$$K^{(r_1+1)}, \; K^{(r_1+2)}, \; \ldots, \; K^{(r_1+r_2)}, K^{(r_1+r_2+1)}, \; K^{(r_1+r_2+2)}, \; \ldots, \; K^{(n)},$$

are nonreal fields, with $\overline{K^{(r_1+i)}} = K^{(r_1+r_2+i)}$ for $i = 1, 2, \ldots, r_2$; so there are r_2 pairs of complex conjugate nonreal fields and

$$n = r_1 + 2r_2.$$

It is also convenient to introduce the following notation: $l_1 = \cdots = l_{r_1} = 1$, $l_{r_1+1} = l_{r_1+2} = \cdots = l_{r_1+r_2} = 2$, so $\sum_{i=1}^{r_1+r_2} l_i = r_1 + 2r_2 = n$. If $x \in K$, we denote by $x^{(1)}, \ldots, x^{(n)}$ all its conjugates, where $x^{(i)} \in K^{(i)}$ $(i = 1, \ldots, n)$.

Let $\varphi \colon K \to \mathbb{R}^n$ be the isomorphism of $\mathbb{Q}$-vector spaces defined by $\varphi(x) = (\xi_1, \ldots, \xi_n) \in \mathbb{R}^n$, where

$$
\begin{cases}
\xi_i = x^{(i)} & \text{for } i = 1, \ldots, r_1, \\
\xi_j = \Re(x^{(j)}) & \text{for } j = r_1 + 1, \ldots, r_1 + r_2, \\
\xi_{j+r_2} = \Im(x^{(j)}) & \text{for } j = r_1 + 1, \ldots, r_1 + r_2,
\end{cases}
$$

(for every complex number $y = a + b\sqrt{-1}$ we denote $a = \Re(y)$, $b = \Im(y)$).

Let $D_1 = \{(\xi_1, \ldots, \xi_n) \in \mathbb{R}^n \mid \prod_{j=1}^{r_1} |\xi_j| \cdot \prod_{j=r_1+1}^{r_1+r_2} (\xi_j^{\,2} + \xi_{j+r_2}^2) \le 1\}$. For example, if $r_1 = 1$, $r_2 = 0$ then $n = 1$ hence D_1 is the closed interval $[-1, 1]$ and $\mathrm{vol}(D_1) = 2$. If $r_1 = 0$, $r_2 = 1$ then $n = 2$ is the closed disk of radius 1 and center 0, hence $\mathrm{vol}(D_1) = \pi$. If $r_1 = 2$, $r_2 = 0$ then $n = 2$ and D_1 is no more bounded; this is also true for every other case.

We may consider symmetric convex bodies D contained in D_1, and apply to them Minkowski's theorem, as we did in (**C**).

E.   *Let K be an algebraic number field of degree n and let J be a nonzero integral ideal of A. For every symmetric convex body D in $\mathbb{R}^n$, such that $D \subseteq D_1$, there exists an element $a \in J$, $a \neq 0$, such that*

$$
|N_{K|\mathbb{Q}}(a)| \le \frac{2^{r_1+r_2}}{\mathrm{vol}(D)}\, N(J) \cdot \sqrt{|\delta|}.
$$

Proof: Let $\{a_1, \ldots, a_n\}$ be a basis of the free Abelian group J and consider the vectors $\varphi(a_1), \ldots, \varphi(a_n)$ in $\mathbb{R}^n$. We shall show that these vectors are linearly independent over $\mathbb{R}$, by computing the determinant of their coordinates (relative to the canonical basis of $\mathbb{R}^n$).

Let $\varphi(a_i) = (\alpha_{1i}, \ldots, \alpha_{ni})$, therefore

$$
\det(\alpha_{ij}) = \det
\begin{pmatrix}
a_1^{(1)} & \cdots\cdots\cdots\cdots\cdots & a_n^{(1)} \\
\cdots\cdots\cdots\cdots\cdots\cdots\cdots \\
a_1^{(r_1)} & \cdots\cdots\cdots\cdots\cdots & a_n^{(r_1)} \\
\Re\, a_1^{(r_1+1)} & \cdots\cdots\cdots & \Re\, a_n^{(r_1+1)} \\
\cdots\cdots\cdots\cdots\cdots\cdots\cdots \\
\Re\, a_1^{(r_1+r_2)} & \cdots\cdots & \Re\, a_n^{(r_1+r_2)} \\
\Im\, a_1^{(r_1+1)} & \cdots\cdots\cdots & \Im\, a_n^{(r_1+1)} \\
\cdots\cdots\cdots\cdots\cdots\cdots\cdots \\
\Im\, a_1^{(r_1+r_2)} & \cdots\cdots & \Im\, a_n^{(r_1+r_2)}
\end{pmatrix}.
$$

To compute this determinant we make successively the following:

(1) Multiply by $i = \sqrt{-1}$ each of the r_2 last rows; this introduces a factor $1/i^{r_2}$.

(2) Add the $(r_1 + r_2 + j)$th row to the $(r_1 + j)$th row to get a new $(r_1 + j)$th row.

(3) Subtract the $(r_1 + j)$th row from the double of the $(r_1 + r_2 + j)$th row and multiply the result by -1 to get the new $(r_1 + r_2 + j)$th row; this introduces a factor $(-1)^{r_2} \cdot (1/2^{r_2})$.

After these transformations we arrive at $(-1/2i)^{r_2} \cdot \det(a_j^{(k)})$ and therefore the absolute value of the determinant is

$$|\det(\alpha_{ij})| = \frac{1}{2^{r_2}} |\det(a_j^{(k)})| = \frac{1}{2^{r_2}} N(J) \cdot \sqrt{|\delta|} \neq 0,$$

as was computed in Chapter 8, (**E**).

Hence, the vectors $\varphi(a_1)$, ..., $\varphi(a_n)$ are linearly independent over $\mathbb{R}$ and define a lattice Λ, whose fundamental parallelotope has volume $\mu = |\det(\alpha_{ij})| = (1/2^{r_2})N(J) \cdot \sqrt{|\delta|}$.

Let $\rho \in \mathbb{R}$, $\rho > 0$ be such that $\rho^n = [2^{r_1+r_2}/\text{vol}(D)]N(J)\sqrt{|\delta|}$. Thus, if we consider the homothetic image $\rho D = \{\rho\xi \mid \xi \in D\}$, then

$$\text{vol}(\rho D) = \rho^n \text{vol}(D) = 2^{r_1+r_2} N(J) \cdot \sqrt{|\delta|} = 2^n \mu.$$

By Minkowski's theorem, there exists $\xi \neq 0$, $\xi \in D$, such that $\rho\xi \in \Lambda = \sum_{i=1}^{n} \mathbb{Z}\varphi(a_i) = \varphi\left(\sum_{i=1}^{n} \mathbb{Z}a_i\right) = \varphi(J)$; thus $\rho\xi = \varphi(a)$, $a \in J$, $a \neq 0$. We have $\varphi(a) \in \rho D \subseteq \rho D_1$, so $\varphi(a) = (\rho\beta_1, \ldots, \rho\beta_n)$ with $(\beta_1, \ldots, \beta_n) \in D_1$.

Now, let us compute $|N_{K|\mathbb{Q}}(a)| = \prod_{j=1}^{n} |a^{(j)}|$. Since $a^{(r_1+r_2+j)} = \overline{a^{(r_1+j)}}$ for $j = 1, \ldots, r_2$ then $|a^{(r_1+r_2+j)}| \cdot |a^{(r_1+j)}| = (\mathfrak{Re}\, a^{(r_1+j)})^2 + (\mathfrak{Im}\, a^{(r_1+j)})^2 = \rho^2(\beta_{r_1+j}^2 + \beta_{r_1+r_2+j}^2)$; similarly, $|a^{(j)}| = \rho|\beta_j|$ for $j = 1, \ldots, r_1$; thus $|N_{K|\mathbb{Q}}(a)| \leq \rho^n = [2^{r_1+r_2}/\text{vol}(D)]N(J) \cdot \sqrt{|\delta|}$. $\blacksquare$

The statement of (**E**) gives an upper bound for the norms of elements of the ideal J. The larger the volume of the convex body $D \subseteq D_1$, the smaller the upper bound in question.

For example, if $r_1 = 0$, $r_2 = 1$ and if we take $D = D_1$, we have $\text{vol}(D_1) = \pi$, and (**E**) states the existence of $a \in J$, $a \neq 0$, such that

$$|N_{K|\mathbb{Q}}(a)| \leq \frac{2}{\pi} N(J)\sqrt{|\delta|} < N(J)\sqrt{|\delta|};$$

this result was already proved in (**C**) for any extension $K|\mathbb{Q}$ of finite degree.

For larger values of r_1, r_2 we have seen that D_1 itself is not a convex body, for it is unbounded and not convex. In this situation, we reach different estimates by appropriate choices of the convex body D.

F.   *Let K be an algebraic number field of degree n, and let J be a nonzero integral ideal of A. There exists an element $a \in J$, $a \neq 0$, such that*

$$|N_{K|\mathbb{Q}}(a)| \le \left(\frac{4}{\pi}\right)^{r_2} \frac{n!}{n^n} \sqrt{|\delta|} \cdot N(J).$$

Proof: Let

$$D = \{(\xi_1, \ldots, \xi_n) \in \mathbb{R}^n \mid |\xi_1| + \cdots + |\xi_{r_1}|$$
$$+ 2\sqrt{\xi_{r_1+1}^2 + \xi_{r_1+r_2+1}^2} + \cdots + 2\sqrt{\xi_{r_1+r_2}^2 + \xi_n^2} \le n\}.$$

We show that D is a symmetric convex body. Everything is easy to check and only the convexity requires some computation.

Let $\lambda, \mu \in \mathbb{R}$, $0 \le \lambda$, $\mu \le 1$, $\lambda + \mu = 1$. If $(\xi_1, \ldots, \xi_n)$, $(\eta_1, \ldots, \eta_n)$ are in D, then so is $(\lambda\xi_j + \mu\eta_j)_j$, because

$$|\lambda\xi_1 + \mu\eta_1| + \cdots$$
$$+ 2\sqrt{(\lambda\xi_{r_1+1} + \mu\eta_{r_1+1})^2 + (\lambda\xi_{r_1+r_2+1} + \mu\eta_{r_1+r_2+1})^2} + \cdots$$
$$\le \lambda|\xi_1| + \mu|\eta_1| + \cdots + 2\lambda\sqrt{\xi_{r_1+1}^2 + \xi_{r_1+r_2+1}^2}$$
$$+ 2\mu\sqrt{\eta_{r_1+1}^2 + \eta_{r_1+r_2+1}^2} + \cdots$$
$$\le \lambda\{|\xi_1| + \cdots + 2\sqrt{\xi_{r_1+1}^2 + \xi_{r_1+r_2+1}^2} + \cdots\}$$
$$+ \mu\{|\eta_1| + \cdots + 2\sqrt{\eta_{r_1+1}^2 + \eta_{r_1+r_2+1}^2} + \cdots\}$$
$$\le \lambda n + \mu n = (\lambda + \mu)n = n$$

(the above inequalities are straightforward). This shows that D is convex.

Moreover $D \subseteq D_1$, because the geometric mean between positive real numbers is not greater than the arithmetic mean, thus

$$\left\{\prod_{j=1}^{r_1} |\xi_j| \cdot \prod_{j=r_1+1}^{r_1+r_2} (\xi_j^2 + \xi_{j+r_2}^2)\right\}^{1/n}$$

$$= \left\{\prod_{j=1}^{r_1} |\xi_j| \cdot \prod_{j=r_1+1}^{r_1+r_2} |\xi_j + i\xi_{j+r_2}|^2\right\}^{1/n}$$

$$\le \frac{1}{n}\left\{\sum_{j=1}^{r_1} |\xi_j| + 2\sum_{j=r_1+1}^{r_1+r_2} |\xi_j + i\xi_{j+r_2}|\right\}$$

$$= \frac{1}{n}\left\{\sum_{j=1}^{r_1} |\xi_j| + 2\sum_{j=r_1+1}^{r_1+r_2} \sqrt{\xi_j^2 + \xi_{j+r_2}^2}\right\}.$$

We may apply (**E**), so there exists $a \in J$, $a \neq 0$ such that

$$|N_{K|\mathbb{Q}}(a)| \leq \frac{2^{r_1+r_2}}{\text{vol}(D)} N(J) \cdot \sqrt{|\delta|}.$$

It remains to compute $\text{vol}(D)$, which depends on r_1, r_2. More generally, if $\rho > 0$ let $D^{(\rho)}$ be the set of vectors $(\xi_1, \ldots, \xi_n) \in \mathbb{R}^n$ such that

$$|\xi_1| + \cdots + 2\sqrt{\xi_{r_1+1}^2 + \xi_{r_1+r_2+1}^2} + \cdots \leq \rho.$$

We denote by $f_{r_1,r_2}(\rho)$ the volume of $D^{(\rho)}$. For example, $f_{1,0}(\rho) = 2\rho$, $f_{0,1}(\rho) = \pi\rho^2/4$. For larger values of r_1, r_2 the computation is done by induction and we shall omit the details. Let us just note that

$$f_{r_1,r_2}(\rho) = 2\int_0^\rho f_{r_1-1,r_2}(\rho - t)\, dt$$

and

$$f_{r_1,r_2}(\rho) = \iint_{t^2+u^2 \leq \rho^2/4} f_{r_1,r_2-1}(\rho - 2\sqrt{t^2 + u^2})\, dt\, du$$

$$= 2\pi \int_0^{\rho/2} f_{r_1,r_2-1}(\rho - 2t)t\, dt.$$

Performing the induction, we obtain $f_{r_1,r_2}(\rho) = 2^{r_1}(\pi/2)^{r_2}\rho^n/n!$ and taking $\rho = n$ we have $\text{vol}(D) = 2^{r_1}(\pi/2)^{r_2}n^n/n!$.

We conclude that

$$|N_{K|\mathbb{Q}}(a)| \leq \frac{2^{r_1+r_2}}{2^{r_1}}\left(\frac{2}{\pi}\right)^{r_2}\frac{n!}{n^n}N(J)\cdot\sqrt{|\delta|}$$

$$= \left(\frac{4}{\pi}\right)^{r_2}\frac{n!}{n^n}N(J)\cdot\sqrt{|\delta|}. \qquad\blacksquare$$

The preceding result gives the following sharper estimation of the discriminant:

G. *If K is an algebraic number field of degree n then*

$$|\delta| \geq \left(\frac{\pi}{4}\right)^{2r_2}\left(\frac{n^n}{n!}\right)^2 \geq \left(\frac{\pi}{4}\right)^n\left(\frac{n^n}{n!}\right)^2.$$

Proof: We take J equal to the unit ideal; hence, applying (**F**), we deduce

$$1 \leq |N_{K|\mathbb{Q}}(a)| \leq \left(\frac{4}{\pi}\right)^{r_2}\frac{n!}{n^n}\sqrt{|\delta|},$$

therefore,

$$|\delta| \geq \left(\frac{\pi}{4}\right)^{2r_2}\left(\frac{n^n}{n!}\right)^2 \geq \left(\frac{\pi}{4}\right)^n\left(\frac{n^n}{n!}\right)^2,$$

since

$$\frac{\pi}{4} < 1 \qquad \text{and} \qquad 2r_2 \leq n. \qquad \blacksquare$$

We may note that $(\pi/4)^n(n^n/n!)^2$ increases monotonically with n. If we express $n!$ by Stirling's formula

$$n! = \sqrt{2\pi n}\, n^n e^{-n+(\alpha/12n)},$$

where $0 < \alpha < 1$, then $e^{\alpha/12n} < e^{1/12} < \sum_{\nu=0}^{\infty}(1/12^\nu) = 12/11$; hence $|\delta| > (\pi e^2/4)^n(1/2\pi n)(11/12)^2$.

Since $\pi e^2/4 > 1$, the right-hand side in the above inequality tends to infinity with n, hence

$$\lim_{n\to\infty} \min_{[K:\mathbb{Q}]=n} \{|\delta_K|\} = \infty.$$

From this fact we may now deduce the following interesting consequence due to Hermite:

H.   *For every integer d there exist at most finitely many fields K having discriminant $\delta_K = d$.*

Proof: Given d, there exists an integer $n_0 \geq 1$ such that if $n \geq n_0$ then $(\pi e^2/4)^n(1/2\pi n)(11/12)^2 > |d|$, hence if K has degree $n \geq n_0$, then its discriminant is greater than $|d|$. Thus, it is enough to prove that for every integer n, there exists at most finitely many fields K of degree n having discriminant equal to d; this is true for $|d| = 1$, as we have seen in (**D**), so we may assume $|d| > 1$. Hence, $r_1 = 1$, $r_2 = 0$ is impossible. If $r_1 = 0$, $r_2 = 1$ then $n = 2$, so K is an imaginary quadratic extension of $\mathbb{Q}$, $K = \mathbb{Q}(\sqrt{a})$, with $a \in \mathbb{Z}$, a square-free; by Chapter 6, (**P**), $d = \delta_K$ is either $4a$ or a, so there exists at most one quadratic field with discriminant d.

From now on we may assume that $r_1 + r_2 > 1$. Given K, with discriminant d, we shall show the existence of a primitive element a of K which is an algebraic integer and such that $|a^{(1)}| \leq \sqrt{|d|}$ and for $i \neq 1$, $|a^{(i)}| < 1$.

Case 1: $r_1 > 0$.

Let $\{x_1, \ldots, x_n\}$ be an integral basis of K, and let $L_i = \sum_{j=1}^{n} x_j^{(i)} X_j$ for $i = 1, \ldots, n$; these linear forms have determinant $|\det(x_j^{(i)})| = |d|$, since K has discriminant d. Given $\tau_1 = \sqrt{|d|}$, $\tau_i = 1$ for $i \neq 1$, we deduce the existence of integers $m_1, \ldots, m_n \in \mathbb{Z}$ not all equal to 0, such that $|L_i(m_1, \ldots, m_n)| < \tau_i = 1$ for $i \neq 1$, $|L_1(m_1, \ldots, m_n)| \leq \tau_1 = \sqrt{|d|}$. This means that the element $a = \sum_{i=1}^{n} m_i x_i \in A$ is such that $|a^{(1)}| \leq \sqrt{|d|}$, while $|a^{(i)}| < 1$ for $i \neq 1$. From $1 \leq |N_{K|\mathbb{Q}}(a)| < |a^{(1)}|$, we deduce also that all the conjugates of $a^{(1)} = a$ are distinct from a and so a has n distinct conjugates and is a primitive element of K.

Case 2: $r_1 = 0$.

We proceed in a similar manner . Let $\{x_1, \ldots, x_n\}$ be an integral basis of the field K. For each $i = 1, \ldots, n$, let $L_i = \sum_{j=1}^{n} x_j^{(i)} X_j$, the numbering being such that $L_{i+r_2} = \bar{L}_i$; so $|\det(x_j^{(i)})| = \sqrt{|d|}$. We consider the following new linear forms:

$$L_1^* = (L_1 + \bar{L}_1)/2 = \sum_{j=1}^{n} \mathfrak{Re}(x_j^{(1)})X_j,$$

$$L_{1+r_2}^* = ((\bar{L}_1 - L_1)/2) \cdot \sqrt{-1} = \sum_{j=1}^{n} \mathfrak{Im}(x_j^{(1)})X_j,$$

$$L_i^* = L_i \quad \text{for} \quad i \neq 1, 1 + r_2.$$

The absolute value of the determinant of the coefficients of these forms is equal to $\frac{1}{2}\sqrt{|d|}$. Let $\tau_1 = \sqrt{|d|}$, $\tau_{1+r_2} = 1$, $\tau_i = 1$ for $i \neq 1, 1 + r_2$. By (**B**) there exist integers a_1, $\ldots$, a_n, not all equal to 0, such that if $a = \sum_{j=1}^{n} a_j x_j$, then $|a| = |a^{(1)}| \leq \sqrt{|d|}$, $|a^{(i)}| < 1$ for all $i \neq 1$. So, as in the first case, $a \in A$, $K = \mathbb{Q}(a)$, and the required condition is satisfied.

Now we consider the set $\mathcal{S}$ of all fields K of degree n and discriminant d; for each such field, let a be chosen as indicated, so $K = \mathbb{Q}(a)$. We shall associate with K, or better with a, an element of a fixed finite set S, with depends only on n, d; explicitly, S is the set of n-tuples of integers $(m_1, \ldots, m_n)$, where $-\mu \leq m_i \leq \mu$ $(i = 1, \ldots, n)$ and

$$\mu = \max_{1 \leq i \leq n} \left\{ \binom{n}{i} \right\} \cdot \sqrt{|d|}.$$

This is achieved as follows. Let $f = X^n + c_1 X^{n-1} + \cdots + c_n \in \mathbb{Z}[X]$ be the minimal polynomial of the algebraic integer a, thus $c_i = (-1)^i s_i$ where s_i is the ith elementary symmetric polynomial on a and its conjugates; hence $|c_i| = |s_i| = |\sum (\text{products of } i \text{ conjugates of } a)| \leq \sum |\text{products of } i$ conjugates of $a| \leq \binom{n}{i}\sqrt{|d|} \leq \mu$.

With $K = \mathbb{Q}(a)$, we associate the n-tuple $(c_1, \ldots, c_n)$ which belongs to S. There are at most finitely many fields K which give rise to the same element $(c_1, \ldots, c_n)$, namely the conjugates of $K = \mathbb{Q}(a)$. Thus, $\mathcal{S}$ is a finite set, proving our statement. ∎

EXERCISES

1. Construct a symmetric convex body S of volume $2^n\mu$ and having no lattice point in its interior, except the origin.

2. Develop in detail the argument in the proof of Theorem 1, where it is asserted that if $y \in \Lambda$, there exists $y' \in \Lambda \cap C$ such that $\rho_{y'} \leq \rho_y$.

3. Show that the homothetic image of a convex set is convex.

4. Establish the following assertions made in the proof of (**B**):

(a) $f_{r_1,r_2}(\rho) = 2\int_0^\rho f_{r_1-1,r_2}(\rho - t)\, dt$.

(b) $f_{r_1,r_2}(\rho) = \iint\limits_{t^2+u^2\leq\rho^2/4} f_{r_1,r_2-1}(\rho - 2\sqrt{t^2+u^2})\, dt\, du$

$$= 2\pi \int_0^{\rho/2} f_{r_1,r_2-1}(\rho - 2t)t\, dt.$$

(c) $f_{r_1,r_2}(\rho) = 2^{r_1}\, (\pi/2)^{r_2}\, \rho^n/n!$.

5. Show that if K is a field of degree 3 then its discriminant satisfies $\delta < -12$ or $20 < \delta$.

6. Show that in every class of ideals of the algebraic number field K with discriminant δ, there exists a nonzero integral ideal J such that

$$N(J) \leq \left(\frac{4}{\pi}\right)^{r_2} \frac{n!}{n^n} \sqrt{|\delta|}.$$

7. Let K be a quadratic field with discriminant δ. Show that in every class of ideals of K there exists a nonzero integral ideal J such that

$$N(J) \leq \begin{cases} \dfrac{1}{2}\sqrt{\delta} & \text{when } \delta > 0, \\[2mm] \dfrac{2}{\pi}\sqrt{|\delta|} & \text{when } \delta < 0. \end{cases}$$

(see Chapter 16, (**N**)).

8. With the notations of (**B**) show that if there exists i_0, $1 \leq i_0 \leq n$, such that L_{i_0} does not have real coefficients, then there exist $x_1, \ldots, x_n \in \mathbb{Z}$, not all equal to 0, such that $|L_i(x_1, \ldots, x_n)| < \tau_i$ for all $i = 1, \ldots, n$.

10

Units

As we have said, two elements of a domain are associated precisely when they generate the same ideal. Thus, by considering ideals, we ignore the units. However, it will become apparent that a number of arithmetic properties are intimately tied up with the units of the ring of integers A of the algebraic number field K.

We shall first consider the simplest type of units.

Any root of unity, that is, any root of a polynomial $X^m - 1$ (with $m \geq 1$) is an algebraic integer. If $\zeta^m = 1$ then $\zeta^{-m} = 1$, so ζ^{-1} is also a root of unity. Thus any root of unity in K is a unit of A.

Let $U = U_K$ denote the group of units of A and let $W = W_K$ denote the subgroup of U consisting of roots of unity. W is a nontrivial subgroup of U, since $1, -1 \in W$.

How large is W? Can it be infinite? What is the structure of the group W? Is W necessarily equal to U? If not, how can one determine the structure of U?

These are the main questions we are faced with in trying to study the units.

10.1 Roots of Unity

We shall begin by describing rather accurately the group W, and afterward we shall examine the units of quadratic fields and certain cyclotomic fields, to gain some insight into the possibilities for a reasonable structure theorem.

First, we prove a remarkable fact:

A. *Let c be any positive real number and let K be an algebraic number field. Then there exist only finitely many algebraic integers x in K such that $|x^{(i)}| \leq c$ for all conjugates $x^{(i)}$ of x.*

Proof: We shall determine a finite set S, depending only on the constant c, and prove that if x is an algebraic integer of K, such that $|x^{(i)}| \leq c$ for all its conjugates $x^{(i)}$, then $x \in S$.

Let $[K : \mathbb{Q}] = n$ and let s_1, s_2, ..., s_n be the elementary symmetric polynomials in n variables, namely, $s_1 = X_1 + X_2 + \cdots + X_n$, $s_2 = \sum_{i<j} X_i X_j$, ..., $s_n = X_1 \cdot X_2 \cdots X_n$.

Let c' be a sufficiently large real number, for example

$$c' = \max\left\{ nc, \binom{n}{2}c^2, \ldots, \binom{n}{k}c^k, \ldots, c^n \right\}.$$

Let F be the set of all monic polynomials of degree at most n, whose coefficients are integers a such that $|a| \leq c'$; F is a finite set. Let S be the set of elements of K which are roots of some polynomial belonging to F; S is also a finite set.

If $|x^{(i)}| \leq c$ for all conjugates of $x \in K$, then $|s_k(x^{(1)}, x^{(2)}, \ldots, x^{(n)})| \leq c'$ for every $k = 1, \ldots, n$. Since x is an algebraic integer, then $s_k(x^{(1)}, x^{(2)}, \ldots, x^{(n)}) \in \mathbb{Z}$, and therefore the polynomial $\prod_{i=1}^{n}(X - x^{(i)})$ belongs to F; that is, $x \in S$. ∎

An immediate corollary is a characterization of the roots of unity in K:

B.   *x is a root of unity in K if and only if x is an algebraic integer of K such that $|x^{(i)}| = 1$ for every conjugate of x.*

Proof: One implication is easily established, for if x is a root of unity, the same holds for all its conjugates. From $x^m = 1$ it follows that $|x|^m = 1$, so $|x| = 1$; the same holds for every conjugate of x.

Conversely, by (**A**) there are only finitely many algebraic integers x in K such that $|x^{(i)}| = 1$ for all conjugates of x.

Since x, x^2, x^3, ... all share this property, then there exist integers r, s, $r < s$, such that $x^r = x^s$. Hence $x^{s-r} = 1$, showing that x is a root of unity. ∎

Combining (**A**) and (**B**), we have the structure of the group W:

C.   *The group W of roots of unity in K is a finite multiplicative cyclic group.*

Proof: By (**A**) and (**B**), W must be finite.

Let h be the maximum of the orders of the elements in W. By Chapter 3, Lemma 2, the order of every element in W divides h, so W is contained in the group of hth roots of unity. By Chapter 2, Section 8, this last group is cyclic, and therefore W itself is cyclic. ∎

The number of elements of W will be denoted by w. It is another numerical invariant attached to the field K. Since 1, -1 are roots of unity belonging to any algebraic number field, then w is even. We shall prove in Chapter 13, (**R**), that w divides $2\delta_K$.

Now we pause and determine the units in special cases.

10.2 Units of Quadratic Fields

Let $K = \mathbb{Q}(\sqrt{-d})$, where d is a nonzero square-free integer. If $d < 0$ it is pretty easy to find all units:

D.   *If $d < 0$, $d \neq -1$, $d \neq -3$, then the units of $\mathbb{Q}(\sqrt{d})$ are 1, -1.*
The units of $\mathbb{Q}(\sqrt{-1})$ are 1, -1, i, $-i$.
The units of $\mathbb{Q}(\sqrt{-3})$ are 1, -1, $(1 + \sqrt{-3})/2$, $(1 - \sqrt{-3})/2$, $(-1 + \sqrt{-3})/2$, $(-1 - \sqrt{-3})/2$.
In all cases, every unit is a root of unity.

Proof: If $d \equiv 2 \pmod 4$ or $d \equiv 3 \pmod 4$, then the integers of $\mathbb{Q}(\sqrt{d})$ are of the form $x = a + b\sqrt{d}$ with $a, b \in \mathbb{Z}$; the conjugate of x is $x' = a - b\sqrt{d}$ and the norm is $N(x) = xx' = a^2 - b^2d$. In order that x be a unit, it is necessary and sufficient that $N(x) = \pm 1$ (Chapter 5, (**N**)). Since $d < 0$, this means that $a^2 - b^2d = 1$.

The only possible solutions are $a = \pm 1$, $b = 0$, except when $d = -1$, where we have another solution: $a = 0$, $b = \pm 1$.

If $d \equiv 1 \pmod 4$ then the integers of $\mathbb{Q}(\sqrt{d})$ are of the form $x = (a + b\sqrt{d})/2$, where $a, b \in \mathbb{Z}$ have the same parity. With the same argument, we are led to solve $a^2 - b^2d = 4$. The only possible solutions are $a = \pm 2$, $b = 0$, except when $d = -3$, where we have another solution: $a = \pm 1$, $b = \pm 1$.

It is also quite clear that all units are roots of unity, those of $\mathbb{Q}(\sqrt{-3})$ being sixth roots of unity. ∎

Let us consider now the more interesting case where $d > 0$, so $\mathbb{Q}(\sqrt{d})$ is contained in the field of real numbers. Thus, the only roots of unity in $\mathbb{Q}(\sqrt{d})$ are 1, -1. We shall show that there exist other units in $\mathbb{Q}(\sqrt{d})$, when $d > 0$.

Lemma 1.   *If $\alpha > 0$ is an irrational real number, then for every integer $m > 0$ there exist integers a, b, not both equal to 0, such that $|a| \leq m$, $|b| \leq m$, and $|a + \alpha b| \leq (1 + \alpha)/m$.*

Proof: Let $f = X + \alpha Y$ and consider the set S of values $f(a, b) = a + \alpha b$, when $0 \leq a \leq m$, $0 \leq b \leq m$. Since α is irrational, if $(a, b) \neq (a', b')$ then $f(a, b) \neq f(a', b')$; therefore $\#S = (m + 1)^2$; all elements in S belong to the interval $[0, (1 + \alpha)m]$. We divide this interval into m^2 equal parts; $[0, (1 + \alpha)/m]$, $[(1 + \alpha)/m, 2(1 + \alpha)/m]$, ... Since there are more elements in S than subintervals, there exist at least two elements in S in the same subinterval:

$$\frac{r(1 + \alpha)}{m} \leq a_1 + \alpha b_1 < a_2 + \alpha b_2 \leq \frac{(r + 1)(1 + \alpha)}{m}.$$

Therefore, letting $a = a_2 - a_1$, $b = b_2 - b_1$, we have $|a + \alpha b| \leq (1 + \alpha)/m$, with a, b not both equal to 0, and $|a| \leq m$, $|b| \leq m$.   ∎

E.   *If d is a positive square-free integer, then the group U of units of the field $\mathbb{Q}(\sqrt{d})$ is $U \cong \{1, -1\} \times C$, where C is an infinite multiplicative cyclic group.*

Proof: As we have already said, $W = \{1, -1\}$. To show that there exists another unit in $\mathbb{Q}(\sqrt{d})$, we shall make use of Lemma 1, with $\alpha = \sqrt{d}$. For every m, let S_m be the set of all couples of integers (a, b), with a, b not both 0, such that $|a| \leq m$, $|b| \leq m$, $|a + b\sqrt{d}| \leq (1 + \sqrt{d})/m$. By the lemma, each set S_m is nonempty. Let us write $S_m = S_m^+ \cup S_m^- \cup S_m^0$, where $S_m^+ = \{(a, b) \in S_m \mid a > 0\}$, $S_m^- = \{(a, b) \in S_m \mid a < 0\}$, $S_m^0 = \{(a, b) \in S_m \mid a = 0\}$. If $(a, b) \in S_m^+$ then $(-a, -b) \in S_m^-$ and vice versa; also, if $m = 1$ then $S_1^0 = \{(0, 1), (0, -1)\}$ and if $m \geq 2$ then $S_m^0 = \varnothing$ because $|b|\sqrt{d} \leq (1 + \sqrt{d})/m$ implies $|b| \leq (1/m)(1 + 1/\sqrt{d}) \leq \frac{1}{2}(1 + 1/\sqrt{d}) < 1$ which is impossible since b cannot be zero.

If $\bigcup_{m \geq 1} S_m$ is a finite set then there exists m_0 such that $1/m_0 < |a + b\sqrt{d}|$ for all $(a, b) \in \bigcup_{m \geq 1} S_m$; however, if m is sufficiently large and if $(a, b) \in S_m$ then $|a + b\sqrt{d}| < (1 + \sqrt{d})/m \leq 1/m_0$, which is a contradiction. Thus, $\bigcup_{m \geq 1} S_m$ is an infinite set, hence $\bigcup_{m \geq 1} S_m^+$ is also an infinite set (otherwise from $\#S_m^- = \#S_m^+$ for every m, it would follow that $\bigcup_{m \geq 1} S_m$ is finite).

From $|a| \leq m$, $|b| \leq m$ it follows that $|a - b\sqrt{d}| \leq |a| + |b|\sqrt{d} \leq m(1 + \sqrt{d})$ so

$$0 \neq |a^2 - b^2 d| = |a - b\sqrt{d}| \cdot |a + b\sqrt{d}|$$

$$\leq m(1 + \sqrt{d})\frac{1 + \sqrt{d}}{m} = (1 + \sqrt{d})^2$$

for every $(a, b) \in \bigcup_{m \geq 1} S_m$, hence also for every $(a, b) \in \bigcup_{m \geq 1} S_m^+$.

Therefore, there exists some integer n, $0 < |n| < (1 + \sqrt{d})^2$, such that $a^2 - b^2 d = n$ for infinitely many couples of integers (a, b), where $a > 0$.

Consider $n^2 + 1$ among these couples (a, b). Let us define $(a_1, b_1) \equiv (a_2, b_2)$ when $a_1 \equiv a_2 \pmod{n}$, $b_1 \equiv b_2 \pmod{n}$; thus we have at most n^2 equivalence classes. Since the number of couples is greater than n^2, there exist at least two distinct couples (a_1, b_1), (a_2, b_2) in the same equivalence class.

Let $x_1 = a_1 + b_1\sqrt{d}$, $x_2 = a_2 + b_2\sqrt{d}$, and consider $u = x_1/x_2$. Since $N(x_1) = N(x_2) = n$ then $N(u) = 1$ and $u \neq \pm 1$, because $x_1 \neq x_2$, $x_1 \neq -x_2$ since $a_1 > 0$, $a_2 > 0$.

But

$$u = \frac{x_1}{x_2} = 1 + \frac{x_1 - x_2}{x_2} = 1 + \frac{(x_1 - x_2)x_2'}{N(x_2)}$$

$$= 1 + \left(\frac{a_1 - a_2}{n} + \frac{b_1 - b_2}{n}\sqrt{d}\right)(a_2 - b_2\sqrt{d}).$$

Noting that $(a_1 - a_2)/n$, $(b_1 - b_2)/n$ are integers, and multiplying out, we obtain $u = a + b\sqrt{d}$, with $a, b \in \mathbb{Z}$. Thus, u is a unit, different from 1, -1.

There must exist some unit u of $\mathbb{Q}(\sqrt{d})$ such that $1 < u$; in fact, if u is a unit, then u, $-u$, u^{-1}, $-u^{-1}$ are units and the largest of these real numbers is greater than 1.

Now we shall prove that among all units $u > 1$ there exists a smallest possible; it is enough to show that for every real number $c > 1$ there exist only finitely many units u such that $1 < u < c$. Now, if u is such a unit, then from $N(u) = uu' = \pm 1$, it follows that $1/c < u' < 1$ or $-1 < u' < -1/c$ and at any rate $|u'| < c$. By (**A**) the set of such units must be finite.

Let u_1 be the smallest unit such that $1 < u_1$. We shall prove that every positive unit u is a power of u_1. In fact, there exists an integer m such that $u_1^m \leq u < u_1^{m+1}$; then u/u_1^m is again a unit such that $1 \leq u/u_1^m < u_1$, hence necessarily $u = u_1^m$. Similarly, all negative units are of the form $-u_1^m$, for $m \in \mathbb{Z}$.

Let C be the multiplicative cyclic group generated by u_1.

The mapping $U \to \{1, -1\} \times C$, defined by $u_1^m \mapsto (1, u_1^m)$, $-u_1^m \mapsto (-1, u_1^m)$ is clearly an isomorphism.   ■

The smallest unit $u_1 > 1$ is called the *fundamental unit* of $\mathbb{Q}(\sqrt{d})$.

A crude method of determining the fundamental unit is the following. First let $d \equiv 2 \pmod 4$ or $d \equiv 3 \pmod 4$.

If $u = a + b\sqrt{d}$ is a unit, $u \neq \pm 1$, so are $-u$, u^{-1}, $-u^{-1}$ and only the largest of these numbers is larger than 1; since these numbers are exactly $\pm a \pm b\sqrt{d}$, then $a + b\sqrt{d} > 1$ when $a > 0$, $b > 0$.

If $u_1 = a_1 + b_1\sqrt{d}$ is the fundamental unit, if $u_m = u_1^m = a_m + b_m\sqrt{d}$, then $b_{m+1} = a_1 b_m + a_m b_1$, so we have $b_1 < b_2 < b_3 < \cdots$. From $\pm 1 = N(u_1) = a_1^2 - b_1^2 d$, we have $b_1^2 d = a_1^2 \mp 1$; thus if we write the sequence d, $4d$, $9d$, $16d$, $\ldots$, then b_1 is the smallest integer such that $0 < b_1$, $b_1^2 d$ is a square plus or minus 1.

For example, let $d = 3$, then $b_1 = 1$, $a_1 = 2$, so $2 + \sqrt{3}$ is the fundamental unit of $\mathbb{Q}(\sqrt{3})$.

Similarly, $1 + \sqrt{2}$, $5 + 2\sqrt{6}$, $8 + 3\sqrt{7}$, are the fundamental units of the fields $\mathbb{Q}(\sqrt{2})$, $\mathbb{Q}(\sqrt{6})$, $\mathbb{Q}(\sqrt{7})$, respectively.

Now, if $d \equiv 1 \pmod 4$, by a similar argument $u_1 = (a_1 + b_1\sqrt{d})/2$ with a_1, b_1 positive integers of the same parity; also $\pm 1 = N(u_1) = (a_1^2 - b_1^2 d)/4$ hence $b_1^2 d = a_1^2 \mp 4$ and we have to find the smallest integer $b_1 > 0$ such that $b_1^2 d$ is a square plus or minus 4. For example $(1 + \sqrt{5})/2$, $(3 + \sqrt{13})/2$ are fundamental units of $\mathbb{Q}(\sqrt{5})$, $\mathbb{Q}(\sqrt{13})$, respectively.

For the next considerations we fix the following notation. Let $u = u_d = (a + b\sqrt{d})/2$ be the fundamental unit of $\mathbb{Q}(\sqrt{d})$; we note that $a \equiv b \pmod 2$.

For any $n \geq 1$ let $u^n = (a_n + b_n\sqrt{d})/2$, with $a_n \equiv b_n \pmod 2$. The integers a_n, b_n may be obtained recursively as we now indicate.

Let $P = a$, $Q = N(u) = (a^2 - db^2)/4 = \pm 1$. We define the recurring sequences $(U_n)_{n \geq 0}$, $(V_n)_{n \geq 0}$ as follows:

$$U_0 = 0, \qquad U_1 = 1, \qquad U_n = PU_{n-1} - QU_{n-2} \qquad \text{for} \quad n \geq 2$$

and

$$V_0 = 2, \qquad V_1 = P, \qquad V_n = PV_{n-1} - QV_{n-2} \qquad \text{for} \quad n \geq 2.$$

Let α, β be the roots of the polynomial $f(X) = X^2 - PX + Q$. Thus

$$\alpha = \frac{P + \sqrt{P^2 - 4Q}}{2}, \qquad \beta = \frac{P - \sqrt{P^2 - 4Q}}{2}.$$

Then $\alpha + \beta = P$, $\alpha\beta = Q$, and $P^2 - 4Q = (\alpha - \beta)^2$.

Lemma 2.   *For every $n \geq 0$:*

$$U_n = \frac{\alpha^n - \beta^n}{\alpha - \beta} \qquad and \qquad V_n = \alpha^n + \beta^n.$$

Proof: For $n = 0$ and 1 we have

$$\frac{\alpha^0 - \beta^0}{\alpha - \beta} = 0 = U_0 \qquad \text{and} \qquad \frac{\alpha - \beta}{\alpha - \beta} = 1 = U_1.$$

By induction, if $n \geq 2$ then

$$\frac{\alpha^n - \beta^n}{\alpha - \beta} = (\alpha + \beta)\frac{\alpha^{n-1} - \beta^{n-1}}{\alpha - \beta} - \alpha\beta\frac{\alpha^{n-2} - \beta^{n-2}}{\alpha - \beta}$$

$$= PU_{n-1} - QU_{n-2} = U_n.$$

Similarly, $\alpha^0 + \beta^0 = 2 = V_0$, $\alpha + \beta = P = V_1$, and by induction, if $n \geq 2$ then

$$\alpha^n + \beta^n = (\alpha + \beta)(\alpha^{n-1} + \beta^{n-1}) - \alpha\beta(\alpha^{n-2} + \beta^{n-2})$$

$$= PV_{n-1} - QV_{n-2} = V_n. \qquad \blacksquare$$

Let $P = a$, $Q = N(u) = (a^2 - db^2)/4 = \pm 1$. Now $P^2 - 4Q = b^2d = (\alpha - \beta)^2$. With the above notations, we have:

Lemma 3.   *For each $n \geq 1$, $a_n = V_n$ and $b_n = bU_n$.*

Proof: The assertion is true for $n = 1$ and we proceed by induction. We have

$$\frac{a_n + b_n\sqrt{d}}{2} = \frac{a_{n-1} + b_{n-1}\sqrt{d}}{2} \cdot \frac{a + b\sqrt{d}}{2}$$

hence $a_n = \frac{1}{2}(a_{n-1}a + b_{n-1}bd)$ and $b_n = \frac{1}{2}(a_{n-1}b + b_{n-1}a)$. From the preceding lemma we obtain

$$a_n = \frac{1}{2}\left[(\alpha^{n-1} + \beta^{n-1})(\alpha + \beta) + \frac{\alpha^{n-1} - \beta^{n-1}}{\alpha - \beta}b^2d\right]$$

$$= \alpha^n + \beta^n = V_n.$$

Similarly

$$b_n = \tfrac{1}{2}[a_{n-1}b + b_{n-1}a]$$

$$= \frac{b}{2}\left[(\alpha^{n-1} + \beta^{n-1}) + \frac{\alpha^{n-1} - \beta^{n-1}}{\alpha - \beta} \cdot (\alpha + \beta)\right]$$

$$= b\frac{\alpha^n - \beta^n}{\alpha - \beta} = bU_n. \qquad\blacksquare$$

In the next lemma we indicate the parity of a_n and b_n.

Lemma 4.   *If a is even, then a_n, b_n are even for every $n \geq 1$. If a is odd, then a_n (and also b_n) is even if and only if 3 divides n.*

Proof: Let $a = P$ be even. The sequence $(V_n \pmod 2)_{n \geq 0}$ is seen to be $0, 0, 0, \ldots$, that is, a_n is even for every $n \geq 1$, and so must be b_n, since $a_n \equiv b_n \pmod 2$.

Now let $a = P$ be odd, so b is also odd. The sequence $(V_n \pmod 2)_{n \geq 0}$ is seen to be $0, 1, 1, 0, 1, 1, \ldots$, thus a_n (and also b_n) is even if and only if 3 divides n. $\blacksquare$

There is a narrow connection between units of the real quadratic field $\mathbb{Q}(\sqrt{d})$ and the Diophantine equations

$$X^2 - dY^2 = \pm 1, \ \pm 4 \qquad\qquad (10.1)$$

($d > 0$ and square-free).

These equations were studied by Fermat, but became better known since they were discussed in the textbook by Pell—so today, they are known as *Pell equations.*

For $\varepsilon = 1, -1, 4,$ or -4 let

$$S_{d,\varepsilon} = \{(x, y) \mid x > 0, y > 0 \ \text{ and } \ x^2 - dy^2 = \varepsilon\}.$$

In the next results we determine explicitly the sets $S_{d,\varepsilon}$.

F.   *With the above notations:*
   (i) *The set of solutions of $X^2 - dY^2 = 1$ is the following:*
   (a) *If $N(u) = 1$:*

$$\begin{cases} \textit{If } a \textit{ is even then } S_{d,1} = \{(a_n/2, b_n/2) \mid n \geq 1\}, \\ \textit{If } a \textit{ is odd then } S_{d,1} = \{(a_n/2, b_n/2) \mid 3 \textit{ divides } n\}. \end{cases}$$

   (b) *If $N(u) = -1$:*

$$\begin{cases} \textit{If } a \textit{ is even then } S_{d,1} = \{(a_n/2, b_n/2) \mid n \textit{ is even}\}, \\ \textit{If } a \textit{ is odd then } S_{d,1} = \{(a_n/2, b_n/2) \mid 6 \textit{ divides } n\}. \end{cases}$$

   (ii) *The set of solutions of $X^2 - dY^2 = -1$ is the following:*

(a) *If $N(u) = 1$ then $S_{d,-1} = \varnothing$.*
(b) *If $N(u) = -1$:*

$$\begin{cases} \textit{If } a \textit{ is even then } S_{d,-1} = \{(a_n/2, b_n/2) \mid n \textit{ is odd}\}, \\ \textit{If } a \textit{ is odd then } S_{d,-1} = \{(a_n/2, b_n/2) \mid n \textit{ is odd and } 3 \textit{ divides } n\}. \end{cases}$$

Proof: Let $\varepsilon = 1$ or -1. We have $(x, y) \in S_{d,\varepsilon}$ if and only if $((2x)^2 - d(2y)^2)/4 = \varepsilon$ or, equivalently, $(2x + 2y\sqrt{d})/2$ is a unit with norm ε. By the preceding results, there exists $n \geq 1$ such that $2x = a_n$, $2y = b_n$. If a is even, this holds for all $n \geq 1$, while if a is odd, this holds if and only if 3 divides n.

Moreover, if $\varepsilon = 1$ and $N(u) = -1$ we must have n even; if $\varepsilon = -1$ and $N(u) = -1$ then n must be odd: finally, if $\varepsilon = -1$ and $N(u) = 1$ no n satisfies the required conditions. ∎

G.   *With the above notations:*
   (i) *The set of solutions of $X^2 - dY^2 = 4$ is the following:*
   (a) *If $N(u) = 1$:*

$$\begin{cases} \textit{If } a \textit{ is even then } S_{d,4} = \{(a_n, b_n) \mid n \geq 1\}, \\ \textit{If } a \textit{ is odd then } S_{d,4} = \{(a_n, b_n) \mid 3 \textit{ divides } n\}. \end{cases}$$

   (b) *If $N(u) = -1$:*

$$\begin{cases} \textit{If } a \textit{ is even then } S_{d,4} = (a_n, b_n) \mid n \textit{ is even}\}, \\ \textit{If } a \textit{ is odd then } S_{d,4} = \{(a_n, b_n) \mid 6 \textit{ divides } n\}. \end{cases}$$

   (ii) *The set of solutions of $X^2 - dY^2 = -4$ is the following:*
   (a) *If $N(u) = 1$ then $S_{d,-4} = \varnothing$.*
   (b) *If $N(u) = -1$:*

$$\begin{cases} \textit{If } a \textit{ is even then } S_{d,-4} = \{(a_n, b_n) \mid n \textit{ is odd}\}, \\ \textit{If } a \textit{ is odd then } S_{d,-4} = \{(a_n, b_n) \mid n \textit{ is odd and } 3 \textit{ divides } n\}. \end{cases}$$

Proof: The proof, similar to the one of (**F**), is left to the reader. ∎

We now consider the special case when $d = p$ is a prime number.

H.   *Let p be a prime number and let $u = u_p = (a + b\sqrt{d})/2$ be the fundamental unit of $\mathbb{Q}(\sqrt{p})$.*
   The following statements are equivalent:
   (1) $N(u) = -1$.
   (2) *The equation $X^2 - pY^2 = -1$ has a nontrivial solution.*
   (3) *The equation $X^2 - pY^2 = -4$ has a nontrivial solution.*
   (4) $p = 2$ or $p \equiv 1 \pmod 4$.

Proof: By **(F)** and **(G)**, $S_{d,-1} \neq \varnothing$ if and only if $N(u) = -1$; similarly, $S_{d,-4} \neq \varnothing$ if and only if $N(u) = -1$. This shows that the statements (1), (2), and (3) are equivalent.

(3) $\rightarrow$ (4)   If $p \neq 2$ and if $x^2 - py^2 = -4$, then -1 is a square modulo p, so $p \equiv 1 \pmod 4$.

(4) $\rightarrow$ (1)   If $p = 2$ then $u = 1 + \sqrt{2}$, so $N(u) = -1$. Let p be a prime, $p \equiv 1 \pmod 4$ and assume that $N(u) = 1$, so $a^2 - pb^2 = 4$, hence $(a + 2)(a - 2) = pb^2$.

Case 1: a is odd.

Then $\gcd(a + 2, a - 2) = 1$. Thus there are integers b_1, b_2 with $b_1 b_2 = b$ and either (a) or (b) holds:

$$\text{(a)} \quad \begin{cases} a + 2 = pb_1^2, \\ a - 2 = b_2^2, \end{cases} \qquad \text{(b)} \quad \begin{cases} a + 2 = b_1^2, \\ a - 2 = pb_2^2. \end{cases}$$

So $4 = pb_1^2 - b_2^2$ or $4 = b_1^2 - pb_2^2$. In the first case $(b_2 + b_1\sqrt{p})/2$ is a unit with norm -1; this implies that $N(u) = -1$, a contradiction.

In the second case, since $(b_1 + b_2\sqrt{p})/2$ is a unit then $a \leq b_1$, $b \leq b_2$, so $b_2 = b$, $b_1 = 1$, and $3 = -pb^2$, which is impossible.

Case 2: a is even.

Now $\gcd(a + 2, a - 2) = 2$ or 4. Thus there exist integers b_1, b_2 with $b = 2b_1 b_2$ or $4b_1 b_2$ and either one of the following cases hold:

$$\text{(c)} \quad \begin{cases} a + 2 = 2pb_1^2, \\ a - 2 = 2b_2^2, \end{cases} \qquad \text{(d)} \quad \begin{cases} a + 2 = 2b_1^2, \\ a - 2 = 2pb_2^2, \end{cases}$$

$$\text{(e)} \quad \begin{cases} a + 2 = 4pb_1^2, \\ a - 2 = 4b_2^2, \end{cases} \qquad \text{(f)} \quad \begin{cases} a + 2 = 4b_1^2, \\ a - 2 = 4pb_2^2. \end{cases}$$

In cases (c) and (d) we obtain $2 = pb_1^2 - b_2^2$ or $2 = b_1^2 - pb_2^2$. These relations are impossible, as seen by reducing modulo 4.

In case (e) we obtain $((2b_2)^2 - p(2b_1)^2)/4 = -1$ so $(2b_2 + 2b_1\sqrt{p})/2$ is a unit of norm -1, which contradicts the assumption that $N(u) = 1$.

Finally, in case (f), $(2b_1 + 2b_2\sqrt{p})/2$ is a unit, therefore $a \leq 2b_1$, $b \leq 2b_2$ so $ab \leq 4b_1 b_2 = b$ in this case. Hence $a = 1$ and $-1 = 4pb_2^2$, which is impossible.

This shows that $N(u)$ must be equal to -1, concluding the proof.   ∎

10.3 Units of Cyclotomic Fields

Let p be an odd prime, ζ a primitive pth root of unity, and let $K = \mathbb{Q}(\zeta)$. Hence K has degree $p - 1$ over $\mathbb{Q}$. As we have seen in Chapter 5, **(X)**, the ring of integers of K is $A = \mathbb{Z}[\zeta]$ and $p = u(1 - \zeta)^{p-1}$, where u is a unit of A.

Now we prove the main result about units in $\mathbb{Q}(\zeta)$:

I.   *The multiplicative group W of roots of unity of $\mathbb{Q}(\zeta)$ is*

$$W = \{1, \zeta, \zeta^2, \ldots, \zeta^{p-1}, -1, -\zeta, -\zeta^2, \ldots, -\zeta^{p-1}\},$$

so $w = 2p$. Every unit of $\mathbb{Q}(\zeta)$ may be written as $u = \pm\zeta^k v$ where v is a positive real unit of A.

Proof: By **(C)**, W is a cyclic group of order w. Since $-\zeta \in W$ and $-\zeta$ has order $2p$ (because p is odd) then $2p$ divides w.

Now, let $x \in W$ be an element of order w. Since $x \in K$ then $\mathbb{Q}(x) \subseteq K$ so $\varphi(w) = [\mathbb{Q}(x) : \mathbb{Q}]$ divides $p - 1 = [K : \mathbb{Q}]$. But $w = p^r \cdot m$ where $r \geq 1$, $m \geq 2$, p does not divide m. So $\varphi(w) = p^{r-1}(p - 1) \cdot \varphi(m)$ and since it divides $p - 1$ then $r = 1$, $\varphi(m) = 1$ so $m = 2$ and therefore $w = 2p$. Thus, $W = \{1, \zeta, \zeta^2, \ldots, \zeta^{p-1}, -1, -\zeta, -\zeta^2, \ldots, -\zeta^{p-1}\}$.

Now, let u be any unit of $\mathbb{Q}(\zeta)$, so $u = a_0 + a_1\zeta^k + \cdots + a_{p-2}\zeta^{p-2}$ with $a_i \in \mathbb{Z}$. The complex conjugate of u, which is also a unit (because $uv = 1$ implies $\overline{u}\,\overline{v} = 1$), is given by

$$\overline{u} = a_0 + a_1\zeta^{-1} + a_2\zeta^{-2} + \cdots + a_{p-2}\zeta^{-(p-2)};$$

then $u' = u \cdot \overline{u}^{-1}$ is also a unit. Moreover, if $u^{(k)} = a_0 + a_1\zeta^k + a_2\zeta^{2k} + \cdots + a_{p-2}\zeta^{k(p-2)}$, for $k = 1, 2, \ldots, p - 1$, are the conjugates of u, then $\overline{u^{(k)}} = \overline{u}^{(k)}$ are the conjugates of $\overline{u}$, so those of u' are $u'^{(k)} = u^{(k)} \cdot \overline{u}^{(k)-1}$ and therefore, for every $k = 1, 2, \ldots, p-1$, we have $|u'^{(k)}| = 1$. By **(B)**, the element u' is a root of unity and therefore of the form $u' = \pm\zeta^h$, with $0 \leq h \leq p - 1$, as shown above.

We must have the positive sign; if not, $u' = -\zeta^h$ then $u = -\zeta^h\overline{u}$. Let us consider the ring $R = A/A(1 - \zeta)$, and let $\theta : A \to R$ be the canonical homomorphism. Then $\theta(\zeta) = 1$, so $\theta(\zeta^k) = 1$ for every $k = 1, \ldots, p - 2$, therefore $\theta(u) = a_0 + a_1 + \cdots + a_{p-2} = \theta(\overline{u})$; from $u = -\zeta^h\overline{u}$, we have $\theta(u) = -\theta(\overline{u})$, thus $\theta(2\overline{u}) = 0$, so $2\overline{u} \in A(1 - \zeta)$, that is, $1 - \zeta$ divides $2\overline{u}$, and since $\overline{u}$ is a unit, $1 - \zeta$ divides 2. Since p is associated with $(1 - \zeta)^{p-1}$ then p divides 2, so $p = 2$, contrary to the hypothesis.

Thus, $u = \zeta^h\overline{u}$. Let k be such that $2k \equiv h \pmod{p}$, so $\zeta^h = \zeta^{2k}$, and therefore

$$\frac{u}{\zeta^k} = \zeta^k\overline{u} = \frac{\overline{u}}{\zeta^{-k}} = \overline{\left(\frac{u}{\zeta^k}\right)}.$$

Letting $v = u/\zeta^k = u\zeta^{p-k}$, we see that $v \in \mathbb{R} \cap A$, v is a unit, and $u = \zeta^k v$. Finally, we may take $v > 0$, by multiplying ζ^k with -1, if necessary.   ∎

We may easily exhibit some outstanding real units of $\mathbb{Q}(\zeta)$.

For example, $(1 - \zeta^s)/(1 - \zeta)$ $(1 \leq s \leq p - 1)$ is a unit (since $1 - \zeta^s$, $1 - \zeta$ are associated elements).

By the preceding proof, there exists an even integer $2k$ such that

$$(1 - \zeta^s)/(1 - \zeta) = \zeta^{2k}(1 - \zeta^{-s})/(1 - \zeta^{-1}),$$

so

$$\frac{1 - \zeta^s}{1 - \zeta} \cdot \frac{1 - \zeta^{-s}}{1 - \zeta^{-1}} = \left(\zeta^k \frac{1 - \zeta^{-s}}{1 - \zeta^{-1}} \right)^2$$

and

$$v_s = \sqrt{\frac{1 - \zeta^s}{1 - \zeta} \cdot \frac{1 - \zeta^{-s}}{1 - \zeta^{-1}}}$$

is a real unit in $\mathbb{Q}(\zeta)$ for every $s = 1, 2, \ldots, p - 1$.

Now, if $1 \leq s$, $s' \leq p - 1$, $s + s' = p$ then $v_s = -v_{s'}$, if $s = 1$ then $v_1 = 1$. Thus $v_2, v_3, \ldots, v_g$ ($g = (p - 1)/2$) are $(p - 3)/2$ real units, distinct from 1, -1, thus not roots of unity. They are called the *circular units*. Actually, they are distinct, because if $v_s = v_t$ then

$$\frac{1 - \zeta^s}{1 - \zeta} \cdot \frac{1 - \zeta^{-s}}{1 - \zeta^{-1}} = \frac{1 - \zeta^t}{1 - \zeta} \cdot \frac{1 - \zeta^{-t}}{1 - \zeta^{-1}},$$

so $\zeta^s + \zeta^{-s} = \zeta^t + \zeta^{-t}$.

However, $\zeta = \cos(2\pi/p) + i\sin(2\pi/p)$, so $\zeta^s = \cos(2\pi s/p) + i\sin(2\pi s/p)$, $\zeta^{-s} = \cos(2\pi s/p) - i\sin(2\pi s/p)$, so that the above relation gives

$$2\cos(2\pi s/p) = 2\cos(2\pi t/p).$$

Therefore, either $s = t$ or $(2\pi s/p) + (2\pi t/p) = 2k\pi$, with $k \in \mathbb{Z}$. But then $s + t = kp$, and, if $1 < s$, $t \leq (p - 1)/2$, this is impossible.

At this point we still have no way of determining, even in very simple particular cases, all the units of a cyclotomic field. Soon, we shall prove a theorem of Dirichlet, which indicates the structure of the group of units of an algebraic number field; this result states the existence of fundamental systems of units which, together with the roots of unity, generate all the units of the field. But even for cyclotomic fields, the determination of a fundamental system of units requires deep results and delicate computations.

10.4 Dirichlet's Theorem

Our aim is to establish Dirichlet's fundamental theorem on the structure of the group of units of a number field K. Let $[K : \mathbb{Q}] = n$, let $r_1 \geq 0$ be the number of real conjugates of K and let $r_2 \geq 0$ be the number of pairs of nonreal complex-conjugate conjugates of K. Thus $n = r_1 + 2r_2$.

We follow the convention already used in the preceding chapter: $K^{(1)}$, $K^{(2)}$, $\ldots$, $K^{(r_1)}$ are the real conjugates of K, while $K^{(r_1+1)}$, $\ldots$,

$K^{(n)}$ are the nonreal conjugates, and $\overline{K^{(r_1+j)}} = K^{(r_1+r_2+j)}$ for all $j = 1, \ldots, r_2$.

If $x \in K$ let $x^{(i)} \in K^{(i)}$ $(i = 1, \ldots, n)$ be the conjugates of x, so $\overline{x^{(r_1+j)}} = x^{(r_1+r_2+j)}$ for $j = 1, \ldots, r_2$.

We shall consider in the sequel the mapping $\lambda : U \to \mathbb{R}^r$ where $r = r_1 + r_2 - 1$, defined by $\lambda(u) = (\log|u^{(1)}|, \log|u^{(2)}|, \ldots, \log|u^{(r)}|)$ (where $|u^{(j)}|$ denotes the positive real number which is the absolute value of $u^{(j)} \in \mathbb{C}$ and log is the natural logarithm function).

J.   *Let u be a unit. Then u is a root of unity if and only if $\lambda(u) = (0, 0, \ldots, 0) \in \mathbb{R}^r$.*

Proof: If $u \in W$ then $|u^{(j)}| = 1$ for every conjugate of u, hence $\lambda(u) = (0, 0, \ldots, 0)$.

Conversely, let $u \in U$ be such that $|u^{(1)}| = |u^{(2)}| = \cdots = |u^{(r)}| = 1$. Since $|N_{K|\mathbb{Q}}(u)| = 1$ then $\sum_{i=1}^{n} \log|u^{(i)}| = 0$. We recall that $|u^{(r_1+j)}| = |u^{(r_1+r_2+j)}|$ for $j = 1, \ldots, r_2$. From the above relations it follows that $2 \log|u^{(r_1+r_2)}| = 0$, so $|u^{(r_1+r_2)}| = |u^{(r_1+2r_2)}| = 1$. By (**B**), u is a root of unity. ∎

Let $q \geq 1$ and let $u_1, \ldots, u_q \in U$ be such that $\{\lambda(u_1), \ldots, \lambda(u_q)\}$ is a linearly independent subset of the $\mathbb{R}$-vector space $\mathbb{R}^r$. Let $G = \{(a_1, \ldots, a_q) \in \mathbb{R}^q \mid$ there exists $v \in U$ such that $\lambda(v) = \sum_{j=1}^{q} a_j \lambda(u_j)\}$.

First, we note that $\mathbb{Z}^q \subseteq G$: indeed, if $(a_1, \ldots, a_q) \in \mathbb{Z}^q$ and $v = \prod_{j=1}^{q} u_j^{a_j}$ then $\lambda(v) = \sum_{j=1}^{q} a_j \lambda(u_j)$. Next, we note that G is an additive subgroup of $\mathbb{R}^q$—the verification is trivial and left to the reader. Every coset of G relative to the subgroup $\mathbb{Z}^q$ contains a unique element of the subset $G_1 = \{(a_1, \ldots, a_q) \in G \mid 0 \leq a_j < 1 \text{ for all } j = 1, \ldots, q\}$. Moreover, different elements of G_1 are in different cosets of G by $\mathbb{Z}^q$. We show:

Lemma 5.   *Under the above hypothesis and notations, $G/\mathbb{Z}^q$ is a finite group.*

Proof: It suffices to show that G_1 is a finite set. Let $U_1 = \{v \in U \mid$ there exist $(a_1, \ldots, a_q) \in G_1$ such that $\lambda(v) \in \sum_{j=1}^{q} a_j \lambda(u_j)\}$. We note that if $v \in U_1$ and $(a_1, \ldots, a_q)$, $(b_1, \ldots, b_q) \in G_1$ are such that $\lambda(v) = \sum_{j=1}^{q} a_j \lambda(u_j) = \sum_{j=1}^{q} b_j \lambda(u_j)$, then $\sum_{j=1}^{q} (a_j - b_j) \lambda(u_j) = 0$. By hypothesis, $a_j = b_j$ for all $j = 1, \ldots, q$. Thus we may define the mapping $v \in U_1 \mapsto (a_1, \ldots, a_q) \in G_1$, where $\lambda(v) = \sum_{j=1}^{q} a_j \lambda(u_j)$. By definition, this mapping is surjective. In order to show that G_1 is finite, it suffices to establish that U_1 is finite. If $v \in U_1$ then $\|\log|v^{(i)}|\| = |\sum_{j=1}^{q} a_j \log|u_j^{(i)}|\| \leq \sum_{j=1}^{q} \|\log|u_j^{(i)}|\|$ for all $i = 1, \ldots, r$.

Let $\alpha_i = \sum_{j=1}^{q} \|\log|u_j^{(i)}|\|$ and $\alpha = \max\{\alpha_1, \ldots, \alpha_r\}$. Then $e^{-\alpha} \leq e^{-\alpha_i} \leq |v^{(i)}| \leq e^{\alpha_i} \leq e^{\alpha}$ for $i = 1, \ldots, r$. Since $|N_{K|\mathbb{Q}}(v)| = 1$ and

$$|v^{(r_1+r_2)}|^2 = \frac{1}{|v^{(1)}| \cdots |v^{(r_1)}| \, |v^{(r_1+1)}|^2 \cdots |v^{(r)}|^2}$$

then there exists $\beta > 0$ such that $|v^{(i)}| \leq \beta$ for all $i = 1, \ldots, n$. By (**A**), U_1 is a finite set, concluding the proof.   ∎

Now we introduce the following notion. The units $u_1, \ldots, u_k$ of A are said to be *independent* whenever a relation

$$u_1^{m_1} u_2^{m_2} \cdots u_k^{m_k} = 1 \quad \text{with} \quad m_i \in \mathbb{Z},$$

is only possible when $m_1 = \cdots = m_k = 0$. Therefore, each u_i belonging to an independent set of units is not a root of unity.

Lemma 6.   *Let $u_1, \ldots, u_k$ be units of A. The following conditions are equivalent:*

    (1) $u_1, \ldots, u_k$ *are independent units.*

    (2) $\lambda(u_1), \ldots, \lambda(u_k)$ *are linearly independent over* $\mathbb{Q}$.

    (3) $\lambda(u_1), \ldots, \lambda(u_k)$ *are linearly independent over* $\mathbb{R}$.

Proof: (1) → (2) Assume that $\lambda(u_1), \ldots, \lambda(u_k)$ are linearly dependent over $\mathbb{Q}$. So there exist integers n_j, not all equal to 0, such that $\sum_{j=1}^{k} n_j \lambda(u_j) = 0$, that is, $\lambda(\prod_{j=1}^{k} u_j^{n_j}) = 0$. By (**J**), $\prod_{j=1}^{k} u_j^{n_j}$ is a root of unity, so there exists $h \geq 1$ such that $\prod_{j=1}^{k} u_j^{h n_j} = 1$, and this is contrary to the hypothesis.

(2) → (3) Assume that $\lambda(u_1), \ldots, \lambda(u_k)$ are linearly dependent over $\mathbb{R}$. By hypothesis, $\lambda(u_j) \neq (0, 0, \ldots,)$ for all $j = 1, \ldots, k$. After renumbering, if necessary, we may assume that $\{\lambda(u_1), \ldots, \lambda(u_q)\}$ is $\mathbb{R}$-linearly independent, but each $\lambda(u_s)$ (with $q < s \leq r$) is of the form $\lambda(u_s) = \sum_{j=1}^{q} a_j \lambda(u_j)$ with $a_j \in \mathbb{R}$. With the previous notation $(a_1, \ldots, a_q) \in G$.

By Lemma 5, if $h = \#(G/\mathbb{Z}^q)$ then $h a_j \in \mathbb{Z}$, so $a_j \in \mathbb{Q}$ for all $j = 1, \ldots, q$. Thus $\{\lambda(u_1), \ldots, \lambda(u_k)\}$ would be $\mathbb{Q}$-linearly dependent, which is a contradiction.

(3) → (1) Assume that $u_1, \ldots, u_k$ are dependent. Then there exist integers m_j, not all equal to 0, such that $u_1^{m_1} \cdots u_k^{m_k} = 1$. Hence $\sum_{j=1}^{k} m_j \lambda(u_j) = 0$ and so $\lambda(u_1), \ldots, \lambda(u_k)$ would be linearly dependent over $\mathbb{Q}$.   ∎

Now we are ready to prove Dirichlet's theorem on the structure of the group of units of an algebraic number field.

Theorem 1 (Dirichlet).   *The group U of units of the ring A of algebraic integers of K has the following structure:*

$$U \cong W \times C_1 \times \cdots \times C_r,$$

where W is the cyclic group of order w of roots of unity belonging to K, each C_i is an infinite multiplicative group, and $r = r_1 + r_2 - 1$.

Proof: We first put aside the trivial case where $r = 0$. This means that $r_1 + r_2 = 1$, thus if $K \neq \mathbb{Q}$ then $r_1 = 0$, $r_2 = 1$, $n = 2$, so $K =$

$\mathbb{Q}(\sqrt{-d})$, $d > 0$. By (**D**) we know that every unit of K is a root of unity, that is, $U = W$. From now on we assume that $r \geq 1$.

We divide the proof into three parts.

Part 1.   We show that $U \cong W \times C_1 \times \cdots \times C_k$, where $0 \leq k \leq r$ and each C_i is an infinite cyclic multiplicative group.

For this purpose, we show that U/W is a free multiplicative group of rank k, where $0 \leq k \leq r$. By Chapter 6, Lemma 1, then $U \cong W \times C_1 \times \cdots \times C_k$ as is required.

If $U = W$ we take $k = 0$. If there exists $u \in U$, $u \notin W$, then $\{u\}$ is an independent set of units. By Lemma 6, if $\{u_1, \ldots, u_k\}$ is any independent set of units, then $\{\lambda(u_1), \ldots, \lambda(u_k)\}$ is an $\mathbb{R}$-linearly independent subset of R^r, thus $k \leq r$. So there exists a maximal independent set of units, say $\{u_1, \ldots, u_k\}$ and $1 \leq k \leq r$.

Let $G = \{(a_1, \ldots, a_k) \in \mathbb{R}^k \mid$ there exists $v \in U$ such that $\lambda(v) = \sum_{i=1}^{k} a_i \lambda(u_i)\}$. G is an additive group, containing the subgroup $\mathbb{Z}^k$. By Lemma 2, $G/\mathbb{Z}^k$ is a finite group; let h be the number of its elements. Let F be the subgroup of U generated by $\{u_1, \ldots, u_k\}$. F is a finitely generated torsion-free Abelian multiplicative group, so F is free of rank k (since $\{u_1, \ldots, u_k\}$ is a maximal subset of independent units in U). We show that for every $u \in U$ we have $u^h = vz$ where $v \in F$, $z \in W$. Indeed, if $u \in F$, we take $v = u$, $z = 1$. If $u \notin F$ then $\{u, u_1, \ldots, u_k\}$ is a dependent set of units, hence by Lemma 6 $\{\lambda(u), \lambda(u_1), \ldots, \lambda(u_k)\}$ is $\mathbb{Q}$-linearly dependent. So there exist $b_1, \ldots, b_k \in \mathbb{Q}$ such that $\lambda(u) = \sum_{i=1}^{k} b_i \lambda(u_i)$.

Let $d > 1$ be the smallest integer such that $db_i = a_i \in \mathbb{Z}$ for all i. Then $\lambda(u^d) = \sum_{i=1}^{k} db_i \lambda(u_i)$ with $(db_1, \ldots, db_k) \in \mathbb{Z}^k$. So d divides the order h of $G/\mathbb{Z}^k$, say $h = de$. Thus $\lambda(u^h) = \sum_{i=1}^{h} hb_i \lambda(u_i)$. Letting $v = \prod_{i=1}^{k} u_i^{hb_i} \in F$ (because $hb_i \in \mathbb{Z}$) then $\lambda(u^h) = \lambda(v)$, thus, by (**J**), $u^h = vz$, with $z \in W$.

Let z_1 be a root of the polynomial $X^h - z$; so $z_1^{hw} = 1$, where $w = \#W$. Let t_i be a root of the polynomial $X^h - u_i$ for each $i = 1, \ldots, k$.

We note that z, t_1, $\ldots$, t_k are complex numbers not necessarily in the field K. We have $u^h = (z_1 \prod_{i=1}^{k} t_i^{hb_i})^h$, hence $u = z_2 z_1 \prod_{i=1}^{k} t_i^{hb_i}$, where $z_2^h = 1$, so $(z_2 z_1)^{hw} = 1$. Let z_3 be a primitive (hw)th root of 1.

Let U' be the multiplicative group of complex numbers, which is generated by z_3, t_1, $\ldots$, t_k. Let W' be the group generated by z_3. Since $W' \cap U$ consists of roots of unity, then $W' \cap U \subseteq W$. Conversely, the elements of W are the wth roots of 1, so they are (wh)th roots of 1, hence $W \subseteq U \cap W'$. Thus $U/W = U/W' \cap U \subseteq U'/W'$ because $U \subseteq U'$. But U'/W' is isomorphic to the free group generated by $\{t_1, \ldots, t_k\}$, which has rank k. Hence the subgroup U/W is also free, of rank at most k. On the other hand, since $\{u_1, \ldots, u_k\}$ is an independent set, then U/W has rank at least k, therefore rank equal to k.

Part 2. We still need to show the existence of sufficiently many independent units. This is basically the content of the following statement:

If $c_1, \ldots, c_r \in \mathbb{R}$ are not all equal to zero, then there exists a unit $u \in U$ such that $\sum_{i=1}^{r} c_i \log |u^{(i)}| \neq 0$.

Proof: We shall require Minkowski's theorem (Chapter 9, **(B)**) in the proof of this statement.

Let $\{a_1, \ldots, a_n\}$ be an integral basis of K, let $d = \det(a_j^{(i)})$. Since d^2 is the discriminant of the field then $1 < d^2 \in \mathbb{Z}$ so $1 < |d|$. Let β be a sufficiently large positive real number, for example, $\beta = (\sum_{i=1}^{r} |c_i|) \log |d| + 1$. We consider the n linear forms

$$L_i = \sum_{j=1}^{n} a_j^{(i)} X_j \qquad (i = 1, \ldots, n).$$

Let $\tau_1, \ldots, \tau_n$ be positive real numbers satisfying the following conditions:

$$\begin{cases} \tau_{r_1+i} = \tau_{r_1+r_2+i} & \text{for } i = 1, \ldots, r_2, \\ \tau_1 \cdots \tau_n = |d|. \end{cases} \tag{10.2}$$

We may choose $\tau_1, \ldots, \tau_r$ arbitrarily and the above relations determine $\tau_{r_1+r_2}$ uniquely. By Chapter 9, **(B)**, there exist integers $x_1, \ldots, x_n$, not all equal to zero, such that if $y^{(i)} = L_i(x_1, \ldots, x_n) = \sum_{j=1}^{n} a_j^{(i)} x_j$ then $|y^{(i)}| \leq \tau_i$ for $i = 1, \ldots, n$. In particular, $y \in A$, $y \neq 0$, and $1 \leq |N_{K|\mathbb{Q}}(y)| \leq |d|$ thus, $\tau_i/|d| = \tau_i/(\tau_1 \cdots \tau_n) \leq 1/\prod_{j \neq i} |y^{(j)}| \leq |y^{(i)}| \leq \tau_i \leq \tau_i |d|$ for every $i = 1, \ldots, n$.

Letting $F(y) = \sum_{i=1}^{r} c_i \log |y^{(i)}|$ we deduce that

$$\left| F(y) - \sum_{i=1}^{r} c_i \log \tau_i \right| = \left| \sum_{i=1}^{r} c_i \log \left(\frac{|y^{(i)}|}{\tau_i} \right) \right| \leq \sum_{i=1}^{r} |c_i| \cdot \left| \log \left(\frac{|y^{(i)}|}{\tau_i} \right) \right|$$

$$\leq \left(\sum_{i=1}^{r} |c_i| \right) \log |d| < \beta$$

because

$$-\log |d| \leq \log \left(\frac{|y^{(i)}|}{\tau_i} \right) \leq 0 \leq \log |d|.$$

Suppose now that for every $h = 1, 2, \ldots$ we choose positive real numbers $\tau_{h1}, \ldots, \tau_{hr}$ satisfying the conditions in Chapter 9, **(B)**, and also the following condition:

$$\sum_{i=1}^{r} c_i \log \tau_{hi} = 2\beta h. \tag{10.3}$$

This is possible since there exists an index i, $1 \leq i \leq r$, for which $c_i \neq 0$.

Let $y_h \in A$, $y_h \neq 0$ be obtained from $\tau_{h1}, \ldots, \tau_{hr}$ in the manner indicated above. Then $|F(y_h) - 2\beta h| = |F(y_h) - \sum_{i=1}^{h} c_i \log \tau_{hi}| < \beta$ so $\beta(2h - 1) < F(y_h) < \beta(2h + 1)$ for all indices $h = 1, 2, \ldots$. Therefore $F(y_1) < F(y_2) < F(y_3) < \cdots$. But $N(Ay_h) = |N_{K|\mathbb{Q}}(y_h)| \leq |d|$, so there exist distinct indices $h \neq h'$ such that $Ay_h = Ay_{h'}$ and therefore $u = y_{h'}/y_h$ is a unit of K. Since $F(y_h) \neq F(y_{h'}) = F(uy_h) = F(u) + F(y_h)$ it follows that $F(u) \neq 0$ so

$$\sum_{i=1}^{r} c_i \log |u^{(i)}| \neq 0. \qquad \blacksquare$$

Part 3.   We conclude the proof of the theorem. By Part 2, if $c_1 = 1, c_2 = \cdots = c_r = 0$ there exists a unit $u_1 \in U$ such that $\log |u_1^{(1)}| \neq 0$.

Now, given $c_1 = -\log |u_1^{(2)}|$, $c_2 = \log |u_1^{(1)}| \neq 0$, $c_3 = \cdots = c_r = 0$ there exists a unit $u_2 \in U$ such that $c_1 \log |u_2^{(1)}| + c_2 \log |u_2^{(2)}| \neq 0$; that is,

$$\det \begin{pmatrix} \log |u_1^{(1)}| & \log |u_2^{(1)}| \\ \log |u_1^{(2)}| & \log |u_2^{(2)}| \end{pmatrix} \neq 0.$$

Repeating this argument, given

$$c_1 = \det \begin{pmatrix} \log |u_1^{(2)}| & \log |u_2^{(2)}| \\ \log |u_1^{(3)}| & \log |u_2^{(3)}| \end{pmatrix},$$

$$c_2 = -\det \begin{pmatrix} \log |u_1^{(1)}| & \log |u_2^{(1)}| \\ \log |u_1^{(3)}| & \log |u_2^{(3)}| \end{pmatrix},$$

$$c_3 = \det \begin{pmatrix} \log |u_1^{(1)}| & \log |u_2^{(1)}| \\ \log |u_1^{(2)}| & \log |u_2^{(2)}| \end{pmatrix} \neq 0,$$

there exists a unit $u_3 \in U$ such that

$$c_1 \log |u_3^{(1)}| + c_2 \log |u_3^{(2)}| + c_3 \log |u_3^{(3)}| \neq 0;$$

that is,

$$\det \begin{pmatrix} \log |u_1^{(1)}| & \log |u_2^{(1)}| & \log |u_3^{(1)}| \\ \log |u_1^{(2)}| & \log |u_2^{(2)}| & \log |u_3^{(2)}| \\ \log |u_1^{(3)}| & \log |u_2^{(3)}| & \log |u_3^{(3)}| \end{pmatrix} \neq 0.$$

In this way, we determine r units $u_1, \ldots, u_r$. Since the determinant $\det(\log |u_j^{(i)}|) \neq 0$, then no column is identically zero, so each u_j is not a root of unity (by (**B**)); moreover, the column vectors are linearly independent over $\mathbb{R}$, hence, by Lemma 6, the units $u_1, \ldots, u_r$ are independent. This shows that we have $k = r$ in Part 1, concluding the proof of the theorem. ∎

Explicitly, Dirichlet's theorem on units says that there exist a root of unity ζ and r units of infinite order $u_1, \ldots, u_r$, such that every unit u may be written uniquely in the form $u = \zeta^{e_0} u_1^{e_1} \cdots u_r^{e_r}$ with $0 \leq e_0 < w$ and $e_1, \ldots, e_r \in \mathbb{Z}$.

Definition 1.   Any set of r independent units $\{u_1, \ldots, u_r\}$ of K (with $r = r_1 + r_2 - 1$) for which the above statement holds is called a *fundamental system of units of K.*

Now, we may introduce a new numerical invariant:

K.   *Let $\{u_1, \ldots, u_r\}$, $\{v_1, \ldots, v_r\}$ be any two fundamental systems of units of K. Then*

$$|\det(\log |u_j^{(i)}|)| = |\det(\log |v_j^{(i)}|)|.$$

Proof: By the theorem, we may write $v_j = \zeta^{b_j} u_1^{a_{1j}} u_2^{a_{2j}} \cdots u_r^{a_{rj}}$ for every $j = 1, \ldots, r$ where $b_j, a_{ij} \in \mathbb{Z}$. Similarly, we may write $u_j = \zeta^{b_j'} v_1^{a_{1j}'} v_2^{a_{2j}'} \cdots v_r^{a_{rj}'}$ for every $j = 1, \ldots, r$. Therefore, by the uniqueness of representation of units, the matrix (a_{ij}') is the inverse of (a_{ij}), and so $\det(a_{ij}) \cdot \det(a_{ij}') = 1$, hence $|\det(a_{ij})| = |\det(a_{ij}')| = 1$. But considering the conjugates of the units, their absolute values, and logarithms, we obtain

$$\log |v_j^{(i)}| = \sum_{h=1}^{r} a_{hj} \log |u_h^{(i)}|$$

and therefore

$$|\det(\log |v_j^{(i)}|)| = |\det(\log |u_j^{(i)}|)|. \qquad ∎$$

We introduce therefore the following concept:

Definition 2.   Let $u_1, \ldots, u_r$ be any fundamental system of units of K. The positive real number

$$R = |\det(\log |u_j^{(i)}|)|$$

is called the *regulator* of K.

For example, if $K = \mathbb{Q}(\sqrt{3})$, we have $r = r_1 + r_2 - 1 = 1$, and as computed before, $u_1 = 2 + \sqrt{3}$ is a fundamental unit. Then $R = |\log(2 + \sqrt{3})|$.

The computation of the regulator is difficult in practice, since it requires the knowledge of a fundamental system of units, which is usually hard to determine.

E X E R C I S E S

1. Let K be an algebraic number field of degree 4. Determine the roots of unity of K.

2. Let K be an algebraic number field of odd degree. Show that K has only two roots of unity: 1, -1.

3. Using properties of Euler's function, show that an algebraic number field contains only finitely many roots of unity.

Hint: If $n = [K : \mathbb{Q}]$ show that for every large d we have $\varphi(d) > n$, so K cannot contain a dth root of unity.

4. Let K be an algebraic number field which is different from $\mathbb{Q}$ and not an imaginary quadratic field. Prove that for every real number $\varepsilon > 0$, there exists an algebraic integer $x \in K$ such that $0 < |x| < \varepsilon$.

5. Determine the fundamental units of the following quadratic fields $\mathbb{Q}(\sqrt{d})$, where:
    (a) $d = 10$.
    (b) $d = 14$.
    (c) $d = 19$.
    (d) $d = 23$.

6. Determine the units of the field $\mathbb{Q}(\sqrt[3]{3})$.

7. Show that $u = 1 - 6\sqrt[3]{6} + 3\sqrt[3]{36}$ is a fundamental unit of $\mathbb{Q}(\sqrt[3]{6})$.

8. Let α be a real number. Show that there exists a pair of relatively prime integers x, y such that $|x/y - \alpha| < 1/y^2$. Prove that if α is rational then there are only finitely many pairs (x, y) with the above properties.

9. If α is a real irrational number, show that there exist an infinite number of pairs of relatively prime integers x, y such that $|x/y - \alpha| < 1/y^2$.

10. Let d be a natural number, not a square. Show that there exist infinitely many pairs of natural numbers x, y such that $|x^2 - dy^2| < 1 + 2\sqrt{d}$.

11. Let d be a square-free natural number. Show that if $u = x + y\sqrt{d}$ is a unit of $\mathbb{Q}(\sqrt{d})$ such that $x > \frac{1}{2}y^2 - 1$ then u is the fundamental unit.

12. Given any natural numbers y_1 and a, let $d = a(ay_1^2 + 2)$. Show that $(1 + ay_1^2) + y_1\sqrt{d}$ is the fundamental unit of $\mathbb{Q}(\sqrt{d})$. Deduce that for every natural number y_1 there exist infinitely many real quadratic fields whose fundamental unit is of the form $x + y_1\sqrt{d}$.

13. Let d be a natural number, not a square. Show that if there exist integers x, y such that $x^2 - dy^2 = -1$ then every odd prime factor of d is congruent to 1 modulo 4. However, verify (for $d = 34$) that the converse is not true.

14. Let p be an odd prime. Prove:

(a) $X^2 - pY^2 = 2$ is solvable if and only if $p \equiv 7 \pmod 8$.

(b) $X^2 - pY^2 = -2$ is solvable if and only if $p \equiv 3 \pmod 8$.

15. Let $d > 2$ be a natural number, not a square. Show that at most one of the equations

$$X^2 - dY^2 = -1,$$
$$X^2 - dY^2 = 2,$$
$$X^2 - dY^2 = -2,$$

has solutions in integers.

16. Let $m > 1$, $l > 1$ be relatively prime integers, let ζ be a primitive mth root of unity, and let ξ be a primitive lth root of unity. Show that $1 - \zeta\xi$ is a unit in the ring of integers of the field $\mathbb{Q}(\zeta, \xi)$.

17. Let ζ be a primitive fifth root of unity. Show, with the methods of this chapter, that the class number of $\mathbb{Q}(\zeta)$ is equal to 1.

18. Let $K = \mathbb{Q}(\sqrt{d})$ be a quadratic field. Show that the number h_0 of strictly equivalent classes of ideals of K is finite (see Chapter 8, Exercise 15). Moreover, $h_0 = h$ when $d < 0$ or $d > 0$ and the fundamental unit has negative norm. Otherwise, $h_0 = 2h$.

19. Show that if $m^2 - 1 > 0$ is square-free then the fundamental unit of the real quadratic field $\mathbb{Q}(\sqrt{m^2 - 1})$ is $\varepsilon = m + \sqrt{m^2 - 1}$.

20. Let ζ be a primitive mth root of unity, and let $K = \mathbb{Q}(\zeta)$, $K_0 = K \cap \mathbb{R}$. Show that a fundamental system of units of K_0 is a maximal independent system of units of K.

21. Let p be an odd prime and ζ a primitive pth root of unity. Show that

$$p = (-1)^{(p-1)/2} \prod_{j=1}^{(p-1)/2} (\zeta^j - \zeta^{-j})^2.$$

Hint: Use the fact that ζ^2 is also a primitive pth root of unity and express p in terms of ζ^2 and its powers.

22. Let p, q be distinct odd primes, let ζ be a primitive pth root of unity, and let A be the ring of algebraic integers of $\mathbb{Q}(\zeta)$. Show that:

(a) $p^{(q-1)/2} \equiv (-1)^{\frac{p-1}{2} \cdot \frac{q-1}{2}} \prod_{j=1}^{(p-1)/2} (\zeta^{jq} - \zeta^{-jq})/(\zeta^j - \zeta^{-j})$ (mod Aq).

(b) $\prod_{j=1}^{(p-1)/2}(\zeta^{jq} - \zeta^{-jq})/(\zeta^j - \zeta^{-j}) = (-1)^r$, where r is the number of integers kq in the set $\left\{ q, 2q, \ldots, \dfrac{p-1}{2}q \right\}$ such that $kq \equiv -s$ (mod p), with $0 < s \le (p-1)/2$.

(c) Prove anew Gauss' reciprocity law.

Hint: For (a) note that $(\zeta^j - \zeta^{-j})^q \equiv \zeta^{jq} - \zeta^{-jq}$ (mod Aq); for (c) make use of Euler's and Gauss' criteria (Chapter 2, **(G)** and **(I)**).

23. Show that if the algebraic number field K contains a nonreal root of unity then $N_{K|\mathbb{Q}}(x) > 0$ for every $x \in K$, $x \ne 0$.

24. Let K be an algebraic number field of degree n. Let U_1 be the group of all units of K having norm $N_{K|\mathbb{Q}}(u) = 1$. Prove:

(a) If n is odd there is a fundamental system of units $\{u_1, \ldots, u_r\}$ of K, such that every unit $u \in U_1$ may be written uniquely in the form $u = u_1^{e_1} \cdots u_r^{e_r}$ with $e_i \in \mathbb{Z}$.

(b) If n is even, if $\{u_1, \ldots, u_r\}$ is a fundamental system of units of K and k, $0 \le k \le r$, is such that $N_{K|\mathbb{Q}}(u_i) = 1$ for $i = 1, \ldots, k$, $N_{K|\mathbb{Q}}(u_i) = -1$ for $i = k+1, \ldots, r$, let $v_i = u_i$ for $i = 1, \ldots, k$, $v_i = u_i u_r$ for $i = k+1, \ldots, r$. Then every unit $u \in U_1$ may be written uniquely in the form $u = \zeta v_1^{e_1} \cdots v_r^{e_r}$ ($e_i \in \mathbb{Z}$), where ζ is any root of unity in K.

25. Let S be a finite set of $s \ge 0$ nonzero prime ideals of the ring A of algebraic integers of K. Let S be the multiplicative set, complement in A of the union $\bigcup_{P \in S} P$. The units of the ring A_S (Chapter 12, Section 1) are called the S-*units* of A. Let U_S denote the group of S-units of A. Prove the analogue of Dirichlet's theorem: U_S/W is a free Abelian group of rank $r + s$, where $r = r_1 + r_2 - 1$.

26. Let $n \ge 2$ be a power of a prime, let ζ be a primitive nth root of unity, $K = \mathbb{Q}(\zeta)$, $K_0 = K \cap \mathbb{R}$. Show that every unit of K is the product of a unit of K_0 with a root of unity.

Hint: Generalize the method of proof of **(I)**.

27. With the notation of the text and assuming $r \geq 1$ show:

(a) There exist units $u_1, \ldots, u_r$ such that for all $j = 1, \ldots, r$ and $i = 1, \ldots, r + 1$, $\quad |u_j^{(i)}| < 1$ when $i \neq j$ and $|u_j^{(j)}| > 1$.

(b) Show that $\{u_1, \ldots, u_r\}$ is a fundamental system of units.

11

Extension of Ideals

In this chapter, we begin the study of extensions of ideals. Let K be an algebraic number field, let $L|K$ be an extension of finite degree, and let A (respectively, B), be the rings of algebraic integers of K (respectively, L). Let I be any nonzero fractional ideal of K. The aim of this study is to relate the decomposition of I into prime ideals of A, with the decomposition into prime ideals of B, of the fractional ideal of L generated by I.

We study this problem, whenever feasible, in a more general situation.

11.1 Extension of Ideals

Let A be a Dedekind domain, K its field of quotients; let $L|K$ be a separable extension of degree n and B the integral closure of A in L. By Chapter 7, (**P**), B is a Dedekind domain.

If I is any fractional ideal of A, let $B.I$ (or also BI) denote the ideal of B generated by I; it consists of all sums $\sum_{i=1}^{m} b_i x_i$ with $m \geq 1$, $b_i \in B$, $x_i \in I$ for all $i = 1, \ldots, m$.

We note that if I, J are fractional ideals in K, then $B.(I \cdot J) = (B.I) \cdot (B.J)$—as follows at once from the definitions.

We begin by proving the following easy fact:

A.   *Let I be a fractional ideal of K. Then $B.I \cap K = I$. If $I \subseteq A$, then $BI \cap A = I$.*

Proof: It is trivial if $I = 0$, so let $I \neq 0$. We consider the fractional ideal I^{-1}, then

$$B = B.A = B.(I \cdot I^{-1}) = (B.I) \cdot (B.I^{-1}),$$

hence $A = B \cap K = [(B.I) \cdot (B.I^{-1})] \cap K \supseteq (B.I \cap K) \cdot (B.I^{-1} \cap K)$. We note that $B.I \cap K$ is also a fractional ideal. Therefore $B.I^{-1} \cap K \subseteq (B.I \cap K)^{-1}$. But from $I \subseteq B.I \cap K$, it follows that $(B.I \cap K)^{-1} \subseteq I^{-1} \subseteq B.I^{-1} \cap K$ and therefore $(B.I \cap K)^{-1} = I^{-1}$, so $B.I \cap K = I$. The last assertion is now obvious. ∎

In particular, if $I \neq A$ then $B.I \neq B$.

Let I be any nonzero fractional ideal of A, so there exists $a \in A$, $a \neq 0$, such that $Aa \cdot I$ is an integral ideal. Let $Aa \cdot I = \prod_{i=1}^{r} P_i^{k_i}$, $Aa = \prod_{i=1}^{r} P_i^{h_i}$ where $P_1, \ldots, P_r$ are distinct nonzero prime ideals of A and $k_i \geq 0$, $h_i \geq 0$ for $i = 1, \ldots, r$. Then $Ba \cdot BI = \prod_{i=1}^{r} BP_i^{k_i}$, $Ba = \prod_{i=1}^{r} BP_i^{h_i}$. If one knows the decomposition of BP as a product of prime ideals of B, for each nonzero prime ideal P of A, then the decompositions of $Ba \cdot BI$ and of Ba are also known, hence so is that of BI.

If P is a nonzero prime ideal of A then $B.P$ may be written in a unique way as a product of powers of prime ideals of B:

$$B.P = \prod_{i=1}^{g} Q_i^{e_i}. \tag{11.1}$$

B.   *With the above notations, let Q be a nonzero prime ideal of B. Then $Q \cap A = P$ if and only if $Q \in \{Q_1, \ldots, Q_g\}$.*

Proof: If $Q \cap A = P$ then $Q \supseteq B.P$, so Q divides $B.P$, hence $Q \in \{Q_1, \ldots, Q_g\}$. Conversely, $Q_i \cap A \supseteq B.P \cap A = P$. Since $Q_i \cap A \neq A$ and P is a maximal ideal, then $Q_i \cap A = P$ for every $i = 1, \ldots, g$. ∎

We assume that the relation (11.1) holds and introduce the following terminology:

Definition 1. g is called the *decomposition number* of P in the extension $L|K$. If necessary, we shall use the notation $g_P(L|K)$, or simply g_P.

We have seen in **(A)** that $g \geq 1$, because $B.P \neq B$. If BP is a prime ideal, we say that P is *inert* in $L|K$.

Definition 2.   For every $i = 1, \ldots, g$, e_i is called the *ramification index* of Q_i in $L|K$. If $e_i = 1$ we say that Q_i is *unramified* in $L|K$.

We shall sometimes use the notations $e(Q_i|P)$ or $e_{Q_i}(L|K)$ for the ramification index e_i.

Now we prove:

C.   *B/BP is a vector space over the field A/P of dimension $[B/BP : A/P] \leq [L : K]$.*

Proof: Let $[L : K] = n$ and let $\bar{b}_1, \ldots, \bar{b}_{n+1} \in B/BP$. Then the elements $b_1, \ldots, b_{n+1}$ are linearly dependent over K, hence there exist $a_1, \ldots, a_{n+1} \in A$, not all equal to 0, such that $\sum_{i=1}^{n+1} a_i b_i = 0$. Let J be the ideal of A generated by $a_1, \ldots, a_{n+1}$. So $J \neq 0$, then $J^{-1}J = A$, so $J^{-1}J \not\subseteq P$. Thus there exists $c \in J^{-1}$ such that $cJ \not\subseteq P$, so there exists i, $1 \leq i \leq n + 1$, such that $ca_i \notin P$. Considering the images in A/P and B/BP, we obtain $\sum_{i=1}^{n+1} \bar{c}\bar{a}_i\bar{b}_i = 0$ showing that $\bar{b}_1, \ldots, \bar{b}_{n+1}$ are linearly dependent over A/P, which concludes the proof. ∎

From relation (11.1) and the isomorphism between A/P-vector spaces $(B/BP)/(Q_i/BP) \cong B/Q_i$, it follows that $[B/Q_i : A/P] \leq [L : K]$ for each $i = 1, \ldots, g$.

Definition 3.   The dimension f_i of B/Q_i over A/P is called the *inertial degree* or *residual degree* of Q_i in $L|K$. We use the notation $f_i = f_{Q_i}(L|K) = f(Q_i|P)$.

We shall now study properties of these numbers (decomposition number, ramification index, inertial degree).

First, we establish the transitivity. Let A, B, K, and L be as before. Let $L'|L$ be a separable extension of finite degree, and B' the integral closure of B in L'.

D.   *With the above notations, let Q' be a prime ideal of B', let $Q = Q' \cap B$, $P = Q \cap A$, and assume $P \neq 0$. Then*

$$e(Q'|P) = e(Q'|Q) \cdot e(Q|P),$$

$$f(Q'|P) = f(Q'|Q) \cdot f(Q|P),$$

Proof: For simplicity, let

$$e = e(Q|P), \qquad e' = e(Q'|Q), \qquad e'' = e(Q'|P),$$
$$f = f(Q|P), \qquad f' = f(Q'|Q), \qquad f'' = f(Q'|P).$$

Since Q^e divides BP, but Q^{e+1} does not divide BP, we may write $BP = Q^e \cdot J$ where Q does not divide J. Similarly, $B'Q = Q'^{e'} \cdot J'$ where Q' does not divide J'.

Hence $B'P = B'(BP) = (B'Q)^e \cdot (B'J) = Q'^{e'e} J'^e \cdot (B'J)$ and Q' does not divide $B'J$, otherwise $Q = Q' \cap B \supseteq B'J \cap B \supseteq J$, contrary to the assumption. Therefore $Q'^{e'e}$ is the exact power of Q' dividing $B'P$, so $e'' = ee'$.

Similarly, by definition, $f = [B/Q : A/P]$, $f' = [B'/Q' : B/Q]$, thus

$$f'' = [B'/Q' : A/P] = ff'. \qquad \blacksquare$$

In the same manner, we note that if $BP = \prod_{i=1}^{g} Q_i^{e_i}$, if g_i' is the decomposition number of Q_i in $L'|L$, then the decomposition number of P in $L'|K$ is $g'' = \sum_{i=1}^{g} g_i'$.

A notable simplification arises in the important case of a Galois extension.

Let $L|K$ be a Galois extension of degree n, let G be its Galois group, so the elements of G leave each element of K fixed, and transform any element of L into its conjugates; in particular, $\sigma(B) \subseteq B$ for every $\sigma \in G$, hence also $B = \sigma(\sigma^{-1}(B)) \subseteq \sigma(B)$, showing that $\sigma(B) = B$. If J is any ideal of B then $\sigma(J)$ is an ideal of B and $\sigma(J) \cap A = J \cap A$. Therefore σ induces a ring-isomorphism $\overline{\sigma} : B/J \to B/\sigma(J)$, namely, if $\overline{b} \in B/J$ then $\overline{\sigma}(\overline{b}) = \overline{\sigma(b)}$ (and this is indeed well defined); $\overline{\sigma}$ leaves fixed every element of $A/(J \cap A)$.

In particular, if Q is a nonzero prime ideal of B, then $B/Q \cong B/\sigma(Q)$ and $\sigma(Q)$ is also a nonzero prime ideal of B.

More interesting is the transivity of the action of G on the set of prime ideals of B, having a given intersection with A:

E.   *If Q, Q' are any prime ideals of B such that $Q \cap A = Q' \cap A \neq 0$, there exists $\sigma \in G$ such that $\sigma(Q) = Q'$.*

Proof: Let $G = \{\sigma_1, \dots, \sigma_n\}$ be the Galois group of $L|K$, and let us assume that $Q' \neq \sigma_i(Q)$ for every $\sigma_i \in G$. By Chapter 7, (**L**), there exists an element $x \in B$ such that $x \notin \sigma_i(Q)$ for $i = 1, \dots, n$, $x \in Q'$. Let

$$a = \prod_{i=1}^{n} \sigma_i(x),$$

then $a \in A \cap Q'$; however, $a \notin Q$ since each $\sigma_i(x) \notin Q$ for $i = 1, \dots, n$, (otherwise $x = \sigma_i^{-1}\sigma_i(x) \in \sigma_i^{-1}(Q)$, for some index i). This is a contradiction, hence there exists $\sigma_i \in G$ such that $\sigma_i(Q) = Q'$.   ∎

As a corollary of (**E**), we have:

F.   *If $L|K$ is a Galois extension of degree n, if $BP = \prod_{i=1}^{g} Q_i^{e_i}$, and if $[B/Q_i : A/P] = f_i$, then $e_1 = \cdots = e_g$, $f_1 = \cdots = f_g$, and if A/P is a finite field, then each B/Q_i is isomorphic to the extension of degree f_i of A/P.*

Proof: Let $BP = \prod_{i=1}^{g} Q_i^{e_i}$; for every index j, $1 \leq j \leq g$, by (**E**), there exists $\sigma \in G$ such that $\sigma(Q_1) = Q_j$. Hence from $BP = \sigma(BP) = \prod_{i=1}^{g} \sigma(Q_i)^{e_i}$ and the uniqueness of the decomposition of BP into a product of prime ideals, it follows that $e_j = e_1$ for every j, $1 \leq j \leq g$. Similarly, from $B/Q_j = B/\sigma(Q_1) \cong B/Q_1$ it follows that $f_j = f_1$ for every j, $1 \leq j \leq g$.

If A/P is finite, there exists only one extension of degree f_1, up to isomorphism, hence all the fields B/Q_i are isomorphic.   ∎

The results indicated hold in the particular case of rings of integers of algebraic number fields of finite degree and their extensions.

Thus, if K is an algebraic number field of degree $[K : \mathbb{Q}] = n$, if A is the ring of integers of K, if p is any prime number, then

$$Ap = \prod_{i=1}^{g} P_i^{e_i}, \tag{11.2}$$

where $P_1, \dots, P_g$ are distinct prime ideals of A, $e_i = e_{P_i}$ are the ramification indices, and $f_i = [A/P_i : \mathbb{Z}/\mathbb{Z}p]$ are the inertial degrees (for $i = 1, \dots, g$).

We have the following fundamental relation:

G.   *With the above notations:*

$$n = \sum_{i=1}^{g} e_i f_i.$$

Proof: Taking the norms of the ideals in (11.2), by Chapter 8, (**D**), (**F**), we have

$$|N_{K|\mathbb{Q}}(p)| = \prod_{i=1}^{g} N(P_i)^{e_i},$$

and since $N_{K|\mathbb{Q}}(p) = p^n$, and $N(P_i) = \#(A/P_i) = p^{f_i}$ then $n = \sum_{i=1}^{g} e_i f_i$. ∎

We note that $g \le n$. If $g = n$, we say that the prime p is *totally decomposed* or *splits completely* in the extension $K|\mathbb{Q}$. In this case, $e_i = f_i = 1$ for each $i = 1, \dots, n$.

We note explicitly the particular case where $K|\mathbb{Q}$ is a Galois extension:

$$n = efg,$$

where e is the ramification index and f is the inertial degree of any prime ideal P_i dividing Ap.

We note:

H.   *If Q is a prime ideal of B, $Q \cap A = P$, if $e \ge 1$ then $[B/Q^e : A/P] = e[B/Q : A/P]$.*

Proof: Let $P \cap \mathbb{Z} = \mathbb{Z}p$. Taking norms

$$N(Q^e) = \#(B/Q^e) = p^{[B/Q^e : \mathbb{F}_p]} = p^{f[B/Q^e : A/P]},$$

where $f = [A/P : \mathbb{F}_p]$. But

$$N(Q^e) = N(Q)^e = \#(B/Q)^e = p^{e[B/Q : \mathbb{F}_p]} = p^{ef[B/Q : A/P]}.$$

Hence $[B/Q^e : A/P] = e[B/Q : A/P]$. ∎

Theorem 1.   *With the above notations $[B/BP : A/P] = [L : K]$ and $[L : K] = \sum_{i=1}^{g} e_i f_i$.*

Proof: Let $P \cap \mathbb{Z} = \mathbb{Z}p$ where p is a prime number. Let $Ap = \prod_{i=1}^{g'} P_i^{e_i'}$ where $P_1 = P$, P_2, ..., $P_{g'}$ are distinct prime ideals of A, $e_i' \ge 1$ for all $i = 1, \dots, g'$ and $f_i' = [A/P_i : \mathbb{F}_p]$. From (**G**), $[K : \mathbb{Q}] = \sum_{i=1}^{g'} e_i' f_i'$. We have $Bp = \prod_{i=1}^{g'} BP_i^{e_i'}$, hence by Chapter 7, Theorem 2, $B/Bp = \prod_{i=1}^{g'} B/BP_i^{e_i'}$.

Let $n_i = [B/BP_i : A/P_i]$; by (C), $n_i \leq [L : K]$ for each $i = 1, \ldots, g'$. Counting the number of elements, we have

$$p^{[L:\mathbb{Q}]} = |N_{L|\mathbb{Q}}(p)| = N(Bp) = \#(B/Bp)$$

$$= \prod_{i=1}^{g'} \#(B/BP_i)^{e_i'} = \prod_{i=1}^{g'} \#(A/P_i)^{e_i' n_i} = \prod_{i=1}^{g'} p^{f_i' e_i' n_i},$$

so by (G):

$$[L : \mathbb{Q}] = \sum_{i=1}^{g'} n_i e_i' f_i' \leq [L : K] \sum_{i=1}^{g'} e_i' f_i'$$

$$= [L : K][K : \mathbb{Q}] = [L : \mathbb{Q}].$$

This implies that $n_i = [L : K]$ for each $i = 1, \ldots, g'$; in particular, $[B/BP : A/P] = [L : K]$. But by Chapter 7, Theorem 2, $B/BP \cong \prod_{i=1}^{g} B/Q_i^{e_i}$ hence

$$[B/BP : A/P] = \left[\prod_{i=1}^{g} B/Q_i^{e_i} : A/P \right] = \sum_{i=1}^{g} [B/Q_i^{e_i} : A/P]$$

$$= \sum_{i=1}^{g} e_i[B/Q_i : A/P] = \sum_{i=1}^{g} e_i f_i$$

by (H).   ∎

As before, if $g = [K : L]$ then $e_i = f_i = 1$ for each $i = 1, \ldots, g$, and the prime ideal P is said to be *totally decomposed* in $L|K$, or also P *splits completely* in $L|K$.

Our next aim is to determine the decomposition number, ramification indices, and inertial degrees. We need some preliminary considerations. Let K be an algebraic number field and $L|K$ an extension of degree n; let A (respectively, B), be the ring of algebraic integers of K (respectively, L).

We assume that B *is a free* A-module. This is the case (by Chapter 6, Theorem 2) if A is a principal ideal domain, say when the ground field is $\mathbb{Q}$.

If $\{x_1, \ldots, x_n\}$ is a basis of the free A-module B, then $\mathrm{discr}_{L|K}(x_1, \ldots, x_n) = [\det(\sigma_i(x_j))]^2 \in A$, where $\sigma_1, \ldots, \sigma_n$ are the K-isomorphisms of L into $\mathbb{C}$. Moreover, if $\{x_1', \ldots, x_n'\}$ is any other basis of the A-module B, then the principal ideals coincide

$$A\,\mathrm{discr}_{L|K}(x_1, \ldots, x_n) = A\,\mathrm{discr}_{L|K}(x_1', \ldots, x_n').$$

Let $t \in B$ be a primitive element, so $L = K(t)$ and $A[t] \subseteq B$. If $\{x_1, \ldots, x_n\}$ is an A-basis of B, we may write $t^j = \sum_{i=1}^{n} c_{ij} x_i$ for $j = 0, 1, \ldots, n - 1$ where $c_{ij} \in A$, hence

$$\mathrm{discr}_{L|K}(1, t, \ldots, t^{n-1}) = [\det(c_{ij})]^2 \cdot \mathrm{discr}_{L|K}(x_1, \ldots, x_n).$$

Let $\alpha = \det(c_{ij})$ so $A\alpha$ depends on t but not on the basis $\{x_1, \ldots, x_n\}$ according to the preceding remark. We note that if $B = A[t]$, then $A\alpha = A$, since $\{1, t, \ldots, t^{n-1}\}$ is then an A-basis of B.

I.   *Let A be a principal ideal domain, and let P be a prime ideal of A such that P does not divide $A\alpha$. Then every element $y \in B$ may be written in the form $y = z/a$, where $z \in A[t]$, $a \in A$, $a \notin P$.*

Proof: If $y \in B$ we may write $y = \sum_{j=0}^{n-1} y_j t^j$ with $y_j = a_j/a_j' \in K$ and $a_j, a_j' \in A$. We may also assume that a_j, $a_j{'}$ are relatively prime elements of the principal ideal domain A.

Applying the K-isomorphisms σ_i to the above relation, we obtain

$$\sigma_i(y) = \sum_{j=0}^{n-1} y_j \sigma_i(t^j) \qquad (i = 1, \ldots, n).$$

Let L' be the smallest Galois extension of K containing L. This means that $\{y_0, \ldots, y_{n-1}\}$ is a solution of the system of linear equations in L', having coefficients $\sigma_i(t^j)$. By Cramer's rule, we have

$$y_j = \sqrt{\frac{\operatorname{discr}_{L|K}(1, t, \ldots, y, \ldots, t^{n-1})}{\operatorname{discr}_{L|K}(1, t, \ldots, t^j, \ldots, t^{n-1})}}.$$

But $A + At + \cdots + Ay + \cdots + At^{n-1} \subseteq B = \sum_{i=1}^{n} Ax_i$ and $\sum_{j=0}^{n-1} At^j \subseteq B = \sum_{i=1}^{n} Ax_i$, so, expressing the generators of the smaller module in terms of the integral basis with coefficients in A, we deduce that

$$\operatorname{discr}_{L|K}(1, t, \ldots, y, \ldots, t^{n-1}) = [\det(c_{ik}^{(j)})]^2 \cdot \operatorname{discr}_{L|K}(x_1, \ldots, x_n),$$

$$\operatorname{discr}_{L|K}(1, t, \ldots, t^j, \ldots, t^{n-1}) = [\det(c_{ik})]^2 \cdot \operatorname{discr}_{L|K}(x_1, \ldots, x_n),$$

with $c_{ik}, c_{ik}^{(j)} \in A$. We conclude that

$$\frac{a_j}{a_j'} = y_j = \frac{\det(c_{ik}^{(j)})}{\det(c_{ik})}.$$

So $a_j' \cdot \det(c_{ik}^{(j)}) = a_j \cdot \det(c_{ik})$ and a_j' divides $\det(c_{ij}) = \alpha$. Thus if P is a prime ideal of A and P does not divide $A\alpha$, then P does not divide Aa_j' for $j = 0, 1, \ldots, n - 1$. Hence, $y = \sum_{j=0}^{n-1}(a_j/a_j') \cdot t^j = z/a$ with $z \in A[t]$, $a \in A$, $a \notin P$. $\blacksquare$

The main result below is valid under the hypothesis indicated.

Let K be an algebraic number field and let $L|K$ be an extension of degree n; let A (respectively, B) be the ring of integers of K (respectively, L). Let $L = K(t)$, with $t \in B$, and let $F \in A[X]$ be the minimal polynomial of t over K. If P is a nonzero prime ideal of A and $\overline{K} = A/P$, for each polynomial $H \in A[X]$, let $\overline{H} \in \overline{K}[X]$ be obtained by the canonical homomorphism $A \to A/P = \overline{K}$.

Theorem 2.   *With the above notations, we assume that one of the following conditions is satisfied:*

(a) *A is a principal ideal domain.*

(b) *$B = A[t]$.*

Let P be a nonzero prime ideal of A, such that, in case (a), P does not divide $A\alpha$ (where α was previously defined). Let $\overline{F} = \prod_{i=1}^{g} \overline{G}_i^{e_i}$ where $G_i \in A[X]$, the polynomials $\overline{G}_1, \ldots, \overline{G}_g$ are distinct and irreducible over $\overline{K}$, $\deg(\overline{G}_i) = f_i$ for $i = 1, \ldots, g$. Then $BP = \prod_{i=1}^{g} Q_i^{e_i}$ where $Q_1, \ldots, Q_g$ are distinct nonzero prime ideals of B and $[B/Q_i : A/P] = f_i$ for every $i = 1, \ldots, g$. Moreover, $Q_i \cap A[t] = A[t]P + A[t]G_i(t)$ for $i = 1, \ldots, g$.

Proof: We consider the following sets:

$$\mathcal{G} = \{\overline{G} \in \overline{K}[X] | \overline{G} \text{ is irreducible and divides } \overline{F}\},$$

$$\mathcal{Q}^{(t)} = \{Q^{(t)} | Q^{(t)} \text{ is a prime ideal of } A[t] \text{ and } Q^{(t)} \cap A = P\},$$

$$\mathcal{Q} = \{Q \,|\, Q \text{ is a prime ideal of } B \text{ and } Q \text{ divides } BP\}.$$

By (**B**), the last set is equal to

$$\{Q \,|\, Q \text{ is a prime ideal of } B, \ Q \cap A = P\}.$$

For each $Q \in \mathcal{Q}$, $Q \cap A[t] \in \mathcal{Q}^{(t)}$. We show that $Q \mapsto Q \cap A[t]$ is a bijection from $\mathcal{Q}$ to $\mathcal{Q}^{(t)}$. Indeed, let $Q, Q' \in \mathcal{Q}$. If $y \in Q$, by (**I**), we may write $y = z/a$ where $a \in A$, $a \notin P$, $z \in A[t]$. Then $z = ay \in Q$ so $z \in A[t] \cap Q = A[t] \cap Q' \subseteq Q'$, and from $a \notin P = A \cap Q'$, $ay = z \in Q'$, then $y \in Q'$; and conversely, so $Q = Q'$.

Next we show that if $Q^{(t)} \in \mathcal{Q}^{(t)}$ there exists $Q \in \mathcal{Q}$ such that $Q \cap A = Q^{(t)}$. Indeed, let

$$Q = \left\{ \frac{z}{a} \,\middle|\, z \in Q^{(t)}, \ a \in A, \ a \notin P \right\}.$$

It is immediate to check that Q is a prime ideal of B (using (**I**)) $Q \cap A[t] = Q^{(t)}$. Indeed, if $z/a \in Q \cap A[t]$ with $z \in Q^{(t)}$, $a \in A$, $a \notin P$, then $z = a \cdot z/a \in Q^{(t)}$, $a \notin Q^{(t)}$, so $z/a \in Q^{(t)}$. Thus $Q \cap A = Q^{(t)} \cap A = P$ and $Q \in \mathcal{Q}$.

We define a mapping Φ from $\mathcal{Q}^{(t)}$ to $\mathcal{G}$. Given $Q^{(t)}$ then $A[t]/Q^{(t)} = B/Q$ (where $Q \cap A[t] = Q^{(t)}$) as already seen. By the canonical homomorphism $A[t] \to B/Q$, we have $\overline{F}(\overline{t}) = 0$, where $\overline{F} \in \overline{K}[X]$ and $\overline{t} = t + Q \in B/Q$. Let $\overline{G} \in \overline{K}[X]$ be the minimal polynomial of $\overline{t}$ over $\overline{K}$, so $\overline{G}$ is irreducible over $\overline{K}$ and $\overline{G}$ divides $\overline{F}$, thus $\overline{G} \in \mathcal{G}$. By definition, $\Phi(Q^{(t)}) = \overline{G}$. Now, given $\overline{G} \in \mathcal{G}$, let θ be a root of $\overline{G}$ and let $\Psi : A[t] \to \overline{K}(\theta) = \overline{K}[\theta]$ be defined by $\Psi(H(t)) = \overline{H}(\theta)$ for every $H \in A[X]$. Then Ψ is a ring-homomorphism; let $Q^{(t)}$ be the kernel of Ψ, so $Q^{(t)}$ is a prime ideal of $A[t]$ and $Q^{(t)} \cap A = P$, thus $Q^{(t)} \in \mathcal{Q}^{(t)}$. Since the canonical homomorphism $A[t] \to A[t]/Q^{(t)}$, has also the kernel $Q^{(t)}$, then $A[t]/Q^{(t)} \cong \overline{K}(\theta)$, so $\overline{t} = t + Q^{(t)}$, θ have the same minimal polynomial $\overline{G}$, thus $\Phi(Q^{(t)}) = \overline{G}$ showing that Φ is surjective. If $Q^{(t)}$, $Q'^{(t)}$ are prime ideals in the set $\mathcal{Q}^{(t)}$ and their images are both equal to the polynomial $\overline{G} \in \mathcal{G}$, if $\theta = t + Q^{(t)} \in A[t]/Q^{(t)}$, $\theta' = t + Q'^{(t)} \in A[t]/Q'^{(t)}$, then $\overline{K}(\theta) \cong \overline{K}(\theta')$, hence the kernels $Q^{(t)}$, $Q'^{(t)}$ of the canonical homomorphisms $A[t] \to \overline{K}(\theta)$, $A[t] \to \overline{K}(\theta')$ are equal. This shows that the mapping Φ is injective.

Let $BP = \prod_{i=1}^{g'} Q_i^{e_i'}$ with $[B/Q_i : A/P] = f_i'$ for every $i = 1, \ldots, g'$. It follows from the bijection between $\mathcal{Q}$ and $\mathcal{G}$ that $g' = g$.

Let Q_i correspond to $\overline{G}_i$ by the bijection. The proof also gave $f_i' = f_i$, since $f_i' = [B/Q_i : A/P] = \deg(\overline{G}_i) = f_i$ for every $i = 1, \ldots, g$.

Now we show that $Q_i \cap A[t]$ is the ideal of $A[t]$ generated by P and $G_i(t)$, where $G_i \in A[X]$ and $\overline{G}_i$ is the image of G_i modulo P. Clearly $Q_i \supseteq A[t]P + A[t]G_i(t)$ since $\overline{G}_i(t + Q_i) = 0$. Conversely, let $z \in A[t] \cap Q_i$. Then $z = H(t)$ with $H \in A[X]$ and $0 = z + Q_i = \overline{H}(t + Q_i)$, hence $\overline{G}_i$ divides $\overline{H}$ and therefore $\overline{H} = \overline{G}_i\overline{K}$, so $H = G_iK + M$ with $K \in A[X]$ and $M \in P[X]$. Thus $z = H(t) = G_i(t)K(t) + M(t) \in A[t]G_i[t] + A[t]P$.

We need to show that $e_i' = e_i$ for $i = 1, \ldots, g$. We have

$$n = \sum_{i=1}^{g} e_i'f_i' = \sum_{i=1}^{g} e_i'f_i$$

(by Theorem 1) and also $n = \sum_{i=1}^{g} e_if_i$, since

$$\overline{F} = \prod_{i=1}^{g} \overline{G}_i^{e_i} \quad \text{and} \quad \deg(\overline{G}_i) = f_i \quad (\text{for } i = 1, \ldots, g).$$

It suffices therefore to show that $e_i' \leq e_i$ for $i = 1, \ldots, g$.

First we show that for each $i = 1, \ldots, g$ there exists $G_i' \in A[X]$ such that $\overline{G}_i' = \overline{G}_i$ and $G_i'(t) \in Q_i$, $G_i'(t) \notin Q_i^2$.

Let $y \in Q_i$, $y \notin Q_i^2$. We may write $y = z/a$ with $z \in A[t]$, $a \in A$, $a \notin P$. Then $z = ay \in A[t] \cap Q_i$ but $z \notin A[t] \cap Q_i$. But $A[t] \cap Q_i = A[t]P + A[t]G_i(t)$ so $z = M(t) + H(t)G_i(t)$ where $M \in P[X]$, $H \in A[X]$. By the correspondence indicated, $G_i(t) \in Q_i$. If $G_i(t) \notin Q_i^2$, we take $G_i' = G_i$. If $G_i(t) \in Q_i^2$ since $z \notin Q_i^2$ then $M(t) \notin Q_i^2$. We also have $z = M(t)[1 - H(t)] + [G_i(t) + M(t)]H(t)$ and we take $G_i' = G_i + M$. So $\overline{G}_i' = \overline{G}_i$ and $G_i'(t) \in Q_i$, $G_i'(t) \notin Q_i^2$.

From $\overline{F} = \prod_{i=1}^{g} \overline{G}_i^{e_i}$ then

$$F - \prod_{i=1}^{g} G_i'^{e_i} \in P[X].$$

From $F(t) \in P[t]$ then

$$\prod_{i=1}^{g} G'_i(t)^{e_i} \in P[t] \subseteq BP = \prod_{i=1}^{g} Q_i^{e'_i},$$

so $\prod_{i=1}^{g} Q_i^{e'_i}$ divides the principal ideal $BG'_1(t)^{e_1} \cdots G'_g(t)^{e_g}$. Since $G'_i(t) \in Q_i$, $G'_i(t) \notin Q_i^2$, then

$$BG'_1(t)^{e_1} \cdots G'_g(t)^{e_g} = Q_1^{e_1} \cdots Q_g^{e_g} J,$$

where J is an ideal, not divisible by $Q_1 \cdots Q_g$. Thus $\prod_{i=1}^{g} Q_i^{e'_i}$ divides $\prod_{i=1}^{g} Q_i^{e_i}$, hence $e'_i \leq e_i$ for $i = 1, \ldots, g$, which concludes the proof.   ∎

The theorem holds in particular when $A = \mathbb{Z}$. It should be noted here that there exist number fields $K \subset L$ such that there is no primitive element t with $B = A[t]$.

11.2 Decomposition of Prime Numbers in Quadratic Fields

Let $K = \mathbb{Q}(\sqrt{d})$, where d is a square-free integer and let p be a prime number; we shall determine the decomposition of Ap into prime ideals of A.

We have $n = 2$, thus by (**G**), the only possibilities are the following:

(1) $g = 2$, $e_1 = e_2 = 1$, $f_1 = f_2 = 1$. In this case, $Ap = P_1 P_2$, with $P_1 \neq P_2$, $N(P_1) = N(P_2) = p$. We say that p is *totally* (or *completely*) *decomposed* in $\mathbb{Q}(\sqrt{d})$.

(2) $g = 1$, $e = 1$, $f = 2$. Thus $Ap = P$ is a prime ideal, p is *inert* in $\mathbb{Q}(\sqrt{d})$, and $N(P) = N(Ap) = p^2$.

(3) $g = 1$, $e = 2$, $f = 1$. Thus $Ap = P^2$, $N(P) = p$, and p is *ramified* in $\mathbb{Q}(\sqrt{d})$.

By Chapter 7, Theorem 2, in case (1) $Ap = P_1 P_2$ then $A/Ap \cong A/P_1 \times A/P_2$, so A/Ap is a Cartesian product of two fields; in particular, it has no nilpotent elements, except 0. If $Ap = P$, then A/Ap is a field. Finally, if $Ap = P^2$ then A/Ap is a ring having a nonzero ideal P/P^2, which is nilpotent. Since the above cases are mutually exclusive, we have therefore another description of the possible phenomena in terms of the ring A/Ap.

The main question is now the following: Given the prime p, for which values of d do we have cases (1) (respectively, (2), (3))?

From Chapter 5, (**V**), if $d \equiv 2$ or $3 \pmod{4}$, then the ring A of integers of $\mathbb{Q}(\sqrt{d})$ consists of all $a + b\sqrt{d}$, with $a, b \in \mathbb{Z}$. If $d \equiv 1 \pmod{4}$, then A consists of all $(a + b\sqrt{d})/2$, where $a, b \in \mathbb{Z}$ and $a \equiv b \pmod{2}$ or, equivalently, $\mathbb{Z} = \mathbb{Z} + \mathbb{Z}\omega$ where $\omega = (1 + \sqrt{d})/2$; in this case, let $c = (1 - d)/4$.

Let

$$F(X) = \begin{cases} X^2 - d & \text{if } d \equiv 2 \text{ or } 3 \pmod 4, \\ X^2 - X + c & \text{if } d \equiv 1 \pmod 4, \end{cases}$$

so $F(X)$ is the minimal polynomial of $\sqrt{d}$ (respectively, of ω), according to the case.

Let $\theta : \mathbb{Z}[X] \to A$ be the mapping defined by

$$\theta(H) = \begin{cases} H(\sqrt{d}) & \text{if } d \equiv 2 \text{ or } 3 \pmod 4, \\ H(\omega) & \text{if } d \equiv 1 \pmod 4. \end{cases}$$

Then θ induces a ring-isomorphism $\mathbb{Z}[X]/(F) \cong A$, where (F) denotes the principal ideal generated by the polynomial F. Using the well-known isomorphism theorems for rings, we have:

J. $A/Ap \cong \mathbb{F}_p[X]/(\overline{F})$.

Proof: $A/Ap \equiv \mathbb{Z}[X]/(F, p\mathbb{Z}[X]) \cong (\mathbb{Z}[X]/p\mathbb{Z}[X])/((F, p\mathbb{Z}[X])/p\mathbb{Z}[X])$ $\cong \mathbb{F}_p[X]/(\overline{F})$, where $(F, p\mathbb{Z}[X])$ denotes the ideal of $\mathbb{Z}[X]$ generated by F and p, $\overline{F} \in \mathbb{F}_p[X]$ is obtained from F by reducing its coefficients modulo p. ∎

We describe explicitly the above isomorphism.

Given $\overline{H} \in \mathbb{F}_p[X]$ let $H \in \mathbb{Z}[X]$ be any polynomial which gives $\overline{H}$ when its coefficients are reduced modulo p. Let $\psi : \mathbb{F}_p[X] \to A/Ap$ be defined by $\psi(\overline{H}) = \theta(H) + Ap \in A/Ap$. It may be easily checked that if $\overline{H} = \overline{H}'$ then $\theta(H) - \theta(H') \in Ap$, so the mapping ψ is well defined. It is clear that ψ is a ring-homomorphism. Since F is the minimal polynomial of $\sqrt{d}$ (respectively, ω), according to the case, it follows that the kernel of ψ is the principal ideal generated by $\overline{F}$.

Finally, since $\{1, \sqrt{d}\}$ (respectively, $\{1, \omega\}$) is a basis of the $\mathbb{Q}$−vector space $\mathbb{Q}(\sqrt{d})$, then the image of ψ is equal to A/Ap, hence ψ induces the isomorphism $\overline{\psi} : \mathbb{F}_p[X]/(\overline{F}) \to A/Ap$.

K.   *Let p be an odd prime. Then p is ramified in $\mathbb{Q}(\sqrt{d})$ if and only if p divides d; p is inert if and only if d is not a square modulo p, that is, $(d/p) = -1$; p is totally decomposed if and only if d is a square modulo p, that is $(d/p) = 1$.*

Proof: We apply (**J**), and the remarks preceding it.

If $\overline{F}$ is irreducible in $\mathbb{F}_p[X]$, then A/Ap is isomorphic to a domain, so Ap is a prime ideal, that is, p is inert in $\mathbb{Q}(\sqrt{d})$.

If $\overline{F} = \overline{H}_1 \overline{H}_2$ where $\overline{H}_1, \overline{H}_2$ are distinct irreducible polynomials of degree 1, then from $(\overline{H}_1 \overline{H}_2) = (\overline{H}_1) \cap (\overline{H}_2)$, $(\overline{H}_1) + (\overline{H}_2) = \mathbb{F}_p[X]$, we deduce easily (or by Chapter 7, (**K**)) that

$$A/Ap \cong \mathbb{F}_p[X]/(\overline{H}_1 \overline{H}_2) \cong \mathbb{F}_p[X]/(\overline{H}_1) \times \mathbb{F}_p[X]/(\overline{H}_2) \cong \mathbb{F}_p \times \mathbb{F}_p$$

(because $\mathbb{F}_p[X]/(\overline{H}_i)$ is an algebraic extension of degree 1 over $\mathbb{F}_p$); so p is totally decomposed in $\mathbb{Q}(\sqrt{d})$.

If $\overline{F}$ is the square of an irreducible polynomial, $\overline{F} = \overline{H}^2$, then $A/Ap \cong \mathbb{F}_p[X]/(\overline{H}^2)$ and the ideal $(\overline{H})/(\overline{H}^2)$ is nonzero and nilpotent; therefore p is ramified in $\mathbb{Q}(\sqrt{d})$.

In this last situation, let $\overline{H} = X + \overline{a}$. If $d \equiv 2$ or $3 \pmod 4$, then $X^2 - \overline{d} = (X + \overline{a})^2 = X^2 + 2\overline{a}X + \overline{a}^2$, hence $\overline{a} = \overline{0}$ and $-\overline{d} = \overline{a}^2 = \overline{0}$, so $p|d$. Conversely, if $p|d$, then $X^2 - \overline{d} = X^2$, so p is ramified.

If $d \equiv 1 \pmod 4$ letting $c = (1 - d)/4$, then

$$\overline{F} = X^2 - X + \overline{c} = (X + \overline{a})^2 = X^2 + 2\overline{a}X + \overline{a}^2,$$

so $2\overline{a} = -\overline{1}$, $\overline{a}^2 = \overline{c}$, hence $4\overline{c} = \overline{1}$ and $\overline{d} = \overline{0}$, so $p|d$. Conversely, if $p|d$ then

$$4(X^2 - X + c) \equiv 4X^2 - 4X + 1 = (2X - 1)^2 \pmod p,$$

hence $\overline{F}$ is the square of a polynomial, so p is ramified.

If p does not divide d, then p is totally decomposed exactly when $X^2 - \overline{d} = (X + \overline{a})(X + \overline{b})$ with $\overline{a} \neq \overline{b}$, so $\overline{a} + \overline{b} = 0$, $\overline{a}\overline{b} = -\overline{d}$, hence, $\overline{a}^2 = \overline{d}$, that is, $(d/p) = 1$. Thus, by exclusion, p is inert when p does not divide d, and $(d/p) = -1$.   ∎

Let us observe at this moment that the type of decomposition of the odd prime p in $\mathbb{Q}(\sqrt{d})$ depends only on the residue class of d modulo p.

L.   *The prime 2 is ramified in $\mathbb{Q}(\sqrt{d})$ if and only if $d \equiv 2 \pmod 4$ or $d \equiv 3 \pmod 4$; 2 is inert in $\mathbb{Q}(\sqrt{d})$ if and only if $d \equiv 5 \pmod 8$; 2 is totally decomposed in $\mathbb{Q}(\sqrt{d})$ if and only if $d \equiv 1 \pmod 8$.*

Proof: If $d \equiv 2$ or $3 \pmod 4$, then the ring of algebraic integers is $A = \mathbb{Z} + \mathbb{Z}\sqrt{d}$ and $A/2A \cong \mathbb{F}_2[X]/(X^2 - \overline{d})$ by **(J)**. Since $\overline{d} = \overline{0}$ or $\overline{d} = \overline{1}$ then $X^2 - \overline{d}$ is a square in $\mathbb{F}_2[X]$, $(X^2 - \overline{1}) = (X - \overline{1})^2$. Therefore $A/2A$ has nonzero nilpotent elements, so 2 is ramified in $\mathbb{Q}(\sqrt{d})$.

If $d \equiv 1 \pmod 4$ then

$$A = A + A\omega \cong \mathbb{Z}[X]/(X^2 - X + c),$$

where $c = (1 - d)/4$. Hence, $A/2A \cong \mathbb{F}_2[X]/(X^2 - X + \overline{c})$.

If $c \equiv 1 \pmod 2$ then $d \equiv 5 \pmod 8$; in this case, since $X^2 - X + \overline{1} = X^2 + X + \overline{1} \in \mathbb{F}_2[X]$ is irreducible over $\mathbb{F}_2$, then $A/2A$ is a field, so 2 is inert in $\mathbb{Q}(\sqrt{d})$.

Finally, if $c \equiv 0 \pmod 2$ then $d \equiv 1 \pmod 8$, $X^2 - X + \overline{c} = X^2 + X = X(X + \overline{1})$, so $A/2A$ is a product of two fields and 2 is totally decomposed in $\mathbb{Q}(\sqrt{d})$.   ∎

We observe that the type of decomposition of 2 in $\mathbb{Q}(\sqrt{d})$ depends only on the residue class of d modulo 8.

As an addendum, we recall that if $d \equiv 2 \pmod 4$ or $d \equiv 3 \pmod 4$ then the discriminant of $\mathbb{Q}(\sqrt{d})$ is $\delta = 4d$, and if $d \equiv 1$

(mod 4) then $\delta = d$ (by Chapter 6, **(P)**). Thus, from **(K)**, **(L)** we deduce: *p is ramified in* $\mathbb{Q}(\sqrt{d})$ *if and only if p divides the discriminant* δ *of* $\mathbb{Q}(\sqrt{d})$.

So for each quadratic field there exist only finitely many prime numbers p which are ramified, and these may be determined by computing the discriminant.

By means of the reciprocity law for the Jacobi symbol (see Chapter 4, **(R)**, **(S)**) we deduce the following fact:

M.   *If p, p′ are prime numbers and* $p \equiv p'$ (mod $|\delta|$) *then p, p′ have the same type of decomposition in* $\mathbb{Q}(\sqrt{d})$.

Proof: If p' is ramified then p' divides δ. From $p \equiv p'$ (mod δ) it follows that $p = p'$.

Let p, p' be odd primes. If $d \equiv 1$ (mod 4) then $d = \delta$ and by the reciprocity law of the Jacobi symbol

$$\left(\frac{d}{p}\right) = \left(\frac{\delta}{p}\right) = (-1)^{\frac{\delta-1}{2}\cdot\frac{p-1}{2}}\left(\frac{p}{\delta}\right) = \left(\frac{p}{\delta}\right) = \left(\frac{p'}{\delta}\right) = \left(\frac{d}{p'}\right).$$

Thus by **(K)** p is totally decomposed if and only if p' is totally decomposed in $\mathbb{Q}(\sqrt{d})$.

If $d \equiv 3$ (mod 4) then $\delta = 4d$ and

$$\left(\frac{d}{p}\right) = (-1)^{\frac{d-1}{2}\cdot\frac{p-1}{2}}\left(\frac{p}{d}\right) = (-1)^{(p-1)/2}\left(\frac{p}{d}\right).$$

But $p \equiv p'$ (mod δ) implies $p \equiv p'$ (mod d) and $p \equiv p'$ (mod 4) so

$$(-1)^{(p-1)/2}\left(\frac{p}{d}\right) = (-1)^{(p'-1)/2}\left(\frac{p'}{d}\right) = \left(\frac{d}{p'}\right)$$

and again p, p' are primes with the same type of decomposition in $\mathbb{Q}(\sqrt{d})$.

If $d \equiv 2$ (mod 4), let $d = 2d'$, d' being odd (since d has no square factor).

We have $\delta = 4d = 8d'$ so $p \equiv p'$ (mod 8) and $p \equiv p'$ (mod d) hence

$$\left(\frac{d}{p}\right) = \left(\frac{2}{p}\right)\left(\frac{d'}{p}\right) = (-1)^{\frac{p^2-1}{8}+\frac{p-1}{2}\cdot\frac{d'-1}{2}}\left(\frac{p}{d'}\right)$$

$$= (-1)^{\frac{p'^2-1}{8}+\frac{p'-1}{2}\cdot\frac{d'-1}{2}}\left(\frac{p'}{d'}\right) = \left(\frac{d}{p'}\right).$$

The case where $p' = 2$ remains to be dealt with. From $p \equiv p'$ (mod δ) it follows that δ is odd, so $d \equiv 1$ (mod 4), $d = \delta$. Now 2 is totally decomposed exactly when $\delta \equiv 1$ (mod 8). From

$$\left(\frac{d}{p}\right) = \left(\frac{\delta}{p}\right) = (-1)^{\frac{\delta-1}{2}\cdot\frac{p-1}{2}}\left(\frac{p}{\delta}\right) = \left(\frac{2}{\delta}\right) = (-1)^{(\delta^2-1)/8}$$

we see that $\delta \equiv 1$ (mod 8) if and only if $(d/p) = 1$; that is, p is totally decomposed in $\mathbb{Q}(\sqrt{d})$.   ∎

This result is interesting insofar as it tells the type of decomposition of any prime number in $\mathbb{Q}(\sqrt{d})$ by considering its class with respect to a unique modulus, namely $|\delta|$. So, in some sense the phenomenon of decomposition of primes is already built in the residue classes modulo $|\delta|$.

11.3 Decomposition of Prime Numbers in Cyclotomic Fields

Let p be a prime number, $m = p^k > 2$ (so if $p = 2$ then $k \geq 2$), and let ζ be a primitive mth root of unity, $K = \mathbb{Q}(\zeta)$, A the ring of integers of K, thus $\zeta \in A$.

K is a Galois extension of degree $\varphi(m) = p^{k-1}(p - 1)$ and its Galois group $\mathcal{G}$ is isomorphic to the multiplicative group $P(m)$ of prime residue classes modulo m.

The minimal polynomial of ζ is the mth cyclotomic polynomial

$$\Phi_m = X^{p^{k-1}(p-1)} + X^{p^{k-1}(p-2)} + \cdots + X^{p^{k-1}} + 1,$$

(see Chapter 2, Exercise 5), and we have

$$\Phi_m = \prod_{\overline{a} \in P(m)} (X - \zeta^a).$$

We shall indicate the decomposition into prime ideals of the ideal Aq, where q is any prime number. We first note that if a, b are nonzero integers, relatively prime to m, then $1 - \zeta^a$, $1 - \zeta^b$ are associated elements of A. In fact, we may write $b \equiv aa' \pmod{m}$ and $a \equiv bb' \pmod{m}$, thus

$$\frac{1 - \zeta^b}{1 - \zeta^a} = \frac{1 - \zeta^{aa'}}{1 - \zeta^a} = 1 + \zeta^a + \zeta^{2a} + \cdots + \zeta^{(a'-1)a} \in A$$

and similarly $\left[(1 - \zeta^a)/(1 - \zeta^b) \right] \in A$. In particular, the absolute values of the norms of the elements $1 - \zeta^a$, $1 - \zeta^b$ are equal.

Let $\xi = 1 - \zeta \in A$.

N. $p = u\xi^{\varphi(p^k)}$, where u is a unit of A. The principal ideal $A\xi$ is prime and $Ap = (A\xi)^{\varphi(p^k)}$, $N(A\xi) = p$.

Proof: From $p = \Phi_m(1) = \prod_{\overline{a} \in P(m)} (1 - \zeta^a)$ and the previous observation, it follows that $p = u\xi^{\varphi(p^k)}$, where u is a unit of A. Thus $Ap = (A\xi)^{\varphi(p^k)}$. Taking norms we have $p^{\varphi(p^k)} = N(Ap) = (N(A\xi))^{\varphi(p^k)}$, hence $N(A\xi) = p$. We conclude that $A\xi$ has to be a prime ideal of A.   ∎

Now let q be any prime number different from p. The type of decomposition of Aq may be obtained from Theorem 2 and a method which will be subsequently generalized.

O.   Let q be any prime number distinct from p, let $f \geq 1$ be the smallest integer such that $q^f \equiv 1 \pmod{p^k}$, and let $g = \varphi(p^k)/f$. Then $Aq =$

$Q_1 \cdots Q_g$ *where* $Q_1, \ldots, Q_g$ *are distinct prime ideals of A, and $N(Q_i) = q^f$.*

Proof: By the theory developed so far, $Aq = (Q_1 \cdots Q_g)^e$ where $Q_1, \ldots, Q_g$ are distinct prime ideals, $e \geq 1$, $N(Q_1) = \cdots = N(Q_g) = q^f$ and $efg = \varphi(p^k)$. Thus, $f = [A/Q_i : \mathbb{Z}/\mathbb{Z}q]$ for each $i = 1, \ldots, g$.

We fix our attention on the prime ideal $Q = Q_1$. Let $\mathcal{Z} = \{\sigma \in \mathcal{G} \mid \sigma(Q) = Q\}$. Then $\mathcal{Z}$ is a subgroup of $\mathcal{G}$ and its index is $(\mathcal{G} : \mathcal{Z}) = g$. Indeed, if $\sigma, \tau \in \mathcal{G}$ then the cosets $\sigma\mathcal{Z} = \tau\mathcal{Z}$ if and only if $\sigma(Q) = \tau(Q)$—as can be checked at once. On the other hand, by (**F**), for each $i = 1, \ldots, g$, there exists $\sigma \in \mathcal{G}$ such that $\sigma(Q) = Q_i$. Thus $g = (\mathcal{G} : \mathcal{Z})$, hence $\#(\mathcal{Z}) = ef$.

With every $\sigma \in \mathcal{Z}$ we associate the mapping $\overline{\sigma} : A/Q \to A/Q$ defined by $\overline{\sigma}(\overline{x}) = \overline{\sigma(x)}$ where $\overline{t}$ denotes $t + Q \in A/Q$ for each $t \in A$. It is obvious that $\overline{\sigma} \in G(A/Q|\mathbb{F}_q)$ and the mapping $\mathcal{Z} \to G(A/Q|\mathbb{F}_q)$ so defined is a group-homomorphism. The kernel is clearly the normal subgroup $\mathcal{T} = \{\sigma \in \mathcal{Z} \mid \sigma(x) \equiv x \pmod{Q} \text{ for every } x \in A\}$. We show that $\mathcal{T}$ is reduced to the identity so the mapping $\mathcal{Z} \to G(A/Q|\mathbb{F}_q)$ is one-to-one. Indeed $\sigma(\zeta)$ is also a primitive p^kth root of unity, so $\sigma(\zeta) = \zeta^s$ where $1 \leq s < p^k$, and $\gcd(s, p^k) = 1$; if $\sigma \in \mathcal{T}$ then Q contains the element $\sigma(\zeta) - \zeta = \zeta^s - \zeta = -\zeta(1 - \zeta^{s-1})$. If $\gcd(s - 1, p) = 1$ then we have seen that $1 - \zeta^{s-1}$ and $1 - \zeta$ are associated elements, so $\xi = 1 - \zeta \in Q$, that is, $Q = A\xi$ and Q divides Ap, which is not the case. If, however, $s - 1 = p^l t$, with $1 \leq l < k$, t not a multiple of p, then $\zeta^{p^{k-l}}$ is a primitive root of unity of order p^l, so

$$X^{p^l} - 1 = \prod_{a=0}^{p^l - 1} (X - \zeta^{ap^{k-l}}),$$

therefore,

$$\zeta^{s-1} = \zeta^{p^l t} - 1 = \prod_{a=0}^{p^l - 1} (\zeta^t - \zeta^{ap^{k-l}}) = \zeta^{p^l t} \prod_{a=0}^{p^l - 1} (1 - \zeta^{ap^{k-l} - t});$$

the elements $1 - \zeta^{ap^{k-l} - t}$ are associated with $1 - \zeta$, therefore in this case we conclude also that $Q = A\xi$, which is not true.

Thus, the mapping from $\mathcal{Z}$ to $G(A/Q|\mathbb{F}_q)$ is one-to-one and $ef = \#\mathcal{Z} \leq \#G(A/Q|\mathbb{F}_q) = [A/Q : \mathbb{F}_q] = f$, implying already that $e = 1$. Now we show that $f \geq 1$ is the smallest integer such that $q^f \equiv 1 \pmod{p^k}$.

Let σ_q be the *Frobenius automorphism*, that is, $\overline{\sigma}_q(\overline{x}) = \overline{x}^q$ for each $\overline{x} \in A/Q$; then $(\overline{\sigma}_q)^f$ is the identity automorphism, therefore σ_q^f is the identity automorphism, since the mapping $\mathcal{Z} \to G(A/Q|\mathbb{F}_q)$ is one-to-one.

Thus $\sigma_q^f(\zeta) = \zeta^{q^f}$ is equal to ζ, so $\zeta^{q^f - 1} = 1$ and therefore p^k divides $q^f - 1$.

If $1 \leq f' \leq f$ and p^k divides $q^{f'} - 1$ then $\sigma_q^{f'}$ is the identity, so $\overline{\sigma}_q^{f'}$ is the identity, which forces $f' = f$, because $\overline{\sigma}_q$ is a generator of $G(A/Q|\mathbb{F}_q)$, so it has order f. ∎

By Theorem 2, the canonical image $\overline{\Phi}_{p^k} \in \mathbb{F}_q[X]$ decomposes as $\overline{\Phi}_{p^k} = \overline{H}_1 \cdots \overline{H}_g$ where $\overline{H}_1, \ldots, \overline{H}_g$ are distinct irreducible polynomials, each of degree f with each $H_i \in A[X]$. Then $Q_i = Aq + AH_i(\zeta)$ for $i = 1, \ldots, g$.

Since the group $P(p^k)$ is cyclic of order $\varphi(p^k)$, for each integer f dividing $\varphi(p^k)$ there exists an integer a, $1 \leq a \leq \varphi(p^k)$, such that a has order f in $P(p^k)$.

By the theorem of Dirichlet on primes in arithmetic progressions, there exist infinitely many primes q, such that $q \equiv a \pmod{p^k}$, so q has order f and thus $Aq = Q_1 \cdots Q_g$, where $Q_1, \ldots, Q_g$ are distinct prime ideals and $fg = \varphi(p^k)$.

In particular, there are infinitely many primes q such that $Aq = Q$ (so $f = \varphi(p^k)$, i.e., $q \bmod p$ is a generator of $P(p^k)$); and similarly, there exist infinitely many primes q such that $Aq = Q_1 \cdots Q_{\varphi(p^k)}$ (so $f = 1$, i.e., $q \equiv 1 \pmod{p}$).

EXERCISES

1. Let $K = \mathbb{Q}(\sqrt{2})$. Determine the decomposition into prime ideals of the ideals $A.2$, $A.3$, $A.5$, $A.7$, where A is the ring of integers of K.

2. Let $K = \mathbb{Q}(\sqrt{-5})$, and let A be the ring of integers of K.

(a) Determine the decomposition into prime ideals of the ideals $A.2$, $A.3$, $A.5$, $A.10$.

(b) Determine all primes $p < 100$ which are inert in the extension $K|\mathbb{Q}$.

(c) Determine all primes $p < 100$ which are totally decomposed in $K|\mathbb{Q}$.

3. Let $K = \mathbb{Q}(x)$ where x is a root of $X^3 - 2$, and let A be the ring of integers of K. Determine the decomposition into prime ideals of $A.2$, $A.3$, and $A.5$.

4. Let ζ_5 be a primitive root of 1 of order 5, and let A be the ring of integers of the cyclotomic field $K = \mathbb{Q}(\zeta_5)$. Let $K^+ = \mathbb{Q}(\zeta_5 + \zeta_5^{-1})$ be the maximal subfield of K, and let A^+ be the ring of integers of K^+. Determine the decomposition into prime ideals of $A.2$, $A.3$, $A.5$ as well as of $A^+.2$, $A^+.3$, $A^+.5$.

5. With notations similar to those of the preceding exercise determine the primes $p < 100$ such that:

   (a) Ap splits completely.

   (b) Ap is ramified.

   (c) Ap is inert.

   (d) Ap is the product of three distinct prime ideals.

   (e) Ap is the product of five distinct prime ideals.

6. Let $K = \mathbb{Q}(\sqrt{2}, \sqrt{3})$, and let A be the ring of integers of K. Determine the decomposition into prime ideals of:

   (a) $A.2$, $A.3$, $A.5$.

   (b) $A.\sqrt{2}$, $A.\sqrt{3}$.

   (c) Determine the prime numbers p which are ramified in $K|\mathbb{Q}$.

   (d) Give a congruence characterization of the primes p which are totally decomposed (respectively, inert) in the extension $K|\mathbb{Q}$.

12

Algebraic Interlude

For the convenience of the reader, this chapter is devoted to the detailed presentation of algebraic results, which will be needed in the sequel.

12.1 Rings of Fractions

Let R be any domain, K the field of quotients of R. We recall that K consists of all equivalence classes a/b of pairs (a, b), with $a, b \in R$, $b \neq 0$, where $(a, b) \equiv (a', b')$ when $b'a = ba'$. The operations in K are defined as follows:

$$\frac{a}{b} + \frac{c}{d} = \frac{ad + bc}{bd}, \qquad \frac{a}{b} \cdot \frac{c}{d} = \frac{ac}{bd}.$$

Every element a of R is identified with $a/1$, making R into a subring of K. If $a \in R$, $a \neq 0$, then a is invertible in K, its inverse being $1/a$.

We observe that we only required that the product of two nonzero elements of R is still not equal to zero. This suggests the concept of a *multiplicative subset* S of a commutative ring R. S is a subset containing 1, not containing zero-divisors and such that if $a, b \in S$ then $ab \in S$. Hence $0 \notin S$. In spite of being mainly interested in rings of fractions of domains, it will become necessary later to consider a ring of fractions of a homomorphic image of a domain, which need not be a domain anymore.

To define the ring of fractions of R by S, we consider pairs (a, s), with $a \in R$, $s \in S$, and state that $(a, s) \equiv (a', s')$ when $s'a = sa'$. Since S contains no zero-divisors, this is an equivalence relation in the above set of pairs. The equivalence class of (a, s) is denoted by a/s. The operations between equivalence classes are defined after the model of the field of quotients:

$$\frac{a}{s} + \frac{a'}{s'} = \frac{as' + a's}{ss'}, \qquad \frac{a}{s} \cdot \frac{a'}{s'} = \frac{aa'}{ss'}.$$

It is an easy matter to check that these operations are well defined, and that we obtain a ring, denoted by $S^{-1}R$ or R_S. It is called the *ring of fractions* of R by S.

R may be considered as a subring of $S^{-1}R$ and every element of S becomes invertible in $S^{-1}R$.

If S_0 is the multiplicative set of all elements of R which are not zero-divisors, then $S_0^{-1}R$ is called the *total ring of fractions* of R. In particular, if R is a domain, this ring $S_0^{-1}R$ is the field of quotients of R. Moreover, if R is a domain and S is any multiplicative subset then $S \subseteq S_0$ and $S^{-1}R$ is contained in the field of quotients $S_0^{-1}R$, hence it is a domain.

In the case where R is a domain and P is a nonzero prime ideal of R, the set-complement S of P in R is a multiplicative set. The ring of fractions $S^{-1}R$ is also denoted by R_P and plays an important role in the sequel.

The following proposition indicates the relationship between the ideals of R and of $S^{-1}R$:

A.   *Let $R' = S^{-1}R$.*

(1) *If J' is any ideal of R' then $R'(J' \cap R) = J'$, hence the mapping $J' \to J' \cap R$, is one-to-one and preserves inclusions; if $J' \neq R'$ then $J' \cap R$ is disjoint from S.*

(2) *The mapping $P' \to P' \cap R$ sends the set of prime ideals of R' onto the set of prime ideals P of R, disjoint from S and $R'P \cap R = P$.*

(3) *In particular, if S is the set-complement of the prime ideal P in R, then by the above mapping we obtain all the prime ideals of R contained in P, and $R' = R_P$ has only one maximal ideal, namely $R'P$.*

Proof: (1)   Obviously $J' \supseteq R'(J' \cap R)$. Conversely, if $x \in J'$ then $x = a/s$, with $a \in R$, $s \in S$; hence $a = sx \in R'J' \subseteq J'$, so $a \in J' \cap R$ and

$$x = (1/s) \cdot a \in R'(J' \cap R).$$

This shows that the mapping $J' \to J' \cap R$ is one-to-one and of course it preserves inclusions. If $(J' \cap R) \cap S$ contains an element s, then

$$1 = (1/s) \cdot s \in R'(J' \cap R) = J'.$$

(2)   If P' is a prime ideal of R', then clearly $P' \cap R$ is a prime ideal of R, and by (1), $P' \cap R$ is disjoint from S.

Conversely, let P be a prime ideal of R, $P \cap S = \emptyset$, and let us show that $R'P$ is a prime ideal of R' such that $R'P \cap R = P$. Every element of $R'P$ is of the form $\sum_{i=1}^{h}(a_i/s_i)x_i$ where $a_i \in R$, $s_i \in S$, $x_i \in P$, $h \geq 1$; this may be rewritten with a common denominator $s = s_1 \cdots s_h \in S$, as follows $\sum_{i=1}^{h}(a_i/s_i)x_i = \sum_{i=1}^{h}(b_i/s)x_i = (1/s)(\sum_{i=1}^{h} b_i x_i) \in R'P$ (where each $b_i \in R$); in other words, every element of $R'P$ is of the form x/s, with $x \in P$, $s \in S$.

Now, $R'P$ is a prime ideal, because if $a/s, b/t \in R'$ and $(a/s) \cdot (b/t) \in R'P$ then $(a/s) \cdot (b/t) = x/u$ with $x \in P$, $a, b \in R$, $s, t, u \in S$; thus

$abu = xst \in P$; since $P \cap S = \varnothing$, then $ab \in P$, so either $a \in P$ or $b \in P$, that is, $a/s \in R'P$ or $b/t \in R'P$, showing that $R'P$ is a prime ideal.

Finally, $P \subseteq R'P \cap R$; conversely, if $a \in R'P \cap R$, then $a = x/s$, with $x \in P$, $s \in S$, so $sa = x \in P$, but $s \notin P$ so $a \in P$.

(3)   This assertion follows immediately from the preceding ones.   ∎

As a corollary, we have:

B.   *If R is a Noetherian ring, S a multiplicative subset of R, then $R' = S^{-1}R$ is also a Noetherian ring.*

Proof: By (**A**), Part (1), there is a one-to-one correspondence, preserving inclusions, from the set of ideals of $R' = S^{-1}R$ into the set of ideals of R. Hence, every strictly increasing chain of ideals of R' must be finite.   ∎

It follows from (**A**) that if P is a nonzero prime ideal of the domain R then $R' = R_P$ has only one maximal ideal $R'P$ and $P = R'P \cap R$; the prime ideals of R' correspond to those of R which are contained in P. Thus, to pass from R to the ring R_P amounts essentially to disregarding all prime ideals P' of R which are not contained in P. This process is usually called the *localizaton of R at P*. It is especially important for us in the case where every nonzero prime ideal of R is maximal (for example, when R is a Dedekind domain). Then $R' = R_P$ has only one nonzero prime ideal $R'P$.

A reverse procedure is the *globalization*. We consider the family of all maximal ideals P_i of the domain R; for every ideal I of R we have $I \subseteq R_{P_i}I$; but in fact the following holds:

C. $I = \bigcap R_{P_i}I$ *(intersection over the set of maximal ideals P_i of the domain R).*

Proof: The result is trivial when R is a field, so we assume that this is not the case. Let $x \in \bigcap R_{P_i}I$. For every maximal ideal P_i of R we may write $x = a_i/b_i$ with $a_i \in I$, $b_i \in R$, $b_i \notin P_i$. Let J be the ideal of R generated by the elements b_i. Since $b_i \notin P_i$ then $J \not\subseteq P_i$ for every maximal ideal P_i. Therefore, $J = R$, because as is known, every ideal of R, distinct from R, is contained in a maximal ideal (this follows from Zorn's lemma; in the case where R is a Noetherian ring it is immediate by the maximal condition on ideals). In particular, 1 may be expressed in terms of the generators of $J = R$, that is, there exist elements $c_{i_1}, \ldots, c_{i_m} \in R$ such that $1 = \sum_{k=1}^{m} c_{i_k} b_{i_k}$ and so

$$x = \sum_{k=1}^{m} c_{i_k}(b_{i_k}x) = \sum_{k=1}^{m} c_{i_k}a_{i_k} \in I. \qquad ∎$$

In order to apply this method to rings of algebraic integers, we want to describe the behavior of the integral closure by going to rings of fractions:

D.   *Let A be a domain, let L be a field containing A; let B be the integral closure of A in L. If S is a multiplicative subset of A, then $S^{-1}B$ is the integral closure of $S^{-1}A$ in L.*

Proof: If $x \in L$ is an integral element over $S^{-1}A$, then there exist elements $a_i/s_i \in S^{-1}A$ (with $a_i \in A$, $s_i \in S$, $i = 1, \ldots, n$), such that

$$x^n + \frac{a_1}{s_1} x^{n-1} + \frac{a_2}{s_2} x^{n-2} + \cdots + \frac{a_n}{s_n} = 0.$$

Letting $s = s_1 \cdots s_n \in S$, we may rewrite

$$x^n + \frac{b_1}{s} x^{n-1} + \frac{b_2}{s} x^{n-2} + \cdots + \frac{b_n}{s} = 0,$$

with $b_i \in A$, and therefore

$$(sx)^n + b_1(sx)^{n-1} + b_2 s(sx)^{n-2} + \cdots + b_n s^{n-1} = 0.$$

This shows that $sx \in L$ is integral over A, hence $sx \in B$ and $x \in S^{-1}B$.

On the other hand, every element b/s of $S^{-1}B$ is integral over $S^{-1}A$; indeed, since $b \in B$, there exist elements $a_1, \ldots, a_n \in A$ such that

$$b^n + a_1 b^{n-1} + \cdots + a_n = 0,$$

so

$$\left(\frac{b}{s}\right)^n + \frac{a_1}{s} \left(\frac{b}{s}\right)^{n-1} + \frac{a_2}{s^2} \left(\frac{b}{s}\right)^{n-2} + \cdots + \frac{a_n}{s^n} = 0,$$

thus b/s is integral over $S^{-1}A$.   ∎

E.   *If A is an integrally closed domain, if S is a multiplicative subset of A, then $S^{-1}A$ is an integrally closed domain.*

Proof: This is a particular case of (**D**).   ∎

Combining the previous results, we now prove:

F.   *If A is a Dedekind domain and S is a multiplicative subset of A, then $A' = S^{-1}A$ is a Dedekind domain. If J is an ideal of A, $J = \prod_{i=1}^r P_i^{e_i}$, then the decomposition of $A'J$ into prime ideals of A' is given by*

$$A'J = \prod_{P_i \cap S = \varnothing} (A'P_i)^{e_i}.$$

Proof: By (**B**) and (**E**), A' is also a Noetherian integrally closed domain. Let us show that every nonzero prime ideal of A' is maximal. By Chapter 7, Definition 2, this implies that A' is a Dedekind domain.

Let P' be a nonzero prime ideal of A', so $P' \cap A = P$ is a prime ideal of A such that $P \cap S = \varnothing$. Also $P \neq 0$ since $P' = A'P$. Thus, P is a maximal ideal of A, hence by (**A**), P' is also a maximal ideal of A'.

From $J = \prod_{i=1}^r P_i^{e_i}$ it follows that

$$A'J = A' \left(\prod_{i=1}^r P_i^{e_i}\right) = \prod_{i=1}^r (A'P_i)^{e_i} = \prod_{P_i \cap S = \varnothing} (A'P_i)^{e_i},$$

noting that if $P_i \cap S \neq \varnothing$ then $A'P_i = A'$ and if $P_i \cap S = \varnothing$ then $A'P_i$ is a prime ideal of A'.   ∎

As a corollary we obtain:

G.   *If A is a Dedekind domain, and if P is a nonzero prime ideal of A, then:*

(1) *A_P is a principal ideal domain, with only one nonzero prime ideal which is $A_P P$.*

(2) *Every nonzero fractional ideal of A_P is a power of $A_P P$ and $A_P P^s \cap A = P^s$ for every $s \geq 1$.*

(3) *An element is invertible in A_P if and only if it does not belong to $A_P P$.*

(4) *If R is a subring of the field of quotients K of A and A_P is properly contained in R then $R = K$.*

Proof: (1)   By **(F)** A_P is a Dedekind domain; by **(A)** it has only one nonzero prime ideal, namely $A_P P$, thus by Chapter 7, **(N)**, A_P is a principal ideal domain.

(2)   By Part (1) every nonzero integral ideal of A_P is a power of $A_P P$; the same holds therefore for the nonzero fractional ideals of A_P.

By **(A)** we know that $A_P P \cap A = P$. If $A_P P^s \cap A = P^s$ for $s \geq 1$ then from **(A)** we have $P^s = A_P P^s \cap A \supset A_P P^{s+1} \cap A \supseteq P^{s+1}$. Since A is a Dedekind domain, there exists no ideal J of A such that $P^s \supset J \supset P^{s+1}$; thus $A_P P^{s+1} \cap A = P^{s+1}$, proving the statement.

(3)   The elements of A_P which are not in the only maximal ideal $A_P P$ are precisely those which generate the unit ideal; so they are the invertible elements of A_P.

(4)   Let $x \in R$, $x \notin A_P$, and let $n > 0$ be such that $A_P x = A_P t^{-n}$ where t is a generator of the principal ideal $A_P P$. If $y \in K$, $y \notin A_P$, let $m > 0$ be such that $A_P y = A_P t^{-m}$. If r is a positive integer such that $rn \geq m$, then $A_P x^{-r} = A_P t^{rn} \subseteq A_P t^m = A_P y^{-1}$, thus $y \in A_P x^r \subseteq A_P R \subseteq R$. This proves that $R = K$.   ∎

Another useful property relates the rings of fractions to quotient rings:

H.   *Let A be a commutative ring, let S be a multiplicative subset of A, and $A' = S^{-1}A$. Let J be an ideal of A, distinct from A, such that: if $as \in J$, $a \in A$, $s \in S$ then $a \in J$. Then the image $\overline{S}$ of S by the canonical mapping $A \to A/J$ is a multiplicative subset of A/J (containing no zero divisors) and there exists a canonical isomorphism $\varphi : \overline{S}^{-1}(A/J) \to A'/A'J$. In particular, if all the elements of $\overline{S}$ are invertible in A/J then $A/J = A'/A'J$ (after an identification). This happens when $J = P$ is a maximal ideal and S is the complement of P in A.*

Proof: First we note that $J \cap S = \varnothing$ because if $a = 1 \cdot a \in J \cap S$ then $1 \in J$, contrary to the hypothesis.

Similarly, $\overline{S}$ is a multiplicative subset of A/J, because $\overline{1} \in \overline{S}$, if $a, b \in S$, then $\overline{a}, \overline{b} \in \overline{S}$, $ab \in S$ so $\overline{ab} \in \overline{S}$; also if $\overline{a}, \overline{b} \in \overline{S}$ and $\overline{a}\overline{b} = \overline{0}$ then $ab \in J$, with $a, b \in S$, hence by the condition on J, we must have $a \in S \cap J$, which is a contradiction.

Given any element of $\overline{S}^{-1}(A/J)$, which is written as $\overline{a}/\overline{s}$, with $\overline{a}, \overline{s} \in A/J$, $\overline{s} \in \overline{S}$, we define $\varphi(\overline{a}/\overline{s}) = \overline{(a/s)}$ (image of $a/s \in A'$ by the canonical homomorphism onto $A'/A'J$). First, we note that φ is well defined, in other words, if $\overline{a}/\overline{s} = \overline{a'}/\overline{s'}$ then $\overline{(a/s)} = \overline{(a'/s')}$. In fact, $\overline{s'}\overline{a} = \overline{s}\overline{a'}$ so $s'a - sa' \in J$, hence

$$\frac{a}{s} - \frac{a'}{s'} = \frac{s'a - a's}{ss'} \in A'J.$$

It follows easily that φ is a ring-homomorphism. Clearly, φ maps $\overline{S}^{-1}(A/J)$ onto $A'/A'J$, since every element of this ring is of the type $\overline{(a/s)}$, with $a \in A$, $s \in S$. Finally the kernel of φ is zero, because from $\overline{(a/s)} = \overline{0}$ we deduce $a/s \in A'J$ so we may write $a/s = a'/s'$ with $a' \in J$, $s' \in S$; hence, $s'a = sa' \in J$; by the hypothesis, we have $a \in J$, so $\overline{a}/\overline{s} = 0$.

For the second assertion, if P is a maximal ideal then A/P is a field. Since $A/P \subseteq \overline{S}^{-1}(A/P)$ and every element of $\overline{S}$ is invertible in A/P it follows that $A/P = \overline{S}^{-1}(A/P)$. Thus φ is an isomorphism between A/P and $A'/A'P$, and we may write $A/P = A'/A'P$, after an identification. ∎

Now we show the generalization of the first assertion of Theorem 1 of Chapter 11:

I.   *Let A be a Dedekind domain, K its field of quotients, and let $L|K$ be a separable extension of degree n and B the integral closure of A in L. If P is any nonzero prime ideal of A then B/BP is a vector space of dimension n over A/P.*

Proof: Let S be the set-complement of P in A and let $A' = S^{-1}A$ be the corresponding ring of fractions. By **(G)**, A' is a principal ideal domain. Since B is the integral closure of A, by **(D)**, $B' = S^{-1}B$ is the integral closure of A' in L. By Chapter 6, **(B)**, B' is contained in a free A'-module of rank n. By Chapter 6, **(I)**, B' is itself a free A'-module of rank at most n; from $L = KB'$ it follows that B' has rank n over A'.

From Chapter 11, **(A)**, we have $BP \cap A = P$ and since $B'P = B'(A'P)$ then **(A)** again implies that $B'P \cap A' = A'P$. Therefore, $B'/B'P$ contains the field $A'/A'P$. We now show that it is a vector space of dimension n. In fact, if $\{x_1, \ldots, x_n\}$ is a basis of the A'-module B', if $\overline{x}_i$ denotes the image of x_i in $B'/B'P$ (by the natural mapping), then $\{\overline{x}_1, \ldots, \overline{x}_n\}$ generates $B'/B'P$ over $A'/A'P$. On the other hand, if $\sum_{i=1}^{n} \overline{a}_i\overline{x}_i = \overline{0}$ (with $\overline{a}_i \in A'/A'P$) then

$$\sum_{i=1}^{n} a_i x_i \in B'P = B'(A'P)$$

and so we may write $\sum_{i=1}^{n} a_i x_i = \sum_{j=1}^{m} a'_j y_j$ with $a'_j \in A'P$, $y_j \in B'$; expressing the elements y_j in terms of the generators $x_1, \ldots, x_n$ of the A'-module B' we may write $\sum_{i=1}^{n} a_i x_i = \sum_{i=1}^{n} a''_i x_i$ with all $a''_i \in A'P$ and necessarily $a_i = a''_i$ for every $i = 1, \ldots, n$ (since $x_1, \ldots, x_n$ are linearly independent over A'). Hence $\bar{a}_i = \bar{a}''_i = \bar{0}$.

The image $\bar{S}$ of S by the natural homomorphism $A \rightarrow A/P$ is the set of nonzero elements of the field A/P; a fortiori, the elements of $\bar{S}$ are invertible in $B'/B'P$. By (**H**), $A'/A'P = A/P$, $B'/B'P = B/BP$, and so B/BP is a vector space of dimension n over A/P.   ∎

12.2 Traces and Norms in Ring Extensions

We shall consider the following general situation:

A is a subring of the commutative ring B and B is a free A-module having a basis with n elements.

It is well-known that any other basis of the A-module B also has n elements.

Let $\theta : B \rightarrow B$ be any linear mapping, $\{z_1, \ldots, z_n\}$ any basis of the A-module B. Then

$$\theta(z_j) = \sum_{i=1}^{n} a_{ij} z_i \qquad (j = 1, \ldots, n)$$

with $a_{ij} \in A$. The matrix $M(\theta) = (a_{ij})_{i,j}$ is called the matrix of θ with respect to the basis $\{z_1, \ldots, z_n\}$.

If $\{z'_1, \ldots, z'_n\}$ is any other basis of the A-module B, and $M'(\theta)$ is the corresponding matrix, $M'(\theta) = (a'_{ij})_{i,j}$, if $z_j = \sum_{i=1}^{n} c_{ij} z'_i$ $(j = 1, \ldots, n)$, with each $c_{ij} \in A$, $C = (c_{ij})_{i,j}$, then we have

$$\theta(z_j) = \sum_{i=1}^{n} a_{ij} z_i = \sum_{i=1}^{n} a_{ij} \left(\sum_{k=1}^{n} c_{ki} z'_k \right) = \sum_{k=1}^{n} \left(\sum_{i=1}^{n} c_{ki} a_{ij} \right) z'_k,$$

and, on the other hand,

$$\theta(z_j) = \sum_{i=1}^{n} c_{ij} \theta(z'_i) = \sum_{i=1}^{n} c_{ij} \left(\sum_{k=1}^{n} a'_{ki} z'_k \right) = \sum_{k=1}^{n} \left(\sum_{i=1}^{n} a'_{ki} c_{ij} \right) z'_k.$$

This shows that $M'(\theta) \cdot C = C \cdot M(\theta)$.

Since C is the matrix of a change of basis, it is invertible, hence $M'(\theta) = C \cdot M(\theta) \cdot C^{-1}$. In particular, $\det(C)$ is a unit of the ring A, because

$$\det(C) \cdot \det(C^{-1}) = 1.$$

Following the well-known method of Linear Algebra, we now consider the matrix $XI - M(\theta)$, where X is an indeterminate, I is the unit $n \times n$

matrix; thus, the entries of $XI - M(\theta)$ are elements of A or linear monic polynomials in X:

$$XI - M(\theta) = \begin{pmatrix} X - a_{11} & -a_{12} & \cdots & -a_{1n} \\ -a_{21} & X - a_{22} & \cdots & -a_{2n} \\ \vdots & \vdots & & \vdots \\ -a_{n1} & -a_{n2} & \cdots & X - a_{nn} \end{pmatrix}.$$

The determinant of this matrix remains unchanged when we change the basis of the A-module B. In fact,

$$\begin{aligned} XI - M'(\theta) &= XI - C \cdot M(\theta) \cdot C^{-1} \\ &= XC \cdot I \cdot C^{-1} - C \cdot M(\theta) \cdot C^{-1} \\ &= C(XI - M(\theta))C^{-1}. \end{aligned}$$

Hence

$$\begin{aligned} \det(XI - M'(\theta)) &= \det(C) \cdot \det(XI - M(\theta)) \cdot \det(C^{-1}) \\ &= \det(XI - M(\theta)). \end{aligned}$$

The polynomial $\det(XI - M(\theta))$ depends therefore only on θ. It is called *the characteristic polynomial of* θ and is denoted by

$$F(\theta) = X^n + a_1 X^{n-1} + a_2 X^{n-2} + \cdots + a_n,$$

with each a_i in A. Sometimes we also use the notation $F_{B|A}(\theta)$.

We now define the *trace of* θ as $\mathrm{Tr}_{B|A}(\theta) = -a_1 \in A$ and the *determinant of* θ as $\det_{B|A}(\theta) = (-1)^n a_n \in A$. It is clear that the trace of θ is the sum of the elements in the diagonal of $M(\theta)$, while the determinant of θ is the determinant of the matrix $M(\theta)$.

Now we apply these general notions to the following special situation. If $x \in B$, $x \neq 0$, let $\theta = \theta_x : B \to B$ be the mapping of multiplication by x; that is, $\theta_x(z) = xz$ for every $z \in B$. Then the characteristic polynomial of θ_x is called *the characteristic polynomial of* x *in* $B|A$ and is denoted by $F_{B|A}(x)$. Similarly, the trace of θ_x is called *the trace of* x *in* $B|A$ and the determinant of θ_x is called *the norm of* x *in* $B|A$. They are, respectively, denoted by $\mathrm{Tr}_{B|A}(x)$ and $N_{B|A}(x)$.

It is important to compare the notions of trace and norm with the ones known for field extensions. As a temporary notation, if $L|K$ is a separable field extension of degree n and $x \in L$ let $\mathrm{Tr}'_{L|K}(x)$ denote the sum of all conjugates of x in $L|K$ and, similarly, $N'_{L|K}(x)$ shall denote the product of all conjugates of x in $L|K$.

J.   *Let $x \in L$ and let $f \in K[X]$ be its minimal polynomial over K. Then the characteristic polynomial is $F_{L|K}(x) = f^s$ where $s = [L : K(x)]$ and $\mathrm{Tr}'_{L|K}(x) = \mathrm{Tr}_{L|K}(x)$, $N'_{L|K}(x) = N_{L|K}(x)$.*

Proof: Let $\{x_1, \ldots, x_r\}$ be a basis of the K-vector space $K(x)$, and let $\{y_1, \ldots, y_s\}$ be a basis of the $K(x)$-vector space L. So

$$\{x_1 y_1, x_2 y_1, \ldots, x_r y_1, x_1 y_2, x_2 y_2, \ldots, x_r y_2, \ldots, x_1 y_s, x_2 y_s, \ldots, x_r y_s\}$$

is a basis of the K-vector space L (with $rs = n$). Since $x \in K(x)$ we have

$$x x_j = \sum_{i=1}^{r} a_{ij} x_i \qquad (\text{for } j = 1, \ldots, r).$$

Therefore, $T = (a_{ij})_{i,j}$ is the matrix of $\theta_x : K(x) \to K(x)$ with respect to the basis $\{x_1, \ldots, x_r\}$. It follows that

$$x x_j y_k = \sum_{i=1}^{r} a_{ij} x_i y_k \qquad (\text{for } j = 1, \ldots, r; \ k = 1, \ldots, s).$$

Hence the matrix of $\theta_x : L \to L$, with respect to the basis of the K-vector space L, considered above, is a block diagonal $n \times n$ matrix

$$M(\theta_x) = \begin{pmatrix} T & 0 & \cdots & 0 \\ 0 & T & \cdots & 0 \\ \vdots & \vdots & & \vdots \\ 0 & 0 & \cdots & T \end{pmatrix}.$$

Therefore

$$F_{L|K}(x) = (F_{K(x)|K}(x))^s,$$

hence

$$\mathrm{Tr}_{L|K}(x) = s \, \mathrm{Tr}_{K(x)|K}(x),$$
$$N_{L|K}(x) = (N_{K(x)|K}(x))^s.$$

As we know, we also have

$$\mathrm{Tr}'_{L|K}(x) = s \, \mathrm{Tr}'_{K(x)|K}(x) \qquad \text{and} \qquad N'_{L|K}(x) = (N_{K(x)|K}(x))^s.$$

Thus, it is enough to show that $F_{K(x)|K} = f$, the minimal polynomial of x over K, for this implies that

$$\mathrm{Tr}'_{K(x)|K}(x) = \mathrm{Tr}_{K(x)|K}(x) \qquad \text{and} \qquad N'_{K(x)|K}(x) = N_{K(x)|K}(x).$$

Now, in $K(x)|K$ the matrix of θ_x with respect to the basis

$$\{1, x, x^2, \ldots, x^{r-1}\}$$

is the *companion matrix* of the minimal polynomial

$$f = X^r + a_1 X^{r-1} + \cdots + a_r;$$

that is,

$$M(\theta_x) = \begin{pmatrix} 0 & 0 & \cdots & 0 & -a_r \\ 1 & 0 & \cdots & 0 & -a_{r-1} \\ 0 & 1 & \cdots & 0 & -a_{r-2} \\ \vdots & \vdots & & \vdots & \vdots \\ 0 & 0 & \cdots & 1 & -a_1 \end{pmatrix}.$$

Hence $F_{K(x)|K}(x) = \det(XI - M(\theta_x)) = X^r + a_1 X^{r-1} + \cdots + a_r = f.$ ∎

Let us note the following algebraic properties of the trace and norm:

K.   *If $x_1, x_2 \in B$, $a \in A$ then*

$$\mathrm{Tr}_{B|A}(x_1 + x_2) = \mathrm{Tr}_{B|A}(x_1) + \mathrm{Tr}_{B|A}(x_2),$$
$$\mathrm{Tr}_{B|A}(ax_1) = a\,\mathrm{Tr}_{B|A}(x_1),$$
$$N_{B|A}(x_1 x_2) = N_{B|A}(x_1) \cdot N_{B|A}(x_2).$$

Proof: The proof is straightforward. For example, $\theta_{x_1+x_2} = \theta_{x_1} + \theta_{x_2}$, hence $M(\theta_{x_1+x_2}) = M(\theta_{x_1}) + M(\theta_{x_2})$, and considering the elements in the diagonal of these matrices, we obtain $\mathrm{Tr}_{B|A}(x_1 + x_2) = \mathrm{Tr}_{B|A}(x_1) + \mathrm{Tr}_{B|A}(x_2)$.

Similarly $\theta_{ax_1} = a\theta_{x_1}$, hence $M(\theta_{ax_1}) = aI \cdot M(\theta_{x_1})$ (where I is the identity matrix) so

$$\mathrm{Tr}_{B|A}(ax_1) = a\,\mathrm{Tr}_{B|A}(x_1).$$

Finally, $\theta_{x_1 x_2} = \theta_{x_1} \circ \theta_{x_2}$, then $M(\theta_{x_1 x_2}) = M(\theta_{x_1}) \cdot M(\theta_{x_2})$ and therefore, considering the determinants, we have $N_{B|A}(x_1 x_2) = N_{B|A}(x_1) \cdot N_{B|A}(x_2)$. ∎

Now we study the behavior of the trace and norm when we consider rings of fractions. Let S be a multiplicative subset of $A \subseteq B$ and $A' = A_S$, $B' = B_S$. If $\{z_1, \ldots, z_n\}$ is an A-basis of B then it is still an A'-basis of B'.

L.   *With these notations, for every $x \in B$ we have:*

$$F_{B'|A'}(x) = F_{B|A}(x), \quad \mathrm{Tr}_{B'|A'}(x) = \mathrm{Tr}_{B|A}(x), \quad N_{B'|A'}(x) = N_{B|A}(x).$$

Proof: Let $\theta_x : B \to B$, $\theta_x' : B' \to B'$ be the homomorphisms of multiplication by $x \in B$ and let $M(\theta_x)$, $M(\theta_x')$ be the corresponding matrices with respect to the basis $\{z_1, \ldots, z_n\}$. Obviously these matrices coincide, therefore the same happens with the characteristic polynomials, with the traces and with the norms of the element x in $B|A$ and in $B'|A'$. ∎

In particular, if B, A are domains having fields of quotients L, K, respectively, then

$$F_{B|A}(x) = F_{L|K}(x), \quad \mathrm{Tr}_{B|A}(x) = \mathrm{Tr}_{L|K}(x), \quad \text{and} \quad N_{B|A}(x) = N_{L|K}(x).$$

We shall study the characteristic polynomial, the trace, and the norm when we consider Cartesian products of rings.

Let $B_1, \ldots, B_r$ be commutative rings containing the subring A and such that each ring B_i is a free A-module of finite rank. Let $B = \prod_{i=1}^{r} B_i$ be their Cartesian product and $\pi_i : B \to B_i$ the ith projection, which is a ring homomorphism from B onto B_i. B contains the subring $\{(a, \ldots, a) \in B \mid a \in A\}$, which is naturally isomorphic to A; hence, we may consider A as a subring of B; then B is also a free A-module of finite rank.

We may easily prove:

M.   *With these notations, if $x \in B$ then*

$$F_{B|A}(x) = \prod_{i=1}^{r} F_{B_i|A}(\pi_i(x)),$$

$$\mathrm{Tr}_{B|A}(x) = \sum_{i=1}^{r} \mathrm{Tr}_{B_i|A}(\pi_i(x)),$$

$$N_{B|A}(x) = \prod_{i=1}^{r} N_{B_i|A}(\pi_i(x)).$$

Proof: It is enough to prove the statement when $r = 2$. Let $\{t_1, \ldots, t_n\}$ be a basis of the A-module B_1, and let $\{u_1, \ldots, u_m\}$ be a basis of the A-module B_2. Then $\{(t_1, 0), \ldots, (t_n, 0), (0, u_1), \ldots, (0, u_m)\}$ is a basis of the A-module $B = B_1 \times B_2$. Let $z_i = (t_i, 0)$ for $i = 1, \ldots, n$ and let $z_{n+i} = (0, u_i)$ for $i = 1, \ldots, m$. If $x = (x_1, x_2) \in B_1 \times B_2$ then

$$(x_1, x_2) \cdot (t_j, 0) = (x_1 t_j, 0) = \left(\sum_{i=1}^{n} a_{ij} t_i, 0 \right) = \sum_{i=1}^{n} a_{ij}(t_i, 0),$$

where $(a_{ij})_{i,j} = M(\theta_{x_1})$, the matrix of the A-linear transformation θ_{x_1} from B_1 to B_1 (with respect to the basis $\{t_1, \ldots, t_n\}$). Similarly,

$$(x_1, x_2) \cdot (0, u_j) = \sum_{i=1}^{m} a'_{ij}(0, u_i),$$

where $(a'_{ij})_{i,j} = M(\theta_{x_2})$, the matrix of the A-linear transformation θ_{x_2} from B_2 to B_2 (with respect to the basis $\{u_1, \ldots, u_m\}$).

Thus the matrix of $\theta_x : B_1 \times B_2 \to B_1 \times B_2$ with respect to the basis $\{z_1, \ldots, z_{n+m}\}$ is

$$M(\theta_x) = \left(\begin{array}{c|c} M(\theta_{x_1}) & 0 \\ \hline 0 & M(\theta_{x_2}) \end{array} \right).$$

Hence $\det(XI - M(\theta_x)) = \det(XI - M(\theta_{x_1})) \cdot \det(XI - M(\theta_{x_2}))$; that is, $F_{B|A}(x) = F_{B_1|A}(x_1) \cdot F_{B_2|A}(x_2)$. The assertions about the trace and norm are now immediate.   ∎

We now consider the effect on the characteristic polynomial of certain ring-homomorphisms.

N.   *Let $\psi : B \to \overline{B}$ be a homomorphism from B onto the ring $\overline{B}$, and let $\psi(A) = \overline{A}$. We assume that there exists a basis $\{z_1, \ldots, z_n\}$ of the A-module B such that $\{\overline{z}_1, \ldots, \overline{z}_n\}$ is a basis of the $\overline{A}$-module $\overline{B}$ (where $\overline{z} = \psi(z)$ for every $z \in B$). If $x \in B$ then*

$$\psi(F_{B|A}(x)) = F_{\overline{B}|\overline{A}}(\overline{x}),$$

$$\psi(\mathrm{Tr}_{B|A}(x)) = \mathrm{Tr}_{\overline{B}|\overline{A}}(\overline{x}),$$

$$\psi(N_{B|A}(x)) = N_{\overline{B}|\overline{A}}(\overline{x}).$$

Proof: If $xz_j = \sum_{i=1}^{n} a_{ij} z_i \ (j = 1, \ldots, n)$ with $a_{ij} \in A$ then

$$\overline{x}\,\overline{z}_j = \sum_{i=1}^{n} \overline{a}_{ij}\overline{z}_i \qquad (j = 1, \ldots, n).$$

Thus $M(\theta_x) = (a_{ij})$, $M(\theta_{\overline{x}}) = (\overline{a}_{ij})$ (with respect to the above bases). Applying ψ to the coefficients of the characteristic polynomial $F_{B|A}(x) = \det(XI - M(\theta_x))$ we obtain $\det(XI - M(\theta_{\overline{x}})) = F_{\overline{B}|\overline{A}}(\overline{x})$. The assertions about the trace and norm follow at once.   ∎

Let R be a ring, and K a subfield of R such that R is a vector space of finite dimension over K. Let $\theta : R \to R$ be a K-linear transformation, and consider a strictly decreasing chain of subspaces of R,

$$R = R_0 \supset R_1 \supset R_2 \supset \cdots \supset R_{k-1} \supset R_k = 0$$

such that $\theta(R_i) \subseteq R_i$ for every $i = 1, \ldots, k$.

The elements of the K-vector space R_{i-1}/R_i are the cosets $z + R_i$, where $z \in R_{i-1}$. Then θ induces a linear transformation

$$\theta_i : R_{i-1}/R_i \to R_{i-1}/R_i$$

defined as follows: $\theta_i(z + R_i) = \theta(z) + R_i$ for every $z \in R_{i-1}$. By virtue of the hypothesis on the subspaces R_i it follows that θ_i is well defined.

For each index $i = 1, \ldots, k$ let $B_i = \{z_{i1}, \ldots, z_{im_i}\}$ be a set of elements of R_{i-1} such that the set of cosets $\{z_{i1} + R_i, \ldots, z_{im_i} + R_i\}$ forms a basis of the vector space R_{i-1}/R_i. Then for every $i = 1, \ldots, k$:

$$B_i \cup B_{i+1} \cup \cdots \cup B_k$$

constitutes a basis of the K-vector space R_{i-1}. In particular,

$$B = B_1 \cup \cdots \cup B_k$$

is a basis of R. The verification is standard and therefore omitted.

We shall consider the matrix $M(\theta)$ of θ with respect to the basis B; it may be expressed in terms of the matrices $M(\theta_i)$ of θ_i with respect to B_i:

$$M(\theta) = \begin{pmatrix} M(\theta_1) & 0 & \cdots & 0 \\ M_{21} & M(\theta_2) & \cdots & 0 \\ \vdots & \vdots & & \vdots \\ M_{k1} & M_{k2} & \cdots & M(\theta_k) \end{pmatrix},$$

where M_{ij} $(i > j)$ are matrices with entries in K of the appropriate size.

Indeed, $\theta(z_{ij}) \in R_{i-1}$ hence it may be expressed in terms of the basis $B_i \cup B_{i+1} \cup \cdots \cup B_k$ as follows:

$$\theta(z_{ij}) = \sum_{h=1}^{m_i} a_{ihj} z_{ih} + \sum_{h=1}^{m_{i+1}} a_{i+1,hj} z_{i+1,h} + \cdots + \sum_{h=1}^{m_k} a_{khj} z_{kh}$$

(with coefficients $a_{ihj} \in K$). Then

$$\theta_i(z_{ij} + R_i) = \sum_{h=1}^{m_i} a_{ihj}(z_{ih} + R_i) + R_i = \sum_{h=1}^{m_i} a_{ihj} z_{ih} + R_i$$

and so $M(\theta)$ has the form indicated.

If we consider the characteristic polynomials of the linear transformations θ, θ_1, ..., θ_k then

$$F_{R|K}(\theta) = \prod_{i=1}^{k} F_{(R_{i-1}/R_i)|K}(\theta_i).$$

O.   *With the above hypotheses and notations, we assume further that:*

(1) *Each R_i is an ideal of R.*

(2) *For every $i = 1, 2, \ldots, k$ there exists no ideal R' of R such that $R_{i-1} \supset R' \supset R_i$.*

(3) *If $y \in R_1$, $z \in R_{i-1}$ then $yz \in R_i$.*

If $x \in R$ and $\theta = \theta_x$, then for every $i = 1, \ldots, k$:

$$F_{(R_{i-1}/R_i)|K}(\theta_i) = F_{(R/R_1)|K}(\theta_1),$$

so

$$F_{R|K}(\theta) = \left[F_{(R/R_1)|K}(\theta_1) \right]^k.$$

Proof: First we show that $(1), (2)$, and (3) imply condition:

(4) Let $y, z \in R$ with $yz \in R_i$; if $y \notin R_i$ then $z \in R_i$.

Indeed, by (1) and (2) R_1 is a maximal ideal of R. By (3), if $t \in R_1$, then $t^k \in R_k = 0$, so $(1 - t)(1 + t + \cdots + t^{k-1}) = 1 - t^k = 1$, thus $1 - t$ is an invertible element of R. Now, if I is a maximal ideal distinct from R_1, then $R = R_1 + I$, so there exist $t \in R_1$, $u \in I$ such that $1 = t + u$, hence

$u = 1 - t$ is invertible and $R = Ru \subseteq I$, which is absurd. This shows that if $y \in R$, but $y \notin R_1$ then y is invertible and proves condition (4).

For every $i = 1, \ldots, k$ there exists an isomorphism of R-modules $\lambda_i :$ $R_{i-1}/R_i \to R/R_1$ such that $\theta_1 \circ \lambda_i = \lambda_i \circ \theta_i$. Indeed, let $u \in R_{i-1}$, $u \notin R_i$, hence $R_i \subset R_i + Ru \subseteq R_{i-1}$. By (1) and (2) we have $R_{i-1} = R_i + Ru$. Given any element $y + R_i \in R_{i-1}/R_i$ let $y = y'u + y''$ with $y' \in R$, $y'' \in R_i$; we put $\lambda_i(y + R_i) = y' + R_1$. This defines uniquely the mapping

$$\lambda_i : R_{i-1}/R_i \to R/R_1.$$

In fact, if $y + R_i = z + R_i$ with $z \in R_{i-1}$ and $z = z'u + z''$ with $z' \in R$, $z'' \in R_i$, then $y - z = (y' - z')u + (y'' - z'')$ so $(y' - z')u \in R_i$. Since $u \notin R_i$, it follows from (4) that $y' - z' \in R_1$ so $y' + R_1 = z' + R_1$. It is also obvious that λ_i is a homomorphism of R-modules. Moreover, if $y = y'u + y''$ where $y' \in R_1$ then $y \in R_i$ (by (3)) hence if $\lambda_i(y + R_i) = \overline{0} \in R/R_1$ then $y + R_i = \overline{0} \in R_{i-1}/R_i$. Of course, for every $y' \in R$ if $y = y'u \in R_{i-1}$, then $\lambda_i(y + R_i) = y' + R_1$. Thus λ_i is an isomorphism.

It remains to show that $\theta_1 \circ \lambda_i = \lambda_i \circ \theta_i$. Given $y \in R_{i-1}$, if $y = y'u + y''$ with $y' \in R$, $y'' \in R_i$, then $xy = (xy')u + xy''$ with $xy' \in R$, $xy'' \in R_i$, and so we have

$$\theta_1(\lambda_i(y + R_i)) = \theta_1(y' + R_1) = \theta(y') + R_1 = xy' + R_1$$
$$= \lambda_i(xy + R_i) = \lambda_i(\theta(y) + R_i) = \lambda_i(\theta_i(y + R_i)).$$

If $\overline{B}_i$ is a basis of the K-vector space R_{i-1}/R_i and $\lambda_i(\overline{B}_i)$ is the corresponding basis of the isomorphic vector space R/R_1, then the matrices of θ_i with respect to $\overline{B}_i$ and of θ_1 with respect to $\lambda_i(\overline{B}_i)$ are the same. Hence,

$$F_{(R_{i-1}/R_i)|K}(\theta_i) = F_{(R/R_1)|K}(\theta_1).$$

Therefore,

$$F_{R|K}(\theta) = \left[F_{(R/R_1)|K}(\theta_1)\right]^k. \quad\blacksquare$$

We apply these considerations of Linear Algebra to the following specific situation.

Let A be a Dedekind domain, K its field of quotients, let $L|K$ be a separable extension of degree n, and B the integral closure of A in L, so B is also a Dedekind domain (Chapter 7, (M)). If P is a nonzero prime ideal of A, let $BP = \prod_{i=1}^{g} Q_i^{e_i}$, where each Q_i is a prime ideal of B. We recall (see (I)) that under these hypotheses B/BP is a vector space of dimension n over A/P. Let $\psi : A \to A/P = \overline{K}$, $\psi_0 : B \to B/BP$, $\psi_i : B \to B/Q_i = \overline{L}_i$ be the canonical ring-homomorphisms; for every $i = 1, \ldots, g$ let

$$\pi_i : B/BP \to B/Q_i^{e_i}$$

be the ith projection induced by the natural isomorphism

$$B/BP \xrightarrow{\sim} \prod_{i=1}^{g} B/Q_i^{e_i};$$

explicitly, if $y \in B$ then $\psi_0(y) = y + BP$, $\pi_i(\psi_0(y)) = y + Q_i^{e_i}$. These mappings are naturally extended to the polynomials, by acting on their coefficients.

With these notations and hypotheses, we have the following relations between characteristic polynomials, traces, and norms:

P. *If $x \in B$ then $F_{L|K}(x) \in A[X]$ and*

$$\psi(F_{L|K}(x)) = \prod_{j=1}^{g} [F_{\overline{L}_j|\overline{K}}(\psi_j(x))]^{e_j},$$

$$\psi(\mathrm{Tr}_{L|K}(x)) = \sum_{j=1}^{g} e_j \, \mathrm{Tr}_{\overline{L}_j|\overline{K}}(\psi_j(x)),$$

$$\psi(N_{L|K}(x)) = \prod_{j=1}^{g} [N_{\overline{L}_j|\overline{K}}(\psi_j(x))]^{e_j}.$$

Proof: Since $x \in B$ its minimal polynomial over K has coefficients in A; therefore its characteristic polynomial, which is a power of the minimal polynomial (see **(J)**), also belongs to $A[X]$.

Let S be the multiplicative set, complement of P in A, and let $A' = S^{-1}A$, $B' = S^{-1}B$, $P' = A'P$, so $B'P' = B'(BP) = B'(A'P)$. By **(H)** we have $B'/B'P = B/BP$, $A'/A'P = A/P$.

A' is a principal ideal domain, and B' is its integral closure in L (by **(D)**); moreover, B' is a free A'-module of rank n. By the corollary of **(L)**:

$$F_{B'|A'}(x) = F_{L|K}(x).$$

Since $B'/B'P' = B/BP$ is a vector space of dimension n over $A'/A'P = A/P = \overline{K}$ (by **(I)**), it follows from **(N)** and **(M)** that

$$\psi(F_{L|K}(x)) = \psi(F_{B'|A'}(x)) = F_{(B'/B'P')|(A'/A'P)}(\psi_0(x))$$

$$= F_{(B/BP)|\overline{K}}(\psi_0(x)) = \prod_{j=1}^{g} F_{(B/Q_j^{e_j})|\overline{K}}(\pi_j \psi_0(x)).$$

It remains now to determine these last characteristic polynomials and for this purpose we apply **(O)**, taking $k = e_j$:

$$R = B/Q_j^k, \qquad R_1 = Q_j/Q_j^k, \ldots, \qquad R_i = Q_j^i/Q_j^k, \ldots,$$

$$R_{k-1} = Q_j^{k-1}/Q_j^k, \qquad R_k = 0,$$

we have R is a ring, $\overline{K} = A/P$ is a subfield of R, and R is a $\overline{K}$-vector space of finite dimension (equal to the inertial degree of Q_j in $L|K$). We have the strictly decreasing chain of $\overline{K}$-subspaces $R \supset R_1 \supset R_2 \supset \cdots \supset R_{k-1} \supset R_k = 0$; actually, we may define a scalar multiplication as follows:

$$(b + Q_j^k) \cdot (y + Q_j^k) = by + Q_j^k, \qquad \text{where} \quad b \in B, \quad y \in Q_j^i.$$

Then each R_i becomes an ideal of R. Since B is a Dedekind domain there exists no ideal J such that $Q_j^{i-1} \supset J \supset Q_j^i$; hence condition (2) of (**O**) is satisfied. Condition (3) is obvious and (4) follows from the fact that B is a Dedekind domain: if $y, z \in B$, $\bar{y} = y + Q_j^k \in R$, $\bar{z} = z + Q_j^k \in R$, and $\bar{y} \cdot \bar{z} = yz + Q_j^k \in R_i$, but $\bar{y} \notin R_1$ then $y \notin Q_j$, $yz \in Q_j^i$ so $z \in Q_j^i$ and $\bar{z} \in R_i$.

Thus, if $x \in B$ and $\bar{x} = x + Q_j^k \in R$ then

$$F_{R|\overline{K}}(\bar{x}) = [F_{(R/R_1)|\overline{K}}(\theta_1)]^k,$$

where $\theta_1 : R/R_1 \to R/R_1$ is defined by $\theta_1(\bar{y} + R_1) = \bar{x}\bar{y} + R_1$, so θ_1 is the mapping of multiplication by $\bar{x} = \pi_j \psi_0(x)$.

Now $R/R_1 = (B/Q_j^k)/(Q_j/Q_j^k) \cong B/Q_j = \bar{L}_j$; the isomorphism

$$\eta : R/R_1 \to B/Q_j$$

is given explicitly as follows: If $\bar{y} + R_1 \in R/R_1$, with $y \in B$, then $\eta(\bar{y} + R_1) = \psi_j(y)$. Then we have

$$
\begin{array}{ccc}
R/R_1 & \xrightarrow{\ \eta\ } & \bar{L}_j \\
\theta_1 \downarrow & & \downarrow \theta_{\psi_j(x)} \\
R/R_1 & \xrightarrow{\ \eta\ } & \bar{L}_j
\end{array}
$$

$$\theta_{\psi_j(x)} \circ \eta(\bar{y} + R_1) = \psi_j(x)\psi_j(y) = \psi_j(xy)$$
$$= \eta(\bar{x}\bar{y} + R_1) = \eta \circ \theta_1(\bar{y} + R_1).$$

Therefore, $F_{(R/R_1)|\overline{K}}(\theta_1) = F_{\bar{L}_j|\overline{K}}(\psi_j(x))$. Concluding, we have shown that

$$\psi(F_{L|K}(x)) = \prod_{j=1}^{g} [F_{\bar{L}_j|\overline{K}}(\psi_j(x))]^{e_j}$$

and the relations for the trace and for the norm follow at once.   ∎

Now we shall prove the transitivity of the trace and norm. We have the following situation: C is a commutative ring, B, A are subrings of C, such that $C \supseteq B \supseteq A$, and we assume that B is a free A-module of rank n, while C is a free B-module of rank m. From this it follows that if $\{x_1, \ldots, x_n\}$ is an A-basis of B and $\{y_1, \ldots, y_m\}$ is a B-basis of C, then

$$\{x_1 y_1, x_2 y_1, \ldots, x_n y_1, x_1 y_2, \ldots, x_n y_m\}$$

is an A-basis of C, and so C is a free A-module of rank mn.

Thus, we may consider for every element $y \in C$ the elements: $\mathrm{Tr}_{C|A}(y)$ and $\mathrm{Tr}_{B|A}(\mathrm{Tr}_{C|B}(y))$ as well as the corresponding elements for the norm.

Q.
$$\mathrm{Tr}_{C|A}(y) = \mathrm{Tr}_{B|A}(\mathrm{Tr}_{C|B}(y)),$$
$$N_{C|A}(y) = N_{B|A}(N_{C|B}(y)),$$

for every element $y \in C$.

Proof: Let θ be any endomorphism of the B-module C. Thus, θ also satisfies $\theta(ay) = a\theta(y)$ for every $a \in A$, that is, θ is also an endomorphism of the A-module C and, as such, it will be denoted by θ_A.

To find the matrix of θ with respect to the basis $\{y_1, \ldots, y_m\}$ we write

$$\theta(y_j) = \sum_{i=1}^{m} b_{ij} y_i, \quad \text{with} \quad b_{ij} \in B.$$

So,

$$b_{ij} = \sum_{k=1}^{n} a_{kij} x_k$$

for all indices $i, j = 1, \ldots, m$.

To find the matrix of θ_A with respect to the basis

$$\{x_1 y_1, x_2 y_1, \ldots, x_n y_1, x_1 y_2, \ldots, x_n y_2, \ldots, x_n y_m\}$$

we note that $\theta(x_l y_j) = \sum_{i=1}^{m} x_l b_{ij} y_i$, and we write

$$x_l b_{ij} = \sum_{k=1}^{n} a_{kij} x_l x_k, \qquad x_l x_k = \sum_{h=1}^{n} a'_{lhk} x_h$$

for all $l, k = 1, \ldots, n$, hence

$$x_l b_{ij} = \sum_{k=1}^{n} a_{kij} \sum_{h=1}^{n} a'_{lhk} x_h = \sum_{h=1}^{n} \left(\sum_{k=1}^{n} a_{kij} a'_{lhk} \right) x_h.$$

Thus

$$\theta(x_l y_j) = \sum_{i=1}^{m} \sum_{h=1}^{n} \left(\sum_{k=1}^{n} a_{kij} a'_{lhk} \right) x_h y_i,$$

therefore, the matrix M of θ_A, with respect to the basis considered above, has entry

$$\sum_{k=1}^{n} a_{kij} a'_{lhk}$$

at the row (h, i) and column (l, j).

On the other hand, By_j is a free A-module with basis $\{x_1 y_j, \ldots, x_n y_j\}$ and similarly for By_i. Let $\theta_{ji} : By_j \to By_i$ be the A-linear transformation defined by $\theta_{ji}(xy_j) = x b_{ij} y_i$ for every $x \in B$. With respect to the above bases, the matrix M_{ji} of θ_{ji} is obtained as follows:

$$\theta_{ji}(x_l y_j) = \sum_{h=1}^{n} \left(\sum_{k=1}^{n} a_{kij} a'_{lhk} \right) x_h y_i,$$

thus M_{ji} is a $n \times n$ matrix, with coefficients in A, and its entry in row h, column l is $\sum_{k=1}^{n} a_{kij}a'_{lhk}$.

Therefore, M may be written as a matrix of m^2 blocks M_{ji}, each being a $n \times n$ matrix with coefficients in A:

$$M = \begin{pmatrix} M_{11} & M_{12} & \cdots & M_{1m} \\ M_{21} & M_{22} & \cdots & M_{2m} \\ \vdots & \vdots & & \vdots \\ M_{m1} & M_{m2} & \cdots & M_{mm} \end{pmatrix}.$$

We now prove that the matrices M_{ji} are permutable (by multiplication). Let λ_{ji} be the endomorphism of the A-module B defined by $\lambda_{ji}(x) = xb_{ij}$; thus $\lambda_{ji}(x_l) = x_l b_{ij} = \sum_{h=1}^{n} (\sum_{k=1}^{n} a_{kij}a'_{lhk})x_h$. With respect to the basis $\{x_1, \ldots, x_n\}$ of B, the matrix of λ_{ji} is equal to M_{ji}. Since $\lambda_{ji} \circ \lambda_{kh}(x) = xb_{hk}b_{ij} = xb_{ij}b_{hk} = \lambda_{kh} \circ \lambda_{ji}(x)$ for every $x \in B$, then the corresponding matrices satisfy $M_{ji}M_{kh} = M_{kh}M_{ji}$, as we have claimed.

Now let $\theta = \theta_y$, the B-endomorphism of C defined by multiplication with y; thus θ_A is the induced A-endomorphism of C. By definition, $\mathrm{Tr}_{C|A}(y) = \mathrm{Tr}(\theta_A)$, so it is equal to the sum of the elements in the diagonal of the matrix M which corresponds to θ_A; as we proved, this sum is equal to

$$\sum_{i=1}^{m} (\text{sum of diagonal elements of } M_{ii}) = \sum_{i=1}^{m} \mathrm{Tr}(M_{ii});$$

but M_{ii} is the matrix of λ_{ii} (which is the A-endomorphism of multiplication by b_{ii}), hence $\mathrm{Tr}(M_{ii}) = \mathrm{Tr}_{B|A}(b_{ii})$, so $\mathrm{Tr}_{C|A}(y) = \sum_{i=1}^{m} \mathrm{Tr}_{B|A}(b_{ii}) = \mathrm{Tr}_{B|A}(\sum_{i=1}^{m} b_{ii}) = \mathrm{Tr}_{B|A}(\mathrm{Tr}_{C|B}(y))$, because the matrix of θ with respect to the B-basis $\{y_1, \ldots, y_m\}$ of C, has diagonal elements b_{ii} $(i = 1, \ldots, m)$.

In order to prove the corresponding statement for the norm, we recall that $N_{C|A}(y) = \det(\theta_A) = \det(M)$.

We shall soon establish in a lemma that the computation of the determinant of M may be done as follows: Regard the blocks M_{ji} as if they were elements, compute the determinant obtaining a matrix with coefficients in A, and then compute the determinant of this matrix.

Now, we note the following general fact: With respect to a given A-basis of B, if M', M'' are, respectively, the matrices of $\theta_{b'}$, $\theta_{b''}$ then $M' + M''$, $M'M''$ are the matrices of $\theta_{b'+b''}$, $\theta_{b'b''}$. From this, we deduce that if μ is the A-endomorphism of B of multiplication by $N_{C|B}(y) = \det(b_{ij})_{i,j}$, then the matrix of μ with respect to $\{x_1, \ldots, x_n\}$ is equal to $\det(M_{ji})_{j,i}$. Thus

$$\det(\mu) = \det(\det[(M_{ji})_{j,i}]) = \det(M) = N_{C|A}(y);$$

on the other hand,

$$\det(\mu) = \det(\theta_{N_{C|B}(y)}) = N_{B|A}(N_{C|B}(y)),$$

showing the formula for the norm.   ∎

Now, we have to prove the lemma used above:

Lemma 1.   *Let X_{ij} be m^2 indeterminates, and consider the $m \times m$ matrix $X = (X_{ij})_{i,j}$; let D be the determinant of X, $D \in \mathbb{Z}[X_{11}, \ldots, X_{mm}]$. If A is a commutative ring, if M_{ij} are $n \times n$ matrices with coefficients in A, for $i, j = 1, \ldots, m$, such that $M_{ij}M_{kh} = M_{kh}M_{ij}$ for any indices i, j, k, h, and if*

$$M = \begin{pmatrix} M_{11} & M_{12} & \cdots & M_{1m} \\ M_{21} & M_{22} & \cdots & M_{2m} \\ \vdots & \vdots & & \vdots \\ M_{m1} & M_{m2} & \cdots & M_{mm} \end{pmatrix}$$

is considered as an $mn \times mn$ matrix with elements in A, then

$$\det(M) = \det(D(M_{11}, \ldots, M_{mm})).$$

Proof: The result is true when $m = 1$, and it will be proved by induction on m. In order to include the case where the ring A may have zero-divisors, we make use of the following device. Let T be a new indeterminate, for all indices i, j, let, as usual, δ_{ij} be 0 when $i \neq j$, and $\delta_{ii} = 1$; we denote by N_{ij} the matrix $N_{ij} = M_{ij} + \delta_{ij}TI_n$, where I_n is the unit $n \times n$ matrix.

Computing the determinant of the matrix X by considering cofactors of the elements in any column, we have the well-known relations

$$\sum_{i=1}^{m} X_{ji}D^{ik} = \delta_{jk}D,$$

where D^{ik} is the cofactor of X_{ki} in the matrix X, so $D^{ik} \in \mathbb{Z}[X_{11}, \ldots, X_{mm}]$.

Let $D^{ik}(N_{11}, \ldots, N_{mm}) = N^{ik}$, so N^{ik} is a $n \times n$ matrix with entries in $A[T]$.

If

$$P = \begin{pmatrix} N^{11} & N^{12} & \cdots & N^{1m} \\ 0 & I_n & \cdots & 0 \\ \vdots & \vdots & & \vdots \\ 0 & 0 & \cdots & I_n \end{pmatrix}$$

and

$$N = \begin{pmatrix} N_{11} & N_{12} & \cdots & N_{1m} \\ N_{21} & N_{22} & \cdots & N_{2m} \\ \vdots & \vdots & & \vdots \\ N_{m1} & N_{m2} & \cdots & N_{mm} \end{pmatrix},$$

by multiplication we have

$$PN = \begin{pmatrix} D(N_{11}, N_{12}, \ldots, N_{mm}) & 0 & \cdots & 0 \\ N_{21} & N_{22} & \cdots & N_{2m} \\ \vdots & \vdots & & \vdots \\ N_{m1} & N_{m2} & \cdots & N_{mm} \end{pmatrix}.$$

Let

$$Q = \begin{pmatrix} N_{22} & \cdots & N_{2m} \\ \vdots & & \vdots \\ N_{m2} & \cdots & N_{mm} \end{pmatrix},$$

so Q is an $(m-1)n \times (m-1)n$ matrix with entries in $A[T]$. Since the first row of PN has only one block which is not zero

$$\det(PN) = \det(D(N_{11}, \ldots, N_{mm})) \cdot \det(Q);$$

but, on the other hand, $\det(PN) = \det(P) \cdot \det(N)$, and $\det(P) = \det(N^{11})$.

Applying the induction on Q, we have

$$\det(Q) = \det(D^{11}(N_{11}, \ldots, N_{mm})) = \det(N^{11}).$$

But $\det(N^{11})$ is a monic polynomial in T, having degree $n(m-1)$, so it is not a zero-divisor in the ring $A[T]$. Therefore, we conclude that

$$\det(D(N_{11}, \ldots, N_{mm})) = \det(N).$$

Now, letting $h : A[T] \to A$ be the homomorphism such that $h(T) = 0$ and h leaves fixed every element of A, we deduce that h induces a homomorphism h^* from the associated matrix rings and

$$\begin{aligned} \det(D(M_{11}, \ldots, M_{mm})) &= \det(D(h^*(N_{11}), \ldots, h^*(N_{mm}))) \\ &= \det(h^*(D(N_{11}, \ldots, N_{mm}))) \\ &= h[\det(D(N_{11}, \ldots, N_{mm}))] \\ &= h(\det(N)) = \det(h^*(N)) = \det(M). \quad \blacksquare \end{aligned}$$

For later use, we record the following special case. Let $N = (x_{ij})_{i,j}$ be an $n \times n$ matrix, let L be an $l \times l$ matrix. For each $i, j = 1, \ldots, n$, let $M_{ij} = (x_{ij}I)L$, where I is the unit $l \times l$ matrix. Let M be as in the statement of Lemma 1.

Now $D(M_{11}, \ldots, M_{nn}) = \det(N)I \cdot L^n$ so $\det(M) = \det(N)^l \cdot \det(L)^n$.

12.3 Discriminant of Ring Extensions

Let B be a commutative ring, and A a subring of B such that B is a free A-module of rank n. If $x_1, \ldots, x_n \in B$ we define the *discriminant of* $(x_1, \ldots, x_n)$ (in the ring extension $B|A$) as

$$\text{discr}_{B|A}(x_1, \ldots, x_n) = \det(\text{Tr}_{B|A}(x_i x_j));$$

that is, the determinant of the matrix whose (i, j)-entry is $\text{Tr}_{B|A}(x_i x_j)$. Thus $\text{discr}(x_1, \ldots, x_n) \in A$.

Let us note at once, if $B = L$, $A = K$ where $L|K$ is a separable field extension of degree n, then by (**J**) the new concept of discriminant coincides with the one in Chapter 2, Section 11.

R.   *If* $(x_1', \ldots, x_n')$ *is another* n-tuple *of elements in* B, *and* $x_j' = \sum_{i=1}^{n} a_{ij} x_i$ *for all* $j = 1, \ldots, n$, *with* $a_{ij} \in A$, *then*

$$\text{discr}_{B|A}(x_1', \ldots, x_n') = \left[\det(a_{ij})\right]^2 \cdot \text{discr}_{B|A}(x_1, \ldots, x_n).$$

Proof: The proof is standard. We first note that

$$\text{Tr}_{B|A}(x_i' x_j') = \text{Tr}_{B|A}\left[\left(\sum_{k=1}^{n} a_{ki} x_k\right)\left(\sum_{h=1}^{n} a_{hj} x_h\right)\right]$$

$$= \sum_{k=1}^{n} \sum_{h=1}^{n} a_{ki} a_{hj} \, \text{Tr}(x_k x_h),$$

hence letting $M = (a_{ij})$ and M' denote the transpose matrix of M, then

$$\begin{aligned}
\text{discr}_{B|A}(x_1', \ldots, x_n') &= \det(\text{Tr}_{B|A}(x_i' x_j')) \\
&= \det(M' \cdot (\text{Tr}(x_k x_h)) \cdot M) \\
&= \det(M') \cdot \det(\text{Tr}(x_k x_h)) \cdot \det(M) \\
&= \left[\det(a_{ij})\right]^2 \cdot \text{discr}_{B|A}(x_1, \ldots, x_n). \quad\blacksquare
\end{aligned}$$

From the next result, we deduce that it is only interesting to consider the discriminant of linearly independent n-tuples:

S.   *If* $\{x_1, \ldots, x_n\}$ *is linearly dependent over the domain* A *then*

$$\text{discr}_{B|A}(x_1, \ldots, x_n) = 0.$$

Proof: We assume that there exist elements $a_1, \ldots, a_n \in A$, not all equal to zero, such that $\sum_{j=1}^{n} a_j x_j = 0$. For example, let $a_1 \neq 0$.

Now, we consider the n-tuple $(x_1', \ldots, x_n')$, where $x_1' = 0$, $x_i' = x_i$ for $i = 2, \ldots, n$. Thus, $x_i' = \sum_{j=1}^{n} a_{ji} x_j$ $(i = 1, \ldots, n)$ by letting $a_{j1} = a_j$, and if $i > 1$, then $a_{ji} = 1$ for $j = i$, $a_{ji} = 0$ for $j \neq i$. By (**R**) we have

$$0 = \text{discr}_{B|A}(0, x_2, \ldots, x_n) = \left[\det(a_{ij})\right]^2 \cdot \text{discr}_{B|A}(x_1, \ldots, x_n).$$

Since $\det(a_{ij}) = a_1 \neq 0$ and A is a domain, then

$$\text{discr}_{B|A}(x_1, \ldots, x_n) = 0. \quad\blacksquare$$

T.   *Let A be a domain, and let $\{x_1,\ldots,x_n\}$, $\{x_1',\ldots,x_n'\}$ be any two bases of the A-module B. Then: either*

$$\operatorname{discr}_{B|A}(x_1,\ldots,x_n) = \operatorname{discr}_{B|A}(x_1',\ldots,x_n') = 0,$$

or

$$\operatorname{discr}_{B|A}(x_1,\ldots,x_n), \qquad \operatorname{discr}_{B|A}(x_1',\ldots,x_n')$$

are associated elements of A (see Chapter 1, Section 1).

Proof: By hypothesis there exist elements $a_{ij} \in A$ such that

$$x_j' = \sum_{i=1}^{n} a_{ij}x_i,$$

for every $j = 1,\ldots,n$. By (**R**) we have

$$\operatorname{discr}_{B|A}(x_1',\ldots,x_n') = \big[\det(a_{ij})\big]^2 \cdot \operatorname{discr}_{B|A}(x_1,\ldots,x_n).$$

Since $(a_{ij})_{i,j}$ is an invertible matrix, then $\det(a_{ij})$ is a unit in the ring A; hence either both discriminants are zero or both are associated elements of A. ∎

The preceding result justifies the following definition:

Let A be a domain, let B be a commutative ring, having A as a subring, and such that B is a free A-module of rank n. If $\{x_1,\ldots,x_n\}$ is any basis of the A-module B, the principal ideal $A \cdot \operatorname{discr}_{B|A}(x_1,\ldots,x_n)$ is called the *discriminant of B relative to A*, and denoted by $\operatorname{discr}(B|A)$.

In the case where A is a field K, $\operatorname{discr}(B|K)$ is either 0 or the unit ideal of K (since K has only trivial ideals). Moreover, we shall see in (**W**), that if L is an algebraic number field then $\operatorname{discr}(L|\mathbb{Q}) = \mathbb{Q}$, the unit ideal of $\mathbb{Q}$; so this concept does not constitute an appropriate generalization of δ_L, the discriminant of the field L, introduced in Chapter 6, Definition 4. In the next chapter we shall explain what is the relative discriminant $\delta_{L|K}$ of an algebraic number field L over a subfield K.

One of the tools used in determining the discriminant is the following easy result:

U.   *Let $B_1, \ldots, B_r$ be commutative rings, containing the domain A and such that each ring B_i is a free A-module of finite rank. Then*

$$\operatorname{discr}(B_1 \times \cdots \times B_r|A) = \prod_{i=1}^{r} \operatorname{discr}(B_i|A).$$

Proof: It is enough to prove the statement when $r = 2$.

Let $\{x_1,\ldots,x_n\}$ be a basis of the A-module B_1, let $\{y_1,\ldots,y_m\}$ be a basis of the A-module B_2. Then $\{(x_1,0),\ldots,(x_n,0),(0,y_1),\ldots,(0,y_m)\}$ is a basis of the A-module $B_1 \times B_2$. Letting $z_i = (x_i,0)$ for $i = 1,\ldots,n$, $z_{n+i} = (0,y_i)$ for $i = 1,\ldots,m$, then $\operatorname{discr}(B_1 \times B_2|A)$ is the principal ideal of A generated by $\det(\operatorname{Tr}_{B_1 \times B_2|A}(z_iz_j))$.

Now, if $t \in B_1$ then $\mathrm{Tr}_{B_1 \times B_2 | A}(t, 0) = \mathrm{Tr}_{B_1|A}(t)$, as we deduce by considering the matrices of the endomorphisms $\theta_{(t,0)}$ of $B_1 \times B_2$ and θ_t of B_1, relative to the basis $\{z_1, \ldots, z_{n+m}\}$ and $\{x_1, \ldots, x_n\}$, respectively. In the same way, if $t \in B_2$ then $\mathrm{Tr}_{B_1 \times B_2 | A}(0, t) = \mathrm{Tr}_{B_2|A}(t)$. Thus,

$$\det(\mathrm{Tr}_{B_1 \times B_2 | A}(z_i z_j)) = \det \left(\begin{array}{c|c} (\mathrm{Tr}_{B_1|A}(x_i x_j)) & 0 \\ \hline 0 & (\mathrm{Tr}_{B_2|A}(y_i y_j)) \end{array} \right)$$

$$= \det(\mathrm{Tr}_{B_1|A}(x_i x_j)) \cdot \det(\mathrm{Tr}_{B_2|A}(y_i y_j))$$

and so this element generates the ideal $\mathrm{discr}(B_1|A) \cdot \mathrm{discr}(B_2|A)$.   ∎

V.   *If K is a field, and if B is a commutative algebra of dimension n over K,* then $\mathrm{discr}(B|K) = 0$ if and only if the trace in $B|K$ is degenerate, that is, there exists an element $x \in B$, $x \neq 0$, such that $\mathrm{Tr}_{B|K}(xy) = 0$ for every $y \in B$.*

Proof: Let us assume that the trace is degenerate, with $x \in B$, $x \neq 0$, such that $\mathrm{Tr}_{B|K}(xy) = 0$ for every $y \in B$. Let us consider a basis $\{x_1, \ldots, x_n\}$ of the vector space B over K, such that $x_1 = x$. Then $\mathrm{discr}(B|K)$ is the ideal of K generated by $\mathrm{discr}_{B|K}(x_1, \ldots, x_n) = \det(\mathrm{Tr}_{B|K}(x_i x_j)) = 0$.

Conversely, if $\mathrm{discr}(B|K) = 0$, let $\{x_1, \ldots, x_n\}$ be a K-basis of B, hence $\mathrm{discr}_{B|K}(x_1, \ldots, x_n) = \det(\mathrm{Tr}_{B|K}(x_i x_j)) = 0$; thus, there exist elements $a_i \in K$, not all equal to zero, such that $\sum_{i=1}^{n} a_i \cdot \mathrm{Tr}_{B|K}(x_i x_j) = 0$ for every $j = 1, \ldots, n$. Thus, letting $x = \sum_{i=1}^{n} a_i x_i$, we have $x \neq 0$ and for every element $y = \sum_{j=1}^{n} b_j x_j \in B$ (with $b_j \in K$) we have

$$\mathrm{Tr}_{B|K}(xy) = \sum_{i,j=1}^{n} a_i b_j \, \mathrm{Tr}(x_i x_j) = 0;$$

this shows that the trace is degenerate.   ∎

Let us assume now that K is a *perfect field*, that is, every algebraic extension L of K is separable; we may improve the preceding result, taking into account the fact that if $L|K$ is separable, there exists an element $x \in L$ such that $\mathrm{Tr}_{L|K}(x) \neq 0$ (see Chapter 2, Section 10). We note that every field of characteristic zero is perfect; also, every finite field is perfect.

W.   *Let K be a perfect field, and let B be a commutative K-algebra of finite dimension. Then $\mathrm{discr}(B|K) \neq 0$ if and only if 0 is the only nilpotent element of B.*

Proof: Let us assume that B contains the nilpotent element $x \neq 0$. Let $\{x_1, \ldots, x_n\}$ be a K-basis of B, such that $x_1 = x$. Since B is commutative,

---

* We may therefore identify K with a subring of B.

then xx_j is also nilpotent. The minimal polynomial of the endomorphism θ_{xx_j}, of multiplication by xx_j, is equal to X^r, for some $r > 0$; as is known from the theory of linear transformations of vector spaces, the characteristic polynomial of θ_{xx_j} is a multiple of the minimal polynomial, having the same irreducible factors and degree n; thus, the characteristic polynomial is X^n, and $\mathrm{Tr}_{B|K}(xx_j) = 0$ for every $j = 1, \ldots, n$.

Hence $\mathrm{discr}(x_1, \ldots, x_n) = \det(\mathrm{Tr}_{B|K}(x_i x_j)) = 0$ because the matrix of traces has the first row of zeros. This shows that $\mathrm{discr}(B|K) = 0$.

Conversely, let us assume that 0 is the only nilpotent element of B. We note that since every ideal of B is in particular a subspace of the K-vector space B, from the fact that B has dimension n over K, every chain of subspaces, hence also of ideals of B, must be finite. Thus, B is a Noetherian ring.

We shall require the following lemma:

Lemma 2.   *If B is a Noetherian ring, such that 0 is the only nilpotent element, then the zero-ideal is the intersection of finitely many prime ideals.*

Assuming the lemma true, we may write $0 = P_1 \cap \cdots \cap P_r$, where each P_i is a prime ideal of B. Since $P_i \cap K$ is an ideal of K, distinct from K, then $P_i \cap K = 0$ for $i = 1, \ldots, r$. Thus $K \subseteq B/P_i$ (up to a natural identification), and B/P_i is a finite-dimensional K-vector space which is also a domain. Since every element of B/P_i is integral over K (by Chapter 5, (**A**)) then $B/P_i = L_i$ is a field (by Chapter 5, (**F**)), so P_i is a maximal ideal of B.

Now, we know that the distinct ideals $P_1, \ldots, P_r$ are maximal; hence $P_i + \bigcap_{j \neq i} P_j = B$, otherwise $P_i \supseteq \bigcap_{j \neq i} P_j$, hence $P_i \supseteq P_j$ for some $j \neq i$ and necessarily $P_i = P_j$ which is not true, since these ideals are distinct. By Chapter 7, (**K**), we have $B = B/0 \cong \prod_{i=1}^r B/P_i = \prod_{i=1}^r L_i$. Hence by (**U**):

$$\mathrm{discr}(B|K) = \prod_{i=1}^r \mathrm{discr}(L_i|K).$$

But the field L_i is a finite extension, thus an algebraic extension of K. Since K is a perfect field, L_i is separable over K. As we quoted, there exists an element $x_i \in L_i$ such that $\mathrm{Tr}_{L_i|K}(x_i) \neq 0$; so the trace is not degenerate (because if there exists $x' \in L_i$, $x' \neq 0$, such that $\mathrm{Tr}_{L_i|K}(x'y) = 0$ for every $y \in L_i$, then from $x_i = x'(x'^{-1}x_i)$ we would have $\mathrm{Tr}_{L_i|K}(x_i) = 0$). By (**V**), $\mathrm{discr}(L_i|K) \neq 0$, thus, $\mathrm{discr}(L_i|K) = K$; hence $\mathrm{discr}(B|K) = K$. ∎

Proof of the Lemma: By Chapter 7, (**C**), we have $0 = \prod_{i=1}^r P_i^{e_i}$, where the prime ideals P_i are distinct, $e_i \geq 1$ for $i = 1, \ldots, r$. We show that $P_1 \cap \cdots \cap P_r = 0$. If $x \in P_1 \cap \cdots \cap P_r$ then $x^{e_1 + e_2 + \cdots + e_r} \in P_1^{e_1} P_2^{e_2} \cdots P_r^{e_r} = 0$, hence x is nilpotent and therefore $x = 0$. ∎

The following result is the crucial part of the main theorem to be proved in the next chapter:

X.   *Let A be a principal ideal domain, and K its field of quotients: let $L|K$ be a separable extension of degree n and B the integral closure of A in L. Let P be a nonzero prime ideal of A such that the field A/P is perfect. Then the ring B/BP has nonzero nilpotent elements if and only if $P \supseteq \mathrm{discr}(B|A)$.*

Proof: B is a free A-module of rank n (Chapter 6, **(B)** and Theorem 1) and a Dedekind domain (by Chapter 7, **(P)**). By **(I)**, B/BP is a vector space of dimension n over the field A/P; actually, if $\{x_1, \ldots, x_n\}$ is a basis of the A-module B, then their images in B/BP form a basis $\{\overline{x}_1, \ldots, \overline{x}_n\}$ of the A/P-vector space B/BP. Thus, $\mathrm{discr}(B|A) = \mathrm{discr}_{B|A}(x_1, \ldots, x_n)$, and $\mathrm{discr}((B/BP)|(A/P)) = \mathrm{discr}_{(B/BP)|(A/P)}(\overline{x}_1, \ldots, \overline{x}_n)$.

Since A/P is a perfect field, by **(V)**, B/BP has nonzero nilpotent elements if and only if $\mathrm{discr}((B/BP)|(A/P)) = 0$; by **(N)** this means that

$$0 = \mathrm{discr}_{(B/BP)|(A/P)}(\overline{x}_1, \ldots, \overline{x}_n) = \det(\mathrm{Tr}_{(B/BP)|(A/P)}(\overline{x}_i\overline{x}_j))$$

$$= \overline{\det(\mathrm{Tr}_{B|A}(x_ix_j))} = \overline{\mathrm{discr}_{B|A}(x_1, \ldots, x_n)},$$

that is, $\mathrm{discr}(B|A) \subseteq P$.   ∎

EXERCISES

1. Determine explicitly the ring of fractions $S^{-1}\mathbb{Z}$ in the following cases:

(a) $S = S_1$ is the set of all odd integers.

(b) $S = S_2$ is the set of all powers of 2.

Determine $S_1^{-1}\mathbb{Z} \cap S_2^{-1}\mathbb{Z}$.

2. Let K be a field, $n \geq 1$, $R = K[X_1, \ldots, X_n]$, and let $S = \{f \in R \mid f(0, \ldots, 0) \neq 0\}$. Determine $S^{-1}R$, and its maximal ideals. If M is a maximal ideal of $S^{-1}R$, determine $\bigcap_{n=0}^{\infty} M^n$ and $S^{-1}R/M$.

3. Let K be a field, and let S be the set of all polynomials $f \in K[X_1, X_2]$ such that $f(0,1) \neq 0$ and $f(1,0) \neq 0$. Show that S is a multiplicative subset of $K[X_1, X_2]$, determine the maximal ideals of $K[X_1, X_2]_S$.

4. Let P be the prime ideal of the ring A of Gaussian integers which is generated by $1 + i$. Determine explicitly the ring of fractions A_P. What is its maximal ideal?

5. Let A be the ring of algebraic integers of a field K of algebraic numbers. Show that there exists an infinite sequence of subrings

$$A = R_0 \subset R_1 \subset \cdots$$

of K and also an infinite sequence of subrings

$$K \supset R_1' \supset R_2' \supset \cdots \supset A.$$

6. Let K be a field, $R = K[[X]]$ (the ring of formal power series with coefficients in K). Let S be the set of power series with nonzero constant term. Calculate $S^{-1}R$.

7. Let $R = \mathbb{Z}[[X]]$ and let S be the multiplicative subset of R generated by 2 and X^2. Determine $S^{-1}R$.

8. Let A be a domain, having the following property: Every element of A is the product of finitely many prime elements (for example, this holds when A is a unique factorization domain). Prove that there exists a one-to-one correspondence between the sets of prime elements of A and the rings of fractions of A (with respect to multiplicative subsets).

9. Let A be a Euclidean domain (see Chapter 5, Definition 5), and let K be its field of quotients. Prove that if B is a subring of K containing A then there exists a multiplicative subset S of A such that $B = A_S$.

10. Let $A = \mathbb{Z}[X, Y]$, $B = A[\sqrt{X}]$, and $C = A[\sqrt{X}, \sqrt{Y}]$. Determine the characteristic polynomial, the trace, and the norm of $\sqrt{X} + \sqrt{Y}$ in the extensions C/A and C/B.

11. Let ζ be a primitive cubic root of 1, let

$$V = \begin{pmatrix} 1 & 1 & 1 \\ 1 & \zeta & \zeta^2 \\ 1 & \zeta^2 & \zeta \end{pmatrix}.$$

Let $\sigma(1) = 1$, $\sigma(\zeta) = \zeta^2$, $\sigma(\zeta^2) = \zeta$ and consider the iterates σ^k of σ, namely $\sigma^2 = \sigma \circ \sigma, \ldots, \sigma^k = \sigma^{k-1} \circ \sigma$. Let

$$\sigma^k(V) = \begin{pmatrix} 1 & 1 & 1 \\ 1 & \sigma^k(\zeta) & \sigma^k(\zeta^2) \\ 1 & \sigma^k(\zeta^2) & \sigma^k(\zeta) \end{pmatrix}.$$

Consider the matrix of 3×3 blocks

$$M = \begin{pmatrix} \sigma^2(V) & \sigma^3(V) & \sigma^4(V) \\ \sigma^3(V) & \sigma^4(V) & \sigma^5(V) \\ \sigma^4(V) & \sigma^5(V) & \sigma^6(V) \end{pmatrix}.$$

Calculate the discriminant of V using the method of Lemma 1.

12. Let $B = \mathbb{Q}[X, Y]$ and $A = \mathbb{Q}[X^2, Y^3]$. Calculate $\mathrm{discr}_{B|A}(X, Y)$.

13

The Relative Trace, Norm, Discriminant, and Different

Let K be an algebraic number field, L an extension of finite degree n over K, and let A, B be, respectively, the rings of algebraic integers in K, L. In this situation, A need not be a principal ideal domain and B need not be a free A-module. We shall introduce the relative trace and norm of fractional ideals of L and, in view of characterizing ramified prime ideals, we shall consider the relative discriminant and relative different.

13.1 The Relative Trace and Norm of an Ideal

Let R be a Dedekind domain, let K be its field of quotients, $L|K$ a separable extension of degree n, and T the integral closure of R in L; so T is also a Dedekind domain.

Let J be a fractional ideal of L (relative to T). Then the set $\{\mathrm{Tr}_{L|K}(x) \mid x \in J\}$ is an R-module.

Since J is a finitely generated T-module, say by the elements $x_1/a_1, \ldots,$ x_m/a_m (with $a_1, \ldots, a_m \in R$, all nonzero, and $x_1, \ldots, x_m \in T$) then if $a = a_1 \cdots a_m \in R$ it follows that $aJ \subseteq T$ and $a\{\mathrm{Tr}_{L|K}(x) \mid x \in J\} \subseteq \{\mathrm{Tr}_{L|K}(ax) \mid x \in J\} \subseteq R$ so $\{\mathrm{Tr}_{L|K}(x) \mid x \in J\}$ is a fractional ideal of K (relative to R).

Definition 1. The *relative trace* in $L|K$ of the fractional ideal J of L (relative to T) is

$$\mathrm{Tr}_{L|K}(J) = \{\mathrm{Tr}_{L|K}(x) \mid x \in J\}.$$

If J is an integral ideal, then so is $\mathrm{Tr}_{L|K}(J)$.

We also note the transitivity property. Let R, T, K, L be as before, let $L'|L$ be a separable extension of finite degree, T' the integral closure of T in L'. If J' is a fractional ideal of L' (relative to T'), then

$$\mathrm{Tr}_{L'|K}(J') = \mathrm{Tr}_{L|K}(\mathrm{Tr}_{L'|L}(J')).$$

Now we shall introduce the relative norm of an ideal.

Let $\mathcal{F}_R$, respectively $\mathcal{F}_T$, be the multiplicative group of nonzero fractional ideals of R, respectively T. By Chapter 7, $\mathcal{F}_R$ and $\mathcal{F}_T$ are free Abelian multiplicative groups generated by the sets of prime ideals.

If Q is a nonzero prime ideal of T and $Q \cap R = P$, let $[T/Q : R/P] = f$.

If $\{x_1, \ldots, x_n\}$ is a K-basis of L where each $x_i \in T$, if $\bar{x}_i$ denotes the canonical image of x_i in T/Q, then $\{\bar{x}_1, \ldots, \bar{x}_n\}$ is a set of generators of the R/P-vector space T/Q. Thus $f \leq n$.

Definition 2.   With the above notations, the *relative norm* of Q is $N_{T|R}(Q) = P^f$.

This definition may be extended to each nonzero fractional ideal $J \in \mathcal{F}_T$. We may write, in a unique way, $J = \prod_{i=1}^r Q_i^{e_i}$, where Q_1, $\ldots$, Q_r are distinct prime ideals of T and $e_1, \ldots, e_r$ are nonzero integers.

Definition 2′.   With the above notations, the *relative norm* of J is

$$N_{T|R}(J) = \prod_{i=1}^{r} N_{T|R}(Q_i)^{e_i}.$$

Since $\mathcal{F}_T$ is a free Abelian multiplicative group generated by the prime ideals, the mapping $N_{T|R} : \mathcal{F}_T \to \mathcal{F}_R$ is well defined.

It is also a group-homomorphism: $N_{T|R}(JJ') = N_{T|R}(J) \cdot N_{T|R}(J')$ for all $J, J' \in \mathcal{F}_T$ (multiplicative property of the relative norm).

Moreover, if $\psi : \mathcal{F}_T \to \mathcal{F}_R$ is a group-homomorphism such that $\psi(Q) = N_{T|R}(Q)$ for every nonzero prime ideal Q of T, then $\psi(J) = N_{T|R}(J)$ for every $J \in \mathcal{F}_T$.

We note also that if J is an integral ideal, then so is $N_{T|R}(J)$.

As already seen (Chapter 12, (**D**) and (**G**)) if P is a prime ideal of R, S the set-complement of P in R, then $R' = S^{-1}R$, is a principal ideal domain, $T' = S^{-1}T$ is the integral closure of R' in L, so we may also consider the relative norm of fractional ideals of T', with respect to R'.

Our first result is:

A.   *With the above notations, if $J \in \mathcal{F}_T$ then*

$$N_{T'|R'}(T'J) = R'N_{T|R}(J).$$

Proof: First, let $J = Q$ be a nonzero prime ideal of T. If $Q \cap R = P_1 \neq P$ then $T'Q = T'$, $R'P_1 = R'$, and $N_{T'|R'}(T'Q) = R'$ while $R'N_{T|R}(Q) = R'P_1^f = R'$ (where $1 \leq f$).

Now we assume that $Q \cap R = P$. From the results of Chapter 12, Section 1, we have

$$T'Q \cap R' = R'P$$

and

$$[T'/T'Q : R'/R'P] = [T/Q : R/P],$$

say equal to f. Then $N_{T'|R'}(T'Q) = (R'P)^f = R'P^f = R'N_{T|R}(Q)$.

It follows by the multiplicativity of the norm, that if $J \in \mathcal{F}_T$ then

$$N_{T'|R'}(T'J) = R'N_{T|Q}(J). \qquad \blacksquare$$

The most important case in our considerations is the following. $R = A$ is the ring of algebraic integers of an algebraic number field K and $T = B$ is the ring of algebraic integers of the algebraic number field L, of degree n over K. In this situation, it is customary to use the notation $N_{L|K}(J)$ instead of $N_{B|A}(J)$ for every fractional ideal J of L, with respect to B.

We compare the relative norm with the (absolute) norm of an ideal, as defined in Chapter 8, Definition 1.

B.   *Let K be an algebraic number field, A the ring of algebraic integers of K. For each nonzero fractional ideal J of K, we have*

$$N_{K|\mathbb{Q}}(J) = \mathbb{Z} \cdot N(J).$$

Proof: Let P be any nonzero prime ideal of A, let $P \cap \mathbb{Z} = \mathbb{Z}p$. Let $[A/P : \mathbb{Z}/\mathbb{Z}p] = f$, so the field A/P has p^f elements, that is, $N(P) = \#(A/P) = p^f$. By definition,

$$N_{K|\mathbb{Q}}(P) = \mathbb{Z}p^f = \mathbb{Z} \cdot N(P).$$

It follows at once from the multiplicativity of the relative norm and of the norm (Chapter 8, (**D**)) that $N_{K|\mathbb{Q}}(J) = \mathbb{Z} \cdot N(J)$ for every $J \in \mathcal{F}_K$. $\qquad \blacksquare$

C.   *With the previous notations, $[L : K] = n$ and any nonzero fractional ideal I of A, we have*

$$N_{L|K}(BI) = I^n.$$

Proof: Once again, due to the multiplicativity of the relative norm of an ideal, it suffices to prove the statement when $I = P$ is a nonzero prime ideal of A.

Let $BP = \prod_{i=1}^{g} Q_i^{e_i}$ where $Q_1, \ldots, Q_g$ are distinct prime ideals of B, $e_i \geq 1$ and $[B/Q_i : A/P] = f_i$ for $i = 1, \ldots, g$. Then $N_{L|K}(BP) = \prod_{i=1}^{g} N(Q_i)^{e_i} = \prod_{i=1}^{g} P^{f_i e_i} = P^n$ since $\sum_{i=1}^{g} e_i f_i = n$, by Chapter 11, Theorem 1. $\qquad \blacksquare$

D.   *Let $K \subseteq L \subseteq L'$ be algebraic number fields, and J' any nonzero fractional ideal of L'. Then*

$$N_{L'|K}(J') = N_{L|K}(N_{L'|L}(J')).$$

Proof: By the multiplicativity of the relative norm, it suffices to prove the statement when $J' = Q'$ is a nonzero prime ideal of the ring B' of algebraic integers of L'. Let $Q' \cap B = Q$, $Q \cap A = P$. Then $N_{L'|K}(Q') = P^{f''}$ where $f'' = [B'/Q' : A/P]$. On the other hand, $N_{L'|L}(Q') = Q^{f'}$ where

$[B'/Q' : B/Q] = f'$ and $N_{L|K}(Q) = P^f$ where $[B/Q : A/P] = f$. Thus $f'' = ff'$ and therefore

$$N_{L|K}(N_{L'|L}(Q')) = N_{L|K}(Q^{f'}) = P^{ff'}$$
$$= P^{f''} = N_{L'|K}(Q'). \qquad \blacksquare$$

E.   *Assume that $L|K$ is a Galois extension with Galois group $\mathcal{G} = \{\sigma_1, \ldots, \sigma_n\}$. For each $J \in \mathcal{F}_L$ we have*

$$\prod_{i=1}^{n} \sigma_i(J) = BN_{L|K}(J).$$

Proof: By multiplicativity, it suffices to prove the result when $J = Q$ is a nonzero prime ideal of B.

Let $Q \cap A = P$, then $BP = (\prod_{i=1}^{g} Q_i)^e$ with $e \geq 1$, $[B/Q : A/P] = f$. Moreover, by Chapter 11, **(B)**, $Q \in \{Q_1, \ldots, Q_g\}$ and by Chapter 11, **(E)**, $\mathcal{G}$ acts transitively on the set $\{Q_1, \ldots, Q_g\}$.

Thus $\prod_{i=1}^{n} \sigma_i(Q) = (Q_1 \cdots Q_g)^{ef} = BP^f = BN_{L|K}(Q).$ $\blacksquare$

For principal ideals, we have:

F.   *If $x \in L$, $x \neq 0$, then $N_{L|K}(Bx) = AN_{L|K}(x)$.*

Proof. First Case: Assume that $L|K$ is a Galois extension, with Galois group $\mathcal{G} = \{\sigma_1, \ldots, \sigma_n\}$. We have, by **(E)**,

$$B \cdot N_{L|K}(Bx) = \prod_{i=1}^{n} \sigma_i(Bx) = \prod_{i=1}^{n} B\sigma_i(x)$$
$$= B \prod_{i=1}^{n} \sigma_i(x) = BN_{L|K}(x).$$

By Chapter 11, **(A)**:

$$N_{L|K}(Bx) = B \cdot N_{L|K}(Bx) \cap K$$
$$= B \cdot AN_{L|K}(x) \cap K = AN_{L|K}(x).$$

General Case: Let L' be the smallest Galois extension of K containing L; let $[L' : L] = m$. We have

$$\left[N_{L|K}(Bx)\right]^m = N_{L|K}((Bx)^m) = N_{L|K}(N_{L'|L}(B'x))$$
$$= \left[N_{L'|K}(B'x)\right] = A \cdot N_{L'|K}(x) = A(N_{L|K}(x))^m.$$

as follows from the first case and **(D)**. Since A is a Dedekind domain, then $N_{L|K}(Bx) = A \cdot N_{L|K}(x)$. $\blacksquare$

13.2 Relative Discriminant and Different of Algebraic Number Fields

As before, let K be an algebraic number field, $L|K$ an extension of degree n and A, B the rings of integers of K, L, respectively.

Definition 3.   The *relative discriminant* of $L|K$ is the ideal $\delta_{L|K}$ of A generated by the elements $\mathrm{discr}_{L|K}(x_1, \ldots, x_n)$, for all possible bases $\{x_1, \ldots, x_n\}$ of $L|K$ such that each $x_i \in B$.

G.   *Let $\{x_1, \ldots, x_n\}$ be a basis of $L|K$ such that each $x_i \in B$. Then $\delta_{L|K} = A \cdot \mathrm{discr}_{L|K}(x_1, \ldots, x_n)$ if and only if B is a free A-module and $\{x_1, \ldots, x_n\}$ is an A-basis of B.*

Proof: If $\{x_1, \ldots, x_n\}$ is an A-basis of B then by definition $\delta_{L|K} \supseteq A \cdot \mathrm{discr}_{L|K}(x_1, \ldots, x_n)$. Now, if $\{x_1', \ldots, x_n'\}$ is any K-basis of L, with $x_j' \in B$ for every $j = 1, \ldots, n$, we have

$$x_j' = \sum_{i=1}^{n} a_{ij} x_i, \qquad \text{with} \quad a_{ij} \in A,$$

so

$$\mathrm{discr}_{L|K}(x_1', \ldots, x_n') = \big[\det(a_{ij})\big]^2 \cdot \mathrm{discr}_{L|K}(x_1, \ldots, x_n)$$

hence every generator of the ideal $\delta_{L|K}$ is contained in

$$A \cdot \mathrm{discr}_{L|K}(x_1, \ldots, x_n),$$

which proves the other inclusion.

Conversely, let us assume that $\delta_{L|K} = A \cdot \mathrm{discr}_{L|K}(x_1, \ldots, x_n)$, where $\{x_1, \ldots, x_n\}$ is a K-basis of L contained in B. Let us show that $\{x_1, \ldots, x_n\}$ generates the A-module B.

Let P be any nonzero prime ideal of A, let S be the multiplicative set complement of P in A, $A' = S^{-1}A$, $B' = S^{-1}B$, $P' = A'P$. By Chapter 12, (**G**), (**D**), and Chapter 6, (**I**), A' is a principal ideal domain, B' is a free A'-module of rank n; let $\{x_1', \ldots, x_n'\}$ be a basis of this module. Writing $x_i' = y_i/s_i$ with $y_i \in B$, $s_i \in S$, we deduce that $\{y_1, \ldots, y_n\}$ is a K-basis of L and $\mathrm{discr}_{L|K}(x_1', \ldots, x_n') \in A' \cdot \mathrm{discr}_{L|K}(y_1, \ldots, y_n) \subseteq A' \cdot \delta_{L|K}$. On the other hand, we have

$$x_j = \sum_{i=1}^{n} a_{ij}' x_i', \qquad \text{with} \quad a_{ij}' \in A'.$$

Hence,

$$\mathrm{discr}_{L|K}(x_1, \ldots, x_n) = \big[\det(a_{ij}')\big]^2 \cdot \mathrm{discr}_{L|K}(x_1', \ldots, x_n'),$$

thus from the hypothesis

$$A'\delta_{L|K} = A' \cdot \mathrm{discr}_{L|K}(x_1, \ldots, x_n) \subseteq A' \cdot \mathrm{discr}_{L|K}(x_1', \ldots, x_n'),$$

hence

$$A' \operatorname{discr}_{L|K}(x_1, \ldots, x_n) = A' \operatorname{discr}_{L|K}(x'_1, \ldots, x'_n).$$

Hence, by Chapter 12, $(\mathbf{T})$, these discriminants are associated elements of A', thus $\left[\det(a'_{ij})\right]^2 \in A'$ is a unit of A', and therefore $\det(a'_{ij})$ is also a unit. This means that the inverse of the matrix $(a'_{ij})_{i,j}$ has coefficients in A' and so each element x'_i belongs to the A'-module generated by $x_1, \ldots, x_n$. Since these elements are linearly independent, they constitute a basis of the A'-module B'.

The above considerations hold for every nonzero prime ideal P of A. It follows that $\{x_1, \ldots, x_n\}$ is a basis of the A-module B. Indeed, if $y \in B$, we may write $y = \sum_{i=1}^n c_i x_i$ with $c_i \in K$; this expression is unique. But for every prime ideal $P \neq 0$ we have $c_i \in A_P$ as we have just shown. Thus

$$c_i \in \bigcap A_P = A \text{ (intersection of all nonzero prime ideals of } A\text{)}$$

as follows from Chapter 12, $(\mathbf{C})$, and $\{x_1, \ldots, x_n\}$ generates the A-module B, as we had to prove. ∎

This result may be applied when A is a principal ideal domain or when there exists a primitive element t of $L|K$ such that $B = A[t]$. It follows that $\delta_{L|\mathbb{Q}}$ is the principal ideal generated by the discriminant δ_L, as introduced in Chapter 6, Definition 4.

Now we come to the main theorem, connecting the ramification and the discriminant:

Theorem 1 (Dedekind). *The nonzero prime ideal P of A is ramified in $L|K$ if and only if $P \supseteq \delta_{L|K}$. In particular, there exist only finitely many prime ideals which are ramified in $L|K$.*

Proof: The second assertion follows at once from the first, by Chapter 7, $(\mathbf{F})$ and $(\mathbf{G})$.

We write $BP = \prod_{i=1}^g Q_i^{e_i}$, where Q_i are distinct prime ideals of B and $e_i \geq 1$. From Chapter 7, Theorem 2, we have $B/BP = \prod_{i=1}^g B/Q_i^{e_i}$.

P is ramified when some e_i is greater than 1; that is, $B/Q_i^{e_i}$ has a nonzero nilpotent element; or, equivalently, B/BP has a nonzero nilpotent element. By Chapter 12, $(\mathbf{V})$, this means that $\operatorname{discr}((B/BP)|(A/P)) = 0$.

If S is the set complement of P in A, if $A' = S^{-1}A$, $B' = S^{-1}B$, $P' = A'P$ it follows from Chapter 12, $(\mathbf{A})$, that $A'/A'P = A/P$, $B'/B'P = B/P$ so the above condition is that $\operatorname{discr}((B'/B'P)|(A'/P')) = 0$. We know further that B' is a free module of rank n over the principal ideal domain A'; moreover, if $\{x'_1, \ldots, x'_n\}$ is any basis of the A'-module B' then the images $\overline{x'_i}$ of these elements by the homomorphism $B' \to B'/B'P$ constitute a basis over $A'/A'P$ (as was proved in Chapter 12, $(\mathbf{I})$).

Now, $\operatorname{discr}((B'/B'P)|(A'/P'))$ is the ideal generated by the elements $\operatorname{discr}_{(B'/B'P)|(A'/P')}(\overline{x'_1}, \ldots, \overline{x'_n})$ for all possible bases $\{\overline{x'_1}, \ldots, \overline{x'_n}\}$ of

$B'/B'P$ over A'/P'. So $\mathrm{discr}((B'/B'P)|(A'/P')) = 0$ exactly when

$$\mathrm{discr}_{(B'/B'P)|(A'/P')}(\overline{x_1'}, \ldots, \overline{x_n'}) = \overline{\mathrm{discr}_{B'|A'}(x_1', \ldots, x_n')} = \overline{0};$$

that is, $\mathrm{discr}_{B'|A'}(x_1', \ldots, x_n') \in P'$, for every basis of the A'-module B'.

This last condition is actually equivalent to $\delta_{L|K} \subseteq P$. Indeed, let $\{x_1, \ldots, x_n\}$ be a K-basis of L, where each x_i belongs to B. If $(x_1', \ldots, x_n')$ is a basis of the A'-module B', expressing $x_1, \ldots, x_n$ in terms of $x_1', \ldots, x_n'$ with coefficients in A', gives

$$\mathrm{discr}_{L|K}(x_1, \ldots, x_n) = \mathrm{discr}_{B'|A'}(x_1, \ldots, x_n)$$
$$\in A' \cdot \mathrm{discr}_{B'|A'}(x_1', \ldots, x_n') \subseteq P',$$

but $\mathrm{discr}_{L|K}(x_1, \ldots, x_n) \in A$, because each $x_i \in B$, so

$$\mathrm{discr}_{L|K}(x_1, \ldots, x_n) \in P' \cap A = P$$

and therefore $\delta_{L|K} \subseteq P$. Conversely, if $\delta_{L|K} \subseteq P$ if $\{x_1', \ldots, x_n'\}$ is a basis of the A'-module B', let $x_i' = x_i/s_i$ with $x_i \in B$, $s_i \in S$; thus $\{x_1, \ldots, x_n\}$ is a K-basis of L contained in B, with

$$\mathrm{discr}_{B'|A'}(x_1', \ldots, x_n') = \mathrm{discr}_{L|K}(x_1', \ldots, x_n')$$
$$= \left(\frac{1}{s_1 \cdots s_n}\right)^2 \cdot \mathrm{discr}_{L|K}(x_1, \ldots, x_n)$$
$$\in A'P = P'. \qquad \blacksquare$$

It will be a feature of the theory that unramified prime ideals may be handled without difficulty; thus the preceding theorem asserts that it is necessary to concentrate only on those finitely many prime ideals which ramify. In the next chapter we shall study in more detail the steps of ramification in the case of Galois extensions.

Dedekind's theorem tells which prime ideals P of A are ramified in $L|K$. A more precise problem is to determine the prime ideals Q of B which are ramified in $L|K$. Clearly, if Q is ramified and $P = Q \cap A$ then P is ramified. Conversely, if $L|K$ is a Galois extension and $Q \cap A = P$ then Q is ramified, as follows from Chapter 11, (**F**).

However, if $L|K$ is not a Galois extension, there may well exist different prime ideals Q, Q' of B such that $Q \cap A = Q' \cap A = P$, and Q is ramified while Q' is not ramified in $L|K$.

To find out which prime ideals Q of B are ramified in $L|K$ we shall introduce the relative different of $L|K$. Besides, we shall also consider the different above a given prime ideal of A. We may treat these two cases simultaneously.

Let R be a Dedekind domain, and K its field of quotients; let $L|K$ be a separable field extension of degree n and T the integral closure of R in L; so T is also a Dedekind domain, L is its field of quotients, $T \cap K = R$.

The relative trace $\text{Tr}_{L|K}$ induced a mapping from $L \times L$ into K, which associates with every pair $(x, y) \in L \times L$ the element $\text{Tr}_{L|K}(xy) \in K$. This is a symmetric K-bilinear form, as is easily verified.

If $x \in L$ let $\varphi_x : L \to K$ be the linear form defined by $\varphi_x(y) = \text{Tr}_{L|K}(xy)$ for every $y \in L$. Thus φ_x belongs to L', the dual of the K-vector space L and since $\varphi_{ax} = a\varphi_x$, $\varphi_{x_1+x_2} = \varphi_{x_1} + \varphi_{x_2}$ (for $a \in K$, $x_1, x_2, x \in L$) we have a K-linear mapping $\varphi : L \to L'$. In order that φ_x be the zero mapping we must have $\text{Tr}_{L|K}(xy) = 0$ for every $y \in L$; this means that $x = 0$ since the trace in the separable extension $L|K$ is nondegenerate (see Chapter 2, Section 10). Therefore, φ is an isomorphism between the K-spaces L, L'.

If $\{x_1, \ldots, x_n\}$ is a K-basis of L, let $x_1^*, \ldots, x_n^* \in L$ be elements such that $\{\varphi_{x_1^*}, \ldots, \varphi_{x_n^*}\}$ is the dual basis; that is,

$$\varphi_{x_i^*}(x_j) = \text{Tr}_{L|K}(x_i^* x_j) = \delta_{ij} \qquad (\delta_{ii} = 1, \ \delta_{ij} = 0 \text{ when } i \neq j).$$

Thus $\{x_1^*, \ldots, x_n^*\}$ is also a basis of L, which we call the *complementary basis* of $\{x_1, \ldots, x_n\}$. Let us note here that

$$\text{discr}_{L|K}(x_1, \ldots, x_n) \cdot \text{discr}_{L|K}(x_1^*, \ldots, x_n^*) = 1.$$

Indeed, if σ_1, $\ldots$, σ_n are the K-isomorphisms of L, if $X = (\sigma_i(x_j))_{i,j}$, $X^* = (\sigma_i(x_j^*))_{i,j}$, and if X' denotes the transpose of the matrix X, then $X^{*\prime} \cdot X = (\text{Tr}_{L|K}(x_i^* x_j))_{i,j}$. Therefore,

$$\det(X) \cdot \det(X^*) = 1;$$

but

$$\text{discr}_{L|K}(x_1, \ldots, x_n) = \det(X)^2$$

and

$$\text{discr}_{L|K}(x_1^*, \ldots, x_n^*) = \det(X^*)^2$$

so

$$\text{discr}_{L|K}(x_1, \ldots, x_n) \cdot \text{discr}_{L|K}(x_1^*, \ldots, x_n^*) = 1.$$

Now we define complementary sets in L. Let M be a subset of L, then $M^* = \{x \in L \mid \text{Tr}_{L|K}(xy) \in R \text{ for every } y \in M\}$ is called the *complementary set* of M (with respect to R).

Let us note at once the following properties:

H. *If M is a subset of L and M^* is the complementary set, then:*

(1) *M^* is a module over R; if $T \cdot M \subseteq M$ then M^* is a module over T.*

(2) *If $M_1 \subseteq M_2 \subseteq L$ then $M_2^* \subseteq M_1^* \subseteq L$.*

(3) *$T \subseteq T^*$ and $\text{Tr}_{L|K}(T^*) \subseteq R$.*

(4) *If M is a free R-module with basis $\{x_1, \ldots, x_n\}$ then M^* is a free R-module with basis $\{x_1^*, \ldots, x_n^*\}$ and $M^{**} = M$.*

Proof: (1)   Let $x_1, x_2 \in M^*$; for every $y \in M$ we have

$$\mathrm{Tr}_{L|K}((x_1 + x_2)y) = \mathrm{Tr}_{L|K}(x_1 y) + \mathrm{Tr}_{L|K}(x_2 y) \in R,$$

so $x_1 + x_2 \in M^*$. If $a \in R$, $x \in M^*$, $y \in M$, then

$$\mathrm{Tr}_{L|K}((ax)y) = a\,\mathrm{Tr}_{L|K}(xy) \in R$$

so $ax \in M^*$.

Now, let us assume that $T \cdot M \subseteq M$. If $b \in T$, $x \in M^*$, and $y \in M$, then $\mathrm{Tr}_{L|K}((bx)y) = \mathrm{Tr}_{L|K}(x(by)) \in R$ since $by \in M$; thus $bx \in M$.

(2)   This is obvious.

(3)   Since T is integral over R, which is integrally closed, then

$$\mathrm{Tr}_{L|K}(T) \subseteq R.$$

Now, if $x, y \in T$ then $\mathrm{Tr}_{L|K}(xy) \in R$, hence $x \in T^*$; that is, $T \subseteq T^*$. It is clear that $\mathrm{Tr}_{L|K}(T^*) \subseteq R$.

(4)   We have $\mathrm{Tr}_{L|K}(x_i^* x_j) = 0$ when $i \neq j$ and $\mathrm{Tr}_{L|K}(x_i^* x_i) = 1$ for every index $i = 1, \ldots, n$. Hence each x_i^* belongs to M^* and therefore $\sum_{i=1}^n R x_i^* \subseteq M^*$. Conversely, let $\sum_{i=1}^n a_i x_i^* \in M^*$, with $a_i \in K$. Then for every $j = 1, \ldots, n$ we have

$$a_j = \mathrm{Tr}_{L|K}\left(\left(\sum_{i=1}^n a_i x_i^* \right) x_j \right) \in R, \qquad \text{so} \qquad M^* \subseteq \sum_{i=1}^n R x_i^*.$$

This shows the first assertion of (4). Since $M \subseteq M^{**}$ and by duality, $x_i^{**} = x_i$ for all $i = 1, \ldots, n$, then $M^{**} = M$.   ∎

I. T^* *is a fractional ideal of L (with respect to T).*

Proof: It is enough to show that the T-module T^* is finitely generated, because if $b \in T$ is a nonzero common denominator of the generators of T^* then $bT^* \subseteq T$.

Since $R[t]$, for $t \in T$, is a finitely generated free R-module, it follows from (**H**) that $R[t]^*$ is a finitely generated R-module. From $R[t] \subseteq T$ it follows that $T^* \subseteq R[t]^*$; since R is a Dedekind domain, hence a Noetherian ring, we deduce that T^* is also a finitely generated R-module (see Chapter 6, (**G**) and (**D**)); a fortiori, T^* is a finitely generated T-module.   ∎

Definition 4.   The ideal of T equal to the inverse of the fractional ideal T^* is called the *different of T over R*, and denoted by $\Delta(T|R)$.

Since $T \subseteq T^*$ then $\Delta(T|R)$ is a nonzero integral ideal of T.

Since T is a Dedekind domain, the ideal $\Delta(T|R)$ may be written in a unique way as $\Delta(T|R) = \prod Q^{s_Q}$ where each Q is a nonzero prime ideal of T and $s_Q \geq 0$ is an integer. Moreover, $s_Q > 0$ only for a finite number of prime ideals Q. The integer s_Q is called *the exponent at Q of the different $\Delta(T|R)$.*

Now we give some indications about the computation of the different. We begin with a result, which dates back essentially to Euler:

J.   *Let $L = K(t)$, where t is integral over R and*

$$g = X^n + c_1 X^{n-1} + \cdots + c_n \in R[X]$$

is the minimal polynomial of t over K, let g' denote its derivative. Then:

(1) $\mathrm{Tr}_{L|K} \left(\dfrac{t^i}{g'(t)} \right) = 0$ *when* $i = 0, 1, \ldots, n-2$,

$\mathrm{Tr}_{L|K} \left(\dfrac{t^{n-1}}{g'(t)} \right) = 1.$

(2) $R[t]^* = \dfrac{1}{g'(t)} R[t].$

Proof: (1)   Let $t = t_1, t_2, \ldots, t_n$ be the conjugates of t over K, which are necessarily distinct and belong to a Galois extension of finite degree over K. We shall compute

$$\mathrm{Tr}_{L|K} \left(\frac{t^i}{g'(t)} \right) = \sum_{k=1}^{n} \frac{t_k^i}{g'(t_k)}$$

for $i = 0, 1, \ldots, n-1$.

Since g is the minimal polynomial of t, we have $g = \prod_{k=1}^{n}(X - t_k)$, hence $1/g = \prod_{k=1}^{n} 1/(X - t_k)$ and we may express the above product as a sum $\sum_{k=1}^{n} a_k/(X - t_k)$, where the elements a_k will now be determined: from $1/g = \sum_{k=1}^{n} a_k/(X - t_k)$ we have

$$1 = \sum_{k=1}^{n} \frac{a_k g}{X - t_k} = \sum_{k=1}^{n} a_k \left(\prod_{i \neq k} (X - t_i) \right)$$

hence for every $j = 1, \ldots, n$:

$$1 = \sum_{k=1}^{n} a_k \left(\prod_{i \neq k}(t_j - t_i) \right) = a_j \prod_{i \neq j}(t_j - t_i),$$

thus

$$a_j = \frac{1}{\prod\limits_{i \neq j}(t_j - t_i)} = \frac{1}{g'(t_j)},$$

and we have found that

$$\frac{1}{g} = \sum_{k=1}^{n} \frac{1}{g'(t_k)(X - t_k)}.$$

By long Euclidean division, we may write

$$\frac{1}{g} = \frac{1}{X^n} + b_1 \frac{1}{X^{n+1}} + b_2 \frac{1}{X^{n+2}} + \cdots$$

with $b_1, b_2, \ldots \in R$, while

$$\sum_{k=1}^{n} \frac{1}{g'(t_k)(X - t_k)} = \sum_{k=1}^{n} \frac{1}{g'(t_k)} \left[\frac{1}{X} + \frac{t_k}{X^2} + \frac{t_k^2}{X^3} + \cdots \right].$$

Comparing the two formal power series, we conclude that

$$\sum_{k=1}^{n} \frac{t_k^i}{g'(t_k)} = 0 \qquad \text{for} \quad i = 0, 1, \ldots, n - 2$$

while

$$\sum_{k=1}^{n} \frac{t_k^{n-1}}{g'(t_k)} = 1.$$

(2)   First we show that $t^j/g'(t) \in R[t]^*$ for $j = 0, 1, \ldots, n - 1$. Indeed, if $y = \sum_{i=0}^{n-1} a_i t^i \in R[t]$ then

$$\mathrm{Tr}_{L|K} \left(\frac{t^j}{g'(t)} y \right) = \sum_{i=0}^{n-1} a_i \, \mathrm{Tr}_{L|K} \left(\frac{t^{j+i}}{g'(t)} \right)$$

$$= a_{n-1-j} + \sum_{j=n-i}^{n-1} a_i b_{n-j-i-1} \in R,$$

where elements $b_k \in R$ are defined above.

The elements $t^j/g'(t)$ (for $j = 0, 1, \ldots, n - 1$) form a basis of L over K, since the set of elements t^j ($j = 0, 1, \ldots, n - 1$) is such a basis.

To prove the inclusion $R[t]^* \subseteq (1/g'(t))R[t]$ let $y \in R[t]^*$, hence we may write $y = \sum_{j=0}^{n-1} a_j (t^j/g'(t))$ with each $a_j \in K$. Then

$$\mathrm{Tr}_{L|K}(y) = \sum_{j=0}^{n-1} a_j \, \mathrm{Tr}_{L|K} \left(\frac{t^j}{g'(t)} \right) = a_{n-1},$$

hence $a_{n-1} \in R$ because $y \in R[t]^*$. Similarly,

$$\mathrm{Tr}_{L|K}(yt) = \sum_{j=0}^{n-1} a_j \, \mathrm{Tr}_{L|K} \left(\frac{t^{j+1}}{g'(t)} \right) = a_{n-2} + a_{n-1} \mathrm{Tr}_{L|K} \left(\frac{t^n}{g'(t)} \right)$$

$$= a_{n-2} - a_{n-1} \left(\sum_{i=1}^{n} c_i \, \mathrm{Tr}_{L|K} \left(\frac{t^{n-i}}{g'(t)} \right) \right) = a_{n-2} - a_{n-1} c_1$$

because $t^n = -(c_1 t^{n-1} + c_2 t^{n-2} + \cdots + c_n)$ with $c_i \in R$. Since $a_{n-2} - a_{n-1} c_1 \in R$ then $a_{n-2} \in R$. Proceeding in the same manner, we deduce that $a_i \in R$ for every $i = 0, 1, \ldots, n - 1$ and therefore

$$R[t]^* = \frac{1}{g'(t)} R[t].$$ ∎

Here is an instance when the different may be explicitly determined:

K.   *Let $L = K(t)$, where $t \in T$ and $g \in R[X]$ is the minimal polynomial of t over K. Then, $\Delta(T|R) = T \cdot g'(t)$ if and only if $T = R[t]$.*

Proof: If $T = R[t]$ we have seen in (**J**) that $T^* = (1/g'(t))T$, hence $\Delta(T|K) = T \cdot g'(t)$.

Conversely, from $R[t] \subseteq T$ we have by (**J**) $T^* \subseteq R[t]^*$ so $T = g'(t)T^* \subseteq g'(t)R[t]^* = R[t]$ hence $R[t] = T$.   ∎

The different satisfies the following characteristic property:

L.   *Let J be a fractional ideal of T. Then $\mathrm{Tr}_{L|K}(J) \subseteq R$ if and only if $J \subseteq T^* = \Delta(T|R)^{-1}$.*

Proof: If $J \subseteq T^*$ then $\mathrm{Tr}_{L|K}(J) \subseteq \mathrm{Tr}_{L|K}(T^*) \subseteq R$. Conversely, from $J = T \cdot J$ and $\mathrm{Tr}_{L|K}(J) \subseteq R$ we deduce that $J \subseteq T^*$.   ∎

Another useful property is the transitivity of the different. Let $L'|L$ be a separable extension of finite degree, and let T' be a Dedekind domain, having a field of quotients equal to L' and equal to the integral closure of T in L'. With these notations:

M. $\Delta(T'|R) = T'\Delta(T|R) \cdot \Delta(T'|T)$.

Proof: A fractional ideal J' of T' is such that $J' \subseteq \Delta(T'|T)^{-1}$ if and only if $T \supseteq \mathrm{Tr}_{L'|L}(J')$. This means that

$$\Delta(T|R)^{-1} \supseteq \Delta(T|R)^{-1} \cdot \mathrm{Tr}_{L'|L}(J') = \mathrm{Tr}_{L'|L}(T'\Delta(T|R)^{-1} \cdot J'),$$

so

$$R \supseteq \mathrm{Tr}_{L|K}(\Delta(T|R)^{-1})$$
$$\supseteq \mathrm{Tr}_{L|K}(\mathrm{Tr}_{L'|L}(T'\Delta(T|R)^{-1} \cdot J')) = \mathrm{Tr}_{L'|K}(T'\Delta(T|R)^{-1} \cdot J').$$

Again, this means that $T'\Delta(T|R)^{-1} \cdot J' \subseteq \Delta(T'|R)^{-1}$, that is,

$$J' \subseteq T'\Delta(T|R) \cdot \Delta(T'|R)^{-1}.$$

So, we have shown that $\Delta(T'|T)^{-1} = T'\Delta(T|R) \cdot \Delta(T'|R)^{-1}$; that is, $\Delta(T'|R) = T'\Delta(T|R) \cdot \Delta(T'|T)$.   ∎

We shall apply the theory just developed in two main instances.

Let K be an algebraic number field, $L|K$ an extension of degree n, let $R = A$, and let $T = B$ be the rings of algebraic integers of K and L, respectively.

Definition 5.   The different $\Delta(B|A)$ is also denoted by $\Delta_{L|K}$ and is called *the different of $L|K$.*

In the special case where the ground field is $K = \mathbb{Q}$ the different of $L|\mathbb{Q}$ is also called *the absolute different of* L, and sometimes denoted by Δ_L.

Now, let P be a nonzero prime ideal of A, S the set complement of P in A, and let $A' = S^{-1}A$, $B' = S^{-1}B$, so B' is the integral closure of the principal ideal domain A' in L and we may take $R = A'$, $T = B'$.

Definition 6.   The different $\Delta(B'|A')$ is called *the different of* $L|K$ *above* P.

Sometimes we denote it by $\Delta_P(L|K)$ or simply Δ_P.

We wish to compare these differents.

N.   *With the above notations,* $B' \cdot \Delta_{L|K} = \Delta(B'|A')$.

Proof: Let $x \in B' \cdot \Delta_{L|K}$; it may be written in the form $x = y/s$ with $y \in \Delta_{L|K} = \Delta(B|A)$, $s \in S$. Let $z \in B'^*$ (the complementary module of the A'-module B'), so $\mathrm{Tr}_{L|K}(zB') \subseteq A'$. We know that B is a finitely generated A-module; let $\{t_1, \ldots, t_m\}$ be a system of generators, and let $\mathrm{Tr}_{L|K}(zt_i) = a_i/s_i$ with $a_i \in A$, $s_i \in S$. If $s_0 = s_1 \cdots s_m \in S$ then

$$\mathrm{Tr}_{L|K}(zs_0t_i) = s_0 \, \mathrm{Tr}_{L|K}(zt_i) \in A$$

for every $i = 1, \ldots, m$. Thus $\mathrm{Tr}_{L|K}(zs_0B) \subseteq A$ so $zs_0 \in B^*$ (complementary module of the A-module B); that is, $yzs_0 \in B$ because $y \in \Delta(B|A)$. We deduce that $xz = yzs_0/ss_0 \in B'$ showing that $x \in \Delta(B'|A')$ and the inclusion $B' \cdot \Delta_{L|K} \subseteq \Delta(B'|A')$.

Conversely, let $x \in \Delta(B'|A')$; B^* is a fractional ideal of B, hence a finitely generated A-module. Let $\{z_1, \ldots, z_m\}$ be a system of generators of the A-module B^*. We have $\mathrm{Tr}_{L|K}(z_iB) \subseteq A$, and since $S \subseteq K$ then

$$\mathrm{Tr}_{L|K}(z_iB') \subseteq A',$$

so $z_i \in B'^*$ hence $xz_i \in B' = S^{-1}B$; so we may write $xz_i = b_i/s_i$ with $b_i \in B$, $s_i \in S$. Let

$$s = s_1 \cdots s_m \in S,$$

then $sxz_i \in B$ for every $i = 1, \ldots, m$, hence also $sxB^* \subseteq B$. This proves that $sx \in \Delta(B|A)$ and $x \in B' \cdot \Delta(B|A)$.   ∎

We are able to compute the different $\Delta_{L|K} = \prod Q^{s_Q}$. Let e_Q be the ramification index of Q in $L|K$.

O.   *For every nonzero prime ideal* Q *of* B *we have* $s_Q \geq e_Q - 1$. *Moreover,* $s_Q = e_Q - 1$ *if and only if the characteristic of* B/Q *does not divide the ramification index* e_Q.

Proof: Given the nonzero prime ideal Q_1 of B let $P = Q_1 \cap A$, and let S be the multiplicative set complement of P in A, $A' = S^{-1}A$, $B' = S^{-1}B$. We note $\Delta_{L|K} = \prod Q^{s_Q}$, with $s_Q \geq 0$ integer, and $BP = \prod_{i=1}^{g} Q_i^{e_i}$, hence by Chapter 12, (**F**), $B'P = \prod_{i=1}^{g} B'Q_i^{e_i}$. From (**N**) and Chapter 12, (**G**), we have

$$\Delta(B'|A') = B' \cdot \Delta_{L|K} = \prod_{i=1}^{g} B'Q_i^{s_i},$$

where $s_i = s_{Q_i}$ for $i = 1, \ldots, g$. Thus the complementary module of the A'-module B' is $B'^* = \prod_{i=1}^{g} B'Q_i^{-s_i}$. The inequalities

$$s_i \geq e_i - 1 \qquad (i = 1, \ldots, g)$$

hold if and only if $\prod_{i=1}^{g} B'Q_i^{1-e_i} \subseteq B'^*$.

Thus let $x \in \prod_{i=1}^{g} B'Q_i^{1-e_i}$; we recall that $P' = A'P$ is a principal ideal, so there exists $t \in K$ such that $P' = A't$; since $B't = B'P = \prod_{i=1}^{g} B'Q_i^{e_i}$, then $xt \in \prod_{i=1}^{g} B'Q_i$. We show that $\mathrm{Tr}_{L|K}(xt) \in A'P$. Since B' is a free A'-module, by Chapter 12, (**L**), $\mathrm{Tr}_{L|K}(xt) = \mathrm{Tr}_{B'|A'}(xt)$. By Chapter 12, (**I**), $B'/B'P$ is a vector space of dimension n over A'/P and by Chapter 12, (**N**), taking the images by the canonical homomorphisms

$$\mathrm{Tr}_{(B'/B'P)|(A'/P)}(\overline{xt}) = \overline{\mathrm{Tr}_{B'|A'}(xt)}.$$

But $(xt)^{e_1 + \cdots + e_g} \in \prod_{i=1}^{g} B'Q_i^{e_i} = B'P$, so $\overline{xt}$ is a nilpotent element of the ring $B'/B'P$. Then the characteristic polynomial of the associated linear transformation of $B'/B'P$ is equal to X^n. Hence $\mathrm{Tr}_{(B'/B'P)|(A'/P)}(\overline{xt}) = 0$, that is

$$\mathrm{Tr}_{L|K}(xt) = \mathrm{Tr}_{B'|A'}(xt) \in A'P.$$

We conclude that $t \, \mathrm{Tr}_{L|K}(x) = \mathrm{Tr}_{L|K}(xt) \in A'P = A't$, hence

$$\mathrm{Tr}_{L|K}(x) \in A'.$$

Now, if $y \in B'$ then xy also belongs to $\prod_{i=1}^{g} B'Q_i^{1-e_i}$ so $\mathrm{Tr}_{L|K}(xy) \in A'$. This shows that $x \in B'^*$.

Now we assume that the characteristic of $B/Q_1 = B'/B'Q_1$ divides the ramification index e_1; we wish to show that

$$J = B'Q_1^{-e_1} \cdot \prod_{i=2}^{g} B'Q_i^{-e_i} \subseteq B'^*,$$

hence $s_1 \geq e_1$. Let $x \in J$; by the previous argument $xt \in \bigcap_{i=2}^{g} B'Q_i$. It follows from Chapter 12, (**P**), that if $\psi : A' \to A'/P$ and $\psi_i : B' \to B'/B'Q_i$ then

$$\psi(\mathrm{Tr}_{L|K}(xt)) = \sum_{i=1}^{g} e_i \cdot \mathrm{Tr}_{(B'/B'Q_i)|(A'/P')}(\psi_i(xt))$$

$$= e_1 \cdot \mathrm{Tr}_{(B'/B'Q_1)|(A'/P')}(\psi_1(xt)).$$

But e_1 is a multiple of the characteristic of $B'/B'Q_1$, hence $\psi(\mathrm{Tr}_{L|K}(xt)) = 0$, so $t \cdot \mathrm{Tr}_{L|K}(x) = \mathrm{Tr}_{L|K}(xt) \in P' = A't$ and therefore $\mathrm{Tr}_{L|K}(x) \in A'$.

Now, if $y \in B'$, then $xy \in J$, so $\text{Tr}_{L|K}(xy) \in A'$, and this shows that $x \in B'^*$.

Conversely, if the characteristic of $B/Q_1 = B'/B'Q_1$ does not divide the ramification index e_1 we proceed as follows. Let $x \in B'$ be an element such that its image $\psi_1(x) \in B'/B'Q_1$ has a nonzero trace. By Chapter 7, (L), (applied to the Dedekind domain B') there exists an element $y \in B'$ such that $y - x \in B'Q_1$, and $y \in B'Q_i^{e_i}$ for $i = 2, \dots, g$. Then by Chapter 12, (P):

$$\psi(\text{Tr}_{L|K}(y)) = \sum_{i=1}^{g} e_i \cdot \text{Tr}_{(B'/B'Q_i)|(A'/P')}(\psi_i(y))$$

$$= e_1 \cdot \text{Tr}_{(B'/B'Q_1)|(A'/P')}(\psi_1(x)) \neq 0$$

since e_1 is not a multiple of the characteristic of $B'/B'Q_1$. Therefore

$$\text{Tr}_{L|K}(y) \notin P' = A't$$

and so

$$\text{Tr}_{L|K}\left(\frac{y}{t}\right) = \frac{1}{t} \cdot \text{Tr}_{L|K}(y) \notin A'.$$

This shows that $y/t \notin B'^*$; since $B't = B'P$ and $y \in B'Q_i^{e_i}$ ($i = 2, \dots, g$), then $y/t \in B'Q_1^{-e_1}$ hence $B'Q_1^{-e_1}$ is not contained in B'^*, it is not true that $-e_1 \geq -s_1$, so $e_1 > s_1 \geq e_1 - 1$; therefore, $s_1 = e_1 - 1$. ∎

Theorem 2.   *An ideal Q of B is ramified in $L|K$ if and only if Q divides the different $\Delta_{L|K}$.*

Proof: We assume that Q is ramified in $L|K$, that is, $e_Q \geq 2$, hence $s_Q \geq 1$ and so Q divides the different $\Delta_{L|K}$.

Conversely, if $e_Q = 1$, since the characteristic of B/Q cannot divide e_Q, we conclude that $s_Q = e_Q - 1 = 0$, so Q does not divide the different. ∎

From Theorems 1 and 2 we expect a relationship between the relative different $\Delta_{L|K}$ and the relative discriminant $\delta_{L|K}$. In fact, we have

P. $N_{L|K}(\Delta_{L|K}) = \delta_{L|K}$.

Proof: Let P be any nonzero prime ideal of A, and let S be the multiplicative set, complement of P in A, $A' = S^{-1}A$, $B' = S^{-1}B$. Then B' is a Dedekind domain, having only finitely many prime ideals (by Chapter 12, (A)), so B' is a principal ideal domain (by Chapter 7, (N)). Let B'^* be the complementary module of the free A'-module B'; it is a fractional ideal of B', hence there exists $y \in L$, $y \neq 0$ such that $B'^* = B'y$, and therefore $B'y^{-1} = \Delta(B'|A') = B' \cdot \Delta_{L|K}$ (by (N)). If $\{x_1', \dots, x_n'\}$ is any basis of the A'-module B' then

$$\{yx_1', \dots, yx_n'\}$$

is a basis of B'^*. By $(\mathbf{H})$, B'^* has also the complementary basis

$$\{x_1'^*, \ldots, x_n'^*\}$$

over A'. By Chapter 12, $(\mathbf{L})$ and $(\mathbf{T})$:

$$\mathrm{discr}_{L|K}(x_1'^*, \ldots, x_n'^*) = \mathrm{discr}_{B'^*|A'}(x_1'^*, \ldots, x_n'^*)$$

and

$$\mathrm{discr}_{L|K}(yx_1', \ldots, yx_n') = \mathrm{discr}_{B'^*|A'}(yx_1', \ldots, yx_n')$$

are associated elements of A'. But

$$\mathrm{discr}_{L|K}(x_1', \ldots, x_n') \cdot \mathrm{discr}_{L|K}(x_1'^*, \ldots, x_n'^*) = 1$$

and

$$\mathrm{discr}_{L|K}(yx_1', \ldots, yx_n') = \left[N_{L|K}(y)\right]^2 \cdot \mathrm{discr}_{L|K}(x_1', \ldots, x_n');$$

therefore, $\left[\mathrm{discr}_{L|K}(x_1', \ldots, x_n')\right]^2 \cdot \left[N_{L|K}(y)\right]^2$ is a unit of A'; so

$$A' \cdot \mathrm{discr}_{L|K}(x_1', \ldots, x_n') = A' \cdot N_{L|K}(y^{-1}).$$

Now $y^{-1} = z/a$ with $z \in B$, $a \in S$, so $B'z = B'a \cdot B'\Delta_{L|K}$. Taking the relative norms (for the ring extension $B'|A'$) it follows from $(\mathbf{A})$ that

$$\begin{aligned}
A'N_{L|K}(z) = A'N_{L|K}(Bz) &= N_{B'|A'}(B'z)\\
&= N_{B'|A'}(B'a \cdot B'\Delta_{L|K})\\
&= N_{B'|A'}(B'a) \cdot N_{B'|A'}(B'\Delta_{L|K})\\
&= A'N_{L|K}(Ba) \cdot A'N_{L|K}(\Delta_{L|K})\\
&= A'N_{L|K}(a) \cdot A'N_{L|K}(\Delta_{L|K}),
\end{aligned}$$

hence $A'N_{L|K}(y^{-1}) = A'N_{L|K}(\Delta_{L|K})$.

We have shown that for every basis $\{x_1', \ldots, x_n'\}$ of the A'-module B' we have $A' \cdot \mathrm{discr}_{L|K}(x_1', \ldots, x_n') = A' \cdot N_{L|K}(\Delta_{L|K})$.

This being seen, we show the inclusion $\delta_{L|K} \subseteq N_{L|K}(\Delta_{L|K})$. Let $\{x_1, \ldots, x_n\}$ be any K-basis of L such that each element x_i belongs to B. For every prime ideal P, let $\{x_1', \ldots, x_n'\}$ be any basis of the A'-module B'. We may write

$$x_j = \sum_{i=1}^n a_{ij}' x_i' \quad (j = 1, \ldots, n) \qquad \text{with} \quad a_{ij}' \in A'.$$

Thus

$$\mathrm{discr}_{L|K} \in A' \cdot \mathrm{discr}_{L|K}(x_1', \ldots, x_n') = A_P \cdot N_{L|K}(\Delta_{L|K}).$$

This shows that $\delta_{L|K} \subseteq \bigcap_P A_P \cdot N_{L|K}(\Delta_{L|K}) = N_{L|K}(\Delta_{L|K})$ (see Chapter 12, $(\mathbf{C})$).

Conversely, let P be any nonzero prime ideal of A, let $\delta_{L|K} = P^s \cdot J$, and $N_{L|K}(\Delta_{L|K}) = P^{s'} \cdot J'$, where J, J' are ideals of A, not multiples of

P. We shall see that $s \leq s'$; since this holds for every prime ideal $P \neq 0$ of A, then $\delta_{L|K}$ divides, hence contains $N_{L|K}(\Delta_{L|K})$.

Now, for every prime ideal $P \neq 0$ of A we choose a basis $\{x_1', \ldots, x_n'\}$ of the A'-module B'; after multiplication with an element of S we may assume that each $x_i' \in B$. Let $A \cdot \mathrm{discr}_{L|K}(x_1', \ldots, x_n') = P^r I$, where I is an ideal of A not a multiple of P. By Chapter 12, (F), $A' \, \mathrm{discr}_{L|K}(x_1', \ldots, x_n') = (A'P)^r$ and since $A' \cdot \mathrm{discr}_{L|K}(x_1', \ldots, x_n') = A' \cdot N_{L|K}(\Delta_{L|K}) = (A'P)^{s'}$, then $s' = r$. But $\delta_{L|K}$ contains, hence divides, $A \cdot \mathrm{discr}_{L|K}(x_1', \ldots, x_n')$, so $s \leq r = s'$. This concludes the proof.   ∎

Using this result and the transitivity of the different, we obtain the transitivity of the discriminant:

Q.   *Let $K \subseteq L \subseteq L'$ be algebraic number fields. Then*

$$\delta_{L'|K} = (\delta_{L|K})^{[L':L]} \cdot N_{L|K}(\delta_{L'|L}).$$

Proof: From (M) we have the following relation between the differents: $\Delta_{L'|K} = B' \Delta_{L|K} \cdot \Delta_{L'|L}$, where B' is the ring of integers of L'. Taking norms, it follows from Chapter 12, (H), (J), and (P) of this chapter, that

$$N_{L'|K}(\Delta_{L'|K}) = N_{L'|K}(B' \Delta_{L|K}) \cdot N_{L'|K}(\Delta_{L'|L})$$

so

$$\begin{aligned}
\delta_{L'|K} &= N_{L|K}\left[N_{L'|L}(B'\Delta_{L|K})\right] \cdot N_{L|K}\left[N_{L'|L}(\Delta_{L'|L})\right] \\
&= N_{L|K}\left(\Delta_{L|K}^{[L':L]}\right) \cdot N_{L|K}(\delta_{L'|L}) \\
&= (\delta_{L|K})^{[L':L]} \cdot N_{L|K}(\delta_{L'|L}).
\end{aligned}$$
    ∎

As an application of these computations, we may establish the following general result, which is due to Ore:

R.   *Let K be an algebraic number field, let W be the multiplicative group of roots of unity in K, and w the number of elements in W. Then w divides $2\delta_K$.*

Proof: We have seen in Chapter 10, (C), that W is a cyclic group. Let $w = \prod_{i=1}^{s} p_i^{k_i}$, where $k_i \geq 1$ and $p_1, \ldots, p_s$ are distinct prime numbers.

Let η be a primitive root of unity of order w, let $w_i = \prod_{j \neq i} p_j^{k_j}$; then $\zeta_i = \eta^{w_i}$ is a primitive root of unity of order $p_i^{k_i}$ belonging to K for every $i = 1, \ldots, s$.

It is enough to show that if K contains a primitive p^kth root of unity ζ then p^k divides $2\delta_K$; this will imply that $w = \prod_{i=1}^{s} p_i^{k_i}$ divides $2\delta_K$.

By (Q) the discriminant $\delta_{\mathbb{Q}(\zeta)}$ divides δ_K, so it is sufficient to prove that p^k divides $2\delta_{\mathbb{Q}(\zeta)}$.

We have seen in Chapter 11, (N) and (O), that p is the only ramified prime in $\mathbb{Q}(\zeta)$ and $Ap = (A\xi)^{\varphi(p^k)}$, thus the different of $\mathbb{Q}(\zeta)|\mathbb{Q}$ is $\Delta_{\mathbb{Q}(\zeta)} =$

$(A\xi)^s$ and by (O) $\varphi(p^k) - 1 \leq s$. Taking norms we have $|\delta_{\mathbb{Q}(\varsigma)}| = p^s$ (by (P) and Chapter 11, (N)).

But

$$s \geq \varphi(p^k) - 1 = p^{k-1}(p-1) - 1 \geq [1 + (k-1)(p-1)](p-1) - 1;$$

this last quantity is at least equal to k when $p > 2$ and equal to $k - 1$ when $p = 2$. In any case p^k divides $2p^s = 2|\delta_{\mathbb{Q}(\varsigma)}|$. ∎

If $t \in L$ let $g(x) = \prod_{i=1}^{n}(X - \sigma_i(t))$ where $\sigma_1 = \varepsilon$, σ_2, ..., σ_n are the K-isomorphisms of L into $\mathbb{C}$. So $g \in K[X]$ and $g(t) = 0$. Thus the minimal polynomial of t over K divides g. Then $g'(t) = \prod_{i=2}^{n}(t - \sigma_i(t))$ and $g'(t) \neq 0$ if and only if t is a primitive element of the extension $L|K$, in other words, t is different from its conjugates over K. For this reason, it is customary to call $g'(t)$ the different of t in $L|K$.

We assume henceforth that t is a primitive element of $L|K$ and $t \in B$; so $g'(t) \in A$, $g'(t) \neq 0$.

We want to compare the fractional ideals B^* and $A[t]^*$. As we know $B^* \subseteq A[t]^* = A[t]/g'(t)$. Let $F_t = \{x \in B \mid x \cdot A[t]^* \subseteq B^*\}$. Then F_t is an ideal of B called the conductor of $A[t]$ in B.

S.   Let $t \in B$ be a primitive element of $L|K$:

(1) The conductor of $A[t]$ in B is equal to $F_t = g'(t)B^*$.

(2) F_t is the largest ideal of B contained in $A[t]$.

Proof: (1)   Indeed $g'(t) \cdot B^* \subseteq g'(t) \cdot A[t]^* = A[t] \subseteq B$, hence

$$(g'(t) \cdot B^*) \cdot A[t]^* = g'(t) \cdot A[t]^* \cdot B^* \subseteq B^*.$$

Conversely, if $x \in B$, $x \cdot A[t]^* \subseteq B^*$ then $x \in x \cdot A[t] \subseteq g'(t) \cdot B^*$.

(2)   Since $B^* \subseteq A[t]^* = A[t]/g'(t)$, then $F_t = g'(t) \cdot B^* \subseteq A[t]$.

Now, let J be an ideal of B which is contained in $A[t]$ and $x \in J$. By (J), $\text{Tr}_{L|K}(x/g'(t)) \in A$. For every $y \in B$, we have $xy \in J$ so $\text{Tr}(yx/g'(t)) \in A$, hence $x/g'(t) \in B^*$ by definition of B^*. Therefore, $x \in g'(t) \cdot B^* = F_t$. ∎

T. $\Delta_{L|K}$ is the ideal of B generated by the differents $g'(t)$ of all the primitive elements $t \in B$.

Proof: Let t be a primitive element of $L|K$. If $t \in B$ then, by (K), $B^* \subseteq A[t]^* = A[t]/g'(t) \subseteq B \cdot (1/g'(t))$ hence

$$g'(t) \cdot B^* \subseteq B \qquad \text{so} \qquad g'(t) \in \Delta_{L|K}.$$

To show the converse, it is enough to establish that for every nonzero prime ideal Q of B there exists a primitive element $t \in B$ such that Q does not contain F_t. If this has been shown, the ideal $\sum_{t \in B} F_t$ generated by $\bigcup_{t \in B} F_t$ is equal to B. Hence

$$\Delta_{L|K} = \Delta_{L|K} \cdot B = \Delta_{L|K} \left(\sum_{t \in B} F_t \right) = \sum_{t \in B} \Delta_{L|K} \cdot F_t$$

$$= \sum_{t \in B} \Delta_{L|K} \cdot g'(t) \cdot B^* = \sum_{t \in B} B \cdot g'(t).$$

So, let us prove the above assertion. Let $Q \cap A = P$ and $B \cdot P = Q^e \cdot J$, where $e \geq 1$, Q does not divide J.

We divide the proof into several steps.

1°.   First we show that there exists $t \in L$ such that:
   (1) $t \in J$, $t \notin Q$.
   (2) The image $\bar{t}$ of t in B/Q is a generator of the multiplicative cyclic group of nonzero elements of the finite field B/Q.
   (3) $t^{N(Q)} - t \notin Q^2$.
   (4) $L = K(t)$.

To choose t let $u \in B$ be such that its image $\bar{u} \in B/Q$ generates the multiplicative group of nonzero elements of B/Q (a finite field), which is known to be cyclic (see Chapter 2, Section 9).

Let $q = N(Q)$, so by Chapter 8, (G), $q \in Q$. Then $u^q - u \in Q$. If $u^q - u \notin Q^2$ let $u_1 = u$. If $u^q - u \in Q^2$, let $v \in Q$, $v \notin Q^2$ and let $u_1 = u + v$. Then $\bar{u}_1 = \overline{u + v} = \bar{u}$ and $u_1^q - u_1 \in Q$ but $u_1^q - u_1 = (u^q - u) + qu^{q-1}v + \binom{q}{2}u^{q-1}v^2 + \cdots + v^q - v \notin Q^2$, since $v \notin Q^2$ but every other summand is in Q^2 (since $q \in Q$).

Since Q does not divide J, then $B = J + Q^2$ and we may write $u_1 = t + v_1$ with $t \in J$, $v_1 \in Q^2$; then $t \notin Q$ because $u_1 \notin Q$. We also have $\bar{t} = \bar{u}_1$ so $t^q - t \in Q$ and

$$t^q - t = (u_1 - v_1)^q - (u_1 - v_1) \equiv u_1^q - u_1 \pmod{Q^2}.$$

Hence $t^q - t \notin Q^2$.

Now we show that $L = K(t)$. First we note that if $t' \in t + JQ^2$ then t' also satisfies conditions (1), (2), (3)—this is trivial to verify. We choose t' in the set $t + JQ^2$ such that $[K(t') : K]$ is maximal. If $JQ^2 \subseteq K(t')$ let $x \in JQ^2$, $x \neq 0$, so $Bx \subseteq JQ^2$. But L is generated by Bx over K, hence by JQ^2, so $L = K(JQ^2) \subseteq K(t') \subseteq L$, thus $K(t') = L$.

If $JQ^2 \not\subseteq K(t')$, let $z \in JQ^2$, $z \notin K(t')$. We consider the family $\mathcal{K} = \{K(t' + az) \mid a \in A\}$ of subfields of L. Since $L|K$ is separable, this family of subfields must be finite. But A is infinite, so there exist $a_1, a_2 \in A$, $a_1 \neq a_2$, such that $K(t' + a_1 z) = K(t' + a_2 z) = K'$. Then $(a_1 - a_2)z \in K'$ so $z \in K'$, $t' \in K'$. Thus $K(t')$ is properly contained in K', because

$z \notin K(t')$. But $K(t' + a_1 z)$ is in the family $\mathcal{K}$, which is a contradiction, and proves (4).

2°.   For every integer $l \geq 1$ and element $b \in B$ there exists $c \in A[t]$ such that $b - c \in Q^l$, where t is chosen as in (1°).

Indeed, let S be a system of representatives of B modulo Q. If $w = t^q - t$ then for every $l > 0$:

$$R = \{s_0 + s_1 w + s_2 w^2 + \cdots + s_{l-1} w^{l-1} \mid s_0, s_1, \ldots, s_{l-1} \in S\}$$

is a system of representatives of B modulo Q^l (see Chapter 8, (**B**)). In particular, we may take $S = \{0, 1, t, t^2, \ldots, t^{q-1}\}$. Then for every element $b \in B$ there exists a unique element $c = s_0 + s_1 w + \cdots + s_{l-1} w^{l-1} \in A[t]$, with each $s_i \in \{0, 1, t, \ldots, t^{q-1}\}$, such that $b - c \in Q^l$.

3°.   There exists $l \geq 0$ such that $Bt^l \subseteq A[t]$.

Indeed, let $a \in B \cdot g'(t) \cap A$, $a \neq 0$; let $Aa = P^r \cdot I$, where $r \geq 0$ and P does not divide the ideal I. If h is the class number of K, we obtain principal ideals by considering hth powers of ideals; let $Aa_1 = Aa^h$, $P^{rh} = Aa_2$, hence $a_1 = a_2 a_3$, where $a_3 \in I^h$. Moreover, $a_3 \notin P$, since the exact power of P which divides a_1 is rh.

Let us show that the principal ideal $B(a_3 t^{rh})$ is contained in $A[t]$. Given $b \in B$, let $c \in A[t]$ be an element such that $b - c \in Q^{erh}$. Then $b a_3 t^{rh} = (b - c)a_3 t^{rh} + c a_3 t^{rh}$. As $c a_3 t^{rh} \in A[t]$ it is enough to show that

$$(b - c)a_3 t^{rh} \in A[t].$$

Now we have

$$B(b - c)a_3 t^{rh} = \frac{B a_2 a_3 (b - c) t^{rh}}{B \cdot P^{rh}} \subseteq \frac{B a_1 \cdot Q^{erh} \cdot B t^{rh}}{Q^{erh} \cdot J^{rh}}$$

$$\subseteq B a^h \cdot B \subseteq g'(t) \cdot B \subseteq A[t]$$

because $t \in J$ and $g'(t)B^* \subseteq A[t]$. Therefore, $B(a_3 t^{rh}) \in A[t]$.

4°.   End of the proof. It follows from (**S**) that $B(a_3 t^{rh}) \subseteq F_t$. But $a_3 \notin P$, so $a_3 \notin Q$ and $t \notin Q$, hence $a_3 t^{rh} \notin Q$, showing that F_t is not contained in Q.   ∎

One might ask whether a similar result holds for the relative discriminant. More precisely, for every primitive integral element x of $L|K$, $x \in B$, we may consider the discriminant $\delta(x) = \mathrm{discr}_{L|K}(1, x, x^2, \ldots, x^{n-1})$.

By definition $\delta(x) \in \delta_{L|K}$ and we wish to compare the ideal $\delta_{L|K}$ of A with the ideal $\delta^*_{L|K}$ which is generated by all the above elements $\delta(x)$. We quote without proof the following results due to Hensel:

For every x as above we have an integral ideal I_x such that $\delta(x) = I_x^2 \cdot \delta_{L|K}$.

A nonzero prime ideal P of A, which divides I_x for every primitive integral element x, is called an *inessential factor of the discriminant*. In order that $\delta_{L|K}^* = \delta(x)$ for some primitive integral element x it is necessary and sufficient that there exists no inessential factor for the discriminant.

A necessary and sufficient condition for a prime ideal P of A not to be an inessential factor of the discriminant $\delta_{L|K}$ is the following:

$$g(f) \leq \frac{1}{f} \sum_{d|f} \mu\left(\frac{f}{d}\right) N(P)^d$$

(for any natural number $f \geq 1$) where $g(f)$ denotes the number of prime ideals Q of B such that $Q \cap A = P$ and Q has inertial degree f over P and μ is the Möbius function (see Chapter 3, Exercise 45).

It follows that if P is an inessential factor of $\delta_{L|K}$ then $N(P) < [L : K] = n$. The converse is also true when there exist n distinct prime ideals Q_i of B such that $Q_i \cap A = P$.

So, if $[L : K] = 2$, then there are no inessential factors for $\delta_{L|K}$. However, Dedekind has shown the existence of inessential prime factors for the discriminant of a cubic field (see Chapter 16, Example 3).

We conclude this section with results relating the degree, rings of integers, and discriminants of fields K_1, K_2 and their compositum $L = K_1 K_2$.

U.   *Let K_1, K_2 be two algebraic number fields, extensions of the field K, let $L = K_1 K_2$ be their compositum; let P be a nonzero prime ideal of the ring of integers A of K. Then:*

(1) *$\Delta_{L|K_2}$ divides $B\Delta_{K_1|K}$, $\Delta_{L|K_1}$ divides $B\Delta_{K_2|K}$ where B is the ring of integers of L.*

(2) *$N_{K_2|K}(\delta_{L|K_2})$ divides $\delta_{K_1|K}^{[L:K_1]}$ and $N_{K_1|K}(\delta_{L|K_1})$ divides $\delta_{K_2|K}^{[L:K_2]}$.*

(3) *P is unramified in $L|K$ if and only if it is unramified in $K_1|K$ and in $K_2|K$.*

Proof: (1)   We show that $\Delta_{L|K_2}$ divides $B\Delta_{K_1|K}$, that is, $\Delta_{K_1|K} \subseteq \Delta_{L|K_2}$. Let A_i be the ring of integers of K_i for $i = 1, 2$.

By (**T**), it suffices to show that if $t \in A_1$, $K_1 = K(t)$, if $g \in A[X]$ is the minimal polynomial of t over K, then $g'(t) \in \Delta_{L|K_2}$. We have $L = K_2(t)$. Let h be the minimal polynomial of t over K_2, so h divides g in $K_2[X]$, thus $g = hk$; since h is monic, then $k \in A_2[X]$. It follows that $g'(t) = h'(t)k(t) \in Bh'(t) \subseteq \Delta_{L|K_2}$.

(2)   We use the various properties for the discriminant and the different which were established in this section. We have

$$N_{K_2|K}(\delta_{L|K_2}) = N_{K_2|K}(N_{L|K_2}(\Delta_{L|K_2})) = N_{L|K}(\Delta_{L|K_2})$$

divides (by (1)):

$$N_{L|K}(B\Delta_{K_1|K}) = N_{K_1|K}(N_{L|K_1}(B\Delta_{K_1|K}))$$

$$= N_{K_1|K}\left(\Delta_{K_1|K}^{[L:K_1]}\right) = \delta_{K_1|K}^{[L:K_1]}.$$

(3)   Clearly, if P is unramified in $L|K$ then P is also unramified in $K_1|K$ and in $K_2|K$.

Conversely, we have (by (**Q**)):

$$\delta_{L|K} = \delta_{K_2|K}^{[L:K_2]} \cdot N_{K_2|K}(\delta_{L|K_2})$$

and

$$N_{K_2|K}(\delta_{L|K_2}) = N_{K_2|K}(N_{L|K_2}(\Delta_{L|K_2})) = N_{L|K}(\Delta_{L|K_2})$$

and this ideal divides

$$N_{L|K}(B\Delta_{K_1|K}) = N_{K_1|K}(N_{L|K_1}(B\Delta_{K_1|K})$$

$$= N_{K_1|K}\left(\Delta_{K_1|K}^{[L:K_1]}\right) = \delta_{K_1|K}^{[L:K_1]}.$$

Thus, if P is unramified in $K_1|K$ and in $K_2|K$, by Theorem 1, P does not divide $\delta_{K_1|K}$ nor $\delta_{K_2|K}$. Hence P does not divide $\delta_{L|K}$, so P is unramified in $L|K$.   ∎

As a corollary we have:

V.   Let $K \subseteq L \subseteq L'$ be algebraic number fields such that L' is the smallest field containing L for which $L'|K$ is a Galois extension. Let P be a nonzero prime ideal of the ring of integers of K. Then P is unramified in $L|K$ if and only if P is unramified in $L'|K$.

Proof: We apply the preceding result for the fields $L = L_1$, L_2, ..., L_m, where each L_i is a conjugate of L over K, noting that L' is the compositum of these fields.   ∎

For algebraic number fields with relatively prime absolute discriminants, we have:

W.   Let K_1, K_2 be algebraic number fields of degree n_1, n_2, respectively, and such that δ_{K_1}, δ_{K_2} are relatively prime. Let $L = K_1K_2$. Then:

(1) $[L : \mathbb{Q}] = n_1 n_2$.

(2) $\delta_L = \delta_{K_1}^{n_2} \delta_{K_2}^{n_1}$.

(3) If A_1, A_2, B are the rings of integers of K_1, K_2, L, respectively, then $B = A_1 A_2$; if $\{x_1, \ldots, x_{n_1}\}$ is an integral basis of K_1, and if $\{y_1, \ldots, y_{n_2}\}$ is an integral basis of K_2 then $\{x_1 y_1, \ldots, x_{n_1} y_{n_2}\}$ is an integral basis of L.

Proof: (1)   We have $[L : \mathbb{Q}] = [K_2 : \mathbb{Q}] \cdot [L : K_2] = n_2[L : K_2]$. If $[L : \mathbb{Q}] < n_1 n_2$ then $[L : K_2] < n_1$. Let $K_1 = \mathbb{Q}(t)$ so $L = K_2(t)$; let $g \in \mathbb{Q}[X]$ be the minimal polynomial of t over $\mathbb{Q}$, so $\deg(g) = n_1$. Since $[L : K_2] < n_1$, the minimal polynomial h of t over K_2 has degree smaller than n_1 and it divides g; let K' be the subfield of K_2 generated by the coefficients of h, so K' is not equal to $\mathbb{Q}$ because g is irreducible over $\mathbb{Q}$. From $K' \subseteq K_2$ we deduce (by (**Q**)) that $\delta_{K'}$ divides δ_{K_2}.

On the other hand, the coefficients of h are elementary symmetric functions of the roots of h, which are among the conjugates of t; thus $h \in K_1'[X]$, where K_1' is the smallest Galois extension of $\mathbb{Q}$ containing K_1. Thus $K' \subseteq K_1'$ and again $\delta_{K'}$ divides $\delta_{K_1'}$.

If p is a prime number dividing $\delta_{K'}$, it divides δ_{K_2} and $\delta_{K_1'}$; by Theorem 1 and (V), p divides δ_{K_1}, which is contrary to the hypothesis.

Thus $|\delta_{K'}| = 1$ and by Chapter 9, (D), we conclude that $K' = \mathbb{Q}$, a contradiction. This proves (1).

(2)   By (Q), we have $\delta_L = N_{K_1|\mathbb{Q}}(\delta_{L|K_1}) \cdot \delta_{K_1}^{n_2} = N_{K_2|\mathbb{Q}}(\delta_{L|K_2}) \cdot \delta_{K_2}^{n_1}$. Hence $\delta_{K_1}^{n_2}$ and $\delta_{K_2}^{n_1}$ divide δ_L and, by hypothesis, $\delta_{K_1}^{n_2}\delta_{K_2}^{n_1}$ divides δ_L.

On the other hand, from (U), we know that $N_{K_2|\mathbb{Q}}(\delta_{L|K_2})$ divides $\delta_{K_1}^{[L:K_1]} = \delta_{K_1}^{n_2}$ by (1); hence $\delta_L = N_{K_2|\mathbb{Q}}(\delta_{L|K_2}) \cdot \delta_{K_2}^{n_1}$ divides $\delta_{K_1}^{n_2}\delta_{K_2}^{n_1}$ and this establishes the equality.

(3)   Let A_1A_2 denote the smallest subring of L containing A_1 and A_2; so $A_1A_2 \subseteq B$.

We compute the discriminant of the set $\{x_1y_1, \ldots, x_{n_1}y_{n_2}\}$. First, we observe that if σ is any isomorphism of L, if σ_{K_1}, σ_{K_2} denote the restrictions of σ to K_1 and K_2, respectively, then the mapping $\sigma \to (\sigma_{K_1}, \sigma_{K_2})$ is injective, because $L = K_1K_2$, and also surjective since $[L : \mathbb{Q}] = n_1n_2$. Hence,

$$\mathrm{discr}_{L|\mathbb{Q}}(x_1y_1, \ldots, x_{n_1}y_{n_2}) = \left[\det(\sigma_i\tau_j(x_ky_l))\right]^2,$$

where $\sigma_1, \ldots, \sigma_{n_1}$ are the isomorphisms of K_1 and $\tau_1, \ldots, \tau_{n_2}$ are the isomorphisms of K_2.

We have to compute the determinant of a matrix, which is the Kronecker product of the matrices

$$(\sigma_i(x_k))_{i,k=1,\ldots,n_1} \quad \text{and} \quad (\tau_j(y_l))_{j,l=1,\ldots,n_2}$$

and for this purpose we use the special case stated after Lemma 1 of Chapter 12.

Thus

$$\mathrm{discr}_{L|\mathbb{Q}}(x_1y_1, \ldots, x_{n_1}y_{n_2}) = \left[\det(\sigma_i(x_j))\right]^{2n_2} \cdot \left[\det(\tau_i(y_j))\right]^{2n_1}$$

$$= \delta_{K_1}^{n_2}\delta_{K_2}^{n_1} = \delta_L.$$

Since $x_iy_j \in B$, by Chapter 6, (M) $\{x_1y_1, \ldots, x_{n_1}y_{n_2}\}$ is an integral basis of B and, in particular, $B = A_1A_2$.   ∎

E X E R C I S E S

1. Let $K = \mathbb{Q}(x)$, where $x^3 - x^2 - 2x + 8 = 0$. If $y = (x^2 - x)/2$, compute its characteristic polynomial in $K|\mathbb{Q}$ and its minimal polynomial over $\mathbb{Q}$.

2. Give an independent proof of Theorem 1 for the case of a quadratic field $K = \mathbb{Q}(\sqrt{d})$.

3. Let p, q be distinct prime numbers, ζ a primitive pth root of unity, and η a primitive qth root of unity:
(a) Find an integral basis for $K = \mathbb{Q}(\zeta, \eta)$ and the discriminant of this field.
(b) Let K' be the maximal real subfield of K. Show that the relative different $\Delta_{K|K'}$ is the unit ideal.

4. Let p, q be distinct prime numbers, $p \equiv 1 \pmod 4$, $q \equiv 1 \pmod 4$. Let $K = \mathbb{Q}(\sqrt{p}, \sqrt{q})$:
(a) Find an integral basis and the discriminant of K.
(b) Let $K' = \mathbb{Q}(\sqrt{pq})$; show that the relative different $\Delta_{K|K'}$ is the unit ideal.

5. Applying Hensel's criterion for the existence of an inessential factor of the discriminant, show the following assertions of the text:
(a) If P is an inessential factor of $\delta_{L|K}$ then $N(P) < [L : K] = n$.
(b) If $N(P) < n$ and BP is decomposed into the product of n distinct prime ideals of B then P is an inessential factor.

6. Let K be an algebraic number field, A the ring of integers, and R a subring of A. Show that there exists an ideal F of A such that:
(a) $F \subseteq R$.
(b) If I is an ideal of A such that $I \subseteq R$ then $I \subseteq F$.

Then F is called the *conductor* of R in A.

7. Let K be an algebraic number field, A the ring of integers, R a subring of A, and F the conductor of R in A. If I is an ideal of R let AI denote the ideal of A generated by I. I is said to be a *regular ideal* of R when $\gcd(AI, F) = A$. Show:
(a) If J is an ideal of A such that $\gcd(J, F) = A$ there exists an ideal I of R such that $J = AI$.

(b) If I, I' are regular ideals of R then $A(I \cdot I') = AI \cdot AI'$.

(c) The number of congruence classes modulo F of elements $a \in R$ such that $\gcd(Aa, F) = A$ divides $\varphi(F)$ (see Chapter 8, Exercise 2).

14

The Decomposition of Prime Ideals in Galois Extensions

14.1 Decomposition and Inertia

Let K be an algebraic number field, $L|K$ a finite extension of degree n, and, as before, let A, B be, respectively, the rings of integers of K, L. Let $\underline{P}$ be a prime ideal of A, and let $B\underline{P} = \prod_{i=1}^{g} P_i^{e_i}$ be the decomposition of $B\underline{P}$ into a product of prime ideals, with $f_i = [B/P_i : A/\underline{P}]$. We shall study in more detail how this decomposition takes place. This has been done by Hilbert, assuming that $L|K$ is a Galois extension.

Accordingly, let $\mathcal{K} = G(L|K)$ be the Galois group of $L|K$, so $\mathcal{K}$ has n elements, the K-automorphisms of L. We shall make appeal to the discussion in Chapter 11, preceding and including (E), (F), and Theorem 1; in particular, we write $e = e_1 = \cdots = e_g$, $f = f_1 = \cdots = f_g$.

We shall also adopt the following notation: $\overline{K} = A/\underline{P}$, $\overline{L}_i = B/P_i$, each field $\overline{L}_i$ is isomorphic with the extension of degree f_i of the finite field $\overline{K}$. Since $f_1 = \cdots = f_g$, all the fields $\overline{L}_i$ are actually isomorphic; however, we shall not identify them.

Definition 1. With the preceding notations, the subgroup $\mathcal{Z}_i$ of $\mathcal{K}$, defined by $\mathcal{Z}_i = \{\sigma \in \mathcal{K} \mid \sigma(P_i) = P_i\}$ is called the *decomposition group of P_i in the extension $L|K$*. The field of invariants of $\mathcal{Z}_i$ is denoted by Z_i and is called *the decomposition field of P_i in the extension $L|K$*.

If necessary, we may also use the following notations:

$$\mathcal{Z}_i = \mathcal{Z}(P_i | \underline{P}) = \mathcal{Z}_{P_i}(L|K),$$
$$Z_i = Z(P_i | \underline{P}) = Z_{P_i}(L|K).$$

A. *The subgroups $\mathcal{Z}_1, \ldots, \mathcal{Z}_g$ of $\mathcal{K}$ are conjugate (by inner automorphisms of $\mathcal{K}$). In particular, if $\mathcal{K}$ is an Abelian group, then $\mathcal{Z}_1 = \cdots = \mathcal{Z}_g$.*

Proof: Let P_i, P_j be distinct prime ideals of B such that $P_i \cap A = P_j \cap A = \underline{P}$. Since $\mathcal{K}$ acts transitively on the set of prime ideals $\{P_1, \ldots, P_g\}$, there exists $\sigma \in \mathcal{K}$ such that $\sigma(P_i) = P_j$. Then $\mathcal{Z}_i = \sigma^{-1}\mathcal{Z}_j\sigma$. ∎

B.   *For every integer* $i = 1, \ldots, g$ *we have:* $[Z_i : K] = (\mathcal{K} : \mathcal{Z}_i) = g$.

Proof: This proof has already been in Chapter 11, (**O**), while discussing the cyclotomic field. We repeat it for the convenience of the reader. We have $\sigma \mathcal{Z}_i = \tau \mathcal{Z}_i$ (with $\sigma, \tau \in \mathcal{K}$) if and only if $\sigma(P_i) = \tau(P_i)$. In fact, if $\sigma \mathcal{Z}_i = \tau \mathcal{Z}_i$, then $\sigma^{-1}\tau \in \mathcal{Z}_i$, so $\sigma^{-1}\tau(P_i) = P_i$, hence $\tau(P_i) = \sigma(P_i)$. Conversely, if $\sigma(P_i) = \tau(P_i)$, then $\sigma^{-1}\tau \in \mathcal{Z}_i$ hence $\tau \mathcal{Z}_i = \sigma \mathcal{Z}_i$.

Thus, the number g of distinct prime ideals P_i is the same as the number of distinct cosets modulo $\mathcal{Z}_i$; that is, the index $(\mathcal{K} : \mathcal{Z}_i)$. From Galois theory, we have $(\mathcal{K} : \mathcal{Z}_i) = [Z_i : K]$.   ■

The decomposition field has the following minimality property:

C.   *If* $Q_i = P_i \cap Z_i$ *(prime ideal of the ring of integers* $B \cap Z_i$ *of* Z_i*), then* P_i *is the only ideal of the ring of integers of* L *which extends* Q_i*. Conversely, if* Z_i' *is a field,* $K \subseteq Z_i' \subseteq L$*, if* $Q_i' = P_i \cap Z_i'$ *and* P_i *is the only extension of* Q_i' *to* L*, then* $Z_i \subseteq Z_i'$*.*

Proof: We have $\mathcal{Z}_i = G(L|Z_i)$; by Chapter 11, (**E**), $\mathcal{Z}_i$ acts transitively on the set of prime ideals of B extending Q_i; but, by definition, $\sigma(P_i) = P_i$ for every $\sigma \in \mathcal{Z}_i$, thus P_i is the only extension of Q_i.

Next, if P_i is the only extension of $Q_i' = P_i \cap Z_i'$, then every element of the Galois group $\mathcal{Z}_i' = G(L|Z_i')$ fixes P_i, hence belongs to $\mathcal{Z}_i$; thus, we have the opposite inclusion for the fixed fields: $Z_i' \supseteq Z_i$.   ■

We fix our attention on one of the prime ideals P_i, which we shall denote by P for simplicity.

Let $Z = Z_P(L|K)$, $\mathcal{Z} = \mathcal{Z}_P(L|K)$, $\overline{L} = B/P$, $\overline{\mathcal{K}} = G(\overline{L}|\overline{K})$. We denote also by $B_Z = B \cap Z$ the ring of integers of Z, by $P_Z = P \cap Z$ the prime ideal of B_Z defined by P, and by $\overline{Z} = B_Z/P_Z$ the corresponding residue class field; accordingly, let e be the ramification index and f the inertial degree of P over $\underline{P}$.

D.   (1) $\overline{Z} = \overline{K}$, *so the inertial degrees are* $f(P_Z|\underline{P}) = 1$, $f(P|P_Z) = f$; *the ramification indices are* $e(P_Z|\underline{P}) = 1$, $e(P|P_Z) = e$.

(2)   *The mapping* $\sigma \in \mathcal{Z} \to \overline{\sigma} \in G(\overline{L}|\overline{K})$ *is a group-homomorphism onto* $G(\overline{L}|\overline{K})$*, having kernel equal to the normal subgroup*

$$\mathcal{T} = \{\sigma \in \mathcal{Z} \mid \sigma(x) \equiv x \pmod{P} \text{ for every element } x \in B\}.$$

Proof: (1)   By the fundamental relation of Theorem 1, Chapter 11, and (**C**), we have

$$[L : Z] = e(P|P_Z) \cdot f(P|P_Z).$$

On the other hand, $[L : K] = efg$ and by (**B**), $[Z : K] = g$. Therefore, $ef = e(P|P_Z) \cdot f(P|P_Z)$. By the transitivity of the ramification index and inertial degree, we must have $e(P|P_Z) = e$, $f(P|P_Z) = f$ and $e(P_Z|\underline{P}) =$

1, $f(P_Z|\underline{P}) = 1$. Therefore

$$[\overline{L} : \overline{Z}] = f(P|P_Z) = f = [\overline{L} : \overline{K}] \quad \text{implies that} \quad \overline{Z} = \overline{K}.$$

(2)   If $\sigma \in \mathcal{Z}$ then $\sigma(P) = P$, hence σ induces the mapping $\overline{\sigma} : \overline{L} \to \overline{L}$, defined by $\overline{\sigma}(\overline{x}) = \overline{\sigma(x)}$ for every $x \in B$. It is immediate that $\overline{\sigma}$ is a $\overline{Z}$-automorphism.

Now, we shall prove that the image is equal to $G(\overline{L}|\overline{K})$. Since $\overline{K}$ is a finite field, there exists $b \in B$ such that $\overline{L} = \overline{K}(\overline{b})$. If $\xi \in G(\overline{L}|\overline{K})$ then $\xi(\overline{b})$ is a conjugate of $\overline{b}$ over $\overline{K}$.

Let h be the minimal polynomial of b over Z; since $L|Z$ is a Galois extension, all the conjugates of b over Z are still in L, and in fact in B; thus h decomposes as $h = \prod_{\sigma \in \mathcal{Z}}(X - \sigma(b))$; considering the images of the coefficients by the canonical mapping $B \to \overline{L}$ (which extends $B_Z \to \overline{Z} = \overline{K}$), we have $\overline{h} = \prod(X - \overline{\sigma(b)}) \in \overline{K}[X]$; of course, $\overline{b}$ is among the roots of $\overline{h}$; the conjugates of $\overline{b}$ over $\overline{K}$ are the roots of its minimal polynomial, which divides $\overline{h}$, thus the conjugates of $\overline{b}$ are among the elements $\overline{\sigma(b)} \in \overline{L}$. In particular, $\xi(\overline{b}) = \overline{\sigma(b)} = \overline{\sigma}(\overline{b})$, for some $\sigma \in \mathcal{Z}$, and therefore ξ and $\overline{\sigma}$ must coincide on every element of $\overline{L}$.

The kernel of the group-homomorphism is obviously the set of all $\sigma \in \mathcal{Z}$ such that $\overline{\sigma}(\overline{x}) = \overline{x}$ for every $\overline{x} \in \overline{L}$, that is, $\sigma(x) \equiv x \pmod{P}$ for every $x \in B$.   ∎

Thus, we have the group-isomorphism $\mathcal{Z}/\mathcal{T} \cong G(\overline{L}|\overline{K})$ for every prime ideal P.

It is convenient to remark that $\mathcal{T}$ is also equal to the set of all $\sigma \in \mathcal{Z}$ such that $\sigma(x) \equiv x \pmod{B_P P}$ for every $x \in B_P$. For if σ satisfies this latter condition, if $x \in B$ then $\sigma(x) - x \in B \cap B_P P = P$. Conversely, if $\sigma \in \mathcal{T}$, if $x \in B_P$, we may write $x = b/s$, with $b, s \in B$, $s \notin P$; let $a = N_{L|Z}(s)$, the product of the conjugates of s over Z, so $a = ss' \in B \cap Z$, $a \notin P$ (because if $\sigma \in \mathcal{Z}$ then $\sigma(P) = P$, so $\sigma(s) \notin P$) and $x = bs'/a$, with $bs' \in B$, $a \in B \cap Z$, $a \notin P$; thus $\sigma(x) - x = (\sigma(bs') - bs')/a \in B_P P$ for $\sigma \in \mathcal{T}$.

Definition 2.   For every prime ideal P_i of B, $\mathcal{T}_i$ is called the *inertial group of P_i* in the extension $L|K$. The field of invariants of $\mathcal{T}_i$ is denoted by T_i and called the *inertial field of P_i in $L|K$*.

We may also adopt the following notations:

$$\mathcal{T}_i = \mathcal{T}(P_i|\underline{P}) = \mathcal{T}_{P_i}(L|K),$$
$$T_i = T(P_i|\underline{P}) = T_{P_i}(L|K).$$

When we fix our attention on one of the prime ideals P_i, which we denote by P for simplicity, then we write $T = T_P(L|K)$ and $\mathcal{T} = \mathcal{T}_P(L|K)$. We also denote by $B_T = B \cap T$ the ring of integers of T, P_T the prime ideal $P \cap T = P_T$, and by $\overline{T}$ the corresponding residue class field.

With these notations, we have the first important result:

Theorem 1.   (1) $T|Z$ *is a Galois extension and* $G(T|Z) = \mathcal{Z}/\mathcal{T} \cong \overline{K}$.

(2) $[T : Z] = f$, $[L : T] = e$.

(3) $\overline{L} = \overline{T}$, *so the inertial degrees are*

$$f(P_T|P_Z) = f, \qquad f(P|P_T) = 1.$$

(4) *The ramification indices of the ideals in question are*

$$e(P_T|P_Z) = 1, \qquad e(P|P_T) = e.$$

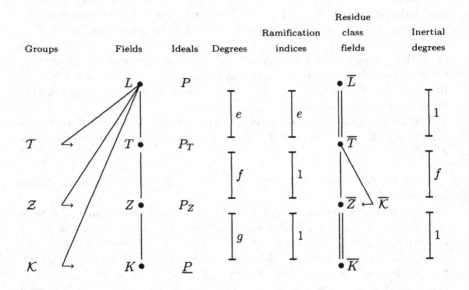

Groups	Fields	Ideals	Degrees	Ramification indices	Residue class fields	Inertial degrees

Proof: The assertion (1) is now obvious.

By (1), we have $[T : Z] = \#(\mathcal{Z}/\mathcal{T}) = \#(\overline{K}) = [\overline{L} : \overline{K}] = f$.

Since $n = efg$, $[Z : K] = g$ (by (**B**)) and $[T : Z] = f$, then $[L : T] = e$.

To show that $\overline{L} = \overline{T}$ we consider the extension $L|T$. Then $\mathcal{Z}_P(L|T) = \mathcal{T}$ and $\mathcal{T}_P(L|T) = \mathcal{T}$, as is obvious. Hence, by (1), $G(\overline{L}|\overline{T}) = \mathcal{T}/\mathcal{T}$ so $\overline{L} = \overline{T}$. Therefore, $[\overline{T} : \overline{Z}] = [\overline{L} : \overline{K}] = f$.

Considering the Galois extension $L|T$, we have

$$[L : T] = e = e(P|P_T) \cdot f(P|P_T);$$

but $f(P|P_T) = 1$, so $e(P|P_T) = e$ and by transitivity of ramification, $e(P_T|P_Z) = 1$.   ∎

It is also useful to know the relative behavior of these groups.

For the next proposition, we consider algebraic number fields K, F, F', L with $K \subseteq F \subseteq L$, $K \subseteq F' \subseteq L$; we assume that $L|K$, $F'|K$, $F|K$ are Galois extensions. Let P, P_F, $P_{F'}$, $\underline{P}$ be prime ideals of the rings of integers of L, F, F', K, such that $P_F = P \cap F$, $P_{F'} = P \cap F'$, $\underline{P} = P_F \cap K = P_{F'} \cap K$. We also denote by $g_{\underline{P}}(L|K)$ the number of prime ideals P of L dividing $\underline{P}$ and by $f_P(L|K)$ the inertial degree of P in $L|K$. The notations

$$g_{\underline{P}}(F|K), \ g_{\underline{P}}(F'|K), \ g_{P_F}(L|F), \ g_{P_{F'}}(L|F'),$$
$$f_P(L|F), \ f_P(L|F'), \ f_{P_F}(F|K), \ f_{P_{F'}}(F'|K),$$

have similar meanings.

E. *With the above notations:*

(1) $\mathcal{Z}_{P_F}(F|K) \cong \mathcal{Z}_P(L|K)/\mathcal{Z}_P(L|F)$,

   $\mathcal{T}_{P_F}(F|K) \cong \mathcal{T}_P(L|K)/\mathcal{T}_P(L|F)$.

(2) *If $F \cap F' = K$ and $L = FF'$ then*

$$\mathcal{Z}_P(L|K) \cong \mathcal{Z}_{P_F}(F|K) \times \mathcal{Z}_{P_{F'}}(F'|K),$$
$$\mathcal{T}_P(L|K) \cong \mathcal{T}_{P_F}(F|K) \times \mathcal{T}_{P_{F'}}(F'|K),$$

and

$$g_{\underline{P}}(L|K) = g_{\underline{P}}(F|K) \cdot g_{\underline{P}}(F'|K),$$
$$f_P(L|K) = f_P(L|F) \cdot f_P(L|F').$$

Proof: (1)   If $\sigma \in \mathcal{Z}_P(L|K)$ let $\sigma|F$ denote the restriction of σ to the field F; thus $\sigma|F \in G(F|K)$ and actually $\sigma|F \in \mathcal{Z}_{P_F}(F|K)$. The mapping $\sigma \to \sigma|F$ is obviously a group-homomorphism and its kernel is $G(L|F) \cap \mathcal{Z}_P(L|K) = \mathcal{Z}_P(L|F)$. It remains to show that the image of the mapping is $\mathcal{Z}_{P_F}(F|K)$. Given $\tau \in \mathcal{Z}_{P_F}(F|K)$ there exists an extension σ of τ to a K-automorphism of L, so $\sigma|F = \tau$; then $\sigma(P)$ is such that $\sigma(P) \cap F = P_F$, because $\sigma|F = \tau$ leaves P_F fixed. By Chapter 11, (**E**), there exists $\sigma' \in G(L|F)$ such that $\sigma'(\sigma(P)) = P$. It follows that $\sigma'\sigma \in \mathcal{Z}_P(L|K)$ and $\tau = (\sigma'\sigma)|F \in \mathcal{Z}_{P_F}(F|K)$.

The second assertion is proved in the same way. If $\sigma \in \mathcal{T}_P(L|K)$ then $\sigma|F \in \mathcal{T}_{P_F}(F|K)$, the kernel of the homomorphism in question is

$$\mathcal{T}_P(L|F) = \mathcal{T}_P(L|K) \cap G(L|F).$$

Finally, the mapping has image equal to $\mathcal{T}_{P_F}(K|F)$, as we may easily see: $\mathcal{T}_P(L|K)/\mathcal{T}_P(L|F) \subseteq \mathcal{T}_{P_F}(F|K)$ (up to isomorphism), hence, considering the orders of these groups, which are ramification indices (by Theorem 1, (2)), we have $e_P(L|K)/e_P(L|F) \leq e_{P_F}(F|K)$; from Chapter 11, (**D**), we must have equality, thereby proving the second assertion.

(2)   In this situation, $G(L|K) \cong G(F|K) \times G(F'|K)$. By the canonical restriction mapping $G(L|K) \to G(F|K)$, the image of $\mathcal{Z}_P(L|K)$

is $\mathcal{Z}_P(L|K)/\mathcal{Z}_P(L|F) \cong \mathcal{Z}_{P_F}(F|K)$ by (1), since $\mathcal{Z}_P(L|K) \cap G(L|F) = \mathcal{Z}_P(L|F)$. The same holds with F' instead of F, hence $\mathcal{Z}_P(L|K) \cong \mathcal{Z}_{P_F}(F|K) \times \mathcal{Z}_{P_{F'}}(F'|K)$. By the same argument we prove the assertion for the inertial subgroups. We use these facts and the homomorphism theorems for groups to establish the relation for the number of extensions of the ideals under consideration. $G(L|K)/\mathcal{Z}_P(L|K)$ contains the subgroup $G(L|F)/\mathcal{Z}_P(L|F)$, noting that $\mathcal{Z}_P(L|F) = \mathcal{Z}_P(L|K) \cap G(L|F)$. This subgroup is isomorphic to $G(L|F)\mathcal{Z}_P(L|K)/\mathcal{Z}_P(L|K)$. The quotient group is isomorphic to

$$G(L|K)/G(L|F)\mathcal{Z}_P(L|K)$$
$$\cong (G(L|K)/G(L|F)) \, / \, (G(L|F)\mathcal{Z}_P(L|K)/G(L|F))$$
$$\cong G(F|K)/\mathcal{Z}_{P_F}(F|K)$$

by (1), since $G(L|F)\mathcal{Z}_P(L|K)$ has the same image as $\mathcal{Z}_P(L|K)$ by the homomorphism considered above.

It follows that $g_{\underline{P}}(L|K) = g_{\underline{P}}(F|K)g_{P_F}(L|F)$.

As was shown, the image of $\mathcal{Z}_P(L|F)$ by the map from $G(L|F)$ to $G(F'|K)$ is equal to $\mathcal{Z}_{P_{F'}}(F'|K)$ hence

$$g_{P_F}(L|F) = \big(G(F'|K) : \mathcal{Z}_{P_{F'}}(F'|K)\big)$$
$$= g_{\underline{P}}(F'|K),$$

showing the required equality.

Since $\mathcal{Z}_P(L|K) \cong \mathcal{Z}_{P_F}(F|K) \times \mathcal{Z}_{P_{F'}}(F'|K)$ and $\mathcal{T}_P(L|K) \cong \mathcal{T}_{P_F}(F|P) \times \mathcal{T}_{P_{F'}}(F'|P)$, then

$$\mathcal{Z}_{P_F}(L|K)/\mathcal{T}_P(L|K) \cong \mathcal{Z}_{P_F}(F|K)/\mathcal{T}_{P_F}(F|K)$$
$$\times \mathcal{Z}_{P_{F'}}(F'|K)/\mathcal{T}_{P_{F'}}(F'|K).$$

So by Theorem 1, $f_P(L|K) = f_P(L|F) \cdot f_P(L|F')$.   ∎

14.2 The Ramification

Now we shall study the ramification, which by Theorem 1, (2), occurs in the extension $L|T$. Since $[L : T] = e_P(L|T)$, we say that P_T is *totally ramified* in $L|T$.

We have to consider the following situation:

$L|T$ is a Galois extension of degree e, with Galois group $\mathcal{T}$, the prime ideal $P_T = P \cap T$ of the ring B_T of integers of T has only one extension P to the ring B of L, $BP_T = P^e$, and the residue class fields are $\overline{L} = \overline{T}$. Thus $\mathcal{Z}_P(L|T) = \mathcal{T}_P(L|T) = \mathcal{T}$.

The following lemma will be useful:

Lemma 1.   *Let R be a Dedekind domain, K its field of quotients, $L|K$ a Galois extension of finite degree, and T the integral closure of R in L. Let P be a prime ideal of R and assume that there exists only one prime ideal Q in T such that $Q \cap R = P$. Let S be the set-complement of P in R and $R' = S^{-1}R$, $T' = S^{-1}T$. Then $T_Q = T'$.*

Proof: Since S is contained in the set-complement of Q in T, then $T' \subseteq T_Q$. Conversely, to show that $T_Q \subseteq T'$, by Chapter 12, (**D**), it suffices to show that every element of T_Q is integral over R'. Since the elements of T are integral over R, hence over R', it suffices to show that for each $t \in T$, $t \notin Q$, $1/t$ is integral over R'.

Let $X^m + a_1 X^{m-1} + \cdots + a_m \in R[X]$ be the minimal polynomial of t over K, so it has coefficients in R. Since $L|K$ is a Galois extension and Q is the only prime ideal of T with $Q \cap R = P$ by Chapter 11, (**E**), all the conjugates of t are in T but not in Q, hence $a_m \in R$, $a_m \notin P$.

Thus $1/a_m \in R'$. But $1/t$ is a root of the polynomial

$$X^m + \frac{a_{m-1}}{a_m} X^{m-1} + \cdots + \frac{a_1}{a_m} X + \frac{1}{a_m} = 0,$$

therefore $1/t$ is integral over R'.   ∎

F.   *Under the hypothesis stated above, let $t \in B_P$ be a generator of the prime ideal $B_P P$ of B_P, that is, $B_P P = B_P t$. Then $\{1, t, \ldots, t^{e-1}\}$ is a basis of the free module B_P over $(B_T)_{P_T}$; in particular, $B_P = (B_P)_{P_T}[t]$. Moreover, t is the root of an Eisenstein polynomial with coefficients in $(B_T)_{P_T}$.*

Proof: First we show that if $a \in T$, $a \neq 0$, then $B_P a = B_P P^{se}$ (for some $s \in \mathbb{Z}$). Indeed, since $(B_P)_{P_T}$ is a principal ideal domain, by Chapter 12, (**G**), there exists $s \in \mathbb{Z}$ such that $(B_T)_{P_T} a = (B_T)_{P_T} P_T^s$, then $B_P a = B_P (B_T)_{P_T} a = B_P (B_T)_{P_T} P_T^s = B_P P_T^s = B_P B P_T^s = B_P P^{se}$.

Next, we prove that if $x = \sum_{i=1}^{e-1} a_i t^i$ with $a_i \in T$, if $0 \leq i$, $j \leq e - 1$ and $a_i \neq 0$, $a_j \neq 0$ then $s_i e + i \neq s_j e + j$ (where $B_P a_i = B_P P^{s_i e}$); this is clear since otherwise we would have $0 \neq i - j = (s_j - s_i)e$ and $|i - j| < e$, which is impossible.

So, if $x = \sum_{i=0}^{e-1} a_i t^i$, with $a_i \in T$, and some $a_i \neq 0$, if

$$m = \min\{s_i e + i \mid a_i \neq 0, \ B_P a_i = B_P P^{s_i e}\}$$

then $B_P x = B_P P^m$.

Indeed, let i be such that $m = s_i e + i$; then for all j such that $a_j \neq 0$, we have $m < s_i e + j$, $x \in B_P P^m$ and if $x \in B_P P^{m+1}$ then $a_i t^i = x - \sum_{j \neq i} a_j t^j \in B_P P^{m+1}$, hence $B_P a_i t^i = B_P P^m \subseteq B_P P^{m+1}$, which is not true.

This implies that $\{1, t, \ldots, t^{e-1}\}$ are linearly independent over T, because if $\sum_{i=0}^{e-1} a_i t^i = 0$ with $a_i \in T$ and some $a_i \neq 0$, then $0 = B_P P^m$, with $m \in \mathbb{Z}$, which is not possible.

Since $[L : T] = e$, the set $\{1, t, \ldots, t^{e-1}\}$ is a T-basis of L. These elements generate the $(B_T)_{P_{T'}}$-module B_P, hence form a basis. Indeed, if $x \in B_P$ we may already write $x = \sum_{i=0}^{e-1} a_i t^i$ with coefficients $a_i \in T$. We have to show that $a_i \in (B_T)_{P_{T'}}$, and we assume that $x \neq 0$, so some $a_i \neq 0$. From $x \in B_P$, if $B_P x = B_P P^m$ then $m \geq 0$. However, as we have seen

$$m = \min\{s_i e + i \mid a_i \neq 0, \; B_P a_i = B_P P^{s_i e}\}$$

so $s_i e + i \geq 0$, $s_i \geq -i/e > -1$; therefore $s_i \geq 0$ since it is an integer; thus, $a_i \in B_P \cap T = (B_T)_{P_{T'}}$.

Let $g = X^e + a_1 X^{e-1} + \cdots + a_e \in T[X]$ be the minimal polynomial of t over T; we shall show that $a_i \in (B_T)_{P_{T'}} P_T$ for $i = 1, \ldots, e$, but $a_e \notin (B_T)_{P_{T'}} P_T^2$; in such a case, g is called an *Eisenstein polynomial*. From $g(t) = 0$, we deduce that $-t^e = a_1 t^{e-1} + \cdots + a_e$; therefore,

$$e = \min\{s_i e + (e - i) \mid a_i \neq 0, \; B_P a_i = B_P P^{s_i e}\};$$

so $e \leq s_i e + (e - i)$, $0 < i/e \leq s_i$; therefore, $a_i \in (B_T)_{P_{T'}} P_T$ for $i = 1, \ldots, e$. On the other hand, if $i \neq e$ then $s_i e + e - i \geq e + e - i > e$; thus the above minimum is attained when $i = e$, that is, $s_e e = e$, hence $(B_T)_{P_{T'}} P_T = (B_T)_{P_{T'}} a_e$ and $a_e \notin (B_T)_{P_{T'}} P_T^2$. ∎

We note at this point that $B_P P$ has a generator $t' \in B$, for if $B_P P = B_P t$, with $t = t'/s$, $t' \in B$, $s \in B$, $s \notin P$ then $B_P t = B_P t'$.

G.   *For every $i = 0, 1, 2, \ldots$ let $\mathcal{V}_i = \{\sigma \in \mathcal{T} \mid \sigma(x) \equiv x \pmod{P^{i+1}}$ for every $x \in B\}$. Then:*

    (1) *Each $\mathcal{V}_i$ is a normal subgroup of $\mathcal{T}$ and*

$$\mathcal{T} = \mathcal{V}_0 \supseteq \mathcal{V}_1 \supseteq \mathcal{V}_2 \supseteq \cdots.$$

    (2) *There exists an index r such that $\mathcal{V}_r$ is the trivial group.*

    (3) *If $t \in B$ is an element such that $B_P P = B_P t$ then*

$$\mathcal{V}_i = \{\sigma \in \mathcal{T} \mid \sigma(t) \cdot t^{-1} \equiv 1 \pmod{B_P P^i}\}$$

    for every $i = 0, 1, 2, \ldots$.

Proof: (1)   We consider the ring B/P^{i+1}; if $\sigma \in \mathcal{Z}$ then $\sigma(P) = P$, hence $\sigma(P^{i+1}) = P^{i+1}$; then each σ acts on B/P^{i+1} in a natural way: $\overline{\sigma}(\overline{x}) = \overline{\sigma(x)}$. Thus, $\sigma \in \mathcal{V}_i$ if and only if σ acts trivially on B/P^{i+1}; that is, $\overline{\sigma}$ is the identity mapping; therefore $\mathcal{V}_i$ is the kernel of the group-homomorphism $\sigma \to \overline{\sigma}$, so $\mathcal{V}_i$ is a normal subgroup of $\mathcal{T}$. Obviously, $\mathcal{T} = \mathcal{V}_0 \supseteq \mathcal{V}_1 \supseteq \mathcal{V}_2 \supseteq \cdots$.

    (2) $\bigcap_{i=0}^{\infty} \mathcal{V}_i$ is the trivial group, for if σ belongs to this intersection then $\sigma(x) - x \in \bigcap_{i=0}^{\infty} P^{i+1}$, that is, $\sigma(x) = x$ for every $x \in B$.

Since T is a finite group, then there exists r such that $\mathcal{V}_r$ consists only of the trivial automorphism.

(3)   If $\sigma \in \mathcal{V}_i$ then $\sigma(t) - t \in B_P P^{i+1}$, hence $\sigma(t)/t - 1 \in B_P P^i$. Conversely, this implies that $\sigma(t) - t \in B_P P^{i+1}$. Next, if $x \in B \subseteq B_P$ we may write $x = \sum_{i=0}^{e-1} a_i t^i$ with $a_i \in (B_T)_{P_r} \subseteq T$, hence

$$\sigma(x) - x = \sum_{i=0}^{e-1} a_i(\sigma(t)^i - t^i);$$

but

$$\sigma(t)^i - t^i = [\sigma(t) - t] \cdot [\sigma(t)^{i-1} + \sigma(t)^{i-2}t + \cdots + t^{i-1}] \in B_P P^{i+1},$$

hence $\sigma(x) - x \in B \cap B_P P^{i+1} = P^{i+1}$.   ∎

Definition 3.   For every $i = 0, 1, 2, \ldots$, $\mathcal{V}_i$ is called the ith *ramification group of P in $L|K$*. The field of invariants of $\mathcal{V}_i$ is denoted by V_i and is called the ith *ramification field of P in $L|K$*.

If necessary, we may use the following more precise notations: $\mathcal{V}_i = \mathcal{V}_i(P|\underline{P}) = \mathcal{V}_{i_r}(L|K)$ and $V_i = V_i(P|\underline{P}) = V_{i_r}(L|K)$.

Thus, in our case

$$K \subseteq Z \subseteq T = V_0 \subseteq V_1 \subseteq V_2 \subseteq \cdots \subseteq V_r = L$$

and each extension $V_i|V_0$ is Galoisian, with Galois group $G(V_i|V_0) = T/\mathcal{V}_i$.

As for the inertial group, let us note that $\mathcal{V}_i$ is also equal to $\{\sigma \in T \mid \sigma(x) \equiv x \pmod{B_P P^{i+1}}$ for every $x \in B_P\}$; the proof is the same, therefore will be omitted.

It is our purpose now to study the structure of the group $T = \mathcal{V}_0$.

Theorem 2.   (1) *There exists a natural group-isomorphism θ from $T/\mathcal{V}_1$ into $\overline{L}^{\cdot}$ (multiplicative group of nonzero elements of $\overline{L}$), hence $T/\mathcal{V}_1$ is a cyclic group whose order is prime to p, where $\mathbb{Z}p = P \cap \mathbb{Z}$.*

(2) *For every $i = 0, 1, 2, \ldots$ there exists an isomorphism θ_i from the group $\mathcal{V}_i/\mathcal{V}_{i+1}$ into the additive group of $\overline{L}$, hence $\mathcal{V}_i/\mathcal{V}_{i+1}$ is an elementary Abelian p-group (that is, a finite-dimensional vector space over the field $\mathbb{F}_p$).*

(3) $\mathcal{V}_1$ *is a p-group. T is a solvable group.*

(4) *If $m = [V_1 : T] = \#(T/\mathcal{V}_1)$ then p does not divide m and $e = mp^s$ for some $s \geq 0$, $p^s = [L : V_1] = \#(\mathcal{V}_1)$.*

Proof: (1) Let $t \in B$ be a generator of the principal ideal $B_P P$ so $t \in B \cap B_P P = P$. If $\sigma \in T$, since $\sigma(P) = P$, then $\sigma(t) \in P \subseteq B_P P = B_P t$, hence there exists $c_\sigma \in B_P$ such that $\sigma(t) = c_\sigma t$.

We show that $c_\sigma \notin B_P P$, because considering $\sigma^{-1} \in T$, $\sigma^{-1}(t) = c_{\sigma^{-1}} t$ with $c_{\sigma^{-1}} \in B_P$, $t = \sigma(\sigma^{-1}(t)) = \sigma(c_{\sigma^{-1}})\sigma(t) = [\sigma(c_{\sigma^{-1}}) \cdot c_\sigma]t$ and

therefore

$$\sigma(c_{\sigma^{-1}}) \cdot c_\sigma = 1,$$

c_σ is invertible in B_P, thus $c_\sigma \notin B_P P$. By Chapter 12, (**H**), $B_P/B_P P \cong B/P = \bar{L}$ and the image of c_σ in $\bar{L}$ is $\bar{c}_\sigma \neq \bar{0}$.

We define the mapping $\bar{\theta} : \mathcal{T} \to \bar{L}^\cdot$ by $\bar{\theta}(\sigma) = \bar{c}_\sigma$. To be convinced that $\bar{\theta}$ is natural, we have to show that $\bar{\theta}$ is independent of the element $t \in B$. If $t' \in B$ is also such that $B_P P = B_P t'$, then $t' = ut$ where $u \in B_P$ is invertible in B_P, hence $u \notin B_P P$; let $\sigma(t') = c'_\sigma t'$, hence $\sigma(ut) = c'_\sigma ut$. But $\sigma \in \mathcal{T}$, therefore $\sigma(u) \equiv u \pmod{B_P P}$, so $\sigma(u) = u + vt$ with $v \in B_P$, and therefore $(u + vt)c_\sigma t = c'_\sigma ut$, $uc_\sigma + vc_\sigma t = c'_\sigma u$ and, considering the images in $\bar{L}$, we have $\bar{u} \cdot \bar{c}_\sigma = \bar{c'_\sigma} \cdot \bar{u}$, so $\bar{c}_\sigma = \bar{c'_\sigma}$. This shows that $\bar{\theta}$ is independent of the choice of t.

$\bar{\theta}$ is a group-homomorphism: $\bar{\theta}(\sigma\tau) = \bar{\theta}(\sigma)\bar{\theta}(\tau)$. In fact, if $\sigma(t) = c_\sigma t$, $\tau(t) = c_\tau t$ then $\sigma\tau(t) = \sigma(c_\tau)\sigma(t) = (c_\tau + vt)c_\sigma t = (c_\tau c_\sigma + vc_\sigma t)t$ where $v \in B_P$; thus $\bar{\theta}(\sigma\tau) = \overline{c_\tau c_\sigma + vc_\sigma t} = \overline{c_\tau c_\sigma} = \bar{\theta}(\sigma)\bar{\theta}(\tau)$.

The kernel of $\bar{\theta}$ is $\mathcal{V}_1$, hence $\bar{\theta}$ induces an isomorphism θ from $\mathcal{T}/\mathcal{V}_1$ into $\bar{L}^\cdot$. In fact, if $\sigma \in \mathcal{V}_1$ then $\sigma(t) \equiv t(P^2)$ hence $\sigma(t) = (1 + bt)t$ with $b \in B$, thus $\bar{\theta}(\sigma) = \overline{1 + bt} = \bar{1}$. Conversely, if $\sigma \in \mathcal{T}$ is such that $\bar{\theta}(\sigma) = \bar{1}$, that is $\bar{c}_\sigma = \bar{1}$, then $\sigma(t) - t = (c_\sigma - 1)t \in B_P t^2$, hence $\sigma(t)/t \equiv 1(B_P P)$ and $\sigma \in \mathcal{V}_1$ (by (**G**)).

Thus $\mathcal{T}/\mathcal{V}_1$ is isomorphic to a subgroup of $\bar{L}^\cdot$. If $\mathbb{Z}p = \mathbb{Z} \cap \underline{P}$, then $\bar{L}$ is a finite field containing $\mathbb{F}_p$, so its nonzero elements form a cyclic group of order $\#(\bar{L}) - 1$, which is not a multiple of p; therefore $\mathcal{T}/\mathcal{V}_1$ is also cyclic of order not a multiple of p.

(2)   Let $i \geq 1$. If $\sigma \in \mathcal{V}_i$ then $\sigma(t) = t + d_\sigma t^{i+1}$ with $d_\sigma \in B$. Let $\tilde{\theta}_i(\sigma) = \bar{d}_\sigma \in \bar{L}$. We shall show that the mapping $\tilde{\theta}_i : \mathcal{V}_i \to \bar{L}$ is a homomorphism into $\bar{L}$. In fact, if $\sigma, \tau \in \mathcal{V}_i$, then

$$\begin{aligned}
\sigma\tau(t) &= \sigma(t + d_\tau t^{i+1}) = \sigma(t) + \sigma(d_\tau) \cdot \sigma(t)^{i+1} \\
&= t + d_\sigma t^{i+1} + (d_\tau + d't^{i+1})(t + d_\sigma t^{i+1})^{i+1} \\
&= t + (d_\sigma + d_\tau)t^{i+1} + ct^{i+2},
\end{aligned}$$

where $d', c \in B$; hence $\tilde{\theta}_i(\sigma\tau) = \overline{d_\sigma + d_\tau} = \tilde{\theta}_i(\sigma) + \tilde{\theta}_i(\tau)$.

The kernel of $\tilde{\theta}_i$ is $\mathcal{V}_{i+1}$, hence $\tilde{\theta}_i$ induces an isomorphism θ_i from $\mathcal{V}_i/\mathcal{V}_{i+1}$ into $\bar{L}$. In fact, if $\sigma \in \mathcal{V}_{i+1}$, then $\sigma(t) = t + ct^{i+2}$, so $d_\sigma = ct$ and $d_\sigma = \bar{0}$. Conversely, if $d_\sigma = \bar{0}$, then $\sigma(t) \equiv t(P^{i+2})$ so $\sigma(t)/t \equiv 1(P^{i+1})$ and therefore $\sigma \in \mathcal{V}_{i+1}$.

Noting that $\bar{L}$ is a finite-dimensional vector space over $\mathbb{F}_p$, the same holds for $\mathcal{V}_i/\mathcal{V}_{i+1}$.

(3)   The group $\mathcal{T}$ has the following sequence of subgroups, each being actually a normal subgroup:

$$\mathcal{T} \supseteq \mathcal{V}_1 \supseteq \mathcal{V}_2 \supseteq \cdots \supseteq \mathcal{V}_r = \{\varepsilon\}$$

(for some r, by (**G**)). Since $\mathcal{T}/\mathcal{V}_1$ is a cyclic group and each group $\mathcal{V}_i/\mathcal{V}_{i+1}$ is an elementary Abelian p-group, then $\mathcal{T}$ is a solvable group and $\mathcal{V}_1$ is a p-group.

(4)   We have $e = [L : V_1] \cdot [V_1 : T]$ (with $T = K$), $[L : V_1] = \#\mathcal{V}_1$ which is a power of p, $[V_1 : T] = \#(\mathcal{T}/\mathcal{V}_1)$ which is relatively prime to p. ∎

In particular, if $\overline{K}$ has characteristic p not dividing the degree $e = [L : T]$, then $\#(\mathcal{V}_1) = 1$, so $V_1 = V_2 = \cdots = L$. Thus, in this case there is no higher ramification present, and $\underline{P}$ is said to be *tamely ramified in* $L|K$. If p divides e, then $\underline{P}$ is said to be *wildly ramified in* $L|K$.

Combining Theorem 1, (1), and Theorem 2, (3), we have:

H.   *If P is any prime ideal of B such that $P \cap A = \underline{P}$, the decomposition group $\mathcal{Z}$ of $\underline{P}$ in $L|K$ is a solvable group.*

Proof: We have $\mathcal{Z} \supseteq \mathcal{T} \supseteq \{\varepsilon\}$, where $\mathcal{T}$ is solvable and $\mathcal{Z}/\mathcal{T}$ is a cyclic group; thus $\mathcal{Z}$ itself is a solvable group. ∎

An interesting application is the following:

I.   *Let K be an algebraic number field, let $f \in K[X]$ be an irreducible polynomial, and let L be the splitting field of f over K. If there exists a prime ideal $\underline{P}$ of the ring of integers A of K which divides only one prime ideal P of the ring of integers of L, then the polynomial f is solvable by radicals.*

Proof: Let $\mathcal{K} = G(L|K)$ be the Galois group of the polynomial f over K. We have $\mathcal{Z}_P(L|K) = \mathcal{K}$, so $\mathcal{K}$ is a solvable group (by (**H**)). By the theorem of Galois, the polynomial f is solvable by radicals. ∎

Later, we shall need the result, which follows. Since $\mathcal{T}$, $\mathcal{V}_1$ are normal subgroups of $\mathcal{Z}$, then $\sigma \mathcal{T} \sigma^{-1} = \mathcal{T}$, $\sigma \mathcal{V}_1 \sigma^{-1} = \mathcal{V}_1$ for every $\sigma \in \mathcal{Z}$, thus considering the cosets of $\mathcal{T}$ by $\mathcal{V}_1$, we have $\sigma(\tau \mathcal{V}_1)\sigma^{-1} = (\sigma\tau\sigma^{-1})\mathcal{V}_1$. Therefore σ acts on the quotient group $\mathcal{T}/\mathcal{V}_1$ by conjugation, defining $\tilde{\sigma}(\tau\mathcal{V}_1) = (\sigma\tau\sigma^{-1})\mathcal{V}_1$. The following proposition due to Speiser describes this action:

J.   *Let $\sigma \in \mathcal{Z}$ be such that its image by the homomorphism $\mathcal{Z} \to \mathcal{Z}/\mathcal{T} \cong G(\overline{L}|\overline{K})$ corresponds to the Frobenius automorphism of $\overline{L}|\overline{K}$. Then:*

(1) *If $\tau \in \mathcal{T}$ then $\tilde{\sigma}(\tau\mathcal{V}_1) = \tau^q \mathcal{V}_1$ where $q = \#(\overline{K})$.*

(2) *If $\mathcal{Z}$ is an Abelian group then $\tau^{q-1} \in \mathcal{V}_1$ for every $\tau \in \mathcal{T}$, and $\mathcal{T}/\mathcal{V}_1$ has order dividing $q - 1$.*

Proof: (1)   Let $B_P P = B_P t$, with $t \in B$ and let us compute $\sigma\tau\sigma^{-1}(t)$. We write $\sigma(t) = c_\sigma t$, $\sigma^{-1}(t) = c_{\sigma^{-1}} t$ with $c_\sigma, c_{\sigma^{-1}} \in B_P$ and as in the proof of Theorem 2, $\sigma(c_{\sigma^{-1}}) \cdot c_\sigma = 1$. If $\tau \in T$ then

$$
\begin{aligned}
\sigma\tau\sigma^{-1}(t) &= \sigma\tau(c_{\sigma^{-1}}t) \\
&= \sigma(\tau(c_{\sigma^{-1}}) \cdot \tau(t)) \\
&= \sigma(c_{\sigma^{-1}} + vt) \cdot \sigma(c_\tau) \cdot \sigma(t) \\
&= (\sigma(c_{\sigma^{-1}}) + \sigma(v) \cdot c_\sigma t)\sigma(c_\tau)c_\sigma t \\
&\equiv \sigma(c_\tau)t \equiv c_\tau^q t \pmod{B_P P^2}
\end{aligned}
$$

(where $v \in B_P$), recalling that the Frobenius automorphism $\tilde{\sigma}$ is defined as being the raising to the power $q = \#(\overline{K})$.

On the other hand, $\tau(t) = c_\tau t$, $\tau^2(t) = \tau(c_\tau) \cdot \tau(t) = (c_\tau + ut)c_\tau t \equiv c_\tau^2 t \pmod{B_P P^2}$ (where $u \in B_P$); similarly $\tau^q(t) \equiv c_\tau^q t$ $\pmod{B_P P^2}$. Thus $\sigma\tau\sigma^{-1} \equiv \tau^q(t)(B_P P^2)$ hence $\tau^{-q}\sigma\tau\sigma^{-1}(t) \equiv t(B_P P^2)$ and $(\tau^{-q}\sigma\tau\sigma^{-1}(t))/t \equiv 1(B_P P)$, so $\tau^{-q}\sigma\tau\sigma^{-1} \in \mathcal{V}_1$ and $\tilde{\sigma}(\tau\mathcal{V}_1) = \sigma\tau\sigma^{-1}\mathcal{V}_1 = \tau^q\mathcal{V}_1$.

(2)   If $\mathcal{Z}$ is an Abelian group, then $(\tau^{-q}\sigma\tau\sigma^{-1})^{-1} = \tau^{q-1} \in \mathcal{V}_1$ for every $\tau \in T$.

Since $T/\mathcal{V}_1$ is a cyclic group and $\tau^{q-1}\mathcal{V}_1 = \mathcal{V}_1$ (unit of $T/\mathcal{V}_1$), then the order of $T/\mathcal{V}_1$ divides $q - 1$.   ∎

In order to have a better insight into the higher ramification, we shall now consider the different above a prime ideal $\underline{P}$ of A.

We assume as before that $L|K$ is a Galois extension. Moreover $B\underline{P} = P^e$, $e \geq 1$. This implies that P is the only prime ideal of B such that $P \cap A = \underline{P}$. If S is the multiplicative set-complement of $\underline{P}$ in A, if $A' = S^{-1}A$, $B' = S^{-1}B$ then $B' = B_P$, as was shown in Lemma 1.

We note $\Delta_P(L|K) = \Delta(B'|A') = \Delta(B_P|A_{\underline{P}})$ the different of $L|K$ above $\underline{P}$. By Chapter 13, (N), and Lemma 1 we have $\Delta_P(L|K) = B' \cdot \Delta_{L|K}$, hence by Chapter 12, (G), we have $\Delta_P(L|K) = B_P P^s$ where $s \geq 0$ is an integer; it is the exponent at P of the different $\Delta_{L|K}$. Sometimes we also denote it by $s_P(L|K)$.

We shall now compute the exponent of the different; it turns out that this expression will involve the orders of the various ramification groups.

Theorem 3.   Let $L|K$ be a Galois extension and P the only extension of $\underline{P}$ to L. We assume that there exists an element t such that $B_P P = B_P t$ and $B_P = A_{\underline{P}}[t]$ (for example, by (**F**), this holds when P is totally ramified over $\underline{P}$, that is, the inertial field T is equal to K). Then the exponent of the different of P in $L|K$ is

$$
s_P(L|K) = \sum_{i=0}^{r-1} [\#(\mathcal{V}_i) - 1],
$$

where $T = \mathcal{V}_0 \supseteq \mathcal{V}_1 \supseteq \cdots \supseteq \mathcal{V}_r = \{\varepsilon\}$ are the ramification groups of P in $L|K$.

Proof: We have seen in Chapter 13, (**T**), that $\Delta_P = B_P \cdot g'(t)$, where g is the minimal polynomial of t over K. We write $g = \prod_{\sigma \in \mathcal{K}}(X - \sigma(t))$, where $\mathcal{K} = G(L|K)$. Then $g'(t) = \prod_{\sigma \neq \varepsilon}(t - \sigma(t))$.

Since P is the only extension of $\underline{P}$, then the decomposition group of P is $\mathcal{Z} = \mathcal{K}$. If $\sigma \in \mathcal{Z}$, $\sigma \notin T$ then $\sigma(t) - t \in B$, but $\sigma(t) - t \notin P$, and similarly, if $\sigma \in \mathcal{V}_i$ but $\sigma \notin \mathcal{V}_{i+1}$ then $\sigma(t) - t \in P^{i+1}$, but $\sigma(t) - t \notin P^{i+2}$ (by (**G**)). If $s = s_P(L|K)$ then $B_P P^s = B_P g'(t) = \prod_{\sigma \neq \varepsilon} B_P(t - \sigma(t))$ and writing $B_P(t - \sigma(t)) = B_P P^{s(\sigma)}$ then

$$s = \sum_{\sigma \neq \varepsilon} s(\sigma) = \sum_{i=0}^{r-1} \sum_{\substack{\sigma \in \mathcal{V}_i \\ \sigma \notin \mathcal{V}_{i+1}}} s(\sigma) = \sum_{i=0}^{r-1} [\#(\mathcal{V}_i) - \#(\mathcal{V}_{i+1})](i+1)$$

$$= [\#(\mathcal{V}_0) - \#(\mathcal{V}_1)] + 2[\#(\mathcal{V}_1) - \#(\mathcal{V}_2)]$$
$$+ 3[\#(\mathcal{V}_2) - \#(\mathcal{V}_3)] + \cdots + r[\#(\mathcal{V}_{r-1}) - 1]$$

$$= \#(\mathcal{V}_0) + \#(\mathcal{V}_1) + \#(\mathcal{V}_2) + \cdots + \#(\mathcal{V}_{r-1}) - r$$

$$= \sum_{i=0}^{r-1} [\#(\mathcal{V}_i) - 1]. \qquad \blacksquare$$

EXERCISES

1. Let K be an algebraic number field, $L|K$ a finite Galois extension, and P a prime ideal of the ring B of integers of L. Show that if K' is a field, $K \subseteq K' \subseteq L$, A' the ring of integers of K' and $P' = P \cap A'$ is unramified and inert over P, then $K' \subseteq Z$ (decomposition field of P in $L|K$).

2. Let K be an algebraic number field, $L|K$ a finite Galois extension, and P a prime ideal of the ring B of integers of L. Show that if K' is a field, $K \subseteq K' \subseteq L$, A' the ring of integers of K', and $P' = P \cap A'$, if P' is unramified over P, then $K' \subseteq T$ (inertial field of P in $L|K$).

3. Let K be an algebraic number field, $L|K$ a finite extension, L' the smallest field containing L and such that $L'|K$ is a Galois extension. Show that if a prime ideal of the ring of integers of K decomposes completely in $L|K$ then it also decomposes completely in $L'|K$.

4. Let K be an algebraic number field, P a prime ideal of its ring of integers. Let $L_1|K$ and $L_2|K$ be finite extensions. Show that if P decomposes completely in $L_1|K$ and $L_2|K$ then it decomposes completely in $L_1 L_2|K$.

5. Let $K|\mathbb{Q}$ be a Galois extension of degree n, let I be an ideal of the ring A of integers of K, such that $\sigma(I) = I$ for every $\sigma \in G(K|\mathbb{Q})$. Show that $I^{n!}$ is generated by some rational integer.

6. Use the previous exercise to show that given the ideal $I \neq 0$ of A there exists an ideal $J \neq 0$ of A such that IJ is a principal ideal (see Chapter 7, (**J**)).

7. Let K be an algebraic number field, and let $L|K$ be a Galois extension of degree n. Show that the relative different $\Delta_{L|K}$ is invariant by the Galois group. Hence $\Delta_{L|K}^n$ is the ideal generated by the relative discriminant $\delta_{L|K}$.

8. Let $n \geq 1$ be an integer. Show that there exists an integer $l(n)$ (depending only on n) such that: If $K|\mathbb{Q}$ is a Galois extension of degree n, if p is a prime number, and p^s divides the discriminant δ_K, then $s \leq l(n)$ (compare this statement with Chapter 9, (**H**)).

Hint: Apply Theorem 3.

9. Let $K|\mathbb{Q}$ be a quadratic extension, let A be the ring of algebraic integers of K, and let P be a nonzero prime ideal of A. Discuss in all cases the decomposition, inertia, and ramification groups and fields of P.

10. Do the previous exercise in the case where $K|\mathbb{Q}$ is a Galois extension of degree p, an odd prime number.

11. Do Exercise 9 in the case where $K|\mathbb{Q}$ is a Galois extension with Galois group equal to the Klein group $\mathcal{K} = \{\varepsilon, \sigma, \tau, \sigma\tau\}$, $\sigma^2 = \tau^2 = \varepsilon$, $\sigma\tau = \tau\sigma$.

15

The Fundamental Theorem
of Abelian Extensions

15.1 The Theorem of Kronecker and Weber

In Chapter 2, Section 8, we have stated that every cyclotomic field $\mathbb{Q}(\zeta)$ (where ζ is a primitive nth root of 1, $n > 2$) is an Abelian extension of $\mathbb{Q}$.

From the arithmetical point of view, the cyclotomic fields have been fairly well studied. In order to investigate other algebraic number fields, it is reasonable to consider first the case of Galois extensions of $\mathbb{Q}$, for which we have already indicated in the preceding chapter a rather elaborate theory of decomposition of ideals. In fact, a whole branch of the theory of algebraic numbers, called *class field theory*, is devoted to the study of Abelian extensions.

Therefore the natural question to ask is the following: Which are the possible Abelian extensions of $\mathbb{Q}$? We certainly cannot attempt to solve this problem in such an elementary text. However, we shall prove a rather interesting theorem about Abelian extensions and indicate without proof the main theorems for Abelian number fields.

Already in Chapter 4, (**P**), we have shown that every quadratic extension of $\mathbb{Q}$ (which is an Abelian extension) is contained in a cyclotomic field. We have also indicated its generalization, by Kronecker and Weber. It is our intention now to prove this theorem; this will provide excellent grounds for the application of the concepts and techniques developed in the preceding chapters.

Theorem 1.   *If L is an algebraic number field, an Abelian extension of $\mathbb{Q}$, then there exists a root of unity ζ, such that $L \subseteq \mathbb{Q}(\zeta)$.*

Proof: We shall consider two crucial particular cases first and then show how to reduce the general case to these two special ones. We shall require several steps.

Case 1: $[L : \mathbb{Q}] = p^m$, $\delta_L = p^k$ where p is an odd prime, $m, k \geq 1$.

A.   *There exists only one prime ideal P in L such that $P \cap \mathbb{Z} = \mathbb{Z}p$; moreover, P is totally ramified in $L|\mathbb{Q}$, and the inertial field T and the first ramification field of P in $L|\mathbb{Q}$ are $V_1 = T = \mathbb{Q}$.*

Proof: Let P, P' be prime ideals such that $P \cap \mathbb{Z} = P' \cap \mathbb{Z} = \mathbb{Z}p$. Since $\mathcal{K} = G(L|\mathbb{Q})$ is an Abelian group, then the decomposition groups $\mathcal{Z}$, $\mathcal{Z}'$ of P, P', respectively, must coincide (Chapter 14, (**A**)).

By Chapter 11, (**E**), there exists $\sigma \in \mathcal{K}$ such that $\sigma(P) = P'$, hence considering the inertial groups of P, P', we have $\mathcal{T}' = \sigma \mathcal{T} \sigma^{-1}$ and therefore $\mathcal{T}' = \mathcal{T}$, because $\mathcal{K}$ is Abelian.

Let T be the inertial field of P and of P'. We show that $\delta_T = 1$. By Chapter 14, Theorem 1, (4), p is unramified in $T|\mathbb{Q}$, since $e(P_T|p) = 1$, so every other prime ideal of T extending p (if it exists) must also be unramified in $T|\mathbb{Q}$, because $T|\mathbb{Q}$ is a Galois extension (by Chapter 11, (**F**)). By Chapter 13, Theorem 1, p does not divide the discriminant δ_T.

Now, if q is a prime, $q \neq p$, and q divides δ_T, then again by the same theorem, q is ramified in $T|\mathbb{Q}$, hence in $L|\mathbb{Q}$, so q divides $\delta_L = p^k$, a contradiction. Thus, δ_T has no prime factor; therefore, $|\delta_T| = 1$. In view of Chapter 9, (**D**), $T = \mathbb{Q}$. If Z is the decomposition field of any prime ideal P containing p, since $T \supseteq Z \supseteq \mathbb{Q}$, then $Z = \mathbb{Q}$. By Chapter 14, (**C**), there exists only one prime ideal P in L, containing p.

Since $[T : \mathbb{Q}] = f(P|p)$, then $f(P|p) = 1$. Finally, from $[L : \mathbb{Q}] = efg$, with $f = g = 1$, we deduce that p is totally ramified in $L|\mathbb{Q}$.

Also, since $[V_1 : T]$ has degree prime to p (characteristic of the residue class field), by Chapter 14, Theorem 2, (4), $V_1 = T$.   ∎

B.   *Let H be a field, $\mathbb{Q} \subseteq H \subseteq L$, such that $[H : \mathbb{Q}] = p$. If C is the ring of integers of H, $Q = P \cap C$, then the exponent of the different of Q in $H|\mathbb{Q}$ is equal to $2(p - 1)$; this value is therefore independent of the field H, provided $[H : \mathbb{Q}] = p$.*

Proof: Since p is totally ramified in $L|\mathbb{Q}$ by transitivity of the ramification index, p is totally ramified in $H|\mathbb{Q}$. If z is a generator of the ideal $C_Q Q$, that is, $C_Q Q = C_Q z$, then by Chapter 14, (**F**), $C_Q = \mathbb{Z}_p[z]$, $\{1, z, \ldots, z^{p-1}\}$ are linearly independent over $\mathbb{Q}$ and z is a root of an Eisenstein polynomial $g = X^p + a_1 X^{p-1} + \cdots + a_p$ with $a_i \in \mathbb{Z}$, p dividing each a_i, but p^2 not dividing a_p. Then $g = \prod_{\sigma \in \mathcal{G}}(X - \sigma(z))$ where $\mathcal{G}$ is the Galois group of $H|\mathbb{Q}$ and $g'(z) = \prod_{\sigma \neq \varepsilon}(z - \sigma(z))$; by Chapter 13, (**T**), $\Delta_Q(H|\mathbb{Q}) = C_Q g'(z)$. We have to compute the largest power of $C_Q Q$ dividing the principal ideal generated by

$$g'(z) = pz^{p-1} + (p-1)a_1 z^{p-2} + \cdots + a_{p-1}.$$

Since p is totally ramified in $H|\mathbb{Q}$, then $C_Q P = C_Q z^p$ (p being the degree of $H|\mathbb{Q}$); on the other hand, if p' is any prime different from p, then $p' \notin Q$ so $C_Q p' = C_Q = C_Q z^0$; hence, for every integer $a \in \mathbb{Z}$, if p^m divides a, but p^{m+1} does not divide a, then $C_Q a = C_Q z^{pm}$, with $m \geq 0$. In particular, let $C_Q(p-i)a_i = C_Q z^{ps_i}$, so $s_i \geq 1$ for every i and $s_0 = 1$. Hence,

$$C_Q[(p-i)a_i z^{p-i-1}] = C_Q z^{ps_i+(p-i-1)}.$$

By the argument in the proof of Chapter 14, (**F**), if i, j are distinct indices, $0 \leq i, j \leq p-1$ then $ps_i + (p-i-1) \neq ps_j + (p-j-1)$; hence, if $C_Q g'(z) = C_Q z^s$ then

$$s = \min_{0 \leq i \leq p-1} \{ps_i + (p-i-1)\}.$$

It follows that $p \leq s$ since $s_i \geq 1$, so $p \leq ps_i + (p-i-1)$ for $i = 0, 1, \ldots, p-1$. On the other hand, $C_Q pz^{p-1} = C_Q z^{p+(p-1)} = C_Q z^{2p-1}$; therefore, we have the inequalities $p \leq s \leq 2p-1$.

But, by Chapter 14, Theorem 3, the exponent of the different is

$$s = \sum_{i=0}^{r-1} \left[\#(V_i') - 1 \right],$$

where V_i' denotes the ith ramification group of Q in $H|\mathbb{Q}$. Since $[H : \mathbb{Q}] = p$ then $\#(V_i')$ is either 1 or p, thus $p-1$ divides s, and therefore

$$1 < \frac{p}{p-1} \leq \frac{s}{p-1} \leq \frac{2p-1}{p-1} = 2 + \frac{1}{p-1} < 3,$$

(because $p \neq 2$). We conclude that $s = 2(p-1)$. ∎

C.   *Let i be the smallest index such that $V_i \neq \mathcal{K} = G(L|\mathbb{Q})$ (hence $i > 1$ by (**A**)). Then $[V_i : \mathbb{Q}] = p$ and V_i is the only field of degree p over $\mathbb{Q}$, contained in L.*

Proof: We have $[V_i : \mathbb{Q}] = (V_{i-1} : V_i) = \#(V_{i-1}/V_i)$. By Chapter 14, Theorem 2, (2), V_{i-1}/V_i is isomorphic to a subgroup of the additive group $\overline{L}$. Since $f(L|\mathbb{Q}) = 1$ then $\overline{L} = \mathbb{F}_p$, so from $V_{i-1} \neq V_i$ it follows that $\#(V_{i-1}/V_i) = p$, and therefore $[V_i : \mathbb{Q}] = p$.

Now, let H be any field such that $\mathbb{Q} \subseteq H \subseteq L$, $[H : \mathbb{Q}] = p$ and assume that $H \neq V_i$; we shall arrive at a contradiction. For this purpose we compute the differents $\Delta_P(L|V_i)$ and $\Delta_P(L|H)$, using Theorem 3 of Chapter 14.

Let $\mathcal{H} = G(L|H)$, then $V_j(L|H) = V_j(L|\mathbb{Q}) \cap \mathcal{H}$ and, similarly, $V_j(L|V_i) = V_j(L|\mathbb{Q}) \cap V_i$ for every $j \geq 0$.

Thus, $V_0(L|V_i) = \cdots = V_i(L|V_i) = V_i$ while $V_j(L|V_i) = V_j$ for $j \geq i+1$ (as before $V_j = V_j(L|\mathbb{Q})$ for every $j \geq 0$). Similarly, $V_0(L|H) = \cdots = V_{i-1}(L|H) = \mathcal{H}$ (since $V_{i-1} = \mathcal{K}$) while $V_i(L|H)$ is properly contained in V_i (otherwise $V_i(L|H) = \mathcal{H}$ hence $V_i = H$ contrary to the hypothesis) and $V_j(L|H) \subseteq V_j$ for $j \geq i+1$. Therefore

$$s_P(L|V_i) = \sum_{j=0}^{r-1} \left[\#V_j(L|V_i) - 1 \right] > \sum_{j=0}^{r-1} \left[\#V_j(L|H) - 1 \right] = s_P(L|H).$$

However, from the transitivity of the different we have $\Delta_P(L|\mathbb{Q}) = B_P \Delta_{P \cap V_i}(V_i|\mathbb{Q}) \cdot \Delta_P(L|V_i)$ and also $\Delta_P(L|\mathbb{Q}) = B_P \Delta_{P \cap H}(H|\mathbb{Q}) \cdot \Delta_P(L|H)$; since P is totally ramified in $L|\mathbb{Q}$, and $[H : \mathbb{Q}] = [V_i : \mathbb{Q}] = p$ by (**B**), we deduce that the exponents of the differents $\Delta_P(L|V_i)$ and $\Delta_P(L|H)$ must coincide, and this is a contradiction. ∎

D. $\mathcal{K} = G(L|\mathbb{Q})$ *is a cyclic group.*

Proof: By hypothesis, $\mathcal{K}$ is an Abelian group of order p^m. By (**C**), $\mathcal{K}$ has only one subgroup of order p^{m-1}. This implies necessarily that $\mathcal{K}$ is a cyclic group: it is a well-known fact in the theory of finite Abelian groups, which may be proved either directly or else by means of the structure theorem of finite Abelian groups (see Lemma 1 below, or Chapter 3, Theorem 3). ∎

For the convenience of the reader, we shall establish this and another easy fact about finite Abelian groups.

First we recall that if p is a prime number dividing the order n of the finite Abelian group G, then G has an element of order p (see Chapter 3, (**M**)).

Lemma 1.   *Let G be a finite Abelian group:*
  (1) *If G has order p^m, where $m \geq 1$ and p is a prime, if H is a subgroup of G of order p^h, if $h < h' \leq m$, then there exists a subgroup H' of G, having order $p^{h'}$ and containing H.*
  (2) *If G has order p^m, $m \geq 2$, p a prime, if G has only one subgroup of order p^{m-1}, then G is a cyclic group.*

Proof: (1)   It is enough to assume that $h' = h + 1 \leq m$ and then repeat the argument. Let $\overline{G} = G/H$ be the quotient group so $\#\overline{G} = p^{m-h}$; thus there exists an element $\overline{x} \in G/H$ of order p. Let H' be the subgroup generated by H and x, so H' contains H properly (since $x \notin H$); but $H' = H \cup Hx \cup \cdots \cup Hx^{p-1}$, because $x^p \in H$; thus $\#H' = p^{h+1}$.

(2)   Let H be the only subgroup of order p^{m-1} of G, let $x \in G$, $x \notin H$, and assume that x has order less than p^m. By (1), the cyclic group generated by x is contained in a subgroup of order p^{m-1} of G, which must be equal to H, by hypothesis, so $x \in H$, a contradiction. This means that x has order p^m and G is a cyclic group. ∎

E.   *In Case 1, $L \subseteq \mathbb{Q}(\zeta)$, where ζ is a root of unity.*

Proof: Let $R = \mathbb{Q}(\zeta)$ where ζ is a primitive root of unity of order p^{m+1}. Thus, $R|\mathbb{Q}$ is an extension of degree $\varphi(p^{m+1}) = p^m(p-1)$, with Galois group isomorphic to the group $P(p^{m+1})$ of prime residue classes modulo p^{m+1} (see Chapter 2, Section 8); by Chapter 3, (**L**), $R|\mathbb{Q}$ is a cyclic extension (since $p \neq 2$). By Chapter 16, (**A**), the discriminant $\delta_{R|\mathbb{Q}} = \delta_R$ is a power of p.

The cyclic group $G(R|\mathbb{Q})$ has a subgroup of order $p-1$ (if σ is a generator then σ^{p^m} has order $p-1$), whose field of invariants we denote by R', so $[R' : \mathbb{Q}] = p^m$. Thus, $R'|\mathbb{Q}$ is a cyclic extension and the discriminant $\delta_{R'}$ is again a power of p (if q is a prime, dividing $\delta_{R'}$ then q is ramified in $R'|\mathbb{Q}$ hence also ramified in $R|\mathbb{Q}$, hence q divides δ_R and so $q = p$, using Theorem 1 of Chapter 13.

Let LR' be the compositum field of L and R'; by Chapter 2, Section 7,

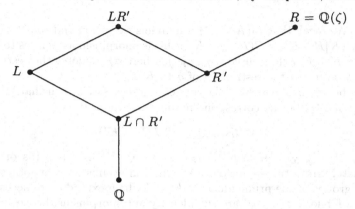

$LR'|\mathbb{Q}$ is also an Abelian extension, with degree $[LR' : \mathbb{Q}] = [LR' : R'] \cdot [R' : \mathbb{Q}] = [L : L \cap R'] \cdot [R' : \mathbb{Q}]$ which is a power of p. Now we show that the discriminant $\delta_{LR'}$ is also a power of p. In fact, if q is a prime dividing $\delta_{LR'}$ then by Chapter 13, Theorem 1, q is ramified in LR'. By Chapter 13, (U), or Lemma 3 below, q is ramified in $L|\mathbb{Q}$ or q is ramified in $R'|\mathbb{Q}$. Hence q divides δ_L or q divides $\delta_{R'}$. In both cases, $q = p$ and therefore $\delta_{LR'}$ is a power of p.

We may now apply (D) to the Abelian extension $LR'|\mathbb{Q}$ with degree and discriminant powers of p; it follows that $LR'|\mathbb{Q}$ is a cyclic extension, and by Galois theory, $G(LR'|L \cap R') \cong G(L|L \cap R') \times G(R'|L \cap R')$. Now, it is quite obvious that such a decomposition as a Cartesian product of cyclic groups of orders powers of p has to be trivial, namely, one of the groups $G(L|L \cap R')$ or $G(R'|L \cap R')$ has to be trivial (see Lemma 2 below). If $L = L \cap R'$ then $L \subseteq R'$, if $R' = L \cap R'$ then $R' \subseteq L$ and since L, R' have the same degree over $\mathbb{Q}$, then in both cases $R' = L$ and so $L = R' \subseteq \mathbb{Q}(\zeta)$.   ∎

This lemma is again included for the convenience of the reader:

Lemma 2.   *Let G be a cyclic group of order p^m, where p is any prime number. If $G \cong H \times H'$, then H, H' are cyclic groups of orders p^h, $p^{h'}$, respectively, with either $h = 1$ or $h' = 1$ (hence $G \cong H'$ or $G \cong H$, respectively).*

Proof: This follows at once from the uniqueness asserted in Theorem 3 of Chapter 3.   ∎

The following lemma has already been proved in Chapter 13, (U). For the case of Galois extensions we have however a simpler proof, independent of the theory of the different:

Lemma 3.   *Let K, K' be algebraic number fields, which are Galois extensions of $\mathbb{Q}$, let $L = K \cdot K'$ be the compositum of these fields. If q is a prime number, unramified in $K|\mathbb{Q}$ and in $K'|\mathbb{Q}$, then q is also unramified in $L|\mathbb{Q}$.*

Proof: We recall that $L|(K \cap K')$ is a Galois extension and that $G(L|K \cap K') \cong G(K|K \cap K') \times G(K'|K \cap K')$; this isomorphism associates to every $\sigma \in G(L|K \cap K')$ the couple $(\sigma_K, \sigma_{K'})$, where σ_K denotes the restriction of σ to K and $\sigma_{K'}$ the restriction of σ to K'.

Let Q be any prime ideal of the ring C of integers of L such that $Q \cap \mathbb{Z} = \mathbb{Z}q$; let $T_Q(L|\mathbb{Q})$ be the corresponding inertial group. If

$$\sigma \in T_Q(L|\mathbb{Q}) \cap G(L|K \cap K')$$

then $\sigma_K \in T_{Q\cap K}(K|K \cap K')$, $\sigma_{K'} \in T_{Q\cap K'}(K'|K \cap K')$ (as one sees immediately from the definition of the inertial groups). By hypothesis the inertial groups of the prime ideals of K, K' which extend q are necessarily trivial; a fortiori, σ_K, $\sigma_{K'}$ are the identity automorphisms, hence σ is the identity automorphism. This proves that Q is unramified in $L|(K \cap K')$. But $Q \cap (K \cap K')$ is also unramified in $(K \cap K')|\mathbb{Q}$ because of the hypothesis. Therefore Q is unramified in $L|\mathbb{Q}$, showing that q is unramified in $L|\mathbb{Q}$. ∎

The proof of Case 1 of the theorem is now complete. We shall continue, considering the case where $p = 2$.

Case 2: $[L : \mathbb{Q}] = 2^m$, $\delta_L = 2^k$, where $m, k \geq 1$.

F.   *Given $m \geq 1$ there exists a real field K such that $[K : \mathbb{Q}] = 2^m$, δ_K is a power of 2, and $K \subseteq \mathbb{Q}(\xi)$ for some root of unity ξ.*

Proof: Let ξ be a primitive root of unity of order 2^{m+2}, and let $K' = \mathbb{Q}(\xi)$. Then $[K' : \mathbb{Q}] = \varphi(2^{m+2}) = 2^{m+1}$ and hence $i = \sqrt{-1} \in K'$.

Let $K = K' \cap \mathbb{R}$, so $K' = K(i)$; in fact, the conjugates of ξ belong to K', and are either real or appear in pairs of complex conjugates $a + bi$, $a - bi$ with $a, b \in \mathbb{R}$; then $2a, 2b \in K' \cap \mathbb{R} = K$, so $a, b \in K$ and $K' = K(i)$. It follows that $[K' : K] = 2$, hence $[K : \mathbb{Q}] = 2^m$.

We shall show that the discriminant δ_K is a power of 2. If q is a prime dividing δ_K then q is ramified in $K|\mathbb{Q}$, hence also ramified in $K'|\mathbb{Q}$; therefore q divides $\delta_{K'}$ (by Chapter 13, Theorem 1). But, if $K' = \mathbb{Q}(\xi)$, then by Chapter 16, (A), $\delta_{K'}$ is a power of 2, so $q = 2$ and δ_K is a power of 2. ∎

G.   *Given $m \geq 1$, there exists only one real field K such that $K|\mathbb{Q}$ is an Abelian extension, $[K : \mathbb{Q}] = 2^m$, and δ_K is a power of 2.*

Proof: If $m = 1$ and $[F : \mathbb{Q}] = 2$, $F \subseteq \mathbb{R}$ then $F = \mathbb{Q}(\sqrt{d})$ with $d > 0$, d square-free. But $\delta_F = d$ when $d \equiv 1 \pmod 4$, $\delta_F = 4d$ when $d \equiv 2$ or 3 (mod 4). So if δ_F is a power of 2, then $d = 2$, $F = \mathbb{Q}(\sqrt{2})$.

Thus, we may assume that $m \geq 2$. If F is an Abelian extension of $\mathbb{Q}$, $F \subseteq \mathbb{R}$, $[F : \mathbb{Q}] = 2^m$, and δ_F is a power of 2, then the Galois group $G(F|\mathbb{Q})$ contains a subgroup of order 2^{m-1} (by Lemma 1), hence F contains a subfield H such that $[H : \mathbb{Q}] = 2$ and the discriminant of H must be a power of 2 (by the same argument); so $H = \mathbb{Q}(\sqrt{2})$. Thus, $G(F|\mathbb{Q})$ contains only one subgroup of order 2^{m-1} and by Lemma 1, it must be cyclic.

If F is different from the field K obtained in **(F)**, we consider the compositum FK; thus $FK \subseteq \mathbb{R}$, $FK|\mathbb{Q}$ is again an Abelian extension of degree a power of 2 and with discriminant a power of 2 (see the argument in **(E)** and Lemma 3); hence by our proof just above (considering FK in place of F), $G(FK|\mathbb{Q})$ is a cyclic group, and

$$G(FK|F \cap K) \cong G(F|F \cap K) \times G(K|F \cap K).$$

By Lemma 2, either $F \subseteq K$ or $K \subseteq F$ and since both fields have the same degree 2^m then $F = K$. ■

H.   *If $L|\mathbb{Q}$ is an Abelian extension of degree 2^m and the discriminant δ_L is a power of 2, then there exists a root of unity ζ such that $L \subseteq \mathbb{Q}(\zeta)$.*

Proof: Since $\mathbb{Q}(i)$ and L are Abelian extensions of $\mathbb{Q}$ then the compositum $L(i)$ is also an Abelian extension of $\mathbb{Q}$. By previous arguments, $L(i)|\mathbb{Q}$ has degree and discriminant powers of 2.

Let $K = L(i) \cap \mathbb{R}$, hence K is a real Abelian extension of $\mathbb{Q}$, with degree and discriminant powers of 2. By **(G)** and **(F)** there exists a root of unity ξ such that $K \subseteq \mathbb{Q}(\xi)$.

Let $L(i) = K(a + bi)$, where $a, b \in \mathbb{R}$. The complex conjugate $a - bi$, which is a conjugate of $a + bi$ over $\mathbb{Q}$, still belongs to $L(i)$; hence $a \in L(i) \cap \mathbb{R} = K$, and $bi \in L(i)$, thus $b^2 \in L(i) \cap \mathbb{R} = K$; it follows that $a + bi$ is a root of the polynomial $X^2 - 2aX + (a^2 + b^2)$ with coefficients in K, so $[L(i) : K] = 2$; since $i \notin K$ then $L \subseteq L(i) = K(i) \subseteq \mathbb{Q}(\xi, i) \subseteq \mathbb{Q}(\zeta)$, where ζ is a root of unity. ■

It remains now to show how it is possible to reduce the general case of the theorem to the preceding ones.

Reduction to Cases 1 and 2:

I.   *If the theorem is true for Abelian extensions having degree a power of a prime, then it is true for any finite Abelian extension of $\mathbb{Q}$.*

Proof: Let L be an algebraic number field, which is an Abelian extension of degree n over $\mathbb{Q}$. We prove now that L is the compositum of finitely many fields $L = L_1 \cdots L_s$, where each L_i is an extension having degree a power of a prime.

By Chapter 3, (**O**), the Abelian Galois group $G(L|\mathbb{Q})$ is isomorphic to the Cartesian product of p_i-groups: $G(L|\mathbb{Q}) \cong \prod_{i=1}^{s} \mathcal{H}_i$, $\#(\mathcal{H}_i) = p_i^{h_i}$, $[L : \mathbb{Q}] = n = \prod_{i=1}^{s} p_i^{h_i}$. Let $\mathcal{L}_i = \prod_{j \neq i} \mathcal{H}_j$ for every $i = 1, \ldots, s$, and let L_i denote the fixed field of the subgroup $\mathcal{L}_i$ of $G(L|\mathbb{Q})$; then $[L_i : \mathbb{Q}]$ is a power of p_i, since $G(L_i|\mathbb{Q}) \cong \mathcal{H}_i$ $(i = 1, \ldots, s)$. Moreover, if $L_1 \cdots L_s$ denotes the compositum of the fields $L_1, \ldots, L_s$ then the Galois group $G(L|L_1 \cdots L_s) \subseteq \bigcap_{i=1}^{s} \mathcal{L}_i = \{\varepsilon\}$ thus $L = L_1 \cdots L_s$.

Assuming the theorem true for each of the Abelian extensions L_i of $\mathbb{Q}$, we may write $L_i \subseteq \mathbb{Q}(\xi_i)$ where ξ_i is a primitive root of unity; let ζ be a primitive root of unity of order equal to the least common multiple of the orders of $\xi_1, \ldots, \xi_s$; then

$$L = L_1 L_2 \cdots L_s \subseteq \mathbb{Q}(\xi_1, \ldots, \xi_s) \subseteq \mathbb{Q}(\zeta). \qquad \blacksquare$$

J.   *If the theorem is true for Abelian extensions having degree and discriminant which are powers of the same prime p, then it is also true for Abelian extensions of degree a power of p.*

Proof: In order to establish (**J**), we shall need to prove the following reduction step:

K.   *Let $L|\mathbb{Q}$ be an Abelian extension of degree n. For every prime q dividing δ_L but not dividing n, there exists an Abelian extension $L'|\mathbb{Q}$ such that $[L' : \mathbb{Q}]$ divides n, $L \subseteq L'(\xi)$ where ξ is a qth root of unity, q does not divide $\delta_{L'}$, and if q' is a prime dividing $\delta_{L'}$ then q' divides δ_L too.*

Assuming (**K**), we may proceed as follows: If $L|\mathbb{Q}$ is an Abelian extension of degree p^m (where p is a prime number), if δ_L is also a power of p, then we are already in the first or second case and the theorem is true.

If there exists a prime q, different from p, such that q divides δ_L, by (**K**) there exists an Abelian extension $L_1|\mathbb{Q}$ and a qth root of unity ξ_1 such that $L \subseteq L_1(\xi_1)$, $[L_1 : \mathbb{Q}]$ is still a power of p, q does not divide the discriminant δ_{L_1} and if q' is any prime dividing δ_{L_1} then already q' divides δ_L; thus, δ_{L_1} has fewer prime factors than δ_L.

If δ_{L_1} is not a power of p, we repeat the same argument; hence there exists an Abelian extension $L_2|\mathbb{Q}$ and a root of unity ξ_2 such that $L_1 \subseteq L_2(\xi_2)$, $[L_2 : \mathbb{Q}]$ is a power of p, and δ_{L_2} has fewer prime factors than δ_{L_1}.

After a finite number of steps, we arrive at an Abelian extension $L_r|\mathbb{Q}$ such that $[L_r : \mathbb{Q}]$ is a power of p, $L_{r-1} \subseteq L_r(\xi_r)$, where ξ_r is a root of unity, and finally δ_{L_r} is now a power of p (perhaps equal to 1, in which case $L_r = \mathbb{Q}$). At worst, by the first or second case, $L_r \subseteq \mathbb{Q}(\xi_{r+1})$, where ξ_{r+1} is a root of unity.

Then $L \subseteq L_1(\xi_1)$, $L_1 \subseteq L_2(\xi_2)$, $\ldots$, $L_{r-1} \subseteq L_r(\xi_r)$, $L_r \subseteq \mathbb{Q}(\xi_{r+1})$ and so $L \subseteq \mathbb{Q}(\xi_1, \ldots, \xi_{r+1}) \subseteq \mathbb{Q}(\zeta)$ where ζ is a root of unity of order equal to the least common multiple of the orders of the roots $\xi_1, \ldots, \xi_{r+1}$. This proves the theorem, except for the need to establish (**K**).

Proof of (**K**). *Case* (i) : L contains a primitive qth root of unity ξ.

Then $L \supseteq \mathbb{Q}(\xi) \supseteq \mathbb{Q}$. Let Q be a prime ideal of the ring of integers of L such that $Q \cap \mathbb{Z} = \mathbb{Z}q$. Since q does not divide $n = [L : \mathbb{Q}]$ then q does not divide $e = e_Q(L|\mathbb{Q})$. By Chapter 13, Theorem 2, if $V_1 = V_{1Q}(L|\mathbb{Q})$ then $[L : V_1]$ is a power of q and divides n, so $L = V_1$. By Chapter 14, (**J**), the ramification index $e = \#(T/V_1) = \#(T)$ divides $q - 1$.

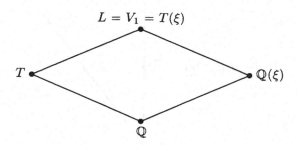

$$L = V_1 = T(\xi)$$

$$T \qquad\qquad \mathbb{Q}(\xi)$$

$$\mathbb{Q}$$

On the other hand, $e = e_Q(L|\mathbb{Q}) = e_Q(L|\mathbb{Q}(\xi)) \cdot e_{Q'}(\mathbb{Q}(\xi)|\mathbb{Q})$, where $Q' = Q \cap \mathbb{Q}(\xi)$. By Chapter 11, (**N**), $e_{Q'}(\mathbb{Q}(\xi)|\mathbb{Q}) = q - 1$, thus $q - 1$ divides e, therefore $e = q - 1$, thus $e_Q(L|\mathbb{Q}(\xi)) = 1$.

Let $L' = T$, the inertial field of Q in $L|\mathbb{Q}$, and we shall prove that it satisfies the required conditions.

Of course, $T|\mathbb{Q}$ is an Abelian extension, and its degree divides n. The inertial group of Q in $L|\mathbb{Q}(\xi)$ is $T \cap G(L|\mathbb{Q}(\xi))$, hence the inertial field is $T(\xi)$; similarly $V_{1Q}(L|\mathbb{Q}(\xi)) = V_{1Q}(L|\mathbb{Q}) \cdot \mathbb{Q}(\xi) = V_1 = L$. Thus $[L : T(\xi)] = e_Q(L|\mathbb{Q}(\xi)) = 1$, hence $L = T(\xi) = L'(\xi)$.

Next, we note that q does not divide δ_T because q is unramified in T (this being the inertial field). If q' is any prime, $q' \neq q$, and q' divides δ_T, then q' is ramified in $T|\mathbb{Q}$ hence also in $L|\mathbb{Q}$, thus q' divides δ_L (by Chapter 13, Theorem 1).

Case (ii) :   General.

We adjoin a primitive qth root of unity ξ to L, obtaining the Abelian extension $L(\xi)|\mathbb{Q}$. Let $F = L \cap \mathbb{Q}(\xi)$, so

$$G(L(\xi)|F) \cong G(L|F) \times G(\mathbb{Q}(\xi)|F).$$

Then

$$[L(\xi) : \mathbb{Q}] = [L(\xi) : F] \cdot [F : \mathbb{Q}] = [L : F] \cdot [\mathbb{Q}(\xi) : F] \cdot [F : \mathbb{Q}]$$

divides $n(q - 1)$ since $[\mathbb{Q}(\xi) : \mathbb{Q}] = q - 1$.

We may apply Case (i) to the Abelian extension $L(\xi)|\mathbb{Q}$. Let q be a prime dividing δ_L but not dividing n; then q is ramified in $L|\mathbb{Q}$, hence also in

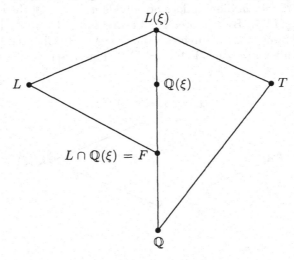

$L(\xi)|\mathbb{Q}$, thus q divides $\delta_{L(\xi)}$ (Chapter 13, Theorem 1). Also q does not divide $[L(\xi) : \mathbb{Q}]$, because this degree divides $n(q - 1)$. By Case (i), if Q is a prime ideal of the ring of integers of $L(\xi)$, $Q \cap \mathbb{Z} = \mathbb{Z}q$, if T is the inertial field of Q in $L(\xi)|\mathbb{Q}$, then $T(\xi) = L(\xi)$ and $[L(\xi) : T] = e_Q(L(\xi)|\mathbb{Q}) = q - 1$, $[L(\xi) : \mathbb{Q}] = [L(\xi) : T] \cdot [T : \mathbb{Q}] = (q - 1) \cdot [T : \mathbb{Q}]$ hence $[T : \mathbb{Q}]$ divides n. Since q is unramified in $T|\mathbb{Q}$ then q does not divide δ_T.

Now, if q' is a prime different from q, and dividing δ_T, then q' is ramified in $T|\mathbb{Q}$, hence also in $L(\xi)|\mathbb{Q}$. By Lemma 3, either q' is ramified in $L|\mathbb{Q}$ or in $\mathbb{Q}(\xi)|\mathbb{Q}$. Since $q' \neq q$, it is not ramified in $\mathbb{Q}(\xi)|\mathbb{Q}$, and q' divides the discriminant of $L|\mathbb{Q}$. Thus, we only need to take $L' = T$. This concludes the proof of the theorem. ∎

15.2 Class Field Theory

In this section we wish to indicate the main results from class field theory. This is appropriate, since the theorem of Kronecker and Weber is better understood when viewed as a theorem from class field theory.

We must of course refrain from entering into any details. This would require means far beyond the level of this book.

15.2.1 The Theory of Hilbert

Hilbert endeavored to relate the decomposition of prime ideals with some form of reciprocity law. He studied quadratic extensions and more generally

Abelian extensions of algebraic number fields K (not assumed equal to $\mathbb{Q}$). This led him to formulate the concept of a class field and to establish various important theorems.

Definition 1.   Let K be an algebraic number field, and let $L|K$ be a Galois extension of finite degree. L is said to be a *class field* of K when the following condition is satisfied: the only prime ideals of K, which are completely decomposed into prime ideals of inertial degree 1 over $\mathbb{Q}$, are the principal prime ideals of K with inertial degree 1 over $\mathbb{Q}$.

With this definition, Hilbert proved several theorems:

Theorem 1 (Existence and uniqueness).   *For every algebraic number field K, there exists one and only one (up to K-isomorphism) class field of K.*

The class field of K is usually called the *Hilbert class field of K*.

Theorem 2 (Isomorphism).   *If L is the Hilbert class field of K, then $G(L|K) = \mathcal{F}/\mathcal{Pr}$ (where $\mathcal{F}$ is the multiplicative group of nonzero fractional ideals of K and $\mathcal{Pr}$ is the subgroup of $\mathcal{F}$ of principal fractional nonzero ideals).*

In particular, $[L : K] = h$ (the class number of K) and $L|K$ is an Abelian extension.

Theorem 3 (The discriminant).   *The relative discriminant $\delta_{L|K}$ of the class field extension $L|K$ is the unit ideal.*

Thus, every prime ideal of K is unramified in the class field extension. It follows from Chapter 9, (**D**), that the Hilbert class field of $\mathbb{Q}$ is equal to $\mathbb{Q}$. Thus, the theory of Hilbert is trivial over the ground field $\mathbb{Q}$.

Theorem 4 (Decomposition).   *The prime ideals of K are decomposed in the Hilbert class field extension $L|K$ according to the following rule: if $f \geq 1$ is the smallest integer such that P^f is a principal ideal of K, then P is decomposed into the product of h/f distinct prime ideals of L, each with inertial degree f over K.*

Therefore, the type of decomposition of a prime ideal P depends only on the class of ideals which contains P—this is the justification for the name "class field," which was given to L.

15.2.2 The Theory of Takagi

Takagi generalized the theory of Hilbert by considering admissible groups of ideals, which had been considered by Weber.

Definition 2.   Let K be an algebraic number field and let A be the ring of algebraic integers of K. For each nonzero integral ideal J of K, let $\mathcal{F}^{(J)}$ be the multiplicative group of nonzero fractional ideals, which are relatively prime to J. Let $\mathcal{R}^{(J)}$ be the subgroup of $\mathcal{F}^{(J)}$ consisting of the principal fractional ideals Aa, such that a is totally positive (that is, all real conjugates of a are positive) and $a \equiv 1 \pmod{J}$. $\mathcal{R}^{(J)}$ is called the *ray defined* by J.

Definition 3.   Every subgroup $\mathcal{H}$ of $\mathcal{F}^{(J)}$ such that $\mathcal{R}^{(J)} \subseteq \mathcal{H} \subseteq \mathcal{F}^{(J)}$ is called an *admissible group of ideals associated to* J.

It is useful to state that for each integral ideal J, the ray $R^{(J)}$ is a subgroup of finite index in $\mathcal{F}^{(J)}$. It is also a fact that a multiplicative group of ideals $\mathcal{H}$ may be admissible with respect to two distinct integral ideals J, J'. This leads to the following considerations:

If $\mathcal{H}$ and $\mathcal{H}'$ are admissible groups of fractional ideals associated to J and J', respectively, let $\mathcal{H} \approx \mathcal{H}'$ if there exists a nonzero integral ideal I such that $\mathcal{H} \cap \mathcal{F}^{(I)} = \mathcal{H}' \cap \mathcal{F}^{(I)}$.

This is an equivalence relation on the set of admissible subgroups of $\mathcal{F}$. It may be shown that in each equivalence class there is an admissible group $\mathcal{H}$ associated to an ideal F such that if $\mathcal{H} \approx \mathcal{H}'$ and $\mathcal{H}'$ is associated to J', then F divides J'. The ideal F is called the *conductor* of the equivalence class of the admissible group $\mathcal{H}$. We denote by $[\mathcal{H}]$ the equivalence class of $\mathcal{H}$.

If $\mathcal{H}$ is an admissible group with conductor F and $\mathcal{R}^{(F)} \subseteq \mathcal{H} \subseteq \mathcal{F}^{(F)}$, and if $\mathcal{H}'$ is an admissible group with $\mathcal{R}^{(JF)} \subseteq \mathcal{H}' \subseteq \mathcal{F}^{(JF)}$, then $\mathcal{H}' = \mathcal{H} \cap \mathcal{F}^{(JF)}$, hence $\mathcal{F}^{(F)}/\mathcal{H} \cong \mathcal{F}^{(JF)}/\mathcal{H}'$.

We define now an order relation on the set of equivalence classes of admissible groups. Let $\mathcal{H}$, $\mathcal{H}'$ be admissible groups, with conductors F, F', respectively. Assume that there exist integral ideals J, J' such that $JF = J'F'$ and $\mathcal{H}_1 \subseteq \mathcal{H}'_1$ where $\mathcal{H}_1 \approx \mathcal{H}$, $\mathcal{H}'_1 \approx \mathcal{H}'$, and $\mathcal{H}_1$, $\mathcal{H}'_1$ are associated to $JF = J'F'$. In this situation, we write $[\mathcal{H}] \leq [\mathcal{H}']$, after verifying that the property depends only on the equivalence class of admissible groups. The above relation is an order relation: $[\mathcal{H}] \leq [\mathcal{H}]$; if $[\mathcal{H}] \leq [\mathcal{H}']$ and $[\mathcal{H}'] \leq [\mathcal{H}]$ then $[\mathcal{H}] = [\mathcal{H}']$; if $[\mathcal{H}] \leq [\mathcal{H}']$ and $[\mathcal{H}'] \leq [\mathcal{H}'']$ then $[\mathcal{H}] \leq [\mathcal{H}'']$.

Let $\mathcal{Pr}_+$ denote the multiplicative group of principal fractional ideals Aa, where a is totally positive. Thus $\mathcal{Pr}_+ = \mathcal{R}^{(A)}$, so the admissible groups with conductor $F = A$ are equivalent to the groups $\mathcal{H}$ such that $\mathcal{Pr}_+ \subseteq \mathcal{H} \subseteq \mathcal{F}$. We recall that the index h_0 of $\mathcal{Pr}_+$ in $\mathcal{F}$ is the number of restricted classes of ideals.

Definition 4.   Let K be an algebraic number field, let J be an integral ideal, and let $\mathcal{H}$ be an admissible group of nonzero fractional ideals of K associated to J. The Galois extension $L|K$ is said to be a *class field of K associated to the admissible group* $\mathcal{H}$ when the following property is satisfied: a prime ideal P of K, having inertial degree 1 over $\mathbb{Q}$, belongs to $\mathcal{H}$ if and only if P is relatively prime to J and P is decomposed into distinct prime ideals of L, each having inertial degree 1 over $\mathbb{Q}$.

If $\mathcal{H} = \mathcal{Pr}$ (with conductor the unit ideal A) the class field of K associated to $\mathcal{Pr}$ is the Hilbert class field of K. If $\mathcal{H} = \mathcal{Pr}_+$ (with conductor also the unit ideal A), then the class field of K associated to $\mathcal{P}r_+$ is called the *absolute class field of K*.

Takagi proved the following theorems:

Theorem 5 (Existence and uniqueness).   *For each admissible group $\mathcal{H}$ of nonzero fractional ideals of K, there exists a class field L of K associated to $\mathcal{H}$. If $\mathcal{H}$, $\mathcal{H}'$ are equivalent admissible groups of nonzero fractional ideals of K, the associated class fields L, L' coincide.*

Theorem 6 (Isomorphism).   *If L is the class field of K associated to the admissible group $\mathcal{H}$ with conductor F, and if $\mathcal{H} \approx \mathcal{H}^{(F)}$ where $\mathcal{R}^{(F)} \subseteq \mathcal{H}^{(F)} \subseteq \mathcal{F}^{(F)}$, then $G(L|K) \cong \mathcal{F}^{(F)}/\mathcal{H}^{(F)}$.*

In particular, $L|K$ is an Abelian extension and $[L : K] = \#(\mathcal{F}^{(F)}/\mathcal{H}^{(F)})$.

A fundamental feature is that the isomorphism between the Galois group $G(L|K)$ and the quotient group is canonical. It is embodied by the *general reciprocity law* of Artin, which is conveniently stated appealing to the *idèles* introduced by Chevalley. We refer the reader to books on class field theory, like Weil [29], Iyanaga [14] or Neukirch [22].

Theorem 7 (The discriminant).   *Let L be the class field extension of K associated to the admissible group $\mathcal{H}$ with conductor F. If P is a prime ideal of K which divides the relative discriminant $\delta_{L|K}$, then P also divides the conductor F.*

So the only prime ideals P which ramify in the extension $L|K$ must divide the conductor F of the admissible group $\mathcal{H}$ to which $L|K$ is the associated class field extension.

Theorem 8 (Decomposition).   *Every prime ideal P of A, which does not divide the conductor F of the admissible group $\mathcal{H}$, decomposes in the extension $L|K$ (where L is the class field of K associated to $\mathcal{H}$) according to the following rule: let $\mathcal{H}^{(F)}$ be such that $\mathcal{H}^{(F)} \approx \mathcal{H}$, $\mathcal{R}^{(F)} \subseteq \mathcal{H}^{(F)} \subseteq \mathcal{F}^{(F)}$, let $f \geq 1$ be the smallest integer such that $P^f \in \mathcal{H}^{(F)}$, then P is decomposed as the product of $[L : K]/f$ distinct prime ideals of L, and each one has inertial degree f in the extension $L|K$.*

Besides the above theorems, which extended the ones by Hilbert, Takagi also proved the following new theorem:

Theorem 9 (Order-reversing correspondence).   *Let $\mathcal{H}$, $\mathcal{H}'$ be admissible groups of nonzero fractional ideals of K, and let L, L', respectively, be the associated class fields. Then $L \subseteq L'$ if and only if $[\mathcal{H}'] \leq [\mathcal{H}]$.*

But the most important theorem is the characterization of class field extensions.

Theorem 10 (Characterization).   *If $L|K$ is an Abelian extension, there exists an admissible group $\mathcal{H}$ of nonzero fractional ideals of K, such that*

L is the class field of K associated to $\mathcal{H}$. Moreover, if $\mathcal{H}'$ has the same property as $\mathcal{H}$, then $\mathcal{H} \approx \mathcal{H}'$.

As a consequence, we have:

Theorem 11 (Conductor and discriminant).   If L is the class field of K associated to the admissible group $\mathcal{H}$ with conductor F, then a prime ideal of K divides the relative discriminant $\delta_{L|K}$ if and only if P divides the conductor.

Altogether, if P is a prime ideal not dividing the ideal F, the type of decomposition of P in the class field extension L, associated to an admissible group $\mathcal{H}$ of conductor F, depends only on the order of the coset of P by $\mathcal{H}^{(F)}$, where $\mathcal{R}^{(F)} \subseteq \mathcal{H}^{(F)} \subseteq \mathcal{F}^{(F)}$, $\mathcal{H} \approx \mathcal{H}^{(F)}$.

For the finitely many prime ideals P dividing the conductor F, Takagi proved:

Theorem 12 (Ramification).   Let $L|K$ be an Abelian extension, let F be the integral ideal of K, and let $\mathcal{H}$ be the admissible group of nonzero fractional ideals and conductor F, $\mathcal{R}^{(F)} \subseteq \mathcal{H} \subseteq \mathcal{F}^{(F)}$, such that L is the class field of K associated to $\mathcal{H}$ (by Theorem 10). Let P be a prime ideal of K such that P^r divides F (with $r \geq 1$) but P does not divide $F' = FP^{-r}$. Let $\mathcal{H}'$ be the smallest admissible group with conductor F', $\mathcal{R}^{(F')} \subseteq \mathcal{H}' \subseteq \mathcal{F}^{(F')}$ such that $\mathcal{H} \subseteq \mathcal{H}'$. Let $f \geq 1$ be the smallest integer such that $P^f \in \mathcal{H}'$. Then P decomposes as the product of $[L : K]/f(\mathcal{H}' : \mathcal{H})$ distinct prime ideals of L raised to the power $e = (\mathcal{H}' : \mathcal{H})$.

For admissible groups which are rays, and the ground field $K = \mathbb{Q}$, we have:

L.   The class field extension of $\mathbb{Q}$ associated to the ray $\mathcal{R}^{(J)}$, where $J = \mathbb{Z}m$, $m \geq 1$, is the field $\mathbb{Q}(\zeta)$, where ζ is a primitive mth root of unity.

Proof: Let p be a prime number, $p \equiv 1 \pmod{m}$. By Chapter 16, (**D**), p decomposes in $\mathbb{Q}(\zeta)$ as the product of distinct prime ideals, each of inertial degree equal to 1. Conversely, if p is a prime number, which decomposes in $\mathbb{Q}(\zeta)$ into the product of distinct prime ideals of inertial degree 1, then p is unramified, so it does not divide the discriminant, hence p does not divide m; moreover, the order of p modulo m is the inertial degree of the prime ideals of $\mathbb{Q}(\zeta)$ dividing p, thus it is equal to 1, that is, $p \equiv 1 \pmod{m}$.

By definition and the uniqueness of the class field, $\mathbb{Q}(\zeta)$ is the class field of $\mathbb{Q}$ associated to the ray $\mathcal{R}^{(\mathbb{Z}m)}$.   ∎

With this property, it is easy to deduce the theorem of Kronecker and Weber from the theorems of Takagi:

Proof: Let $L|\mathbb{Q}$ be an Abelian extension, so by Theorem 10 there exists an admissible group $\mathcal{H}$ of nonzero fractional ideals of $\mathbb{Q}$, such that $\mathcal{R}^{(\mathbb{Z}m)} \subseteq$

$\mathcal{H} \subseteq \mathcal{F}^{(\mathbb{Z}m)}$, where $\mathbb{Z}m$ $(m \geq 1)$ is the conductor of $\mathcal{H}$, such that L is the class field of $\mathbb{Q}$ associated to $\mathcal{H}$. By Theorem 9 and (**L**), $L \subseteq \mathbb{Q}(\zeta)$, where ζ is a primitive mth root of 1. ∎

Another important theorem was proved in 1930 by Furtwängler:

Theorem 13 (Principal ideal theorem).   *Let L be the class field of K associated to the ray $\mathcal{R}^{(A)} = \mathcal{P}r_+$ of conductor equal to the unit ideal. Then every ideal of K generates a principal ideal of L.*

It should be noted that this does not imply that the ring of algebraic integers of L is a principal ideal domain, since not every ideal of L is generated by some ideal of K. A natural question arises. Let $K = K_0 \subset K_1 \subset K_2 \subset \cdots$ be the *tower of absolute class fields* of K, i.e., for each $i \geq 1$, K_i is the absolute class field of K_{i-1} (class field associated to the ray associated to the unit ideal). The question is whether there always exists an index i, depending on K, such that $K_i = K_{i+1} = \cdots$, in other words, K_i has restricted class number $h_0 = 1$, hence also $h = 1$. This would imply that K is a subfield of a field of algebraic numbers in which every ideal is principal.

This problem remained open for a long time. Using methods of Galois cohomology, Golod and Shafarevich provided a negative answer by exhibiting a criterion and an explicit counterexample.

Exercises

1. Let I be a set of indices, let $\leq$ be an order relation on I, and let $(S_i)_{i \in I}$ be a family of sets. For every pair of indices $(i, j) \in I \times I$, such that $i \leq j$, let $\pi_{ji} : S_j \to S_i$ be a mapping such that:

   (1) π_{ii} is the identity map.

   (2) If $i \leq j \leq k$ then $\pi_{ki} = \pi_{ji} \circ \pi_{kj}$.

Show:

   (a) There exists a set S and a family of maps $\pi_i : S \to S_i$ with the following properties:

   (1) If $i \leq j$ then $\pi_i = \pi_{ji} \circ \pi_j$.

   (2) If S' is a set, if for every $i \in I$, $\pi_i' : S' \to S_i$ is a map such that $\pi_i' = \pi_{ji} \circ \pi_j'$ when $i \leq j$ then there exists a unique map $\theta : S' \to S$ such that $\pi_i \circ \theta = \pi_i'$ for every $i \in I$.

   (b) The set S and maps $(\pi_i)_{i \in I}$ are uniquely defined, in the following sense: if $\widetilde{S}$, $(\widetilde{\pi}_i)_{i \in I}$ satisfy the properties of (a) then there exists a bijection $\sigma : S \to \widetilde{S}$ such that $\widetilde{\pi}_i \circ \sigma = \pi_i$ for every $i \in I$.

$(S, (\pi_i)_{i\in I})$ is called the *inverse limit* or *projective limit* of the family $(S_i)_{i\in I}$, with respect to the maps π_{ji} for $i \leq j$. It is denoted by $S = \varprojlim S_i$.

2. In the previous exercise, assume that each set is a group and that each map is a group-homomorphism. Show that S is a group and the maps π_i are group-homomorphisms. Explicitly, S is the subgroup of the product $\prod_{i\in I} S_i$, consisting of the families $(s_i)_{i\in I}$ such that if $i \leq j$ then $\pi_{ji}(s_j) = s_i$.

3. Let $K|K_0$ be a Galois extension of infinite degree, and let $\mathcal{G}$ be the family of fields F, $K_0 \subseteq F \subseteq K$ such that $F|K_0$ is a Galois extension of finite degree. Show:

(a) If $F, F' \in \mathcal{G}$ then $FF' \in \mathcal{G}$.

If $F|K_0$, $F'|K0$ are Galois Txtensions, $K_0 \subseteq F' \subseteq F \subseteq K$, let

$$\rho_{FF'} : G(F|K_0) \to G(F'|K_0)$$

be the group-homomorphism which associates with every K_0-automorphism of F its restriction to F'.

(b) If $F'' \subseteq F' \subseteq F$ belong to $\mathcal{G}$ then $\rho_{FF''} = \rho_{F'F''} \circ \rho_{FF'}$ and ρ_{FF} is the identity.

(c) If $F' \subseteq F$ belong to $\mathcal{G}$ then $\rho_{KF'} = \rho_{FF'} \circ \rho_{KF}$.

(d) Let H be a group, for every $F \in \mathcal{G}$ let $\theta_F : H \to G(F|K_0)$ be a group-homomorphism such that if $F' \subseteq F$ then $\rho_{FF'} \circ \theta_F = \theta_{F'}$. Show that there is a unique group-homomorphism $\theta : H \to G(K|K_0)$ such that $\rho_{KF} \circ \theta = \theta_F$ for every $F \in \mathcal{G}$. Conclude that $G(K|K_0)$ is the inverse limit of the family of finite groups $G(F|K_0)$ (for $F \in \mathcal{G}$) with respect to the group-homomorphisms $\rho_{FF'}$.

4. Let $\xi_1, \ldots, \xi_k$ be roots of unity, and let m_i be the order of ξ_i. Let $m = \mathrm{lcm}(m_1, \ldots, m_k)$ and let ζ be a primitive mth root of unity. Show that $\mathbb{Q}(\xi_1, \ldots, \xi_k) = \mathbb{Q}(\zeta)$.

5. Let p be a prime number and let T_p be the field generated over $\mathbb{Q}$ by all p^mth roots of unity (for $m = 1, 2, \ldots$). Show:

(a) $T_p|\mathbb{Q}$ is a Galois extension of infinite degree.

(b) If T_p' denotes the field generated by $\bigcup_{q \neq p} T_q$, then $T_p \cap T_p' = \mathbb{Q}$ (for every prime number p).

(c) The Abelian closure of $\mathbb{Q}$, denoted by $\mathbb{A}b$, is the field generated by $\bigcup T_p$ (for all prime numbers p).

Hint: Apply the theorem of Kronecker and Weber.

6. Let $K|K_0$ be a Galois extension. For every $i = 1, 2, \ldots$ let $L_i|K_0$ be a Galois extension, $L_i \subseteq K$, such that:

(a) If L_i' is the subfield of K generated by $\bigcup_{j \neq i} L_j$ then $L_i \cap L_i' = K_0$.

(b) K is generated by $\bigcup_{i=1}^{\infty} L_i$.

Show that

$$G(K|K_0) \cong \prod_{i=1}^{\infty} G(L_i|K_0).$$

7. Show that $G(\mathbb{A}b|\mathbb{Q}) \cong \prod_p G(T_p|\mathbb{Q})$.

Hint: Apply the two previous exercises.

8. Let p be an odd prime and $G_p = G(T_p|\mathbb{Q})$. Prove that

$$G_p \cong \mathbb{Z}/(p-1) \times \mathbb{Z}_p,$$

where $\mathbb{Z}_p = \varprojlim \mathbb{Z}/\mathbb{Z}p^m$ for $m \geq 1$, with respect to the canonical group-homomorphisms $\mathbb{Z}/\mathbb{Z}p^m \rightarrow \mathbb{Z}/\mathbb{Z}p^n \cong (\mathbb{Z}/\mathbb{Z}p^n)/(\mathbb{Z}p^m/\mathbb{Z}p^n)$ when $m \leq n$.

Hint: Apply the previous exercises and Chapter 3, (**J**).

9. Show that $G_2 = G(T_2|\mathbb{Q})$ is isomorphic to $\mathbb{Z}/2 \times \mathbb{Z}_2$, where $\mathbb{Z}_2 = \varprojlim \mathbb{Z}/\mathbb{Z}2^m$ for $m \geq 1$.

Hint: Apply Chapter 3, (**K**).

16

Complements and Miscellaneous Numerical Examples

In this chapter we shall work out some numerical examples to illustrate the theories developed so far. We shall also add some complements.

The handling of a definite example may offer many difficulties. These may be due to the noneffectiveness of the methods of proof of the theorems; that is, no algorithm in a finite number of steps may be seen from the proof indicated. Sometimes, even though there is a theoretically finite algorithm to solve the problem, it may be too long to perform and therefore shortcuts have to be found.

Accordingly, in the first section we point out the existence of certain algorithms.

16.1 Some Algorithms

We begin by pointing out that the matters discussed here are the object of much attention and research, especially after the advent of computers, but they had been often expressly considered by the classical mathematicians.

We should stress that much more is known about computational methods than we are able to explain in this text. We suggest therefore that the reader consult more specialized books about these problems.

The basic problems are the following: the determination of the ring of integers and of an integral basis; the computation of the different and the discriminant; the decomposition of a natural number into a product of prime ideals of the ring of integers of the field; the determination of a fundamental system of units and the computation of the regulator. Concerning the class number, there exist classical analytical methods for its determination; however, these will not be dealt with in this book.

We recall that there exist algorithms, that is, procedures with finitely many steps:

(a) to recognize if a natural number is a prime;

(b) to find the complete factorization into primes of any given natural number;

(c) to find if a polynomial with coefficients in a finite field is irreducible;

(d) to express any polynomial with coefficients in a finite field as the product of irreducible polynomials with coefficients in the given field;

(e) to find if a polynomial with coefficients in $\mathbb{Z}$ is irreducible over $\mathbb{Z}$, or equivalently by Gauss' lemma, over $\mathbb{Q}$;

(f) to express any polynomial of $\mathbb{Z}[X]$ as the product of irreducible polynomials of $\mathbb{Z}[X]$; and

(g) to determine all the subgroups of a finite Abelian group.

To these basic algorithms, we add the following ones, more closely related to the subject of this book.

16.1.1 Calculation of the Minimal Polynomial, Trace and Norm of an Element

Let $F = X^n + a_1 X^{n-1} + \cdots + a_n \in \mathbb{Z}[X]$ be an irreducible polynomial, let t be a root of F and $K = \mathbb{Q}(t)$, so $K = Q[t] \cong \mathbb{Q}[X]/(F)$ (where (F) denotes the principal ideal of all multiples of F in $\mathbb{Q}[X]$). Thus t may be viewed as a symbol, the elements of K are written in a unique way in the form $x = b_0 + b_1 t + \cdots + b_{n-1} t^{n-1}$. The operations are performed as if t were an indeterminate, by then t^n is identified with $-(a_1 t^{n-1} + a_2 t^{n-2} + \cdots + a_n)$; similarly, t^{n+1}, t^{n+2}, ... are expressed in terms of the lower powers t^j $(0 \le j \le n-1)$ by the above relation.

Thus, given $x = b_{10} + b_{11} t + \cdots + b_{1,n-1} t^{n-1}$ we may write x^2, ..., x^{n-1}, x^n as linear combinations of 1, t, ..., t^{n-1}, say

$$x^i = \sum_{j=0}^{n-1} b_{ij} t^j, \qquad i = 1, 2, \ldots, n.$$

The search of a polynomial $G(X) = X^n + c_1 X^{n-1} + \cdots + c_n$ with $G(x) = 0$ requires the solution of a system of n linear equations in the n unknowns c_1, c_2, ..., c_n using Cramer's rule, for example, or Gauss' elimination. This leads in a finite number of steps to a solution in $\mathbb{Q}$, thus giving $G \in \mathbb{Q}[X]$ such that $G(x) = 0$. By determining the irreducible factors of G, we obtain one which has the root x. It is therefore the minimal polynomial of x over $\mathbb{Q}$.

The trace and norm of x are the appropriate coefficients in the minimal polynomial of x.

16.1.2 Calculation of the Discriminant of a Set $\{x_1, \ldots, x_n\}$

By definition, $\mathrm{discr}_{K|\mathbb{Q}}(x_1, \ldots, x_n) = \det(\mathrm{Tr}(x_i x_j))$. To compute the discriminant one needs to compute the traces, which is possible as indicated above, and then the determinant.

For the special case of the discriminant of $\{1, t, \ldots, t^{n-1}\}$ there is a more expeditious method. By Chapter 2, Section 11, $\mathrm{discr}(F)$ is the discriminant of a matrix whose entries are the sums of like powers of the roots of F, namely $p_k = t_1^k + \cdots + t_n^k$ where $t_1, \ldots, t_n$ are the roots of F. By Newton's formulas $p_0 = n$, p_1, p_2, $\ldots$, p_{2n-2} are computed recursively in terms of the coefficients of F, without requiring the actual computation of the roots of F.

It is also sometimes feasible to use the norm of the derivative of $F \in \mathbb{Z}[X]$, according to the formula in Chapter 2, Section 11.

16.1.3 Determination of an Integral Basis, Ring of Integers and Discriminant

Let $F \in \mathbb{Z}[X]$ be an irreducible monic polynomial of degree n, let t be a root of F, and let $K = \mathbb{Q}(t)$, A the ring of integers of K, so $t \in A$, hence $\mathbb{Z}[t] \subseteq A$. Let $\mathbb{Z}[t]^*$ denote the complementary $\mathbb{Z}$-module. By Chapter 13, (J) and (H), $\mathbb{Z}[t]^* = (1/F'(t)) \mathbb{Z}[t]$ and $\mathbb{Z}[t]^*$ is a free Abelian group of the same rank n. Therefore $\mathbb{Z}[t]^*/\mathbb{Z}[t]$ is a finitely generated torsion Abelian group, hence it is a finite Abelian group. According to (g) it is possible to determine in finitely many steps all the subgroups of $\mathbb{Z}[t]^*/\mathbb{Z}[t]$, hence also all the subgroups M such that $\mathbb{Z}[t] \subseteq M \subseteq \mathbb{Z}[t]^*$. By using the result in Chapter 6, (L), it is possible to find a basis of M and by multiplication, it is possible to identify those $\mathbb{Z}[t]$-submodules M which are actually subrings of K.

We show that there is the largest subring which is equal to A. Indeed, if there is a subring M as above, not contained in A, let MA be the set of finite sums of elements of the form xy, where $x \in M$, $y \in A$. Then MA is a subring of K and A is properly contained in MA (if $x \in M$, $x \notin A$, then $x = x \cdot 1 \in MA$). Moreover, MA is a finitely generated $\mathbb{Z}$-module. Since every element of K is of the form a/m, where $a \in A$, $m \in \mathbb{Z}$, then there exists $r \in \mathbb{Z}$ such that $MA \subseteq (1/r)A$. By Chapter 7, Exercise 31, (f), there exists a prime ideal P of A such that the subring MA contains the set

$$\{x \in K \mid \text{there exists } m \geq 0 \text{ such that } P^m A x \subseteq A\} = \bigcup_{m \geq 0} P^{-m}.$$

So $P^{-m} \subseteq (1/r)A$ and P^m divides Ar for every $m \geq 0$. This is impossible, showing that every subring M, $\mathbb{Z}[t] \subseteq M \subseteq \mathbb{Z}[t]^*$ must be contained in A. Therefore A is determined as the largest of the subrings M which correspond to subgroups H of $G = \mathbb{Z}[t]^*/\mathbb{Z}[t]$.

So it is possible to determine the ring A and an integral basis in finitely many steps.

One may wish to find out if $\mathbb{Z}[t] = A$. If $\{x_1, \ldots, x_n\}$ is an integral basis, since $t^j = \sum_{i=1}^{n} a_{ij} x_i$ (with $a_{ij} \in \mathbb{Z}$), then $d = \operatorname{discr}_{K|\mathbb{Q}}(1, t, \ldots, t^{n-1}) = \left[\det(a_{ij})\right]^2 \delta_K$. Thus, if d is square-free, then $d = \delta_K$ so $A = \mathbb{Z}[t]$. However, it is important to note that it may well happen that the discriminant has a square factor $m^2 > 1$ which divides $\operatorname{discr}_{K|\mathbb{Q}}(1, u, \ldots, u^{n-1})$ for every integer u of K such that $K = \mathbb{Q}(u)$. This possibility was indicated at the end of Chapter 13.

16.1.4 Decomposition into Prime Ideals

If p is any prime number, then $Ap = P_1^{e_1} \cdots P_g^{e_g}$ where the prime ideals $P_1, \ldots, P_g$, and the various integers g, e_i, f_i, may be computed in finitely many steps. This follows from Chapter 11, Theorem 2. Indeed, let $F \in \mathbb{Z}[X]$ be the minimal polynomial of the primitive element $t \in A$ of K. Let $\overline{F} \in \mathbb{F}_p[X]$ be its canonical image. If $\overline{F} = \overline{H}_1^{e_1} \cdots \overline{H}_g^{e_g}$, where $g \geq 1$, $e_1, \ldots, e_g \geq 1$, $H_1, \ldots, H_g \in A[X]$, $\overline{H}_i \in \mathbb{F}_p[X]$ are distinct irreducible polynomials—this decomposition may be performed in finitely many steps; let $f_i = \deg(\overline{H}_i)$, then $P_i = Ap + AH_i(t)$, $N(P_i) = p^{f_i}$.

16.2 Complements on Cyclotomic Fields

In Chapter 5, Section 5, and Chapter 6, Section 5, we have determined the ring of algebraic integers and the discriminant of the cyclotomic field $K = \mathbb{Q}(\zeta)$, where ζ is a primitive pth root of unity and p is an odd prime. Namely $A = \mathbb{Z}[\zeta]$, $\delta_{\mathbb{Q}(\zeta)} = (-1)^{(p-1)/2} p^{p-2}$.

Now we shall extend these results. First, let $m = p^k > 2$, where p is a prime, $k \geq 1$, let $K = \mathbb{Q}(\zeta)$, where ζ is a primitive mth root of unity, and let A denote the ring of integers of K.

We begin by noting that if $\mu = \varphi(p^k)$ then $\{1, \zeta, \ldots, \zeta^{\mu-1}\}$ is a $\mathbb{Q}$-basis of K and of course $\mathbb{Z}[\zeta] \subseteq A$. We shall prove:

A. $A = \mathbb{Z}[\zeta]$, so $\{1, \zeta, \ldots, \zeta^{\mu-1}\}$ is an integral basis. The discriminant is

$$\delta_{\mathbb{Q}(\zeta)} = (-1)^{\varphi(p^k)/2} p^{p^{k-1}(k(p-1)-1)}.$$

Proof: First we compute the discriminant; $d = \operatorname{discr}_{K|\mathbb{Q}}(1, \zeta, \ldots, \zeta^{\mu-1})$. By Chapter 2, Section 11,

$$d = (-1)^{\mu(\mu-1)/2} N_{K|\mathbb{Q}}(\Phi'_m(\zeta)),$$

where $\Phi_m \in \mathbb{Z}[X]$ is the mth cyclotomic polynomial. But by Chapter 2, Section 11,

$$X^m - 1 = (X^{m/p} - 1)\Phi_m$$

hence

$$mX^{m-1} = \frac{m}{p}X^{(m/p)-1}\Phi_m + (X^{m/p} - 1)\Phi_m',$$

so

$$m\zeta^{m-1} = (\zeta^{m/p} - 1)\Phi_m'(\zeta).$$

Considering the conjugates of this expression and taking their product, we arrive at

$$m^\mu \left(\prod_{\overline{a} \in P(m)} \zeta^a\right)^{m-1} = \prod_{\overline{a} \in P(m)} (\zeta^{ap^{k-1}} - 1) \prod_{\overline{a} \in P(m)} \Phi_m'(\zeta^a).$$

But

$$\prod_{\overline{a} \in P(m)} \zeta^a = N_{K|\mathbb{Q}}(\zeta) = (-1)^\mu \cdot 1 = 1$$

and

$$\prod_{\overline{a} \in P(m)} (\zeta^{ap^{k-1}} - 1) = \prod_{\overline{a} \in P(m)} (1 - \zeta^{ap^{k-1}}) = p^{p^{k-1}},$$

because $\zeta^{p^{k-1}}$ is a primitive pth root of unity, $\prod_{\overline{a} \in P(p)}(1 - \zeta^a) = \Phi_p(1) = p$ and a system of representatives of the prime residue classes modulo $m = p^k$ gives rise to p^{k-1} systems of representatives of prime residue classes modulo p. Thus

$$m^\mu = p^{p^{k-1}} N_{K|\mathbb{Q}}(\Phi_m'(\zeta))$$

and so

$$d = (-1)^{\mu(\mu-1)/2} p^{p^{k-1}(k(p-1)-1)}.$$

Since $\mu(\mu - 1)/2 \equiv \mu/2 \pmod{2}$, then

$$d = (-1)^{\mu/2} p^{p^{k-1}(k(p-1)-1)}.$$

In order to show that $\mathbb{Z}[\zeta] = A$ we consider an arbitrary element $x \in A$. It may be written in a unique way in the form

$$x = x_0 + x_1\zeta + x_2\zeta^2 + \cdots + x_{\mu-1}\zeta^{\mu-1},$$

where each $x_i \in \mathbb{Q}$. It is our purpose to show that in fact each x_i is an integer. We shall prove that if q is a prime number and $x \in Aq$ then also $x_i \in \mathbb{Z}q$ $(i = 0, 1, \ldots, \mu - 1)$. If this is established we are able to conclude that each $x_i \in \mathbb{Z}$. In fact, let $x_i = a_i/b_i$ with $a_i, b_i \in \mathbb{Z}$ relatively prime integers. Let $l = \operatorname{lcm}(b_0, b_1, \ldots, b_{\mu-1})$ and $l = b_i l'_i$. Assume that there exists a prime number q such that q^r is the highest power of q dividing b_j; so q does not divide l'_j. From $l = q^r l'$ we have

$$lx = \sum_{i=0}^{\mu-1}(a_i l'_i)\zeta^i = (l'x)q^r \in Aq^r.$$

If we show that q divides each coefficient $a_i l'_i$, in particular, q divides $a_j l'_j$, but q does not divide l'_j, then q divides a_j, a contradiction. Thus $l = 1$ and each $x_i \in \mathbb{Z}$.

Considering the conjugates in $K|\mathbb{Q}$ we obtain the μ relations

$$\sigma_i(x) = \sum_{j=0}^{\mu-1} x_j \sigma_i(\zeta^j), \qquad i = 1, \ldots, \mu.$$

Thus $(x_0, \ldots, x_{\mu-1})$ is a solution of the system of linear equations with coefficients $\sigma_i(\zeta^j) \in A$ and determinant whose square is

$$(\det(\sigma_i(\zeta^j)))^2 = \operatorname{discr}_{K|\mathbb{Q}}(1, \zeta, \ldots, \zeta^{\mu-1}) = d.$$

Let α_j be the determinant of the matrix obtained from $(\sigma_i(\zeta_j))$ by replacing the jth column by that formed with elements $\sigma_1(x), \ldots, \sigma_\mu(x) \in A$. Thus $\alpha_j \in A$. By Cramer's rule, $dx_j = \alpha_j \sqrt{d} \in A \cap \mathbb{Q} = \mathbb{Z}$ for $j = 0, 1, \ldots, \mu - 1$.

If $q \neq p$ is a prime number such that $x \in Aq$ then $x = qy$, $y = y_0 + y_1\zeta + \cdots + y_{\mu-1}\zeta^{\mu-1}$ with $y_i \in \mathbb{Q}$, $x_j = qy_j$ so $dx_j = dy_jq$, with $dy_j \in \mathbb{Z}$. Thus q divides $dx_j = d(a_j/b_j)$ $(a_j, b_j \in \mathbb{Z})$, hence q divides da_j. But $q \neq p$, hence q does not divide d, so q divides a_j; that is, q divides x_j for $j = 0, 1, \ldots, \mu - 1$.

Now we shall prove that if $x \in Ap$ then p also divides x_j for every $j = 0, 1, \ldots, \mu - 1$. Let $h = x_0 + x_1 X + \cdots + x_{\mu-1}X^{\mu-1}$ so if $\xi = 1 - \zeta$ then

$$x = h(\zeta) = h(1 - \xi)$$

$$= h(1) - \xi \cdot h'(1) + \xi^2 \cdot \frac{h''(1)}{2!} - \xi^3 \frac{h'''(1)}{3!} + \cdots$$

$$+ (-1)^{\mu-1}\xi^{\mu-1} \cdot \frac{h^{(\mu-1)}(1)}{(\mu-1)!}.$$

The coefficients $h^{(k)}(1)/(k!)$ are integers which may easily be computed:

$$\frac{h^{(\mu-1)}(1)}{(\mu-1)!} = x_{\mu-1},$$

$$\frac{h^{(\mu-2)}(1)}{(\mu-2)!} = x_{\mu-2} + (\mu-1)x_{\mu-1},$$

$$\frac{h^{(\mu-3)}(1)}{(\mu-3)!} = x_{\mu-3} + (\mu-2)x_{\mu-2} + \binom{\mu-1}{2}x_{\mu-1},$$

$$\vdots$$

$$\frac{h''(1)}{2!} = x_2 + \binom{3}{2}x_3 + \cdots + \binom{k}{2}x_k + \cdots + \binom{\mu-1}{2}x_{\mu-1},$$

$$\frac{h'(1)}{1!} = x_1 + 2x_2 + 3x_3 + \cdots + (\mu-1)x_{\mu-1},$$

$$h(1) = x_0 + x_1 + x_2 + \cdots + x_{\mu-1}.$$

Since $p = u\xi^\mu$ and p divides x then ξ divides x, hence ξ divides $h(1)$. Thus $h(1) \in A\xi \cap \mathbb{Z} = Ap$, so p divides $h(1)$. But $\mu > 1$, thus ξ divides $p/\xi \in A$, which divides

$$-\frac{x - h(1)}{\xi} = h'(1) - \xi\frac{h''(1)}{2!} + \cdots + (-1)^{\mu-1}\xi^{\mu-2}\frac{h^{(\mu-1)}(1)}{(\mu-1)!}.$$

Thus ξ divides $h'(1)$ and so p divides $h'(1)$. We may continue in this manner showing successively that p divides

$$\frac{h''(1)}{2!}, \quad \frac{h'''(1)}{3!}, \quad \ldots, \quad \frac{h^{(\mu-1)}(1)}{(\mu-1)!}.$$

Taking into account the values of these elements, we deduce that p divides $x_{\mu-1}$, p divides $x_{\mu-2} + (\mu-1)x_{\mu-1}$, hence p divides $x_{\mu-2}$. Similarly, p divides $x_{\mu-3}$, $\ldots$, p divides x_0.

Thus we have established that $A = \mathbb{Z}[\zeta]$ and therefore the discriminant of K is $\delta_K = d = (-1)^{\mu/2}p^{p^{k-1}(k(p-1)-1)}$. ∎

We consider now the cyclotomic field $K = \mathbb{Q}(\zeta)$ generated by an mth root of unity ζ, where $m > 2$ is any integer. We may assume that if m is even then 4 divides m. Indeed, if $m = 2m'$, where m' is odd, if ζ' is a primitive (m')th root of unity then $\zeta'^m = 1$, so $\zeta' \in \mathbb{Q}(\zeta)$. On the other hand, $(\zeta^{m'})^2 = 1$, so $\zeta^{m'} = 1$, or $\zeta^{m'} = -1$, and in this case $(-\zeta)^{m'} = -\zeta^{m'} = 1$. So ζ or $-\zeta$ belongs to $\mathbb{Q}(\zeta')$ and therefore $\mathbb{Q}(\zeta) = \mathbb{Q}(\zeta')$ with m' odd.

B. *We have, with the above hypothesis:*

$$(1) \qquad \delta_{\mathbb{Q}(\zeta)} = (-1)^{s\varphi(m)/2}\frac{m^{\varphi(m)}}{\prod_{q|m}q^{\varphi(m)/(q-1)}},$$

where s is the number of distinct prime factors of m.

(2) $A = \mathbb{Z}[\zeta]$.

Proof: (1)   The proof is by induction. By $(\mathbf{A})$, it is true when $s = 1$. We assume it is true for $s - 1$. Let p be a prime dividing m, and $m = p^k m'$, where $k \geq 1$, p does not divide m'. Then there exist integers a, b such that $1 = ap^k + bm'$ and $\zeta = \zeta^{ap^k} \cdot \zeta^{bm'}$, with $(\zeta^{ap^k})^{m'} = 1$, $(\zeta^{bm'})^{p^k} = 1$.

Thus if ξ is a primitive (m')th root of unity and η is a primitive (p^k)th root of unity, then ζ^{ap^k} is a power of ξ, $\zeta^{bm'}$ is a power of η, and we conclude that $\mathbb{Q}(\zeta) = \mathbb{Q}(\xi) \cdot \mathbb{Q}(\eta)$.

By induction $\delta_{\mathbb{Q}(\xi)}$ and $\delta_{\mathbb{Q}(\eta)}$ are relatively prime. By Chapter 13, $(\mathbf{W})$:

$$
\begin{aligned}
\delta_{\mathbb{Q}(\zeta)} &= \delta_{\mathbb{Q}(\xi)}^{\varphi(p^k)} \cdot \delta_{\mathbb{Q}(\eta)}^{\varphi(m')} \\
&= (-1)^{(s-1)[\varphi(m')/2]\varphi(p^k)} \cdot \frac{m'^{\varphi(m')\varphi(p^k)}}{\prod_{q|m'} q^{[\varphi(m')/(q-1)]\varphi(p^k)}} \\
&\quad \times (-1)^{[\varphi(p^k)/2]\cdot\varphi(m')} \cdot \frac{p^{k\varphi(p^k)\varphi(m')}}{p^{[\varphi(p^k)/(p-1)]\varphi(m')}} \\
&= (-1)^{[\varphi(m)/2]s} \frac{m^{\varphi(m)}}{\prod_{q|m} q^{\varphi(m)/(q-1)}} .
\end{aligned}
$$

(2)   If B is the ring of algebraic integers of $\mathbb{Q}(\xi)$ and C is the ring of algebraic integers of $\mathbb{Q}(\eta)$, then by induction, $B = \mathbb{Z}[\xi]$, $C = \mathbb{Z}[\eta]$. Again, by Chapter 13, $(\mathbf{W})$, $A = BC = \mathbb{Z}[\xi, \eta] = \mathbb{Z}[\zeta]$. ∎

We applied Eisenstein's irreducibility criterion to show that $\Phi_p(X) \in \mathbb{Z}[X]$ is irreducible.

In the same way, we may show that for each $k \geq 2$, the polynomial $\Phi_{p^k}(X)$ is irreducible. It is equivalent to showing that $\Phi_{p^k}(X + 1)$ is irreducible. But

$$
\begin{aligned}
\Phi_{p^k}(X + 1) &= \frac{(X + 1)^{p^k} - 1}{(X + 1)^{p^{k-1}} - 1} = \frac{\{[(X + 1)^{p^{k-1}} - 1] + 1\}^p - 1}{(X + 1)^{p^{k-1}} - 1} \\
&= [(X + 1)^{p^{k-1}} - 1]^{p-1} + \binom{p}{1}[(X + 1)^{p^{k-1}} - 1]^{p-2} + \cdots \\
&\quad + \binom{p}{p - 2}[(X + 1)^{p^{k-1}} - 1] + p.
\end{aligned}
$$

Developing, we obtain a polynomial with leading coefficient 1, constant term p, and all other coefficients multiples of p; hence by Eisenstein's criterion, it is irreducible.

More generally we have the following corollary:

C.   For every $m > 2$ the cyclotomic polynomial $\Phi_m \in \mathbb{Z}[X]$ is irreducible in $\mathbb{Q}[X]$.

Proof: By Chapter 13, (**W**):

$$[\mathbb{Q}(\zeta) : \mathbb{Q}] = [\mathbb{Q}(\xi) : \mathbb{Q}] \cdot [\mathbb{Q}(\eta) : \mathbb{Q}]$$
$$= \varphi(m') \cdot \varphi(p^k) = \varphi(m),$$

thus Φ_m is necessarily the minimal polynomial of ζ over $\mathbb{Q}$, so Φ_m is irreducible in $\mathbb{Q}[X]$. ∎

Now we shall extend the results of Chapter 11, Section 3, about the decomposition of Ap as a product of prime ideals.

Let $m > 2$ and assume that 4 divides m if m is even. Let ζ be a primitive mth root of 1.

D.   *Let p be a prime.*
  (1) *If $m = p^k m'$, with $k \geq 1$, and p does not divide m', let f be the order of p modulo m', let $g = \varphi(m')/f$. Then $Ap = (P_1 \cdots P_g)^{\varphi(p^k)}$ where P_1, ..., P_g are distinct prime ideals.*
  (2) *If p does not divide m, let f be the order of p modulo m, $g = \varphi(m)/f$. Then $Ap = P_1 \cdots P_g$, where P_1, ..., P_g are distinct prime ideals.*

Proof: Let s be the number of distinct prime factors of m. If $s = 1$, the assertions are true and have been proved in Chapter 11, (**N**), (**O**).

We proceed by induction on s, assuming the results true for $s - 1$. Let q be a prime dividing m, let $m = q^k m'$ where $k \geq 1$ and q does not divide m'. Let ξ be a primitive (m')th root of 1 and η a primitive (q^k)th root of 1. As seen in the previous proof, $\mathbb{Q}(\zeta) = \mathbb{Q}(\xi)\mathbb{Q}(\eta)$ and also $A = BC$, where $A = \mathbb{Z}[\zeta]$, $B = \mathbb{Z}[\xi]$, $C = \mathbb{Z}[\eta]$ are rings of integers of $\mathbb{Q}(\zeta)$, $\mathbb{Q}(\xi)$, $\mathbb{Q}(\eta)$.

We prove (1), taking $q = p$. By Chapter 11, (**N**), $Cp = R^{\varphi(p^k)}$ where R is a prime ideal of C. By induction $Bp = Q_1 \cdots Q_{g'}$ where Q_1, ..., $Q_{g'}$ are prime ideals of B, $f'g' = \varphi(m')$, and f' is the order of p modulo m'. Let $Ap = (P_1 \cdots P_g)^e$ where $efg = \varphi(m)$, f being the inertial degree of each prime ideal P_i. By the transitivity of the inertial degrees, ramification indices, and decomposition numbers:

$$efg = \varphi(m) = \varphi(p^k)\varphi(m') \leq e(f'g') \leq efg,$$

then $e = \varphi(p^k)$, $f = f'$, $g = g'$, so $Ap = (P_1 \cdots P_g)^{\varphi(p^k)}$, $fg = \varphi(m')$, and f is the order of p modulo m'.

To prove (2) we proceed similarly; now p does not divide m, so $p \neq q$ and p is unramified in $\mathbb{Q}(\xi)$ and in $\mathbb{Q}(\eta)$. By Chapter 13, (**W**), p is unramified in $\mathbb{Q}(\zeta)$. By induction we have $Bp = Q_1 \cdots Q_{g'}$, with Q_1, ..., $Q_{g'}$ distinct prime ideals of B, $f'g' = \varphi(m')$, and f' is the order of p modulo m'. Also, by induction, $Cp = R_1 \cdots R_{g''}$, with distinct prime ideals R_1, ..., $R_{g''}$ of C, $f''g'' = \varphi(q^k)$, and f'' is the order of p modulo q^k. Finally, we have $Ap = P_1 \cdots P_g$, where P_1, ..., P_g are distinct prime ideals, $fg = \varphi(m)$, and f is the inertial degree of each ideal P_i in $\mathbb{Q}(\zeta)|\mathbb{Q}$. We need to show that f is equal to the order of p modulo m.

By Chapter 14, (E), $g = g'g''$. Since $fg = \varphi(m) = \varphi(m')\varphi(q^k) = f'g'f''g''$ then $f = f'f''$.

If $f_0 = \gcd(f', f'')$, let F be the unique extension of $\mathbb{F}_p$ with $[F : \mathbb{F}_p] = f_0$. We have $F \subseteq B/Q$, $F \subseteq C/R$ where $Q = P \cap \mathbb{Q}(\eta)$, $R = P \cap \mathbb{Q}(\xi)$, P is a prime ideal of $\mathbb{Q}(\zeta)$. Since $\gcd(f'/f_0, f''/f_0) = 1$ and $[B/Q : F] = f'/f_0$, $[C/R : F] = f''/f_0$ then $B/Q \cap C/R = F$ and we have

$$f = f_0[A/P : F] = f_0[B/Q : F][C/R : F] = f_0 \frac{f'}{f_0} \cdot \frac{f''}{f_0} = \frac{f}{f_0}$$

thus $f_0 = 1$, that is, $\gcd(f', f'') = 1$, so $f = f'f'' = \mathrm{lcm}(f', f'')$. Since f' is the order of p modulo m' and f'' is the order of p modulo q^k, then $f = f'f''$ is the order of p modulo $m'q^k = m$. ∎

16.3 Some Cubic Fields

Example 1:   Let $K = \mathbb{Q}(t)$, where t is a root of $f = X^3 - X - 1$. This is an irreducible polynomial, because by Gauss' lemma the only roots in $\mathbb{Q}$ could be 1, -1.

The discriminant of f is equal to $-(4(-1)^3 + 27(-1)^2) = -23$ (see Chapter 2, Exercise 48).

Let δ_K be the discriminant of K. We have $-23 = m^2 \delta_K$, where $m \in \mathbb{Z}$; hence $m^2 = 1$ and $\delta_K = -23$. It follows also that $\{1, t, t^2\}$ is an integral basis. The only ramified prime is $p = 23$.

To see, for example, how the primes 2, 3, 5, 7, 23 are decomposed, we consider the decomposition into irreducible polynomials of the images of f in the fields $\mathbb{F}_2$, $\mathbb{F}_3$, $\mathbb{F}_5$, $\mathbb{F}_7$, $\mathbb{F}_{23}$.

Over $\mathbb{F}_2$, $\overline{f}$ is irreducible, hence $A \cdot 2 = Q_2$ where Q_2 has inertial degree 3.

Over $\mathbb{F}_3$, $\overline{f}$ is irreducible, hence $A \cdot 3 = Q_3$ where Q_3 has inertial degree 3.

Over $\mathbb{F}_5$, $\overline{f} = (X - \overline{2})(X^2 + \overline{2}X + \overline{3})$, hence $A \cdot 5 = Q_5 \cdot Q_5'$, Q_5 has inertial degree 1, Q_5' has inertial degree 2. We have homomorphisms ψ_5 from A onto $\mathbb{F}_5$ and ψ_5' from A onto $\mathbb{F}_{25}$ with kernels, respectively, equal to Q_5, Q_5'. $\psi_5(t) = \overline{2}$, $\psi_5(t^2) = \overline{4}$, hence $Q_5 \supseteq \mathbb{Z} \cdot 5 \oplus \mathbb{Z}(t-2) \oplus \mathbb{Z}(t^2-4)$. On the other hand, if $a, b, c \in \mathbb{Z}$ and $\psi_5(a + bt + ct^2) = 0$ then $\overline{a} + \overline{2}\,\overline{b} + \overline{4}\,\overline{c} = \overline{0}$, so there exists $m \in \mathbb{Z}$ such that $a = 5m - 2b - 4c$, thus

$$a + bt + ct^2 = 5m + b(t - 2) + c(t^2 - 4).$$

This shows that $Q_5 = \mathbb{Z} \cdot 5 \oplus \mathbb{Z}(t - 2) \oplus \mathbb{Z}(t^2 - 4)$. From

$$t^2 - 4 = (t + 2)(t - 2)$$

and

$$N_{K|\mathbb{Q}}(t - 2) = (-1)^3 f(2) = -5$$

it follows that 5, $t^2 - 4$ belong to the principal ideal $A(t - 2)$, hence $Q_5 = A(t - 2)$.

From $A \cdot 5 = Q_5 \cdot Q_5'$, it also follows that Q_5' is the principal ideal generated by $5/(t - 2) = -(t' - 2)(t'' - 2)$ where t', t'' are the conjugates of t. But $tt't'' = 1$, $t + t' + t'' = 0$, hence

$$(t' - 2)(t'' - 2) = t't'' - 2(t' + t'') + 4 = 1/t + 2t + 4 = (2t^2 + 4t + 1)/t.$$

From $N_{K|\mathbb{Q}}(t) = 1$, t is a unit, thus $Q_5' = A(2t^2 + 4t + 1)$.

Over $\mathbb{F}_7$, $\bar{f} = (X - \bar{5})(X^2 + \bar{5}X + \bar{3})$, hence $A \cdot 7 = Q_7 \cdot Q_7'$, Q_7 has inertial degree 1, Q_7' has inertial degree 2. We have homomorphisms ψ_7 from A onto $\mathbb{F}_7$ and ψ_7' from A onto $\mathbb{F}_{49}$, with kernels Q_7, Q_7'. $\psi_7(t) = \bar{5}$, $\psi_7(t^2) = \bar{4}$, hence by a calculation already explained $Q_7 = \mathbb{Z} \cdot 7 \oplus \mathbb{Z}(t - 5) \oplus \mathbb{Z}(t^2 - 4)$. But $t^2 - 4 = (t + 5)(t - 5) + 3 \times 7$ and $N_{K|\mathbb{Q}}(t - 5) = (-1)^3 f(5) = -119 = -7 \times 17$, hence $7 \notin A(t - 5)$; otherwise, $7 = x(t - 5)$, $x \in A$, and taking norms, $343 = -N_{K|\mathbb{Q}}(x) \times 7 \times 17$, which is impossible because $N_{K|\mathbb{Q}}(x) \in \mathbb{Z}$. Therefore Q_7 is the ideal generated by 7, $t - 5$ (since the decomposition $A \cdot 7 = Q_7 \cdot Q_7'$ implies that $Q_7 \neq A \cdot 7$).

Next, $\psi_7'(7t) = 0$, $\psi_7'(t^2 + 5t + 3) = 0$ then $Q_7' = \mathbb{Z} \cdot 7 \oplus \mathbb{Z} \cdot 7t \oplus \mathbb{Z}(t^2 + 5t + 3)$. In fact, the generators are linearly independent over $\mathbb{Q}$ and if $\psi_7'(a + bt + ct^2) = 0$ then $\bar{a} + \bar{b}\bar{t} + \bar{c}(-\bar{5}\bar{t} - \bar{3}) = \bar{0}$ so

$$a = 7m + 3c, \qquad b = 7l + 5c \qquad \text{with} \quad l, m \in \mathbb{Z},$$

and

$$a + bt + ct^2 = 7m + 7lt + c(t^2 + 5t + 3).$$

We conclude that Q_7' is the ideal generated by 7 and $t^2 + 5t + 3$.

Over $\mathbb{F}_{23}$, we know already that 23 is ramified, hence either $A \cdot 23 = Q_{23}^2 \cdot Q_{23}'$ or $A \cdot 23 = Q_{23}^3$. To decide what actually happens, we factorize $X^3 - X - \bar{1}$ into irreducible polynomials modulo 23. Since there will be a root of multiplicity at least 2, this will be a common root of $\bar{f} = X^3 - X - \bar{1}$, and of its derivative $\bar{f}' = \bar{3}X^2 - \bar{1}$. Multiplying $\bar{f}'$ by X and subtracting $\bar{3}\bar{f}$ we have $2X + 3 \equiv 0 \pmod{23}$ hence $t \equiv 10 \pmod{23}$ is a double root and this yields the congruence $X^3 - X - 1 \equiv (X - 10)^2(X - 3) \pmod{23}$. Therefore the decomposition is $A \cdot 23 = Q_{23}^2 \cdot Q_{23}'$ where Q_{23}, Q_{23}' have inertial degree equal to 1.

By a similar argument, we show that $Q_{23} = \mathbb{Z} \cdot 23 \oplus \mathbb{Z}(t - 10) \oplus \mathbb{Z}(t^2 - 8)$, $t^2 - 8 = (t + 10)(t - 10) + 4 \times 23$. Q_{23} is the ideal generated by 23 and $t - 10$, because $N_{K|\mathbb{Q}}(t - 10) = (-1)^3 f(10) = -989 = -23 \times 43$, $N_{K|\mathbb{Q}}(23) = 23^3$, hence $23 \notin A(t - 10)$ and $t - 10 \notin A \cdot 23$.

In the same way, $Q_{23}' = \mathbb{Z} \cdot 23 \oplus \mathbb{Z}(t - 3) \oplus \mathbb{Z}(t^2 - 9)$, $t^2 - 9 = (t + 3)(t - 3)$ and $N_{K|\mathbb{Q}}(t - 3) = (-1)^3 f(3) = -23$, thus Q_{23}' is the principal ideal generated by $t - 3$. In particular, Q_{23}^2 is the principal ideal generated by $3t^2 + 9t + 1$, because $-23 = (t - 3)(t' - 3)(t'' - 3)$, $tt't'' = 1$, $t + t' + t'' = 0$,

$N_{K|\mathbb{Q}} = 1$, so t is a unit and

$$\frac{23}{t-3} = \frac{1}{t}(3t^2 + 9t + 1).$$

Example 2:   Let $K = \mathbb{Q}(t)$, where t is a root of $f = X^3 - 3X + 9$.

This is an irreducible polynomial (by Gauss' lemma the only roots in $\mathbb{Q}$ could be ± 1, ± 3, ± 9 and none of these numbers is a root).

The discriminant of f is $d = -(4 \times (-3)^3 + 27 \times 9^2) = -27 \times 7 \times 11$.

We shall determine a subring A_1 of A which properly contains $\mathbb{Z}[t]$. From

$$t^3 - 3t + 9 = 0 \tag{16.1}$$

we have $1 - 3/t^2 + 9/t^3 = 0$ and multyplying by 3, $(3/t)^3 - (3/t)^2 + 3 = 0$. Let $u = 3/t$, so

$$u^3 - u^2 + 3 = 0 \tag{16.2}$$

and $u \in A$. The $\mathbb{Z}$-module A_1 generated by $\{1, t, u\}$ is actually a subring of A. In fact, dividing (16.1) by t and (16.2) by u, we have

$$\begin{cases} t^2 = 3 - 3u, \\ u^2 = u - t, \\ tu = 3, \end{cases}$$

and this provides the multiplication table of A_1. Also $u = 1 - \frac{1}{3}t^2$ and since this expression is unique (because $\{1, t, t^2\}$ is a $\mathbb{Q}$-basis of K) then $u \notin \mathbb{Z}[t]$. Therefore, $\mathbb{Z}[t]$ is properly contained in A_1. Considering the discriminants, we have $d = m^2 d_1$, where $d_1 = \text{discr}_{K|\mathbb{Q}}(1, t, u)$, with $1 < m^2$. Hence $m^2 = 9$ (this is the only square dividing d) and $d_1 = -3 \times 7 \times 11$. Now, from $A_1 \subseteq A$ follows $d_1 = r^2 \delta_K$. But d_1 has no square-factors, thus $r^2 = 1$, $d_1 = \delta_K$, $A_1 = A$, and $\{1, t, u\}$ is an integral basis. The only ramified primes are therefore 3, 7, 11.

We shall describe in detail the decomposition of some primes p in A. Referring to our discussion in Chapter 11, preceding Theorem 2, we have

$$\alpha = \sqrt{\frac{\text{discr}_{K|\mathbb{Q}}(1, t, t^2)}{\text{discr}_{K|\mathbb{Q}}(1, t, u)}} = 3.$$

Over $\mathbb{F}_2$, $X^3 - 3X + 9 \equiv X^2 + X + 1$ (mod 2) and this polynomial is irreducible over $\mathbb{F}_2$. Hence $A \cdot 2 = Q_2$ where the inertial degree of Q_2 in $K|\mathbb{Q}$ is equal to 3.

Over $\mathbb{F}_{17}$, $X^3 - 3X + 9$ has the root 5 mod 17 and we have $X^3 - 3X + 9 \equiv (X - 5)(X^2 + 5X + 5)$ (mod 17) where $X^2 + 5X + 5$ is irreducible modulo 17. Thus $A \cdot 17 = Q_{17} \cdot Q'_{17}$, where Q_{17} has inertial degree 1, Q'_{17} has inertial degree 2.

Over $\mathbb{F}_7$ we know that $X^3 - 3X + 9$ must have at least a double root. So, we look for the roots common to $X^3 - 3X + 9$ and its derivative $3X^3 - 3$. 1 mod 7 is such a root and we have the decomposition $X^3 - 3X + 9 \equiv$

$(X-1)^2(X+2)$ (mod 7). Thus $A \cdot 7 = Q_7^2 \cdot Q_7'$, where Q_7, Q_7' have inertial degree equal to 1.

The prime ideal Q_7 is the kernel of the homomorphism $\psi_7 : A \to \mathbb{F}_7$ such that $\psi_7(t) = \bar{1}$, while Q_7' is the kernel of ψ_7', with $\psi_7'(t) = -\bar{2}$. From the relations between u, t we deduce that $3\psi_7(u) = \bar{3} - \psi_7(t^2) = \bar{2}$, hence $\psi_7(u) = \bar{3}$. Similarly $\psi_7'(u) = \bar{2}$. By the computation explained in Example 1, $Q_7 = \mathbb{Z} \cdot 7 \oplus \mathbb{Z}(t-1) \oplus \mathbb{Z}(u-3)$. Since $u - 3 = -u(t-1)$ and $N_{K|\mathbb{Q}}(t-1) = (-1)^3 f(1) = -7$, then Q_7 is the principal ideal generated by $t - 1$. Hence Q_7^2 is the principal ideal generated by $(t-1)^2$ and Q_7' is the principal ideal generated by $7/(t-1)^2$. If t', t'' are the conjugates of t, then $t + t' + t'' = 0$, $tt't'' = -9$ so $-7 = (t-1)(t'-1)(t''-1)$, $(t'-1)(t''-1) = -9/t + t + 1$:

$$\frac{7}{(t-1)^2} = \tfrac{1}{7}(t'-1)^2(t''-1)^2 = \frac{(t^2 + t - 9)^2}{7t^2}$$

$$= \frac{[(t^2 + t - 9)u]^2}{63} = \frac{9(1 + t - 3u)^2}{63} = -(2 + t),$$

so $Q_7' = A(t + 2)$. We could also see this directly, noting that $Q_7' = \mathbb{Z} \cdot 7 \oplus \mathbb{Z}(t+2) \oplus \mathbb{Z}(u-2)$, $N_{K|\mathbb{Q}}(t+2) = (-1)^3 f(-2) = -7$. Thus $7 \in A(t+2)$ and $u(t+2) = 3 + 2u = 2(u-2) + 7$ therefore $4u(t+2) = (u-2) + 7(u-2) + 2$, and $u - 2 = 4u(t+2) - (u-1) \cdot 7 \in A(t+2)$. This shows again that $Q_7' = A(t+2)$.

Over $\mathbb{F}_{11}$ we have, similarly, $X^3 - 3X + 9 \equiv (X+1)^2(X-2)$ (mod 11) so $A \cdot 11 = Q_{11}^2 \cdot Q_{11}'$, where Q_{11}, Q_{11}' have inertial degree equal to 1.

Q_{11} is the kernel of the homomorphism $\psi_{11} : A \to \mathbb{F}_{11}$ such that $\psi_{11}(t) = -\bar{1}$. Similarly, $\psi_{11}' : A \to \mathbb{F}_{11}$ has kernel Q_{11}' and $\psi_{11}'(t) = \bar{2}$. Then $3\psi_{11}(u) = \bar{3} - \psi_{11}(t^2) = \bar{3} - \bar{1} = \bar{2}$, hence $\psi_{11}(u) = \bar{8}$. Similarly, $\psi_{11}'(u) = \bar{7}$. Thus

$$Q_{11} = \mathbb{Z} \cdot 11 \oplus \mathbb{Z}(t+1) \oplus \mathbb{Z}(u+3),$$

$$Q_{11}' = \mathbb{Z} \cdot 11 \oplus \mathbb{Z}(t-2) \oplus \mathbb{Z}(u+4).$$

Since $u + 3 = u(t+1)$ and $N_{K|\mathbb{Q}}(t+1) = (-1)^3 f(-1) = -11$ then $11 \in A(t+1)$ therefore $Q_{11} = A(t+1)$. Next, $N_{K|\mathbb{Q}}(t-2) = (-1)^3 f(2) = -11$, so $11 \in A(t-2)$. Also $u(t-2) = 3 - 2u = -2(u+4) + 11$:

$$5u(t-2) = (u+4) - 11(u+4) + 11$$

hence $u + 4 \in A(t-2)$, showing that $Q_{11}' = A(t-2)$.

Now we describe the decomposition of $A \cdot 3$. The method indicated in Chapter 11, Theorem 2, cannot be applied to the prime number 3.

From the relations satisfied by t, u it follows that if $\psi : A \to \mathbb{F}_3$ is any homomorphism then $\psi(t) = \bar{t}$, $\psi(u) = \bar{u}$ satisfy $\bar{t}^2 = \bar{0}$, $\bar{u}^2 = \bar{u}$, $\bar{t}\bar{u} = \bar{0}$. The only possibilities ψ_3, ψ_3' are

$$\psi_3(t) = \bar{0}, \qquad \psi_3(u) = \bar{0},$$

and

$$\psi_3'(t) = \bar{0}, \qquad \psi_3'(u) = \bar{1}.$$

If $Q_3 = \ker(\psi_3)$, $Q_3' = \ker(\psi_3')$ then Q_3, Q_3' have inertial degree 1. We have $Q_3 = \mathbb{Z} \cdot 3 \oplus \mathbb{Z}t \oplus \mathbb{Z}u$, $Q_3' = \mathbb{Z} \cdot 3 \oplus \mathbb{Z}t \oplus \mathbb{Z}(u-1)$. From $t = -(u-1)u$ and $3 = N_{K|\mathbb{Q}}(u)$ it follows that Q_3 is the principal ideal generated by u. Similarly, $-3 = N_{K|\mathbb{Q}}(u-1)$, so Q_3' is the principal ideal generated by $u-1$. Thus $At = Q_3 \cdot Q_3'$ and $A \cdot 3 = Atu = Q_3^2 \cdot Q_3'$.

Example 3: Now we discuss a classical example of Dedekind. Let $K = \mathbb{Q}(t)$, where t is a root of $f = X^3 + X^2 - 2X + 8$.

f is irreducible over $\mathbb{Q}$, because if $f = f_1 f_2$, with $f_1, f_2 \in \mathbb{Z}[X]$, $\deg(f_1) > 0$, $\deg(f_2) > 0$, then reducing the coefficients modulo 2 we would have $\bar{f} = X^3 + X^2 = X^2(X + \bar{1})$. The constant term of f_2 is congruent to 1 mod 2. Since it divides 8 it must be 1 or -1. But $f(1) \neq 0$, $f(-1) \neq 0$, so f is irreducible.

The discriminant of f is equal to $d = 4 - 4 \times 8 + 18 \times (-2) \times 8 - 4 \times (-2)^3 - 27 \times 8^2 = -2012 = -4 \times 503$ (see Chapter 2, Exercise 48).

We shall determine a subring A_1 of A which properly contains $\mathbb{Z}[t]$. From

$$t^3 + t^2 - 2t + 8 = 0 \tag{16.3}$$

we have $1 + 1/t - 2/t^2 + 8/t^3 = 0$ and, multilpying by 8, $8 + 8/t - 16/t^2 + 64/t^3 = 0$. Letting $u = 4/t$ then

$$u^3 - u^2 + 2u + 8 = 0, \tag{16.4}$$

hence $u \in A$.

The $\mathbb{Z}$-module A_1 generated by $\{1, t, u\}$ is a subring of A. In fact, dividing (16.3) by t and (16.4) by u we have

$$\begin{cases} t^2 = 2 - t - 2u, \\ u^2 = -2 - 2t + u, \\ ut = 4. \end{cases}$$

These relations provide the multiplication table in A_1 and show that A_1 is a subring of A. Moreover, $u \notin \mathbb{Z}[t]$ since $u = 1 - (1/2)t - (1/2)t^2$ (and the expression of u in terms of the $\mathbb{Q}$-basis $\{1, t, t^2\}$ is unique). Thus $\mathbb{Z}[t]$ is properly contained in A_1. If $d_1 = \mathrm{discr}_{K|\mathbb{Q}}(1, t, u)$ then $d = m^2 d_1$ with $1 < m^2$. Hence $m^2 = 4$ and $d_1 = -503$. Since 503 is prime then $A = A_1$ and the discriminant of K is $\delta_K = d_1 = -503$.

The only ramified prime is 503. In our discussion in Chapter 11, before Theorem 2, we have

$$\alpha = \sqrt{\frac{\mathrm{discr}_{K|\mathbb{Q}}(1, t, t^2)}{\mathrm{discr}_{K|\mathbb{Q}}(1, t, u)}} = 2.$$

Let us study the decomposition of the primes 2 and 503 in the ring A.

From the relations satisfied by t, u we see that if $\psi : A \to \mathbb{F}_2$ is any homomorphism then $\psi(\bar{t}) = \bar{t}$, $\psi(u) = \bar{u}$ satisfy $\bar{t}^2 = \bar{t}$, $\bar{u}^2 = \bar{u}$, $\bar{t}\bar{u} = \bar{0}$. The only possibilities ψ_2, ψ_2', ψ_2'' are

$$\begin{aligned}
\psi_2(t) &= \bar{0}, & \psi_2(u) &= \bar{0}, \\
\psi_2'(t) &= \bar{1}, & \psi_2'(u) &= \bar{0}, \\
\psi_2''(t) &= \bar{0}, & \psi_2''(u) &= \bar{1}.
\end{aligned}$$

If $Q_2 = \ker(\psi_2)$, $Q_2' = \ker(\psi_2')$, $Q_2'' = \ker(\psi_2'')$ then Q_2, Q_2', Q_2'' have inertial degree 1 and $A \cdot 2 = Q_2 \cdot Q_2' \cdot Q_2''$. We have

$$\begin{aligned}
Q_2 &= \mathbb{Z} \cdot 2 \oplus \mathbb{Z}t \oplus \mathbb{Z}u, \\
Q_2' &= \mathbb{Z} \cdot 2 \oplus \mathbb{Z}(t-1) \oplus \mathbb{Z}u, \\
Q_2'' &= \mathbb{Z} \cdot 2 \oplus \mathbb{Z}t \oplus \mathbb{Z}(u-1).
\end{aligned}$$

Now we show that these prime ideals are principal. $N_{K|\mathbb{Q}}(t) = -8$ and similarly $N_{K|\mathbb{Q}}(u) = -8$ since $ut = 4$. t divides $N_{K|\mathbb{Q}}(t)$, hence the only prime ideals appearing in the decomposition of At are those which divide 2. But $t \notin Q_2'$ and $t \in Q_2$, $t \in Q_2''$. Similarly, $u \notin Q_2''$ but $u \in Q_2$, $u \in Q_2'$. So $A \cdot tu = A \cdot 4 = Q_2^2 \cdot Q_2'^2 \cdot Q_2''^2$. We show that Q_2^2 does not divide At. Otherwise, either $At = Q_2^2 \cdot Q_2''$, thus $Au = Q_2'^2 \cdot Q_2''$ which is impossible; or $At = Q_2^2 \cdot Q_2''^2$, $Au = Q_2'^2$, then $8 = |N_{K|\mathbb{Q}}(u)| = N(Au) = (N(Q_2'))^2 = 4$, which is absurd. With the same argument, we see that $Q_2'^2$ does not divide Au. Therefore $At = Q_2 \cdot Q_2''^2$ and $Au = Q_2 \cdot Q_2'^2$.

Let us note that if $a \in \mathbb{Q}$ and if t, t', t'' are conjugate over $\mathbb{Q}$ then

$$N_{K|\mathbb{Q}}(t - a) = (t - a)(t' - a)(t'' - a) = -8 + 2a - a^2 - a^3 = -f(a).$$

If $a \in \mathbb{Z}$ is odd then $t - a \in Q_2'$, $t - a \notin Q_2$, $t - a \notin Q_2''$ and $u - a \in Q_2''$, $u - a \notin Q_2$, $u - a \notin Q_2'$. Thus $N_{K|\mathbb{Q}}(t-1) = -8$, therefore the prime ideals dividing $A(t-1)$ must be among Q_2, Q_2', Q_2''. From the above we know that $A(t-1)$ must be a power of Q_2' and taking norms we conclude that $A(t-1) = Q_2'^3$. In the same manner, we see from $N_{K|\mathbb{Q}}(t+3) = -8 - 6 - 9 + 27 = 4$ that $A(t+3) = Q_2'^2$. Hence

$$Q_2' = A\left(\frac{t-1}{t+3}\right)$$

is a principal ideal. In terms of the integral basis, we may write $(t-1)/(t+3) = a + bt + cu$ where $a, b, c \in \mathbb{Z}$ are easily determined taking into account the multiplication table; namely $(t-1)/(t+3) = -5 + 3t + 2u$.

If $a \in \mathbb{Z}$ is even but not a multiple of 4 then $t - a \notin Q_2'$, and $t - a \in Q_2$, $t - a \in Q_2''$, $t - a \notin Q_2''^2$ (as may be seen from the norms). Similarly, $u - a \notin Q_2''$ and $u - a \in Q_2$, $u - a \in Q_2'$, $u - a \notin Q_2'^2$. Thus $N_{K|\mathbb{Q}}(t-2) = -16$, $N_{K|\mathbb{Q}}(t+2) = -8$ and the decomposition of $A(t-2)$, $A(t+2)$ is easily seen: $A(t-2) = Q_2^3 \cdot Q_2''$, $A(t+2) = Q_2^2 \cdot Q_2''$

(because of the norms). Hence

$$Q_2 = A \left(\frac{t-2}{t+2} \right) = A(2 - t - u).$$

Finally, $A \cdot 2 = Q_2 \cdot Q_2' \cdot Q_2''$, thus Q_2'' is also a principal ideal, namely

$$Q_2'' = A \left(\frac{2(t+2)(t+3)}{(t-1)(t-2)} \right) = A(-5 - 2t + u).$$

Now we study the decomposition of other primes p into prime ideals of A.

If $p = 3$ we have $A \cdot 3 = Q_3$ since $\overline{f} = X^3 + X^2 + X + \overline{2}$ is irreducible over $\mathbb{F}_3$.

Let $p = 5$. Then $\overline{f} = X^3 + X^2 + 3X + \overline{3} = (X + \overline{1})(X^2 + \overline{3})$ over $\mathbb{F}_5$. Hence $A \cdot 5 = Q_5 \cdot Q_5'$, where Q_5 has inertial degree 1 and Q_5' has inertial degree 2. If ψ_5, ψ_5' are homomorphisms from A with kernels Q_5, Q_5', respectively, then $\psi_5(t)\psi_5(u) = \overline{4}$ hence $\psi_5(u) = \overline{1}$ and by a computation already explained, we see that $Q_5 = \mathbb{Z} \cdot 5 \oplus \mathbb{Z}(t+1) \oplus \mathbb{Z}(u-1)$. Similarly, $Q_5' = \mathbb{Z} \cdot 5 \oplus \mathbb{Z}(t^2 + 3) \oplus \mathbb{Z}(u^2 + 2)$. From $N_{K|\mathbb{Q}}(t+1) = (-1)^3 f(-1) = 10$ we see that the prime ideals dividing $t + 1$ are among those dividing $A \cdot 2$, $A \cdot 5$. We have seen that $t + 1 \in Q_2'$. Taking the norms into account, we must have $A(t + 1) = Q_2' \cdot Q_5$. So

$$Q_5 = A \left(\frac{(t+1)(t+3)}{t-1} \right)$$

and this generator of Q_5 may be easily expressed in terms of the integral basis $\{1, t, u\}$. From $A \cdot 5 = Q_5 \cdot Q_5'$ we deduce also that Q_5' is a principal ideal generated by

$$\frac{5(t-1)}{(t+1)(t+3)} \in A.$$

For the primes $p = 7$, $p = 11$, we see with some computation that $\overline{f}$ is irreducible over $\mathbb{F}_7$, respectively, over $\mathbb{F}_{11}$. Hence $A \cdot 7 = Q_7$, $A \cdot 11 = Q_{11}$.

We conclude the study of this example by noting the following facts.

For every integer $v \in A$ we have $A \neq \mathbb{Z} \oplus \mathbb{Z}v \oplus \mathbb{Z}v^2$. Indeed, $A \cdot 2 = Q_2 \cdot Q_2' \cdot Q_2''$, so the prime ideals Q_2, Q_2', Q_2'' have inertial degree equal to 1. Thus $\psi_2(A) = \psi_2'(A) = \psi_2''(A) = \mathbb{F}_2$. If we had $A = \mathbb{Z}[v]$ then the homomorphisms would be determined by the image of v. The only possibilities are $\overline{0}, \overline{1} \in \mathbb{F}_2$, so there would only exist two homomorphisms from A onto $\mathbb{F}_2$, a contradiction.

This tells us that the discriminant of K has an inessential factor, namely 2. Indeed $A \cdot 2 = Q_2 \cdot Q_2' \cdot Q_2''$, $N(A \cdot 2) = 2 < 3 = [K : \mathbb{Q}]$ (see Chapter 13, after $(\mathbf{T})$).

The class number of K is 1; that is, every ideal of A is principal. It is enough to show that every prime ideal is principal. By Chapter 9, $(\mathbf{F})$, in every class of ideals of K there exists an integral nonzero ideal J such that

$$N(J) \leq \left(\frac{4}{\pi}\right)^{r_2} \frac{n!}{n^n} \sqrt{|\delta_K|}.$$

In our case, $n = 3$, $r_2 \leq 1$, $|\delta_K| = 503$, hence $N(J) < 7$. Thus it suffices to prove that every prime ideal of A having norm less than 7 is principal. This has already been established.

Example 4:   Let $K_1 = \mathbb{Q}(t)$ be a field of degree 3 over $\mathbb{Q}$ (where t is an algebraic integer). Let $t_1 = t$, t_2, t_3 be the conjugates of t over $\mathbb{Q}$. We assume that $K_1 = \mathbb{Q}(t_1)$, $K_2 = \mathbb{Q}(t_2)$, $K_3 = \mathbb{Q}(t_3)$ are distinct fields. Let $K = K_1 K_2 = K_1 K_3 = K_2 K_3$ (since $t_1 + t_2 + t_3 \in \mathbb{Z}$) so $K|\mathbb{Q}$ is a Galois extension of degree 6. We denote by A the ring of integers of K and by B_i the ring of integers of K_i $(i = 1, 2, 3)$.

The Galois group of $K|\mathbb{Q}$ is the symmetric group on three letters $\mathfrak{K} = \mathfrak{S}_3$. Moreover, $K = K_i(\sqrt{\delta})$ for $i = 1, 2, 3$ where $\delta = \delta_{K_1} = \delta_{K_2} = \delta_{K_3}$. Indeed, $\mathbb{Z}[t_i] \subseteq B_i$, hence $d_i = \mathrm{discr}_{K_i|\mathbb{Q}}(t_i) = (t_1 - t_2)^2(t_1 - t_3)^2(t_2 - t_3)^2 = m_i^2 \delta$ with $m_i \in \mathbb{Z}$; thus $K \supseteq K_1(\sqrt{\delta}) = K_1(\sqrt{d_1}) \supseteq K_1$. But $\sqrt{d_1} \notin K_1$ hence $K = K_1(\sqrt{\delta})$. In fact, if $\sqrt{d_1} \in K_1$ from $[K_1 : \mathbb{Q}] = 3$ it would follow that $\sqrt{d_1} \in \mathbb{Q}$; however, for the permutation

$$\sigma = \begin{pmatrix} t_1 & t_2 & t_3 \\ t_1 & t_3 & t_2 \end{pmatrix} \in \mathfrak{K}$$

we have $\sigma(\sqrt{d_1}) = (t_1 - t_3)(t_1 - t_2)(t_3 - t_2) = -\sqrt{d_1}$.

Let $L = \mathbb{Q}(\sqrt{\delta})$ thus $[K : L] = 3$, $[L : \mathbb{Q}] = 2$, $[K : K_i] = 2$, $[K_i : \mathbb{Q}] = 3$ for $i = 1, 2, 3$. We denote by C the ring of integers of L.

The nontrivial subgroups of $\mathfrak{K} = \mathfrak{S}_3$ are $\mathfrak{A} = G(K|L)$, the alternating group on three letters, $\mathfrak{B}_i = G(K|K_i)$, group of order 2 generated by the transposition

$$\tau_i = \begin{pmatrix} t_i & t_j & t_k \\ t_i & t_k & t_j \end{pmatrix}, \qquad \text{for} \quad i = 1, 2, 3.$$

Clearly $\mathfrak{A} \cap \mathfrak{B}_i = \{\varepsilon\}$; $\mathfrak{A}\mathfrak{B}_i = \mathfrak{K}$; $\mathfrak{A}$ is a normal subgroup of $\mathfrak{K}$; $\mathfrak{B}_1, \mathfrak{B}_2, \mathfrak{B}_3$ are conjugate subgroups.

We shall discuss all possible types of decomposition of an arbitrary prime number p in $K|\mathbb{Q}$.

The following notations will be used:

$$P = P_1, P_2, \ldots \quad \text{denote prime ideals of } A,$$
$$Q_{i1}, Q_{i2}, \ldots \quad \text{denote prime ideals of } B_i,$$
$$R_1, R_2, \ldots \quad \text{denote prime ideals of } C.$$

Case 1: p is unramified in $K|\mathbb{Q}$.

The inertial group and the inertial field of P in $K|\mathbb{Q}$ are, respectively, equal to $T_P(K|\mathbb{Q}) = \{\varepsilon\}$, $T_P(K|\mathbb{Q}) = K$. The possibilities for the decomposition group of P in $K|\mathbb{Q}$ are the following:

(a) $\mathcal{Z}_P(K|\mathbb{Q}) = \{\varepsilon\}$

(b) $e_P(K|\mathbb{Q}) = \mathfrak{A}$; and

(c) $\mathcal{Z}_P(K|\mathbb{Q}) = \mathfrak{B}_i$ $(i = 1, 2, 3)$.

(a)   In this case, $Ap = P_1 P_2 P_3 P_4 P_5 P_6$, $B_i p = Q_{i1} Q_{i2} Q_{i3}$ $(i = 1, 2, 3)$, $Cp = R_1 R_2$, where the above prime ideals are distinct. Each prime ideal P_i, Q_{ij}, R_i has degree 1 over $\mathbb{Q}$.

In fact, since p is unramified, by the fundamental relation $n = efg$ we have:

(1) in the extension $K|\mathbb{Q} : n = g = 6$, $e = f = 1$;

(2) in the extension $K_i|\mathbb{Q} : n = g = 3$, $e = f = 1$; and

(3) in the extension $L|\mathbb{Q} : n = g = 2$, $e = f = 1$.

(b)   In this case, $Ap = P_1 P_2$, $B_i p = Q_{i1}$ $(i = 1, 2, 3)$, $Cp = R_1 R_2$.

In fact, $g = (\mathfrak{K} : \mathfrak{A}) = 2$ so $Ap = P_1 P_2$. Also $\mathcal{Z}_P(K|K_i) = \mathcal{Z}_P(K|\mathbb{Q}) \cap \mathfrak{B}_i = \{\varepsilon\}$ hence $P \cap B_i$ decomposes into the product of two prime ideals of A, thus necessarily $A(P_1 \cap B_i) = P_1 P_2$. Therefore $B_i p$ must be a prime ideal of B_i, $B_i p = Q_{i1}$. On the other hand, by Chapter 14, (**E**):

$$\mathcal{Z}_{P \cap C}(L|\mathbb{Q}) = \mathcal{Z}_P(K|\mathbb{Q})/\mathcal{Z}_P(K|L) = \mathfrak{A}/\mathfrak{A} = \{\varepsilon\},$$

thus Cp is the product of two prime ideals, $Cp = R_1 R_2$.

(c)   In this case, $Ap = P_1 P_2 P_3$, $B_i p = Q_{i1} Q_{i2}$ $(i = 1, 2, 3)$, $Cp = R_1$. Moreover, $AQ_{i1} = P_1$, $f(P_1|q_{i1}) = 2$, $AQ_{i2} = P_2 P_3$, $f(Q_{i1}|\mathbb{Z}p) = 1$, $f(Q_{i2}|\mathbb{Z}p) = 2$, $f(R_1|\mathbb{Z}p) = 2$.

In fact, $g = (\mathfrak{K} : \mathfrak{B}_i) = 3$ so Ap is the product of three prime ideals. Since $\mathcal{Z}_P(K|K_i) = \mathcal{Z}_P(K|\mathbb{Q}) \cap \mathfrak{B}_i$ then $P \cap B_i$ (denoted by Q_{i1}) generates a prime ideal of A, that is, $AQ_{i1} = P_1$. From the fundamental relation in $K|K_i$ we have $f(P_1|Q_{i1}) = 2$. Let $Q_{i2} = P_2 \cap B_i$, so we know that $Q_{i2} \neq Q_{i1}$. Since P_2 is conjugate to P_1 by some $\sigma_2 \in \mathfrak{K}$ then $\sigma_2 \notin \mathfrak{B}_i$ and $\mathcal{Z}_{P_2}(K|\mathbb{Q}) = \sigma_2^{-1} \mathcal{Z}_P(K|\mathbb{Q})\sigma_2 = \mathfrak{B}_j$ $(j \neq i)$. Hence

$$\mathcal{Z}_{P_2}(K|K_i) = \mathcal{Z}_{P_2}(K|\mathbb{Q}) \cap \mathfrak{B}_i = \{\varepsilon\},$$

thus $P_2 \cap B_i = Q_{i2}$ decomposes into the product of two prime ideals of A (which are distinct from P_1), hence $AQ_{i2} = P_2 P_3$.

Since $f(P_2|\mathbb{Z}p) = f(P_1|\mathbb{Z}p) = f(P_1|Q_{i1}) \cdot f(Q_{i1}|\mathbb{Z}p) = 2$ and $f(P_2|Q_{i2}) = f(P_3|Q_{i2})$ cannot be 2 then $f(Q_{i2}|\mathbb{Z}p) = 2$.

From $\mathcal{Z}_P(K|L) = \mathcal{Z}_P(K|\mathbb{Q}) \cap \mathfrak{A} = \mathfrak{B}_i \cap \mathfrak{A} = \{\varepsilon\}$ it follows that if $R_1 = P \cap C$ then AR_1 is the product of three prime ideals of A, thus necessarily $AR_1 = P_1 P_2 P_3$, $Cp = R_1$, and $f(R_1|\mathbb{Z}p) = 2$.

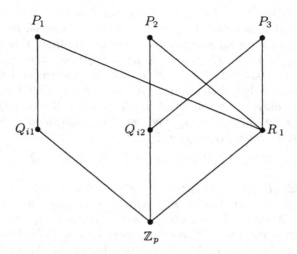

Case 2: p is ramified in $K|\mathbb{Q}$.

The possibilities for the inertial group and the decomposition group of P in $K|\mathbb{Q}$ are the following:

(a) $\mathcal{T}_P(K|\mathbb{Q}) = \mathcal{Z}_P(K|\mathbb{Q}) = \mathfrak{K}$;
(b) $\mathcal{T}_P(K|\mathbb{Q}) = \mathfrak{A}$, $\mathcal{Z}_P(K|\mathbb{Q}) = \mathfrak{K}$;
(c) $\mathcal{T}_P(K|\mathbb{Q}) = \mathcal{Z}_P(K|\mathbb{Q}) = \mathfrak{A}$; and
(d) $\mathcal{T}_P(K|\mathbb{Q}) = \mathcal{Z}_P(K|\mathbb{Q}) = \mathfrak{B}_i$ for $i = 1, 2, 3$.

It suffices to recall that $\mathcal{T}_P(K|\mathbb{Q}) \neq \{\varepsilon\}$ and that $\mathcal{T}_P(K|\mathbb{Q})$ is a normal subgroup of $\mathcal{Z}_P(K|\mathbb{Q})$.

(a)   This case may only happen when $p = 3$. Then $A \cdot 3 = P_1^6$, $B_i \cdot 3 = Q_{i1}^3$, $C \cdot 3 = R_1^2$.

Let $\mathcal{V}_1$ be the first ramification group of P in $K|\mathbb{Q}$. $\mathcal{V}_1$ is a normal subgroup of $\mathcal{T} = \mathcal{T}_P(K|\mathbb{Q})$ and $\mathcal{T}/\mathcal{V}_1$ is a cyclic group (Chapter 14, Theorem 2); thus $\mathcal{V}_1 \neq \{\varepsilon\}$ since $\mathfrak{K} = \mathfrak{S}_3$. But the order of $\mathcal{V}_1$ is a power of p and $\#\mathfrak{S}_3 = 6$ thus $\mathcal{V}_1 \neq \mathfrak{K}$. Hence $\mathcal{V}_1 = \mathfrak{A}$, it has order 3, so necessarily $p = 3$.

From $\mathcal{T}_P(K|\mathbb{Q}) = \mathcal{Z}_P(K|\mathbb{Q}) = \mathfrak{K}$ it follows that the decomposition number and inertial degree of P in $K|\mathbb{Q}$ are equal to 1, hence $A \cdot 3 = P_1^6$.

By transitivity of the decomposition number and inertial degree, those of $P_1 \cap B_i$, $P_1 \cap C$ are also equal to 1, hence $B_i \cdot 3 = Q_{i1}^3$, $C \cdot 3 = R_1^2$.

The case in question may actually arise; for example, when $K_1 = \mathbb{Q}(\sqrt[3]{3})$ (the reader should verify this statement).

(b)   In this case, $Ap = P_1^3$, $B_ip = Q_{i1}^3$ $(i = 1, 2, 3)$, $Cp = R_1$.

We have $\mathcal{T}_P(K|L) = \mathcal{Z}_P(K|L) = \mathfrak{A}$, therefore if $P_1 \cap C = R_1$ then the decomposition number of P_1 in $K|L$ is 1, the inertial degree is also 1, so by the fundamental relation R_1 is totally ramified, that is, $AR_1 = P_1^3$. Also $\mathcal{T}_{R_1}(L|\mathbb{Q}) = \mathcal{T}_P(K|\mathbb{Q})/\mathcal{T}_P(K|L) = \{\varepsilon\}$ hence the inertial degree $f(R_1|\mathbb{Z}p) = 2$. Thus $f(P_1|\mathbb{Z}p) = 2$ and by the fundamental relation for $K|\mathbb{Q}$, P_1 is the only prime ideal of A dividing $\mathbb{Z}p$; that is, $Ap = P_1^3$. Then there is only one prime ideal in B_i dividing B_ip, say Q_{i1}; by the fundamental relation, $B_ip = Q_{i1}^3$ because $f(Q_{i1}|\mathbb{Z}p)$ divides the degree 3 and the inertial degree $f(P_1|\mathbb{Z}p) = 2$.

(c)   In this case, $p \neq 2$ and $Ap = P_1^3 P_2^3$, $B_ip = Q_{i1}^3$ $(i = 1, 2, 3)$, $Cp = R_1 R_2$.

We have $\mathcal{Z}_P(K|\mathbb{Q}) = \mathfrak{A}$, hence $\mathcal{Z}_P(K|L) = \mathfrak{A}$ and $\mathcal{Z}_P(L|\mathbb{Q}) = \{\varepsilon\}$. Thus $Cp = R_1 R_2$ and $AR_1 = P_1^3$, $AR_2 = P_2^3$ hence $Ap = P_1^3 P_2^3$. Also $\mathcal{Z}_P(K|K_i) = \mathfrak{A} \cap \mathfrak{B}_i = \{\varepsilon\}$ thus if $Q_{i1} = P_1 \cap B_i$ then AQ_{i1} is the product of two distinct prime ideals, thus necessarily $AQ_{i1} = P_1 P_2$. Hence B_ip has only one prime factor Q_{i1}. But the inertial degrees of P_1, P_2 in $K|\mathbb{Q}$ are equal to 1, so $f(Q_{i1}|\mathbb{Z}p) = 1$ and by the fundamental relation $B_ip = Q_{i1}^3$.

We show now that $p \neq 2$. If $p = 2$ let $\mathcal{V}_1$ be the first ramification group of P_1 in $K|\mathbb{Q}$; by Chapter 14, Theorem 2, $\#(\mathcal{T}_{P_1}(K|\mathbb{Q})/\mathcal{V}_1)$ divides $\#(\mathbb{F}_2^{\cdot}) = 1$, so $\mathcal{V}_1 = \mathfrak{A}$ has order 3. But $\#(\mathcal{V}_1)$ is a power of p, thus $p = 3$, a contradiction.

(d)   In this case, $Ap = P_1^2 P_2^2 P_3^2$, $B_ip = Q_{i1} Q_{i2}^2$ $(i = 1, 2, 3)$, $Cp = R_1^2$.

In fact, $g = (\mathfrak{K} : \mathfrak{B}_i) = 3$ so Ap has three different prime factors. Since p is ramified then $Ap = P_1^2 P_2^2 P_3^2$. Let $Q_{i1} = B_i \cap P_1$. Since $\mathcal{Z}_P(K|K_i) = \mathfrak{B}_i$ then $AQ_{i1} = P_1^2$. On the other hand, $\mathcal{Z}_{P_2}(K|\mathbb{Q}) = \mathfrak{B}_j$ $(j \neq i)$ hence $\mathcal{Z}_{P_2}(K|K_i) = \{\varepsilon\}$. If $Q_{i2} = B_i \cap P_2 (\neq Q_{i1})$ then AQ_{i2} is the product of two different prime ideals; that is, $AQ_{i2} = P_2 P_3$ and therefore $B_ip = Q_{i1} Q_{i2}^2$. From $\mathcal{Z}_P(K|L) = \mathfrak{B}_i \cap \mathfrak{A} = \{\varepsilon\}$, if $R_1 = P_1 \cap C$ then AR_1 has three different prime factors; that is, $AR_1 = P_1 P_2 P_3$ and so $Cp = R_1^2$.

From this discussion we deduce:

If $A \cdot 2 = P_1 P_2 P_3 P_4 P_5 P_6$ then 2 is unramified in $K_1|\mathbb{Q}$; however, it divides the discriminant of every primitive integral element b of K_1. Thus 2 is an inessential factor of the discriminant δ_{K_1}.

Indeed, 2 is unramified in $K|\mathbb{Q}$, hence also in $K_1|\mathbb{Q}$. On the other hand, if $b = b_1$, b_2, b_3 are the conjugates of b over $\mathbb{Q}$, then

$$\text{discr}_{K_1|\mathbb{Q}}(b) = (b_1 - b_2)^2 (b_1 - b_3)^2 (b_2 - b_3)^2.$$

But $N(P_1) = 2$ (since P_1 has inertial degree 1), so b_1, b_2, b_3 are not all in different residue classes modulo P_1. Hence P_1^2 divides $\text{discr}_{K_1|\mathbb{Q}}(b) \in \mathbb{Z}$, so 4 divides $\text{discr}_{K_1|\mathbb{Q}}(b)$.

16.4 Biquadratic Fields

Example 5:   Let $K = \mathbb{Q}(\sqrt{7}, i)$, let A be the ring of integers of K, B the ring of integers of $\mathbb{Q}(\sqrt{7})$, C the ring of integers of $\mathbb{Q}(i)$.

$K|\mathbb{Q}$ is a Galois extension, its Galois group is $\{\varepsilon, \sigma, \tau, \sigma\tau\}$ where $\sigma(\sqrt{7}) = \sqrt{7}$, $\sigma(i) = -i$, $\tau(\sqrt{7}) = -\sqrt{7}$, $\tau(i) = i$, and $\sigma\tau(\sqrt{7}) = -\sqrt{7}$, $\sigma\tau(i) = -i$. Thus the fixed field of $\{\varepsilon, \sigma\}$ is $\mathbb{Q}(\sqrt{7})$, while the fixed field of $\{\varepsilon, \tau\}$ is $\mathbb{Q}(i)$.

Every element of K may be written uniquely in the form

$$x = a + b\sqrt{7} + ci + d\sqrt{7}i \qquad \text{with} \quad a, b, c, d \in \mathbb{Q}.$$

Then

$$\sigma(x) = a + b\sqrt{7} - ci - d\sqrt{7}i,$$

$$\tau(x) = a - b\sqrt{7} + ci - d\sqrt{7}i,$$

$$\sigma\tau(x) = a - b\sqrt{7} - ci + d\sqrt{7}i.$$

If $x \in A$ then $\mathrm{Tr}_{K|\mathbb{Q}(\sqrt{7})}(x) = x + \sigma(x) \in B$, $N_{K|\mathbb{Q}(\sqrt{7})}(x) = x \cdot \sigma(x) \in B$ and, similarly, $\mathrm{Tr}_{K|\mathbb{Q}(i)}(x) = x + \tau(x) \in C$, $N_{K|\mathbb{Q}(i)}(x) = x \cdot \tau(x) \in C$. We express these conditions in terms of the coefficients of x:

$$\begin{cases} 2a + 2b\sqrt{7} \in B, \\ (a+b\sqrt{7})^2 + (c+d\sqrt{7})^2 = [(a^2 + c^2) + 7(b^2 + d^2)] + 2(ab+cd)\sqrt{7} \in B, \\ 2a + 2ci \in C, \\ (a + ci)^2 - 7(b + di)^2 = [(a^2 - c^2) - 7(b^2 - d^2)] + 2(ac - 7bd)i \in C. \end{cases}$$

Taking into account that $B = \mathbb{Z} \oplus \mathbb{Z}\sqrt{7}$, $C = \mathbb{Z} \oplus \mathbb{Z}i$, then

$$2a \in \mathbb{Z}, \qquad 2b \in \mathbb{Z}, \qquad 2c \in \mathbb{Z},$$

$$(a^2 + c^2) + 7(b^2 + d^2) \in \mathbb{Z}, \qquad 2(ab + cd) \in \mathbb{Z},$$

$$(a^2 - c^2) - 7(b^2 - d^2) \in \mathbb{Z}, \qquad 2(ac - 7bd) \in \mathbb{Z}.$$

From these relations we deduce: $2c^2 + 14b^2 \in \mathbb{Z}$, $2a^2 + 14d^2 \in \mathbb{Z}$. Letting $a = \frac{1}{2}a'$, $b = \frac{1}{2}b'$, $c = \frac{1}{2}c'$ with $a', b', c' \in \mathbb{Z}$, then $12b^2 = 3b'^2 \in \mathbb{Z}$, so $(c'^2 + b'^2)/2 \in \mathbb{Z}$, and therefore b', c' have the same parity. Since $14d^2$ has a denominator at most equal to 2 then $d = \frac{1}{2}d'$, $d' \in \mathbb{Z}$, and again $12d^2 = 3d'^2 \in \mathbb{Z}$ so $(a'^2 + d'^2)/2 \in \mathbb{Z}$ and a', d' have the same parity. But

$$\frac{a'}{2} + \frac{d'}{2}\sqrt{7}i = \frac{a' - d'}{2} + d'\left(\frac{1 + \sqrt{7}i}{2}\right) \qquad \text{with} \quad \frac{a' - d'}{2} \in \mathbb{Z},$$

and, similarly,

$$\frac{b'}{2}\sqrt{7} + \frac{c'}{2}i = \frac{b' - c'}{2}\sqrt{7} + c'\left(\frac{\sqrt{7} + i}{2}\right) \qquad \text{with} \quad \frac{b' - c'}{2} \in \mathbb{Z}.$$

Thus every element $x \in A$ is a linear combination with coefficients in $\mathbb{Z}$ of 1, $\sqrt{7}$, $(\sqrt{7}+i)/2$, $(1+\sqrt{7}i)/2$, and since these elements are integers and are linearly independent, they constitute an integral basis of $K|\mathbb{Q}$.

Let us note here that $t = (\sqrt{7}+i)/2$ is a primitive integral element of K. Its minimal polynomial is easily computed and equal to $f = X^4 - 3X^2 + 4$. Its discriminant is $d = 16 \times (-3)^4 \times 4 - 128 \times (-3)^2 \times 4^2 + 256 \times 4^3 = 4^3 \times 7^2$ (see Chapter 2, Exercise 48). The fact that d has square factors does not allow us to decide at once whether $\mathbb{Z}[t]$ is equal to A. But computing the discriminant of the integral basis 1, $\sqrt{7}$, $(\sqrt{7}+i)/2$, $(1+\sqrt{7}i)/2$ we arrive at

$$\delta_K = \det \begin{pmatrix} 1 & \sqrt{7} & \dfrac{\sqrt{7}+i}{2} & \dfrac{1+\sqrt{7}i}{2} \\[2mm] 1 & -\sqrt{7} & \dfrac{-\sqrt{7}+i}{2} & \dfrac{1-\sqrt{7}i}{2} \\[2mm] 1 & \sqrt{7} & \dfrac{\sqrt{7}-i}{2} & \dfrac{1-\sqrt{7}i}{2} \\[2mm] 1 & -\sqrt{7} & \dfrac{-\sqrt{7}-i}{2} & \dfrac{1+\sqrt{7}i}{2} \end{pmatrix}^2 = 4^2 \times 7^2.$$

Thus $\mathbb{Z}[t] \neq A$. In the notation of the proof of Chapter 11, (I), $\alpha = 2$. The only ramified prime ideals are $2, 7$.

Since $f \equiv (X^2 + 2)^2 \pmod 7$ and $X^2 + \bar{2}$ is irreducible over $\mathbb{F}_7$ it follows that $A \cdot 7 = Q_7^2$ where Q_7 has inertial degree 2.

Let ψ_7 be the homomorphism from A onto $\mathbb{F}_{7^2}$ having kernel equal to Q_7. From $\psi_7(7) = 0$ it follows that $\psi_7(\sqrt{7}) = 0$. From $i^2 = -1$ it follows that $\psi_7(i)$ is the square root of -1 over $\mathbb{F}_7$ ($X^2 + \bar{1}$ is irreducible over $\mathbb{F}_7$ because -1 is not a square modulo 7, see Chapter 4, (H)). So $\psi_7(t) = \frac{1}{2}\sqrt{-1} = \gamma \in \mathbb{F}_{7^2}$. Finally, $\psi_7[(1 + \sqrt{7}i)/2] = \psi_7(1/2) = \bar{4} \in \mathbb{F}_7$.

If $x = a + b\sqrt{7} + ct + du \in Q_7$, with $u = (1 + \sqrt{7}i)/2$ then $\bar{a} + \bar{c}\gamma + \bar{d}\bar{4} = \bar{0}$, and since $\gamma \notin \mathbb{F}_7$ then $\bar{a} + \bar{4}\bar{d} = \bar{0}$, $\bar{c} = \bar{0}$, so there exist integers $l, m \in \mathbb{Z}$ such that $a = 7l - 4d$, $c = 7m$, and $x = 7l + b\sqrt{7} + 7mt + d(u - 4)$. This shows that $Q_7 = \mathbb{Z} \cdot 7 \oplus \mathbb{Z}\sqrt{7} \oplus \mathbb{Z} \cdot 7t \oplus \mathbb{Z}(u - 4)$. Now we conclude that Q_7 is the principal ideal generated by $\sqrt{7}$, because $u - 4 = -[(\sqrt{7} - i)/2]\sqrt{7}$ and $(\sqrt{7} - i)/2 \in A$, being a conjugate of t.

We now consider the decomposition of 2. We need to describe the possible homomorphisms ψ from A onto a field extension of $\mathbb{F}_2$. From $[\psi(\sqrt{7})]^2 = \psi(7) = \psi(1) = \bar{1}$ it follows that $\psi(\sqrt{7}) = \bar{1}$. Similarly $[\psi(i)]^2 = \psi(-1) = \bar{1}$, hence $\psi(i) = \bar{1}$.

Let $t' = (-\sqrt{7} + i)/2$ so $t + t' = i$, $tt' = -2$, therefore, $\psi(t) + \psi(t') = \bar{1}$, $\psi(t)\psi(t') = \bar{0}$. Let $u' = (1 - \sqrt{7}i)/2$ so $u + u' = 1$, $uu' = 2$, hence $\psi(u) + \psi(u') = \bar{1}$, $\psi(u)\psi(u') = \bar{0}$. From $tu = ((\sqrt{7}+i)/2) \cdot ((1+\sqrt{7}i)/2) = 2i$ and $t'u' = ((-\sqrt{7} + i)/2) \cdot ((1 - \sqrt{7}i)/2) = 2i$ we have $\psi(t)\psi(u) = \bar{0}$, $\psi(t')\psi(u') = \bar{0}$. Then $\psi(t)$, $\psi(u)$ are either $\bar{0}$ or $\bar{1}$ and there are only two possible homomorphisms ψ_2, ψ_2', namely:

$$\psi_2(\sqrt{7}) = \bar{1}, \qquad \psi_2(t) = \bar{0}, \qquad \psi_2(u) = \bar{1},$$
$$\psi_2'(\sqrt{7}) = \bar{1}, \qquad \psi_2'(t) = \bar{1}, \qquad \psi_2'(u) = \bar{0}.$$

Letting Q_2, Q_2' be their kernels, then Q_2, Q_2' have inertial degree equal to 1. But $K|\mathbb{Q}$ is a Galois extension, so the ramification indices of Q_2, Q_2' are equal, hence $A \cdot 2 = Q_2^2 \cdot Q_2'^2$. Now, it is easily seen that

$$Q_2 = \mathbb{Z} \cdot 2 \oplus \mathbb{Z}(1 - \sqrt{7}) \oplus \mathbb{Z}t \oplus \mathbb{Z}(u - 1),$$
$$Q_2' = \mathbb{Z} \cdot 2 \oplus \mathbb{Z}(1 - \sqrt{7}) \oplus \mathbb{Z}(t - 1) \oplus \mathbb{Z}u.$$

But $2 = -iut \in At$, $u - 1 = (1 + \sqrt{7}i)/2 - 1 = (-1 + \sqrt{7}i)/2 = it \in At$. Thus Q_2 is the ideal generated by $1 - \sqrt{7}$ and t. We note that $N_{K|\mathbb{Q}}(1 - \sqrt{7}) = (1 - \sqrt{7})^2(1 + \sqrt{7})^2 = 36$ and

$$N_{K|\mathbb{Q}}(t) = \left(\frac{\sqrt{7} + i}{2}\right) \cdot \left(\frac{\sqrt{7} - i}{2}\right) \cdot \left(\frac{-\sqrt{7} + i}{2}\right) \cdot \left(\frac{-\sqrt{7} - i}{2}\right) = 4,$$

hence $t \notin A(1 - \sqrt{7})$. It is easily seen that $1 - \sqrt{7} \notin At$ (for if $1 - \sqrt{7} = xt$ with $x \in A$, and if we express x in terms of the integral basis, we arrive at an impossibility).

For Q_2' we observe that $u - 1 = it$ implies that $u = i(t - i) \in A(t - i)$, $2 = -iut = t(t - i) \in A(t - i)$, thus Q_2' is the ideal generated by $1 - \sqrt{7}$, $t - i$. We note also that

$$N_{K|\mathbb{Q}}(t - i) = \left(\frac{\sqrt{7} - i}{2}\right) \cdot \left(\frac{\sqrt{7} + i}{2}\right) \cdot \left(\frac{-\sqrt{7} - i}{2}\right) \cdot \left(\frac{-\sqrt{7} + i}{2}\right) = 4$$

hence $t - i \notin A(1 - \sqrt{7})$ and it is also easily seen that $1 - \sqrt{7} \notin A(t - i)$.

Let us now compute the decomposition of 3. $f = X^4 - 3X^2 + 4 \equiv X^4 + 1 \equiv (X^2 + X + 2)(X^2 - X + 2) \pmod 3$, these factors being irreducible over $\mathbb{F}_3$. Then $A \cdot 3 = Q_3 \cdot Q_3'$, where Q_3, Q_3' have inertial degree equal to 2. Let ψ be a homomorphism from A onto a field extension of $\mathbb{F}_3$. Then $\psi(\sqrt{7})^2 = \psi(7) = \bar{1}$ so $\psi(\sqrt{7}) = \bar{1}$ or $\psi(\sqrt{7}) = \bar{2}$. From $\psi(i)^2 = \psi(-1) = -\bar{1}$ and the fact that -1 is not a square modulo 3, then $\psi(i) = \gamma$, where $\gamma \in \mathbb{F}_{3^2}$ is a root of $X^2 + \bar{1}$. then $\psi(t) = -(\psi(\sqrt{7}) + \gamma)$, $\psi(u) = -(1 + \psi(\sqrt{7})\gamma)$, and we have the following homomorphisms ψ_3, ψ_3' defined by

$$\psi_3(\sqrt{7}) = \bar{1}, \qquad \psi_3(t) = -(\gamma + \bar{1}), \qquad \psi_3(u) = -(\gamma + \bar{1}),$$
$$\psi_3'(\sqrt{7}) = -\bar{1}, \qquad \psi_3'(t) = -(\gamma + \bar{2}), \qquad \psi_3'(u) = \gamma + \bar{2}.$$

If Q_3 is the kernel of ψ_3 and Q_3' is the kernel of ψ_3' then by a computation already explained

$$Q_3 = \mathbb{Z} \cdot 3 \oplus \mathbb{Z}(\sqrt{7} - 1) \oplus \mathbb{Z} \cdot 3t \oplus \mathbb{Z}(u - t),$$
$$Q_3' = \mathbb{Z} \cdot 3 \oplus \mathbb{Z}(\sqrt{7} + 1) \oplus \mathbb{Z} \cdot 3t \oplus \mathbb{Z}(u + t).$$

Since

$$(\sqrt{7} - 1)\left(\frac{\sqrt{7} + i}{2}\right) = \frac{7 + \sqrt{7}i - \sqrt{7} - i}{2} = 3 + (u - t)$$

then Q_3 is the ideal generated by 3 and $\sqrt{7} - 1$. Moreover, $3 \notin A(\sqrt{7} - 1)$ and $\sqrt{7} - 1 \notin A \cdot 3$, as one sees taking norms. We also deduce that Q_3' is the ideal generated by 3 and $\sqrt{7} + 1$, since Q_3' is conjugate to Q_3.

16.5 Binomial Extensions

Example 6:   * Let $f = X^p - a \in \mathbb{Z}[X]$, where $a = q_1 q_2 \cdots q_r$, and $p, q_1, \ldots, q_r$ are distinct prime numbers, $p \neq 2$.

The roots of $f = X^p - a$ are $t, t\zeta, \ldots, t\zeta^{p-1}$ where ζ is a primitive pth root of unity, $t^p = a$. Thus

$$f = \prod_{i=0}^{p-1}(X - t\zeta^i).$$

Now we show that f is irreducible over $\mathbb{Q}$. It has no linear factor, otherwise there would exist a rational number whose pth power is equal to a.

If the minimal polynomial $g \in \mathbb{Q}[X]$ of t has degree less than p, it is of the form

$$g = \prod_{j=1}^{k}(X - t\zeta^{i_j}) \in \mathbb{Z}[X]$$

with $1 < k < p$ and $0 = i_1 < i_2 < \cdots < i_k \leq p - 1$. Hence, $t(\zeta^{i_1} + \zeta^{i_2} + \cdots + \zeta^{i_k}) \in \mathbb{Q}$ and $t \in \mathbb{Q}(\zeta)$. But $\mathbb{Q}(t\zeta^{i_2})$ and $\mathbb{Q}(t)$ are conjugate over $\mathbb{Q}$, hence the Galois groups $G(\mathbb{Q}(\zeta)|\mathbb{Q}(t\zeta^{i_2}))$ and $G(\mathbb{Q}(\zeta)|\mathbb{Q}(t))$ are conjugate subgroups of $G(\mathbb{Q}(\zeta)|\mathbb{Q})$. Since this is an Abelian group then $\mathbb{Q}(t) = \mathbb{Q}(t\zeta^{i_2})$ hence $\zeta^{i_2} \in \mathbb{Q}(t)$; taking j such that $ji_2 \equiv 1 \pmod{p}$ we have $\zeta \in \mathbb{Q}(t)$, that is, $\mathbb{Q}(\zeta) = \mathbb{Q}(t)$. Thus g has degree $p - 1$ and f would have a linear factor over $\mathbb{Q}$, which is impossible.

It follows that $K = \mathbb{Q}(t)$ has degree p over $\mathbb{Q}$. Let A be the ring of integers of K. A contains $\mathbb{Z}[t]$ and we know that A, $\mathbb{Z}[t]$ are free Abelian additive groups of rank p.

E.   *We have $Ap \subset \mathbb{Z}[t]$ and the Abelian group $A/\mathbb{Z}[t]$ is isomorphic to $(\mathbb{Z}/\mathbb{Z}p)^j$ for some j, $0 \leq j < p$.*

---

* See Gautheron, V. and Flexor, M., Un Exemple de Détermination des Entiers d'un Corps de Nombres, *Bull. Sci. Math.*, (2) **93** (1969), 3–13. See also "Rectificatif," *ibid.*, (2) **96** (1972), 172–179.

Proof: By Chapter 6, (**B**), $\mathbb{Z}[t] \subseteq A \subseteq [1/f'(t)]\mathbb{Z}[t]$, where $f'(t) = pt^{p-1}$. Then $Apt^{p-1} \subseteq \mathbb{Z}[t]$, hence multiplying by t, $Apa \subseteq \mathbb{Z}[t]t \subseteq \mathbb{Z}[t]$. Let us note that $\mathbb{Z}[t]$ is not contained in Ap, because $1 \notin Ap$ (otherwise $1/p \in A \cap \mathbb{Q} = \mathbb{Z}$).

In order to show that $Ap \subseteq \mathbb{Z}[t]$ it suffices to prove that if $\psi : A \to A/At$ is the canonical homomorphism then $\psi(\mathbb{Z}) = \psi(A)$. In fact, this means that given any element of A there exists an element of $\mathbb{Z}$ in the same residue class modulo At, so $A = \mathbb{Z} + At$. Repeating, $A = \mathbb{Z} + (\mathbb{Z} + At)t = \mathbb{Z} + \mathbb{Z}t + At^2$, and in this manner, $A \subseteq \mathbb{Z}[t] + At^p = \mathbb{Z}[t] + Aa$ hence $Ap \subseteq \mathbb{Z}[t]p + Aap \subseteq \mathbb{Z}[t]$.

We shall compare the decomposition into prime ideals of $\mathbb{Z}a = \mathbb{Z}q_1 \cdot \mathbb{Z}q_2 \cdots \mathbb{Z}q_r$ and At. Let $At = Q_1^{l_1} \cdot Q_2^{l_2} \cdots Q_s^{l_s}$, where each Q_i is a prime ideal of A and $1 \leq l_i$. We shall prove that $s = r$, $l_i = 1$, and $Aq_i = Q_i^p$ for every index $i = 1, \ldots, r$.

Indeed, $Aa = (At)^p = Q_1^{pl_1} \cdot Q_2^{pl_2} \cdots Q_s^{pl_s}$; on the other hand, $Aa = Aq_1 \cdot Aq_2 \cdots Aq_r$. Hence we have $r \leq s$, because of the uniqueness of the decomposition into prime ideals. Each Q_i divides one (and only one) ideal Aq_j. For example, the decomposition into prime ideals of Aq_1 is (after renumbering) of the type $Aq_1 = Q_1^{h_1} \cdot Q_2^{h_2} \cdots Q_k^{h_k}$, where the exponents h_j satisfy the fundamental relation $\sum_{j=1}^{k} f_{Q_j}(K|\mathbb{Q}) \cdot h_j = p$. So $pl_1 = h_1$ and from $h_1 \leq p$ it follows that $h_1 = p$ and $l_1 = 1$. Therefore, by the fundamental relation $k = 1$, so $Aq_1 = Q_1^p$. By the same token $Aq_i = Q_i^p$ for $i = 1, \ldots, r$ and we conclude that $s = r$.

So, each prime ideal q_i is totally ramified in $K|\mathbb{Q}$ and therefore $A/Q_i \cong \mathbb{Z}/\mathbb{Z}q_i$ $(i = 1, \ldots, r)$.

Now we have $A/At \cong \prod_{i=1}^{r} A/Q_i \cong \prod_{i=1}^{r} \mathbb{Z}/\mathbb{Z}q_i \cong \mathbb{Z}/\mathbb{Z}a$. If $\psi : A \to A/At$ denotes the canonical homomorphism then $\psi(\mathbb{Z})$ is a subring of the ring $A/At \cong \mathbb{Z}/\mathbb{Z}a$. Since this ring has no nontrivial subring then $\psi(\mathbb{Z}) = \psi(A)$ as we intended to show.

From $Ap \subset \mathbb{Z}[t] \subseteq A$ it follows that $A/\mathbb{Z}[t] \cong (A/Ap)/(\mathbb{Z}[t]/Ap)$ and since A/Ap is a vector space of dimension p over $\mathbb{F}_p$ (see Chapter 11, Theorem 1), and $\mathbb{Z}[t]/Ap \neq 0$, then $A/\mathbb{Z}[t]$ has dimension j, $0 \leq j < p$. ∎

F.   *The discriminant of $K|\mathbb{Q}$ is $\delta_K = \mathbb{Z}p^{p-2j}a^{p-1}$, where j is as in (**E**).*

Proof: The discriminant of t is

$$d = \operatorname{discr}_{K|\mathbb{Q}}(1, t, \ldots, t^{p-1}) = (-1)^{(1/2)p(p-1)} N_{K|\mathbb{Q}}(f'(t)).$$

But $f'(t) = pt^{p-1}$. The conjugates of t are $t, t\zeta, \ldots, t\zeta^{p-1}$ (where ζ is a primitive pth root of unity), so

$$N_{K|\mathbb{Q}}(f'(t)) = p^p t^{p(p-1)} \prod_{i=0}^{p-1} \zeta^i = p^p a^{p-1},$$

and therefore $|d| = p^p a^{p-1}$.

We have seen that $\mathbb{Z}[t]$ is contained in A and the Abelian group $A/\mathbb{Z}[t]$ is a vector space of dimension j over $\mathbb{F}_p$. By Chapter 6, (L), there exists an integral basis $\{x_1, \ldots, x_p\}$ of the Abelian group A and integers $f_1, \ldots, f_p$ such that if $y_i = f_i x_i$ $(i = 1, \ldots, p)$ then $\{y_1, \ldots, y_p\}$ is a basis of the Abelian group $\mathbb{Z}[t]$. Then $\mathbb{Z}d = (f_1 \cdots f_p)^2 \delta_K$. But

$$\mathbb{F}_p^j \cong A/\mathbb{Z}[t] \cong \bigoplus_{i=1}^{p} (\mathbb{Z}x_i/\mathbb{Z}f_i x_i) \cong \bigoplus_{i=1}^{p} \mathbb{Z}/\mathbb{Z}f_i,$$

thus $\prod_{i=1}^{p} f_i = p^j$. We conclude that $\delta_K = \mathbb{Z}p^{p-2j}a^{p-1}$. ∎

Since p is odd then $2j < p$. From this result, we know that the only ramified primes are p and $q_1, \ldots, q_r$. As we saw in (E), each prime q_i is totally ramified in $K|\mathbb{Q}$. Now we study the ramification of p.

G.   *The following facts are equivalent:*

(1) $A = \mathbb{Z}[t]$;

(2) $j = 0$; *and*

(3) p *is totally ramified (that is, $Ap = P^p$ where P is a prime ideal of A).*

If these conditions are not satisfied then $Ap = P_1^{l_1} \cdots P_s^{l_s}$, *with* $l_i < p$, $2j = \sum_{i=1}^{s} f_i$, *where f_i is the inertial degree of P_i over p.*

Proof: The equivalence of (1) and (2) has been seen in (E).

(1) → (3)   Let $A = \mathbb{Z}[t]$. By reducing modulo p the coefficients of $f = X^p - a$, we obtain $\overline{f} = X^p - \overline{a} = (X - \overline{a})^p$, because $\overline{a} \in \mathbb{F}_p$ hence $\overline{a}^p = \overline{a}$.

From Chapter 11, Theorem 2, we know that there exists only one prime ideal P of A above p; that is, $Ap = P^e$; if f is the inertial degree then $ef = p$. But p is ramified in $K|\mathbb{Q}$, hence $e > 1$, so $e = p$, $f = 1$, and p is therefore totally ramified.

(3) → (2)   Let $Ap = P^p$ so the inertial degree of P over p is equal to 1. Since the ramification index of P over p is the characteristic of the residue class field, it follows from Chapter 13, (O), that the exponent of the different at P is $s_P \geq e_P = p$; that is, P^p divides the different Δ_K. Taking norms, we conclude from Chapter 13, (P), that $p^p = N(P^p)$ divides $N(\Delta_K) = \delta_K = \mathbb{Z}p^{p-2j}a^{p-1}$. Hence $j = 0$.

Now we assume that p is not totally ramified in $K|\mathbb{Q}$, so $Ap = P_1^{l_1} \cdots P_s^{l_s}$, with each $l_i < p$ and $\sum_{i=1}^{s} l_i f_i = p$, where f_i is the inertial degree of P_i in $K|\mathbb{Q}$. Hence p does not divide l_i and therefore by Chapter 13, (O), the exponent of the different at each P_i is $s_i = l_i - 1$.

We have seen in the proof of (E) that $At = Q_1 Q_2 \cdots Q_r$, where each prime ideal Q_i of A is totally ramified over q_i; that is, $Aq_i = Q_i^p$. Since $p \neq q_i$, by Chapter 13, (O), the exponent of the different Δ_K at Q_i is equal to $p - 1$. Since $P_1, \ldots, P_s$ and $Q_1, \ldots, Q_r$ are only ramified prime ideals,

then

$$\Delta_K = P_1^{l_1-1} \cdots P_s^{l_s-1} \cdot Q_1^{p-1} \cdots Q_r^{p-1}.$$

Taking norms we have

$$\delta_K = N_{K|\mathbb{Q}}(\Delta_K) = \mathbb{Z}p^{\sum_{i=1}^{s} f_i(l_i-1)} \cdot (q_1 \cdots q_r)^{p-1}.$$

But $\delta_K = \mathbb{Z}p^{p-2j}a^{p-1}$ and, comparing, we get $p - 2j = \sum_{i=1}^{s} f_i l_i - \sum_{i=1}^{s} f_i$, so $2j = \sum_{i=1}^{s} f_i$. ∎

H.   *If p is not totally ramified then $j = 1$, $Ap = P_1^e P_2^{p-e}$, and the inertial degrees of P_1, P_2 are equal to 1.*

Proof: We shall find a new relation between j and the inertial degrees $f_1, \ldots, f_s$. This will be done considering the conductor of A into $\mathbb{Z}[t]$ and localizing above p.

Let S be the multiplicative set-complement of $\mathbb{Z}p$ in $\mathbb{Z}$, let $\mathbb{Z}' = S^{-1}\mathbb{Z}$, $\mathbb{Z}[t]' = S^{-1}\mathbb{Z}[t]$, $A' = S^{-1}A$. The ring $\mathbb{Z}'$ has only one nonzero prime ideal which is $\mathbb{Z}'p$ and $\mathbb{Z}'/\mathbb{Z}'p \cong \mathbb{Z}/\mathbb{Z}p = \mathbb{F}_p$ (by Chapter 12, (**H**)).

Reducing the coefficients of $f = X^p - a$ modulo p we obtain $\overline{f} = X^p - \overline{a} = (X - \overline{a})^p$ since $\overline{a}^p = \overline{a}$ in $\mathbb{F}_p$. By Chapter 11, Theorem 2, there exists only one prime ideal $\underline{P}$ of $\mathbb{Z}[t]$ such that $\underline{P} \cap \mathbb{Z} = \mathbb{Z}p$ and $\mathbb{Z}[t]p = \underline{P}^p$. Hence the inertial degree of $\underline{P}$ over p is equal to 1. It follows that $\mathbb{Z}[t]'\underline{P}$ is the only nonzero prime ideal of $\mathbb{Z}[t]'$ and $\mathbb{Z}[t]'/\mathbb{Z}[t]'\underline{P} \cong \mathbb{Z}[t]/\underline{P} = \mathbb{F}_p$. We note also that $\underline{P}$ is the ideal of $\mathbb{Z}[t]$ generated by $t - a$ and p.

Since p is not totally ramified then $A \neq \mathbb{Z}[t]$. Let F_t be the conductor of A in $\mathbb{Z}[t]$. Thus $F_t = Af'(t)\Delta_K^{-1}$. We shall show that $A'F_t = \mathbb{Z}[t]'\underline{P}$. By Chapter 13, (**N**):

$$\Delta(A'|\mathbb{Z}') = A' \cdot \Delta_K = A'(Af'(t) \cdot F_t^{-1}) = A'p \cdot A't^{p-1} \cdot (A'F_t)^{-1}$$

$$= A'p \cdot (A'F_t)^{-1} = \prod_{i=1}^{s} A'P_i^{l_i} \cdot (A'F_t)^{-1}.$$

On the other hand, since p is not totally ramified, each $l_i < p$, so the exponent of the different at P_i is $l_i - 1$, hence $\Delta(A'|\mathbb{Z}') = \prod_{i=1}^{s} A'P_i^{l_i-1}$. Comparing these expressions

$$A'F_t = \prod_{i=1}^{s} A'P_i = \bigcap_{i=1}^{s} A'P_i.$$

But by Chapter 13, (**S**), $F_t \subseteq \mathbb{Z}[t]$ so $A'F_t \subseteq \mathbb{Z}[t]'$ and so

$$A'F_t = A'F_t \cap \mathbb{Z}[t]' = \bigcap_{i=1}^{s}(A'P_i \cap \mathbb{Z}[t]') = \mathbb{Z}[t]'\underline{P}$$

because each prime ideal $A'P_i \cap \mathbb{Z}[t]'$ lies over $\mathbb{Z}'p$, therefore $A'P_i \cap \mathbb{Z}[t]' = \mathbb{Z}[t]'\underline{P}$.

We shall prove that $j = \left(\sum_{i=1}^{s} f_i\right) - 1$, where f_i is the inertial degree of P_i. We have $A'/A'F_t \cong \prod_{i=1}^{s} A'/A'P_i$, hence $\#(A'/A'F_t) = p^{\sum_{1}^{s} f_i}$, because $A'/A'P_i = A/P_i$; $\mathbb{Z}[t]'/A'F_t = \mathbb{Z}[t]'/\mathbb{Z}[t]'\underline{P} = \mathbb{F}_p$, hence $\#(\mathbb{Z}[t]'/A'F_t) = p$. So the Abelian group

$$A'/\mathbb{Z}[t]' \cong (A'/A'F_t)/(\mathbb{Z}[t]'/A'F_t)$$

has $p^{\sum_{1}^{s} f_i - 1}$ elements. But $A'/\mathbb{Z}[t]' \cong A/\mathbb{Z}[t] \cong \mathbb{F}_p{}^j$ and so $j = \sum_{1}^{s} f_i - 1$.

From $2j = \sum_{1}^{s} f_i$ we deduce that $2j = j + 1$, hence $j = 1$, $\sum_{1}^{s} f_i = 2$ and $s \leq 2$.

If we had $s = 1$ then $f_1 = 2$ and from the fundamental relation $e_1 f_1 = p$, so p would be even, which is not true. Thus $s = 2$ and $f_1 = f_2 = 1$, $e_1 + e_2 = p$ and we have concluded the proof.   ∎

It remains to indicate the conditions to decide whether $j = 0$ or $j = 1$.

I.   *The following conditions are equivalent:*

(1) $A = \mathbb{Z}[t]$.

(2) $a^{p-1} \not\equiv 1 \pmod{p^2}$.

If $A \neq \mathbb{Z}[t]$ then

$$\left\{1, t, \ldots, t^{p-2}, \frac{1 + ta^{p-2} + t^2 a^{p-3} + \cdots + t^{p-1}}{p}\right\}$$

is an integral basis.

Proof: (1) → (2)   By (**G**) we know that $Ap = P^p$. In the proof of (**H**) we have seen that P is generated by the elements $t - a$, p. Since $p \in P^p$ then $t - a \notin P^2$. Thus $A(t - a) = PJ$ where P does not divide the ideal J of A. Taking norms, $|N_{K|\mathbb{Q}}(t - a)| = N(A(t - a)) = N(P) \cdot N(J) = p \cdot N(J)$. Since P is the only prime ideal of A whose norm is a power of p, from the decomposition of J into a product of prime ideals, it follows that p does not divide $N(J)$.

But the minimal polynomial of $t - a$ over $\mathbb{Q}$ is $(X + a)^p - a$ (this polynomial is irreducible because $\mathbb{Q}(t - a) = \mathbb{Q}(t)$). So $|N_{K|\mathbb{Q}}(t - a)| = a^p - a = a(a^{p-1} - 1)$. We conclude that $a(a^{p-1} - 1) \not\equiv 0 \pmod{p^2}$ and since p does not divide a, $a^{p-1} \not\equiv 1 \pmod{p^2}$.

(2) → (1)   Now we assume that $A \neq \mathbb{Z}[t]$, hence $j \neq 0$; therefore $j = 1$ and $Ap = P_1^e P_2^{p-e}$, with $P_1 \cap \mathbb{Z}[t] = P_2 \cap \mathbb{Z}[t] = \underline{P}$, the only prime ideal of $\mathbb{Z}[t]$ above p. Since $t - a \in \underline{P}$ then $A(t - a)$ is a multiple of P_1 and of P_2, so we may write $A(t - a) = P_1^{m_1} P_2^{m_2} J$ where J is an ideal of A not containing p and $m_1 \geq 1$, $m_2 \geq 1$. Taking norms we have

$$|a(a^{p-1} - 1)| = N(A(t - a)) = p^{m_1 + m_2} \cdot N(J),$$

with $m_1 + m_2 \geq 2$, hence $a^{p-1} \equiv 1 \pmod{p^2}$ since p does not divide a.

Now we shall prove that when $A \neq \mathbb{Z}[t]$ there exists an integral basis of the type indicated. Let us begin by showing that

$$z = (1 + ta^{p-2} + t^2 a^{p-3} + \cdots + t^{p-1})/p$$

belongs to A. We have

$$\begin{aligned}
(t^{p-1})^p - 1 &= (t^{p-1} - 1)(1 + t^{p-1} + t^{2(p-1)} + \cdots + t^{(p-1)^2}) \\
&= (t^{p-1} - 1)(1 + t^{p-1} + at^{p-2} + a^2 t^{p-3} + \cdots + a^{p-2}t) \\
&= (t^{p-1} - 1)pz.
\end{aligned}$$

From $(t^{p-1})^p = a^{p-1}$, we deduce that $pzt^{p-1} = a^{p-1} - 1 + pz$, and therefore $p^p z^p a^{p-1} = (a^{p-1} - 1 + pz)^p$. Thus z is a root of the polynomial

$$p^p a^{p-1} X^p - (a^{p-1} - 1 + pX)^p$$

which has leading coefficient $p^p(a^{p-1} - 1) \in \mathbb{Z}$. The coefficient of X^i $(0 \leq i < p)$ is equal to $c_i = \binom{p}{p-i} p^i (a^{p-1} - 1)^{p-i}$. But $a^{p-1} \equiv 1 \pmod{p^2}$ because $A \neq \mathbb{Z}[t]$, hence for $i = 1, \ldots, p-1$:

$$\binom{p}{p-i} p^i (a^{p-1} - 1)^{p-i} = p^{1+i+2(p-i-1)} (a^{p-1} - 1) b \qquad \text{with} \quad b \in \mathbb{Z},$$

therefore, $1 + i + 2(p - i - 1) = 2p - 1 - i \geq p$; showing that $p^p(a^{p-1} - 1)$ divides c_i. Similarly, $p^p(a^{p-1} - 1)$ divides $c_0 = (a^{p-1} - 1)^p$ because $(a^{p-1} - 1)^{p-1} = p^{2(p-1)} b$, $b \in \mathbb{Z}$, and $p \leq 2(p-1)$. Hence z is an algebraic integer.

Since $1, t, \ldots, t^{p-2}, t^{p-1}$ are linearly independent over $\mathbb{Z}$, then the same holds obviously for the elements $1, t, \ldots, t^{p-2}, z$. It remains to show that these elements generate A. We have $A/\mathbb{Z}[t] \cong \mathbb{Z}/\mathbb{Z}p$ since $j = 1$, hence $A/\mathbb{Z}[t]$ is a cyclic Abelian additive group, generated by every element different from 0. Since $z \in A$, $z \notin \mathbb{Z}[t]$ then $A = \mathbb{Z}z + \mathbb{Z}[t]$. Finally, t^{p-1} belongs to the Abelian group generated by $1, t, \ldots, t^{p-2}, z$ because

$$t^{p-1} = -1 - a^{p-2}t - a^{p-3}t^2 - \cdots - at^{p-2} + pz. \qquad \blacksquare$$

We conclude with the following observation. The results indicated above do not hold when some square divides a. For example, let $f = X^3 - 4$, $K = \mathbb{Q}(t)$, where $t^3 = 4$. The discriminant of t is $d = -27 \times 16$. Letting $u = 2/t$, from $t^3 - 4 = 0$ we have $2 - 8/t^3 = 0$, so $u^3 - 2 = 0$, therefore $u \in A$. The $\mathbb{Z}$-module A_1 generated by $1, t, u$ is a subring of A, because $t^2 - 4/t = 0$ implies $t^2 = 2u$. Similarly, $u^2 = t$ and $ut = 2$. Since $\{1, t, t^2\}$ is a $\mathbb{Q}$-basis of K and $u = t^2/2$ then $u \notin \mathbb{Z}[t]$. Hence $\mathbb{Z}[t] \neq A$. We have $3u = (3/2)t^2 \notin A$, thus pA is not contained in $\mathbb{Z}[t]$.

Let $d_1 = \operatorname{discr}_{K|\mathbb{Q}}(1, t, u)$. From the expression of $1, t, t^2$ in terms of $1, t, u$ it follows that $d = 4d_1$, hence $d_1 = -27 \times 4$. Since $A_1 \subseteq A$ then $d_1 = m^2 \delta_K$, $m^2 \geq 1$. It follows that $|\delta_K| \neq 3^{3-2j} \times 4^2$.

The prime ideals of A dividing 3 are the kernels of the homomorphisms $\psi : A \to \mathbb{F}_3$ such that $\bar{u}^2 = \bar{t}$, $\bar{t}^2 = 2\bar{u}$, $\bar{u}\bar{t} = \bar{2}$. The only possibility is

$\bar{t} = \bar{1}$, $\bar{u} = \bar{2}$. So there is only one prime ideal P dividing 3 and $A \cdot 3 = P^3$. Thus 3 is totally ramified, yet $\mathbb{Z}[t] \neq A$.

16.6 Relative Binomial Extensions

Example 7:   Let p be a prime number, K an algebraic number field containing a primitive pth root of unity ζ, and A the ring of algebraic integers of K. Let $a \in K$ be an element which is not the pth power of an element of K and consider the polynomial $F = X^p - a \in K[X]$. Let t be a root of F, $L = K(t)$ and B the ring of integers of L. We shall determine the decomposition of the prime ideals of A in $L|K$.

First we note that $L|K$ is a Galois extension. In fact, the roots of F are t, $t\zeta$, $\ldots$, $t\zeta^{p-1}$ and they belong to L.

Next, we show that $[L : K] = p$. Indeed, by hypothesis the minimal polynomial G of t over K cannot be linear (since a is not the pth power of an element of K). Thus $G = \prod_{j=1}^{k}(X - t\zeta^{i_j})$ with $0 = i_1 < i_2 < \cdots < i_k \leq p - 1$ and $1 < k \leq p$. It follows that $t(\sum_{j=1}^{k} \zeta^{i_j}) \in K$. If $k < p$ then $\sum_{j=1}^{k} \zeta^{i_j} \neq 0$ so $t \in K(\zeta) = K$ which contradicts the hypothesis about a. Hence $k = p$; that is, G has degree p, $G = F$, and $[L : K] = p$.

We deduce that the Galois group $G(L|K)$ is the cyclic group with p elements.

Let $\underline{P}$ be any nonzero prime ideal of A, then $B\underline{P} = \prod_{i=1}^{g} P_i^e$ and all the prime ideals P_i have the same inertial degree f over K. From the fundamental relation $p = efg$ we deduce the following possibilities:

(α) $e = p$, $f = g = 1$: $B\underline{P} = P^p$, $\underline{P}$ is totally ramified,
(β) $f = p$, $e = g = 1$: $B\underline{P} = P$, $\underline{P}$ is inert,
(γ) $g = p$, $e = f = 1$: $B\underline{P} = P_1 \cdots P_p$, $\underline{P}$ is decomposed.

Now we shall indicate which case holds for any given prime ideal $\underline{P}$ of A.

J.   Let $Aa = \underline{P}^h J$, where J is an ideal of A not a multiple of $\underline{P}$ and $h > 0$ is an integer not a multiple of p. Then $\underline{P}$ is ramified.

Proof: We may assume without loss of generality that $\underline{P}$ divides Aa, but $\underline{P}^2$ does not divide Aa. Indeed, let $b \in \underline{P}$, $b \notin \underline{P}^2$, let l, l' be integers such that $lh + l'p = 1$. Taking $a' = a^l b^{pl'}$ then $X^p - a'$ generates the same field extension $L|K$. In fact, if $t'^p = a' = a^l b^{pl'} = t^{pl}b^{pl'}$ then $t' = \zeta^i t^l b^{l'} \in L$ (for some pth root of unity ζ^i); conversely, $a'^h = a^{lh}b^{pl'h} = a(b^h a^{-1})^{l'p}$, hence $a = a'^h(ab^{-h})^{l'p}$ and in the same way we see that $t \in K(t')$. From the choice of b we deduce that the exact power of $\underline{P}$ which divides $Aa' = Aa^l \cdot Ab^{pl'}$ is $\underline{P}^{hl+pl'} = \underline{P}$.

Thus $Aa = \underline{P} \cdot J$ where J is not a multiple of $\underline{P}$. Now let P be the ideal of B generated by $B\underline{P}$ and Bt; that is, $P = \gcd(B\underline{P}, Bt)$. Then

$$P^p = \gcd(B\underline{P}^p, Ba) = \gcd(B\underline{P}, B\underline{P} \cdot BJ) = B\underline{P}.$$

It follows from the fundamental relation that P is necessarily a prime ideal of B and $\underline{P}$ is totally ramified in $L|K$.   ∎

In the second case, P is a prime ideal of A such that $Aa = \underline{P}^h J$, where J is an ideal of A not a multiple of $\underline{P}$ and p divides the integer $h \geq 0$. Actually, we may assume without loss of generality that $\underline{P}$ does not divide Aa. Indeed, we choose an element $b \in \underline{P}$, $b \notin \underline{P}^2$; if $h > 0$ and if $a' = a(b^{-h/p})^p$ then $X^p - a'$ generates the same field extension $L|K$ (because a root t' of this equation satisfies $t'^p = a' = t^p (b^{-h/p})^p$, so $t' \in K(t)$ and, conversely, $t \in K(t')$). Moreover, the exact power of $\underline{P}$ dividing Aa' is now equal to $h - (h/p)p = 0$. In particular, $\gcd(Aa, \underline{P}) = A$.

K. *Assume that $\underline{P}$ divides neither Aa nor Ap.*

    (1) *If the congruence $X^p \equiv a \pmod{\underline{P}}$ has a solution in A then $\underline{P}$ decomposes in $L|K$.*

    (2) *If the above congruence has no solution in A then $\underline{P}$ is inert in $L|K$.*

Proof: (1) Let us assume that there exists $x \in A$ such that $x^p - a \in \underline{P}$. We have $x^p - a = (x - t)(x - \zeta t) \cdots (x - \zeta^{p-1} t)$.

Let $P_i = B\underline{P} + B(x - \zeta^{i-1} t) = \gcd(B\underline{P}, B(x - \zeta^{i-1} t))$ for $i = 1, \ldots, p$. These ideals of B are conjugate over K.

We have $P_1 P_2 \cdots P_p = B\underline{P}$ because every element of $P_1 P_2 \cdots P_p$ is a sum of elements of the form

$$\prod_{i=1}^{p}(y_i + z_i(x - \zeta^{i-1} t)) = y + z(x^p - a) \in B\underline{P},$$

where $y_i, y \in B\underline{P}$, $z_i, z \in B$. Hence each P_i is different from B (for these ideals are conjugate, so if some $P_i = B$ then $P_1 = \cdots = P_p = B$ and $P_1 P_2 \cdots P_p = B$, contradiction). It follows that $A \neq P_i \cap A \supseteq \underline{P}$ and therefore $P_i \cap A = \underline{P}$ for $i = 1, 2, \ldots, p$.

Taking into account the fundamental relation, from $B\underline{P} = P_1 P_2 \cdots P_p$, with each $P_i \neq B$, we see that each P_i must already be a prime ideal of B.

It remains to show that these ideals P_i are all distinct. By Galois theory, the decomposition group of P_i in $L|K$ has order 1 or p, so either $P_1 = P_2 = \cdots = P_p$ or these ideals are all distinct. In the first case, we would have $x - t$, $x - t\zeta \in P_1 = \cdots = P_p$, hence $t(1 - \zeta) \in P_1$; since $\gcd(Aa, \underline{P}) = A$ then $\gcd(Bt, P_1) = B$ so there exist elements $y \in P_1$, $z \in B$ such that $1 = y + zt$ and $1 - \zeta = y(1 - \zeta) + zt(1 - \zeta) \in P_1 \cap A = \underline{P}$. Now $p = u(1 - \zeta)^{p-1}$ where u is a unit, so $p \in \underline{P}$, that is, $\underline{P}$ divides Ap which contradicts the hypothesis. We conclude that $\underline{P}$ decomposes in $L|K$.

(2) Let us assume now that $\underline{P}$ is not inert in $L|K$. So either $B\underline{P} = P_1^p$ or $B\underline{P} = P_1 P_2 \cdots P_p$. At any rate, P_1 is a prime ideal of B with inertial degree over A equal to 1, thus $B/P_1 \cong A/\underline{P}$ and $N_{L|K}(P_1) = \underline{P}$. In particular, there exists $x \in A$ such that $x - t \in P_1$. So P_1 divides $B(x - t)$

and therefore $\underline{P} = N_{L|K}(P_1)$ divides $N_{L|K}(B(x-t)) = A(x^p - a)$ because $N_{L|K}(x-t) = \prod_{i=1}^{p}(x - \zeta^{i-1}t) = x^p - a$. This means that the congruence $X^p \equiv a \pmod{\underline{P}}$ has the solution $x \in A$. ∎

It remains now to study the case where $\underline{P}$ divides Ap but does not divide Aa. Since $p = u(1-\zeta)^{p-1}$ then $\underline{P}$ divides $A(1-\zeta)$. We write $A(1-\zeta) = \underline{P}^m \cdot \underline{J}$ where $\underline{P}$ does not divide the ideal $\underline{J}$ and $m \geq 1$.

L.   *With these notations we have:*

(1) *If the congruence $X^p \equiv a \pmod{\underline{P}^{mp+1}}$ has a solution in A then $\underline{P}$ decomposes in L.*

(2) *If the above congruence has no solution in A, but the congruence $X^p \equiv a \pmod{\underline{P}^{mp}}$ has a solution in A, then $\underline{P}$ is inert in L.*

(3) *If the congruence $X^p \equiv a \pmod{\underline{P}^{mp}}$ has no solution in A, then $\underline{P}$ is ramified in L.*

Proof: (1)   We first show that the congruence $X^p \equiv a \pmod{\underline{P}^{mp+1}}$ has a solution if and only if $B\underline{P} = P_1 P_2 \cdots P_p$, where $P_1, \ldots, P_p$ are distinct prime ideals of B. This will establish (1).

In fact, if $\underline{P}$ decomposes in $L|K$ then the inertial degree of each P_i is equal to 1. Hence $P_i^s \cap A \neq \underline{P}^{s-1}$ for $s \geq 1$; otherwise, P_i^s divides $(B\underline{P})^{s-1} = B\underline{P}^{s-1}$ and the exponent of P_i in the decomposition of $B\underline{P}$ into prime ideals could not be equal to 1. Now we show that $P_i^s \cap A = \underline{P}^s$ for every $s \geq 1$; this is true for $s = 1$ and if it is true for $s - 1$ then $\underline{P}^s \subseteq P_i^s \cap A \subseteq P_i^{s-1} \cap A = \underline{P}^{s-1}$ with $P_i^s \cap A \neq \underline{P}^{s-1}$, hence also $P_i^s \cap A = \underline{P}^s$. We deduce that $A/\underline{P}^s$ is a subring of B/P_i^s. But these rings have the same number of elements, namely $N(\underline{P})^s = N(P_i)^s$, so $B/P_i^s = A/\underline{P}^s$.

Thus there exists $x \in A$ such that $t \equiv x \pmod{P_1^{mp+1}}$; that is, P_1^{mp+1} divides $B(x-t)$. Taking norms and noting that $N_{L|K}(x-t) = x^p - a$, we deduce that $\underline{P}^{mp+1}$ divides $x^p - a$, so the congruence $X^p \equiv a \pmod{\underline{P}^{mp+1}}$ has a solution in A.

Conversely, let $x \in A$ be such that $x^p \equiv a \pmod{\underline{P}^{mp+1}}$. Let $u \in \underline{P}^{-m}$, $u \notin \underline{P}^{-m+1}$, hence $Au \cdot \underline{P}^m = I$ is an integral ideal of A. Consider $v = u(x-t)$. This element is a root of

$$(X - ux)^p + u^p a = X^p - \binom{p}{1}uxX^{p-1} + \binom{p}{2}u^2x^2X^{p-2} - \cdots$$

$$+ \binom{p}{p-1}u^{p-1}x^{p-1}X - u^p(x^p - a).$$

The coefficients of this polynomial belong to A as we show now. We know that $\underline{P}^{m(p-1)}$ divides $A(1-\zeta)^{p-1} = Ap$. For every j such that $1 \leq j \leq p-1$ we have $m(p-1) - mj \geq 0$, hence $\binom{p}{j}u^jx^j \in A$. Moreover, $x^p - a \in \underline{P}^{mp+1}$, thus $u^p(x^p - a) \in \underline{P}$.

Therefore, $v \in B$ and the same holds for the conjugates $u(x - \zeta^{i-1}t)$ for $i = 1, \ldots, p - 1$. We consider the integral ideals

$$P_i = B\underline{P} + Bu(x - \zeta^{i-1}t) = \gcd(B\underline{P}, Bu(x - \zeta^{i-1}t)) \text{ for } i = 1, \ldots, p-1,$$

conjugate over K. We have $P_1 P_2 \cdots P_p = B\underline{P}$, because every element of $P_1 P_2 \cdots P_p$ is a sum of elements of the form

$$\prod_{i=1}^{p}(y_i + z_i u(x - \zeta^{i-1}t)) = y + zu^p(x^p - a) \in B\underline{P},$$

where $y_i, y \in B\underline{P}$, $z_i, z \in B$. Hence each ideal P_i is different from B, since these ideals are conjugate and $P_1 P_2 \cdots P_p = B\underline{P}$. Therefore $P_i \cap A = \underline{P}$ for $i = 1, \ldots, p$.

Taking into account the fundamental relation, from $B\underline{P} = P_1 P_2 \cdots P_p$, with each $P_i \neq B$, we see that each P_i must already be a prime ideal of B.

It remains to show that all these conjugate ideals P_i are distinct. Considering the decomposition group of P_i in $L|K$, we see that either all ideals P_i are distinct or $P_1 = P_2 = \cdots = P_p$, because the Galois group of $L|K$ has order p. In the latter case, $u(x - t)$ and $u(x - \zeta t)$ belong to P_1 so $ut(1 - \zeta) \in P_1$ and, taking the pth power, $u^p a(1 - \zeta)^p \in P_1 \cap A = \underline{P}$. Now, the exact power of $\underline{P}$ dividing $A(u^p a(1 - \zeta)^p)$ is $\underline{P}^{-mp+mp}$, so $\underline{P}$ would not divide this ideal, a contradiction.

(2)   Now we assume that the congruence $X^p \equiv a \pmod{\underline{P}^{mp}}$ has a solution $x \in A$. We choose $u \in \underline{P}^{-m}$, $u \notin \underline{P}^{-m+1}$, and $v = u(x - t)$. As before, the minimal polynomial of v oer K is $g(X - ux)^p - u^p a$, hence $v \in B$ because $x^p a \in \underline{P}^{mp}$. The different of the element v in $L|K$ is $g'(v) = p(ut)^{p-1}$, o $\gcd(P, Bg'(v)) = B$ for every prime ideal P dividing $B\underline{P}$. Indeed, if P contains $g'(v)$ then $P \cap A = \underline{P}$ contains $g'(v)^p = p^p u^{p(p-1)} a^{p-1}$. But this is not true, because the exact power of $\underline{P}$ dividing this element is $\underline{P}^{p(p-1)m - p(p-1)m}$.

We conclude from Chapter 13, (**T**), that $\Delta_{L|K}$ does not divide any of the prime ideals P such that $P \cap A = \underline{P}$. This shows that $\underline{P}$ is unramified in $L|K$.

If we also assume that the congruence $X^p \equiv a \pmod{\underline{P}^{mp+1}}$ has no solution in A, then from (1) we deduce that $\underline{P}$ is not decomposed. Therefore, $\underline{P}$ is inert in $L|K$.

(3)   We assume that the congruence $X^p \equiv a \pmod{\underline{P}^{mp}}$ has no solution in A.

However, the congruence $X^p \equiv a \pmod{\underline{P}}$ has a solution in A. Indeed, $A/\underline{P}$ is a finite field with p^r elements and the multiplicative group of $A/\underline{P}$ is cyclic. Let ω be a generator, so there exists an integer α such that ω^α is the residue class of a modulo $\underline{P}$. Since p, $p^r - 1$ are relatively prime, let c, $c' \in \mathbb{Z}$ be such that $cp + c'(p^r - 1) = 1$. Hence $\alpha \equiv c\alpha p \pmod{p^r - 1}$. Taking $\beta = c\alpha$ we have $\omega^{\beta p} = \omega^\alpha$ and if $x \in A$ has residue class modulo $\underline{P}$ equal to ω^β then $x^p \equiv a \pmod{\underline{P}}$.

Let $l \geq 1$ be the largest integer such that the congruence $X^p \equiv a$ (mod $\underline{P}^l$) has a solution in A. We show that l is not a multiple of p. For suppose that $l = ph$, that $x \in A$ is a solution of $X^p \equiv a$ (mod $\underline{P}^{ph}$), and that $X^p \equiv a$ (mod $\underline{P}^{ph+1}$) has no solution in A. Let $y \in \underline{P}^h$, $y \notin \underline{P}^{h+1}$, hence $y^p \in \underline{P}^{ph}$, $y^p \notin \underline{P}^{ph+1}$. Similarly, by hypothesis, $a - x^p \in \underline{P}^{ph}$, $a - x^p \notin \underline{P}^{ph+1}$. But $\underline{P}^{ph}/\underline{P}^{ph+1}$ is a vector space of dimension 1 over $A/\underline{P}$ (see Chapter 8, (**A**)), hence there exists $z \in A$ such that $zy^p \equiv a - x^p$ (mod $\underline{P}^{ph+1}$). Since $A/\underline{P}$ is a finite field of characteristic p, the raising to the pth power is an automorphism, so there exists $w \in A$ such that $z \equiv w^p$ (mod $\underline{P}$). Therefore, $x^p + y^p w^p \equiv a$ (mod $\underline{P}^{ph+1}$). But then $x + yw \in A$ would satisfy the congruence $(x+yw)^p \equiv a$ (mod $\underline{P}^{ph+1}$), which is a contradiction. Indeed, $(x+yw)^p \equiv x^p + y^p w^p$ (mod $\underline{P}^{ph+1}$) since the exact power of $\underline{P}$ dividing each of the terms $\binom{p}{i} x^{p-i} y^i w^i$ is at least $m(p - 1) + ih \geq ph + 1$, because $h \leq m - 1$, $1 \leq i \leq p - 1$.

This shows that l is not a multiple of p. We write $l = ph + k$, with $1 \leq k < p$ and $h \leq m - 1$.

Let $u \in \underline{P}^h$, $u \notin \underline{P}^{h+1}$, and $x^p \equiv a$ (mod $\underline{P}^l$), and consider $v = u(x - t)$, which is a root of $(X - ux)^p + u^p a$. As before, we deduce that $v \in B$ and $\underline{P}^k$ is the exact power of $\underline{P}$ dividing $u^p(x^p - a) = N_{L|K}(v)$. Therefore $v \notin B\underline{P}$, otherwise, $B\underline{P}$ divides Bv, hence $\underline{P}^p = N_{L|K}(B\underline{P})$ divides $A \cdot N_{L|K}(v)$, which is false since $k < p$.

Let $P = B\underline{P} + Bv = \gcd(B\underline{P}, Bv)$. So $B\underline{P} \neq P$. Also $P \neq B$. This may be seen by considering the ideals $P_1 = P$, P_2, ..., P_p which are the conjugates of P in $L|K$. Then $P_1 P_2 \cdots P_p = B\underline{P}$, by the argument already used: If $y_i \in B\underline{P}$, $z_i \in B$, then

$$\prod_{i=1}^{p} (y_i + z_i u(x - \zeta^{i-1} t)) = y + z u^p (x^p - a) \in B\underline{P}$$

with $y \in B\underline{P}$, $z \in B$. Since the ideals P_i are all equal or all distinct, $P \neq B$.

We conclude that $B\underline{P}$ is not a prime ideal of B, hence $\underline{P}$ is not inert in $L|K$. By (1), $\underline{P}$ is not decomposed, hence $\underline{P}$ is ramified in $L|K$. ∎

From the study of the decomposition of the prime ideals of A in $L|K$ we may infer the following result about the relative discriminant:

M.   *The relative discriminant $\delta_{L|K}$ is the unit ideal of A if and only if the following conditions hold:*

(1) *Aa is the pth power of an ideal of A.*

(2) *In the case $\gcd(Aa, Ap) = A$, there exists $x \in A$ which satisfies the congruence $X^p \equiv a$ (mod $A(1 - \zeta)^p$).*

Proof: We assume that $\delta_{L|K} = A$, so by Chapter 13, Theorem 1, no prime ideal of A is ramified in $L|K$.

Let $\underline{P}$ be a prime ideal of A dividing Aa, and let $Aa = \underline{P}^m \cdot \underline{J}$, where $\underline{J}$ is an ideal of A not a multiple of $\underline{P}$. By ($\mathbf{J}$), m is a multiple of p. This being true for every prime ideal $\underline{P}$ dividing Aa, it follows that Aa is the pth power of an ideal of A.

Now we assume that $\gcd(Aa, Ap) = A$, and let $A(1 - \zeta) = \prod_{i=1}^{r} \underline{P}_i^{m_i}$ be the decomposition of $A(1 - \zeta)$ into prime ideals of A. Thus $\underline{P}_i$ divides Ap, hence $\underline{P}_i$ does not divide Aa. Since $\underline{P}_i$ is not ramified, by ($\mathbf{L}$) there exists $x_i \in A$ such that $x_i^p \equiv a \pmod{\underline{P}_i^{m_i p}}$. By the Chinese remainder theorem, there exists $x \in A$ such that $x \equiv x_i \pmod{\underline{P}_i^{m_i}}$ for $i = 1, \ldots, r$, so $x^p \equiv a \pmod{A(1 - \zeta)^p}$.

Conversely, we assume now that conditions (1) and (2) hold and we shall show that every prime ideal $\underline{P}$ of A is unramified in $L|K$. If $Aa = \underline{P}^m \cdot \underline{J}$, where $\underline{J}$ is an ideal of A, not a multiple of $\underline{P}$, and $m > 0$, then by (1) m is a multiple of p. Replacing a by another element a', we may assume without loss of generality that $\underline{P}$ does not divide Aa.

If $\underline{P}$ does not divide Ap then by ($\mathbf{K}$), $\underline{P}$ is not ramified in $L|K$. If $\underline{P}$ divides Ap then necessarily Ap does not divide Aa, because $\underline{P}$ does not divide Aa. Since $Ap = A(1 - \zeta)^{p-1}$ then $\underline{P}$ divides $A(1 - \zeta)$. Let $A(1 - \zeta) = \underline{P}^m \cdot \underline{J}$, with $m \geq 1$, where $\underline{J}$ is an ideal of A, not a multiple of $\underline{P}$. By (2) there exists $x \in A$ such that $x^p - a \in A(1 - \zeta)^p \subseteq \underline{P}^{mp}$. By ($\mathbf{L}$) it follows that $\underline{P}$ is not ramified. By Chapter 13, Theorem 1, $\delta_{L|K}$ is the unit ideal. ∎

In Chapter 9, ($\mathbf{D}$), we showed that if K is any algebraic number field then its absolute discriminant $\delta_{K|\mathbb{Q}}$ is different from the unit ideal. In the preceding result we saw that this fact need no more be true for the relative discriminant.

In class field theory, it is important to consider field extensions with relative discriminant equal to 1, as we saw in Chapter 15, Section 2.

16.7 The Class Number of Quadratic Extensions

In this section we indicate a method to compute the class number of quadratic extensions; we follow the exposition of Hasse [7] or [8, Chapter III, §29.3]. The procedure is appropriate for small values of the discriminant. There exist more efficient methods using the theory of characters and transcendental arguments; the interested reader should consult the literature (see [9], [3, Chapter V, §4]).

Let $K = \mathbb{Q}(\sqrt{d})$ where d is a square-free integer, let $\delta = \delta_K$ be the discriminant, and A the ring of integers of K. We recall that if $d \equiv 2, 3 \pmod 4$ then $\delta = 4d$, and if $d \equiv 1 \pmod 4$ then $\delta = d$.

Also, if $d \equiv 2, 3 \pmod 4$ then $\{1, \sqrt{d}\}$ is an integral basis of K. If $d \equiv 1 \pmod 4$ let $\omega = (1 + \sqrt{d})/2$, then $\{1, \omega\}$ is an integral basis. In this latter case, the elements of A are of the form $(a + b\sqrt{d})/2$, where $a, b \in \mathbb{Z}$, $a \equiv b$

(mod 2). In all cases, the elements of A are of the form $(u + v\sqrt{d})/2$ with $u, v \in \mathbb{Z}$, $u^2 \equiv v^2 d$ (mod 4).

Let $\mathbb{D}$ be the set of prime numbers p which are decomposed in K, that is, $Ap = PP'$, with P, P' distinct prime ideals of A.

Let $\mathcal{J}$ be the set of prime numbers q which are inert in K; that is, $Aq = Q$, where Q is a prime ideal of A.

Let $\mathcal{R}$ be the set of prime numbers r which are ramified in K; that is, $Ar = R^2$, where R is a prime ideal of A.

A nonzero integral ideal I of A is said to be *normalized* when its norm satisfies the following conditions:

$$(1) \qquad N(I) = \prod_{r \in \mathcal{R}} r^{e_r} \prod_{p \in \mathbb{D}} p^{e_p}$$

with $e_r = 0$ or 1, $e_p \geq 0$.

(2) If $\delta > 0$ then $N(I) \leq \frac{1}{2}\sqrt{|\delta|}$; if $\delta < 0$ then $N(I) \leq (2/\pi)\sqrt{|\delta|}$.

A nonzero integral ideal I of A is *primitive* when there exists no integer $m \in \mathbb{Z}$, $m \neq 1$, such that Am divides I.

Let $\mathcal{N}$ be the set of normalized primitive ideals of A.

N. *Every nonzero fractional ideal of K is equivalent to some ideal belonging to $\mathcal{N}$.*

Proof: From Chapter 9, (**E**), it follows that every nonzero fractional ideal of K is equivalent to an ideal J such that $N(J) \leq \left[2^{r_1+r_2}/\mathrm{vol}(D)\right]\sqrt{|\delta|}$, where D is a symmetric convex body in $\mathbb{R}^2$ such that $D \subseteq D_1$ and D_1 is defined as follows. If $\delta > 0$ then $r_1 = 2$, $r_2 = 0$, and $D_1 = \{(\xi_1, \xi_2) \mid |\xi_1|\,|\xi_2| \leq 1\}$. If $\delta < 0$ then $r_1 = 0$, $r_2 = 1$, and $D_1 = \{(\xi_1, \xi_2) \mid \xi_1^2 + \xi_2^2 \leq 1\}$.

Thus if $\delta > 0$ we may choose D to be the square with vertices $(2,0)$, $(0,2)$, $(-2,0)$, $(0,-2)$; therefore $\mathrm{vol}(D) = 8$, hence $N(J) \leq \frac{1}{2}\sqrt{|\delta|}$. If $\delta < 0$ we take $D = D_1$ hence $\mathrm{vol}(D) = \pi$ and $N(J) \leq (2/\pi)\sqrt{|\delta|}$.

Noting that $Ar = R^2$ for every $r \in \mathcal{R}$, $Aq = Q$ for every $q \in \mathcal{J}$, $Ap = PP'$ for every $p \in \mathbb{D}$ are principal ideals, then the integral ideal J is equivalent to an integral ideal I such that

$$I = \prod_{r \in \mathcal{R}} R^{e_r} \cdot \prod_{p \in \mathbb{D}} P^{e_p} P'^{e'_p}$$

with $e_r = 0$ or 1, e_p, $e'_p \geq 0$, and $e_p = 0$ or $e'_p = 0$. Thus I is a normalized ideal. But I is also primitive, as follows from the limitation on the exponents e_r $(r \in \mathcal{R})$, e_p $(p \in \mathbb{D})$. ∎

Let $N(\mathcal{N})$ denote the set of integers $N(I)$, where $I \in \mathcal{N}$. Thus if $m \in N(\mathcal{N})$ then

$$m = \prod_{r \in \mathcal{R}} R^{e_r} \prod_{p \in \mathbb{D}} p^{e_p}$$

with $e_r = 0$ or 1, $e_p \geq 0$, and $m \leq \frac{1}{2}\sqrt{|\delta|}$, when $\delta > 0$ or $m \leq (2/\pi)\sqrt{|\delta|}$ when $\delta < 0$.

Given the field K it is possible to determine all integers $m \in N(\mathcal{N})$ by the computation of a finite number of Legendre symbols.

Once the set of integers $N(\mathcal{N})$ is known, the question is to find all possible ideals in $\mathcal{N}$ and to decide when $I, I' \in \mathcal{N}$ are equivalent.

In $m \in N(\mathcal{N})$, let $k_m = \#\{p \in \mathbb{D} \mid p|m\}$; then there exists 2^{k_m} ideals $I \in \mathcal{N}$ such that $N(I) = m$. Indeed, if p^{e_p} is the exact power of $p \in \mathbb{D}$ dividing m, if $Ap = PP'$, we may choose an ideal $I \in \mathcal{N}$ such that P^{e_p} divides I or such that P'^{e_p} divides I. This gives rise to 2^{k_m} possible ideals $I \in \mathcal{N}$ such that $N(I) = m$.

O.   *Let $m \in N(\mathcal{N})$ and let $x = (u + v\sqrt{d})/2$, with $u, v \in \mathbb{Z}$, $u \equiv v$ (mod 2), and u, v even when $d \equiv 2, 3$ (mod 4). The following conditions are equivalent:*

(1) $N(Ax) = m$.

(2) $(u^2 - v^2d)/4 = \pm m$ and $\gcd(u/2, v/2) = 1$ when $d \equiv 2, 3$ (mod 4), or $\gcd((u - v)/2, v) = 1$ when $d \equiv 1$ (mod 4).

Proof: (1) $\rightarrow$ (2) $m = N(Ax) = |N_{K|\mathbb{Q}}(x)| = |(u^2 - v^2d)/4|$. Moreover, since $m \in N(\mathcal{N})$, then $\gcd(u/2, v/2) = 1$ when $d \equiv 2, 3$ (mod 4), or $\gcd((u - v)/2, v) = 1$ when $d \equiv 1$ (mod 4).

(2) $\rightarrow$ (1)   Let $I = Ax$ so $N(I) = m$ and I is a primitive ideal, as follows from the hypothesis on u, v.   ∎

We need therefore to find for which integers $m \in N(\mathcal{N})$, m or $-m$ admits a primitive representation as indicated in (2) of (**O**).

Of course, if $d < 0$ then necessarily $-m$ has no such representation.

If $d \equiv 2, 3$ (mod 4) then we have to consider representations of the type $\pm m = a^2 - b^2d$, with $\gcd(a, b) = 1$.

As a corollary, we deduce:

P.   *Assume that the class number of K is $h = 1$ and that $m \in N(\mathcal{N})$.*

If $d < 0$ then m admits a primitive representation.

If $d > 0$ and the fundamental unit ε of K has norm $N(\varepsilon) = 1$ then m or $-m$ admits a primitive representation.

If $d > 0$ and $N(\varepsilon) = -1$ then m and $-m$ both admit a primitive representation.

Proof: In view of (**O**) and the hypothesis that $h = 1$ for every $m \in N(\mathcal{N})$, m or $-m$ admits a primitive representation. If $d < 0$, $-m$ has no primitive representation, as already said.

If $d > 0$ and $N(\varepsilon) = -1$, if m (or $-m$) has a primitive representation, so also has $-m$ (respectively, m).   ∎

Summarizing, to determine the class number of K, we proceed as follows:

Step 1:   To determine the set $\mathcal{N}$ of normalized primitive ideals and the set $N(\mathcal{N})$ of integers which are their norms.

Step 2:   For every pair of ideals $I, J \in \mathcal{N}$, to decide whether I, J are in the same class; it is equivalent to decide whether $I \cdot J^{-1}$ is a principal ideal.

Now we shall consider numerical examples.

Case 1: $d > 0$.

Let n be the largest integer such that $n \leq \frac{1}{2}\sqrt{|\delta|}$.

(a) <u>$n = 1$.</u> The fields with discriminant $\delta = 5, 8, 12, 13$ are the only ones such that $n = 1$. In this case $\mathcal{N}$ contains only one element and therefore $h = 1$.

(b) <u>$n = 2$.</u> The fields with discriminant $\delta = 17, 21, 24, 28, 29, 33$ are the only ones such that $n = 2$. Let us consider some of these values.

If $\delta = 17$ then $N(\mathcal{N}) = \{1, 2\}$ since $17 \equiv 1 \pmod 8$ so $A \cdot 2 = P_2 \cdot P_2'$. But $-2 = (3^2 - 17 \cdot 1^2)/4$ is a primitive representation of -2, so $P_2 = Ax$, $x = (3 + \sqrt{17})/2$, $P_2' = Ax'$, $x' = (3 - \sqrt{17})/2$. Thus $h = 1$.

If $\delta = 24$ then $d = 6$ and $N(\mathcal{N}) = \{1, 2\}$, since 2 is ramified, so $A \cdot 2 = R_2^2$. But $-2 = 2^2 - 6 \cdot 1^2$ is a primitive representation of -2, hence $R_2 = Ax$, $x = 2 + \sqrt{6}$. Thus $h = 1$.

(c) <u>$n = 3$.</u> The fields with discriminant $\delta = 37, 40, 41, 44, 53, 56, 57, 60, 61$ are the only ones such that $n = 3$. We consider one of these values.

If $\delta = 40$ then $d = 10$ and $N(\mathcal{N}) = \{1, 2, 3\}$ since 2 is ramified and $\left(\frac{10}{3}\right) = \left(\frac{1}{3}\right) = 1$; hence $A \cdot 2 = R_2^2$ and $A \cdot 3 = P_3 \cdot P_3'$.

But ± 2 have no primitive representation, otherwise $\pm 2 = a^2 - 10b^2$ (with $a, b \in \mathbb{Z}$) hence $a^2 = 10b^2 \pm 2$; this is impossible since the last digit of a square is not equal to 2 or to 8. For the same reason ± 3 has no primitive representation.

Finally, $-2 \cdot 3 = 2^2 - 10 \cdot 1^2$, hence $R_2 \cdot P_3$ is principal and so is $R_2 \cdot P_3'$. Therefore $h = 2$.

It should be noted that when the discriminant is positive, the question of primitive representation of a prime number may not be as immediate to answer as in the above example. In general, it requires the theory of the Hilbert symbol.

From the tables of class numbers of real quadratic fields, one sees that there exist 142 square-free integers d, $2 \leq d < 500$, such that the class number of $\mathbb{Q}(\sqrt{d})$ is equal to 1. It is not known whether there exists an infinite number of real quadratic fields with class number 1.

Case 2: $d < 0$.

Let n be the largest integer such that $n \leq (2/\pi)\sqrt{|\delta|}$.

(a) <u>$n = 1$.</u> The fields with discriminant $\delta = -3, -4, -7, -8$ are the only ones such that $n = 1$. In this case, $\mathcal{N}$ contains only one element and therefore $h = 1$.

(b) <u>$n = 2$.</u> The fields with discriminant $\delta = -11, -15, -19, -20$ are the only ones such that $n = 2$.

We consider some of these values.

If $\delta = -11$ then $N(\mathcal{N}) = \{1\}$ since $-11 \equiv 5 \pmod 8$ so $A \cdot 2$ is a prime ideal. Hence $h = 1$.

If $\delta = -15$ then $N(\mathcal{N}) = \{1, 2\}$ since $-15 \equiv 1 \pmod 8$, so $A \cdot 2 = P_2 \cdot P_2'$. Now $2 \neq (u^2 + 15v^2)/4$ for $u, v \in \mathbb{Z}$, $\gcd(u, v) = 1$. Hence P_2, P_2' are not principal ideals; but $P_2 \cdot P_2'$ is principal, hence $h = 2$.

(c) $\underline{n = 3}$. The fields with discriminant $\delta = -23, -24, -31, -35, -39$ are the only ones such that $n = 3$.

We consider one of these values.

If $\delta = -31$ then $N(\mathcal{N}) = \{1, 2\}$ since $-31 \equiv 1 \pmod 8$ so $A \cdot 2 = P_2 \cdot P_2'$; on the other hand, $(-31/3) = -1$ hence 3 is inert in $\mathbb{Q}(\sqrt{-31})$ and $3 \notin N(\mathcal{N})$. Since $2 \neq (u^2 + 31v^2)/4$ for $u, v \in \mathbb{Z}$, $\gcd(u, v) = 1$ then P_2, P_2' are not principal ideals. Similarly $2^2 \neq (u^2 + 31v^2)/4$ for $u, v \in \mathbb{Z}$, but $2^3 = (1^2 + 31 \cdot 1^2)/4$, hence P_2^2 is not a principal ideal, but P_2^3 is a principal ideal. From $P_2 \cdot P_2' = A \cdot 2$ it follows that P_2', $P_2'^2$ are not principal ideals, but $P_2'^3$ is a principal ideal; actually, $P_2^2 P_2'^{-1} = P_2^3 \cdot (A \cdot 2)^{-1}$ is a principal ideal, so P_2' is equivalent to P_2^2. We conclude that $h = 3$, and the class of the ideal P_2 is a generator of the class group.

The question of determination of the imaginary quadratic fields with class number 1 may be tackled as follows:

Q.   *If $\mathbb{Q}(\sqrt{d})$, $d < 0$, has class number 1 then $\mathcal{N}$ contains only the unit ideal.*

Proof: This is true when $|\delta| < 7$, so let $|\delta| \geq 7$.

If $I \in \mathcal{N}$, $I \neq A$, there exists a prime ideal P dividing I, so $N(P) = p \leq (2/\pi)\sqrt{|\delta|}$. If P is a principal ideal generated by $(u + v\sqrt{d})/2$, with u, v integers, $u \equiv v \pmod 2$ and u, v even when $d \not\equiv 1 \pmod 4$, then

$$p = N(P) = \left| N_{K|\mathbb{Q}}\left(\frac{u + v\sqrt{d}}{2}\right) \right| = \left| \frac{u^2 - v^2 d}{4} \right|.$$

So $v \neq 0$ and therefore $(2/\pi)\sqrt{|\delta|} \geq p \geq |\delta|/2$, hence $|\delta| < 7$, a contradiction. Then P is not a principal ideal and $h > 1$.   ■

Gauss developed a theory of genera of binary quadratic forms. It implies:

If $h = 1$ then $\delta = -4, -8$, or $-p$, where p is a prime, $p \equiv 3 \pmod 4$.

We shall discuss this topic more amply in Chapter 28, Section 1.

If $\delta \neq -3, -4, -7, -8$, in order for $\mathcal{N}$ to contain only the unit ideal it is necessary and sufficient that $-p \equiv 5 \pmod 8$ and if q is an odd prime number, $q \leq n$ (largest integer such that $n \leq (2/\pi)\sqrt{|\delta|}$) then $(-p/q) = -1$ (this means that q is inert).

$\underline{\text{Let } n = 2}$. Then $-20 \leq \delta \leq -11$. From $\delta = -p$, $p \equiv 3 \pmod 4$, then $\delta \in \{-11, -19\}$. But 2 should be inert, thus $\delta \equiv 5 \pmod 8$, so $\delta = -11, -19$ satisfy the required conditions.

Let $n = 3$. Then $-39 \leq \delta \leq -23$. From $\delta = -p$, $p \equiv 3$ (mod 4), then $\delta \in \{-23, -31\}$. However, -23, -31 do not satisfy the condition $\delta \equiv 5$ (mod 8).

Let $n = 4$. Then $-59 \leq \delta \leq -40$. Again $\delta \in \{-41, -43, -47, -53\}$. From $\delta \equiv 5$ (mod 8) it follows that $\delta = -43$. Also $(-43/3) = -1$, hence $\delta = -43$ satisfies the required conditions.

Let $n = 5$. Then $-88 \leq \delta \leq -62$ so $\delta \in \{-67, -71, -73, -79, -83\}$. But $\delta \equiv 5$ (mod 8), so $\delta \in \{-67, -83\}$. From $(-67/3) = -1$, $(-67/5) = -1$, and $(-83/3) = 1$ we deduce that $\delta = -67$ is the only possibility.

If $n = 6$ then $-120 \leq \delta \leq -89$ so

$$\delta \in \{-89, -97, -101, -103, -107, -109, -113, -117\}.$$

From $\delta \equiv 5$ (mod 8) it follows that $\delta = -107$. But $(-107/3) = 1$ so no value of δ is possible.

If $n = 7$ then $-157 \leq \delta \leq -121$, so

$$\delta \in \{-131, -137, -139, -149, -151, -157\}.$$

From $\delta \equiv 5$ (mod 8) it follows that $\delta \in \{-131, -139\}$. But $(-131/3) = 1$ and $(-139/5) = 1$ so no value of δ is possible.

If $n = 8$ then $-199 \leq \delta \leq -158$ so

$$\delta \in \{-163, -167, -173, -181, -191, -193, -197, -199\}.$$

From $\delta \equiv 5$ (mod 8) it follows that $\delta = -163$ is the only possibility.

Altogether, we have shown that if $-200 \leq \delta < 0$ and $\mathbb{Q}(\sqrt{d})$ has class number 1 then $\delta = -3, -4, -7, -8, -11, -19, -43, -163$. As we have mentioned in Chapter 5, Section 4, these are the only imaginary quadratic fields with class number equal to 1.

16.8 Prime Producing Polynomials

Euler discovered that if $q = 2, 3, 5, 11, 17$, or 41, then $f_q(X) = X^2 + X + q$ has prime values for $k = 0, 1, \ldots, q - 2$.

Thus $f_{41}(X)$ assumes 40 successive initial prime values: 41, 43, 47, 53, 61, 71, 83, 97, 113, 131, 151, 173, 197, 223, 251, 281, 313, 347, 383, 421, 461, 503, 547, 593, 641, 691, 743, 797, 853, 911, 971, 1033, 1097, 1163, 1231, 1301, 1373, 1447, 1523, 1601.

Note that

$$f_q(q-1) = (q-1)^2 + (q-1) + q = (q-1)[q - 1 + 1] + q$$
$$= [(q-1) + 1]q = q^2.$$

It is an interesting fact that this property is intimately connected with the class number of imaginary quadratic fields, as discovered by Rabinowitsch in 1912 (see Ribenboim [22, Chapter III, Section IV B]).

First we need the following fact. We simply denote by (x, y) the ideal generated by the elements x, y of K.

R.   *Let p be an odd prime, $d \equiv 1 \pmod 4$.*

    (1) *If a is any integer, then $(p, a + \sqrt{d}) = (p, \ell(a-1)+\omega)$ where $2\ell \equiv 1 \pmod p$.*

    (2) *If p is not inert in $\mathbb{Q}(\sqrt{d})$, there exists an integer b, $0 \le b \le p-1$, such that p divides $N_{K|\mathbb{Q}}(b + \omega)$.*

Proof: (1)   Since

$$a + \sqrt{d} = a - 1 + 2\omega \equiv 2\ell(a - 1) + 2\omega \pmod p$$

and since p is odd, there exists s such that $2s \equiv 1 \pmod p$. It follows that

$$(p, a + \sqrt{d}) = (p, (a - 1) + 2\omega) = (p, 2\ell(a - 1) + 2\omega) = (p, \ell(a - 1) + \omega).$$

(2)   If p is not inert in $\mathbb{Q}(\sqrt{d})|\mathbb{Q}$ then $Ap = P_1 P_2$, where P_1, P_2 are (not necessarily distinct) prime ideals of A. More precisely, from Chapter 11, Theorem 2, if

$$X^2 - d \equiv (X - a)(X - a') \pmod p,$$

then

$$P_1 = (p, a + \sqrt{d}) = (p, a - 1 + 2\omega).$$

Since p is odd, then 2 is invertible modulo p and there exists b such that $0 \le b \le p-1$, $2b \equiv a-1 \pmod p$. Then $P_1 = (p, b+\omega)$. Then $p = N(P_1)$ divides $N(A(b + \omega)) = N_{K|\mathbb{Q}}(b + \omega)$. ∎

The main result is the following:

S.   *Let q be a prime number, and let $f_q(X) = X^2 + X + q$. Then the following conditions are equivalent:*

    (1) $q = 2, 3, 5, 11, 17, 41$.

    (2) *$f_q(k)$ is a prime number for every $k = 0, 1, \ldots, q - 2$.*

    (3) *The class number of $\mathbb{Q}(\sqrt{1 - 4q})$ is equal to 1.*

Proof: (1) $\to$ (2)   This is a simple numerical verification.

(2) $\to$ (3)   Let $1 - 4q = u^2 d$, where d is square-free, so $d \equiv 1 \pmod 4$, $\mathbb{Q}(\sqrt{1 - 4q}) = \mathbb{Q}(\sqrt{d})$, and its discriminant is $\delta = d$. If $q = 2$ or 3, then $d = -7$ or -11; as was computed, $h_{-7} = 1$, $h_{-11} = 1$. Thus, we may assume that $q \ge 5$. As follows from the discussion in Section 7, in particular (**Q**), it suffices to show that if p is any prime, $p \le (2/\pi)\sqrt{|\delta|}$, then p is inert in $\mathbb{Q}(\sqrt{1 - 4q})|\mathbb{Q}$.

If $p = 2$, since $q = 2t - 1$ and $u^2 \equiv 1 \pmod 8$, then $du^2 = 1 - 4q \equiv 5 \pmod 8$ so $d \equiv 5 \pmod 8$, hence $(2/d) = -1$ and 2 is inert.

Let p be an odd prime. If it is not inert, then by $(\mathbf{R})$, there exists an integer b such that $0 \leq b \leq p - 1$ and p divides

$$N_{K|\mathbb{Q}}(b + \omega) = \left(b + \frac{1 + \sqrt{d}}{2}\right)\left(b + \frac{1 - \sqrt{d}}{2}\right) = b^2 + b + q = f_q(b).$$

Now we note that $b \neq p - 1$. Otherwise p divides $f_p(p - 1) = (p - 1)^2 + (p - 1) + q \equiv q \pmod{p}$, so $q = p \leq (2/\pi)\sqrt{|d|} < \sqrt{|d|} \leq \sqrt{4q - 1}$ so $q = 2$ or 3, which is a contradiction.

So $b \neq p - 1$ and from the hypothesis $f_q(b)$ is a prime number, so it must be equal to p.

We conclude that $\sqrt{4q - 1} > p = f_q(b) \geq f_q(0) = q$, hence again $q = 2$ or 3, a contradiction.

(3) $\rightarrow$ (1)   It is known that if the class number of $\mathbb{Q}(\sqrt{d})$ (with $d < -3$, d square-free) is equal to 1, then $d = 1 - 4q = -7, -11, -19, -43, -67, -163$. This implies that $q = 2, 3, 5, 11, 17,$ or 41.   ∎

In the implication (3) $\rightarrow$ (1) we used the determination of all imaginary quadratic fields with class number 1. But it is possible to give a direct proof that (3) $\rightarrow$ (2)

For this proof, the theory of genera—more precisely, the results quoted in Section 7—will be needed.

(3) $\rightarrow$ (2)   Let $d = 1 - 4q$ and assume that the class number of $\mathbb{Q}(\sqrt{d})$ is equal to 1. Then either $d = -1, -2, -3, -7,$ or $d < -7$. Since the class number is 1, by the theory of genera, $d = -p$, where p is a prime, $p \equiv 3 \pmod{4}$; also $q > 2$.

As noted before, 2 is inert in $\mathbb{Q}(\sqrt{d}) = \mathbb{Q}(\sqrt{-p})$, so $p \equiv 3 \pmod{8}$.

We show that if ℓ is an odd prime, $\ell < q$, then $(\ell/p) = -1$. Indeed, if $(\ell/p) = +1$ then ℓ is decomposed in $\mathbb{Q}(\sqrt{-p})$. Since the class number is 1, there exists an algebraic integer $\alpha = (a + b\sqrt{-p})/2$ with $a \equiv b \pmod{2}$, such that $A\ell = A\alpha \cdot A\alpha'$, where α' is the conjugate of α. Taking norms

$$\ell^2 = N(A\ell) = N(A\alpha)N(A\alpha') = N(A\alpha)^2,$$

hence $\ell = N(A\alpha) = (a^2 + b^2 p)/4$, so $p + 1 = 4q > 4\ell = a^2 + b^2 p$. Hence $1 > a^2 + (b^2 - 1)p$, so $a = 0$, $b^2 = 1$, and $4\ell = p$, which is absurd.

Now assume that there exists k, $0 \leq k \leq q - 2$, such that $f_q(k) = k^2 + k + q$ is not a prime. Then there exists a prime ℓ such that $\ell^2 \leq k^2 + k + q = a\ell$, with $a \geq 1$. Since $k^2 + k + q$ is odd, then $\ell \neq 2$. Also

$$4\ell^2 \leq (2k + 1)^2 + p < \left(\frac{p - 1}{2}\right)^2 + p = \left(\frac{p + 1}{2}\right)^2.$$

Hence $\ell < (p + 1)/4 = q$. As was shown, $(\ell/p) = -1$. However,

$$4a\ell = (2k + 1)^2 + 4q - 1 = (2k + 1)^2 + p,$$

hence $-p$ is a square modulo ℓ and using Gauss' reciprocity law

$$1 = \left(\frac{-p}{\ell}\right) = \left(\frac{-1}{\ell}\right)\left(\frac{p}{\ell}\right) = (-1)^{(\ell-1)/2}\left(\frac{\ell}{p}\right)(-1)^{\frac{\ell-1}{2}\cdot\frac{p-1}{2}} = \left(\frac{\ell}{p}\right)$$

and this is absurd.   ∎

EXERCISES

1. Let $K_1 = \mathbb{Q}(t)$, $K_2 = \mathbb{Q}(u)$, $K_3 = \mathbb{Q}(v)$, where t, u, v satisfy the equations

$$t^3 - 18t - 6 = 0,$$
$$u^3 - 36u - 78 = 0,$$
$$v^3 - 54v - 150 = 0.$$

Show that these fields have the same discriminant $22\,356 = 2^2 \times 3^5 \times 23$.

2. Let K_1, K_2, K_3 be the fields of the previous exercise. Show that they are distinct, by considering the decomposition of the prime numbers 5, 11.

3. Let $g = X^3 - 7X - 7$.
   (a) Show that g is irreducible over $\mathbb{Q}$.

Let t be a root of g, $K = \mathbb{Q}(t)$.
   (b) Find the discriminant of K and the ring of integers A of K.
   (c) Determine the decomposition into prime ideals of the ideals of A generated by 2, 3, 5, 7, 11.
   (d) Does there exist an inessential factor of the discriminant?

4. Let $K = \mathbb{Q}(\sqrt[3]{175})$ and let A be the ring of integers of K.
   (a) Show that $\{1, \sqrt[3]{175}, \sqrt[3]{245}\}$ is an integral basis, and that the discriminant of K is $\delta = -3^3 \times 5^2 \times 7^2$.
   (b) Show that 3, 5, 7 are completely ramified in $K|\mathbb{Q}$.
   (c) Show that δ has no inessential factor and find the decomposition of 2 into prime ideals of A.
   (d) Show that A has no integral basis of the form $\{1, x, x^2\}$.

5. Determine the ring of integers and the discriminant of the field $K = \mathbb{Q}(t)$, where $t^3 - t - 4 = 0$. Which prime numbers are ramified in K? Determine the decomposition into prime ideals of A of the ideals generated by 2, 3, 5. Determine whether there is an inessential factor of the discriminant.

6. Let a, b be square-free positive integers, $\gcd(a, b) = 1$. Let $c = ab$ when $a^2 \equiv b^2 \pmod{9}$ or $c = 3ab$ when $a^2 \not\equiv b^2 \pmod 9$. Show that the discriminant of $\mathbb{Q}(\sqrt[3]{ab^2})$ is $\delta = -3c^2$.

7. Determine the ring of integers, the discriminant, and the decomposition of prime numbers in the field $\mathbb{Q}(\sqrt{2}, i)$.

8. Determine the ring of integers, the discriminant, and the decomposition of prime numbers in the field $\mathbb{Q}(\zeta_{12})$, where ζ_{12} is a primitive twelfth root of unity.

9. Let ζ_7 be a primitive seventh root of unity. Determine the minimal polynomial of $\zeta_7 + \zeta_7^{-1}$ and show that the discriminant of $\mathbb{Q}(\zeta_7 + \zeta_7^{-1})$ is equal to 49.

10. Let q be a prime number not dividing n. In order that there exists $x \in \mathbb{Z}$ such that $\Phi_n(x) \equiv 0 \pmod q$ it is necessary and sufficient that $q \equiv 1 \pmod n$. In this case the solutions are the integers x such that $x^n \equiv 1 \pmod q$. The number of pairwise incongruent solutions modulo q is $\varphi(n)$.

Hint: Use the results about the decomposition of primes in cyclotomic fields.

11. Let q be a prime factor of n, and let $n = q^a n_1$, n_1 not divisible by q. Show that there exists an integer x such that $\Phi_n(x) \equiv 0 \pmod q$ if and only if $q \equiv 1 \pmod{n_1}$. In this case, the solutions are the integers x such that $x^{n_1} \equiv 1 \pmod q$. The number of pairwise incongruent solutions modulo q is $\varphi(n_1)$.

Hint: Use the results about the decomposition of primes in cyclotomic fields.

12. Determine the integers x such that $\Phi_{20}(x) \equiv 0 \pmod{41}$.

13. Show that there exists no integer x such that $\Phi_{15}(x) \equiv 0 \pmod 5$.

14. Prove the following particular case of Dirichlet's theorem: for every natural number n there exist infinitely many prime numbers which are congruent to 1 modulo n.

Hint: Let m be the product of all primes congruent to 1 modulo $n > 2$; show that every prime factor of $\Phi_n(nm)$ is congruent to 1 modulo n, and note that this is impossible.

15. Let $m = p^k > 2$, p a prime number, $k \geq 1$; let ζ be a primitive mth root of unity. Show that if s is any integer, such that $1 \leq s \leq$

m, $\gcd(s, m) = 1$, then

$$v_s = \sqrt{\frac{1 - \zeta^s}{1 - \zeta} \cdot \frac{1 - \zeta^{-s}}{1 - \zeta^{-1}}}$$

is a real unit of $\mathbb{Q}(\zeta)$ (see Chapter 10, Section 3).

16. Let $m = p_1^{k_1} \cdots p_r^{k_r}$ with $r > 1$, p_i prime numbers, $k_i \geq 1$, and let ζ be a primitive mth root of unity. Show that if s is any integer such that $1 \leq s \leq m$, $\gcd(s, m) = 1$, then $v_s = \sqrt{(1 - \zeta^s)(1 - \zeta^{-s})}$ is a real unit of $\mathbb{Q}(\zeta)$ (see Chapter 10, Section 3, and the previous exercise).

17. Let L be the algebraic number field of Example 7 with $K = \mathbb{Q}(\zeta)$, and let σ be a generator of the Galois group of $\mathbb{Q}(\zeta)|\mathbb{Q}$, $\sigma(\zeta) = \zeta^s$, where $1 \leq s \leq p - 1$, $\gcd(s, p) = 1$. Show:

(a) $L|\mathbb{Q}$ is a Galois extension if and only if there exists r, $1 \leq r \leq p - 1$, such that $\sigma(a)/a^r$ is the pth power of an element of $\mathbb{Q}(\zeta)$.

(b) $L|\mathbb{Q}$ is an Abelian extension if and only if $\sigma(a)/a^s$ is the pth power of an element of $\mathbb{Q}(\zeta)$.

(c) If $L|\mathbb{Q}$ is a Galois extension then $L = \mathbb{Q}(\zeta) \cdot M$, where $M|\mathbb{Q}$ is an extension of degree p.

18. Let $n > 2$ and let h be an integer, $\gcd(h, n) = 1$. Show:

(a) $\mathbb{Q}(\cos(2\pi h/n))|\mathbb{Q}$ has degree $\varphi(n)/2$.

(b) If $n \neq 4$ then

$$\mathbb{Q}(\sin(2\pi h/n))|\mathbb{Q} \quad \text{has degree} \quad \begin{cases} \varphi(n) & \text{when } \gcd(n, 8) < 4, \\ \varphi(n)/4 & \text{when } \gcd(n, 8) = 4, \\ \varphi(n)/2 & \text{when } \gcd(n, 8) > 4. \end{cases}$$

(c) If $n > 4$ then

$$\mathbb{Q}(\tan(2\pi h/n))|\mathbb{Q} \quad \text{has degree} \quad \begin{cases} \varphi(n) & \text{when } \gcd(n, 8) < 4, \\ \varphi(n)/2 & \text{when } \gcd(n, 8) = 4, \\ \varphi(n)/4 & \text{when } \gcd(n, 8) > 4. \end{cases}$$

19. Let ζ be a primitive nth root of unity, where $n > 2$. Show that $\mathbb{Q}(\zeta + \zeta^{-1})|\mathbb{Q}$ has degree $\varphi(n)/2$.

20. Let $K|\mathbb{Q}$ be a real quadratic extension, and let ε_0 be a fundamental unit of K, having norm equal to 1. Let x be an algebraic integer of K such that $N_{K|\mathbb{Q}}(x) < 0$ and $Ax = J^2$, where J is an ideal of the ring of integers of K. Show that J is not a principal ideal.

21. Show that the class number of $\mathbb{Q}(\sqrt{34})$ is $h = 2$.

Hint: By a result of Chapter 9, reduce to the consideration of the principal ideals generated by 2, 3, 5; study their prime ideal decompositions in $\mathbb{Q}(\sqrt{34})$ and, by the previous exercise, show that the ideal generated by 3 and $1 - \sqrt{34}$ is not a principal ideal, and also show that the ideal generated by 5 and $3 - \sqrt{34}$ is not principal; conclude by showing that these ideals are equivalent.

22. Show that the class number of $\mathbb{Q}(\sqrt{21})$ is $h = 1$.

23. Show that the class number of $\mathbb{Q}(\sqrt{37})$ is $h = 1$.

24. Show that the class number of $\mathbb{Q}(\sqrt{65})$ is $h = 2$.

25. Show that the class number of $\mathbb{Q}(\sqrt{-19})$ is $h = 1$.

26. Show that the class number of $\mathbb{Q}(\sqrt{-163})$ is $h = 1$.

27. Show that the class number of $\mathbb{Q}(\sqrt{-23})$ is $h = 3$.

28. Show that the class number of $\mathbb{Q}(\sqrt{-14})$ is $h = 4$.

29. Show that the class number of $\mathbb{Q}(\sqrt{-127})$ is $h = 5$.

30. Show that the class number of $\mathbb{Q}(\sqrt{-39})$ is $h = 4$.

31. Let p be a prime number, $p \equiv 1 \pmod 4$. Show that the ideal class group of $\mathbb{Q}(\sqrt{-p})$ has an element of order 2.

Part Three

Part Three

17

Local Methods for Cyclotomic Fields

In his investigations about Fermat's last theorem, Kummer developed the theory of cyclotomic fields. His ideas and results were extended by his illustrious contemporaries or successors, among them Dedekind, Hermite, Hurwitz, and Hensel.

Some of the ideas and theorems proved by Kummer for cyclotomic fields, and later extended for all number fields, are the concepts of an ideal (*ideal number* in Kummer's terminology), the unique factorization of ideals into a product of prime ideals, the classes of ideals, the finiteness of the class number, the finite generation of the group of units, and the type of decomposition of prime numbers into prime ideals of the cyclotomic field. These topics have already been dealt with in this book.

As a preparation presenting a proof of Kummer's result on Fermat's last theorem (see next chapter) we shall consider here the methods, which today are called "local."

Let $p > 2$ be a prime number, let $\zeta = \cos(2\pi/p) + i\sin(2\pi/p)$ be a primitive pth root of unity, $K = \mathbb{Q}(\zeta)$ the pth cyclotomic field, and $A = \mathbb{Z}[\zeta]$ the ring of cyclotomic integers. We have $[K : \mathbb{Q}] = p - 1$.

17.1 p-Adic and λ-Adic Numbers

In his research on cyclotomic fields, Kummer worked with λ-adic numbers, which are a generalization of p-adic numbers. In this section we indicate the definitions and a few results. The topic belongs to the Theory of Valuations, and it is fully developed in my book [26].

17.1.1 The p-Adic Numbers

In order to study divisibility properties of a prime p, it is often convenient to consider the development of integers in the base p:

$$a = a_0 + a_1 p + \cdots + a_m p^m$$

with $0 \leq a_i \leq p - 1$, $p^m \leq a < p^{m+1}$.

Numbers defined by infinite p-adic developments are the *p-adic integers*. Hensel described the operations of addition and multiplication of p-adic integers, and proved a very important theorem concerning the existence of p-adic integers which are roots of certain polynomials.

The p-adic numbers may be considered as being the limits of sequences of integers, relative to the p-adic distance. These considerations allowed the introduction of methods of Analysis in the study of questions on divisibility.

We shall describe here very briefly the concepts of p-adic numbers and give a few results which will be needed. The systematic study of these numbers is given in [23, Chapter 2].

Let p be any prime number. For any nonzero integer a let $v_p(a) = m$ if p^m divides a, but p^{m+1} does not divide a. For any nonzero rational number a/b, let $v_p(a/b) = v_p(a) - v_p(b)$, where $a, b \in \mathbb{Z}$, $b \neq 0$. Let $v_p(0) = \infty$.

Then the following properties are satisfied:

(1) $v_p(x) = \infty$ if and only if $x = 0$.

(2) $v_p(xy) = v_p(x) + v_p(y)$.

(3) $v_p(x + y) \geq \min\{v_p(x), v_p(y)\}$.

(By convention $n < \infty$ and $n + \infty = \infty + n = \infty + \infty = \infty$, for every integer n.)

Moreover, we also have

(3′)   If $v_p(x) < v_p(y)$ then $v_p(x + y) = v_p(x)$.

The mapping $v_p : \mathbb{Q} \to \mathbb{Z} \cup \{\infty\}$ is called the *p-adic valuation* of $\mathbb{Q}$.

The set $A_{v_p} = \{x \in \mathbb{Q} \mid v_p(x) \geq 0\}$ is a subring of $\mathbb{Q}$, containing $\mathbb{Z}$ called the *ring of the valuation v_p*. It is easy to see that

$$A_{v_p} = \mathbb{Z}_{p\mathbb{Z}} = \left\{ \frac{a}{b} \in \mathbb{Q} \,\middle|\, a, b \in \mathbb{Z},\ b \neq 0,\ \gcd(a, b) = 1,\ p \nmid b \right\}.$$

The ring $\mathbb{Z}_{p\mathbb{Z}}$ has the unique maximal ideal $\mathbb{Z}_{p\mathbb{Z}}p$, and the residue field $\mathbb{Z}_{p\mathbb{Z}}/\mathbb{Z}_{p\mathbb{Z}}p \cong \mathbb{F}_p$.

Let

$$d_p : \mathbb{Q} \times \mathbb{Q} \to \mathbb{R}_{\geq 0}$$

be defined by $d_p(x, y) = p^{-v_p(x-y)}$ where $x \neq y$ and $d_p(x, x) = 0$. Then d_p satisfies the following properties:

(1) $d_p(x, y) = 0$ if and only if $x = y$.

(2) $d_p(x, y) = d_p(y, x)$.

(3) $d_p(x, y) \leq \max\{d_p(x, z), d_p(z, y)\}$.

(4) $d_p(x + z, y + z) = d_p(x, y)$.

So d_p is a distance function compatible with the operation of addition, thus $\mathbb{Q}$ becomes a metric space; d_p is called the *p-adic distance*.

The completion of $\mathbb{Q}$ relative to the p-adic distance is again a field, denoted $\mathbb{Q}_p$ and its elements are called p-adic numbers. The nonzero elements α of $\mathbb{Q}_p$ are represented by p-adic developments

$$\alpha = \sum_{i=m}^{\infty} a_i p^i$$

with $0 \leq a_i \leq p - 1$, $m \in \mathbb{Z}$, and $a_m \neq 0$.

If

$$\alpha_n = \sum_{i=m}^{n} a_i p^i$$

(for each $n \geq m$) then $\alpha = \lim_{n \to \infty} \alpha_n$ (the limit is relative to the p-adic distance).

The p-adic valuation may be extended by continuity to a valuation of the field $\mathbb{Q}_p$ (still denoted by v_p), which is defined as follows:

$$v_p \left(\sum_{i=m}^{\infty} a_i p^i \right) = m \qquad \text{if} \quad a_m \neq 0.$$

Thus, the values of v_p are also integers or infinity.

The topological closure of $\mathbb{Z}_p\mathbb{Z}$ in the field $\mathbb{Q}_p$ is a ring, still denoted by $\mathbb{Z}_p$. Its elements are called the p-adic integers. Thus $\alpha \in \mathbb{Q}_p$ is a p-adic integer exactly when $v_p(\alpha) \geq 0$. It is also clear that $\mathbb{Z}_p \cap \mathbb{Q} = \mathbb{Z}_p\mathbb{Z}$.

The only nonzero prime ideal of $\mathbb{Z}_p$ is $\mathbb{Z}_p p$, consisting of the multiples of p. The residue field of v_p is $\mathbb{Z}_p / \mathbb{Z}_p p$, which is isomorphic to the field $\mathbb{F}_p$.

If $\alpha, \beta \in \mathbb{Q}_p$, we say that α divides β if there exists $\gamma \in \mathbb{Z}_p$ such that $\alpha\gamma = \beta$; this means that $v_p(\alpha) \leq v_p(\beta)$.

The element $\alpha \in \mathbb{Z}_p$ is a unit in $\mathbb{Z}_p$ when α divides 1, i.e., $v_p(\alpha) = 0$. The set U_p of units of $\mathbb{Z}_p$ is a multiplicative group.

If $\alpha, \beta, \gamma \in \mathbb{Q}_p$, $\gamma \neq 0$, we write $\alpha \equiv \beta \pmod{\gamma}$ if γ divides $\alpha - \beta$. Similarly, if $\gamma \in \mathbb{Q}_p$, $\gamma \neq 0$ and $F(X), G(X) \in \mathbb{Q}_p[X]$ we write $F(X) \equiv G(X) \pmod{\gamma}$ when γ divides each coefficient of $F(X) - G(X)$.

These congruence relations satisfy the usual properties of congruences of integers.

Hensel proved, in 1908, what today is known as Hensel's lemma:

A.   *Let $F(X)$ be a monic polynomial with coefficients in $\mathbb{Z}_p$. If $a \in \mathbb{Z}$ is a simple root of the congruence*

$$F(X) \equiv 0 \pmod{p},$$

then there exists a p-adic integer $\alpha \in \mathbb{Z}_p$ such that $\alpha \equiv a \pmod{p}$ and $F(\alpha) = 0$.

The proof of this result can be found in [26, Chapter 3].

We apply Hensel's lemma to the polynomial $X^{p-1} - 1$:

B. $\mathbb{Z}_p$ *contains* $p - 1$ $(p-1)$*th roots of unity. More precisely, for every* $j = 1, 2, \ldots, p-1$, *there exists a unique element* $\omega_j \in \mathbb{Z}_p$ *such that* $\omega_j^{p-1} = 1$ *and* $\omega_j \equiv j \pmod{p}$.

Proof: For every $j = 1, \ldots, p-1$, $j^{p-1} \equiv 1 \pmod{p}$, so

$$X^{p-1} - 1 \equiv \prod_{j=1}^{p-1} (X - j) \pmod{p}.$$

Thus $1, 2, \ldots, p - 1$ are all the roots of this congruence, and they are simple. By (**A**), for every j there exists $\omega_j \in \mathbb{Z}_p$ such that $\omega_j^{p-1} = 1$ and $\omega_j \equiv j \pmod{p}$.

For the uniqueness, we observe that if $\omega \in \mathbb{Z}_p$, $\omega^{p-1} = 1$, and $\omega \equiv k \pmod{p}$, then ω must coincide with one of the roots of $X^{p-1} - 1$, say $\omega = \omega_j$; then $j \equiv \omega_j = \omega \equiv k \pmod{p}$, so $j = k$, i.e., $\omega = \omega_k$.   ∎

Let $P(p) = (\mathbb{Z}/p\mathbb{Z})^{\cdot}$ denote the multiplicative group of nonzero residue classes modulo p. Let Ω denote the multiplicative group of $(p - 1)$th roots of unity in $\mathbb{Z}_p$.

As a corollary, we have:

C.   *The mapping which associates to each nonzero residue class* j *modulo* p *the* $(p-1)$*th root of unity* ω_j *in* $\mathbb{Z}_p$, *such that* $\omega_j \equiv j \pmod{p}$, *establishes an isomorphism between the multiplicative groups* $P(p)$ *and* Ω. *Moreover,* ω_g *is a generator of* Ω *if and only if* g *is a primitive root modulo* p.

Proof: Indeed, if $1 \leq j, k, h \leq p - 1$ and $jk \equiv h \pmod{p}$, by (**B**) it follows that $\omega_j \omega_k \equiv \omega_h \pmod{p}$. Since $\omega_j \equiv j \pmod{p}$, the mapping $j \pmod{p} \mapsto \omega_j$ is an isomorphism. The last assertion is trivial.   ∎

D.   *With the above notations:*
   (1) *If* $p - 1 \nmid r$ *then* $\sum_{\omega \in \Omega} \omega^r = 0$.
   (2) *If* $p - 1 \mid r$ *then* $\sum_{\omega \in \Omega} \omega^r = p - 1$.

Proof: (1)   Let g be a primitive root modulo p, so ω_g is a generator of the multiplicative group Ω. Then

$$\sum_{\omega \in \Omega} \omega^r = \sum_{j=0}^{p-2} \omega_g^{jr} = \frac{1 - \omega_g^{(p-1)r}}{1 - \omega_g^r} = 0 \qquad \text{when} \quad p - 1 \nmid r.$$

   (2)   If $p - 1 \mid r$ then $\omega^r = 1$ for every $\omega \in \Omega$, hence $\sum_{\omega \in \Omega} \omega^r = p - 1$.   ∎

17.1.2 The λ-Adic Numbers

We now consider the pth cyclotomic field $K = \mathbb{Q}(\zeta)$. Let $\lambda = 1 - \zeta$, then $P = A\lambda$ is a prime ideal and $Ap = P^{p-1} = A\lambda^{p-1}$.

We introduce the λ-*adic valuation* v_λ on K, by defining: for $\alpha \neq 0$, $v_\lambda(\alpha) = m$ if $A\lambda^m$ divides $A\alpha$ but $A\lambda^{m+1}$ does not divide $A\alpha$; moreover, $v_\lambda(0) = \infty$.

Then $v_\lambda : K \to \mathbb{Z} \cup \{\infty\}$ satisfies the following properties (for $\alpha, \beta \in K$):

(1) $v_\lambda(\alpha) = \infty$ if and only if $\alpha = 0$.

(2) $v_\lambda(\alpha\beta) = v_\lambda(\alpha) + v_\lambda(\beta)$.

(3) $v_\lambda(\alpha + \beta) \geq \min\{v_\lambda(\alpha), v_\lambda(\beta)\}$.

Moreover:

(3′) If $v_\lambda(\alpha) < v_\lambda(\beta)$ then $v_\lambda(\alpha + \beta) = v_\lambda(\alpha)$.

We also note that $v_\lambda(p) = p-1$ and, more generally, $v_\lambda(x) = (p-1)v_p(x)$ for every $x \in \mathbb{Q}$.

Let $d_\lambda : K \times K \to \mathbb{R}_{\geq 0}$ be the mapping defined by $d_\lambda(\alpha, \beta) = e^{-v_\lambda(\alpha-\beta)}$ when $\alpha \neq \beta$ and $d_\lambda(\alpha, \alpha) = 0$.

Then d_λ satisfies the same properties indicated for the p-adic distance; d_λ is called the λ-*adic distance function* and K becomes a metric space.

Let $\widehat{K}$ denote the completion of the metric space K. The operations of addition and multiplication extend by continuity from K to $\widehat{K}$, which is a commutative ring. But, in fact, every $\alpha \in \widehat{K}$, $\alpha \neq 0$ is invertible, so $\widehat{K}$ is a topological field, whose elements are called the λ-*adic numbers*.

The λ-adic valuation and the λ-adic distance extend canonically by continuity to $\widehat{K}$, by letting

$$\widehat{v}_\lambda \left(\lim_{n \to \infty} \alpha_n \right) = \lim_{n \to \infty} v_\lambda(\alpha_n),$$

when $(\alpha_n)_{n \geq 0}$ is any Cauchy sequence in K.

We define $\widehat{d}_\lambda(\alpha, \beta) = e^{-\widehat{v}_\lambda(\alpha-\beta)}$ for $\alpha, \beta \in \widehat{K}$. It is immediate that $\widehat{v}_\lambda$ and $\widehat{d}_\lambda$ satisfy the same properties already indicated for v_λ, d_λ.

For simplicity, we shall use the notations v_λ, d_λ, instead of $\widehat{v}_\lambda$, $\widehat{d}_\lambda$.

The set

$$\widehat{A} = \{x \in \widehat{K} \mid v_\lambda(x) \geq 0\}$$

is a subring of $\widehat{K}$, $\widehat{K}$ is the field of fractions of $\widehat{A}$, and $\widehat{A} \cap K = A_{v_\lambda}$ (the ring of the valuation v_λ).

The elements of $\widehat{A}$ are called λ-*adic integers*. The unique maximal ideal of $\widehat{A}$ is $\widehat{A}\lambda$, and $\widehat{A}/\widehat{A}\lambda \cong \mathbb{F}_p$.

Hensel's lemma, given in (**A**), is still valid with $\widehat{A}$ instead of $\mathbb{Z}_p$:

E.   *Let $F(X)$ be a monic polynomial with coefficients in $\widehat{A}$. If $a \in A$ is a simple root of the congruence*

$$F(X) \equiv 0 \pmod{\widehat{A}\lambda},$$

then there exists a λ-adic integer $\alpha \in \widehat{A}$ such that $\alpha \equiv a \pmod{\widehat{A}\lambda}$ and $F(\alpha) = 0$.

The proof can be found in [26, Chapter 3].

$\widehat{K}$ contains the subfield $\mathbb{Q}_p$ of p-adic numbers and $[\widehat{K} : \mathbb{Q}_p] = [K : \mathbb{Q}] = p - 1$.

$\widehat{K}|\mathbb{Q}_p$ is a Galois extension with a Galois group canonically isomorphic to $G(K|\mathbb{Q}) = P(p)$. Explicitly, every $\sigma \in G(K|\mathbb{Q})$ may be extended in a unique way by continuity to an element of $G(\widehat{K}|\mathbb{Q}_p)$, still denoted by σ.

Indeed, if

$$\alpha \in \widehat{K}, \quad \alpha = \lim_{n \to \infty} \alpha_n \quad (\text{with } \alpha_n \in K) \qquad \text{then} \quad \sigma(\alpha) = \lim_{n \to \infty} \sigma(\alpha_n),$$

because $(\sigma(\alpha_n))_{n \geq 0}$ is still a Cauchy sequence in K. Thus, if $\alpha \in K$ then

$$\mathrm{Tr}_{\widehat{K}|\mathbb{Q}_p}(\alpha) = \mathrm{Tr}_{K|\mathbb{Q}}(\alpha), \qquad N_{\widehat{K}|\mathbb{Q}_p}(\alpha) = N_{K|\mathbb{Q}}(\alpha).$$

17.2 The λ-Adic Exponential and Logarithm

We begin this section with a brief discussion of formal power series. Other aspects of formal power series are discussed in [23, Chapter 7].

17.2.1 Formal Power Series

Let A be a commutative ring, let $X_1, \ldots, X_r$ $(r \geq 1)$ be the indeterminates. A (formal) power series in $X_1, \ldots, X_r$, with coefficients in A, is a formal sum

$$S = S(X_1, \ldots, X_r) = \sum_{m=0}^{\infty} S_m,$$

where each S_m is zero or a homogeneous polynomial of degree m, in the indeterminates $X_1, \ldots, X_r$, with coefficients in A.

S_0 is the constant term of S; $S_0 \in A$.

If

$$S = \sum_{n=0}^{\infty} S_n, \qquad T = \sum_{n=0}^{\infty} T_n$$

we define

$$S + T = \sum_{n=0}^{\infty} (S_n + T_n) \quad \text{and} \quad ST = \sum_{n=0}^{\infty} \left(\sum_{i+j=n} S_i T_j \right).$$

With these operations, the set $A[[X_1, \ldots, X_r]]$ of power series is a commutative ring. The ring A is naturally identified with a subring of $A[[X_1, \ldots, X_r]]$.

The elements 0 and 1 of A are also the zero and unit elements of $A[[X_1, \ldots, X_r]]$.

The *order* $\omega(S)$ of S is defined to be the smallest integer $n \geq 0$ such that $S_n \neq 0$. By convention, the order of zero series is ∞.

It is clear that $\omega(S + T) \geq \min\{\omega(S), \omega(T)\}$ and $\omega(ST) \geq \omega(S) + \omega(T)$. We assume henceforth that A is an integral domain. Then $\omega(ST) = \omega(S) + \omega(T)$ and $A[[X_1, \ldots, X_r]]$ is an integral domain.

In the case of one indeterminate X, each power series is of the form

$$\sum_{n=0}^{\infty} s_n X^n;$$

each s_n is called a coefficient of $S(X)$.

If

$$T^{(j)} = \sum_{m=0}^{\infty} T_m^{(j)} \in A[[X_1, \ldots, X_r]]$$

and $\omega(T^{(j)}) \geq j$ for every $j \geq 0$, then for every $n \geq 0$ the following sum is finite:

$$U_n = \sum_{j=0}^{\infty} T_n^{(j)} = \sum_{j=0}^{n} T_n^{(j)}.$$

Let

$$U = \sum_{n=0}^{\infty} U_n \in A[[X_1, \ldots, X_r]].$$

In this case, we write $U = \sum_{j=0}^{\infty} T^{(j)}$.

A series $S \in A[[X_1, \ldots, X_r]]$ is invertible if there exists a series $T \in A[[X_1, \ldots, X_r]]$ such that $ST = 1$. Then $S_0 T_0 = 1$ so S_0 is an invertible element of A. Conversely, if S_0 is invertible, let $S_0^* \in A$ be such that $S_0^* S_0 = 1$, let $T = -S_0^*(S - S_0)$ so $\omega(T) \geq 1$ hence $\omega(T^j) \geq j$. Then

$$S_0^* \left(\sum_{j=0}^{\infty} T^j \right) \in A[[X_1, \ldots, X_r]]$$

is the inverse of S:

$$S_0^* \left(\sum_{j=0}^{\infty} T^j \right) S = (1 - T) \left(\sum_{j=0}^{\infty} T^j \right) = 1.$$

We write S^{-1} for the inverse of S.

Let

$$S = \sum_{n=0}^{\infty} s_n X^n \in A[[X]], \qquad T = \sum_{n=0}^{\infty} T_n \in A[[X_1, \ldots, X_r]].$$

If $\omega(T) \geq 1$ or if S is a polynomial, it makes sense to consider the power series

$$\sum_{n=0}^{\infty} s_n T^n.$$

We write

$$S(T) = S(T(X_1, \ldots, X_r)) = \sum_{n=0}^{\infty} s_n T^n.$$

$S(T)$ is called the power series obtained by substituting T for X in S.

If

$$S(X) = \sum_{n=0}^{\infty} s_n X^n,$$

its derivative is

$$S'(X) = \sum_{n=1}^{\infty} n s_n X^{n-1}.$$

Iterating, we have the higher derivatives

$$S''(X) = \sum_{n=2}^{\infty} n(n - 1) s_n X^{n-2},$$

etc.

If $S, T \in A[[X]]$, $\omega(T) \geq 1$, and $S_1 = S(T(X))$ then $S_1'(X) = S'(T(X)) \cdot T'(X)$.

If A has characteristic 0, i.e., $\mathbb{Z} \subseteq A$, if $S \in A[[X]]$ and its derivative is 0, then $S = S_0$ is a constant.

With the same hypothesis, if $S(X)$, $T(X)$ have the same derivative $S'(X) = T'(X)$ and the same constant term, then $S(X) = T(X)$.

Let $d \geq 0$ and let D_d be the set of all power series $S \in A[[X_1, \ldots, X_r]]$ such that $\omega(S) \geq d$. Then D_d is an ideal of the ring $A[[X_1, \ldots, X_r]]$.

If $S, T \in A[[X_1, \ldots, X_r]]$ we write $S \equiv T \ (\text{ord } d)$, when $\omega(S - T) \geq d$; this is an equivalence relation. Moreover, if S, T, U, V are power series, if $S \equiv T$ (ord d), $U \equiv V$ (ord d), then $S \pm U \equiv T \pm V$ (ord d), $SU \equiv TV$ (ord d).

Let $S, T \in A[[X]]$, $U, V \in A[[X_1, \ldots, X_r]]$; assume that S is a polynomial or $\omega(U) \geq 1$, and that T is a polynomial or $\omega(V) \geq 1$; if $S \equiv T$ (ord d), $U \equiv V$ (ord d), then $S(U) \equiv T(V)$ (ord d).

From now on, we assume that A is an integral domain containing $\mathbb{Q}$, and we shall introduce the exponential and logarithmic series. Let $r \geq 1$ and

$$\mathcal{A} = \{S \in A[[X_1, \ldots, X_r]] \mid \omega(S) \geq 1\},$$
$$\mathcal{M} = \{1 + S \in A[[X_1, \ldots, X_r]] \mid \omega(S) \geq 1\}.$$

From the above considerations, it follows that $\mathcal{A}$ is an additive group and $\mathcal{M}$ is a multiplicative group. Let $\exp : \mathcal{A} \to \mathcal{M}$ be the mapping defined by

$$\exp(S) = \sum_{n=0}^{\infty} \frac{1}{n!} S^n \qquad \text{for every} \quad S \in \mathcal{A}. \tag{17.1}$$

We also write $e^S = \exp(S)$. In particular, if $r = 1$, $X = X_1$:

$$e^X = \exp(X) = \sum_{n=0}^{\infty} \frac{1}{n!} X^n \tag{17.2}$$

is called the *exponential series*.

On the other hand, let $\log : \mathcal{M} \to \mathcal{A}$ be the mapping defined by

$$\log(1 + S) = \sum_{n=1}^{\infty} \frac{(-1)^{n-1}}{n} S^n \tag{17.3}$$

for every $1 + S \in \mathcal{M}$. In particular, if $r = 1$, $X = X_1$:

$$\log(1 + X) = \sum_{n=1}^{\infty} \frac{(-1)^{n-1}}{n} X^n \tag{17.4}$$

is called the *logarithmic series*.

We note that the derivatives of these series are

$$\exp'(X) = \exp(X),$$

$$\log'(1 + X) = \frac{1}{1 + X},$$

because $\log'(1 + X) = 1 - X + X^2 - X^3 + \cdots$.

Lemma 1.   *The mapping* $\exp$ *is an isomorphism from the additive group* $\mathcal{A}$ *onto the multiplicative group* $\mathcal{M}$; *the mapping* $\log$ *is the inverse isomorphism.*

Explicitly we have:

    (1) $\exp(\log(1 + S)) = 1 + S$;

    (2) $\log(\exp(S)) = S$;

    (3) $\exp(S + T) = \exp(S)\exp(T)$; *and*

(4) $\log((1 + S)(1 + T)) = \log(1 + S) + \log(1 + T)$;

for any series S, T of order at least 1.

Proof: (1)   We show that in $A[[X]]$ we have $\exp(\log(1 + X)) = 1 + X$. Indeed, $\log(1 + X)$ has order 1, so we may write

$$\exp(\log(1 + X)) = \sum_{n=0}^{\infty} \frac{[\log(1 + X)]^n}{n!} = 1 + \sum_{n=1}^{\infty} A_n X^n,$$

where each A_n is given by a finite sum. Taking derivatives

$$\frac{\exp(\log(1 + X))}{1 + X} = \sum_{n=1}^{\infty} nA_n X^{n-1},$$

hence

$$1 + \sum_{n=1}^{\infty} A_n X^n = \exp(\log(1 + X)) = (1 + X)\left(\sum_{n=1}^{\infty} nA_n X^{n-1}\right).$$

Comparing the coefficients on both sides, it follows that $A_1 = 1$, $A_2 = A_3 = \cdots = 0$, hence $\exp(\log(1 + X)) = 1 + X$. By substitution, the relation holds for every $S \in \mathcal{A}$.

(2)   We show that in $A[[X]]$ we have $\log(\exp(X)) = X$. Writing $\exp(X) = 1 + T(X)$, then

$$\log(\exp(X)) = \sum_{n=1}^{\infty} \frac{(-1)^{n-1}}{n}[T(X)]^n = \sum_{n=1}^{\infty} A_n X^n$$

where each A_n is given by a finite sum.

Taking derivatives

$$\frac{\exp'(X)}{\exp(X)} = \sum_{n=1}^{\infty} nA_n X^{n-1}.$$

Since $\exp'(X) = \exp(X)$ then $A_1 = 1$, $A_2 = A_3 = \cdots = 0$, so $\log(\exp(X)) = X$. By substitution, the relation holds for every $S \in \mathcal{A}$.

(3)

$$\exp(S + T) = \sum_{n=0}^{\infty} \frac{1}{n!}(S + T)^n = \sum_{n=0}^{\infty} \frac{1}{n!}\left[\sum_{a+b=n} \binom{n}{a} S^a T^b\right]$$

$$= \sum_{n=0}^{\infty}\left[\sum_{a+b=n} \frac{S^a}{a!} \frac{T^b}{b!}\right] = \left(\sum_{a=0}^{\infty} \frac{S^a}{a!}\right)\left(\sum_{b=0}^{\infty} \frac{T^b}{b!}\right)$$

$$= \exp(S)\exp(T).$$

(4)   By (1), $\exp(\log(1+X)) = 1 + X$. Substituting $S + T + ST$ for X we have

$$\exp(\log((1+S)(1+T))) = (1+S)(1+T)$$
$$= \exp(\log(1+S)) \cdot \exp(\log(1+T))$$
$$= \exp(\log(1+S) + \log(1+T)),$$

hence by (2):

$$\log((1+S)(1+T)) = \log(1+S) + \log(1+T). \qquad \blacksquare$$

17.2.2 The λ-Adic Exponential and Logarithm.

Let p be an odd prime; we use the same notations as in Section 17.1, (**B**).

A series $\sum_{n=0}^{\infty} \alpha_n$ of elements $\alpha_n \in \widehat{K}$ is said to be *convergent* when the sequence of partial sums $\sum_{n=0}^{N} \alpha_n$ (for all $N \geq 0$) is a convergent sequence in $\widehat{K}$. We write

$$\alpha = \sum_{n=0}^{\infty} \alpha_n \qquad \text{if} \quad \alpha = \lim_{N \to \infty} \left(\sum_{n=0}^{N} \alpha_n \right).$$

It is easy to see that the series $\sum_{n=0}^{\infty} \alpha_n$ is convergent if and only if $\lim_{n \to \infty} \alpha_n = 0$ and, in turn, this is equivalent to $\lim_{n \to \infty} v_\lambda(\alpha_n) = \infty$.

The formal power series

$$S = \sum_{n=0}^{\infty} S_n \in \widehat{K}[[X_1, \ldots, X_r]]$$

(with $r \geq 1$) is said to be *convergent at* $(\xi_1, \ldots, \xi_r)$, where each $\xi_i \in \widehat{K}$, when the series $\sum_{n=0}^{\infty} S_n(\xi_1, \ldots, \xi_r)$ is convergent in $\widehat{K}$; in this case, we write

$$S(\xi_1, \ldots, \xi_r) = \sum_{n=0}^{\infty} S_n(\xi_1, \ldots, \xi_r).$$

The set of all $(\xi_1, \ldots, \xi_r)$, such that the series S is convergent at $(\xi_1, \ldots, \xi_r)$, is called the *domain of convergence* of S. We are especially interested in series $S(X) \in \widehat{K}[[X]]$.

Let $S \in \widehat{K}[[X]]$ and $T \in \widehat{K}[[X_1, \ldots, X_r]]$ where $\omega(T) \geq 1$ or $S \in \widehat{K}[X]$. Let $U = S(T(X_1, \ldots, X_r))$. If T is convergent at $(\xi_1, \ldots, \xi_r)$ and S is convergent at $T(\xi_1, \ldots, \xi_r)$ then U is convergent at $(\xi_1, \ldots, \xi_r)$ and

$$U(\xi_1, \ldots, \xi_r) = S(T(\xi_1, \ldots, \xi_r)).$$

In the book [23, Chapter 7] we have studied in greater detail power series in valued fields. The present situation of $\widehat{K}$, endowed with v_λ, is just a special case.

Now, we determine the domains of convergence of the series $\exp(X)$, $\log(1 + X)$, when considered as series with coefficients in $\hat{K}$.

We need the following easy lemmas. If x is any real number, $[x]$ denotes the only integer such that $[x] \leq x < [x] + 1$.

Lemma 2.   *Let x be any real number, and let $a \geq 1$ be any integer. Then*

$$\left[\frac{x}{a}\right] = \left[\frac{[x]}{a}\right].$$

Proof: $[x]/a - 1 \leq x/a - 1 < [x/a]$ Also $a[x/a] \leq a(x/a) < [x] + 1$, hence $a[x/a] \leq [x]$, so $[x/a] \leq [x]/a$. This proves the lemma.   ∎

Legendre proved:

Lemma 3.   *If $n \geq 1$ and p is a prime, the exponent of the exact power of p dividing $n!$ is*

$$e = \left[\frac{n}{p}\right] + \left[\frac{n}{p^2}\right] + \left[\frac{n}{p^3}\right] + \cdots = \frac{n - s_n}{p - 1},$$

where

$$n = a_0 + a_1 p + \cdots + a_k p^k, \qquad 0 \leq a_i \leq p - 1,$$

and $s_n = a_0 + a_1 + \cdots + a_k$.

Proof: Let $n! = p^e m$, where $p \nmid m$, thus $e = v_p(n!)$. Let $n = n_1 p + r_1$, $0 \leq n_1$, $0 \leq r_1 < p$, so $n_1 = [n/p]$. The multiples of p no bigger than n are $p, 2p, \ldots, n_1 p \leq n$, so

$$p^{n_1} \cdot n_1! = p \cdot 2p \cdots n_1 p = p^e m', \qquad p \nmid m'.$$

Thus $n_1 + e_1 = e$ where $e_1 = v_p(n_1!)$. Since $n_1 < n$, by induction

$$e_1 = \left[\frac{n_1}{p}\right] + \left[\frac{n_1}{p^2}\right] + \cdots.$$

By Lemma 2:

$$\left[\frac{n_1}{p^i}\right] = \left[\frac{[n/p]}{p^i}\right] = \left[\frac{n}{p^{i+1}}\right]$$

so

$$e = \left[\frac{n}{p}\right] + \left[\frac{n}{p^2}\right] + \left[\frac{n}{p^3}\right] + \cdots.$$

Now we note that if

$$n = a_0 + a_1 p + \cdots + a_k p^k \qquad \text{with} \quad 0 \leq a_i \leq p - 1,$$

then

$$\left[\frac{n}{p}\right] = a_1 + a_2 p + \cdots + a_k p^{k-1},$$

$$\left[\frac{n}{p^2}\right] = a_2 + a_3 p + \cdots + a_k p^{k-2},$$

. . . .

$$\left[\frac{n}{p^k}\right] = a_k.$$

so

$$\sum_{i=1}^{k} \left[\frac{n}{p^i}\right] = a_1 + a_2(p+1) + a_3(p^2 + p + 1) + \cdots + a_k(p^{k-1} + p^{k-2}$$

$$+ \cdots + p + 1) = \frac{1}{p-1}[a_1(p-1) + a_2(p^2 - 1) + \cdots + a_k(p^k - 1)]$$

$$= \frac{1}{p-1}(n - s_n). \qquad \blacksquare$$

With these lemmas we show:

F.   (1)   *The domain of convergence of the exponential series in $\widehat{K}$ is*

$$\{\xi \in \widehat{K} \mid v_\lambda(\xi) \geq 2\} = \widehat{A}\lambda^2.$$

(2)   *The domain of convergence of the logarithmic series in $\widehat{K}$ is*

$$\{1 + \xi \in \widehat{K} \mid v_\lambda(\xi) \geq 1\} = 1 + \widehat{A}\lambda.$$

Proof: (1)   We have

$$v_\lambda\left(\frac{\xi^n}{n!}\right) = n v_\lambda(\xi) - v_\lambda(n!) = n v_\lambda(\xi) - (p-1)v_p(n!),$$

where $\xi \in \widehat{K}$.

By Lemma 3, $v_p(n!) = (n - s_n)/(p-1)$ where

$$n = a_0 + a_1 p + \cdots + a_k p^k, \qquad 0 \leq a_i \leq p - 1,$$

is the expression of n in the basis p and $s_n = a_0 + a_1 + \cdots + a_k$.
   Hence

$$v_\lambda\left(\frac{\xi^n}{n!}\right) = n(v_\lambda(\xi) - 1) + s_n \geq n(v_\lambda(\xi) - 1).$$

So if $v_\lambda(\xi) \geq 2$ then $\lim v_\lambda(\xi^n/n!) = \infty$ and the series converges at ξ. But if $v_\lambda(\xi) \leq 1$, since there exist infinitely many integers n such that $s_n = 1$ (namely all the powers of p), then the limit of $v_\lambda(\xi^n/n!)$ when n tends to infinity does not exist, hence the exponential series is not convergent for such ξ.

(2) If $\xi \in \widehat{K}$ then

$$v_\lambda \left(\frac{(-1)^{n-1}\xi^n}{n} \right) = nv_\lambda(\xi) - v_\lambda(n) = nv_\lambda(\xi) - (p-1)v_p(n).$$

Since $p^{v_p(n)}$ divides n then $v_p(n) \log p \le \log n$ and

$$v_\lambda \left(\frac{(-1)^{n-1}\xi^n}{n} \right) \ge nv_\lambda(\xi) - \frac{p-1}{\log p} \log n.$$

Again if $v_\lambda(\xi) \ge 1$ then the limit of the general term is ∞ and the series converges at ξ. However, if $v_\lambda(\xi) \le 0$ then

$$v_\lambda \left(\frac{(-1)^{n-1}\xi^n}{n} \right) \le -v_\lambda(n)$$

and considering integers of the form $n = p^k$ we see that the general term has no limit. ∎

We define the λ-*adic exponential* and *logarithmic functions* as follows:

$$\exp(\xi) = \sum_{n=0}^{\infty} \frac{\xi^n}{n!} \qquad \text{when} \quad v_\lambda(\xi) \ge 2$$

and

$$\log(1 + \eta) = \sum_{n=1}^{\infty} \frac{(-1)^{n-1}}{n} \eta^n \qquad \text{when} \quad v_\lambda(\eta) \ge 1.$$

Now we indicate several properties of the λ-adic exponential and λ-adic logarithm.

First we note that for every $m \ge 1$, $\widehat{A}\lambda^m$ is obviously an additive group. Similarly, $1 + \widehat{A}\lambda^m$ is a multiplicative group. Indeed, it suffices to note that if $\alpha \in \widehat{A}$ then

$$\frac{1}{1 - \alpha\lambda^m} = 1 + \alpha\lambda^m + \alpha^2\lambda^{2m} + \cdots + \alpha^k\lambda^{km} + \cdots = 1 + \beta\lambda^m$$

since

$$\alpha + \alpha^2\lambda^m + \cdots + \alpha^k\lambda^{(k-1)m} + \cdots$$

converges to some element $\beta \in \widehat{A}$, because $v_\lambda(\alpha^k\lambda^{(k-1)m}) \ge (k-1)m$.

G.   *The exponential function defines an isomorphism from the additive group $\widehat{A}\lambda^2$ onto the multiplicative group $1 + \widehat{A}\lambda^2$. The inverse isomorphism is defined by the logarithmic function (restricted to $1 + \widehat{A}\lambda^2$).*

Proof: First we show that if $\xi \in \widehat{A}\lambda^2$ then $v_\lambda(\exp(\xi) - 1) = v_\lambda(\xi)$.

Indeed, let

$$\exp(\xi) = 1 + \eta, \qquad \eta = \sum_{n=1}^{\infty} \frac{\xi^n}{n!}.$$

If $n \geq 2$ then

$$v_\lambda\left(\frac{\xi^n}{n!}\right) = n v_\lambda(\xi) - n + s_n \geq (n-1)(v_\lambda(\xi) - 1) + v_\lambda(\xi) > v_\lambda(\xi)$$

because $v_\lambda(\xi) \geq 2$, $s_n \geq 1$. So for every $m > 1$:

$$v_\lambda\left(\sum_{n=1}^{m} \frac{\xi^n}{n!}\right) = v_\lambda(\xi)$$

as follows from (3'). At the limit we have $v_\lambda(\eta) = v_\lambda(\xi) \geq 2$. Clearly $\exp(0) = 1$. From the above, it follows that if $\xi \in \hat{A}\lambda^2$, $\xi \neq 0$, then $\exp(\xi) \neq 1$.

Taking $S(X) = \exp(X)$, $T(X_1, X_2) = X_1 + X_2$, $U(X_1, X_2) = \exp(X_1) \cdot \exp(X_2)$ by Lemma 1, $S(T(X_1, X_2)) = U(X_1, X_2)$.

Thus, if $\xi_1, \xi_2 \in \hat{A}\lambda^2$, since $\xi_1 + \xi_2 \in \hat{A}\lambda^2$, then $\exp(\xi_1 + \xi_2) = \exp(\xi_1)\exp(\xi_2)$. If $\xi \in \hat{A}\lambda^2$ then $1 = \exp(0) = \exp(\xi - \xi) = \exp(\xi)\exp(-\xi)$, so $\exp(-\xi) = (\exp(\xi))^{-1}$. It follows that if $\xi_1 \neq \xi_2$ then $\exp(\xi_1) \neq \exp(\xi_2)$.

So the exponential function defines an injective homomorphism from $\hat{A}\lambda^2$ into $1 + \hat{A}\lambda^2$.

Now we consider the restriction of the logarithmic function to the subgroup $1 + \hat{A}\lambda^2$, and we show that if $\eta \in \hat{A}\lambda^2$ then $v_\lambda(\log(1 + \eta)) = v_\lambda(\eta)$. Since

$$\log(1 + \eta) = \sum_{n=1}^{\infty} \frac{(-1)^{n-1}}{n} \eta^n$$

we need to compute $v_\lambda\left(((-1)^{n-1}/n)\eta^n\right)$.

Let $n = p^k n' \geq 2$, where $k \geq 0$ and p does not divide n'. Then $v_p(n) = k$ and

$$v_\lambda\left(\frac{(-1)^{n-1}}{n}\eta^n\right) = n v_\lambda(\eta) - (p-1)k > v_\lambda(\eta).$$

Indeed, this is true if $k = 0$ because $n \geq 2$. If $k \geq 1$ it follows from $n - 1 \geq p^k - 1$ and $(p^k - 1)/(p - 1) = p^{k-1} + \cdots + p + 1 > k$.

Hence if $m \geq 1$ we have

$$v_\lambda\left(\sum_{n=1}^{m} \frac{(-1)^{n-1}}{n}\eta^n\right) = v_\lambda(\eta)$$

and taking the limit, $v_\lambda(\log(1 + \eta)) = v_\lambda(\eta)$.

Thus the logarithmic function maps $1 + \widehat{A}\lambda^2$ into $\widehat{A}\lambda^2$.

Moreover, from Lemma 1, it follows that

$$\log(\exp(\xi)) = \xi \quad \text{and} \quad \exp(\log(1 + \eta)) = 1 + \eta \quad \text{for} \quad \xi, \eta \in \widehat{A}\lambda^2.$$

Therefore the exponential function is an isomorphism from $\widehat{A}\lambda^2$ onto $1 + \widehat{A}\lambda^2$ and the logarithmic function is the inverse isomorphism. ∎

We shall also work with the polynomials

$$E_p(X) = \sum_{n=0}^{p-1} \frac{X^n}{n!} \quad \text{and} \quad L_p(1 + X) = \sum_{n=0}^{p-1} \frac{(-1)^{n-1}}{n} X^n.$$

They have p-integral coefficients and degree $p - 1$.

We note that $\exp(X) \equiv E_p(X)$ (ord p) and $\log(1 + X) \equiv L_p(1 + X)$ (ord p). It follows that if $T(X)$ is a power series and $w(T(X)) \geq 1$, then $\exp(T(X)) \equiv E_p(T(X))$ (ord p) and $\log(1 + T(X)) \equiv L_p(1 + T(X))$ (ord p). In particular, $\exp(\log(1 + X)) \equiv E_p(\log(1 + X))$ (ord p), $\log(\exp(X)) \equiv L_p(\exp(X))$ (ord p).

Lemma 4.   *Let X, Y be indeterminates.*

(1) $E_p(X) \cdot E_p(Y) \equiv E_p(X + Y)$ (ord p).

(2) $[E_p(X)]^k \equiv E_p(kX)$ (ord p) *for $k \geq 1$.*

(3) $L_p((1 + X)(1 + Y)) \equiv L_p(1 + X) + L_p(1 + Y)$ (ord p).

(4) $L_p((1 + X)^k) \equiv kL_p(1 + X)$ (ord p) *for $k \geq 1$.*

(5) $L_p(E_p(X)) \equiv X$ (ord p).

(6) $E_p(L_p(1 + X)) \equiv 1 + X$ (ord p).

Proof: (1)   Since $E_p(X) \equiv \exp(X)$ (ord p), $E_p(Y) \equiv \exp(Y)$ (ord p), then by Lemma 1:

$$E_p(X)E_p(Y) \equiv \exp(X)\exp(Y) = \exp(X + Y) \equiv E_p(X + Y) \text{ (ord } p).$$

(2)   This follows at once from (1).

(3)   From $L_p(1 + X) \equiv \log(1 + X)$ (ord p) and $L_p(1 + Y) \equiv \log(1 + Y)$ (ord p) it follows from Lemma 1 that

$$L_p(1 + X) + L_p(1 + Y) \equiv \log(1 + X) + \log(1 + Y)$$
$$= \log((1 + X)(1 + Y)) \equiv L_p((1 + X)(1 + Y)) \text{ (ord } p).$$

(4)   This follows at once from (3).

(5)   Since $E_p(X) \equiv \exp(X)$ (ord p), then

$$L_p(E_p(X)) \equiv L_p(\exp(X)) \equiv \log(\exp(X)) = X \text{ (ord } p).$$

(6)   From $L_p(1 + X) \equiv \log(1 + X)$ (ord p) it follows that

$$E_p(L_p(1 + X)) \equiv E_p(\log(1 + X)) \equiv \exp(\log(1 + X)) = 1 + X \text{ (ord } p).$$

∎

To conclude we note a general fact: let A be any commutative ring, let I be any ideal of A, and let $P \in A[X]$. If $\alpha, \beta \in A$ and $\alpha \equiv \beta \pmod{I}$, then $P(\alpha) \equiv P(\beta) \pmod{I}$.

From this it follows that if $m \geq 1$, $\alpha, \beta \in \widehat{A}$, and $\alpha \equiv \beta \pmod{\widehat{A}\lambda^m}$ then

$$E_p(\alpha) \equiv E_p(\beta) \pmod{\widehat{A}\lambda^m} \qquad (17.5)$$

and

$$L_p(1 + \alpha) \equiv L_p(1 + \beta) \pmod{\widehat{A}\lambda^m}. \qquad (17.6)$$

17.3 The λ-Adic Integers

In this section we shall study in more detail the ring of λ-adic integers.

We recall that $A = \mathbb{Z}[\zeta]$, i.e., $\{1, \zeta, \zeta^2, \ldots, \zeta^{p-2}\}$ is a basis of the $\mathbb{Z}$-module A.

From this we easily obtain:

H. *The ring $\widehat{A}$ of λ-adic integers is a free module over the ring $\mathbb{Z}_p$ of p-adic integers, having the basis $\{1, \zeta, \zeta^2, \ldots, \zeta^{p-2}\}$.*

Proof: A is the direct sum

$$A = \bigoplus_{j=0}^{p-2} \mathbb{Z}\zeta^j.$$

Let M be the multiplicative set of integers not multiples of p. Then $M^{-1}\mathbb{Z} = \mathbb{Z}_{\mathbb{Z}p}$ (the ring of p-integral rational numbers) and

$$M^{-1}A = \bigoplus_{j=0}^{p-2} \mathbb{Z}_{\mathbb{Z}p}\zeta^j.$$

But $M^{-1}A = A_{A\lambda}$. Indeed, if $\alpha \in A$ and $m \in M$, then $m \notin A\lambda$ since $A\lambda \cap \mathbb{Z} = \mathbb{Z}_{\mathbb{Z}p}$. So $\alpha/m \in A_{A\lambda}$, showing that $M^{-1}A \subseteq A_{A\lambda}$. Conversely, let $\alpha \in A$, let $\beta \in A$, and let $\beta \notin A\lambda$. The conjugates $\sigma^j(\beta)$ cannot belong to $\sigma^j(A\lambda)$, but this is a prime ideal containing p, hence equal to $A\lambda$. So $\sigma^j(\beta) \notin A\lambda$ and therefore

$$N(\beta) = \prod_{j=0}^{p-2} \sigma^j(\beta) \notin A\lambda.$$

But $N(\beta) \in \mathbb{Z}$ hence $N(\beta) \notin \mathbb{Z}p$, so

$$\frac{\alpha}{\beta} = \frac{\alpha\sigma(\beta) \cdots \sigma^{p-2}(\beta)}{N(\beta)} \in M^{-1}A,$$

proving the equality.

Therefore

$$A_{A\lambda} = \bigoplus_{j=0}^{p-2} \mathbb{Z}_{\mathbb{Z}_p}\zeta^j.$$

Taking the closure in the completion $\widehat{K}$, relative to the valuation v_λ, we deduce that

$$\widehat{A} = \bigoplus_{j=0}^{p-2} \mathbb{Z}_p\zeta^j,$$

which was to be proved.   ∎

Our aim is to indicate another basis of the $\mathbb{Z}_p$-module $\widehat{A}$, which behaves nicely with respect to the trace.

First we show:

I.   If $\alpha \in \widehat{A}$, $v_\lambda(\alpha) = 1$, and $\{1, \alpha, \alpha^2, \ldots, \alpha^{p-2}\}$ are linearly independent over $\mathbb{Q}_p$, then this set is a $\mathbb{Z}_p$-basis of $\widehat{A}$.

Proof: Since $[\widehat{K} : \mathbb{Q}_p] = p - 1$ then $\{1, \alpha, \alpha^2, \ldots, \alpha^{p-2}\}$ is a basis of the $\mathbb{Q}_p$-vector space $\widehat{K}$.

Given $\beta \in \widehat{A} \subseteq \widehat{K}$, we may write, in unique way,

$$\beta = \sum_{j=0}^{p-2} c_j\alpha^j,$$

with each $c_j \in \mathbb{Q}_p$; we shall prove that each $c_j \in \mathbb{Z}_p$.

Let

$$k = \min_{0 \le j \le p-2}\{v_p(c_j)\} \qquad \text{then} \qquad \min_{0 \le j \le p-2}\{v_p(p^{-k}c_j)\} = 0,$$

so each $d_j = p^{-k}c_j \in \mathbb{Z}_p$.

We note that if $0 \le i < j \le p - 2$ then $v_\lambda(d_i\alpha^i) \ne v_\lambda(d_j\alpha^j)$. Otherwise $v_\lambda(d_i) + iv_\lambda(\alpha) = v_\lambda(d_j) + jv_\lambda(\alpha)$ so

$$(p - 1)[v_p(d_i) - v_p(d_j)] = (j - i)v_\lambda(\alpha) = j - i,$$

that is, $j - i$ is a multiple of $p - 1$; but this is impossible, since $0 < j - i < p - 1$.

Let j_0 be the smallest index such that $v_p(d_{j_0}) = 0$. Then $v_\lambda(d_{j_0}\alpha^{j_0}) < v_\lambda(d_j\alpha^j)$ when $j \ne j_0$. Otherwise

$$j_0 = v_\lambda(d_{j_0}\alpha^{j_0}) \ge v_\lambda(d_j\alpha^j) = (p - 1)v_p(d_j) + j;$$

if $v_p(d_j) \ge 1$ this implies that $j_0 \ge p-1$, which is impossible; if $v_p(d_j) = 0$ then $j_0 > j$, contrary to the choice of j_0. Therefore

$$v_\lambda(p^{-k}\beta) = v_\lambda\left(\sum_{j=0}^{p-2} d_j\alpha^j\right) = v_\lambda(d_{j_0}\alpha^{j_0}) = j_0 < p - 1;$$

on the other hand,

$$v_\lambda(p^{-k}\beta) = -kv_\lambda(p) + v_\lambda(\beta) \geq -k(p-1).$$

Therefore $k \geq 0$ so each $c_j = p^k d_j \in \mathbb{Z}_p$.   ∎

We now show the existence of a special element ρ, which will play an important role.

J. $\widehat{A}$ *contains one element ρ which is unique, satisfying the following two conditions:*

(1) $\rho^{p-1} = -p$; *and*

(2) $\rho \equiv -\lambda \pmod{\widehat{A}\lambda^2}$.

Hence $\widehat{A}\rho = \widehat{A}\lambda$.

Proof: First we establish the uniqueness. If $\rho, \rho_1 \in \widehat{A}$ satisfy the above properties then

$$\rho^{p-1} = -p = \rho_1^{p-1} \qquad \text{so} \qquad (\rho_1 \rho^{-1})^{p-1} = 1,$$

i.e., $\eta = \rho_1 \rho^{-1}$ is a $(p-1)$th root of 1. But

$$\rho_1 \equiv -\lambda \equiv \rho \pmod{\widehat{A}\lambda^2} \qquad \text{so} \qquad \rho\eta \equiv \rho \pmod{\widehat{A}\lambda^2}.$$

Since $\rho = -\lambda + \alpha\lambda^2 = \lambda(-1 + \alpha\lambda)$ with $\alpha \in \widehat{A}$ then $\widehat{A}\rho = \widehat{A}\lambda$. Thus $\eta \equiv 1 \pmod{\widehat{A}\lambda}$. If $\eta \neq 1$ then $X - \eta$ divides

$$\frac{X^{p-1} - 1}{X - 1} = X^{p-2} + \cdots + X + 1$$

and computing these polynomials at 1, we deduce that $1 - \eta$ divides $p - 1$. But λ divides $1 - \eta$, hence λ divides both p and $p - 1$, which is impossible. Thus $\eta = 1$ and this proves that $\rho_1 = \rho$.

To prove the existence of ρ, let $\alpha = -p/(1 - \zeta)^{p-1}$.

First we show that $\alpha \equiv 1 \pmod{\widehat{A}\lambda}$.

We write

$$\frac{p}{(1 - \zeta)^{p-1}} = \frac{\Phi_p(1)}{(1 - \zeta)^{p-1}} = \frac{(1 - \zeta)(1 - \zeta^2) \cdots (1 - \zeta^{p-1})}{(1 - \zeta)(1 - \zeta) \cdots (1 - \zeta)}$$

$$= (1 + \zeta)(1 + \zeta + \zeta^2) \cdots (1 + \zeta + \zeta^2 + \cdots + \zeta^{p-2}).$$

Since $\lambda = 1 - \zeta$ then $\zeta \equiv 1 \pmod{\widehat{A}\lambda}$, hence

$$\begin{cases} 1 + \zeta \equiv 2 \pmod{\widehat{A}\lambda}, \\ 1 + \zeta + \zeta^2 \equiv 3 \pmod{\widehat{A}\lambda}, \\ \cdots\cdots\cdots\cdots\cdots\cdots\cdots \\ 1 + \zeta + \cdots + \zeta^{p-2} \equiv p - 1 \pmod{\widehat{A}\lambda}. \end{cases}$$

Hence, from Wilson's congruence

$$\frac{p}{(1 - \zeta)^{p-1}} \equiv 1 \cdot 2 \cdots (p-1) = (p-1)! \equiv -1 \pmod{\widehat{A}\lambda}.$$

Therefore $\alpha \equiv 1 \pmod{\widehat{A}\lambda}$.

We shall show that α is a $(p-1)$th power in $\widehat{A}$. Let

$$F(X) = X^{p-1} - \alpha \qquad \text{then} \quad F(1) \equiv 0 \pmod{\widehat{A}\lambda}.$$

But the derivative

$$F'(X) = (p-1)X^{p-2} \qquad \text{is such that} \quad F'(1) \not\equiv 0 \pmod{\widehat{A}\lambda}.$$

Thus the image of 1 modulo $\widehat{A}\lambda$ is a simple root of $\overline{F}(X) = X^{p-1} - \overline{\alpha}$ (polynomial with coefficients reduced modulo $\widehat{A}\lambda$). According to Hensel's lemma of Section 17.1 there exists a root β of $F(X)$ in $\widehat{A}$, such that $\beta \equiv 1 \pmod{\widehat{A}\lambda}$, so $\beta^{p-1} = \alpha$.

Let $\rho = -\beta(1 - \zeta)$, then

$$\rho^{p-1} = \beta^{p-1}(1-\zeta)^{p-1} = \alpha(1-\zeta)^{p-1} = -p$$

and

$$\rho = -\beta(1-\zeta) \equiv -(1-\zeta) = -\lambda \pmod{\widehat{A}\lambda^2}. \qquad \blacksquare$$

In the next few propositions we derive some congruences satisfied by expressions involving this element ρ.

K.   (1) $[E_p(\rho)]^p \equiv 1 \pmod{\widehat{A}\lambda^{2p-1}}$.

(2) *For every integer $k \geq 1$: $E_p(k\rho) \equiv \zeta^k \pmod{\widehat{A}\lambda^p}$.*

Proof: (1)   We write $E_p(X) = 1 + XG(X)$, where

$$G(X) = 1 + \frac{X}{2!} + \cdots + \frac{X^{p-2}}{(p-1)!} \in \mathbb{Z}_p[X].$$

So

$$[E_p(X)]^p = 1 + pH(X) + X^p[G(X)]^p,$$

where $H(X) \in \mathbb{Z}_p[X]$. By Lemma 4:

$$[E_p(X)]^p = E_p(pX) + X^pT(X),$$

where $T(X) \in \mathbb{Z}_p[X]$. We show first that $pH(\rho) \equiv p\rho \pmod{\widehat{A}\lambda^{2p-1}}$. Indeed,

$$pH(X) = [E_p(X)]^p - 1 - X^p[G(X)]^p$$
$$= \{E_p(pX) - 1\} + X^p(T(X) - [G(X)]^p).$$

Since

$$E_p(pX) - 1 = \frac{pX}{1!} + \frac{p^2X^2}{2!} + \cdots + \frac{p^{p-1}X^{p-1}}{(p-1)!} \in \mathbb{Z}_p[X]$$

then $T(X) - [G(X)]^p \in p\mathbb{Z}_p[X]$. Hence $pH(\rho) \equiv p\rho \pmod{\widehat{A}\lambda^{2p-1}}$, because $\widehat{A}\rho = \widehat{A}\lambda$ and $\widehat{A}p = \widehat{A}\lambda^{p-1}$.

Next we show that

$$\rho^p \equiv -p\rho \pmod{\widehat{A}\lambda^{2p-1}}.$$

Indeed, since $G(\rho) \equiv 1 \pmod{\widehat{A}\lambda}$ then $[G(\rho)]^p \equiv 1 \pmod{\widehat{A}\lambda^p}$ hence $\rho^p[G(\rho)]^p \equiv \rho^p \equiv -p\rho \pmod{\widehat{A}\lambda^{2p}}$.

Therefore

$$\left[E_p(\rho)\right]^p = 1 + pH(\rho) + \rho^p[G(\rho)]^p \equiv 1 + p\rho - p\rho = 1 \pmod{\widehat{A}\lambda^{2p-1}}.$$

(2)   We first show that $E_p(\rho) \equiv \zeta \pmod{\widehat{A}\lambda^p}$.
We have

$$\rho \equiv \zeta - 1 \pmod{\widehat{A}\lambda^2} \qquad \text{then} \qquad E_p(\rho) \equiv 1 + \rho \equiv \zeta \pmod{\widehat{A}\lambda^2},$$

so $\zeta^{-1}E_p(\rho) \equiv 1 \pmod{\widehat{A}\lambda^2}$ and there exists an element $\alpha \in \widehat{A}$ such that $\zeta^{-1}E_p(\rho) = 1 + \alpha\lambda^2$. Raising to the pth power

$$[\zeta^{-1}E_p(\rho)]^p = \left[E_p(\rho)\right]^p \equiv 1 \pmod{\widehat{A}\lambda^{2p-1}}$$

by the first part of the proof. On the other hand,

$$(1 + \alpha\lambda^2)^p = 1 + p\alpha\lambda^2 + \binom{p}{2}\alpha^2\lambda^4 + \cdots + \alpha^p\lambda^{2p}$$

and comparing

$$\alpha\left(p\lambda^2 + \binom{p}{2}\alpha\lambda^4 + \cdots + \alpha^{p-1}\lambda^{2p}\right) \equiv 0 \pmod{\widehat{A}\lambda^{2p-1}}.$$

Since $v_\lambda(p\lambda^2) = (p-1) + 2 = p + 1$ and

$$v_\lambda\left[\binom{p}{k}\alpha^{k-1}\lambda^{2k}\right] \geq (p-1) + 2k > p + 1, \qquad k = 2, \ldots, p-1,$$

$$v_\lambda(\alpha^{p-1}\lambda^{2p}) \geq 2p > p + 1,$$

then

$$v_\lambda\left(p\lambda^2 + \binom{p}{2}\alpha\lambda^4 + \cdots + \alpha^{p-1}\lambda^{2p}\right) = p + 1,$$

therefore $v_\lambda(\alpha) \geq (2p - 1) - (p + 1) = p - 2$, i.e., $\alpha \equiv 0 \pmod{\widehat{A}\lambda^{p-2}}$ and, consequently, $\zeta^{-1}E_p(\rho) \equiv 1 \pmod{\widehat{A}\lambda^p}$, so $E_p(\rho) \equiv \zeta \pmod{\widehat{A}\lambda^p}$.

Now if $k > 1$, it follows from Lemma 4 that $E_p(k\rho) \equiv \left[E_p(\rho)\right]^k \pmod{\widehat{A}\lambda^p}$. But $E_p(\rho) \equiv \zeta \pmod{\widehat{A}\lambda^p}$ hence $E_p(k\rho) \equiv \zeta^k \pmod{\widehat{A}\lambda^p}$. ∎

L.   (1) If $\alpha \in \widehat{A}\lambda^2$ then $L_p(1 + \alpha) \equiv \log(1 + \alpha) \pmod{\widehat{A}\lambda^p}$.
    (2) If $\alpha_1, \alpha_2 \in \widehat{A}\lambda$ then

$$L_p((1 + \alpha_1)(1 + \alpha_2)) \equiv L_p(1 + \alpha_1) + L_p(1 + \alpha_2) \pmod{\widehat{A}\lambda^p}$$

and

$$L_p((1 + \alpha_1)^{-1}) \equiv -L_p(1 + \alpha_1) \pmod{\widehat{A}\lambda^p}.$$

(3) $L_p(\zeta) \equiv \rho \pmod{\widehat{A}\lambda^p}$.

Proof: (1)   We have

$$\log(1 + \alpha) - L_p(1 + \alpha) = \sum_{n=p}^{\infty} \frac{(-1)^n}{n} \alpha^n$$

and we shall determine the value $v_\lambda(\alpha^n/n)$ for $n \geq p$.

We have $v_\lambda(\alpha^n/n) = nv_\lambda(\alpha) - v_\lambda(n) \geq 2n - v_\lambda(n)$. Since $v_\lambda(n) = (p-1)v_p(n)$ and since $p^{v_\lambda(n)}$ divides n then $v_p(n) \log p \leq \log n$. Therefore

$$v_\lambda\left(\frac{\alpha^n}{n}\right) \geq 2n - (p-1)\frac{\log n}{\log p}.$$

We shall show that this last quantity is greater than p. For this purpose we consider the function of t, $t \geq 2$:

$$F(t) = \frac{\log t}{t - 1};$$

since

$$F'(t) = \frac{1 - 1/t - \log t}{(t-1)^2} < 0$$

for $t \geq 2$, $F(t)$ is monotone decreasing. In particular, if $p \leq n$ then $\log p/(p-1) \geq \log n/(n-1)$.

Therefore

$$2n - (p-1)\frac{\log n}{\log p} \geq 2n - (n-1) = n + 1 > p.$$

This proves that $v_\lambda(\alpha^n/n) > p$, and establishes the statement (1).

(2)   By Lemma 4:

$$L_p((1 + \alpha_1)(1 + \alpha_2)) \equiv L_p(1 + \alpha_1) + L_p(1 + \alpha_2) \pmod{\widehat{A}\lambda^p},$$

since $\alpha_1, \alpha_2 \in \widehat{A}\lambda$.

The second formula follows at once.

(3)   By Lemma 4:

$$L_p(E_p(\rho)) \equiv \rho \pmod{\widehat{A}\lambda^p}.$$

By (K), $E_p(\rho) \equiv \zeta \pmod{\widehat{A}\lambda^p}$ hence by (17.6):

$$L_p(E_p(\rho)) \equiv L_p(\zeta) \pmod{\widehat{A}\lambda^p}$$

and therefore $L_p(\zeta) \equiv \rho \pmod{\widehat{A}\lambda^p}$.   ∎

With the element ρ we build a basis:

M. $\{1, \rho, \rho^2, \ldots, \rho^{p-2}\}$ *is a basis of the* $\mathbb{Z}_p$-*module* $\widehat{A}$.

Proof: Since $\widehat{A}\rho = \widehat{A}\lambda$ then $v_\lambda(\rho) = 1$. Now we show that the elements 1, ρ, ρ^2, ..., ρ^{p-2} are linearly independent over $\mathbb{Q}_p$.

Indeed, if $c_0 + c_1\rho + \cdots + c_{p-2}\rho^{p-2} = 0$ with $c_i \in \mathbb{Q}_p$, not all zero, then there must exist two indices i, j such that $0 \leq i < j \leq p - 2$ and $v_\lambda(c_i) + iv_\lambda(\rho) = v_\lambda(c_j) + jv_\lambda(\rho)$ (as follows from (3')). So $j - i = (j - i)v_\lambda(\rho) = v_\lambda(c_i) - v_\lambda(c_j) = (p - 1)[v_p(c_i) - v_p(c_j)]$.

This is not possible since $j - i < p - 1$.

It follows from (I) that $\{1, \rho, \rho^2, \ldots, \rho^{p-2}\}$ is a basis of the $\mathbb{Z}_p$-module $\widehat{A}$. ∎

So every element of $\widehat{A}$ (respectively, $\widehat{K}$) may be written in a unique way in the form $a_0 + a_1\rho + \cdots + a_{p-2}\rho^{p-2}$ with $a_i \in \mathbb{Z}_p$ (respectively, $a_i \in \mathbb{Q}_p$).

N. (1) *For every* $i = 1, 2, \ldots, p - 2$, *the trace of* ρ^i *in the extension* $\widehat{K}|\mathbb{Q}_p$ *is 0.*

   (2) *An element of* $\widehat{K}$ *has trace 0 if and only if it is of the form*

$$\sum_{i=1}^{p-2} a_i\rho^i, \qquad a_i \in \mathbb{Q}_p.$$

Proof: (1) By definition $\mathrm{Tr}_{\widehat{K}|\mathbb{Q}_p}(\rho^i)$ is equal to the trace of the endomorphism of $\widehat{K}$ of multiplication with ρ^i. This is equal to the trace of the matrix of this endomorphism with respect to any basis, say $\{1, \rho, \rho^2, \ldots, \rho^{p-2}\}$. As seen immediately, all elements in the diagonal of this matrix are zero, hence the trace of ρ^i is also 0.

If

$$\xi = \sum_{i=0}^{p-2} a_i\rho^i$$

then

$$\mathrm{Tr}_{\widehat{K}|\mathbb{Q}_p}(\xi) = \mathrm{Tr}_{\widehat{K}|\mathbb{Q}_p}(a_0) = (p - 1)a_0.$$

So $\mathrm{Tr}_{\widehat{K}|\mathbb{Q}_p}(\xi) = 0$ exactly when $a_0 = 0$. ∎

Now we consider the real cyclotomic field $K^+ = \mathbb{Q}(\zeta + \zeta^{-1})$, which is the field fixed by the automorphism of complex-conjugation: $\sigma_{-1}(\zeta) = \zeta^{-1} = \overline{\zeta}$. The extension of σ_{-1} to $\widehat{K}$ leaves invariant a subfield, denoted $\widehat{K}^+$. It is easily seen that $\widehat{K}^+$ coincides with the closure of K^+ in $\widehat{K}$. The elements of $\widehat{K}^+$ are the *real λ-adic numbers*. The elements of the ring $\widehat{A}^+$, the closure of A^+ in $\widehat{K}^+$ are the *real λ-adic integers*.

O. (1) $\sigma_{-1}(\rho) = -\rho$, $\sigma_{-1}(\rho^2) = \rho^2$, $\widehat{K}^+$ *is the set of all elements*

$$\sum_{i=0}^{(p-3)/2} a_i\rho^{2i},$$

with $a_i \in \mathbb{Q}_p$; $\widehat{A}^+$ *is the set of all elements of the above form,*
with $a_i \in \mathbb{Z}_p$; $[\widehat{K}^+ : \mathbb{Q}_p] = (p-1)/2$.

(2) *Let*

$$S = \{\alpha \in \widehat{A}^+ \mid \mathrm{Tr}_{\widehat{K}|\mathbb{Q}_p}(\alpha) = 0\}.$$

Then S is a $\mathbb{Z}_p$-module equal to the set of all elements of the form

$$\sum_{i=1}^{(p-3)/2} a_i \rho^{2i} \quad with \quad a_i \in \mathbb{Z}_p.$$

Proof: (1)   Since $\rho^{p-1} = -p$ then $[\sigma_{-1}(\rho)]^{p-1} = -p = \rho^{p-1}$ thus
$\sigma_{-1}(\rho) = \eta\rho$ where $\eta \in \widehat{K}$, $\eta^{p-1} = 1$.

But there are at most $p-1$ $(p-1)$th roots of 1 in $\widehat{K}$. By (**B**), they all
belong to $\mathbb{Z}_p$. In particular, $\eta \in \mathbb{Q}_p$.

It follows that

$$\rho = \sigma_{-1}(\sigma_{-1}(\rho)) = \eta\sigma_{-1}(\rho) = \eta^2\rho$$

hence $\eta^2 = 1$. If $\eta = 1$ then σ_{-1} leaves invariant every element of $\widehat{K}$
(since these are combinations of $1, \rho, \ldots, \rho^{p-2}$), and σ_{-1} would be the
identity automorphism. In particular, every element of $\mathbb{Q}(\zeta)$ would be real,
a contradiction. This proves that $\eta = -1$, so $\sigma_{-1}(\rho) = -\rho$.

It follows that $\sigma_{-1}(\rho^i) = \rho^i$ exactly when i is even. Hence the subfield
$\widehat{K}^+$, invariant by σ_{-1}, is the one indicated.

(2)   Combining (**N**) with what we have just proved immediately yields
the present statement.   ∎

We conclude this section with a result about units:

P.   (1)   *If ε is a unit of K then $\varepsilon^{p-1} \equiv 1 \pmod{\widehat{A}\lambda}$ and*

$$\mathrm{Tr}_{\widehat{K}|\mathbb{Q}_p}(\log(\varepsilon^{p-1})) = 0.$$

If, moreover, ε is a real unit then $\log(\varepsilon^{p-1}) \in S$.

(2)   *If $\delta_2, \ldots, \delta_{(p-1)/2}$ are the circular units of K then $\log(\delta_k^{p-1}) \in S$
for $k = 2,, \ldots, (p-1)/2$.*

Proof: (1)   Let ε be a unit of K, hence also of $\widehat{K}$; thus $\varepsilon = a_0 +$
$a_1\rho + \cdots + a_{p-2}\rho^{p-2}$, with $a_i \in \mathbb{Z}_p$ and actually $v_p(a_0) = 0$. Thus $\varepsilon \equiv a_0$
$\pmod{\widehat{A}\lambda}$ hence $\varepsilon^{p-1} \equiv a_0^{p-1} \pmod{\widehat{A}\lambda}$. But $a_0 = m_0 + a'p$ with $m_0 \in$
$\mathbb{Z}$, $a' \in \mathbb{Z}_p$, so $a_0^{p-1} \equiv m_0^{p-1} \equiv 1 \pmod{\mathbb{Z}_p p}$. We conclude that $\varepsilon^{p-1} \equiv 1$
$\pmod{\widehat{A}\lambda}$ and the λ-adic logarithm is defined for ε^{p-1}.

Since each $\mathbb{Q}_p$-automorphism of $\widehat{K}$ is continuous, then

$$\operatorname{Tr}_{\widehat{K}|\mathbb{Q}_p}(\log \varepsilon^{p-1}) = \sum_{j=1}^{p-1} \sigma_j(\log \varepsilon^{p-1}) = \sum_{j=1}^{p-1} \log(\sigma_j(\varepsilon^{p-1}))$$

$$= \log \left(\prod_{j=1}^{p-1} \sigma_j(\varepsilon^{p-1}) \right) = \log(N_{\widehat{K}|\mathbb{Q}_p}(\varepsilon^{p-1}))$$

$$= \log(N_{K|\mathbb{Q}}(\varepsilon^{p-1})) = \log((\pm 1)^{p-1})$$

$$= \log 1 = 0.$$

Finally, if ε is a real unit so is ε^{p-1} and, therefore, $\log(\varepsilon^{p-1})$ is also a real λ-adic number, hence it belongs to S.

(2)   This follows from (1), because the circular units are real positive units of K.   ∎

Exercises

1. Calculate the 7-adic developments of the integers $328, 171$.

2. Calculate the 7-adic developments of $-1, -2, -3, \ldots, -6, -7$.

3. Calculate the 7-adic developments of $\frac{1}{2}, -\frac{1}{2}, \frac{3}{5}, -\frac{3}{5}$.

4. Which of the following integers have a square-root in the field of 5-adic numbers $2, -2, 5, -1, 25$? In the affirmative, write the 5-adic development of the square-root.

5. Is $\frac{2}{3}$ a square in $\mathbb{Q}_5$? If so, find the 5-adic development of its two square-roots.

6. Let $p = 5, 7$, or 11. For which values of p does the field $\mathbb{Q}_p$ contain a primitive cubic root of 1? In the affirmative, calculate the p-adic development.

7. Which of the numbers $2, -2, 3, \frac{1}{3}, -1$ have a cubic root in $\mathbb{Q}_5, \mathbb{Q}_7$? In the affirmative, calculate the 5-adic or 7-adic development of the cubic root.

8. Does $\mathbb{Q}_7$ contain a fifth root of 2? If so, calculate its 7-adic development.

9. Let $p = 5$. Calculate the 5-adic developments of the four fourth roots of 1.

10. Let p be a prime. Show:

    (a) Each p-adic series $\sum_{i=m}^{\infty} a_i p^i$ (with $m \in \mathbb{Z}$, $a_m \neq 0$, $0 \leq a_i < p$) converges to a unique element x of $\mathbb{Q}_p$ (with respect to the p-adic distance d_p); in this situation, write $x = \sum_{i=m}^{\infty} a_i p^i$ and show that $v_p(x) = m$.

    (b) Distinct p-adic series converge to distinct elements of $\mathbb{Q}_p$.

    (c) Every x of the completion $\mathbb{Q}_p$ of $\mathbb{Q}$ is the sum of a uniquely defined p-adic series.

11. A p-adic development $\sum_{i=0}^{\infty} a_i p^i$ is said to be *finite* if there exists $r \geq 0$ such that $a_i = 0$ for all $i \geq r$. Show that $x \in \mathbb{Q}_p$ has a finite p-adic development if and only if $x \in \mathbb{N}$.

12. A p-adic development $\sum_{i=0}^{\infty} a_i p^i$ is said to be *periodic infinite* if there exists $r \geq 0$ and $k \geq 1$ such that

$$a_r = a_{r+k} = a_{r+2k} = \cdots,$$

$$a_{r+1} = a_{r+1+k} = a_{r+1+2k} = \cdots,$$

$$\cdots \cdots \cdots \cdots \cdots \cdots \cdots \cdots \cdots \cdots$$

$$a_{r+k-1} = a_{r+2k-1} = a_{r+3k-1} = \cdots,$$

and $a_r, a_{r+1}, \ldots, a_{r+k-1}$ are not all 0. Show that $x \in \mathbb{Q}_p$ has an infinite periodic p-adic development if and only if $x \in \mathbb{Q}$, $x \notin \mathbb{N}$.

13. Determine the rational numbers with 7-adic developments:

    (a) $3 + 6 \times 7 + 6 \times 7^2 + 6 \times 7^3 + \cdots$, and

    (b) $2 + 7^2 + 3 \times 7^3 + 2 \times 7^4 + 7^6 + 3 \times 7^7 + 2 \times 7^8 + \cdots$.

14. Prove *Hensel's lemma* (statement (**A**)). With the notation of the statement, show:

    (a) There exists $a_1 \in \mathbb{Z}$, $0 \leq a_1 < p$, such that $F(a + a_1 p) \in \mathbb{Z}_p p^2$.

    (b) Show by induction that there exist $a_1, a_2, \ldots, a_n \in \mathbb{Z}$, $0 \leq a_i < p$, for each i, such that $F(a + a_1 p + \cdots + a_n p^n) \in \mathbb{Z}_p p^{n+1}$.

    (c) Let $\alpha = \sum_{i=0}^{\infty} a_i p^i \in \mathbb{Z}_p$ with $a_0 = a$. Show that $F(\alpha) = 0$ and that $\alpha \equiv a \pmod{\mathbb{Z}_p p}$.

15. Let $F \in \mathbb{Z}_p[X]$ be a nonconstant polynomial, and let F' denote its derivative. Assume that there exists $a \in \mathbb{Z}$ such that $2v_p(F'(a)) < v(F(a))$. Show that there exists $\alpha \in \mathbb{Z}_p$ such that $F(\alpha) = 0$ and $\alpha \equiv a \pmod{\mathbb{Z}_p p}$.

16. *q-adic solutions of Fermat's equation.* Let p, q be prime numbers (not assumed to be necessarily distinct). Show that the equation $X^p + Y^p = Z^p$ has a solution in nonzero integral q-adic numbers.

    Hint: For $p \neq q$ use Exercise 14, for $p = q$ use Exercise 15.

17. Prove the λ-adic *Hensel's lemma* (statement (**E**)) for the field $\widehat{K}$ and the λ-adic valuation.

    Hint: Follow the method indicated in Exercise 14.

18. Let $n \geq 1$ and let $a_0, a_1, \ldots, a_{n-1}$ be complex numbers. The *circulant* of $(a_0, a_1, \ldots, a_{n-1})$ is the determinant of the matrix

$$\begin{pmatrix} a_0 & a_1 & \cdots & a_{n-1} \\ a_{n-1} & a_0 & \cdots & a_{n-2} \\ \cdots & \cdots & \cdots & \cdots \\ a_1 & a_2 & \cdots & a_0 \end{pmatrix}.$$

Let $G(X) = a_0 + a_1 X + \cdots + a_{n-1} X^{n-1}$ and for $j = 0, 1, \ldots, n-1$ let $\zeta_j = \cos(2\pi j/n) + i \sin(2\pi j/n)$.

    Show that the circulant is equal to

$$\prod_{j=0}^{n-1} G(\zeta_j)$$

and also equal to $\text{Res}(G(X), X^n - 1)$ (the resultant of the polynomials—see Chapter 2, Exercises 53, 54).

19. Let $p = 5$. Express the inverse of $1 - \zeta + \zeta^2 + 3\zeta^3$ as a $\mathbb{Q}$-linear combination of $1, \zeta, \zeta^2, \zeta^3$. Do the same for $(3 - 2\zeta^2)/(1 - \zeta + \zeta^2 + 3\zeta^3)$.

20. Let $p = 5$. Express the periodic infinite λ-adic development

$$3 - \lambda + 2\lambda^2 - 4\lambda^3 - \lambda^4 + 2\lambda^5 - 4\lambda^6 - \lambda^7 + \cdots$$

as a $\mathbb{Q}$-linear combination of $1, \zeta, \zeta^2, \zeta^3$.

21. Calculate the inverse of the series $1 - 2X + 3X^2 - 4X^3 + 5X^4 - \cdots$.

22. We consider the power series in one indeterminate. Let $f : \mathcal{A} \to \mathcal{A}$ be the mapping defined by $f(S) = \exp(S) - 1$. Calculate the terms of degree at most 4 in the series $f^2(S)$ where $f^2 = f \circ f$.

23. We consider the power series in one indeterminate. Let $g : \mathcal{M} \to \mathcal{M}$ be the mapping defined by $g(1 + S) = 1 + \log(1 + S)$. Calculate the terms of degree at most 4 in the series $g^2(1 + S)$ where $g^2 = g \circ g$.

24. Let p be a prime number. Following as a model the text about the λ-adic exponential and logarithmic functions on $\widehat{K}$, define the exponential and logarithmic p-adic functions on $\mathbb{Q}_p$.

    Determine the domains of convergence and study the properties of the p-adic exponential and logarithmic functions.

25. Let $p = 5$. Calculate explicitly the λ-adic development of the element ρ from statement (**J**).

26. Let $p = 5$. Use the explicit λ-adic development of ρ (obtained in the preceding exercise) to verify the congruences in the statement (**K**).

27. Let $p = 5$. Calculate explicitly the expressions of λ, λ^2, λ^3 as $\mathbb{Z}_p$-linear combinations of $1, \rho, \rho^2, \rho^3$.

28. Let $p = 5$. Express $\zeta + \zeta^{-1}$ as a linear combination of $1, \rho, \rho^2, \rho^3$.

18

Bernoulli Numbers

Bernoulli numbers appear in the expression of sums of a fixed power of consecutive integers. These sums are intimately connected with the class number of cyclotomic fields.

18.1 Algebraic Properties

18.1.1 Recurrence for the Bernoulli Numbers

The series $(\exp(X) - 1)/X$ has order 0 and constant term 1, so it is invertible. We write its inverse in the form

$$\frac{X}{\exp(X) - 1} = \sum_{n=0}^{\infty} \frac{B_n}{n!} X^n. \tag{18.1}$$

The numbers B_n are called the *Bernoulli numbers*.

A. *The Bernoulli numbers are rational numbers; $B_0 = 1$, $B_1 = -\frac{1}{2}$. For every $k \geq 1$ the following recurrence relation is satisfied*

$$\binom{k+1}{1} B_k + \binom{k+1}{2} B_{k-1} + \cdots + \binom{k+1}{k} B_1 + 1 = 0. \tag{18.2}$$

Proof:

$$X = [\exp(X) - 1]\left[\frac{X}{\exp(X) - 1}\right] = \left(X + \frac{1}{2!}X^2 + \frac{1}{3!}X^3 + \cdots\right)$$
$$\times \left(B_0 + \frac{B_1}{1!}X + \frac{B_2}{2!}X^2 + \frac{B_3}{3!}X^3 + \cdots\right).$$

Comparing the coefficients of the powers of X, we deduce that $B_0 = 1$, $B_1 = -\frac{1}{2}$ and if $k \geq 2$ then

$$\frac{B_k}{k!} + \frac{B_{k-1}}{2!(k-1)!} + \frac{B_{k-2}}{3!(k-2)!} + \cdots + \frac{B_1}{k!} + \frac{1}{(k+1)!} = 0.$$

Multiplying with $(k + 1)!$ we have

$$\binom{k+1}{1} B_k + \binom{k+1}{2} B_{k-1} + \binom{k+1}{3} B_{k-2} + \cdots + \binom{k+1}{k} B_1 + 1 = 0.$$

It follows, by induction on k, that each B_k is a rational number.   ∎

Moreover, we have:

B.   (1) *If $k \geq 3$ and k is odd, then $B_k = 0$.*

   (2) *If $k \geq 1$ then $(-1)^{k-1} B_{2k} > 0$.*

Proof: (1)   We consider the series

$$S(X) = \frac{X}{2} + \frac{X}{\exp(X) - 1} = 1 + \sum_{k=2}^{\infty} \frac{B_k}{k!} X^k.$$

We have

$$S(-X) = -\frac{X}{2} - \frac{X}{\exp(-X) - 1} = \frac{X \exp(X)}{\exp(X) - 1} - \frac{X}{2}$$

hence

$$S(-X) - S(X) = \frac{X}{\exp(X) - 1} (\exp(X) - 1) - \frac{2X}{2} = 0.$$

Therefore $B_k = 0$ for every odd index $k \geq 3$.

   (2)   This proof is due to Mordell. We have

$$\frac{X}{\exp(X) + 1} = \frac{X}{\exp(X) - 1} - \frac{2X}{\exp(2X) - 1}$$

$$= \sum_{n=0}^{\infty} (1 - 2^n) \frac{B_n}{n!} X^n.$$

Multiplying with $X/(\exp(X) - 1)$ gives

$$\frac{X}{2} \cdot \frac{2X}{\exp(2X) - 1} = \left(\sum_{n=0}^{\infty} \frac{B_n}{n!} X^n \right) \left(\sum_{n=0}^{\infty} (1 - 2^n) \frac{B_n}{n!} X^n \right).$$

The left-hand side is equal to

$$\frac{X}{2} \left(\sum_{k=0}^{\infty} \frac{2^k B_k}{k!} X^k \right).$$

Since $B_k = 0$ when k is odd, $k > 1$, comparing coefficients of X^{2k} on both sides, yields (for $k \geq 2$):

$$0 = \sum_{m+n=2k} \frac{1 - 2^n}{m! n!} B_m B_n.$$

In the right-hand side, if $m = 1$ or $n = 1$ the summand is 0; it suffices to consider the summands with even indices and $2n \geq 2$. Thus

$$0 = \sum_{\substack{2m+2n=2k \\ 0 < m,n}} \frac{1 - 2^{2n}}{(2m)!(2n)!} B_{2m} B_{2n} + \frac{1 - 2^{2k}}{(2k)!} B_{2k}.$$

The proof of the proposition is by induction. The result is assumed true for B_{2i} for $i = 1, \ldots, k-1$. By induction $(-1)^{m-1} B_{2m} > 0$, $(-1)^{n-1} B_{2n} > 0$, for $1 \leq m, n$, hence

$$(-1)^{m+n-1} \frac{1 - 2^{2n}}{(2m)!(2n)!} B_{2m} B_{2n} = (-1)^{k-1} \frac{1 - 2^{2n}}{(2m)!(2n)!} B_{2m} B_{2n} > 0$$

so $(-1)^{k-1} B_{2k} > 0$.   ∎

In Subsection 18.1.3 we shall give another proof that $(-1)^{k-1} B_{2k} > 0$. Here are some Bernoulli numbers:

$$B_1 = -\tfrac{1}{2}, \qquad\qquad B_{18} = \tfrac{43867}{798},$$

$$B_2 = \tfrac{1}{6}, \qquad\qquad B_{20} = -\tfrac{174611}{330},$$

$$B_4 = -\tfrac{1}{30}, \qquad\qquad B_{22} = \tfrac{854513}{138},$$

$$B_6 = \tfrac{1}{42}, \qquad\qquad B_{24} = -\tfrac{236364091}{2730},$$

$$B_8 = -\tfrac{1}{30}, \qquad\qquad B_{26} = \tfrac{8553103}{6},$$

$$B_{10} = \tfrac{5}{66}, \qquad\qquad B_{28} = -\tfrac{23749461029}{870},$$

$$B_{12} = -\tfrac{691}{2730}, \qquad\qquad B_{30} = \tfrac{8615841276005}{14322},$$

$$B_{14} = \tfrac{7}{6}, \qquad\qquad B_{32} = -\tfrac{7709321041217}{510},$$

$$B_{16} = -\tfrac{3617}{510}, \qquad\qquad B_{34} = \tfrac{2577687858367}{6},$$

etc.

We shall also need the following formal power series expressions:

C.

$$\frac{\exp(X)}{\exp(X) - 1} = \frac{1}{X} + \frac{1}{2} + \sum_{k=1}^{\infty} \frac{B_{2k}}{(2k)!} X^{2k-1}$$

and

$$\log\left(\frac{\exp(X) - 1}{X}\right) = \frac{X}{2} + \sum_{k=1}^{\infty} \frac{B_{2k}}{(2k)!2k} X^{2k}.$$

Proof: We have

$$\frac{\exp(X)}{\exp(X) - 1} = \frac{\exp(X)}{X} \cdot \frac{X}{\exp(X) - 1}$$

$$= \left[\frac{1}{X} + \sum_{k=0}^{\infty} \frac{X^k}{(k+1)!}\right]\left[1 + \sum_{k=1}^{\infty} \frac{B_k}{k!} X^k\right]$$

$$= \frac{1}{X} + \sum_{k=1}^{\infty} \frac{B_k}{k!} X^{k-1} + \sum_{k=0}^{\infty} \frac{X^k}{(k+1)!}$$

$$+ \sum_{k=1}^{\infty} \left(\sum_{h=1}^{k} \frac{B_h}{h!(k-h+1)!}\right) X^k$$

$$= \frac{1}{X} - \frac{1}{2} + \sum_{k=1}^{\infty} \frac{B_{2k}}{(2k)!} X^{2k-1} + 1$$

$$+ \frac{1}{(k+1)!} \sum_{k=1}^{\infty} \left\{1 + \sum_{h=1}^{k} \binom{k+1}{h} B_h\right\} X^k.$$

It follows from (**A**) that the expression in the bracket is equal to 0. Hence the first formula is established.

For the second formula we consider the derivatives of the expressions in the right-hand side and in the left-hand side of the identity. We note that

$$\frac{X}{2} + \sum_{k=1}^{\infty} \frac{B_{2k}}{(2k)!2k} X^{2k}$$

has derivative equal to

$$\frac{1}{2} + \sum_{k=1}^{\infty} \frac{B_{2k}}{(2k)!} X^{2k-1}.$$

Similarly, the derivative of $\log((\exp(X) - 1)/X)$ is equal to

$$\frac{X}{\exp(X) - 1} \cdot \frac{X \exp(X) - \exp(X) + 1}{X^2} = \frac{\exp(X)}{\exp(X) - 1} - \frac{1}{X}.$$

By virtue of the first formula these derivatives are equal. But both expressions have constant term equal to 0, so they are equal, and this proves the second formula. ∎

18.1.2 Relations of Bernoulli Numbers with Trigonometric Functions

The Bernoulli numbers were introduced as coefficients in the power series expansion of the function $z/(e^z - 1)$. Since the trigonometric functions may be defined in terms of the exponential functions, there are also relations between certain trigonometric functions and the Bernoulli numbers.

D.   *For $|z| < \pi$:*

$$z \cot z = \sum_{k=0}^{\infty} (-1)^k \frac{2^{2k} B_{2k}}{(2k)!} z^{2k}. \qquad (18.3)$$

Proof: From

$$e^{iz} = \cos z + i \sin z \qquad (18.4)$$

and

$$e^{-iz} = \cos z - i \sin z$$

it follows that

$$\sin z = \frac{e^{iz} - e^{-iz}}{2i} \qquad (18.5)$$

$$\cos z = \frac{e^{iz} + e^{-iz}}{2} \qquad (18.6)$$

here, for $0 < |z| < \pi$, $\sin z \neq 0$, and

$$\cot z = \frac{\cos z}{\sin z} = i \frac{e^{iz} + e^{-iz}}{e^{iz} - e^{-iz}} = \frac{i(e^{2iz} + 1)}{e^{2iz} - 1} = i + \frac{1}{z} \cdot \frac{2iz}{e^{2iz} - 1}.$$

Expanding $2iz/(e^{2iz} - 1)$ as a power series

$$\frac{2iz}{e^{2iz} - 1} = \sum_{k=0}^{\infty} \frac{B_k}{k!} (2iz)^k = 1 - iz + \sum_{k=1}^{\infty} \frac{B_{2k}}{(2k)!} 2^{2k} (-1)^k z^{2k}.$$

Substituting

$$\cot z = \frac{1}{z} + \sum_{k=1}^{\infty} (-1)^k \frac{B_{2k} 2^{2k}}{(2k)!} z^{2k-1}. \qquad ∎$$

E.   *For $|z| < \pi/2$:*

$$\tan z = \sum_{k=1}^{\infty} (-1)^{k-1} \frac{2^{2k}(2^{2k} - 1) B_{2k}}{(2k)!} z^{2k-1}. \qquad (18.7)$$

Proof:

$$\tan z = \cot z - 2 \cot 2z \qquad (18.8)$$

as is well known. So for $|z| < \pi/2$, using (**D**):

$$\tan z = \sum_{k=1}^{\infty} (-1)^k \frac{B_{2k} 2^{2k}}{(2k)!} z^{2k-1} - 2 \sum_{k=1}^{\infty} (-1)^k \frac{B_{2k} 2^{4k-1}}{(2k)!} z^{2k-1}$$

$$= \sum_{k=1}^{\infty} (-1)^{k-1} \frac{2^{2k}(2^{2k} - 1) B_{2k}}{(2k)!} z^{2k-1}. \qquad ∎$$

Identity (18.7) shows that the *tangent coefficients* T_{2k-1}, which are defined by

$$\tan z = \sum_{k=1}^{\infty} \frac{T_{2k-1}}{(2k-1)!} z^{2k-1}, \qquad (18.9)$$

satisfy

$$T_{2k-1} = \frac{2^{2k}(2^{2k}-1)}{2k} (-1)^{k-1} B_{2k}. \qquad (18.10)$$

F. *For $|z| < \pi/2$:*

$$z \, \operatorname{cosec} z = \sum_{k=0}^{\infty} (-1)^k \frac{B_{2k}}{(2k)!} 2(1 - 2^{2k-1}) z^{2k}. \qquad (18.11)$$

Proof: The well-known identity

$$\operatorname{cosec} 2z = \cot z - \cot 2z$$

gives

$$2z \, \operatorname{cosec} 2z = 2z \cot z - 2z \cot 2z.$$

Applying (18.3):

$$2z \, \operatorname{cosec} 2z = 2 \sum_{k=0}^{\infty} (-1)^k \frac{2^{2k} B_{2k}}{(2k)!} z^{2k} - \sum_{k=0}^{\infty} (-1)^k \frac{2^{2k} B_{2k}}{(2k)!} 2^{2k} z^{2k}$$

$$= \sum_{k=0}^{\infty} (-1)^k \frac{2(1 - 2^{2k-1}) B_{2k}}{(2k)!} (2z)^{2k},$$

and this proves (18.11). ■

18.1.3 Bernoulli Numbers and the Zeta-Function

It is easy to see—and we shall return to this topic in Chapter 22—that for $s > 1$, the series $\sum_{n=1}^{\infty} 1/n^s$ is convergent; on the other hand, the harmonic series $\sum_{n=1}^{\infty} 1/n$ is divergent.

Let

$$\zeta(s) = \sum_{n=1}^{\infty} \frac{1}{n^s} \qquad \text{for} \quad s > 1. \qquad (18.12)$$

The zeta-function was studied by Euler (for real numbers $s > 1$) and later Riemann (for complex numbers s with real part $\mathfrak{Re}(s) > 1$). (See Chapter 22).

In order to present Euler's beautiful result, we need the expansion in partial fractions of the cotangent function cot z, which was also given by Euler.

G.   *For $|z| < \pi$:*

$$\cot z = \frac{1}{z} - 2z \sum_{n=1}^{\infty} \frac{1}{n^2 \pi^2 - z^2}. \qquad (18.13)$$

Proof: The function $\cot z = \cos z / \sin z$ is meromorphic, with period π. By (18.3), for $|z| < \pi$:

$$\cot z = \frac{1}{z} + \sum_{k=1}^{\infty} (-1)^k \frac{2^{2k} B_{2k}}{(2k)!} z^{2k-1}$$

hence cot z has poles of order 1, with residue 1 at the points $k\pi$ for all $k \in \mathbb{Z}$, and no other poles.

Let $f(z) = \pi \cot \pi z$. Then $f(z)$ is a meromorphic function, with period 1, with only poles at k for every $k \in \mathbb{Z}$, having order 1 and residue 1.

Now we introduce the function

$$h(z) = \frac{1}{z} + \sum_{n=1}^{\infty} \frac{2z}{z^2 - n^2} = \frac{1}{z} + \sum_{\substack{n=-\infty \\ n \neq 0}}^{\infty} \left(\frac{1}{z+n} - \frac{1}{n} \right). \qquad (18.14)$$

This function is also meromorphic with period 1. Its only poles are n, for $n \in \mathbb{Z}$; they are of order 1, having residue 1.

We shall compare the functions $f(z)$ and $h(z)$, by considering the difference

$$g(z) = f(z) - h(z),$$

and showing that it is identically zero.

At $z = 0$:

$$\lim_{z \to 0} \left(f(z) - \frac{1}{z} \right) = 0$$

and

$$\lim_{z \to 0} \left(h(z) - \frac{1}{z} \right) = 0,$$

hence

$$\lim_{z \to 0} \left(f(z) - h(z) \right) = g(0) = 0.$$

By differentiating termwise the series for $g(z)$, for $|z| < \pi$:

$$g'(z) = -\frac{\pi^2}{\sin^2 \pi z} + \frac{1}{z^2} + \sum_{n=1}^{\infty} \frac{2(z^2 + n^2)}{(z-n)^2(z+n)^2}$$

$$= -\frac{\pi^2}{\sin^2 \pi z} + \frac{1}{z^2} + \sum_{n=1}^{\infty} \left[\frac{1}{(z+n)^2} + \frac{1}{(z-n)^2} \right]$$

$$= -\frac{\pi^2}{\sin^2 \pi z} + \sum_{n=-\infty}^{\infty} \frac{1}{(z+n)^2} .$$

In particular, for $z/2$ and $(z+1)/2$:

$$g'\left(\frac{z}{2}\right) = -\frac{\pi^2}{\sin^2(\pi z/2)} + 4 \sum_{n=-\infty}^{\infty} \frac{1}{(z+2n)^2} ,$$

$$g'\left(\frac{z+1}{2}\right) = -\frac{\pi^2}{\cos^2(\pi z/2)} + 4 \sum_{n=-\infty}^{\infty} \frac{1}{(z+2n+1)^2} .$$

Hence

$$g'\left(\frac{z}{2}\right) + g'\left(\frac{z+1}{2}\right) = -\frac{4\pi^2}{\sin^2 \pi z} + 4 \sum_{n=-\infty}^{\infty} \frac{1}{(z+n)^2} = 4g'(z).$$

Let

$$M = \sup_{|z|<2} |g'(z)|.$$

If $|z| < 2$ then $|z/2| < 2$ and $|(z+1)/2| < 2$, hence

$$|g'(z)| \le \frac{1}{4} \left\{ \left| g'\left(\frac{z}{2}\right) \right| + \left| g'\left(\frac{z+1}{2}\right) \right| \right\} \le \frac{M}{2} .$$

Thus $M \le M/2$ hence $M = 0$, showing that $g'(z) = 0$ for $|z| < 2$.

Therefore the entire function $g'(z) = 0$ for every $z \in \mathbb{C}$. Thus $g(z)$ is a constant and since $g(0) = 0$, then $g(z) = 0$ for every $z \in \mathbb{C}$. Therefore $f(z) = h(z)$ for every $z \in \mathbb{C}$. Thus

$$\pi \cot \pi z = \frac{1}{z} + \sum_{n=1}^{\infty} \frac{2z}{z^2 - n^2} .$$

Hence, for $|z| < \pi$:

$$\cot z = \frac{1}{z} - 2z \sum_{n=1}^{\infty} \frac{1}{n^2 \pi^2 - z^2} . \qquad \blacksquare$$

In the next result, the function zeta appears:

H.   *For $|z| < \pi$:*

$$\cot z = \frac{1}{z} - 2 \sum_{k=1}^{\infty} \zeta(2k) \frac{z^{2k-1}}{\pi^{2k}}. \tag{18.15}$$

Proof: From

$$\frac{1}{n^2\pi^2 - z^2} = \frac{1}{n^2\pi^2} \times \frac{1}{1 - (z/n\pi)^2} = \sum_{k=0}^{\infty} \frac{z^{2k}}{(n\pi)^{2k+2}}.$$

Hence by (18.13):

$$\cot z = \frac{1}{z} - 2z \sum_{n=1}^{\infty} \sum_{k=0}^{\infty} \frac{z^{2k}}{(n\pi)^{2k+2}}.$$

Since the above double series converges absolutely for $|z| < \pi$, interchanging the order of summation gives

$$\cot z = \frac{1}{z} - 2z \sum_{k=0}^{\infty} \frac{z^{2k}}{\pi^{2k+2}} \left(\sum_{n=1}^{\infty} \frac{1}{n^{2k+2}} \right) = \frac{1}{z} - 2 \sum_{k=1}^{\infty} \zeta(2k) \frac{z^{2k-1}}{\pi^{2k}}. \quad \blacksquare$$

And now we derive Euler's famous expression for $\zeta(2k)$ in terms of Bernoulli numbers.

I.   *For $k \geq 1$:*

$$\zeta(2k) = (-1)^{k-1} \frac{(2\pi)^{2k}}{2(2k)!} B_{2k}. \tag{18.16}$$

Proof: It is enough to compare (18.3) and (18.15). For every $k \geq 1$, the coefficients of z^{2k-1} in both series are equal

$$(-1)^k \frac{2^{2k} B_{2k}}{(2k)!} = -2 \frac{\zeta(2k)}{\pi^{2k}}$$

hence

$$\zeta(2k) = (-1)^{k-1} \frac{(2\pi)^{2k}}{2(2k)!} B_{2k}. \quad \blacksquare$$

Explicitly, taking into account that

$$B_2 = \tfrac{1}{6}, \qquad B_4 = -\tfrac{1}{30}, \qquad B_6 = \tfrac{1}{42}, \quad \text{etc.}$$

Euler gave the values of the sums

$$\zeta(2) = \sum_{n=1}^{\infty} \frac{1}{n^2} = \frac{\pi^2}{6}, \tag{18.17}$$

$$\zeta(4) = \sum_{n=1}^{\infty} \frac{1}{n^4} = \frac{\pi^4}{90}, \tag{18.18}$$

$$\zeta(6) = \sum_{n=1}^{\infty} \frac{1}{n^6} = \frac{\pi^6}{945}, \qquad (18.19)$$

and so on.

It is now easy to determine the signs of the Bernoulli numbers; we give a new proof of part (2) of (**B**): Since $\zeta(2k) > 0$ it follows from (**I**) that $(-1)^{k-1}B_{2k} > 0$.

Concerning the absolute values of the Bernoulli numbers, we prove:

J.   (1) *If $k \geq 2$ then*

$$\frac{2(2k)!}{(2\pi)^{2k}} < |B_{2k}| < \frac{(2k)!}{12(2\pi)^{2k-2}} .$$

(2) *If $k \geq 3$ then*

$$|B_{2k}| > \frac{k(2k-1)}{6(2k+1)}|B_{2k-2}|$$

   and if $k \geq 4$ then $|B_{2k}| > |B_{2k-2}|$.

(3) *Asymptotically, as k tends to infinity*

$$|B_{2k}| \sim \frac{2(2k)!}{(2\pi)^{2k}} \sim 4\sqrt{\pi k}\left(\frac{k}{\pi e}\right)^{2k} .$$

(4) *For every integer $N \geq 1$:*

$$\lim_{k \to \infty} \frac{|B_{2k}|}{(2k)^N} = \infty.$$

Proof: (1)   If $1 < s < s'$ are real numbers then $\zeta(s) > \zeta(s') > 1$ and $\lim_{s \to \infty} \zeta(s) = 1$. Hence from (**I**), if $k \geq 2$:

$$\frac{2(2k)!}{(2\pi)^{2k}} < |B_{2k}| < \frac{2(2k)!}{(2\pi)^{2k}}\zeta(2) = \frac{(2k)!}{12(2\pi)^{2k-2}}$$

because $\zeta(2) = \pi^2/6$.

(2)   By (1), if $k \geq 3$ then

$$\frac{k(2k-1)}{6(2k+1)}|B_{2k-2}| < \frac{k(2k-1)(2k-2)!}{6(2k+1)12(2\pi)^{2k-4}} = \frac{(2\pi)^4(2k)!}{12^2(2k+1)(2\pi)^{2k}}$$

$$< \frac{2(2k)!}{(2\pi)^{2k}} < |B_{2k}|,$$

since $\pi^4/3^2(2k+1) < 2$.

For $k \geq 7$ we have $k(2k-1) > 6(2k+1)$ hence $|B_{2k}| > |B_{2k-2}|$. Also for $k = 3, 4, 5, 6$, we have

$$|B_6| = \tfrac{1}{42}, \qquad |B_8| = \tfrac{1}{30}, \qquad |B_{10}| = \tfrac{5}{66}, \qquad |B_{12}| = \tfrac{691}{2730},$$

with $\frac{1}{42} < \frac{1}{30} < \frac{5}{66} < \frac{691}{2730}$.

(3)   It follows from (**I**) that

$$|B_{2k}| \sim \frac{2(2k)!}{(2\pi)^{2k}}.$$

By Stirling's formula

$$(2k)! \sim \sqrt{2\pi}\sqrt{2k}\left(\frac{2k}{e}\right)^{2k}, \qquad (18.20)$$

$$|B_{2k}| \sim 4\sqrt{\pi k}\left(\frac{k}{\pi e}\right)^{2k}.$$

(4)   By the above, we have

$$\frac{|B_{2k}|}{(2k)^N} \sim \frac{4}{(2\pi)^{2k}}\sqrt{\pi k}\left(\frac{2k}{e}\right)^{2k}\frac{1}{(2k)^N} = \left(\frac{k}{\pi e}\right)^{2k-N}\frac{4\sqrt{\pi k}}{(2\pi e)^N},$$

thus

$$\lim_{k\to\infty}\frac{|B_{2k}|}{(2k)^N} = \infty. \qquad \blacksquare$$

18.1.4 Sums of Equal Powers of Successive Natural Numbers

Now we shall study the sums $\sum_{j=1}^n j^k$.

K.   *For every integer $k \geq 0$ there exists a polynomial $S_k(X) \in \mathbb{Q}[X]$ with the following properties:*

(1) *$S_k(X)$ has degree $k+1$, leading coefficient $1/(k+1)$, and constant term equal to zero.*

(2) *$(k+1)!\,S_k(X) \in \mathbb{Z}[X]$.*

(3) *$S_k(n) = \sum_{j=1}^n j^k$ for every $n \geq 1$.*

These polynomials satisfy the recurrence relation: $S_0(X) = X$ and if $k \geq 1$ then

$$\binom{k+1}{1}S_k(X) + \binom{k+1}{2}S_{k-1}(X) + \cdots + \binom{k+1}{k}S_1(X) + S_0(X)$$

$$= (X+1)^{k+1} - 1. \qquad (18.21)$$

Proof: The statement is true for $k = 0$. For $k \geq 1$ we have

$$(X+1)^{k+1} - X^{k+1} = \binom{k+1}{1}X^k + \binom{k+1}{2}X^{k-1} + \cdots + \binom{k+1}{k}X + 1.$$

Let $n \geq 1$ be an arbitrary integer and let X be successively equal to $1, 2, \ldots, n$. Adding up the relations so obtained, we have, by induction

$(n + 1)^{k+1} - 1$

$$= \binom{k+1}{1} \left(\sum_{j=1}^{n} j^k \right) + \binom{k+1}{2} S_{k-1}(n) + \cdots + \binom{k+1}{k} S_1(n) + S_0(n).$$

If $S_k(X) \in \mathbb{Q}[X]$ is defined by the relation

$(X + 1)^{k+1} - 1$

$$= \binom{k+1}{1} S_k(X) + \binom{k+1}{2} S_{k-1}(X) + \cdots + \binom{k+1}{k} S_1(X) + X,$$

then by induction on k, we see that $S_k(X)$ has degree $k + 1$, leading coefficient $1/(k+1)$, no constant term, $(k+1)! S_k(X) \in \mathbb{Z}[X]$, $S_k(n) = \sum_{j=1}^{n} j^k$ for every $n \geq 1$, and the recurrence relation (18.21) is satisfied. ∎

For example:

$S_0(X) = X,$

$S_1(X) = \frac{1}{2} X^2 + \frac{1}{2} X,$

$S_2(X) = \frac{1}{3} X^3 + \frac{1}{2} X^2 + \frac{1}{6} X,$

$S_3(X) = \frac{1}{4} X^4 + \frac{1}{2} X^3 + \frac{1}{4} X^2,$

$S_4(X) = \frac{1}{5} X^5 + \frac{1}{2} X^4 + \frac{1}{3} X^3 - \frac{1}{30} X,$

$S_5(X) = \frac{1}{6} X^6 + \frac{1}{2} X^5 + \frac{5}{12} X^4 - \frac{1}{12} X^2,$

$S_6(X) = \frac{1}{7} X^7 + \frac{1}{2} X^6 + \frac{1}{2} X^5 - \frac{1}{6} X^3 + \frac{1}{42} X,$

$S_7(X) = \frac{1}{8} X^8 + \frac{1}{2} X^7 + \frac{7}{12} X^6 - \frac{7}{24} X^4 + \frac{1}{12} X^2,$

$S_8(X) = \frac{1}{9} X^9 + \frac{1}{2} X^8 + \frac{2}{3} X^7 - \frac{7}{15} X^5 + \frac{2}{9} X^3 - \frac{1}{30} X,$

etc.

Euler expressed the coefficients of $S_k(X)$ (for $k \leq 16$) in terms of the Bernoulli numbers and he indicated how to compute the coefficients recursively

L.   *For every $k \geq 1$:*

$(k + 1)S_k(X) = X^{k+1}$

$$- \binom{k+1}{1} B_1 X^k + \binom{k+1}{2} B_2 X^{k-1} + \cdots + \binom{k+1}{k} B_k X. \quad (18.22)$$

Proof: Let $n \geq 2$ be an arbitrary integer. We consider the formal power series in the indeterminate T:

$$U(T) = k! T[1 + \exp(T) + \exp(2T) + \cdots + \exp((n-1)T)].$$

The coefficient of T^{k+1} is equal to

$$k! \left[\frac{1}{k!} + \frac{2^k}{k!} + \cdots + \frac{(n-1)^k}{k!} \right] = S_k(n-1).$$

On the other hand

$$U(T) = k! T \left(\frac{1 - \exp(nT)}{1 - \exp(T)} \right) = k! \left(\frac{T}{\exp(T) - 1} \right) [\exp(nT) - 1]$$

$$= k! \left(1 + \frac{B_1}{1!} T + \frac{B_2}{2!} T^2 + \frac{B_3}{3!} T^3 + \cdots \right)$$

$$\times \left(nT + \frac{n^2 T^2}{2!} + \frac{n^3 T^3}{3!} + \cdots \right).$$

By comparing the coefficients of T^{k+1} we have

$$S_k(n-1) = k! \left(\frac{n^{k+1}}{(k+1)!} + \frac{B_1}{1!} \frac{n^k}{k!} + \frac{B_2}{2!} \frac{n^{k-1}}{(k-1)!} + \cdots + \frac{B_k}{k!} n \right)$$

hence

$$(k+1)S_k(n-1)$$

$$= n^{k+1} + \binom{k+1}{1} B_1 n^k + \binom{k+1}{2} B_2 n^{k-1} + \cdots + \binom{k+1}{k} B_k n.$$

Since this holds for every $n \geq 2$, then

$$(k+1)S_k(X-1)$$

$$= X^{k+1} + \binom{k+1}{1} B_1 X^k + \binom{k+1}{2} B_2 X^{k-1} + \cdots + \binom{k+1}{k} B_k X. \tag{18.23}$$

But $S_k(n) = S_k(n-1) + n^k$ for every $n \geq 2$. Thus $S_k(X) = S_k(X-1) + X^k$ and since $B_1 = -\frac{1}{2}$ then

$$(k+1)S_k(X)$$

$$= X^{k+1} - \binom{k+1}{1} B_1 X^k + \binom{k+1}{2} B_2 X^{k-1} + \cdots + \binom{k+1}{k} B_k X. \quad \blacksquare$$

Writing (18.23) with $X = 1$, we obtain the recurrence relation (18.2) for the Bernoulli numbers.

Let us note the following congruences:

$$S_{2k}(p-1) \equiv \begin{cases} 0 \pmod{p} & \text{if } p - 1 \nmid 2k, \\ -1 \pmod{p} & \text{if } p - 1 \mid 2k. \end{cases} \tag{18.24}$$

The first part follows from (18.23), while the second part follows from $j^{p-1} \equiv 1 \pmod{p}$ when $1 \leq j \leq p - 1$.

Formula (18.22) may be rewritten as follows:

$$S_k(X) = \frac{X^{k+1}}{k+1} + \frac{X^k}{2} + \frac{1}{2}\binom{k}{1}B_2X^{k-1} + \frac{1}{4}\binom{k}{3}B_4X^{k-3}$$

$$+ \cdots + \frac{1}{k-1}\binom{k}{k-2}B_{k-1}X^2 + \frac{1}{k}\binom{k}{k-1}B_kX. \quad (18.25)$$

For even integers $2k$, formula (18.25) becomes

$$S_{2k}(X) = \frac{X^{2k+1}}{2k+1} + \frac{X^{2k}}{2} + \left(\sum_{i=1}^{k-1}\binom{2k}{2i}\frac{1}{2i+1}B_{2k-2i}X^{2i+1}\right) + B_{2k}X.$$

$$(18.26)$$

Indeed,

$$\frac{1}{2k+1}\binom{2k+1}{2j}B_{2j}X^{2k+1-2j} = \frac{1}{2k+1}\binom{2k+1}{2k+1-2j}B_{2j}X^{2k+1-2j}$$

$$= \binom{2k}{2j}\frac{1}{2k+1-2j}B_{2j}X^{2k+1-2j} = \binom{2k}{2i}\frac{1}{2i+1}B_{2k-2i}X^{2i+1}$$

with $i = k - j$. Then (18.26) follows at once from (18.22).

18.1.5 Quadratic Identities

We shall give here a quadratic relation satisfied by Bernoulli numbers, which was discovered by Euler.

Lemma 1.   *Let* $n \geq 1$, *and let* $U = \{u_1, u_2, \ldots, u_n\}$, $V = \{v_1, v_2, \ldots, v_n\}$ *be sequences of elements in an integral domain containing* $\mathbb{Z}$. *For every* $k = 1, \ldots, n$, *let*

$$U_k = \sum_{i=1}^{k} u_i, \qquad V_k = \sum_{i=1}^{k} v_i.$$

Then

$$U_nV_n = \sum_{j=1}^{n}(u_jV_j + v_jU_j) - \sum_{j=1}^{n}u_jv_j. \quad (18.27)$$

Proof:

$$U_nV_n = \sum_{i,j=1}^{n} u_iv_j$$

$$= u_1v_1 + (u_2V_2 + v_2U_2 - u_2v_2) + (u_3V_3 + v_3U_3 - u_3v_3)$$

$$+ \cdots + (u_nV_n + v_nU_n - u_nv_n) = \sum_{j=1}^{n}(u_jV_j + v_jU_j) - \sum_{j=1}^{n}u_jv_j.$$

This may be easily seen (for example, when $n = 4$) by arranging the products as indicated below and adding according to the different sectors.

u_1v_1	u_1v_2	u_1v_3	u_1v_4
u_2v_1	u_2v_2	u_2v_3	u_2v_4
u_3v_1	u_3v_2	u_3v_3	u_3v_4
u_4v_1	u_4v_2	u_4v_3	u_4v_4

∎

We give first the following expression for the product of the polynomials $S_k(X)$, $S_h(X)$:

M.   Let $k \geq 1$, $h \geq 1$. Then

$$S_k(X)S_h(X) = \frac{1}{k+1}\left[S_{h+k+1}(X) + \sum_{j=2}^{k}\binom{k+1}{j}B_j S_{h+k+1-j}(X)\right]$$

$$+ \frac{1}{h+1}\left[S_{h+k+1}(X) + \sum_{j=2}^{h}\binom{h+1}{j}B_j S_{h+k+1-j}(X)\right].$$

$$(18.28)$$

Proof: Let $n \geq 1$ and consider the sequences

$$U = \{1^k, 2^k, \ldots, n^k\}, \qquad V = \{1^h, 2^h, \ldots, n^h\}.$$

Then $U_j = S_k(j)$, $V_j = S_h(j)$, and

$$\sum_{j=1}^{n} u_j v_j = S_{k+h}(j) \qquad \text{for} \quad j = 1, 2, \ldots, n.$$

By the above lemma

$$S_k(n)S_h(n) + S_{k+h}(n) = \sum_{j=1}^{n}[j^k S_h(j) + j^h S_k(j)].$$

We compute now the right-hand side.
By (18.22):

$$(k+1)X^h S_k(X) = X^{h+k+1}$$

$$- \binom{k+1}{1}B_1 X^{h+k} + \sum_{j=2}^{k}\binom{k+1}{j}B_j X^{h+k+1-j}.$$

Let X be equal to 1, 2, $\ldots$, n in the above relation; adding up the relations so obtained

$$(k+1)\sum_{\ell=1}^{n}\ell^{h}S_{k}(\ell) = S_{h+k+1}(n) - \binom{k+1}{1}B_{1}S_{k+h}(n)$$

$$+ \sum_{j=2}^{k}\binom{k+1}{j}B_{j}S_{h+k+1-j}(n).$$

By interchanging h and k:

$$(h+1)\sum_{\ell=1}^{n}\ell^{k}S_{h}(\ell) = S_{h+k+1}(n) - \binom{h+1}{1}B_{1}S_{k+h}(n)$$

$$+ \sum_{j=2}^{h}\binom{h+1}{j}B_{j}S_{k+h+1-j}(n).$$

Noting that $B_1 = -\frac{1}{2}$, it follows that

$$S_{k}(n)S_{h}(n) = \frac{1}{k+1}\left[S_{h+k+1}(n) + \sum_{j=2}^{k}\binom{k+1}{j}B_{j}S_{h+k+1-j}(n)\right]$$

$$+ \frac{1}{h+1}\left[S_{h+k+1}(n) + \sum_{j=2}^{h}\binom{h+1}{j}B_{j}S_{h+k+1-j}(n)\right].$$

This holds for every $n \geq 1$, hence the statement is proved.   ∎

Here are some special cases. Taking $h = 1$, $k = 2$ gives the formula of Djamchid ben Massoud (1589) which was rediscovered by Fermat (1636):

$$S_1(X)S_2(X) = \tfrac{1}{6}[5S_4(X) + S_2(X)] \tag{18.29}$$

hence

$$S_4(X) = \tfrac{1}{5}S_2(X)[6S_1(X) - 1]. \tag{18.30}$$

As a particular case, we note

$$[S_k(X)]^2 = \frac{2}{k+1}\left[S_{2k+1}(X) + \sum_{j=2}^{k}\binom{k+1}{j}B_{j}S_{2k+1-j}(X)\right] \tag{18.31}$$

for $k \geq 1$.

Since $B_j = 0$ for j odd, $j \geq 3$, (18.31) means that each $[S_k(X)]^2$ belongs to the $\mathbb{Q}$-vector space generated by $\{S_{2k+1}(X), S_{2k-1}(X), S_{2k-3}(X), \ldots, S_{k_0}(X)\}$ (where $k_0 = k + 1$ or $k + 2$, whichever is odd). Several special cases had been known

$$[S_1(X)]^2 = S_3(X), \tag{18.32}$$

$$[S_2(X)]^2 = \tfrac{1}{3}[2S_5(X) + S_3(X)], \tag{18.33}$$

$$[S_3(X)]^2 = \tfrac{1}{2}[S_7(X) + S_5(X)], \tag{18.34}$$

(this formula is due to Jacobi):

$$[S_4(X)]^2 = \tfrac{1}{15}[6S_9(X) + 10S_7(X) - S_5(X)], \tag{18.35}$$

$$[S_5(X)]^2 = \tfrac{1}{6}[2S_{11}(X) + 5S_9(X) - S_7(X)]. \tag{18.36}$$

From the above relations between the polynomials $S_k(X)$, it is easy to obtain quadratic relations between Bernoulli numbers.

N.   *If* $k, h \geq 1$, *then*

$$
0 = \frac{1}{k+1}\left[B_{h+k+1} + \sum_{j=2}^{k}\binom{k+1}{j}B_j B_{h+k+1-j}\right]
$$
$$
+ \frac{1}{h+1}\left[B_{h+k+1} + \sum_{j=2}^{h}\binom{h+1}{j}B_j B_{h+k+1-j}\right]. \tag{18.37}
$$

and

$$B_k B_h$$

$$
= \frac{1}{2(k+1)}\left[(h+k+1)B_{h+k} + \sum_{j=2}^{k}\binom{k+1}{j}(h+k+1-j)B_j B_{h+k-j}\right]
$$
$$
+ \frac{1}{2(h+1)}\left[(h+k+1)B_{h+k} + \sum_{j=2}^{h}\binom{h+1}{j}(h+k+j-1)B_j B_{h+k-j}\right]. \tag{18.38}
$$

Proof: Equating the coefficients of X on both sides of (18.28) yields the first identity. The second one is obtained by comparing the coefficients of X^2.   ∎

As a special case

$$
B_k^2 = \frac{1}{k+1}\left[(2k+1)B_{2k} + \sum_{j=2}^{k}\binom{k+1}{j}(2k-j+1)B_j B_{2k-j}\right] \tag{18.39}
$$

for $k \geq 1$.

The following relation was found by Euler:

O.   *If* $m \geq 2$, *then*

$$
(2m+1)B_{2m} + \sum_{j=1}^{m-1}\binom{2m}{2j}B_{2j}B_{2m-2j} = 0. \tag{18.40}
$$

Proof: Applying (18.37) with $k = 2m - 2$, $h = 1$, gives

$$0 = \frac{1}{2m-1}\left[B_{2m} + \sum_{j=1}^{m-1} \binom{2m-1}{2j} B_{2j} B_{2m-2j} \right] + \tfrac{1}{2} B_{2m}$$

since $B_s = 0$ for s odd, $s \geq 3$. Thus

$$0 = \frac{2m+1}{2} B_{2m} + \sum_{j=1}^{m-1} \binom{2m-1}{2j} B_{2j} B_{2m-2j}.$$

Multiplying with $2m$ and noting that

$$2m \binom{2m-1}{2j} = \binom{2m}{2j}(2m - 2j),$$

then

$$0 = (2m+1)m B_{2m} + \sum_{j=1}^{m-1} \binom{2m}{2j}(2m - 2j) B_{2j} B_{2m-2j}. \qquad (18.41)$$

Applying again (18.38) with $k = 2m - 1$, $h = 1$ gives, since $B_{2m-1} = 0$:

$$0 = \frac{1}{2 \times 2m}\left[(2m+1)B_{2m} + \sum_{j=1}^{m-1} \binom{2m}{2j}(2m + 1 - 2j) B_{2j} B_{2m-2j} \right]$$

$$+ \frac{1}{2 \times 2}(2m+1)B_{2m},$$

hence

$$0 = (2m+1)(m+1) B_{2m} + \sum_{j=1}^{m-1} \binom{2m}{2j}(2m + 1 - 2j) B_{2j} B_{2m-2j}. \quad (18.42)$$

Subtracting (18.41) from (18.42) gives the relation of the statement. ∎

It is possible to use this identity to derive a new proof that $(-1)^{k-1} B_{2k} > 0$ for $k \geq 1$.

18.2 Arithmetical Properties

In this section we study the arithmetical properties of the rational numbers B_{2k}. We shall prove an important theorem about the denominator. Much less can be said about the numerator.

18.2.1 The Denominator of the Bernoulli Numbers

We recall the following terminology.

If $r = a/b$, with a, b nonzero integers, $\gcd(a, b) = 1$, if m is a nonzero integer, we say that m divides r when m divides a.

If $r, s \in \mathbb{Q}$, $m \in \mathbb{Z}$, $m \neq 0$, then we define the congruence modulo m, by saying that $r \equiv s \pmod{m}$ when m divides $r - s$.

If p is any prime number, we denote by $\mathbb{Z}_{p\mathbb{Z}}$ the ring of all rational numbers $r = a/b$ (as above) which are p-integral, that is, such that p does not divide b. Clearly

$$\mathbb{Z} = \bigcap_{p \text{ prime}} \mathbb{Z}_{p\mathbb{Z}}.$$

For every $k \geq 1$ we write

$$B_{2k} = \frac{N_{2k}}{D_{2k}}$$

where N_{2k}, D_{2k} are relatively prime nonzero integers and $D_{2k} > 0$. N_{2k} is the numerator and D_{2k} is the denominator of the Bernoulli number B_{2k}.

The important theorem of von Staudt and Clausen describes completely the denominator D_{2k}.

We begin by recalling (Chapter 17, (**B**)) that if p is any prime, for every $a = 1, 2, \ldots, p - 1$ there exists ω_a, the unique $(p - 1)$th root of 1 in $\mathbb{Z}_p$ (ring of p-adic integers), such that $\omega_a \equiv a \pmod{p}$.

We write $\omega_a = a + \rho_a p$ with $\rho_a \in \mathbb{Z}_p$.

It was proved in Chapter 17, (**D**), that

$$\sum_{a=1}^{p-1} \omega_a^k = \begin{cases} 0 & \text{when } p - 1 \nmid k, \\ p - 1 & \text{when } p - 1 \mid k. \end{cases} \tag{18.43}$$

For every $k \geq 1$, we define

$$\beta_k = \begin{cases} \dfrac{B_k}{k} & \text{when } p - 1 \nmid k, \\[2ex] \dfrac{pB_k - p + 1}{pk} & \text{when } p - 1 \mid k. \end{cases} \tag{18.44}$$

Lemma 2.   *If $k \geq 1$, and p is any prime then*

$$\beta_k + \sum_{a=1}^{p-1} a^{k-1}\rho_a + \sum_{j=2}^{k} \frac{1}{k}\binom{k}{j}p^{j-1}\left[\frac{B_{k+1-j}}{k+1-j} + \sum_{a=1}^{p-1} a^{k-j}\rho_a^j \right]$$

$$+ \frac{p^k}{(k+1)k} = 0. \tag{18.45}$$

Proof:

$$\sum_{a=1}^{p-1} \omega_a^k = \sum_{a=1}^{p-1} (a + \rho_a p)^k$$

$$= \sum_{a=1}^{p-1} \left[a^k + \binom{k}{1} a^{k-1} \rho_a p + \cdots + \binom{k}{j} a^{k-j} \rho_a^j p^j + \cdots + \rho_a^k p^k \right].$$

By (18.23), we have

$$\sum_{a=1}^{p-1} \omega_a^k = \frac{p^{k+1}}{k+1} + \sum_{j=1}^{k} p^j \binom{k}{j} \left[\frac{B_{k+1-j}}{k+1-j} + \sum_{a=1}^{p-1} a^{k-j} \rho_a^j \right]$$

after noting that

$$\frac{1}{k+1} \binom{k+1}{k+1-j} = \frac{1}{k+1-j} \binom{k}{j}.$$

Using (18.43) and (18.44) and dividing by pk we obtain the relation of the statement. ∎

Before proving the next result, we compute some p-adic values.
If p is a prime, $k \geq 1$, then

$$v_p \left(\frac{p^k}{k(k+1)} \right) \geq 0.$$

If $j \geq 2$, let $j = j_0 + j_1 p + j_2 p^2 + \cdots + j_h p^h$, with $h \geq 0$, $0 \leq j_i \leq p-1$, and let $s_j = j_0 + j_1 + \cdots + j_h \geq 1$.
By Chapter 17, Section 2, Lemma 3, we have, for $j > 2$:

$$v_p \left(\frac{p^{j-1}}{j!} \right) = j - 1 - v_p(j!) = j - 1 - \frac{j - s_j}{p - 1}$$

$$= \frac{p-2}{p-1} j + \frac{s_j}{p-1} - 1 \geq \frac{3p-5}{p-1} - 1 = 2 - \frac{2}{p-1}.$$

Thus if $p > 2$, $j > 2$, then

$$v_p \left(\frac{p^{j-1}}{j!} \right) > 1.$$

On the other hand, if $p > 2$, $j = 2$, then $v_p(p/2!) = 1$.
Now we prove:

P.   (1) *If $p \geq 3$ and $k \geq 1$ then $\beta_k \in \mathbb{Z}_p \mathbb{Z}$. Moreover, if $p - 1$ divides k, then $pB_k + 1 \in p\mathbb{Z}_p \mathbb{Z}$.*

   (2) *If $k \geq 1$ then $2B_k \in \mathbb{Z}_2 \mathbb{Z}$ and $2B_k + 1 \in 2\mathbb{Z}_2 \mathbb{Z}$ when $k = 1$ or k is even.*

Proof: (1)   If $k = 1$ then $\beta_1 = B_1 = -\frac{1}{2} \in \mathbb{Z}_{p\mathbb{Z}}$.

We proceed by induction on k, and we need to show that $v_p(\beta_k) \geq 0$ for $k \geq 2$. For this purpose, we compute the p-adic values of the summands in (18.45).

For each $a = 1, \ldots, p-1$, we have $v_p(a^{k-1}\rho_a) \geq 0$. As already indicated

$$v_p\left(\frac{p^k}{k(k+1)}\right) \geq 0 \quad \text{and} \quad v_p\left(\frac{1}{k}\binom{k}{j}p^{j-1}\right) \geq 1$$

when $p > 2$, $j \geq 2$.

It remains to consider the summands

$$\alpha_j = \frac{1}{k}\binom{k}{j}p^{j-1}\frac{B_{k+1-j}}{k+1-j},$$

where $j = 2, \ldots, k$.

If $p - 1$ does not divide $k + 1 - j$ then by induction

$$\frac{B_{k+1-j}}{k+1-j} = \beta_{k+1-j} \in \mathbb{Z}_{p\mathbb{Z}},$$

so $\alpha_j \in \mathbb{Z}_{p\mathbb{Z}}$.

If $p - 1$ divides $k + 1 - j$ then by induction

$$\frac{B_{k+1-j}}{k+1-j} - \frac{p-1}{p(k+1-j)} = \beta_{k+1-j} \in \mathbb{Z}_{p\mathbb{Z}}.$$

Since $p \nmid (k+1-j)$, if $j > 2$, then

$$\alpha_j = \frac{1}{k}\binom{k}{j}p^{j-1}\left(\frac{p-1}{p(k+1-j)} + \beta_{k+1-j}\right) \in \mathbb{Z}_{p\mathbb{Z}},$$

while

$$\alpha_2 = \frac{1}{2}\binom{k}{2}p\left(\frac{p-1}{p(k-1)} + \beta_{k-1}\right)$$

$$= \frac{(k-1)p}{2}\left(\frac{p-1}{p(k-1)} + \beta_{k-1}\right) \in \mathbb{Z}_{p\mathbb{Z}}.$$

If $p - 1$ divides k, since

$$\beta_k = \frac{pB_k - p + 1}{pk} \in \mathbb{Z}_{p\mathbb{Z}},$$

then $pB_k + 1 \in p\mathbb{Z}_{p\mathbb{Z}}$.

(2)   If $p = 2$ the proof is similar. The result is true when $k = 1$ and we proceed by induction.

In order to show that $2B_k \in \mathbb{Z}_{2\mathbb{Z}}$ it suffices to prove, as in part (1), that each summand

$$\alpha_j = \frac{1}{k}\binom{k}{j}2^j\frac{B_{k+1-j}}{k+1-j}$$

belongs to $\mathbb{Z}_{2\mathbb{Z}}$, for $j = 2, \ldots, k$.

For $j = 2$ we have $\alpha_2 = 2B_{k-1} \in \mathbb{Z}_{2\mathbb{Z}}$, by induction.

Similarly, if $j > 2$ then α_j is a multiple of $(k-1)\cdots(k-j+1)$ $2B_{k+1-j}/(k+1-j)$ hence, by induction it belongs to $\mathbb{Z}_{2\mathbb{Z}}$.

Thus, by (18.45), $2\beta_k \in \mathbb{Z}_{2\mathbb{Z}}$, hence $2B_k = 2k\beta_k + 1 \in \mathbb{Z}_{2\mathbb{Z}}$.

If $k = 1$ or k is even then $2B_k + 1 = 2k\beta_k + 2 \in 2\mathbb{Z}_{2\mathbb{Z}}$. ∎

It is now easy to obtain the theorem of von Staudt and Clausen (1840):

Theorem 1.   *Let $k \geq 1$. Then:*

(1)
$$B_{2k} + \sum_{\substack{p \text{ prime} \\ p-1|2k}} \frac{1}{p} \in \mathbb{Z}.$$

(2) *The denominator of B_{2k} is the product of distinct primes p, namely those such that $p-1$ divides $2k$. In particular, $6|D_{2k}$.*

Proof: (1)   Let q be an arbitrary prime. If $q - 1|2k$ then, by (**P**), $B_{2k} + 1/q \in \mathbb{Z}_{q\mathbb{Z}}$. Since $1/p \in \mathbb{Z}_{q\mathbb{Z}}$ when p is a prime distinct from q, then $B_{2k} + \sum_{p-1|2k} 1/p \in \mathbb{Z}_{q\mathbb{Z}}$ for every prime q such that $q - 1|2k$.

If $q - 1 \nmid 2k$ then $q \neq 2$ and, by (**P**), $\beta_{2k} = B_{2k}/2k \in \mathbb{Z}_{q\mathbb{Z}}$ so $B_{2k} \in \mathbb{Z}_{q\mathbb{Z}}$ and therefore

$$B_{2k} + \sum_{p-1|2k} \frac{1}{p} \in \mathbb{Z}_{q\mathbb{Z}}$$

(because each prime p above is different from q).

Thus

$$B_{2k} + \sum_{p-1|2k} \frac{1}{p} \in \bigcap_q \mathbb{Z}_{q\mathbb{Z}} = \mathbb{Z}.$$

(2)   The second assertion is now immediate. ∎

We have the following congruences:

Q.   *Let $k \geq 1$.*

(1) *If p is a prime and $p - 1$ divides $2k$ then $pB_{2k} \equiv -1 \pmod{p}$.*

(2) *If $m \geq 2$ then*

$$mN_{2k} \equiv D_{2k}S_{2k}(m-1) \pmod{m^2}.$$

In particular, if $p - 1$ does not divide $2k$ then $pB_{2k} \equiv S_{2k}(p-1)$ (mod p^2), hence $S_{2k}(p-1) \equiv 0$ (mod p), while if $p - 1$ divides $2k$ then $pB_{2k} \equiv S_{2k}(p-1)$ (mod p).

Proof: (1)   If $p - 1|2k$ then by Theorem 1:

$$B_{2k} = m - \frac{1}{p} - \sum_{\substack{q \neq p \\ q-1|2k}} \frac{1}{q},$$

where $m \in \mathbb{Z}$. Hence $pB_{2k} + 1 = p[m - (\sum 1/q)]$. The rational number in brackets is p-integral, so $pB_{2k} \equiv -1 \pmod{p}$.

(2)   By (18.23) we have

$$S_{2k}(m-1) = S_{2k}(m) - m^{2k} = \frac{m^{2k+1}}{2k+1} - \frac{m^{2k}}{2}$$

$$+ \left[\sum_{i=1}^{k-1} \binom{2k}{2i} \frac{1}{2i+1} B_{2(k-i)} m^{2i+1} \right] + B_{2k} m,$$

hence

$$D_{2k} S_{2k}(m-1) - mN_{2k} = -\frac{D_{2k}}{2} m^{2k}$$

$$+ \sum_{i=1}^{k} \binom{2k}{2i} \frac{D_{2k}}{2i+1} B_{2(k-i)} m^{2i+1}$$

so

$$D_{2k} S_{2k}(m-1) - mN_{2k} = -m^2 \left(\frac{D_{2k}}{2} m^{2k-2} \right)$$

$$+ m^2 \left[\sum_{i=1}^{k} \binom{2k}{2i} \frac{D_{2k}}{2i+1} B_{2(k-i)} m^{2i-1} \right].$$

We wish to prove that the right-hand side is an integer, multiple of m^2. Since 2 divides D_{2k} and $k \geq 1$ then $(D_{2k}/2)m^{2k-2} \in \mathbb{Z}$.

Let

$$D_{2k} \frac{B_{2(k-i)}}{2i+1} m^{2i-1} = \frac{c_i}{d_i}, \qquad i = 1, \ldots, k,$$

with $c_i, d_i \in \mathbb{Z}$, $\gcd(c_i, d_i) = 1$. We shall prove that $\gcd(d_i, m) = 1$.

We write

$$A = d_i D_{2k} N_{2(k-i)} m^{2i-1} \qquad \text{and} \qquad A' = c_i(2i+1)D_{2(k-i)},$$

so $A = A'$.

If p is a prime dividing both d_i and m, then $p \nmid c_i$. Since $p^2 \nmid D_{2(k-i)}$, by Theorem 1, p divides $2i+1$, and we write $2i+1 = p^a r$, with $p \nmid r$, $a \geq 1$. Taking the p-adic values of A, A' we have, by Theorem 1:

$$2i = 1 + (2i-1) \leq v_p(A) = v_p(A') \leq v_p(2i+1) + 1,$$

hence

$$p^a - 2 \leq p^a r - 2 = 2i - 1 \leq v_p(2i+1) = a.$$

If $p \geq 5$ or if $p = 3$, $a \geq 2$ we have $a + 3 < p^a$, which contradicts the above inequality. If $p = 3$, $a = 1$, noting that 3 divides D_{2k} and $D_{2(k-i)}$ then

$$3 \leq 2i+1 = 1+1+(2i-1) \leq v_3(A) = v_3(A') = v_3(2i+1) + 1 = 2$$

so we again reach a contradiction.

Thus, we have shown that

$$D_{2k}S_{2k}(m-1) - mN_{2k} = m^2 z + m^2 \frac{x}{y},$$

where $x, y, z \in \mathbb{Z}$ and $\gcd(m, y) = 1$, $\gcd(x, y) = 1$. So $m^2(x/y) \in \mathbb{Z}$ and this implies that $y = 1$. So

$$D_{2k}S_{2k}(m-1) \equiv mN_{2k} \pmod{m^2}.$$

The last assertion follows at once, using Theorem 1 and the above congruence. ∎

18.2.2 The Numerator of the Bernoulli Numbers

We shall now consider divisibility properties of the numerator of Bernoulli numbers. The results are much less conclusive than for the denominator. The first fact, which is classical, is a trivial consequence of (**P**):

R.   If $p - 1 \nmid 2k$, if $t \geq 1$, and p^t divides $2k$ then $p^t | N_{2k}$.

Proof: By (**P**), $\beta_{2k} = B_{2k}/2k \in \mathbb{Z}_{p\mathbb{Z}}$. Since $p^t | 2k$ then $p^t | N_{2k}$. ∎

The next result is due to Carlitz (1953):

S.   If $t \geq 1$, $p - 1 | 2k$ and $p^t | 2k$ then p^t divides the numerator of $B_{2k} + 1/p - 1$, except when $p = 2$, $k = 1$.

Proof: If $p = 2$, $k = 1$ then $t = 1$ and 2 does not divide the numerator of $B_2 + \frac{1}{2} - 1 = -\frac{1}{3}$.

Now let $p \geq 3$ or $p = 2$, $k \geq 2$.

By Lemma 2:

$$\beta_{2k} + \sum_{a=1}^{p-1} a^{2k-1}\rho_a + \sum_{j=2}^{2k} \frac{1}{2k}\binom{2k}{j} p^{j-1}\left[\frac{B_{2k+1-j}}{2k+1-j} + \sum_{a=1}^{p-1} a^{2k-j}\rho_a^j\right]$$
$$+ \frac{p^{2k}}{(2k+1)2k} = 0,$$

where $\rho_a \in \mathbb{Z}_{p\mathbb{Z}}$ for $a = 1, \ldots, p-1$.

As was indicated after Lemma 2, $p^{2k}/((2k+1)2k) \in \mathbb{Z}_{p\mathbb{Z}}$ and also

$$\frac{1}{2k}\binom{2k}{j} p^{j-1} = (2k-1) \cdots (2k-j+1)\frac{p^{j-1}}{j!} \in \mathbb{Z}_{p\mathbb{Z}},$$

because

$$v_p\left(\frac{p^{j-1}}{j!}\right) = j - 1 - \frac{j - s_j}{p - 1} \geq 0.$$

Let

$$\alpha_j = \frac{1}{2k} \binom{2k}{j} p^{j-1} \frac{B_{2k+1-j}}{2k+1-j}.$$

If $p \geq 3$, $j \geq 3$ then $v_p(p^{j-1}/j!) \geq 1$, as was shown after Lemma 2, hence

$$\alpha_j = (2k-1)\cdots(2k-j+2) \frac{p^{j-1}}{j!} B_{2k+1-j} \in \mathbb{Z}_{p\mathbb{Z}},$$

by (**P**).

If $p \geq 3$, $j = 2$ then $\alpha_2 = (p/2)B_{2k-1} = 0$ (when $k \geq 2$) or $\alpha_2 = -p/4 \in \mathbb{Z}_{p\mathbb{Z}}$ (when $k = 1$).

If $p = 2$ and $j \geq 4$ then $\alpha_j \in \mathbb{Z}_{2\mathbb{Z}}$ since 2 divides $(2k-1)(2k-2)\cdots(2k-j+2)$. If $j = 3$ then $\alpha_3 = (2k-1)(2^2/3!)B_{2k-2} \in \mathbb{Z}_{2\mathbb{Z}}$ and, finally, if $j = 2$ then $\alpha_2 = B_{2k-1} = 0 \in \mathbb{Z}_{2\mathbb{Z}}$, because $k \geq 2$.

Thus, in all cases, $\alpha_j \in \mathbb{Z}_{p\mathbb{Z}}$.

We deduce that

$$\beta_{2k} = \frac{pB_{2k} - p + 1}{2kp} \in \mathbb{Z}_{p\mathbb{Z}},$$

hence p^t divides $B_{2k} - 1 + 1/p$. ∎

The following result was given explicitly by Frobenius in 1910: We write

$$\frac{B_{2k}}{2k} = \frac{N'_{2k}}{D'_{2k}},$$

where N'_{2k}, D'_{2k} are relatively prime integers, $D'_{2k} > 0$. Then:

T.   *If $k \geq 1$ and p is a prime then $p|D_{2k}$ if and only if $p|D'_{2k}$.*

Proof:

$$\frac{B_{2k}}{2k} = \frac{N_{2k}}{2kD_{2k}} = \frac{N'_{2k}}{D'_{2k}}$$

so $N_{2k}D'_{2k} = 2kD_{2k}N'_{2k}$.

If $p|D_{2k}$ then $p \nmid N_{2k}$ so $p|D'_{2k}$.

Conversely, if $p|D'_{2k}$ then $p \nmid N'_{2k}$. If $p \nmid D_{2k}$ then, by (**P**), $p - 1 \nmid 2k$ and by the above relation $p|2k$, say $2k = p^t r$, with $t \geq 1$, $p \nmid r$. By (**R**), $p^t|N_{2k}$. Hence $p^{t+1}|2kD_{2k}N'_{2k}$, so $p^{t+1}|2k$, which is a contradiction. ∎

The next result is due to von Staudt.

U.   *If $k \geq 1$ consider the decomposition $k = k_1 k_2$ with $k_1 \geq 1$, $k_2 \geq 1$ and such that the prime factors of k dividing k_2 are precisely those which divide D_{2k}. Then k_1 divides N_{2k}.*

Proof: We have $\gcd(k_1, k_2) = \gcd(k_1, D_{2k}) = 1$.

Let p be any prime dividing k_1 and let p^t (with $t \geq 1$) be the exact power of p dividing k_1. Then $p \nmid D_{2k}$ so $p - 1 \nmid 2k$. Since $p^t | 2k$, by (**R**), p^t divides N_{2k}. This shows that $k_1 | N_{2k}$. ∎

For example, taking $k = 17$ then $k_1 = 17$, $k_2 = 1$, and 17 divides $N_{34} = 2\,577\,687\,858\,367$. Similarly, if $k = 22$ then $k_1 = 11$, $k_2 = 2$, and 11 divides $N_{44} = -27\,833\,269\,579\,301\,024\,235\,023$.

18.2.3 The Congruence of Kummer

The following congruence, first proved by Kummer in 1851, is quite useful from the practical point of view, since it allows us to reduce the index of the Bernoulli number by multiples of $p - 1$.

V.   *If p is a prime and $p - 1 \nmid 2k$ then*

$$\frac{B_{2k+p-1}}{2k + p - 1} \equiv \frac{B_{2k}}{2k} \pmod{p}.$$

Proof: The idea of the proof is to consider a formal power series whose coefficients, reduced modulo p, are known to have a period $p - 1$, and which, on the other hand, are related to $B_{2k}/2k$.

For this purpose we consider the set $\mathcal{S}$ of all formal power series

$$S(X) = \sum_{k=0}^{\infty} \frac{c_k}{k!} X^k$$

with the following properties:

(1) each c_k is a p-integral rational number;

(2) $c_{2k+p-1} \equiv c_{2k} \pmod{p}$ for every $k \geq 1$.

We note the following easy facts:

(a) If $a_1, \ldots, a_n$ ($n \geq 1$) are p-integral rational numbers and $S_1(X), \ldots, S_n(X) \in \mathcal{S}$ then $a_1 S_1(X) + \cdots + a_n S_n(X) \in \mathcal{S}$.

(b) If $a_1, a_2 \ldots$ are p-integral rational numbers, if $S_1(X), S_2(X), \ldots$ $\in \mathcal{S}$ and they have order $o(S_h(X)) \geq h$ (for every $h \geq 1$), then $\sum_{h=1}^{\infty} a_h S_h(X)$ is still a power series, which belongs to $\mathcal{S}$.

(c) If a is a p-integral rational number then $a \in \mathcal{S}$.

(d) If $n \geq 1$ then

$$\exp(nX) = \sum_{k=0}^{\infty} \frac{n^k}{k!} X^k$$

belongs to $\mathcal{S}$, because $n^{2k+p-1} \equiv n^{2k} \pmod{p}$ for every $k \geq 1$.

(e) If $m \geq 1$ then

$$(\exp(X) - 1)^m = \exp(mX) - \binom{m}{1} \exp((m-1)X)$$

$$+ \binom{m}{2} \exp((m-2)X) - \cdots + (-1)^m$$

belongs to $\mathcal{S}$.

This follows from (a), (c), and (d).

Now, let g be a primitive root modulo p, $1 < g < p$. Let

$$S(X) = \frac{gX}{\exp(gX) - 1} - \frac{X}{\exp(X) - 1} = \sum_{k=0}^{\infty} \frac{B_k(g^k - 1)}{k!} X^k.$$

Let $\exp(X) - 1 = Y$, so $\exp(gX) = (1+Y)^g$ hence

$$S(X) = \frac{gX}{(1+Y)^g - 1} - \frac{X}{Y} = X\,T(Y)$$

where

$$T(Y) = \frac{g}{(1+Y)^g - 1} - \frac{1}{Y} = \frac{g}{gY + \binom{g}{2}Y^2 + \cdots + Y^g} - \frac{1}{Y}.$$

Since $(1/g)\binom{g}{2}, (1/g)\binom{g}{3}, \ldots, 1/g$ are p-integral, by long Euclidean division

$$T(Y) = \frac{1}{Y + (1/g)\binom{g}{2}Y^2 + \cdots + (1/g)Y^g} - \frac{1}{Y} = \sum_{k=0}^{\infty} c_k Y^k,$$

where each coefficient c_k is p-integral. Hence, by (b) above,

$$T(Y) = \sum_{k=0}^{\infty} c_k(\exp(X) - 1)^k = \sum_{n=0}^{\infty} \frac{a_n}{n!} X^n$$

belongs to $\mathcal{S}$, because $(\exp(X) - 1)^k$ has order k.

Comparing the coefficients of X^{2k} in the two expressions of $S(X) = X \cdot T(Y)$, we have

$$\frac{B_{2k}(g^{2k} - 1)}{(2k)!} = \frac{a_{2k-1}}{(2k-1)!},$$

hence

$$\frac{B_{2k}}{2k}(g^{2k} - 1) = a_{2k-1}.$$

From $a_{2k-1} \equiv a_{2k-1+p-1} \pmod{p}$ it follows that

$$\frac{B_{2k+p-1}}{2k+p-1}(g^{2k+p-1}-1) \equiv \frac{B_{2k}}{2k}(g^{2k}-1) \pmod{p}.$$

But $p-1 \nmid 2k$ hence $g^{2k} \equiv g^{2k+p-1} \not\equiv 1 \pmod{p}$, so

$$\frac{B_{2k+p-1}}{2k+p-1} \equiv \frac{B_{2k}}{2k} \pmod{p}.$$ ■

We easily obtain the corollary:

W.   *If $p \neq 2,3$, then $6N_{2p} \equiv pD_{2p} \pmod{p^2}$.*

Proof: Indeed, since $p-1 \neq 1,2$ then $p-1 \nmid 2p$. Hence, by (**V**):

$$\frac{B_{2p}}{2p} \equiv \frac{B_{p+1}}{p+1} \equiv \frac{B_2}{2} = \frac{1}{12} \pmod{p}$$

so $6N_{2p} \equiv pD_{2p} \pmod{p^2}$. ■

E X E R C I S E S

1. Calculate B_k for $k \leq 20$.

2. Show that for every $k \geq 1$:
$$\binom{2k+2}{2}B_{2k} + \binom{2k+2}{4}B_{2k-2} + \cdots + \binom{2k+2}{2k}B_2 = k.$$

3. Show, for each $k \geq 1$:
$$\binom{2k+1}{2}B_{2k} + \binom{2k+1}{4}B_{2k-2} + \cdots + \binom{2k+1}{2k}B_2 = \frac{1}{2}.$$

4. Show that if $k > 1$ then $2B_{2k} \equiv 1 \pmod 4$.

5. Give a new proof that $(-1)^{k-1}B_{2k} > 0$.

   Hint: Apply (**O**).

6. Calculate $S_k(X)$ for $9 \leq k \leq 12$.

7. Show, for $k \geq 1$:
$$S_k(X) = \frac{X^{k+1}}{k+1} + \frac{X^k}{2} + \frac{1}{2}\binom{k}{1}B_2 X^{k-1} + \frac{1}{4}\binom{k}{3}B_4 X^{k-3}$$
$$+ \cdots + \frac{1}{k-1}\binom{k}{k-2}B_{k-1}X^2 + \frac{1}{k}\binom{k}{k-1}B_k X.$$

8. Show that, for every $k \geq 1$, $S_{2k+1}(X)$ may be written in a unique way as a linear combination with natural coefficients of $[S_1(X)]^2$, $[S_2(X)]^2, \ldots, [S_k(X)]^2$.

9. Let p, q, r, m be positive integers. Show that if

$$\left(\sum_{j=1}^{n} j^k \right)^p = \left(\sum_{j=1}^{n} j^m \right)^q,$$

for all $n \geq 1$, then $(k, p) = (3, 1)$ and $(m, q) = (1, 2)$ or vice versa.

10. Calculate $\zeta(2k)$ for $1 \leq k \leq 10$.

11. Show that

$$\log \left(\frac{\exp(X)}{X} \right) = \frac{X}{2} + \sum_{k=1}^{\infty} \frac{1}{(2k)!} \cdot \frac{B_{2k}}{2k} X^{2k}.$$

12. Prove the recurrence relation

$$\sum_{j=1}^{k} \binom{2k}{2j} 2^{2j} (2^{2j} - 1) B_{2j} - 2k = 0.$$

Hint: Use the fact that $\sin x = \tan x \cdot \cos x$ (for $|x| < \pi/2$) and the Taylor series for these functions.

13. Show that

$$\lim_{k \to \infty} \frac{|B_{2k}|^{1/2k}}{k/\pi e} = 1$$

and that

$$\frac{k}{\pi e} < |B_{2k}|^{1/2k}.$$

Hint: Use Stirling's formula

$$k! = \sqrt{2\pi k} \left(\frac{k}{e} \right)^k e^{\theta/12k},$$

where $0 < \theta < 1$.

14. Show that for every $k \geq 1$ and every prime p:

$$pB_{2k} \equiv S_{2k}(p) \pmod{\mathbb{Z}_{\mathbb{Z}p}}.$$

15. Let p be an odd prime. Show that

$$S_{(p-1)/2} \left(\frac{p-1}{2} \right) \equiv 2 \left(\left(\frac{2}{p} \right) - 2 \right) \times B_{(p-1)/2} \pmod{p}.$$

16. Show that if p is a prime number and $p \equiv 1 \pmod 3$ then B_{2p} has denominator equal to 6.

17. Show that for every even integer k there exist infinitely many even integers h such that B_k and B_h have the same denominator.

   Hint: Use Dirichlet's theorem on primes in arithmetic progressions.

18. Show that if k is odd then $4\left(2^{2k} - 1\right) B_{2k}/2k$ is an integer.

19. Let p be a prime such that $2p + 1$ is composite. Show that the numerator N_{2p} of B_{2p} has a prime factor $p \equiv 3 \pmod 4$.

20. Let $p_1, \ldots, p_r$ be primes greater than 3, let $k = (p_1 - 1)(p_2 - 1)$ $\cdots (p_r - 1)$ and let n be an integer such that $n \equiv 1 \pmod k$. Show that $p_1, \ldots, p_r$ do not divide the numerator of $B_{2n}/2n$.

21. Let $m \geq 3$ be odd and let $k \geq 1$. Show that

$$2(1 - 2^{\varphi(m)-2k})N_{2k} \equiv -2kD_{2k} \sum_{j=1}^{[m/2]} j^{2k-1} \pmod m.$$

22. The *Euler numbers* E_{2n} are defined by the Taylor series of the secant function (for $|x| < \pi/2$):

$$\sec x = 1 - \frac{E_2}{2!} x^2 + \frac{E_4}{4!} x^4 - \cdots + (-1)^n \frac{E_{2n}}{(2n)!} x^{2n} + \cdots.$$

Prove the recurrence relation for Euler numbers

$$E_{2n} + \binom{2n}{2} E_2 + \binom{2n}{4} E_4 + \cdots + \binom{2n}{2n-2} E_{2n-2} + 1 = 0.$$

23. Show that each Euler number E_{2n} is an integer.

24. Show that the Euler numbers are odd and satisfy the congruence

$$E_{2n} \equiv \binom{2n}{2} + \binom{2n}{4} + \cdots + \binom{2n}{2n-2} + 1 \pmod 2.$$

25. Compute E_{2k} for $1 \leq k \leq 10$.

26. Show that

$$E_{2k} = 1 - \sum_{j=1}^{k} \binom{2k}{2j - 1} \times \frac{2^{2j}(2^{2j} - 1)}{2j} \cdot B_{2j}.$$

   Hint: $\sec x = \sin x \cdot \tan x + \cos x$ (for $|x| < \pi/2$).

27. Show that if p is an odd prime and $k \geq 1$ then $E_{2k} \equiv E_{2k+(p-1)}$ $\pmod p$ and $E_{2p} \equiv -1 \pmod p$.

28. Show that for every $k \geq 1$, $E_{2k} \equiv -1 \pmod{6}$ and $E_{2k} \equiv E_{2k+4}$ (mod 10). Conclude that the Euler numbers have a last digit alternatively equal to 1 and 5.

19

Fermat's Last Theorem for Regular Prime Exponents

Around 1636 Fermat conjectured that if $n \geq 3$ and if x, y, z are integers such that $x^n + y^n = z^n$, then x, y, or z is equal to 0. This conjecture, usually called *Fermat's last theorem*, was proved true in 1995 by Wiles.

The classical result of Kummer on Fermat's last theorem is the main object of this chapter.

19.1 Regular Primes and the Lemma of Units

Definition 1.   An odd prime p is said to be *regular* if p does not divide the class number h of $K = \mathbb{Q}(\zeta)$; otherwise, p is called *irregular*.

A.   *The following conditions are equivalent:*
  (1) *p is a regular prime.*
  (2) *If I is any nonzero fractional ideal of K and if I^p is a principal ideal, then I is a principal ideal.*

Proof: (1) $\rightarrow$ (2)   If I is not a principal ideal, the class of I in the class group $Cl(K)$ has order p; so p divides the order h of $Cl(K)$ and therefore p is an irregular prime.

(2) $\rightarrow$ (1)   If p divides h, since $Cl(K)$ is an Abelian group, there exists an ideal I whose class has order p; thus I^p is a principal ideal, but I is not a principal ideal.   ∎

Kummer proved the following facts about the class number h of $K = \mathbb{Q}(\zeta)$:

(a)   The class number h^+ of the real cyclotomic field $K^+ = \mathbb{Q}(\zeta + \zeta^{-1})$ divides h, so we write $h = h^- h^+$, where h^- is an integer called the *relative class number*; note that h^- is not a class number, but just the quotient of h by h^+.

(b) h^- is given by the following expression:

$$h^- = |\gamma|/(2p)^{t-1},$$

where

$$t = \frac{p-1}{2}, \quad \gamma = G(\eta)G(\eta^3)\cdots G(\eta^{p-2}), \quad \eta = \cos\frac{2\pi}{p-1} + i\sin\frac{2\pi}{p-1},$$

$$G(X) = \sum_{j=0}^{p-2} g_j X^j, \quad g_j \equiv g^j \pmod{p}, \quad 1 \le g_j \le p-1,$$

and finally g is a primitive root modulo p, $1 \le g \le p-1$.

We note that since $|\gamma| = h^-(2p)^{t-1}$ then $|\gamma|$ is independent of the choices of η and of g. We may choose a primitive root g modulo p such that $g^{p-1} \equiv 1 \pmod{p^2}$.

(c)   For each $k = 2, \ldots, (p-1)/2$ let

$$\delta_k = \frac{\sin(k\pi/p)}{\sin(\pi/p)},$$

so δ_k is a real positive unit of K, with

$$\delta_k^2 = \frac{1-\zeta^k}{1-\zeta} \cdot \frac{1-\zeta^{-k}}{1-\zeta^{-1}}.$$

The $(p-3)/2$ units $\delta_2, \ldots, \delta_{(p-1)/2}$ are the circular units. They are multiplicatively independent and generate an Abelian free multiplicative group V of rank $(p-3)/2$.

The index of V in the group U^+ of units of K^+ is finite and $(U^+ : V) = h^+$.

We shall prove these results in Part Four, Chapter 27.

Before proceeding, we need the following:

Lemma 1.   *If g is a primitive root modulo p, there exists g' such that $g' \equiv g \pmod{p}$ and $g'^{\,p-1} \equiv 1 \pmod{p^2}$.*

Proof: If $g^{p-1} = 1 + bp$, $b \in \mathbb{Z}$, let $a \in \mathbb{Z}$ and consider the congruence

$$(g + ap)^{p-1} \equiv g^{p-1} + (p-1)g^{p-2}ap \equiv 1 + bp - g^{p-2}ap \pmod{p^2}.$$

Choosing a such that $g^{p-2}a \equiv b \pmod{p}$, then $g + ap$ has the required property. Note also that $\overline{g + ap} = \overline{g}$.   ∎

Let η be a primitive $(p-1)$th root of 1. Let B denote the ring of integers of $\mathbb{Q}(\eta)$. We recall that (see Chapter 16, **(A)**):

$$B = \mathbb{Z}[\eta], \quad [\mathbb{Q}(\eta) : \mathbb{Q}] = \varphi(p-1), \quad \Phi_{p-1}(X) = \prod(X - \eta^k),$$

(product over k, $1 \le k \le p-2$, and $\gcd(k, p-1) = 1$), and $\Phi_{p-1}(X)$ divides $X^{p-1} - 1$. Reducing modulo the ideal Bp, we have $\overline{\Phi}_{p-1}(X) = \prod(X - \overline{\eta}^k)$, where $\overline{\eta}$ denotes the image of η modulo Bp.

We have $\bar{\eta}^{p-1} = \bar{1}$, thus $\bar{\eta}$ belongs to $\mathbb{F}_p$.

B. $\bar{\eta}$ *is a generator of* $\mathbb{F}_p^{\cdot}$.

Proof: We show that $\bar{\eta}^d \neq \bar{1}$ for every d, $1 \leq d < p - 1$. Otherwise, $\eta^d - 1 \in Bp$, i.e., p divides $\eta^d - 1$ for some d, $1 \leq d < p-1$. Let $e \geq 1$ be the largest integer such that p^e divides $\eta^d - 1$, so $\eta^d = 1 + p^e\beta$ (with $\beta \in B$ and p does not divide β). Then $1 = \eta^{(p-1)d} = (1 + p^e\beta)^{p-1} \equiv 1 + (p-1)p^e\beta$ (mod Bp^{e+1}) hence $p^e\beta \equiv 0$ (mod Bp^{e+1}) so p divides β, which is a contradiction. Therefore, $\bar{\eta}^k \neq \bar{\eta}^j$ when $1 \leq k, j \leq p - 2$, and $k \neq j$.   ∎

If η is a primitive root of 1 of order $p - 1$ and g is a primitive root modulo p, it follows from (**B**) that there exists k_0, $1 \leq k_0 \leq p-1$, $\gcd(k_0, p-1) = 1$, such that $\eta \equiv g^{k_0}$ (mod Bp).

We shall now describe the prime ideals of B which divide Bp:

C.   *Let* g *be a primitive root modulo* p, *and* η *any primitive* $(p - 1)th$ *root of* 1.

   (1) *Bp is the product of* $\varphi(p - 1)$ *distinct prime ideals, each with norm equal to* p, *namely* $Bp + B(\eta - g^k)$ *with* $1 \leq k \leq p - 2$, $\gcd(k, p - 1) = 1$.

   (2) *The prime ideals of B dividing Bp may be labeled as follows:*

$$P_k^{(g,\eta)} = Bp + B(1 - g\eta^k)$$

   with $1 \leq k \leq p - 2$, $\gcd(k, p - 1) = 1$. *Moreover, this labeling depends only on the residue class* $\bar{g}$ *of* g *modulo* p *and on* η, *the primitive root of unity.*

   (3) *If P is any prime ideal of B dividing Bp, if g is any primitive root modulo* p, *it is possible to choose* η, *a primitive* $(p - 1)th$ *root of* 1, *such that* $P = P_{p-2}^{(g,\eta)}$.

Proof: (1)   We apply Kummer's theory of decomposition of ideals (see Chapter 11, Theorem 2). From $\eta \equiv g^{k_0}$ (mod Bp) with $\gcd(k_0, p-1) = 1$, it follows that

$$\Phi_{p-1}(X) = \prod_{\bar{k} \in P(p-1)} (X - \eta^k) \equiv \prod_{\bar{k} \in P(p-1)} (X - g^{k_0 k})$$

$$\equiv \prod_{\bar{l} \in P(p-1)} (X - g^l) \text{ (mod } Bp),$$

hence Bp is the product of the $\varphi(p-1)$ distinct prime ideals $Bp + B(\eta - g^l)$, for $1 \leq l \leq p - 2$, $\gcd(l, p - 1) = 1$.

   Moreover, since $p \equiv 1$ (mod $p - 1$), each of these prime ideals has norm equal to p.

   (2)   Given g, η, and k, $1 \leq k \leq p - 2$, $\gcd(k, p - 1) = 1$, let $P_k^{(g,\eta)} = Bp + B(1 - g\eta^k)$. We note that $P_k^{(g,\eta)}$ depends on k, η and only on the residue class $\bar{g}$ of g modulo p.

We show that $P_k = P_k^{(g,\eta)}$ is a prime ideal. Indeed, if $\eta \equiv g^{k_0} \pmod{p}$, with $1 \le k_0 \le p - 2$, $\gcd(k_0, p - 1) = 1$, then

$$1 - g\eta^k = \eta^{-1}(\eta - g\eta^{k+1}) \equiv \eta^{-1}(\eta - g^{1+k_0(k+1)})$$
$$\equiv \eta^{-1}(\eta - g^l) \pmod{Bp},$$

where $l \equiv 1 + k_0(k+1) \pmod{p-1}$, $1 \le l \le p-2$. So $P_k = Bp + B(\eta - g^l)$ is a prime ideal of B.

If $1 \le k, l \le p - 2$, $\gcd(k, p - 1) = \gcd(l, p - 1) = 1$, and $P = P_k^{(g,\eta)} = P_l^{(g,\eta)}$ then $(1 - g\eta^k) - (1 - g\eta^l) = g(\eta^l - \eta^k) \in P$ and since g is a unit, then $\overline{\eta}^l = \overline{\eta}^k$ in $\mathbb{F}_p$. This implies that $k = l$.

So the mapping $k \mapsto P_k^{(g,\eta)}$ is one-to-one. But there are exactly $\varphi(p - 1)$ prime ideals dividing Bp, so the mapping is onto the set of these prime ideals, as was stated.

(3)   Given P and g, let η be a primitive $(p - 1)$th root of 1, and let $P = P_k^{(g,\eta)}$, with $1 \le k \le p - 2$, $\gcd(k, p - 1) = 1$. Then $1 - g\eta^k \in P$. Let $\eta_1 = \eta^{-k}$ then $1 - g\eta_1^{p-2} = 1 - g\eta^k \in P$ so $P = P_{p-2}^{(g,\eta_1)}$.   ∎

Our aim now is to characterize the regular primes. Here is a first result.

D. p *divides* h^- *if and only if there exists an integer* k, $1 \le k \le (p - 3)/2$, *such that* p^2 *divides the sum*

$$S_{2k}(p - 1) = \sum_{j=1}^{p-1} j^{2k}.$$

Proof: We choose a primitive root g modulo p, such that $g^{p-1} \equiv 1 \pmod{p^2}$; this is possible by Lemma 1.

Since $h^- = |\gamma|/(2p)^{t-1}$ as indicated before, p divides h^- if and only if Bp divides the ideal generated by γ/p^{t-1}. Since Bp is the product of distinct prime ideals, then p divides h^- if and only if each prime ideal P which divides Bp also divides $B(\gamma/p^{t-1})$.

Given any such prime ideal P, by (C), there is η, a $(p - 1)$th root of 1, such that $P = P_{p-2}^{(g,\eta)}$, that is, P divides $B(1 - g\eta^{p-2})$. But, as indicated before, $\gamma = G(\eta)G(\eta^3) \cdots G(\eta^{p-2})$, so

$$B\left(\frac{\gamma}{p^{t-1}}\right) = B\left(\frac{G(\eta)}{p}\right) \times B\left(\frac{G(\eta^3)}{p}\right) \times \cdots \times B\left(\frac{G(\eta^{p-2})}{p}\right).$$

Thus if P divides $B(\gamma/p^{t-1})$ then there exists k odd, $1 \le k \le p - 2$, such that P^2 divides $BG(\eta^k)$. Conversely, if P^2 divides $BG(\eta^k)$ then P^2/Bp divides $B(G(\eta^k)/p)$, so if $J = Bp/P$ then P divides $JB(\gamma/p^{t-1})$, and since P does not divide J, then P divides $B(\gamma/p^{t-1})$.

Actually $k \leq p - 4$, because P does not divide $BG(\eta^{p-2})$. Indeed, $P = P_{p-2}^{(g,\eta)}$ contains $1 - g\eta^{p-2}$, therefore

$$G(\eta^{p-2}) = \sum_{j=0}^{p-2} g_j \eta^{(p-2)j} \equiv \sum_{j=0}^{p-2} (g\eta^{p-2})^j \equiv p - 1 \equiv -1 \pmod{P}$$

since $g_j \equiv g^j \pmod{p}$, $g\eta^{p-2} \equiv 1 \pmod{P}$. This shows that P does not divide $BG(\eta^{p-2})$.

We shall express the above condition in a different way involving only rational integers.

Since $1 - g^{p-1} = \prod_{j=0}^{p-2}(1 - g\eta^j) \equiv 0 \pmod{p^2}$ and since $P = P_{p-2}^{(g,\eta)}$ does not contain the elements $1 - g\eta^j$ for $j < p - 2$, then P^2 divides $1 - g\eta^{p-2}$; thus $g\eta^{-1} \equiv g\eta^{p-2} \equiv 1 \pmod{P^2}$ hence $g \equiv \eta \pmod{P^2}$. Therefore

$$G(\eta^k) = \sum_{j=0}^{p-2} g_j \eta^{jk} \equiv \sum_{j=0}^{p-2} g_j g^{jk} \pmod{P^2}$$

and P^2 divides $BG(\eta^k)$ exactly when P^2 divides $B(\sum_{j=0}^{p-2} g_j g^{jk})$.

Taking conjugates, if P' is any prime ideal of B dividing Bp then P^2 divides $B(\sum_{j=0}^{p-2} g_j g^{jk})$ exactly when P'^2 divides $B(\sum_{j=0}^{p-2} g_j g^{jk})$. Since p is unramified in B, this means that p^2 divides $\sum_{j=0}^{p-2} g_j g^{jk}$.

Now we express this sum in a different way.

Let $g_j \equiv g^j + a_j p \pmod{p^2}$ with $a_j \in \mathbb{Z}$. So

$$g_j^{k+1} \equiv g^{j(k+1)} + (k+1)g^{jk}pa_j \equiv g^{j(k+1)} + (k+1)g^{jk}(g_j - g^j)$$
$$\equiv (k+1)g_j g^{jk} - kg^{j(k+1)} \pmod{p^2}.$$

Adding these relations for $j = 0, 1, \ldots, p - 2$ we have

$$\sum_{j=0}^{p-2} g_j^{k+1} \equiv (k+1)\left[\sum_{j=0}^{p-2} g_j g^{jk}\right] - k\left[\sum_{j=0}^{p-2} g^{j(k+1)}\right] \pmod{p^2}.$$

But

$$\sum_{j=0}^{p-2} g^{j(k+1)} = \frac{g^{(p-1)(k+1)} - 1}{g^{k+1} - 1} \equiv 0 \pmod{p^2}$$

because $g^{k+1} \not\equiv 1 \pmod{p}$ when $k \leq p - 4$ and $g^{p-1} \equiv 1 \pmod{p^2}$.

Hence

$$\sum_{j=0}^{p-2} g_j^{k+1} \equiv (k+1)\left[\sum_{j=0}^{p-2} g_j g^{jk}\right] \pmod{p^2}$$

and since $k + 1 \not\equiv 0 \pmod{p}$ then p^2 divides $\sum_{j=0}^{p-2} g_j g^{jk}$ if and only if p^2 divides $\sum_{j=0}^{p-2} g_j^{k+1}$. But the sets of integers $\{g_0, g_1, \ldots, g_{p-2}\}$ and

404   19. Fermat's Last Theorem for Regular Prime Exponents

$\{1, 2, \ldots, p-1\}$ coincide because g is a primitive root modulo p. So the condition becomes: p^2 divides $\sum_{j=1}^{p-1} j^{k+1} = S_{k+1}(p-1)$ where $k \in \{1, 3, \ldots, p-4\}$ or, still, p^2 divides $S_k(p-1)$ where $k \in \{2, 4, \ldots, p-3\}$. ∎

In Chapter 18, (**L**), we have expressed the sums $S_k(n) = \sum_{j=1}^{n} j^k$ in terms of Bernoulli numbers.

We may now derive easily:

E.   *p divides h^- if and only if p divides the numerator of at least one of the Bernoulli numbers B_2, B_4, $\ldots$, B_{p-3}.*

Proof: By (**D**), p divides h^- if and only if there exists k, $2 \le k \le p-3$, such that p^2 divides $S_{2k}(p-1)$. By Chapter 18, (**Q**), we have the congruence

$$D_{2k} S_{2k}(p-1) \equiv pN_{2k} \pmod{p^2},$$

where $B_{2k} = N_{2k}/D_{2k}$, N_{2k}, D_{2k} are integers, $D_{2k} > 0$, and $\gcd(N_{2k}, D_{2k}) = 1$.

Hence p divides h^- if and only if p^2 divides pN_{2k}, that is, p divides N_{2k}, for some k, $2 \le 2k \le p-3$. ∎

Now we shall study the divisibility of h^+ by the prime p, as well as the units.

We recall from Chapter 10, Section 3, that if $k = 2, \ldots, (p-1)/2$ then

$$\delta_k = \sqrt{\frac{1-\zeta^k}{1-\zeta} \cdot \frac{1-\zeta^{-k}}{1-\zeta^{-1}}}$$

are the circular units. Let S be the $\mathbb{Z}_p$-module of all real λ-adic integers with trace zero (see Chapter 17, (**O**)). We recall that $\log \delta_k^{p-1} \in S$ for $k = 2, \ldots, (p-1)/2$ (see Chapter 17, (**P**)).

F.   *If p does not divide h^- then*

$$\{\log \delta_2^{p-1}, \log \delta_3^{p-1}, \ldots, \log \delta_{(p-1)/2}^{p-1}\}$$

is a basis of the $\mathbb{Z}_p$-module S.

Proof: By Chapter 17, (**O**), we may write

$$\log(\delta_k^{p-1}) = \sum_{i=1}^{(p-3)/2} a_{ik} \rho^{2i} \quad (k = 2, \ldots, (p-1)/2),$$

where each $a_{ik} \in \mathbb{Z}_p$. It suffices to show that $\det((a_{ik})_{i,k})$ is invertible in $\mathbb{Z}_p$.

From Chapter 17, (**P**), $\delta_k^{p-1} \equiv 1 \pmod{\widehat{A}\lambda}$. Since $\widehat{A}\lambda = \widehat{A}\rho$ it follows from Chapter 17, (**O**), that $\delta_k^{p-1} = 1 + \alpha$ where $\alpha \in \widehat{A}\lambda^2$. By Chapter 17, (**L**), $\log(\delta_k^{p-1}) \equiv L_p(\delta_k^{p-1}) \pmod{\widehat{A}\lambda^p}$.

But

$$\delta_k^2 = \frac{1 - \zeta^k}{1 - \zeta} \cdot \frac{1 - \zeta^{-k}}{1 - \zeta^{-1}}$$

$$= (1 + \zeta + \cdots + \zeta^{k-1})(1 + \zeta^{-1} + \cdots + \zeta^{-(k-1)}) \equiv k^2 \pmod{\widehat{A}\lambda},$$

since $\lambda = 1 - \zeta$, $\zeta^{-1}\lambda = \zeta^{-1} - 1$.

Hence $\delta_k^{2p} \equiv k^{2p} \pmod{\widehat{A}\lambda^p}$ since $p \equiv 0 \pmod{\widehat{A}\lambda^{p-1}}$. But $k^p \equiv k \pmod{\widehat{A}\lambda^{p-1}}$ since $p \equiv 0 \pmod{\widehat{A}\lambda^{p-1}}$ and

$$\delta_k^{2(p-1)} \equiv \delta_k^{-2}k^2 \equiv \frac{(1 - \zeta)(1 - \zeta^{-1})k^2}{(1 - \zeta^k)(1 - \zeta^{-k})}$$

$$\equiv \frac{\zeta - 1}{\rho} \times \frac{\zeta^{-1} - 1}{-\rho} \times \left(\frac{\zeta^k - 1}{k\rho}\right)^{-1} \times \left(\frac{\zeta^{-k} - 1}{-k\rho}\right)^{-1}$$

$$\equiv \frac{\zeta - 1}{\rho} \times \frac{\zeta^{p-1} - 1}{(p-1)\rho} \times \left(\frac{\zeta^k - 1}{k\rho}\right)^{-1}$$

$$\times \left(\frac{\zeta^{p-k} - 1}{(p-k)\rho}\right)^{-1} \pmod{\widehat{A}\lambda^{p-1}}.$$

Since $\zeta - 1 = -\lambda \equiv \rho \pmod{\widehat{A}\lambda^2}$ then $(\zeta - 1)/\rho \equiv 1 \pmod{\widehat{A}\lambda}$, hence for any integer $j = 1, \ldots, p - 1$, we have also

$$\frac{\zeta^j - 1}{j\rho} = \frac{(\zeta - 1)(\zeta^{j-1} + \cdots + \zeta + 1)}{j\rho} \equiv \frac{\zeta - 1}{\rho} \equiv 1 \pmod{\widehat{A}\lambda}.$$

It follows from Chapter 17, (**L**), and Chapter 17, Lemma 4, that

$$L_p(\delta_k^{2(p-1)}) \equiv L_p\left(\frac{\zeta - 1}{\rho}\right) + L_p\left(\frac{\zeta^{p-1} - 1}{(p-1)\rho}\right)$$

$$- L_p\left(\frac{\zeta^k - 1}{k\rho}\right) - L_p\left(\frac{\zeta^{p-k} - 1}{(p-k)\rho}\right) \pmod{\widehat{A}\lambda^{p-1}}.$$

By Chapter 17, (**K**):

$$\frac{\zeta^j - 1}{j\rho} \equiv \frac{E_p(j\rho) - 1}{j\rho} \pmod{\widehat{A}\lambda^{p-1}}.$$

Therefore

$$L_p(\delta_k^{2(p-1)}) \equiv L_p\left(\frac{E_p(\rho) - 1}{\rho}\right)$$

$$+ L_p\left(\frac{E_p((p-1)\rho) - 1}{(p-1)\rho}\right) - L_p\left(\frac{E_p(k\rho) - 1}{k\rho}\right)$$

$$- L_p\left(\frac{E_p((p-k)\rho) - 1}{(p-k)\rho}\right) \pmod{\widehat{A}\lambda^{p-1}}.$$

Since $(E_p(X) - 1)/X \equiv (\exp(X) - 1)/X$ (ord $p - 1$) then

$$L_p\left(\frac{E_p(X) - 1}{X}\right) \equiv L_p\left(\frac{\exp(X) - 1}{X}\right) \equiv \log\frac{\exp(X) - 1}{X} \text{ (ord } p - 1\text{).}$$

It follows from Chapter 18, (C), that

$$L_p\left(\frac{E_p(X) - 1}{X}\right) \equiv \frac{X}{2} + \sum_{i=1}^{(p-3)/2} \frac{B_{2i}}{(2i)!\, 2i} X^{2i} \text{ (ord } p - 1\text{).}$$

Hence

$$L_p\left(\frac{E_p(j\rho) - 1}{j\rho}\right) \equiv \frac{j\rho}{2} + \sum_{i=1}^{(p-3)/2} \frac{B_{2i}}{(2i)!\, 2i} (j\rho)^{2i} \pmod{\widehat{A}\lambda^{p-1}}.$$

Therefore

$$L_p(\delta_k^{2(p-1)}) \equiv \frac{\rho + (p-1)\rho + k\rho + (p-k)\rho}{2}$$

$$+ \sum_{i=1}^{(p-3)/2} \frac{B_{2i}}{(2i)!\, 2i} \{\rho^{2i} + [(p-1)\rho]^{2i} - (k\rho)^{2i} - [(p-k)\rho]^{2i}\}$$

$$\equiv \sum_{i=1}^{(p-3)/2} \frac{2B_{2i}}{(2i)!\, 2i} (1 - k^{2i})\rho^{2i} \pmod{\widehat{A}\lambda^{p-1}}.$$

Since $\delta_k^{p-1} \equiv 1 \pmod{\widehat{A}\lambda^2}$, by Chapter 17, Lemma 4, and Chapter 17, (L):

$$\log(\delta_k^{p-1}) \equiv L_p(\delta_k^{p-1}) \equiv \sum_{i=1}^{(p-3)/2} \frac{B_{2i}}{(2i)!\, 2i} (1 - k^{2i})\rho^{2i} \pmod{\widehat{A}\lambda^{p-1}}.$$

From the unique expression of $\log(\delta_k{}^{p-1})$ it follows that

$$a_{ik} \equiv \frac{B_{2i}(1 - k^{2i})}{(2i)!\, 2i} \pmod{p},$$

for $i = 1, \ldots, (p - 3)/2$ and $k = 2, 3, \ldots, (p - 1)/2$.
    Hence

$$\det((a_{ik})) = \prod_{i=1}^{(p-3)/2} \frac{B_{2i}}{(2i)!\, 2i} D,$$

where

$$D = \det \begin{pmatrix} 1 - 2^2 & 1 - 3^2 & \cdots & 1 - \left(\dfrac{p-1}{2}\right)^2 \\ 1 - 2^4 & 1 - 3^4 & \cdots & 1 - \left(\dfrac{p-1}{2}\right)^4 \\ \cdots\cdots\cdots\cdots\cdots\cdots\cdots\cdots \\ 1 - 2^{p-3} & 1 - 3^{p-3} & \cdots & 1 - \left(\dfrac{p-1}{2}\right)^{p-3} \end{pmatrix}.$$

But D may be easily computed: from

$$(1 - k^{2i}) - (1 - k^{2(i-1)}) = k^{2(i-1)}(1 - k^2)$$

then

$$D = (1 - 2^2)(1 - 3^2)\cdots\left(1 - \left(\dfrac{p-1}{2}\right)^2\right)$$

$$\times \det \begin{pmatrix} 1 & 1 & \cdots & 1 \\ 2^2 & 3^2 & \cdots & \left(\dfrac{p-1}{2}\right)^2 \\ 2^4 & 3^4 & \cdots & \left(\dfrac{p-1}{2}\right)^4 \\ \cdots\cdots\cdots\cdots\cdots\cdots\cdots\cdots \\ 2^{p-5} & 3^{p-5} & \cdots & \left(\dfrac{p-1}{2}\right)^{p-5} \end{pmatrix}$$

$$= \prod_{1 \le l < m \le (p-1)/2} (l^2 - m^2) = \prod_{1 \le l < m \le (p-1)/2} (l + m)(l - m) \not\equiv 0 \pmod{p}.$$

Since p does not divide h^-, by (**E**), p does not divide the numerators of the Bernoulli numbers B_2, B_4, $\ldots$, B_{p-3}. Hence $\det(a_{ik}) \not\equiv 0 \pmod{p}$. Therefore it is invertible in $\mathbb{Z}_p$. It follows that

$$\{\log \delta_k^{p-1} \mid k = 2, \ldots, (p-1)/2\}$$

also constitutes a basis of the $\mathbb{Z}_p$-module S.   ∎

Kummer proved the following important and rather surprising result:

G.   *If p divides h^+, then p divides h^-.*

Proof: Let U^+ be the group of real positive units of K, let V be the subgroup generated by the circular units δ_k, for $k = 2, \ldots, (p-1)/2$. Then $h^+ = (U^+ : V)$, as was already mentioned.

If p divides h^+ there exists $\varepsilon \in U^+$ such that $\varepsilon \notin V$ but $\varepsilon^p \in V$. So

$$\varepsilon^p = \prod_{k=2}^{(p-1)/2} \delta_k^{e_k}$$

with integers e_k not all multiples of p, because $\varepsilon \notin V$.

Then

$$\varepsilon^{(p-1)p} = \prod_{k=2}^{(p-1)/2} (\delta_k^{p-1})^{e_k}.$$

We have $p \log(\varepsilon^{p-1}) = \sum_{k=2}^{(p-1)/2} e_k \log(\delta_k^{p-1})$ in K. Let S be the $\mathbb{Z}_p$-module of all real λ-adic integers with trace zero.

We have seen that $\log(\varepsilon^{p-1}) \in S$ and $\log(\delta_k^{p-1}) \in S$ for $k = 2, \ldots,$ $(p-1)/2$.

If p does not divide h^-, by (**F**):

$$\{\log \delta_2^{p-1}, \log \delta_3^{p-1}, \ldots, \log \delta_{(p-1)/2}^{p-1}\}$$

is a basis of S. Hence

$$\log(\varepsilon^{p-1}) = \sum_{k=2}^{(p-1)/2} a_k \log(\delta_k^{p-1}),$$

with $a_k \in \mathbb{Z}_p$.

Comparing the above expressions, $pa_k = e_k$ for $k = 2, \ldots, (p-1)/2$. This is a contradiction. ∎

The statements (**E**) and (**G**) together give *Kummer's regularity criterion*:

Theorem 1.   *For a prime $p > 2$ the following statements are equivalent:*

(1) *p is a regular prime;*

(2) *p does not divide h_p^-; and*

(3) *p does not divide the numerators of the Bernoulli numbers $B_2, B_4, \ldots, B_{p-3}$.*

Proof: As stated above, this is a combination of (**E**) and (**G**). ∎

Now we prove Kummer's lemma on units:

H.   *If p is a regular prime, if ε is a unit of K such that $\varepsilon \equiv m$ (mod $\widehat{A}\lambda^{p-1}$), where $m \in \mathbb{Z}$, then $\varepsilon = \varepsilon_1^p$, where ε_1 is a unit of K.*

Proof: By Chapter 7, (**F**), we have $\varepsilon = \zeta^j \varepsilon'$ where ε' is a real positive unit, $0 \le j \le p-1$. By Chapter 17, (**O**), $\varepsilon' \equiv a_0 \pmod{\widehat{A}\lambda^2}$ with $a_0 \in \mathbb{Z}_p$, so $a_0 \equiv m' \pmod{\mathbb{Z}_p p}$ with $m' \in \mathbb{Z}$. Hence $\varepsilon' \equiv m' \pmod{\widehat{A}\lambda^2}$ and necessarily $m' \ne 0$ since ε' is a unit. Since $\rho \equiv -\lambda \equiv \zeta - 1 \pmod{\widehat{A}\lambda^2}$ then $\zeta^j \equiv 1 + j\rho \pmod{\widehat{A}\lambda^2}$ and therefore $m \equiv \varepsilon \equiv m'(1 + j\rho) \pmod{\widehat{A}\lambda^2}$.

By Chapter 17, (**P**), we must have $j = 0$. showing that ε is a real positive unit.

From $\varepsilon^{p-1} \equiv m^{p-1} \equiv 1 \pmod{\widehat{A}\lambda^{p-1}}$ we deduce that $\log(\varepsilon^{p-1}) \equiv 0 \pmod{\widehat{A}\lambda^{p-1}}$. Thus there exists $\alpha \in \widehat{A}$ such that $p\alpha = \log(\varepsilon^{p-1})$. Taking the traces, $p\operatorname{Tr}_{K|\mathbb{Q}_p}(\alpha) = \operatorname{Tr}_{K|\mathbb{Q}_p}(\log(\varepsilon^{p-1})) = 0$ by Chapter 17, (**P**). So $\operatorname{Tr}_{K|\mathbb{Q}_p}(\alpha) = 0$. But ε is a real unit, so $\log(\varepsilon^{p-1}) \in S \subseteq \widehat{A}^+$ by Chapter 17, (**P**). Hence α is real. Thus $\alpha \in \widehat{A}^+$, so $\alpha \in S$.

Since $p \nmid h^-$, by (**F**), $\alpha = \sum_{k=2}^{(p-1)/2} c_k \log(\delta_k^{p-1})$, with $c_k \in \mathbb{Z}_p$.

Let n be the order of εV in the quotient group U^+/V (where U^+ is the group of real units and V the subgroup generated by the circular units). Since $(U^+ : V) = h^+$ as was mentioned before, then n divides h^+. Since p is regular then p does not divide h^+, hence it does not divide n.

From $\varepsilon^n \in V$ we may write

$$\varepsilon^n = \prod_{k=2}^{(p-1)/2} \delta_k^{d_k} \qquad \text{with} \quad d_k \in \mathbb{Z}.$$

So

$$\varepsilon^{n(p-1)} = \prod_{k=2}^{(p-1)/2} \delta_k^{(p-1)d_k}$$

hence

$$np\alpha = n\log(\varepsilon^{p-1}) = \sum_{k=2}^{(p-1)/2} d_k \log(\delta_k^{p-1}).$$

By (**F**), $npc_k = d_k$ for every k.

Hence $c_k \in \mathbb{Z}_p \cap \mathbb{Q} = \mathbb{Z}_{p\mathbb{Z}}$, so $c_k = c_k'/c_k''$, with $c_k', c_k'' \in \mathbb{Z}$, p not dividing c_k''. Thus p divides d_k for every k.

So $\varepsilon^n = \varepsilon_0^p$ where $\varepsilon_0 \in V \subseteq U^+$.

Finally, since p does not divide n, there exist integers s, t such that $1 = sp + tn$; then $\varepsilon = \varepsilon^{sp}\varepsilon^{nt} = (\varepsilon^s\varepsilon_0^t)^p$, proving the statement. ∎

19.2 Kummer's Theorem

Before the statement and proof of Kummer's theorem, we need to consider certain special elements of the cyclotomic field.

Definition 2.   An integer $\alpha \in A \setminus A\lambda$ is said to be *semi-primary* if there exists $m \in \mathbb{Z}$ such that $\alpha \equiv m \pmod{A\lambda^2}$.

I.   *With this definition we have:*
   (1) *If $m \in \mathbb{Z}$ then m is semi-primary if and only if $p \nmid m$.*

(2) *If $\alpha \in A \backslash A\lambda$ and $\alpha \equiv m + n\lambda$ (mod $A\lambda^2$) with $m, n \in \mathbb{Z}$, $m \not\equiv$ 0 (mod p), if $\ell \in \mathbb{Z}$ is such that $\ell m \equiv n$ (mod p), then $\zeta^\ell \alpha \equiv m$ (mod $A\lambda^2$), so $\zeta^\ell \alpha$ is semi-primary.*

(3) *If $\alpha, \beta \in A \backslash A\lambda$ are semi-primary, then $\alpha\beta$ is semi-primary and there exists $\ell \in \mathbb{Z}$ such that $0 \leq \ell \leq p - 1$, $\ell\beta \equiv \alpha$ (mod $A\lambda^2$).*

(4) *If α, β are semi-primary, if $\gamma \in A$ is such that $\alpha\gamma = \beta$, then γ is semi-primary.*

(5) *If ζ^k (with $k \geq 1$) is semi-primary, then $\zeta^k = 1$.*

Proof: (1)   If m is semi-primary, then $m \notin A\lambda$; since $A\lambda \cap \mathbb{Z} = \mathbb{Z}p$, then $p \nmid m$; and conversely.

(2)   Now $\zeta^\ell = (1 - \lambda)^\ell \equiv 1 - \ell\lambda$ (mod $A\lambda^2$); from $\ell m \equiv n$ (mod p) then

$$\zeta^\ell \alpha \equiv (1 - \ell\lambda)(m + n\lambda) \equiv m + (n - m\ell)\lambda \equiv m \pmod{A\lambda^2},$$

so $\zeta^\ell \alpha$ is semi-primary.

(3)   If $\alpha \equiv m$ (mod $A\lambda^2$), $\beta \equiv n$ (mod $A\lambda^2$) with $\alpha, \beta \notin A\lambda$, then $\alpha\beta \notin A\lambda$ (since $A\lambda$ is a prime ideal) and $\alpha\beta \equiv mn$ (mod $A\lambda^2$), so $\alpha\beta$ is semi-primary. If $\ell \in \mathbb{Z}$ is such that $\ell n \equiv m$ (mod p), then $\ell\beta \equiv \ell n \equiv m \equiv \alpha$ (mod $A\lambda^2$), because $Ap = (A\lambda)^{p-1}$.

(4)   Clearly $\gamma \notin A\lambda$. If $\alpha \equiv m$ (mod $A\lambda^2$), $\beta \equiv n$ (mod $A\lambda^2$), then $p \nmid m$; let $m' \in \mathbb{Z}$ be such that $m'm \equiv 1$ (mod p). Then $m'n \equiv m'\beta \equiv m'\alpha\gamma \equiv m'm\gamma \equiv \gamma$ (mod $A\lambda^2$), hence γ is semi-primary.

(5)   Let $\zeta^k \equiv m$ (mod $A\lambda^2$), with $m \in \mathbb{Z}$. Then $m \equiv (1 - \lambda)^k \equiv 1 - k\lambda$ (mod $A\lambda^2$), so $m \equiv 1$ (mod $A\lambda$), hence $m \equiv 1$ (mod p). Therefore $m \equiv 1$ (mod $A\lambda^{p-1}$) and $k\lambda \equiv 0$ (mod $A\lambda^2$), so $k \equiv 0$ (mod $A\lambda$) hence $k \equiv 0$ (mod p) and $\zeta^k = 1$. ∎

The following preliminary result will play a key role in the proof of Kummer's theorem:

J.   *Let $\alpha, \beta, \gamma \in A$, $\alpha\beta\gamma \neq 0$ and assume that $\alpha^p + \beta^p + \gamma^p = 0$.*

(1) *If $\lambda \nmid \gamma$, then for all $k = 0, 1, \ldots, p - 1$ there exists an ideal J_k of A such that*

$$A(\alpha + \zeta^k\beta) = J_k^p I,$$

where $I = \gcd(A(\alpha + \zeta^j\beta) \mid j = 0, 1, \ldots, p - 1)$.

(2) *If $\alpha^p + \beta^p = \varepsilon\delta^p\lambda^{mp}$, where $\varepsilon \in U$, $m \geq 1$, $\delta \in A$, $\lambda \nmid \alpha\beta\delta$, then $m \geq 2$.*

(3) *If $\gamma = \delta\lambda^m$, $m \geq 1$, $\delta \in A$, $\lambda \nmid \alpha\beta\delta$, if $I' = \gcd(A\alpha, A\beta)$, then there exists an index j_0, $0 \leq j_0 \leq p - 1$, and there exist ideals J_0, J_1, $\ldots$, J_{p-1} of A such that*

$$A(\alpha + \zeta^{j_0}\beta) = (A\lambda)^{p(m-1)+1} I' J_{j_0}^p,$$
$$A(\alpha + \zeta^j\beta) = (A\lambda)I' J_j^p \qquad \text{for } j \neq j_0,$$

and the ideals $J_0, J_1, \ldots, J_{p-1}$ are pairwise relatively prime and not multiples of $A\lambda$.

Proof: (1) Write

$$-\gamma^p = \alpha^p + \beta^p = \prod_{k=0}^{p-1}(\alpha + \zeta^k\beta).$$

First we show:

If $j < k$ then $\gcd(A(\alpha + \zeta^j\beta), A(\alpha + \zeta^k\beta)) = I$.

Indeed, if P is any prime ideal, $e \geq 1$, and

$$P^e|A(\alpha + \zeta^j\beta), \qquad P^e|A(\alpha + \zeta^k\beta),$$

then

$$\alpha + \zeta^j\beta \in P^e, \qquad \alpha + \zeta^k\beta \in P^e,$$

so

$$(\zeta^j - \zeta^k)\beta = \zeta^j(1 - \zeta^{k-j})\beta \in P^e.$$

But $1 - \zeta^{k-j} \sim 1 - \zeta = \lambda$, so $P^e|A\lambda \cdot A\beta$.

Also $(\zeta^k - \zeta^j)\alpha = \zeta^j(\zeta^{k-j} - 1)\alpha \in P^e$, so $P^e|A\lambda \cdot A\alpha$.

But $P \neq A\lambda$, otherwise $A\lambda$ divides $\prod_{i=0}^{p-1}(\alpha + \zeta^i\beta) = A\gamma^p$, contrary to the hypothesis.

So $P^e|A\alpha$, $P^e|A\beta$ hence $P^e|A(\alpha + \zeta^i\beta)$ for every $i = 0, 1, \ldots, p - 1$. Hence

$$\gcd(A(\alpha + \zeta^j\beta), A(\alpha + \zeta^k\beta)) = I.$$

From

$$\left(\frac{A\gamma}{I}\right)^p = \prod_{i=0}^{p-1}\frac{A(\alpha + \zeta^i\beta)}{I}$$

with pairwise relatively prime ideals $\dfrac{A(\alpha + \zeta^i\beta)}{I}$, it follows that

$$\frac{A(\alpha + \zeta^i\beta)}{I} = J_i^p \qquad \text{for} \quad i = 0, 1, \ldots, p - 1,$$

with $J_0, J_1, \ldots, J_{p-1}$ pairwise relatively prime ideals and $A\lambda \nmid J_i$, because $\lambda \nmid \gamma$.

(2) Multiplying α, β with roots of unity, the new elements still satisfy the same relation. So by (I) there is no loss of generality to assume that α, β are semi-primary.

Suppose $m = 1$.

By hypothesis there exist $a, b \in \mathbb{Z}$ such that $\alpha \equiv a \pmod{A\lambda^2}$ and $\beta \equiv b \pmod{A\lambda^2}$. Then $\alpha^p \equiv a^p \pmod{A\lambda^{p+1}}$ and $\beta^p \equiv b^p \pmod{A\lambda^{p+1}}$. Then $a^p + b^p = \alpha^p + \beta^p + \mu\lambda^{p+1} = \lambda^p(\varepsilon\delta^p + \mu\lambda)$, with $\mu \in A$, $\lambda \nmid \delta$.

Since $Ap = A\lambda^{p-1}$ then p divides $a^p + b^p$, $p^2 \nmid a^p + b^p$, so $a^p + b^p = ps$, with $p \nmid s$. Hence $A(a^p + b^p) = Ap \cdot As = A\lambda^{p-1} \cdot As$, with $A\lambda \nmid As$; but $A\lambda^p | A(a^p + b^p)$, and this is a contradiction.

(3) Write

$$-\delta^p \lambda^{mp} = \alpha^p + \beta^p = \prod_{j=0}^{p-1} (\alpha + \zeta^j \beta),$$

hence there exists j such that $A\lambda | A(\alpha + \zeta^j \beta)$. If $j \neq k$, then $\alpha + \zeta^k \beta = (\alpha + \zeta^j \beta) + \beta \zeta^j (\zeta^{k-j} - 1)$ with $1 - \zeta^{k-j} \sim 1 - \zeta = \lambda$; then $A\lambda | A(\alpha + \zeta^k \beta)$. Thus

$$\frac{\alpha + \beta}{\lambda}, \frac{\alpha + \zeta\beta}{\lambda}, \ldots, \frac{\alpha + \zeta^{p-1}\beta}{\lambda} \in A.$$

But these p elements are pairwise incongruent modulo $A\lambda$, otherwise there exist $j < k$ such that

$$\alpha + \zeta^j \beta \equiv \alpha + \zeta^k \beta \pmod{A\lambda^2},$$

so $\zeta^j (1 - \zeta^{k-j})\beta \equiv 0 \pmod{A\lambda^2}$ and $A\lambda^2 | A(1 - \zeta^{k-j})A\beta$, and finally $A\lambda | A\beta$, which contradicts the hypothesis.

Since $A/A\lambda \cong \mathbb{F}_p$ then $\#(A/A\lambda) = p$. So there exists j_0 such that $\alpha + \zeta^{j_0}\beta \equiv 0 \pmod{A\lambda^2}$, and if $j \neq j_0$ then $\alpha + \zeta^j \beta \not\equiv 0 \pmod{A\lambda^2}$. Hence $(A\lambda)^{mp-(p-1)}$ divides $A(\alpha + \zeta^{j_0}\beta)$ with $mp - (p-1) = (m-1)p + 1 \geq p - 1 > 1$, by part (2) of this lemma.

Next, $\alpha, \beta \in I'$ so $\alpha + \zeta^k \beta \in I'$ for all $k = 0, 1, \ldots, p-1$ and $A\lambda \nmid I'$ because $\lambda \nmid \alpha$, $\lambda \nmid \beta$.

Hence

$$A(\alpha + \zeta^{j_0}\beta) = (A\lambda)^{p(m-1)+1} I' J'_{j_0}$$
$$A(\alpha + \zeta^j \beta) = A\lambda \cdot I' J'_j \quad \text{for } j \neq j_0,$$

where $A\lambda \nmid J'_k$ for all $k = 0, 1, \ldots, p-1$. Now, we show:

The ideals J'_k are pairwise relatively prime. Indeed, if P is a prime ideal and $P | J'_j$, $P | J'_k$, with $j < k$, then $P \neq A\lambda$, so $A\lambda \cdot I'P$ divides $A(\alpha + \zeta^j \beta)$ and $A(\alpha + \zeta^k \beta)$. Hence $\alpha + \zeta^j \beta$, $\alpha + \zeta^k \beta \in A\lambda \cdot I'P$, so $\zeta^j (1 - \zeta^{k-j})\beta \in A\lambda \cdot I'P$, hence $A\lambda \cdot I'P | A\lambda \cdot A\beta$, so $I'P | A\beta$. Similarly $I'P | A\alpha$, then $I'P | \gcd(A\alpha, A\beta) = I'$, which is impossible.

Since

$$\left(\frac{A\delta}{I'}\right)^p = \prod_{j=0}^{p-1} J'_j$$

then for every j there exists an ideal J_j such that $J'_j = J_j^p$, with $A\lambda \nmid J_j$, and the ideals J_j are pairwise relatively prime. ∎

We are ready to prove the famous theorem of Kummer which may be called "Kummer's Monumental Theorem."

Theorem 2.   *If p is a regular odd prime, if $\alpha, \beta, \gamma \in K$, and $\alpha^p + \beta^p + \gamma^p = 0$, then $\alpha\beta\gamma = 0$.*

Proof: Assume that there exist $\alpha, \beta, \gamma \in K$, $\alpha\beta\gamma \neq 0$, such that $\alpha^p + \beta^p + \gamma^p = 0$. It may be also assumed, without loss of generality, that $\alpha, \beta, \gamma \in \mathbb{Z}[\zeta]$, after multiplying with a common denominator.

Case 1: $\lambda \nmid \alpha\beta\gamma$.

Let $p = 3$: $A/A\lambda \cong \mathbb{F}_3$ so $\alpha, \beta, \gamma \equiv \pm 1 \pmod{A\lambda}$; then $\alpha^3, \beta^3, \gamma^3 \equiv \pm 1 \pmod{A\lambda^3}$ and $0 = \alpha^3 + \beta^3 + \gamma^3 \equiv \pm 1, \pm 3 \pmod{A\lambda^3}$, hence necessarily $A\lambda^3$ divides $A3 = A\lambda^2$, which is impossible.

Let $p = 5$: $A/A\lambda \cong \mathbb{F}_5$, so $\alpha, \beta, \gamma \equiv \pm 1, \pm 2 \pmod{A\lambda}$ and $\alpha^5, \beta^5, \gamma^5 \equiv \pm 1 \pm 32 \pmod{A\lambda^5}$. Then $0 = \alpha^5 + \beta^5 + \gamma^5 \equiv \pm 1, \pm 3, \pm 30, \pm 34, \pm 63, \pm 65, \pm 96 \pmod{A\lambda^5}$. Since $\lambda^5 \sim 5\lambda$ the above congruences are obviously impossible.

Now let $p \geq 7$. It may be assumed without loss of generality that α, β are semi-primary, after multiplication with roots of unity.

Write

$$-\gamma^p = \alpha^p + \beta^p = \prod_{j=0}^{p-1}(\alpha + \zeta^j\beta).$$

Since $\lambda \nmid \gamma$ then $\lambda \nmid \alpha + \zeta^j\beta$ for all $j = 0, 1, \ldots, p-1$. Hence there exists a root of unity ζ^h such that $\zeta^h(\alpha + \zeta^{p-1}\beta)$ is semi-primary.

By **(J)**, $A(\alpha + \zeta^j\beta) = J_j^p$ for all $j = 0, 1, \ldots, p-1$, where $J_0, J_1, \ldots, J_{p-1}$ are pairwise relatively prime ideals not multiples of $A\lambda$. Then

$$A\left(\frac{\alpha + \zeta^j\beta}{\alpha + \zeta^{p-1}\beta}\right) = \left(\frac{J_j}{J_{p-1}}\right)^p \qquad \text{for all} \quad j = 0, 1, \ldots, p-1.$$

Since p is regular, by **(A)**,

$$\frac{J_j}{J_{p-1}}$$

is a principal ideal

$$\frac{J_j}{J_{p-1}} = A\left(\frac{\mu_j}{\nu_j}\right) \qquad \text{with} \quad \mu_j, \nu_j \in A, \quad \nu_j \neq 0.$$

So

$$J_j \cdot A\nu_j = J_{p-1} \cdot A\mu_j \qquad \text{for all} \quad j = 0, 1, \ldots, p-1.$$

Since $A\lambda \nmid J_j, J_{p-1}$, if $e > 0$, then $A\lambda^e | A\mu_j$ if and only if $A\lambda^e | A\nu_j$. Dividing by λ^e, it may also be assumed that $\lambda \nmid \mu_j, \nu_j$.

From the above,

$$\frac{\alpha + \zeta^j \beta}{\alpha + \zeta^{p-1}\beta} = \omega_j \left(\frac{\mu_j}{\nu_j}\right)^p, \quad \text{with} \quad \omega_j \text{ a unit.}$$

By Kummer's result on units (see Chapter 10, (**F**)), there exists a real unit ε_j and an integer c_j, $0 \le c_j \le p-1$, such that

$$\zeta^{-h}\omega_j = \varepsilon_j \zeta^{c_j}.$$

Thus

$$\nu_j^p(\alpha + \zeta^j \beta) = \varepsilon_j \zeta^{c_j} \zeta^h (\alpha + \zeta^{p-1}\beta)\mu_j^p.$$

Let

$$\alpha' = \frac{\alpha}{\zeta^h(\alpha + \zeta^{p-1}\beta)}, \qquad \beta' = \frac{\beta}{\zeta^h(\alpha + \zeta^{p-1}\beta)},$$

hence $\zeta^h(\alpha' + \zeta^{p-1}\beta') = 1$ and

$$\nu_j^p(\alpha' + \zeta^j \beta') = \varepsilon_j \zeta^{c_j} \mu_j^p.$$

But $A/A\lambda \cong \mathbb{F}_p$ so there exist $m_j, n_j \in \mathbb{Z}$ such that $\mu_j \equiv m_j$ (mod $A\lambda$), $\nu_j \equiv n_j$ (mod $A\lambda$), hence $\mu_j^p \equiv m_j^p$ (mod $A\lambda^p$), $\nu_j^p \equiv n_j^p$ (mod $A\lambda^p$). So

$$n_j^p(\alpha' + \zeta^j \beta') \equiv \varepsilon_j \zeta^{c_j} m_j^p \pmod{A\lambda^p}.$$

Taking the complex conjugate

$$n_j^p(\overline{\alpha'} + \zeta^{-j}\overline{\beta'}) \equiv \varepsilon_j \zeta^{-c_j} m_j^p \pmod{A\lambda^p}$$

(note that $\overline{\lambda} = 1 - \zeta^{-1} \sim \lambda$).
   So

$$n_j^p \zeta^{-c_j}(\alpha' + \zeta^j \beta') \equiv \varepsilon_j m_j^p \equiv n_j^p \zeta^{c_j}(\overline{\alpha'} + \zeta^{-j}\overline{\beta'}) \pmod{A\lambda^p}.$$

Since $\lambda \nmid n_j$ then

$$\alpha' + \zeta^j \beta' \equiv \zeta^{2c_j}(\overline{\alpha'} + \zeta^{-j}\overline{\beta'}) \pmod{A\lambda^p}. \qquad (19.1)$$

Evaluation of c_j:
   Since α, β, $\zeta^h(\alpha + \zeta^{p-1}\beta)$ are semi-primary, by (**I**), there exist integers a, b, $0 \le a, b \le p-1$, such that

$$\alpha \equiv a\zeta^h(\alpha + \zeta^{p-1}\beta) \pmod{A\lambda^2},$$
$$\beta \equiv b\zeta^h(\alpha + \zeta^{p-1}\beta) \pmod{A\lambda^2},$$

so

$$\begin{cases} \alpha' \equiv a \pmod{A\lambda^2}, \\ \beta' \equiv b \pmod{A\lambda^2}, \end{cases}$$

and from (19.1):

$$a + \zeta^j b \equiv \zeta^{2c_j}(a + \zeta^{-j}b) \pmod{A\lambda^2}.$$

But $\zeta^t = (1 - \lambda)^t \equiv 1 - t\lambda \pmod{A\lambda^2}$ for every $t \in \mathbb{Z}$, so

$$a + b - jb\lambda \equiv (1 - 2c_j\lambda)(a + b + jb\lambda) \pmod{A\lambda^2}.$$

It follows that

$$2c_j(a + b)\lambda \equiv 2jb\lambda \pmod{A\lambda^2}$$

hence

$$c_j(a + b) \equiv jb \pmod{A\lambda}.$$

Since $c_j, a, b, j \in \mathbb{Z}$ then

$$c_j(a + b) \equiv jb \pmod{p} \qquad \text{for} \quad j = 0, 1, \ldots, p - 2.$$

Note that $a + b \equiv 1 \pmod{p}$, because $\alpha + \zeta^{p-1}\beta = (a + \zeta^{p-1}b)\zeta^h(\alpha + \zeta^{p-1}\beta) \pmod{A\lambda^2}$, so $1 \equiv \zeta^h(a + \zeta^{p-1}b) \pmod{A\lambda^2}$. But $\zeta^h \equiv 1 \pmod{A\lambda}$, $\zeta^{p-1} \equiv 1 \pmod{A\lambda}$ hence $1 \equiv a + b \pmod{A\lambda}$ and therefore $1 \equiv a + b \pmod{p}$.

Thus $c_j \equiv jb \pmod{p}$ for every $j = 0, 1, \ldots, p - 2$. Since $p \geq 5$, then

$$\begin{cases} c_0 \equiv 0 \pmod{p}, \\ c_1 \equiv b \pmod{p}, \\ c_2 \equiv 2b \pmod{p}, \\ c_3 \equiv 3b \pmod{p}, \end{cases}$$

and from (19.1):

$$\begin{cases} \alpha' + \beta' - \overline{\alpha'} - \overline{\beta'} = \rho_0\lambda^p, \\ \alpha' + \zeta\beta' - \zeta^{2b}\overline{\alpha'} - \zeta^{2b-1}\overline{\beta'} = \rho_1\lambda^p, \\ \alpha' + \zeta^2\beta' - \zeta^{4b}\overline{\alpha'} - \zeta^{4b-2}\overline{\beta'} = \rho_2\lambda^p, \\ \alpha' + \zeta^3\beta' - \zeta^{6b}\overline{\alpha'} - \zeta^{6b-3}\overline{\beta'} = \rho_3\lambda^p, \end{cases}$$

with $\rho_0, \rho_1, \rho_2, \rho_3 \in A$.

Let

$$M = \begin{pmatrix} 1 & 1 & -1 & -1 \\ 1 & \zeta & -\zeta^{2b} & -\zeta^{2b-1} \\ 1 & \zeta^2 & -\zeta^{4b} & -\zeta^{4b-2} \\ 1 & \zeta^3 & -\zeta^{6b} & -\zeta^{6b-3} \end{pmatrix}$$

then

$$\det(M) = (1 - \zeta)(1 - \zeta^{2b})(1 - \zeta^{2b-1})(\zeta - \zeta^{2b})(\zeta - \zeta^{2b-1})(\zeta^{2b} - \zeta^{2b-1}).$$

If $\det(M) = 0$ then $\det(M) \equiv 0 \pmod{A\lambda^p}$. If $\det(M) \neq 0$ then by Cramer's rule

$$\alpha' = \frac{\det(M_1)}{\det(M)}, \qquad \beta' = \frac{\det(M_2)}{\det(M)}, \qquad \overline{\alpha'} = \frac{\det(M_3)}{\det(M)}, \qquad \overline{\beta'} = \frac{\det(M_4)}{\det(M)},$$

where M_i is the matrix obtained from M by replacing the ith column by the column

$$(\rho_0\lambda^p, \rho_1\lambda^p, \rho_2\lambda^p, \rho_3\lambda^p)^T.$$

Thus $\det(M_i) \in A\lambda^p$; since $\lambda \nmid \alpha'$, β', $\overline{\alpha'}$, $\overline{\beta'}$ then $\lambda^p \mid \det(M)$. Thus, in all cases

$$(1 - \zeta)(1 - \zeta^{2b})(1 - \zeta^{2b-1})(\zeta - \zeta^{2b})(\zeta - \zeta^{2b-1})(\zeta^{2b} - \zeta^{2b-1})$$
$$\equiv 0 \pmod{A\lambda^p}.$$

Consider the following cases:

(a) $b \equiv 0 \pmod{p}$; then $\beta \equiv b\zeta^h(\alpha + \zeta^{p-1}\beta) \equiv 0 \pmod{A\lambda}$, which is impossible.

(b) $b \equiv 1 \pmod{p}$; then $\beta \equiv b\zeta^h(\alpha + \zeta^{p-1}\beta) \equiv \alpha + \beta \pmod{A\lambda}$, so $\alpha \equiv 0 \pmod{A\lambda}$, again absurd.

(c) $b \not\equiv 0, 1 \pmod{p}$, $2b \not\equiv 1 \pmod{p}$. Then all factors in $\det(M)$ are associated with λ (note that $1 \leq b \leq p - 1$, so $2b \not\equiv 2$ $\pmod{p}$. Thus λ^p divides λ^6 hence $p \leq 6$, which is absurd, because it was assumed that $p \geq 7$.

(d) Since cases (a), (b), (c) cannot happen, then $b \not\equiv 0, 1 \pmod{p}$ and $2b \equiv 1 \pmod{p}$; but $a + b \equiv 1 \pmod{p}$, so $a \equiv b \pmod{p}$ hence $\alpha \equiv \beta \pmod{A\lambda}$.

By the symmetry relative to α, β, γ, then also $\alpha \equiv \gamma \pmod{A\lambda}$. In conclusion, $0 = \alpha^p + \beta^p + \gamma^p \equiv \alpha + \beta + \gamma \equiv 3\alpha \pmod{A\lambda}$. Since $p \neq 3$, then $\lambda \mid \alpha$, absurd.

Case 2: $\lambda \mid \alpha\beta\gamma$.

Assume, for example, $\lambda \mid \gamma$ and write $\gamma = \delta\lambda^m$, with $m \geq 1$, $\lambda \nmid \delta$, so

$$\alpha^p + \beta^p = -\delta^p\lambda^{mp}.$$

Thus there exists a relation of the form

$$\alpha^p + \beta^p = \varepsilon\delta^p\lambda^{mp}. \tag{19.2}$$

with ε a unit, $\lambda \nmid \delta$, m minimal, $m \geq 1$. By (**J**), $m \geq 2$.

Also $\lambda \nmid \alpha$, otherwise, $\lambda \mid \alpha$ hence $\lambda \mid \beta$; writing $\alpha = \lambda\alpha_1$, $\beta = \lambda\beta_1$, then

$$\alpha_1^p + \beta_1^p = \varepsilon\delta^p\lambda^{(m-1)p},$$

which is contrary to the choice of m minimal. By (**J**), there exists j_0 with the properties indicated. Replacing β by $\zeta^{j_0}\beta$, and changing notations, from (**J**) it may be written

$$\begin{cases} A(\alpha + \beta) = (A\lambda)^{p(m-1)+1}I'J_0^p, \\ A(\alpha + \zeta^k\beta) = (A\lambda)I'J_k^p & \text{for } 1 \leq k \leq p - 1, \end{cases}$$

where $I' = \gcd(A\alpha, A\beta)$, $J_0, J_1, \ldots, J_{p-1}$ are pairwise relatively prime ideals, and $A\lambda \nmid J_0 \cdots J_{p-1}$.

Then

$$(A\lambda)^{p(m-1)} \cdot A\left(\frac{\alpha + \zeta^k\beta}{\alpha + \beta}\right) = \left(\frac{J_k}{J_0}\right)^p \quad \text{for } 1 \leq k \leq p - 1.$$

So $\left(\dfrac{J_k}{J_0}\right)^p$ is a principal ideal. Since p is regular, then $\dfrac{J_k}{J_0}$ is a principal ideal. Hence there exist $\mu_k, \nu_k \in A$, $\nu_k \neq 0$, such that

$$\frac{J_k}{J_0} = A\left(\frac{\mu_k}{\nu_k}\right).$$

It may be assumed that $\lambda \nmid \mu_k \nu_k$, because $\lambda \nmid J_0 J_k$. So there exists a unit ε_k such that

$$(\alpha + \zeta^k \beta)\lambda^{p(m-1)} = \varepsilon_k(\alpha + \beta)\left(\frac{\mu_k}{\nu_k}\right)^p \quad \text{for} \quad 1 \leq k \leq p-1.$$

For $k = 1, 2$:

$$(\alpha + \zeta\beta)\lambda^{p(m-1)} = \varepsilon_1(\alpha + \beta)\left(\frac{\mu_1}{\nu_1}\right)^p,$$

$$(\alpha + \zeta^2\beta)\lambda^{p(m-1)} = \varepsilon_2(\alpha + \beta)\left(\frac{\mu_2}{\nu_2}\right)^p.$$

Multiplying the first relation by $1 + \zeta$ and subtracting the second:

$$\zeta(\alpha + \beta)\lambda^{p(m-1)} = (\alpha + \beta)\left[\varepsilon_1(1 + \zeta)\left(\frac{\mu_1}{\nu_1}\right)^p - \varepsilon_2\left(\frac{\mu_2}{\nu_2}\right)^p\right],$$

hence

$$(\mu_1\nu_2)^p - \frac{(\mu_2\nu_1)^p \varepsilon_2}{\varepsilon_1(1 + \zeta)} = \frac{\zeta}{\varepsilon_1(1 + \zeta)}\lambda^{p(m-1)}(\nu_1\nu_2)^p.$$

Since $1 + \zeta$ is a unit, the above relation is of the form

$$(\alpha')^p + \varepsilon'(\beta')^p = \varepsilon''(\delta')^p \lambda^{p(m-1)}$$

with $\alpha' = \mu_1\nu_2$, $\beta' = \mu_2\nu_1$, $\delta' = \nu_1\nu_2$.

Since $m \geq 2$ then λ^p divides $(\alpha')^p + \varepsilon'(\beta')^p$. But $\lambda \nmid \beta' = \mu_2\nu_1$, so $A\beta' + A\lambda = A$ and there exists $\kappa \in A$ such that $\kappa\beta' \equiv 1 \pmod{A\lambda}$, so $\kappa^p(\beta')^p \equiv 1 \pmod{A\lambda^p}$. Hence $(\kappa\alpha')^p + \varepsilon' \equiv 0 \pmod{A\lambda^p}$; thus, there exists $\rho \in A$ such that $\varepsilon' \equiv \rho^p \pmod{A\lambda^p}$.

But $A/A\lambda \cong \mathbb{F}_p$, so there exists $r \in \mathbb{Z}$ such that $\rho \equiv r \pmod{A\lambda}$ and $\varepsilon' \equiv \rho^p \equiv r^p \pmod{A\lambda^p}$.

By Kummer's lemma on units (**H**), there exists ε_1', a unit of K, such that $\varepsilon' = (\varepsilon_1')^p$, hence

$$(\alpha')^p + (\varepsilon_1'\beta')^p = \varepsilon''(\delta')^p \lambda^{p(m-1)}.$$

This is a relation of the form (19.2) with $m - 1$ instead of m—which is contrary to the minimality of m. ∎

As a special case, we mention explicitly:

K.   *If p is a regular odd prime, if x, y, z are integers, and $x^p + y^p = z^p$, then $xyz = 0$.*

19.3 Irregular Primes

It is not yet established and appears to be difficult to prove, that there are infinitely many regular primes.

On the other hand, 37 is an irregular prime, since 37 divides the numerator of B_{32} (see Chapter 18, examples before (**C**)).

The other irregular primes less than 100 are 59, 67.

Numerical evidence indicates that the proportion of regular primes among all primes is clearly to be $1/\sqrt{3} = 0.61\ldots$, indicating that not only are there infinitely many regular primes, but also that they are much more numerous than the irregular primes. Yet, it is fairly easy to show (Jensen, 1915) that there are infinitely many irregular primes. First we prove a preliminary result. In Chapter 18, (**U**), we considered the decomposition $k = k_1 k_2$, where $k_1 \geq 1$, $k_2 \geq 1$, and the prime factors of k_2 are exactly the primes dividing the denominator D_{2k} of B_{2k}. In this case, k_1 divides the numerator N_{2k}.

L.   *Let p be a prime. The following conditions are equivalent:*

   (1) *p is an irregular prime;*

   (2) *there exists an integer $k \geq 1$ such that p divides $N_{2k}/2k$; and*

   (3) *there exists an integer $k \geq 1$ such that p divides N_{2k}/k_1 (where k_1 was defined above).*

Proof: (1) $\rightarrow$ (2) If p is irregular, by (**E**) there exists k, $1 \leq k \leq (p-3)/2$, such that $p|N_{2k}$. Since $2|D_{2k}$ then N_{2k} is odd, so $p \neq 2$.

Also p does not divide k, hence p divides $N_{2k}/2k$.

(2) $\rightarrow$ (3)   This is trivial, because $N_{2k}/k_1 = 2k_2 \times N_{2k}/2k$.

(3) $\rightarrow$ (1)   Since p divides N_{2k}/k_1 then p divides N_{2k}, so p does not divide D_{2k}, therefore $p - 1 \nmid 2k$. Let $2k = m(p-1) + n$, with $1 \leq n < p - 1$. Since n is even, then $n = 2h$, with $2 \leq 2h \leq p - 3$. By Kummer's congruence, Chapter 18, (**V**), $B_{2k}/2k \equiv B_{2h}/2h \pmod{p}$. So $h(N_{2k}/k_1)D_{2h} \equiv k_2 N_{2h} D_{2k} \pmod{p}$.

By hypothesis, p divides N_{2k}/k_1. Since $p \nmid D_{2k}$ then $p \nmid k_2$, hence $p|N_{2h}$. By (**E**), p is an irregular prime.   ∎

Now, we give Carlitz' short proof (1954) of the existence of infinitely many irregular primes:

M.   *There exist infinitely many irregular primes.*

Proof: Let $p_1 = 37, \ldots, p_m$ be irregular primes. Due to the growth of $|B_{2k}|$ with k, see Chapter 18, (**J**), there is an index k such that $2k$ is a multiple of $(p_1 - 1)(p_2 - 1)\cdots(p_m - 1)$ and $|B_{2k}| > 2k$.

Let $|B_{2k}|/2k = a/b$ with $a > b \geq 1$, $\gcd(a, b) = 1$. Since $a > 1$, it has a prime factor p. So $p \nmid b$, and therefore $N_{2k}/2k = \pm D_{2k}a/b$ is a p-integral multiple of p. By (**L**), p is an irregular prime.

But $p\nmid b$ so $p|N_{2k}$ hence $p\nmid D_{2k}$. By the theorem of von Staudt and Clausen (Chapter 18, Theorem 1), $p - 1\nmid 2k$. Therefore $p \neq p_1, \ldots, p_m$, and this concludes the proof. ∎

In his paper, Jensen proved more:

N.   *There exists an infinite number of irregular primes p such that $p \equiv 3$* (mod 4).

Proof: 59 is the smallest irregular prime congruent to 3 modulo 4. Assume that $p_1 = 59$, $p_2, \ldots, p_m$ are irregular primes, congruent to 3 modulo 4.

By Dirichlet's theorem on primes in arithmetic progressions (see Chapter 24) there exists a prime q such that

$$q \equiv 1 \left(\bmod 12 \prod_{i=1}^{m} p_i(p_i - 1) \right).$$

In particular, $q \neq p_i$ $(i = 1, \ldots, m)$. Then $D_{2q} = 6$, as follows from von Staudt and Clausen's theorem (Chapter 18, Theorem 1): a prime ℓ divides D_{2q} exactly if $\ell - 1|2q$, so $\ell = 2$, 3, and ℓ cannot be any other factor ($\ell - 1 \neq q$, $\ell - 1 \neq 2q$; otherwise, since $q \equiv 1$ (mod 6) then $\ell = 2q + 1 = 2(6t + 1) + 1 = 12t + 3$, which is composite).

Now we use Kummer's congruence (Chapter 18, (**V**)). Since $q \equiv 1$ (mod $p_i - 1$) then $2q \equiv 2$ (mod $p_i - 1$) and therefore $B_{2q}/2q \equiv B_2/2$ (mod p_i). Since $B_2 = \frac{1}{6}$ and $D_{2q} = 6$ then $N_{2q} \equiv q$ (mod p_i) hence $p_i \nmid N_{2q}$.

From Chapter 18, (**Q**), we have

$$4N_{2q} \equiv D_{2q}S_{2q}(3) \equiv 6S_{2q}(3) \pmod{16}$$

so

$$2N_{2q} \equiv 3(1^{2q} + 2^{2q} + 3^{2q}) \equiv 3(1 + 0 + 1) \equiv 6 \pmod{8},$$

hence

$$N_{2q} \equiv 3 \pmod 4.$$

Thus there exists a prime factor p of N_{2q} such that $p \equiv 3$ (mod 4). Since $p_i \nmid N_{2q}$ then $p \neq p_i$ $(i = 1, \ldots, m)$. We shall prove that p is irregular, using (**L**).

Since $q \equiv 1$ (mod 4), then $q \neq p$. We have $p|N_{2q}$, $p\nmid 2q$, hence p divides $N_{2q}/2q$, so p is an irregular prime. This concludes the proof. ∎

The above result was successfully extended by various authors; we shall present here Metsänkylä's result.

We begin with two lemmas.

Let $m > 2$ be an integer, let $m = 2^{h_0}p_1^{h_1} \cdots p_s^{h_s}$ be the decomposition of m into prime factors ($s \geq 0$, $h_0 \geq 0$, $h_1 \geq 1, \ldots, h_s \geq 1$, $p_i \neq 2$);

let $h = \max(h_0, h_1, \ldots, h_s) \geq 1$. Recall also the notation $S_{2k}(m - 1) = \sum_{j=1}^{m-1} j^{2k}$.

Lemma 2.   *There exists an integer $a \geq 1$, such that $\gcd(a, m) = 1$ and if $k \geq h$ and $k \equiv 1 \pmod{\varphi(m^2)}$, then $6S_{2k}(m - 1) \equiv am \pmod{m^2}$.*

Proof: Let $q = p_i$ (or 2), $\ell = h_i$ (or h_0). We shall prove that if $k \geq h$, $k \equiv 1 \pmod{\varphi(m^2)}$ then

$$6S_{2k}(m - 1) \equiv (1 - q)m \pmod{q^{2\ell}}. \tag{19.3}$$

By hypothesis $k \geq \ell$, $k \equiv 1 \pmod{\varphi(q^{2\ell})}$ since $\varphi(q^{2\ell})$ divides $\varphi(m^2)$.

If j is an integer and q divides j then $j^{2k} \equiv 0 \pmod{q^{2\ell}}$. If q does not divide j then $j^{\varphi(q^{2\ell})} \equiv 1 \pmod{q^{2\ell}}$ hence from $k \equiv 1 \pmod{\varphi(q^{2\ell})}$ we deduce that $j^k \equiv j \pmod{q^{2\ell}}$. So

$$6S_{2k}(m - 1) = 6 \sum_{j=1}^{m-1} j^{2k} \equiv 6 \sum_{\substack{j=1 \\ q \nmid j}}^{m-1} j^2 \pmod{q^{2\ell}}.$$

But

$$6 \sum_{\substack{j=1 \\ q \nmid j}}^{m-1} j^2 = 6 \left(S_2(m-1) - \sum_{j'=1}^{m/q-1} (qj')^2 \right) = 6 \left(S_2(m - 1) - q^2 S_2 \left(\frac{m}{q} - 1 \right) \right).$$

By Chapter 18, (18.23), $S_2(m - 1) = \frac{1}{3}m^3 - \frac{1}{2}m^2 + \frac{1}{6}m$. So the above sum is equal to

$$2m^3 - 3m^2 + m - q^2 \left(2 \left(\frac{m}{q} \right)^3 - 3 \left(\frac{m}{q} \right)^2 + \frac{m}{q} \right)$$

$$= 2m^2 \left(m - \frac{m}{q} \right) + m(1 - q).$$

So $6S_{2k}(m - 1) \equiv (1 - q)m \pmod{q^{2\ell}}$.

By the Chinese remainder theorem there exists an integer a (depending only on m) such that $a \equiv 1 - p_i \pmod{p_i^{2h_i}}$ (where $p_0 = 2$) for $i = 0, 1, \ldots, s$. Hence $6S_{2k}(m - 1) \equiv am \pmod{p_i^{2h_i}}$ for $i = 0, 1, \ldots, s$, and therefore $6S_{2k}(m - 1) \equiv am \pmod{m^2}$.   ∎

Lemma 3.   *Let $m > 2$, let $G = (\mathbb{Z}/m)^{\cdot}$ be the group of invertible residue classes modulo m, and let H be a proper subgroup of G. Then:*

(1) *If m is odd, there exists an integer f such that $f \pmod m = \overline{f} \in G \setminus H$ and $f \not\equiv 1 \pmod{p_i}$ for each prime factor p_i of m.*

(2) *If 4 divides m and 3 does not divide m, there exists an f satisfying the additional congruence $f \not\equiv 1 \pmod 4$.*

(3) *If 12 divides m and $H \neq H_0 = \{\overline{x} \mid x \equiv \pm 1 \pmod{12}\}$ then there exists an f satisfying the same conditions as in (2).*

Proof: (1)   Let m be odd, and let $T = \{\overline{x} \in G \mid x \in \mathbb{Z} \text{ is such that } x \not\equiv 1$ (mod p_i) for every prime factor p_i of $m\}$.

$T \neq \varnothing$, since by the Chinese remainder theorem, there exists $x \in \mathbb{Z}$ such that $x \equiv 2$ (mod p_i) for every p_i dividing m.

We shall prove that the subgroup generated by T is G, hence T is not contained in H.

If $3 \nmid m$ let $\overline{x} \in G$; if $3 \mid m$ let $\overline{x} \in G$ where $x \equiv 1$ (mod 3). So $x \not\equiv 0$ (mod p_i) for every p_i dividing m.

Let $y \in \mathbb{Z}$ be defined as follows: if $p_i \mid m$ and $x \equiv -1$ (mod p_i), let $y_i \equiv x - 1$ (mod p_i); however, if $x \not\equiv -1$ (mod p_i), let $y_i \equiv x + 1$ (mod p_i). Then let $y \in \mathbb{Z}$ be such that $y \equiv y_i$ (mod p_i) for every p_i dividing m.

Then $\gcd(y, m) = 1$, hence there exists $z \in \mathbb{Z}$ such that $yz \equiv x$ (mod m). Therefore $z \not\equiv 1$ (mod p_i) for every p_i dividing m, and so $\overline{z} \in T$.

If $3 \nmid m$ then $y \not\equiv 1$ (mod p_i) when $p_i \mid m$; so $\overline{y} \in T$ and $\overline{x} = \overline{y}\,\overline{z}$ belongs to the subgroup generated by T, showing that this subgroup is equal to G.

If $3 \mid m$ then $y \equiv x + 1 \equiv 2$ (mod 3), so again $y \not\equiv 1$ (mod p_i) for every prime p_i dividing m. Hence $\overline{y} \in T$ and $\overline{x}$ belongs to the subgroup generated by T. Moreover, $-\overline{1} \in T$ and $-\overline{1} \neq \overline{x}$ for every $\overline{x} \in G$, chosen as indicated. So the subgroup generated by T contains at least $\varphi(m)/2 + 1$ elements. It must therefore be equal to G.

(2)   If 4 divides m, but 3 does not divide m, we proceed in the same way, with the set

$$T_0 = \{\overline{x} \in G \mid x \in \mathbb{Z} \text{ is such that } x \not\equiv 1 \text{ (mod 4) and } x \not\equiv 1 \text{ (mod } p_i)$$

for every odd prime factor of p_i of $m\}$.

Again, the subgroup generated by T_0 is G, hence T_0 is not contained in H.

(3)   If 12 divides m, we consider again the set T_0 of (2). Then $T_0 \subseteq H_0 = \{\overline{x} \in G \mid x \equiv \pm 1 \text{ (mod 12)}\}$, because if $x \in T_0$ then $x \in \overline{G}$, so x must be odd, and since $x \not\equiv 1$ (mod 4) then $x \equiv -1$ (mod 4); similarly, from $x \not\equiv 1$ (mod 3), and $x \not\equiv 0$ (mod 3), then $x \equiv -1$ (mod 3); hence $x \equiv -1$ (mod 12), showing that $T_0 \subseteq H_0$.

The subgroup H_0 has order $\varphi(m)/2$. Indeed, by the surjective homomorphism

$$\theta : G \to (\mathbb{Z}/12)^{\cdot} = \{\pm\overline{1}, \pm\overline{5}\}$$

the image of H_0 has index 2 in $(\mathbb{Z}/12)^{\cdot}$, so H_0 has index 2 in $G = (\mathbb{Z}/m)^{\cdot}$, thus H_0 has order $\varphi(m)/2$.

Now we show that the subgroup generated by T_0 is equal to H_0. In fact, $-\overline{1} \in H_0$ and if $\overline{a} \in G$ with $a \equiv 1$ (mod 4) and $a \equiv 1$ (mod p_i) for every odd prime factor of m, then $-\overline{a} \in T_0$ and $\overline{a} = (-\overline{1})(-\overline{a})$ belongs to the subgroup generated by T_0. So this subgroup has at least $\varphi(m)/4 + 1$ elements hence it must be equal to H_0.

We show now that T_0 is not contained in H. Indeed, if $T_0 \subseteq H$ then $H_0 \subseteq H$, so $H = \theta^{-1}(\theta(H))$, with $\{\pm 1\} \subseteq \theta(H) \subseteq (\mathbb{Z}/12)^{\cdot}$.

If $\theta(H) = (\mathbb{Z}/12)^{\cdot}$ then $H = G$, contrary to the hypothesis. If $\theta(H) = \{\pm\bar{1}\}$ then $H = H_0$ again, contrary to the hypothesis.

This proves that T_0 is not contained in H, concluding the demonstration. ∎

We now prove Metsänkylä's theorem:

O.   *Let $m > 2$, let G be the group of invertible residue classes modulo m, and let H be a proper subgroup of G. Then there exist infinitely many irregular primes p such that p modulo m is not in H.*

Proof: It is sufficient to consider the case where m is odd or 4 divides m. If the theorem has already been proved in this case and if m' is even but not a multiple of 4, then $m' = 2m$, where m is odd. Then $\mathbb{Z}/m' \xrightarrow{\sim} \mathbb{Z}/2 \times \mathbb{Z}/m$ by the map x (mod m') $\mapsto$ (x (mod 2), x (mod m)). Then $\gcd(x, m') = 1$ if and only if x is odd and $\gcd(x, m) = 1$. Hence $G' = (\mathbb{Z}/m')^{\cdot} \xrightarrow{\sim} (\mathbb{Z}/m)^{\cdot} = G$ (by the above isomorphism), so the subgroups of G' and G correspond to each other by the above isomorphism, and may be identified.

Let $m = 2^{h_0} p_1^{h_1} \cdots p_s^{h_s}$ with p_i odd primes, $s \geq 0$, $h_0 = 0$ or $h_0 \geq 2$, and $h_1 \geq 1, \ldots, h_s \geq 1$. Let $h = \max(h_0, h_1, \ldots, h_s)$. We note that 2 and m divide $\varphi(m^2)$. Suppose that $S = \{q_1, \ldots, q_r\}$ (with $r \geq 0$) is a finite set of primes, with $q_i > 3$. It is enough to show the existence of a prime q_{r+1} such that q_{r+1} is irregular, $q_{r+1} \notin S$, $\bar{q}_{r+1} \in G \setminus H$. Repeating this procedure, the theorem will be established.

Let $M = 3\varphi(m^2)(q_1 - 1) \cdots (q_r - 1)$, hence $6|M$, $m|M$.

By Lemma 2 there exists an integer a such that $\gcd(a, m) = 1$ and if $k \geq h$, $k \equiv 1$ (mod $\varphi(m^2)$) then $6S_{2k}(m - 1) \equiv am$ (mod m^2). We distinguish two cases.

Case 1: $\bar{a} \notin H$.

By Dirichlet's theorem on primes in arithmetic progression (see Chapter 24), there exists a prime q, such that $q > h$, $q \equiv 1$ (mod M). By von Staudt and Clausen's theorem (Chapter 18, Theorem 1), the only primes ℓ dividing D_{2q} are those such that $\ell - 1$ divides $2q$, that is, 2, 3, and $2q + 1$ (if it is prime); however, $2q + 1 \equiv 3$ (mod M) and $3|M$ so 3 would divide $2q + 1$. This shows that $D_{2q} = 6$.

It follows from Chapter 18, (**Q**), that $mN_{2q} \equiv 6S_{2q}(m - 1)$ (mod m^2).

By Lemma 2, $6S_{2q}(m - 1) \equiv am$ (mod m^2) hence $N_{2q} \equiv a$ (mod m), in particular, $\gcd(N_{2q}, m) = 1$.

Since q is odd, then $N_{2q} > 0$ by Chapter 18, (**I**). From the hypothesis that $\bar{a} \notin H$ it follows that $N_{2q} \neq 1$, so N_{2q} has a prime factor p such that $\bar{p} \in G \setminus H$.

Now we prove that p is an irregular prime. Since $q \equiv 1$ (mod m) then $\bar{q} \in H$ so $p \neq q$. So p divides $N_{2q}/2q$ and, by (**L**), p is irregular.

We have also $p \notin S$. Indeed, from $q_i > 3$ and $q \equiv 1$ (mod M), it follows that $2q \equiv 2 \not\equiv 0$ (mod $q_i - 1$) for every $i = 1, \ldots, r$. By Kummer's congruence (Chapter 18, (**V**)):

$$\frac{B_{2q}}{2q} \equiv \frac{B_2}{2} \equiv \frac{1}{12} \not\equiv 0 \pmod{q_i}$$

for $i = 1, \ldots, r$.
   But

$$\frac{B_{2q}}{2q} \equiv \frac{N_{2q}}{2q} \cdot \frac{1}{D_{2q}} \equiv 0 \pmod{p}$$

(since p divides N_{2q} then p does not divide D_{2q}) so $p \notin S$. Thus, in this case, we take $q_{r+1} = p$.

Case 2: $\bar{a} \in H$.
   We begin by noting that if 12 divides m then $H \neq H_0 = \{\bar{x} \mid x \equiv \pm 1 \pmod{12}\}$. Indeed, it follows from the relation (19.3) that $6S_{2k}(m - 1) \equiv (1 - 2)m = -m \pmod{2^{2h_0}}$, so from Lemma 2 we have $am \equiv -m \pmod{2^{2h_0}}$ and therefore $a \equiv -1 \pmod{2^{2h_0}}$ so $a \equiv -1 \pmod{4}$. Similarly, since $3|m$ and 3^{h_1} (with $h_1 \geq 1$) is the exact power of 3 dividing m, then

$$am \equiv 6S_{2k}(m - 1) \equiv (1 - 3)m = -2m \pmod{3^{2h_1}}$$

hence $a \equiv -2 \pmod{3}$. Therefore $a \equiv 7 \pmod{12}$, so $\bar{a} \notin H_0$ and therefore $H \neq H_0$.
   It follows from Lemma 3 that in all cases (even when 12 divides m), there exists an integer f such that $\bar{f} \in G \setminus H$, $f \not\equiv 1 \pmod{p_i}$, $i = 1, \ldots, s$, and if $4|m$ also $f \not\equiv 1 \pmod{4}$.
   Let $\ell_1, \ldots, \ell_t$ ($t \geq 0$) be the odd prime factors of M distinct from those dividing m.
   By the Chinese remainder theorem there exists an integer g_1' satisfying the following congruences:

$$\begin{cases} g_1' \equiv -1 \pmod{4}, \\ g_1' \equiv f \pmod{m} \text{ (if } 4|m \text{ the first congruence follows from this one)}, \\ g_1' \equiv -1 \pmod{\ell_i} \text{ for } i = 1, \ldots, t. \end{cases}$$

So $\gcd(g_1', M) = 1$. By Dirichlet's theorem on primes in arithmetic progression, there exists a prime g', $g' > 3$, $g' \equiv g_1' \pmod{M}$. Since $g' \equiv 3 \pmod{4}$ then $g' = 2n' + 1$ where n' is odd. Also $\bar{g'} = \bar{f} \in G \setminus H$. By Chapter 18, Theorem 1, 2, 3, and g' divide $D_{2n'}$. Let g be the smallest prime, $3 < g$, such that $\bar{g} \in G \setminus H$ and g divides $D_{2n'}$ (hence $g \leq g'$). We write $g = 2n + 1$ so 2, 3, g divide D_{2n}, thus $D_{2n} = 6cg$, where 2, 3, g do not divide c. Since g divides $D_{2n'}$ then $2n = g - 1$ divides $2n'$ so n divides n', hence n is odd.
   We now show: if c_j is a prime factor of c then $\bar{c}_j \in H$. Indeed, if $\bar{c}_j \notin H$ since $c_j|c$, then $c_j \neq 2, 3, g$ and $c_j|D_{2n}$ so $c_j - 1$ divides $2n$, hence also $2n'$, so $c_j|D_{2n'}$. Therefore $c_j > g$ (by the choice of g), and $2n = g - 1 < c_j - 1$, a contradiction.

We also have $\gcd(n', M) = 1$. Indeed, n' is odd. Moreover, $2n' = g' - 1 \equiv f - 1 \pmod{p_j}$ and $2n' = g' - 1 \equiv -2 \pmod{\ell_k}$ so ℓ_k does not divide n'. Hence $\gcd(n, M) = 1$ and therefore $\gcd(n, gM) = 1$.

Let $d_1 = 1, d_2, \ldots, d_u$ be the factors of n, so each d_k is odd. Let $e_1, \ldots, e_u$ be distinct primes, each such that $e_k > gM$.

By the Chinese remainder theorem there exists an integer q' satisfying the following congruences:

$$\begin{cases} nq' \equiv 1 \pmod{gM}, \\ 2d_k q' \equiv -1 \pmod{e_k^2} \quad (k = 1, \ldots, u). \end{cases}$$

By Dirichlet's theorem there exists a prime q, $q \geq h$, $q \equiv q'$ $\pmod{gMe_1^2 \cdots e_u^2}$, hence q satisfies also the above congruences.

Let $Q = nq$. We show that D_{2Q} and D_{2n} have the same prime factors, hence by Chapter 18, Theorem 1, $D_{2Q} = D_{2n} = 6cg$. Indeed, if a prime ℓ divides D_{2n}, then $\ell - 1 | 2n$ hence $\ell - 1 | 2Q$, so $\ell | D_{2Q}$. Conversely, if $\ell | D_{2Q}$ then $\ell - 1 | 2nq$. But the factors of $2nq$ are either d_k, $2d_k$, $d_k q$, $2d_k q$ (where $d_1, d_2, \ldots, d_u$ are the factors of n). So $\ell - 1$ must be of type d_k or $2d_k$, hence $\ell = 2$ or $\ell = 3$ or $\ell - 1$ divides $2n$, because all the other cases are impossible: $\ell - 1 = d_k q$ implies $\ell = d_k q + 1$ is even, so $\ell = 2$; $\ell - 1 = 2d_k q$ implies $\ell = 2d_k q + 1 \equiv 0 \pmod{e_k^2}$, which is impossible.

From $D_{2Q} = 6cg$, it follows from Chapter 18, $(\mathbf{Q})$, and from Lemma 2 that $mN_{2Q} \equiv 6cgS_{2Q}(m - 1) \equiv cgam \pmod{m^2}$, so $N_{2Q} \equiv cga \pmod{m}$.

We have $\gcd(c, m) = 1$, because if c_j is a prime factor of c and c_j divides m then c_j divides N_{2Q}; but c_j divides D_{2Q}, which is a contradiction.

From a previous observation, $\overline{c}_j \in H$ (for every prime factor of c), hence $\overline{c} \in H$.

Let $Q = Q_1 Q_2$ be the unique decomposition of Q indicated before (Chapter 18, $(\mathbf{U})$). The prime factors of Q_2 divide D_{2Q}, hence they are among $2, 3, g, c_j$. But $Q \equiv 1 \pmod{6g}$ since $6 | M$; so $2, 3, g$ do not divide Q. Hence these numbers do not divide Q_2, so the prime factors of Q_2 are among the c_j, hence $\overline{Q}_2 \in H$. From $Q \equiv 1 \pmod{m}$ we have $\overline{Q} = \overline{1} \in H$, hence $\overline{Q}_1 \in H$.

We have

$$\frac{N_{2Q}}{Q_1} \equiv \frac{c}{Q_1} ga = c'ga \pmod{m}.$$

Since $\overline{c'} \in H$, $\overline{a} \in H$, $\overline{g} \notin H$, then the residue class of N_{2Q}/Q_1 is not in H. Noting that $N_{2Q} > 0$ (since Q is odd), it follows that $N_{2Q}/Q_1 \neq 1$, so N_{2Q}/Q_1 contains a prime factor p such that $\overline{p} \in G \setminus H$. We also have $p \neq Q$ because $\overline{Q} = \overline{1} \in H$. From $(\mathbf{L})$, p is an irregular prime.

We also have $p \notin S$, the verification being the same as in Case 1. We take $q_{r+1} = p$ and the proof is complete. ∎

It is interesting to note that much less is known about the irregular primes in the residue class 1 modulo m ($m > 2$), even though there is numerical evidence in support of the existence of infinitely many irregular primes p such that $p \equiv 1 \pmod{m}$.

EXERCISES

1. Let $p = 7$. Verify with explicit calculations the facts proved in **(C)**.

2. Let $p = 11$. Verify by explicit calculations the facts proved in **(C)**.

3. Let $p = 7$. With the notation of **(F)** verify by explicit calculation that $\{\log \delta_2^6, \log \delta_3^6\}$ is a basis of the $\mathbb{Z}_7$-module S.

4. Let $p = 5$ and $\delta = ((1 - \zeta^2)/(1 - \zeta)) \times ((1 - \zeta^{-2})/(1 - \zeta^{-1}))$. Show with explicit calculations that $\delta^5 \equiv m \pmod{A \times 5}$, where $m \in \mathbb{Z}$.

5. Let $p = 7$ and

$$\delta_2 = \frac{1 - \zeta^2}{1 - \zeta} \times \frac{1 - \zeta^{-2}}{1 - \zeta^{-1}}, \qquad \delta_3 = \frac{1 - \zeta^3}{1 - \zeta} \times \frac{1 - \zeta^{-3}}{1 - \zeta^{-1}}.$$

Calculate $m_2, m_3 \in \mathbb{Z}$ such that $\delta_2^6 \equiv m_2 \pmod{A \times 7}$ and $\delta_3^6 \equiv m_3 \pmod{A \times 7}$.

6. Write up in detail, as simple as possible, a proof that the equation $X^3 + Y^3 + Z^3 = 0$ has only trivial solution in the field $\mathbb{Q}(\sqrt{-3})$.

7. Write up in detail a proof that the equation $X^5 + Y^5 + Z^5 = 0$ has only trivial solution in the field $\mathbb{Q}(\zeta)$, where $\zeta = \cos(2\pi/5) + i\sin(2\pi/5)$.

8. Let m be a square-free integer, $m \neq 0, 1$. Show that there exist nonzero integers a, b, c such that $(a + b\sqrt{m})^2 + (a - b\sqrt{m})^2 = c^2$ if and only if m has no prime factor p such that $p \equiv 3$ or $5 \pmod{8}$.

9. Let $m \geq 1$, m square-free. Show that $X^2 + Y^2 + Z^2 = 0$ has a nontrivial solution in $\mathbb{Q}(\sqrt{-m})$ if and only if $m \not\equiv 7 \pmod{8}$.

Hint: Use the theorem of Gauss: a natural number n is the sum of three squares if and only if n is not of the form $4^e(8k + 7)$, with $e \geq 0$, $k \geq 0$. Use also the fact that the product of two sums of two squares is the sum of two squares.

10. Let m be a square-free integer, $m \neq 0, 1$. The equation $X^4 + Y^4 = Z^4$ has a nontrivial solution in $\mathbb{Q}(\sqrt{m})$ if and only if $m = -7$. In this case, every nontrivial solution is proportional to

$$x = \pm(1 + \sqrt{-7}), \qquad y = \pm(1 - \sqrt{-7}), \qquad z = \pm 2.$$

(with arbitrary sign combinations).

Hint: First note that a solution is proportional to $x = a_1 + b_1\sqrt{m}$, $y = a_2 + b_2\sqrt{m}$, $z = c$; next remark that $a_1b_1a_2b_2 \neq 0$. Let $e = a_1^2b_1^2 + a_2^2b_2^2$. Obtain a quadratic relation for m, consider its discriminant Δ, which must be a square. Obtain new relations leading to an equation $X^4 + Y^4 = 2Z^2$, and invoke that its rational solutions (x, y, z) have $x^2 = y^2$; continue the analysis in the same vein, to reach the required conclusion.

11. Show:

    (a) For every rational number $k \in \mathbb{Q}$, $k \neq 0, 1$, let

$$\begin{cases} x_k = 3 + \sqrt{-3(1 + 4k^3)}, \\ y_k = 3 - \sqrt{-3(1 + 4k^3)}, \\ z_k = 6k. \end{cases}$$

    Show that $x_k^3 + y_k^3 + z_k^3 = 0$.

    (b) Assume that x, y, z are nonzero elements of the field $\mathbb{Q}(\sqrt{m})$, where m is a square-free integer, $m \neq 0, 1$, and that $x^3 + y^3 = z^3$. Then there exists $c \in \mathbb{Q}(\sqrt{m})$ and $k \in \mathbb{Q}$, $k \neq 0, -1$, such that $x = cx_k$, $y = cy_k$, $z = cz_k$.

    (c) Let $k \in \mathbb{Q}$, $k \neq 0, -1$, let $a \in \mathbb{Q}$, $a \neq 0$ with $ak \neq -1$. If there exists a square-free integer m, $m \neq 0, 1$, and there exists $c \in \mathbb{Q}(\sqrt{m})$, $c \neq 0$, such that $x_{ak} = cx_k$, $y_{ak} = cy_k$, $z_{ak} = cz_k$, then $a = 1$.

12. Let M be a square-free integer, $m \neq 0$, and let p be an odd prime such that $(m/p) \neq -1$. Assume also that $1 + k^p \not\equiv (1 + k)^p \pmod{p^2}$ for all $k = 1, 2, \ldots, p - 2$. Show that if $\alpha, \beta, \gamma \neq 0$ and $\alpha^p + \beta^p\gamma^p = 0$ then p divides $\alpha\beta\gamma$.

13. Let p be any odd prime. Show that the equation $X^p + Y^p + Z^p = 0$ has only trivial solutions in $\mathbb{Q}(\sqrt{(-1)^{(p-1)/2}p}\,)$.

Hint: Apply the result in Chapter 4, **(N)**.

14. Let m be a square-free integer, $m \neq 0, 1$. Show that if (x, y, z) is a solution of $X^3 + Y^3 + Z^3 = 0$ in $\mathbb{Q}(\sqrt{m})$, there exists a solution (x', y', z') and $\alpha \in \mathbb{Q}(\sqrt{m})$, $\alpha \neq 0$, such that $x' = \alpha x$, $y' = \alpha y$, $z' = \alpha z$, and

$$\begin{cases} x' = a + b\sqrt{m}, \\ y' = a - b\sqrt{m}, \\ z' = c, \end{cases}$$

where $a, b, c \in \mathbb{Z}$.

A solution like (x', y', z') is called a *conjugate solution.*

15. Let m be a square-free integer, $m \neq 0, 1$. Show that the equation $X^3 + Y^3 + Z^3 = 0$ has a conjugate solution in $\mathbb{Q}(\sqrt{m})$ if and only if it has one conjugate solution in $\mathbb{Q}(\sqrt{-3m})$.

16. Let m be a square-free integer, $m \neq 0, 1$. Show that the following statements are equivalent:

(i) The equation $X^3 + Y^3 + Z^3 = 0$ has a nontrivial solution in $\mathbb{Q}(\sqrt{m})$.

(ii) There exist nonzero integers x, y, z such that

$$3x(x^3 + 4y^3) + mz^2 = 0$$

and $\gcd(x, y) = 1$.

(iii) There exist nonzero integers x, y, z such that

$$x(x^3 + 4y^3) - mz^2 = 0$$

and $\gcd(x, y) = 1$.

17. Show that the equation $X^3 + Y^3 + Z^3 = 0$ has infinitely many nontrivial pairwise nonproportional solutions in $\mathbb{Q}(\sqrt{-2})$.

18. Let m be a square-free integer, $m \neq 0, 1$. Show that if $X^3 + Y^3 + Z^3 = 0$ has a nontrivial solution in $\mathbb{Q}(\sqrt{m})$ then it has infinitely many nontrivial solutions in $\mathbb{Q}(\sqrt{m})$ (which are pairwise nonproportional).

19. Show that 37 is an irregular prime.

Hint: Compute by recursion N_{2k} modulo 37 and apply Theorem 1.

20

More on Cyclotomic Extensions

In this chapter we shall describe the work of Gauss and Lagrange on the resolution by radicals of cyclotomic polynomials. Then we will describe some of the work of Jacobi and Kummer on the ideal theory of rings of cyclotomic integers.

These theories are classical and we shall give a presentation close to the original, even when there are more modern and sophisticated treatments available.

20.1 Resolution by Radicals of the Cyclotomic Equation

Our aim is to give Gauss' method to express the roots of unity by radicals.

Let $n \geq 2$ and let $\Phi_n(X)$ be the nth cyclotomic polynomial.

We recall the formulas

$$\Phi_{q^e}(X) = \Phi_q(X^{q^{e-1}})$$

and

$$\Phi_{q^e m}(X) = \frac{\Phi_m(X^{q^e})}{\Phi_m(X^{q^{e-1}})},$$

where $e \geq 1$, q is a prime, and q does not divide m. Thus the determination of the roots of an arbitrary cyclotomic polynomial is reduced to the determination of the roots of $\Phi_q(X)$ for every prime q.

From an expression of the qth roots of unity by radicals (for every prime q), one obtains the expression of nth roots of unity by radicals.

Henceforth, let q be an odd prime.

It is illuminating to present the ideas of the method before we develop the details.

(i)   The method of resolution of $\Phi_q(X)$ is by induction on q; Gauss assumed known how to solve by radicals the equations $\Phi_p(X) = 0$ for every prime $p < q$.

(ii)   Let ρ be a primitive qth root of 1, $L = \mathbb{Q}(\rho)$. Gauss constructed a sequence of fields

$$L = \mathbb{Q}(\rho) \supset L_{r-1} \supset L_{r-2} \supset \cdots \supset L_1 \supset L_0 = \mathbb{Q}$$

such that ρ is expressible by radicals in terms of the elements of L_{r-1} and a pth root of 1 (where p divides $q - 1$); similarly, each element of L_{r-1} is expressible by radicals in terms of L_{r-2} and a (p')th root of 1 (where p' divides $q - 1$), and so on.

By superposition of these formulas, and taking into account the induction hypothesis, then ρ is expressible by means of radicals of rational numbers.

(iii)   The fields L_1, L_2, ... are successively obtained as follows. Since $[L : \mathbb{Q}] = q - 1$, let p_1 be a prime dividing $q - 1$, $q - 1 = f_1 p_1$. L_1 is an appropriate subfield of L of degree $[L_1 : \mathbb{Q}] = p_1$; moreover, L_1 has to be defined so that it has a generator, which is expressible by radicals in terms of elements of $\mathbb{Q}$ and of ζ_{p_1} (a primitive (p_1)th root of 1); thus every element of L_1 is expressible by radicals by virtue of the induction assumption.

Next, let p_2 be a prime dividing f_1, $f_1 = f_2 p_2$. L_2 is an appropriate field such that $L_1 \subset L_2 \subset L$, $[L_2 : L_1] = p_2$; moreover, L_2 has to be defined so that it has a generator, which is expressible by radicals in terms of elements of L_1 and of ζ_{p_2} (a primitive (p_2)th root of 1); thus every element of L_2 is expressible by radicals in terms of elements of $\mathbb{Q}$, in view of the inductive assumption. This construction is repeated until $L_r = L$ for some $r \geq 1$. Eventually, ρ and its conjugates are expressible by superposed radicals of rational numbers.

(iv)   The typical step in the construction is of the following type:

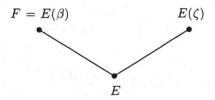

Let $\zeta = \zeta_p$ be a primitive pth root of 1, where p is a prime (dividing $q - 1$) and $F | E$ is a cyclic extension of degree p, β a primitive element of the extension $F | E$, and τ a generator of the Galois group of $F | E$.

Gauss expressed β and its conjugates as linear combinations, with coefficients powers of ζ, of the Lagrange resolvents $\langle \zeta^j, \beta \rangle_\tau$ $(j = 0, 1, \ldots, p-1)$, which are elements (to be defined below) of the field $F(\zeta)$. The Lagrange resolvents have the important property that $\langle \zeta^j, \beta \rangle_\tau^p \in E(\zeta)$, so β and its conjugates are expressible in terms of elements of $E(\zeta)$, using rational operations and pth roots.

(v)   The choice of the appropriate extension $E(\beta)|E$ such that $\langle \zeta^j, \beta \rangle_\tau^p \in E(\zeta)$ is possible by taking β to be one of the Gaussian periods with p terms, to be defined below.

Thus we need to introduce the Lagrange resolvents and the Gaussian periods.

The *Lagrange resolvents* (Lagrange, 1767) may be defined in the following general situation.

Let E be a number field, let $F|E$ be a cyclic extension of degree n, and let τ be a generator of the Galois group G of this extension.

Let $\widetilde{E}|E$ be a given field extension. If $\alpha \in \widetilde{E}$, $\alpha \neq 0$, and $\beta \in F$, the *Lagrange resolvent* (defined by α, β, τ) is the element

$$\langle \alpha, \beta \rangle_\tau = \beta + \alpha\tau(\beta) + \alpha^2\tau^2(\beta) + \cdots + \alpha^{n-1}\tau^{n-1}(\beta). \qquad (20.1)$$

If there is no ambiguity, we simply write $\langle \alpha, \beta \rangle$.

The Lagrange resolvent is an element of the field $F\widetilde{E}$, compositum of F and $\widetilde{E}$.

$F\widetilde{E}|\widetilde{E}$ is a cyclic extension of degree dividing $[F : E] = n$; the Galois group of $F\widetilde{E}|\widetilde{E}$ is isomorphic to the Galois group of $F|(F \cap \widetilde{E})$. τ admits a natural extension $\widetilde{\tau}$ to an $\widetilde{E}$-automorphism of $F\widetilde{E}$, thus $\widetilde{\tau}(\langle \alpha, \beta \rangle_\tau) = \langle \alpha, \tau(\beta) \rangle_\tau$, and $\widetilde{\tau}$ is a generator of $\mathrm{Gal}(F\widetilde{E}|\widetilde{E})$.

We have:

A.   *With above hypotheses and notations:*

(1) $\alpha \cdot \widetilde{\tau}\langle \alpha, \beta \rangle = \langle \alpha, \beta \rangle + (\alpha^n - 1)\beta$.

(2) *If $\alpha^n = 1$ then $\alpha\widetilde{\tau}(\langle \alpha, \beta \rangle) = \langle \alpha, \beta \rangle$ and $\langle \alpha, \beta \rangle^n \in \widetilde{E}$.*

(3) *If $n = p$ is a prime, $\zeta = \zeta_p$ is a primitive pth root of 1, and $F = E(\beta)$ then β and its conjugates are linear combinations, with coefficients in $E(\zeta)$ of the Lagrange resolvents $\langle \zeta^j, \beta \rangle$ ($j = 0, 1, \ldots, p - 1$):*

$$\tau^j(\beta) = \frac{1}{p}[\langle 1, \beta \rangle + \zeta^{-j}\langle \zeta, \beta \rangle + \zeta^{-2j}\langle \zeta^2, \beta \rangle + \cdots + \zeta^{-(p-1)j}\langle \zeta^{p-1}, \beta \rangle].$$

Proof: (1)

$$\alpha\widetilde{\tau}(\langle \alpha, \beta \rangle) = \alpha\tau(\beta) + \alpha^2\tau^2(\beta) + \cdots + \alpha^{n-1}\tau^{n-1}(\beta) + \alpha^n\beta$$
$$= \langle \alpha, \beta \rangle + (\alpha^n - 1)\beta.$$

(2)   If $\alpha^n = 1$ then $\alpha\widetilde{\tau}(\langle \alpha, \beta \rangle) = \langle \alpha, \beta \rangle$.

Raising to the nth power $(\widetilde{\tau}(\langle \alpha, \beta \rangle))^n = \langle \alpha, \beta \rangle^n$ so $\langle \alpha, \beta \rangle^n$ is invariant by $\widetilde{\tau}$. Therefore $\langle \alpha, \beta \rangle^n \in \widetilde{E}$.

(3)   Now let $n = p$, $F = E(\beta)$, then using $\alpha = \zeta^j$:

$$
\begin{aligned}
\beta + \; & \tau(\beta) \; + \cdots + \; & \tau^{p-1}(\beta) & = \langle 1, \beta \rangle, \\
\beta + \; & \zeta\tau(\beta) \; + \cdots + \; & \zeta^{p-1}\tau^{p-1}(\beta) & = \langle \zeta, \beta \rangle, \\
\beta + \; & \zeta^2\tau(\beta) \; + \cdots + \; & \zeta^{2(p-1)}\tau^{p-1}(\beta) & = \langle \zeta^2, \beta \rangle, \\
& \qquad \cdots \cdots \cdots \cdots \cdots \cdots \\
\beta + \; & \zeta^{p-1}\tau(\beta) + \cdots + \; & \zeta^{(p-1)^2}\tau^{p-1}(\beta) & = \langle \zeta^{p-1}, \beta \rangle.
\end{aligned}
$$

Multiplying the ith equation with ζ^{-ij} (for $i = 0, 1, \ldots, p-1$) and adding up these equations we obtain

$$
\tau^j(\beta) = \frac{1}{p}[\langle 1, \beta \rangle + \zeta^{-j}\langle \zeta, \beta \rangle + \zeta^{-2j}\langle \zeta^2, \beta \rangle + \cdots + \zeta^{-(p-1)j}\langle \zeta^{p-1}, \beta \rangle]. \quad \blacksquare
$$

In view of this proposition, β and its conjugates are expressible in terms of elements of $E(\zeta)$ and their pth roots.

Later, we shall study the resolvents in more detail.

Now, we turn to the *Gaussian periods*.

Let E be a number field and let $F|E$ be a cyclic extension of degree n, $F = E(\beta)$. Let τ be a generator of the Galois group of $F|E$. If $f, r \geq 1$ are any integers such that $fr = n$, we define the r *periods of* f *terms* (relative to τ and β):

$$
\begin{aligned}
\mu_0 = \; & \beta \; + \; \tau^r(\beta) \; + \; \tau^{2r}(\beta) \; + \cdots + \tau^{(f-1)r}(\beta), \\
\mu_1 = \; & \tau(\beta) \; + \; \tau^{r+1}(\beta) \; + \tau^{2r+1}(\beta) + \cdots + \tau^{(f-1)r+1}(\beta), \\
& \cdots \cdots \cdots \cdots \cdots \cdots \cdots \cdots \cdots \cdots \cdots \cdots \\
\mu_{r-1} = \; & \tau^{r-1}(\beta) + \tau^{2r-1}(\beta) + \tau^{3r-1}(\beta) + \cdots + \tau^{n-1}(\beta).
\end{aligned}
\qquad (20.2)
$$

We note that $\sum_{j=0}^{r-1} \mu_j = \mathrm{Tr}_{F|E}(\beta)$.

For convenience, if j is any integer, if $j \equiv j_0 \pmod{r}$ with $0 \leq j_0 \leq r - 1$, we define $\mu_j = \mu_{j_0}$.

The periods are conjugate to each other, since $\tau^i(\mu_j) = \mu_{i+j}$ for every $i = 0, 1, \ldots, n - 1$ and every index j. Thus $F(\mu_0) = F(\mu_1) = \cdots = F(\mu_{r-1}) = F(\mu_0, \ldots, \mu_{r-1})$. In particular, $\tau^r(\mu_j) = \mu_j$ for every index $j = 0, 1, \ldots, r - 1$.

Let F' be the subfield of E of all elements which are invariant by τ^r. So $F(\mu_0, \ldots, \mu_{r-1}) \subseteq F'$. The Galois group of $F|F'$ is generated by τ^r and $[F : F'] = f$, $[F' : \mathbb{Q}] = r$.

We shall see later more properties of the Gaussian periods.

Putting together all the steps developed so far, we have the method of Gauss to solve the cyclotomic equation $\Phi_q(X) = 0$ by radicals.

Let ρ be a primitive qth root of 1, $L = \mathbb{Q}(\rho)$, so $L|\mathbb{Q}$ is a cyclic extension of degree $q - 1$. Let h be a primitive root modulo q and let τ be the generator of the Galois group of $L|\mathbb{Q}$ defined by $\tau(\rho) = \rho^h$.

Let p_1 be a prime dividing $q - 1$: $q - 1 = f_1 p_1$. Let ζ_{p_1} be a primitive (p_1)th root of 1 and let $\mu_0, \ldots, \mu_{p_1-1}$ be the p_1 periods with f_1 terms (relative to τ, ρ). Let L_1 be the subfield of all elements fixed by $\tau_1 = \tau^{p_1}$, so $L_1 \supseteq \mathbb{Q}(\mu_0, \ldots, \mu_{p_1-1}) \supseteq \mathbb{Q}$; since $[L_1 : \mathbb{Q}] = p_1$ is a prime number and

$\mu_0 \notin \mathbb{Q}$ then $L_1 = \mathbb{Q}(\mu_0, \ldots, \mu_{p_1-1}) = \mathbb{Q}(\mu_0) = \cdots = \mathbb{Q}(\mu_{p_1-1})$. Using the Lagrange resolvents $\langle \zeta_{p_1}^i, \mu_j \rangle_{\tilde{\tau}}$ ($\tilde{\tau}$ denotes the restriction of τ to L_1), then each μ_j is expressible in terms of ζ_{p_1}, elements of $\mathbb{Q}$, and their (p_1)th roots.

In the second step, we consider the extension $L|L_1$, with $L = L_1(\rho)$, $[L : L_1] = f_1$. Let p_2 be a prime dividing f_1: $f_1 = f_2 p_2$. Let ζ_{p_2} be a primitive (p_2)th root of 1 and let $\mu'_0, \ldots, \mu'_{p_2-1}$ be the p_2 periods with f_2 terms (relative to τ_1, ρ).

We need to observe now that $\mu'_0 \notin L_1$. Indeed, otherwise, $\mu'_0 = \tau_1(\mu'_0)$, that is

$$\rho + \tau_1^{p_2}(\rho) + \tau_1^{2p_2}(\rho) + \cdots + \tau_1^{(f_2-1)p_2}(\rho)$$
$$- \tau_1(\rho) - \tau_1^{p_2+1}(\rho) - \cdots - \tau_1^{(f_2-1)p_2+1}(\rho) = 0,$$

that is

$$\rho + \rho^{h^{p_1 p_2}} + \rho^{h^{2p_1 p_2}} + \cdots + \rho^{h^{(f_2-1)p_1 p_2}}$$
$$- \rho^{h^{p_1}} - \rho^{h^{p_1(p_2+1)}} - \cdots - \rho^{h^{p_1[(f_2-1)p_2+1]}} = 0.$$

All the exponents are congruent modulo q to numbers between 1 and $q - 1$, moreover, all the exponents are pairwise incongruent modulo q.

Thus, we have a nontrivial linear combination of $\rho, \rho^2, \ldots, \rho^{q-1}$ which is equal to 0—but this is impossible (noting that $\rho^{q-1} = -(1 + \rho + \cdots + \rho^{q-2})$ this would lead to a nontrivial combination of $1, \rho, \ldots, \rho^{q-2}$). So we have indeed $L_1(\mu'_0) = L_1(\mu'_1) = \cdots = L_1(\mu'_{p_2-1}) = L_1(\mu'_0, \ldots, \mu'_{p_2-1}) \neq L_1$.

Let L_2 be the subfield of all elements fixed by $\tau_2 = \tau_1^{p_2} = \tau^{p_1 p_2}$, so $L_2 \supseteq L_1(\mu'_0, \ldots, \mu'_{p_2-1})$ and since $[L_2 : L_1] = p_2$ is a prime number then $L_2 = L_1(\mu'_0, \ldots, \mu'_{p_2-1})$. Using the Lagrange resolvents $\langle \zeta_{p_2}^i, \mu'_j \rangle_{\tilde{\tau}_1}$ ($\tilde{\tau}_1$ denotes the restriction of τ_1 to L_2) then each μ'_j is expressible in terms of ζ_{p_2}, elements of L_1, and their (p_2)th roots.

This process may now be continued, until some field $L_r = L = \mathbb{Q}(\rho)$.

For small values of q, say $q \leq 17$, the calculations are quite manageable in size.

Concerning segments constructible by ruler and compass we note that they are obtainable as intersections of straight lines and/or circles. So they are measured by numbers, which are solutions of systems of equations of degree at most equal to 2. As Gauss noted, the side of a regular polygon with q sides may be constructed by ruler and compass, exactly when 2 is the only prime factor of $q - 1$, that is, q is a prime of the form $2^m + 1$. In this event, it follows easily that m itself must be a power of 2, so $q = 2^{2^n} + 1$.

The numbers $F_n = 2^{2^n} + 1$ (for $n \geq 0$) are called *Fermat numbers*.

So, for example, the sides of the regular polygons with $q = 3, 5, 17, 257$ sides may be constructed with ruler and compass.

Gauss gave the expression by radicals of the primitive seventeenth root of 1, by 1832, Richelot computed explicitly the root of 1 of order 257 by means of radicals.

20.2 The Gaussian Periods

We shall now indicate some properties of the Gaussian periods in a special situation.

Let q be an odd prime, ρ a primitive qth root of 1, h a primitive root modulo q, $L = \mathbb{Q}(\rho)$, $B = \mathbb{Z}[\rho]$, and let τ be the generator of the Galois group of $L|\mathbb{Q}$ defined by $\tau(\rho) = \rho^h$.

Every element $\alpha \in L$ may be indifferently written in a unique way as $\alpha = \sum_{i=0}^{q-2} a_i\rho^i$ or as $\alpha = \sum_{j=0}^{q-2} a_j'\rho^{h^j}$ with $a_i, a_j' \in \mathbb{Q}$; moreover, $\alpha \in B$ if and only if each $a_i, a_j' \in \mathbb{Z}$.

Comparing these two representations, and noting that $\rho^{h^{(q-1)/2}} = \rho^{q-1} = -(1 + \rho + \cdots + \rho^{q-2})$, it follows that $a_0 = -a_{(q-1)/2}'$ and $a_i = a_j' - a_{(q-1)/2}'$, where $i \equiv h^j \pmod{q}$ for $i = 1, \ldots, q-2$.

In the present situation, if $q - 1 = fr$, the r periods with f terms (relative to ρ and τ, or h) are

$$\mu_0 = \rho + \rho^{h^r} + \rho^{h^{2r}} + \cdots + \rho^{h^{(f-1)r}},$$

$$\mu_1 = \rho^h + \rho^{h^{r+1}} + \rho^{h^{2r+1}} + \cdots + \rho^{h^{(f-1)r+1}}, \qquad (20.3)$$

$$\cdots\cdots\cdots\cdots\cdots\cdots\cdots\cdots\cdots$$

$$\mu_{r-1} = \rho^{h^{r-1}} + \rho^{h^{2r-1}} + \rho^{h^{3r-1}} + \cdots + \rho^{h^{q-2}}.$$

We have

$$\sum_{j=0}^{r-1} \mu_j = -1.$$

As before, for every j, we write $\mu_j = \mu_{j_0}$ if $0 \le j_0 \le r-1$ and $j \equiv j_0 \pmod{r}$.

The periods μ_j are conjugate to each other: $\tau^i(\mu_j) = \mu_{i+j}$ for $i = 0, 1, \ldots, q-2$, and any j. In particular, $\tau^r(\mu_j) = \mu_j$ for $j = 0, 1, \ldots, r-1$.

Let L' denote the subfield of L which is fixed by τ^r, so $[L : L'] = f$, $[L' : \mathbb{Q}] = r$, the Galois group of $L|L'$ is generated by τ^r, and the Galois group of $L'|\mathbb{Q}$ is generated by the restriction τ' of τ to L'. Let B' denote the ring of integers of L'.

B.   (1) $\{\mu_0, \mu_1, \ldots, \mu_{r-1}\}$ *is a basis of the $\mathbb{Z}$-module B'.*

(2) $L' = \mathbb{Q}(\mu_0, \ldots, \mu_{r-1})$, $B' = \mathbb{Z}[\mu_0, \ldots, \mu_{r-1}]$.

(3) $\{1, \rho, \rho^2, \ldots, \rho^{f-1}\}$ *is a basis of the B'-module B.*

(4) *The polynomial of periods*

$$F_{\mu_0}(X) = \prod_{i=0}^{r-1}(X - \mu_i) \tag{20.4}$$

has coefficients in $\mathbb{Z}$ and is irreducible.

Proof: (1) The elements μ_0, μ_1, ..., μ_{r-1} are linearly independent over $\mathbb{Z}$: if $\sum_{i=0}^{r-1} a_i\mu_i = 0$ (with $a_i \in \mathbb{Z}$) replacing each μ_i by its expression from (20.3), we have a linear combination of ρ, ρ^2, ..., ρ^{q-1} which is equal to 0, and with coefficients $a_0, a_1, \ldots, a_{r-1} \in \mathbb{Z}$. So each $a_i = 0$.

On the other hand, if $\alpha \in B' \subseteq B$, we may write $\alpha = \sum_{i=0}^{q-2} a_i\rho^{h^i}$ with $a_i \in \mathbb{Z}$. Since $\tau^r(\alpha) = \alpha$ then

$$\sum_{i=0}^{q-2} a_i\rho^{h^{i+r}} = \sum_{i=0}^{q-2} a_i\rho^{h^i}$$

and from the uniqueness of the expression, we deduce that

$$a_0 = a_r = \cdots = a_{(f-1)r},$$
$$a_1 = a_{r+1} = \cdots = a_{(f-1)r+1},$$
$$\cdots\cdots\cdots\cdots\cdots\cdots\cdots$$
$$a_{r-1} = a_{2r-1} = \cdots = a_{q-2}.$$

Hence

$$\alpha = \sum_{j=0}^{r-1} a_j\mu_j.$$

(2) Clearly $L' \supseteq \mathbb{Q}(\mu_0,\ldots,\mu_{r-1})$ and $B' \supseteq \mathbb{Z}[\mu_0,\ldots,\mu_{r-1}]$. The converse follows from (1).

(3) Let

$$G(X) = \prod_{i=0}^{f-1}(X - \rho^{h^{ir}}).$$

be the polynomial whose roots are the summands of the period μ_0. Then each coefficient of $G(X)$ is invariant by τ^r, hence it belongs to $B \cap L' = B'$.

Thus

$$G(X) = X^f + \alpha_1 X^{f-1} + \cdots + \alpha_f$$

and since ρ is a root of $G(X)$, then $\rho^f = -(\alpha_1\rho^{f-1} + \cdots + \alpha_f)$. So ρ^f is a linear combination of 1, ρ, ..., ρ^{f-1} with coefficients in B'. Multiplying the above relation successively by ρ, ρ^2, ..., we deduce that ρ^{f+1}, ρ^{f+2}, ..., ρ^{q-1} are also linear combinations of 1, ρ, ρ^2, ..., ρ^{f-1} with coefficients in B'. Thus every element of $B = \mathbb{Z}[\rho]$ is a linear combination of 1, ρ, ρ^2, ..., ρ^{f-1} with coefficients in B'.

So $\{1, \rho, \rho^2, \ldots, \rho^{f-1}\}$ is a system of generators of the L'-vector space L. Since $[L : L'] = f$ then $\{1, \rho, \rho^2, \ldots, \rho^{f-1}\}$ are linearly independent over L', hence over B'.

(4)   The coefficients of $F_{\mu_0}(X)$ belong to $\mathbb{Q}$, since they are invariant by τ; hence they are in $B' \cap \mathbb{Q} = \mathbb{Z}$.

Since $F_{\mu_0}(\mu_0) = 0$, the minimal polynomial of μ_0 divides $F_{\mu_0}(X)$; its roots are all the conjugates of μ_0, so it must coincide with $F_{\mu_0}(X)$, which is therefore irreducible.   ∎

It is not true in general that

$$\mathbb{Z}[\mu_0, \ldots, \mu_{r-1}] = \mathbb{Z}[\mu_0] = \cdots = \mathbb{Z}[\mu_{r-1}].$$

For example, let $q = 13$, $f = 3$, $r = 4$, and $h = 2$. Then the periods are

$$\mu_0 = \rho + \rho^3 + \rho^{-4},$$
$$\mu_1 = \rho^2 + \rho^6 + \rho^5,$$
$$\mu_2 = \rho^4 + \rho^{-1} + \rho^{-3},$$
$$\mu_3 = \rho^{-5} + \rho^{-2} + \rho^{-6}.$$

We shall show that the unique expressions of μ_1, μ_2, μ_3, as polynomials in μ_0 with rational coefficients, require some nonintegral coefficients. Indeed

$$\mu_0^2 = \mu_1 + 2\mu_2,$$
$$\mu_0\mu_1 = \mu_0 + \mu_1 + \mu_3,$$
$$\mu_0\mu_2 = 3 + \mu_1 + \mu_3,$$

and

$$\mu_0^3 = \mu_0\mu_1 + 2\mu_0\mu_2 = 6 + \mu_0 + 3\mu_1 + 3\mu_3$$
$$= 6 + \mu_0 + 3(-1 - \mu_0 - \mu_2),$$

hence

$$\mu_2 = \tfrac{1}{3}(-\mu_0^3 - 2\mu_0 + 3).$$

And from this we obtain

$$\mu_1 = \mu_0^2 - 2\mu_2 = \tfrac{1}{3}(2\mu_0^3 + 3\mu_0^2 + 4\mu_0 - 6),$$
$$\mu_3 = -1 - \mu_0 - \mu_1 - \mu_2 = \tfrac{1}{3}(-\mu_0^3 - 3\mu_0^2 - 5\mu_0).$$

It follows from (**B**) that given i, j, $0 \le i, j \le r - 1$, there exist integers $n_{ijk} \in \mathbb{Z}$ $(0 \le k \le r - 1)$, which are unique, such that

$$\mu_i\mu_j = \sum_{k=0}^{r-1} n_{ijk}\mu_k.$$

More precisely:

C.   *We have the relations*

$$\sum_{i=0}^{r-1} \mu_i \mu_{i+k} = n_k q - f \qquad for \quad 0 \le k \le r - 1,$$

where

$$n_k = \begin{cases} 1, & when\ f\ is\ even\ and\ k = 0, \\ 1, & when\ f\ is\ odd\ and\ k = 0\ or\ r/2, \\ 0, & otherwise. \end{cases}$$

Proof: First we evaluate the product

$$\mu_0 \mu_k = \left(\sum_{\ell=0}^{f-1} \rho^{h^{\ell r}} \right) \left(\sum_{j=0}^{f-1} \rho^{h^{k+jr}} \right).$$

Writing $j \equiv i + \ell \pmod{q-1}$, then the above product is equal to

$$\mu_0 \mu_k = \sum_{\ell=0}^{f-1} \sum_{i=0}^{f-1} \rho^{h^{\ell r}(1 + h^{k+ir})}.$$

Let

$$\mu'_i = \sum_{\ell=0}^{f-1} \rho^{h^{\ell r}(1 + h^{k+ir})}.$$

If $1 + h^{k+ir} \not\equiv 0 \pmod{q}$ there exists a unique t, $0 \le t \le q - 2$, such that $1 + h^{k+ir} \equiv h^t \pmod{q}$; hence μ'_i is equal to the period μ_t.

If $1 + h^{k+ir} \equiv 0 \pmod{q}$ then $\mu'_i = f$.

Therefore, we may write

$$\mu_0 \mu_k = n_k f + m_{k,0} \mu_0 + m_{k,1} \mu_1 + \cdots + m_{k,r-1} \mu_{r-1} \qquad (20.5)$$

with integers $n_k \ge 0$, $m_{k,0} \ge 0, \ldots, m_{k,r-1} \ge 0$.

Now, we determine n_k.

(I)   If f is even and $k = 0$, let $f = 2f'$, then $1 + h^{f'r} \equiv 0 \pmod{q}$ since $fr = q - 1$. So $\mu'_{f/2} = f$. On the other hand, if $0 \le i < f$ and $\mu'_i = f$ then we have $1 + h^{ir} \equiv 0 \pmod{q}$, hence $2ir \equiv 0 \pmod{q-1}$, that is, $2ir = mrf$; but $mf = 2i < 2f$, so $m = 0$ or 1. If $m = 0$ then $i = 0$, an absurdity, because q is odd. Thus $m = 1$, $i = f/2$. Therefore, in this case, $n_k = 1$.

(II)   If f is odd (hence r is even) and $k = r/2$, let $i = (f-1)/2$. Then $1 + h^{r/2 + (f-1)r/2} \equiv 0 \pmod{q}$, so $\mu'_{(f-1)/2} = f$. On the other hand, if $0 \le i < f$ and $\mu'_i = f$, we have $1 + h^{r/2 + ir} \equiv 0 \pmod{q}$, hence $r + 2ir = mrf$; thus $mf = 1 + 2i < 1 + 2f$; it follows that m is odd, so $m = 1$ and $i = (f-1)/2$.

(III)   We consider the remaining cases. If $1 + h^{k+ir} \equiv 0 \pmod{q}$ then $2k + 2ir = mrf$ and $0 \le r(mf - 2i) = 2k < 2r$, thus $mf - 2i = 0$ or 1, and $k = 0$. If $m = 0$ then $i = 0$, $k = 0$, an absurdity, since q is odd. Thus $m = 1$ and f is even, which is a case already studied.

If $mf = 2i + 1 < 2f + 1$ then m is odd, $m \le 2$, so $m = 1$, f is odd, $i = (f - 1)/2$, and also $k = r/2$, which was the case (II) above.

Therefore, in case (III), $n_k = 0$.

Since $\mu_0 \mu_k$ is the sum of f^2 terms of the form ρ^i and since each period contains f such terms, all appearing with different exponents i, $0 \le i \le q - 1$, it follows that

$$n_k + m_{k,0} + m_{k,1} + \cdots + m_{k,q-1} = f.$$

Applying the automorphisms τ^i, we obtain from (20.3):

$$\mu_i \mu_{k+i} = n_k f + m_{k,0} \mu_i + m_{k,1} \mu_{i+1} + \cdots + m_{k,r-1} \mu_{r-1+i}.$$

Hence, from $\sum_{i=0}^{r-1} \mu_i = -1$ we conclude that

$$\sum_{i=0}^{r-1} \mu_i \mu_{i+k} = n_k(q - 1) - (m_{k,0} + m_{k,1} + \cdots + m_{k,r-1}) = n_k q - f. \qquad \blacksquare$$

20.3 Lagrange Resolvents and the Jacobi Cyclotomic Function

We shall use the following notations: q, p are prime numbers such that $q - 1 = 2kp$.

ζ = primitive pth root of 1,

g = primitive root modulo p,

h = primitive root modulo q,

$K = \mathbb{Q}(\zeta)$, $A = \mathbb{Z}[\zeta]$,

σ = generator of the Galois group of $K|\mathbb{Q}$, defined by $\sigma(\zeta) = \zeta^g$,

ρ = primitive qth root of 1,

$L = \mathbb{Q}(\rho)$, $B = \mathbb{Z}[\rho]$,

τ = generator of Galois group of $L|\mathbb{Q}$, defined by $\tau(\rho) = \rho^h$,

$\mu_0, \ldots, \mu_{p-1}$: the p periods with $2k$ terms (relative to ρ, τ),

$L' = \mathbb{Q}(\mu_0, \ldots, \mu_{p-1}) = \mathbb{Q}(\mu_0) = \cdots = \mathbb{Q}(\mu_{p-1})$,

$B' = \mathbb{Z}[\mu_0, \ldots, \mu_{p-1}]$,

τ' = restriction of τ to L'.

We note that $L \cap K = \mathbb{Q}$. Indeed, the prime q is totally ramified in L and unramified in K, hence it is both totally ramified and unramified in $L \cap K$, so $L \cap K = \mathbb{Q}$.

Thus $\mathbb{Q}(\rho, \zeta)$ is a Galois extension of K with the Galois group isomorphic to the one of $L|\mathbb{Q}$ and generated by the automorphism $\tilde{\tau}$, defined by

$$\begin{cases} \tilde{\tau}(\rho) = \rho^h, \\ \tilde{\tau}(\zeta) = \zeta. \end{cases}$$

Similarly, $\mathbb{Q}(\rho, \zeta)$ is a Galois extension of L with the Galois group isomorphic to the one of $K|\mathbb{Q}$ and generated by the automorphism $\tilde{\sigma}$, defined by

$$\begin{cases} \tilde{\sigma}(\rho) = \rho, \\ \tilde{\sigma}(\zeta) = \zeta^g. \end{cases}$$

It is convenient to work with indices, as we now define.

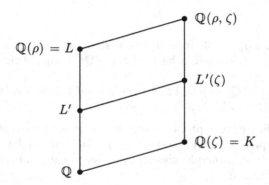

If t is any integer, not a multiple of q, then there exists a unique integer s, $0 \le s \le q - 2$, such that $t \equiv h^s \pmod{q}$. s is called the *index* of t (with respect to h, q), and we write $s = \mathrm{ind}_h(t)$, or simply $s = \mathrm{ind}(t)$ if there is no ambiguity concerning the choice of the primitive root h.

For example, $\mathrm{ind}(1) = 0$, $\mathrm{ind}(-1) = (q - 1)/2$. If $t \equiv t' \pmod{q}$ then $\mathrm{ind}(t) = \mathrm{ind}(t')$ and if t, t' are not multiples of q then $\mathrm{ind}(tt') \equiv \mathrm{ind}(t) + \mathrm{ind}(t') \pmod{q - 1}$. It is also clear that every integer s, $0 \le s \le q - 2$, is an index, namely $s = \mathrm{ind}(h^s)$.

With this notation, if $p \nmid n$ and $q \nmid m$ then the Lagrange resolvents may be written

$$\langle \zeta^n, \rho^m \rangle_\tau = \sum_{u=0}^{q-2} \zeta^{nu} \rho^{mh^u} = \sum_{t=1}^{q-1} \zeta^{n\,\mathrm{ind}_h(t)} \rho^{mt}. \tag{20.6}$$

The complex-conjugate of $\langle \zeta^n, \rho^m \rangle_\tau$ is

$$\overline{\langle \zeta^n, \rho^m \rangle_\tau} = \langle \zeta^{-n}, \rho^{-m} \rangle_\tau.$$

A first result to record is the following:

D.   *With the above notations, for every $n = 1, 2, \ldots, p - 1$:*
   (1) $\langle \zeta^n, \rho \rangle_\tau = \langle \zeta^n, \mu_0 \rangle_{\tau'} \neq 0$ *and it belongs to* $L'(\zeta)$.
   (2) $\langle \zeta^n, \rho \rangle_\tau^p \in \mathbb{Q}(\zeta)$.

Proof: (1)

$$
\begin{aligned}
\langle \zeta^n, \rho \rangle_\tau &= \quad \rho \quad + \zeta^n \rho^h + \zeta^{2n} \rho^{h^2} + \cdots + \zeta^{(q-2)n} \rho^{h^{q-2}} \\
&= \quad \rho \quad + \quad \zeta^n \rho^h \quad + \cdots + \zeta^{(p-1)n} \rho^{h^{p-1}} \\
&\quad + \; \rho^{h^p} \; + \quad \zeta^n \rho^{h^{p+1}} \quad + \cdots + \zeta^{(p-1)n} \rho^{h^{2p-1}} \\
&\qquad \cdots \cdots \cdots \cdots \cdots \cdots \cdots \cdots \cdots \cdots \cdots \cdots \cdots \cdots \cdots \\
&\quad + \rho^{h^{(2k-1)p}} + \zeta^n \rho^{h^{(2k-1)p+1}} + \cdots + \zeta^{(p-1)n} \rho^{h^{q-2}} \\
&= \quad \mu_0 \quad + \quad \zeta^n \mu_1 \quad + \cdots + \zeta^{(p-1)n} \mu_{p-1} \\
&= \langle \zeta^n, \mu_0 \rangle_{\tau'} \in L'(\zeta).
\end{aligned}
$$

Moreover, $\langle \zeta^n, \mu_0 \rangle_{\tau'} \neq 0$. Indeed, the periods $\mu_0, \mu_1, \ldots, \mu_{p-1}$, which are a basis of $L' | \mathbb{Q}$, are still a basis of $L'(\zeta) | \mathbb{Q}(\zeta)$, since this extension has still degree p.
   (2)   Since $[L' : \mathbb{Q}] = p$ and ζ is a pth root of 1, it follows from (**A**) that $\langle \zeta^n, \mu_0 \rangle_{\tau'}^p \in \mathbb{Q}(\zeta)$, so by (1), $\langle \zeta^n, \rho \rangle_\tau^p \in \mathbb{Q}(\zeta)$.   ∎

We shall indicate more properties of the Lagrange resolvents. In particular, we shall express the pth power of $\langle \zeta^n, \rho \rangle_\tau$ in terms of Jacobi cyclotomic functions, which, in principle, should be easier to calculate, since they only involve ζ (and not ρ).

E.   *If $p \nmid n$, $q \nmid m$, then*

$$\langle \zeta^n, \rho^m \rangle = \langle \zeta^n, \rho \rangle \zeta^{-n \, \mathrm{ind}(m)}.$$

Proof:

$$\langle \zeta^n, \rho^m \rangle = \sum_{t=1}^{q-1} \zeta^{n \, \mathrm{ind}(t)} \rho^{mt}.$$

But $\mathrm{ind}(tm) \equiv \mathrm{ind}(t) + \mathrm{ind}(m) \pmod{q - 1}$, hence

$$\langle \zeta^n, \rho^m \rangle = \zeta^{-n \, \mathrm{ind}(m)} \sum_{t=1}^{q-1} \zeta^{n \, \mathrm{ind}(tm)} \rho^{tm}$$

$$= \zeta^{-n \, \mathrm{ind}(m)} \sum_{s=1}^{q-1} \zeta^{n \, \mathrm{ind}(s)} \rho^s = \zeta^{-n \, \mathrm{ind}(m)} \langle \zeta^n, \rho \rangle.$$ ∎

F.   *If $p \nmid n$, $q \nmid m$, then $\langle \zeta^n, \rho^m \rangle^p$ and $\tilde{\sigma}(\langle \zeta^n, \rho^m \rangle)/\langle \zeta^n, \rho^m \rangle^g$ belong to $\mathbb{Q}(\zeta)$.*

Proof: By **(E)** and **(D)** we have

$$\langle \zeta^n, \rho^m \rangle^p = \langle \zeta^n, \rho \rangle^p \zeta^{-pn \,\mathrm{ind}(m)} = \langle \zeta^n, \rho \rangle^p \in \mathbb{Q}(\zeta).$$

Similarly

$$\tilde{\tau}(\tilde{\sigma}(\langle \zeta^n, \rho^m \rangle)) = \langle \zeta^{ng}, \rho^{mh} \rangle = \langle \zeta^{ng}, \rho \rangle \zeta^{-ng \,\mathrm{ind}(mh)}$$
$$= \tilde{\sigma}(\langle \zeta^n, \rho \rangle) \zeta^{-ng \,\mathrm{ind}(mh)},$$

and

$$\tilde{\tau}(\langle \zeta^n, \rho^m \rangle^g) = \langle \zeta^n, \rho^{mh} \rangle^g = \langle \zeta^n, \rho \rangle^g \zeta^{-ng \,\mathrm{ind}(mh)}.$$

Hence the quotient

$$\tilde{\sigma}(\langle \zeta^n, \rho^m \rangle)/\langle \zeta^n, \rho^m \rangle^g$$

is invariant under $\tilde{\tau}$ and therefore belongs to $\mathbb{Q}(\zeta)$. ∎

G.   *If $p \nmid n$, then $\langle \zeta^n, \rho \rangle \langle \zeta^{-n}, \rho \rangle = q$.*

Proof:
$$\langle \zeta^n, \rho \rangle \langle \zeta^{-n}, \rho \rangle = \left(\sum_{t=1}^{q-1} \zeta^{n \,\mathrm{ind}(t)} \rho^t \right) \left(\sum_{s=1}^{q-1} \zeta^{-n \,\mathrm{ind}(s)} \rho^s \right)$$

$$= \sum_{t=1}^{q-1} \sum_{s=1}^{q-1} \zeta^{n[\mathrm{ind}(t)-\mathrm{ind}(s)]} \rho^{t+s}.$$

For each s let r be defined by the congruence $t \equiv rs \pmod{q}$. Since $p \mid q-1$ then $\zeta^{q-1} = 1$, so the above sum is equal to

$$\sum_{s=1}^{q-1} \sum_{r=1}^{q-1} \zeta^{n \,\mathrm{ind}(r)} \rho^{(r+1)s}.$$

But $q - 1 = 2kp$ so

$$\sum_{r=1}^{q-1} \zeta^{n \,\mathrm{ind}(r)} = \sum_{m=1}^{q-1} \zeta^m = 2k \sum_{m=1}^{p} \zeta^m = 0,$$

hence we may add $\sum_{r=1}^{q-1} \zeta^{n \,\mathrm{ind}(r)} = 0$ and write

$$\langle \zeta^n, \rho \rangle \langle \zeta^{-n}, \rho \rangle = \sum_{s=0}^{q-1} \sum_{r=1}^{q-1} \zeta^{n \,\mathrm{ind}(r)} \rho^{(r+1)s}$$

$$= \sum_{r=1}^{q-1} \zeta^{n \,\mathrm{ind}(r)} \sum_{s=0}^{q-1} \rho^{(r+1)s}.$$

But

$$\sum_{s=0}^{q-1} \rho^{(r+1)s} = \begin{cases} q, & \text{when } r = q - 1, \\ 0, & \text{when } 1 \le r \le q - 2, \end{cases}$$

hence

$$\langle\zeta^n,\rho\rangle\langle\zeta^{-n},\rho\rangle = q\zeta^{n\,\text{ind}(q-1)} = q\zeta^{n\,\text{ind}(-1)} = q\zeta^{n(q-1)/2} = q\zeta^{nkp} = q. \quad\blacksquare$$

H.   If $p\nmid n$ then $\langle\zeta^n,\rho\rangle\langle\zeta^{-n},\rho^{-1}\rangle = q$, hence $|\langle\zeta^n,\rho\rangle| = \sqrt{q}$.

Proof: By (**E**) we have

$$\langle\zeta^n,\rho\rangle\overline{\langle\zeta^n,\rho\rangle} = \langle\zeta^n,\rho\rangle\langle\zeta^{-n},\rho^{-1}\rangle = \langle\zeta^n,\rho\rangle\langle\zeta^{-n},\rho\rangle\zeta^{-n\,\text{ind}(-1)}$$
$$= q\zeta^{n(q-1)/2} = q\zeta^{nkp} = q. \quad\blacksquare$$

I.   If $p\nmid nm(n+m)$, then

$$\frac{\langle\zeta^n,\rho\rangle\langle\zeta^m,\rho\rangle}{\langle\zeta^{n+m},\rho\rangle} = \sum_{r=1}^{q-2}\zeta^{n\,\text{ind}(r)-(n+m)\,\text{ind}(r+1)}.$$

Proof:

$$\langle\zeta^n,\rho\rangle\langle\zeta^m,\rho\rangle = \left(\sum_{t=1}^{q-1}\zeta^{n\,\text{ind}(t)}\rho^t\right)\left(\sum_{s=1}^{q-1}\zeta^{m\,\text{ind}(s)}\rho^s\right)$$
$$= \sum_{s=1}^{q-1}\sum_{t=1}^{q-1}\zeta^{n\,\text{ind}(t)+m\,\text{ind}(s)}\rho^{t+s}.$$

For every s, t let r be defined by the congruence $t \equiv rs \pmod{q}$. The above sum is equal to

$$\sum_{s=1}^{q-1}\sum_{r=1}^{q-1}\zeta^{n\,\text{ind}(r)+(n+m)\,\text{ind}(s)}\rho^{(r+1)s}$$

$$= \sum_{r=1}^{q-1}\zeta^{n\,\text{ind}(r)}\left(\sum_{s=1}^{q-1}\zeta^{(n+m)\,\text{ind}(s)}\rho^{(r+1)s}\right)$$

$$= \sum_{r=1}^{q-2}\zeta^{n\,\text{ind}(r)}\left(\sum_{s=1}^{q-1}\zeta^{(n+m)\,\text{ind}(s)}\rho^{(r+1)s}\right)$$

$$+ \zeta^{n\,\text{ind}(q-1)}\left(\sum_{s=1}^{q-1}\zeta^{(n+m)\,\text{ind}(s)}\right).$$

Since p does not divide $n+m$, then ζ^{n+m} is still a primitive pth root of 1, hence, since p divides $q-1$ then $\sum_{s=1}^{q-1}\zeta^{(n+m)\,\text{ind}(s)} = 0$, and we conclude that

$$\langle\zeta^n,\rho\rangle\langle\zeta^m,\rho\rangle = \sum_{r=1}^{q-2}\zeta^{n\,\text{ind}(r)}\left(\sum_{s=1}^{q-1}\zeta^{(n+m)\,\text{ind}(s)}\rho^{(r+1)s}\right).$$

Since q does not divide $r + 1$, letting $u \equiv (r+1)s \pmod{q}$, then the above sum is equal to

$$\sum_{r=1}^{q-2} \zeta^{n\,\mathrm{ind}(r)} \sum_{u=1}^{q-1} \zeta^{(n+m)[\mathrm{ind}(u)-\mathrm{ind}(r+1)]} \rho^u$$

$$= \left(\sum_{u=1}^{q-1} \zeta^{(n+m)\,\mathrm{ind}(u)} \rho^u \right) \left(\sum_{r=1}^{q-2} \zeta^{n\,\mathrm{ind}(r)-(n+m)\,\mathrm{ind}(r+1)} \right)$$

$$= \langle \zeta^{n+m}, \rho \rangle \left(\sum_{r=1}^{q-2} \zeta^{n\,\mathrm{ind}(r)-(n+m)\,\mathrm{ind}(r+1)} \right). \qquad \blacksquare$$

If $p \nmid nm(n+m)$, we define the *Jacobi cyclotomic function* (where $q \equiv 1 \pmod{p}$):

$$\psi_{n,m}(\zeta) = \sum_{r=1}^{q-2} \zeta^{n\,\mathrm{ind}(r)-(n+m)\,\mathrm{ind}(r+1)} \in \mathbb{Q}(\zeta). \qquad (20.7)$$

Thus by **(I)** we have

$$\psi_{n,m}(\zeta) = \frac{\langle \zeta^n, \rho \rangle \langle \zeta^m, \rho \rangle}{\langle \zeta^{n+m}, \rho \rangle}. \qquad (20.8)$$

We define also, for $p \nmid n\ell(\ell + 1)$:

$$\psi_\ell(\zeta^n) = \psi_{n,\ell n}(\zeta) = \sum_{r=1}^{q-2} \zeta^{n\,\mathrm{ind}(r)-n(\ell+1)\,\mathrm{ind}(r+1)}. \qquad (20.9)$$

In particular, if $\ell = 1, 2, \ldots, p - 2$:

$$\psi_\ell(\zeta) = \frac{\langle \zeta, \rho \rangle \langle \zeta^\ell, \rho \rangle}{\langle \zeta^{\ell+1}, \rho \rangle} = \sum_{r=1}^{q-2} \zeta^{\mathrm{ind}(r)-(\ell+1)\,\mathrm{ind}(r+1)}. \qquad (20.10)$$

In view of **(G)**, this is equal to

$$\psi_\ell(\zeta) = \frac{1}{q} \langle \zeta, \rho \rangle \langle \zeta^\ell, \rho \rangle \langle \zeta^{-\ell-1}, \rho \rangle. \qquad (20.11)$$

One of the advantages of introducing the functions $\psi_{n,m}(\zeta)$ is the comparative ease of their computation; it is only necessary to have a table of indices relative to the primitive root h modulo q. To simplify the computations in Gauss' method of solution of the cyclotomic equation $\Phi_q(X) = 0$, Jacobi has also shown:

J.
$$\langle \zeta, \rho \rangle^p = q\psi_1(\zeta)\psi_2(\zeta)\cdots\psi_{p-2}(\zeta).$$

Proof:

$$\prod_{\ell=1}^{p-2} \psi_\ell(\zeta) = \prod_{\ell=1}^{p-2} \frac{\langle \zeta, \rho \rangle \langle \zeta^\ell, \rho \rangle}{\langle \zeta^{\ell+1}, \rho \rangle}$$

$$= \langle \zeta, \rho \rangle^{p-2} \times \frac{\langle \zeta, \rho \rangle}{\langle \zeta^{p-1}, \rho \rangle}.$$

But by (**G**):

$$\frac{1}{\langle \zeta^{p-1}, \rho \rangle} = \frac{1}{\langle \zeta^{-1}, \rho \rangle} = \frac{\langle \zeta, \rho \rangle}{q}$$

so

$$q \prod_{\ell=1}^{p-2} \psi_\ell(\zeta) = \langle \zeta, \rho \rangle^p. \qquad \blacksquare$$

It follows that:

K.   *If $p \nmid nm(n+m)$, then*
$$\psi_{n,m}(\zeta)\psi_{n,m}(\zeta^{-1}) = q.$$

If $\ell = 1, 2, \ldots, p-2$, then
$$\psi_\ell(\zeta)\psi_\ell(\zeta^{-1}) = q.$$

Proof: By (20.8) and (**G**):

$$\psi_{n,m}(\zeta)\psi_{n,m}(\zeta^{-1}) = \frac{\langle \zeta^n, \rho \rangle \langle \zeta^m, \rho \rangle}{\langle \zeta^{n+m}, \rho \rangle}$$

$$\times \frac{\langle \zeta^{-n}, \rho \rangle \langle \zeta^{-m}, \rho \rangle}{\langle \zeta^{-(n+m)}, \rho \rangle} = \frac{q \cdot q}{q} = q.$$

The particular case $\psi_\ell(\zeta)\psi_\ell(\zeta^{-1}) = q$ follows at once, taking $n = 1$, $m = \ell$. $\qquad \blacksquare$

By (20.8) and (20.7) we note also that

$$\psi_{n,m}(\zeta) = \psi_{m,n}(\zeta) = \sum_{r=1}^{q-2} \zeta^{m\,\text{ind}(r)-(n+m)\,\text{ind}(r+1)}. \qquad (20.12)$$

Also if $m \equiv m' \pmod{p}$ or $n \equiv n' \pmod{p}$ then

$$\psi_{n,m}(\zeta) = \psi_{n,m'}(\zeta), \qquad \psi_{n,m}(\zeta) = \psi_{n',m}(\zeta).$$

It follows:

L.   (1) *If $nm \equiv 1 \pmod{p}$ then $\psi_n(\zeta^m) = \psi_m(\zeta)$.*

(2) *If $p \nmid nm(n+m)$ then $\psi_{-(n+m),m}(\zeta) = \psi_{n,m}(\zeta)$.*

(3) *If $p \nmid n(n+1)$ then $\psi_n(\zeta) = \psi_{-(n+1)}(\zeta)$.*

(4) *If $p \nmid nm\ell(n+m)(n+m+\ell)$ then*

$$\psi_{n,m}(\zeta)\psi_{n+m,\ell}(\zeta) = \frac{\langle \zeta^n, \rho \rangle \langle \zeta^m, \rho \rangle \langle \zeta^\ell, \rho \rangle}{\langle \zeta^{n+m+\ell}, \rho \rangle}$$

and the above product is invariant under any permutation of the numbers n, m, ℓ. In particular

$$\psi_{n,m}(\zeta)\psi_{n+m,\ell}(\zeta) = \psi_{m,\ell}(\zeta)\psi_{m+\ell,n}(\zeta).$$

(5) *If $p \nmid 2n(2n+1)(2n+2)$ then*

$$\psi_{2n}(\zeta)\psi_{2n+1}(\zeta) = \psi_1(\zeta)\psi_n(\zeta^2).$$

Proof: (1) $\psi_n(\zeta^m) = \psi_{m,nm}(\zeta) = \psi_{m,1}(\zeta) = \psi_{1,m}(\zeta) = \psi_m(\zeta)$,

by the properties already established.

(2)

$$\frac{\psi_{-(n+m),m}(\zeta)}{\psi_{n,m}(\zeta)} = \frac{\langle \zeta^{-(n+m)}, \rho \rangle \langle \zeta^m, \rho \rangle}{\langle \zeta^{-n}, \rho \rangle} \times \frac{\langle \zeta^{n+m}, \rho \rangle}{\langle \zeta^n, \rho \rangle \langle \zeta^m, \rho \rangle} = \frac{q}{q} = 1$$

by (**G**).

(3)

$$\psi_n(\zeta) = \psi_{1,n}(\zeta) = \psi_{n,1}(\zeta) = \psi_{-(n+1),1}(\zeta)$$
$$= \psi_{1,-(n+1)}(\zeta) = \psi_{-(n+1)}(\zeta),$$

using (2).

(4)   This follows at once from (20.8).

(5)   By (4), we have, when $p \nmid 2n\ell(2n\ell + \ell)(2n\ell + 2\ell)$:

$$\psi_{2n\ell,\ell}(\zeta)\psi_{2n\ell+\ell,\ell}(\zeta) = \psi_{\ell,\ell}(\zeta)\psi_{2\ell,2n\ell}(\zeta).$$

The left-hand side is equal to

$$\psi_{\ell,2n\ell}(\zeta)\psi_{\ell,(2n+1)\ell}(\zeta) = \psi_{2n}(\zeta^\ell)\psi_{2n+1}(\zeta^\ell),$$

while the right-hand side is equal to $\psi_1(\zeta^\ell)\psi_n(\zeta^{2\ell})$. Taking $\ell = 1$, we have $\psi_{2n}(\zeta)\psi_{2n+1}(\zeta) = \psi_1(\zeta)\psi_n(\zeta^2)$.   ∎

Jacobi also considered the following integers $J_{\ell,m}$, where $0 < \ell, m < q - 1$, and $\ell + m \neq q - 1$:

$$J_{\ell,m} = \sum_{t=1}^{q-2} h^{m \, \mathrm{ind}_h(t) + [2(q-1) - \ell - m] \, \mathrm{ind}_h(t+1)}. \tag{20.13}$$

Clearly, we have

$$J_{\ell,m} \equiv \sum_{t=1}^{q-2} t^m (t+1)^{2(q-1)-(\ell+m)} \pmod{q}. \tag{20.14}$$

M.   *With the above notations*

$$
\begin{cases}
J_{\ell,m} \equiv 0 \pmod{q} & \text{when } \ell + m < q - 1, \\
J_{\ell,m} \equiv -\left(\dfrac{2(q-1) - (\ell+m)}{q-1-m}\right) \not\equiv 0 \pmod{q} & \text{when } \ell + m > q - 1.
\end{cases}
$$

Proof: Let

$$
n = \begin{cases}
q - 1 - (\ell + m) & \text{when } \ell + m < q - 1, \\
2(q-1) - (\ell + m) & \text{when } \ell + m > q - 1,
\end{cases}
$$

so $0 < n < q - 1$ and $n \equiv -(\ell + m) \pmod{q-1}$.

From $(t+1)^{q-1} \equiv 1 \pmod{q}$, where $1 \le t \le q - 2$, we deduce

$$
J_{\ell,m} \equiv \sum_{t=1}^{q-2} t^m (t+1)^n \equiv \sum_{t=1}^{q-1} t^m (t+1)^n \pmod{q}.
$$

So

$$
J_{\ell,m} \equiv \sum_{t=1}^{q-1} t^m \sum_{s=0}^{n} \binom{n}{s} t^s \equiv \sum_{s=0}^{n} \binom{n}{s} \sum_{t=1}^{q-1} t^{m+s} \pmod{q}.
$$

If $q - 1$ divides $m + s$ then $\sum_{t=1}^{q-1} t^{m+s} \equiv -1 \pmod{q}$.

If $q - 1$ does not divide $m + s$ then

$$
\sum_{t=1}^{q-1} t^{m+s} \equiv \sum_{t=1}^{q-1} h^{(m+s)\, \mathrm{ind}_h(t)}
$$

$$
\equiv \sum_{r=0}^{q-2} h^{(m+s)r} \equiv \frac{(h^{m+s})^{q-1} - 1}{h^{m+s} - 1} \equiv 0 \pmod{q}.
$$

Now we assume that $\ell + m < q - 1$.

If $s = 0, 1, \ldots, n$ we have

$$
0 < m \le m + s \le m + n = q - 1 - \ell < q - 1
$$

hence

$$
J_{\ell,m} \equiv 0 \pmod{q}.
$$

If $\ell + m > q - 1$ then $m < q - 1 < 2(q-1) - \ell = m + n < 2(q-1)$, so there exists exactly one value of s, namely $s = q - 1 - m$, such that $m + s$ is divisible by $q - 1$. Thus

$$
J_{\ell,m} \equiv -\binom{n}{q-1-m}
$$

$$
\equiv -\frac{n(n-1)\cdots(n-q+m+2)}{(q-1-m)!} \not\equiv 0 \pmod{q},
$$

since $n < q$.   ■

20.4 On the Decomposition into Prime Ideals of the Cyclotomic Field

In this section we follow Kummer and examine in more detail the decomposition into prime ideals of the cyclotomic field. In the main result, we shall show how to obtain principal ideals by multiplying certain sets of prime ideals.

We keep the preceding notations.

If q is a prime different from p, let f denote the order of q modulo p. Let $q^f - 1 = 2kp$ and let $p - 1 = fr$.

We recall that $Aq = Q_0 Q_1 \cdots Q_{r-1}$, where the prime ideals Q_i are pairwise distinct and $Q_i = Aq + AF_i(\zeta)$, where

$$\Phi_p(X) \equiv \prod_{i=0}^{r-1} F_i(X) \pmod{q}$$

with $F_i \in \mathbb{Z}[X]$ pairwise distinct irreducible polynomials, each of degree f.

In particular, if $q \equiv 1 \pmod{p}$, that is, $f = 1$, we have $F_i(X) = X - h^{2ki}$ for $i = 0, \ldots, p - 2$, where h is a primitive root modulo q. Thus $Q_i = Aq + A(\zeta - h^{2ki})$ for $i = 0, \ldots, p - 2$.

Indeed, $(h^{2ki})^p \equiv (h^i)^{q-1} \equiv 1 \pmod{q}$ and for $i \neq j$, $0 \leq i, j \leq p - 2$, necessarily $h^{2ki} \not\equiv h^{2kj} \pmod{q}$; otherwise, $h^{2k(j-i)} \equiv 1 \pmod{q}$, so $q - 1 = 2kp$ divides $2k(j - i)$, hence p divides $j - i$, which is absurd. So

$$\Phi_p(X) = \prod_{i=0}^{p-2} (X - \overline{h}^{2ki})$$

in $\mathbb{F}_q[X]$ and therefore $Q_i = Aq + A(\zeta - h^{2ki})$, as stated.

Thus we also have

$$Aq = \prod_{i=0}^{p-2} \sigma^i(Q), \tag{20.15}$$

where $Q = Q_0$ and $\sigma(\zeta) = \zeta^g$, g a primitive root modulo p.

Since $p - 1 = fr$, we may consider the r periods of f terms of the cyclotomic field $K = \mathbb{Q}(\zeta)$, associated to the primitive root g modulo p:

$$\begin{aligned}
\eta_0 &= \zeta + \zeta^{g^r} + \zeta^{g^{2r}} + \cdots + \zeta^{g^{(f-1)r}}, \\
\eta_1 &= \zeta^g + \zeta^{g^{r+1}} + \zeta^{g^{2r+1}} + \cdots + \zeta^{g^{(f-1)r+1}}, \\
&\cdots\cdots\cdots\cdots\cdots\cdots\cdots\cdots\cdots\cdots\cdots\cdots \\
\eta_{r-1} &= \zeta^{g^{r-1}} + \zeta^{g^{2r-1}} + \zeta^{g^{3r-1}} + \cdots + \zeta^{g^{p-2}}.
\end{aligned}$$

Hence $\sigma^r(\eta_i) = \eta_i$ for $i = 0, 1, \ldots, r - 1$. By (**B**):

$$F_{\eta_0}(X) = \prod_{i=0}^{r-1} (X - \eta_i) \in \mathbb{Z}[X]. \tag{20.16}$$

We now study the congruence

$$F_{\eta_0}(X) \equiv 0 \pmod{q}.$$

N.   *Let Q be any prime ideal of A dividing Aq. With the above notations:*
(1) $\eta_j^q \equiv \eta_j \pmod{Q}$ *for $j = 0, 1, \ldots, r - 1$.*
(2) *The congruence $F_{\eta_0}(X) \equiv 0 \pmod{q}$ has at least one root $u \in \mathbb{Z}$, $0 \le u \le q - 1$.*

Proof: (1)   Since Q divides Aq then for every $j = 0, 1, \ldots, r - 1$, we have

$$\eta_j^q = (\zeta^{g^j} + \zeta^{g^{j+r}} + \cdots + \zeta^{g^{j+(f-1)r}})^q$$
$$\equiv \zeta^{qg^j} + \zeta^{qg^{j+r}} + \cdots + \zeta^{qg^{j+(f-1)r}} \pmod{Q}.$$

We have $q \equiv g^t \pmod{p}$ for some t, $0 \le t \le p - 2$. From $q^f \equiv 1 \pmod{p}$ it follows that $g^{tf} \equiv 1 \pmod{p}$ so $p - 1 = rf$ divides tf, that is, $t = ir$ with $0 \le i \le f - 1$.
Hence

$$\eta_j^q \equiv \zeta^{g^{j+ir}} + \zeta^{g^{j+(i+1)r}} + \cdots + \zeta^{g^{j+(f-1+i)r}} \equiv \eta_j \pmod{Q}.$$

(2)   We have

$$X(X - 1) \cdots (X - q + 1) \equiv X^q - X \pmod{q}. \tag{20.17}$$

For any integer m, replacing X by $m - \eta_j$ we obtain

$$(m - \eta_j)(m - \eta_j - 1) \cdots (m - \eta_j - q + 1) \equiv (m - \eta_j)^q - (m - \eta_j)$$
$$\equiv m^q - \eta_j{}^q - m + \eta_j \pmod{Q}.$$

By (1), for $j = 0, 1, \ldots, r - 1$:

$$(m - \eta_j)(m - \eta_j - 1) \cdots (m - \eta_j - q + 1) \equiv 0 \pmod{Q}.$$

Multiplying all these congruences, we obtain

$$F_{\eta_0}(m)F_{\eta_0}(m - 1) \cdots F_{\eta_0}(m - q + 1) \equiv 0 \pmod{Q}$$

and since the left-hand side is in $\mathbb{Z}$ then

$$F_{\eta_0}(m)F_{\eta_0}(m - 1) \cdots F_{\eta_0}(m - q + 1) \equiv 0 \pmod{q}.$$

So, for some integer m', $0 \le m' \le q$, $F_{\eta_0}(m - m') \equiv 0 \pmod{q}$. If $m - m' \equiv u \pmod{q}$, $0 \le u \le q - 1$, then $F_{\eta_0}(u) \equiv 0 \pmod{q}$.   ■

O.   *Let Q be any prime ideal of A dividing Aq. For every period η_j, $0 \le j \le r - 1$, there exists a unique integer $u_j \in \mathbb{Z}$, $0 \le u_j \le q - 1$, such that $\eta_j \equiv u_j \pmod{Q}$. In particular*

$$F_{\eta_0}(X) \equiv \prod_{j=0}^{r-1}(X - u_j) \pmod{q}.$$

Proof: Let $\{a_0, a_1, \ldots, a_{n-1}\}$ be the set of all integers such that $0 \leq a_i \leq q - 1$, and $F_{\eta_0}(a_i) \equiv 0 \pmod{q}$. By (**N**) this set is not empty, that is, $n \geq 1$.

Let $\{b_0, b_1, \ldots, b_{q-1-n}\} = \{0, 1, \ldots, q - 1\} \setminus \{a_0, a_1, \ldots, a_{n-1}\}$ so $F(b_i) \not\equiv 0 \pmod{q}$ for $i = 0, 1, \ldots, q - 1 - n$.

By (**N**), $\eta_j^q \equiv \eta_j \pmod{Q}$ for $j = 0, 1, \ldots, n-1$. It follows from (20.17) that

$$\eta_j(\eta_j - 1) \cdots (\eta_j - q + 1) \equiv \eta_j^q - \eta_j \equiv 0 \pmod{Q}.$$

This is rewritten as

$$(\eta_j - a_0)(\eta_j - a_1) \cdots (\eta_j - a_{n-1})(\eta_j - b_0)(\eta_j - b_1) \cdots (\eta_j - b_{q-1-n})$$
$$\equiv 0 \pmod{Q}.$$

But $\eta_j - b_i \not\equiv 0 \pmod{Q}$, hence

$$(\eta_j - a_0)(\eta_j - a_1) \cdots (\eta_j - a_{n-1}) \equiv 0 \pmod{Q}. \tag{20.18}$$

For every j let $I_j = \{i \mid 0 \leq i \leq n - 1, \ \eta_j \equiv a_i \pmod{Q}\}$. By (20.18) there exists i such that $\eta_j - a_i \equiv 0 \pmod{Q}$. Thus $I_j \neq \varnothing$.

I_j has at most one element. Indeed, if $\eta_j \equiv a_i \pmod{Q}$ and $\eta_j \equiv a_{i'} \pmod{Q}$ then $a_i \equiv a_{i'} \pmod{Q}$ so $a_i \equiv a_{i'} \pmod{q}$, therefore $a_i = a_{i'}$, and $i = i'$.

Let i_j be the unique element of I_j and, for simplicity, we write $u_j = a_{i_j}$. Then

$$F_{\eta_0}(X) = \prod_{j=0}^{r-1}(X - \eta_j) \equiv \prod_{j=0}^{r-1}(X - u_j) \pmod{Q}.$$

Finally, since the coefficients of $F_{\eta_0}(X)$ are in $\mathbb{Z}$, then the above is a congruence modulo q. ∎

We cannot expect that the integers $u_0, u_1, \ldots, u_{r-1}$ are pairwise incongruent modulo q—just consider the situation where $q < r$ (which is possible when $f > 1$).

Now let K' be the subfield of K fixed by σ^r, let A' denote the ring of integers of K'. By (**B**), we have

$$K' = \mathbb{Q}(\eta_0, \eta_1, \ldots, \eta_{r-1}) = \mathbb{Q}(\eta_0) = \cdots = \mathbb{Q}(\eta_{r-1})$$

and $A' = \mathbb{Z}[\eta_0, \eta_1, \ldots, \eta_{r-1}]$. As we have already indicated, in general $A' \neq \mathbb{Z}[\eta_i]$, $0 \leq i \leq r - 1$. Therefore, the decomposition of $A'q$ into the prime ideals of A' cannot be obtained by the method indicated by Kummer (Chapter 11, Theorem 2) which involved the decomposition of $F_{\eta_0}(X)$ modulo q as a product of irreducible factors over the field $\mathbb{F}_q$.

However, we may still prove:

P. *$A'q$ is the product of r distinct prime ideals:*

$$A'q = Q_0'Q_1' \cdots Q_{r-1}'.$$

Each ideal Q_i' has inertial degree 1 in the extension $K'|\mathbb{Q}$, that is, $A'/Q_i' \cong \mathbb{F}_q$. With appropriate numbering, if $Aq = Q_0 Q_1 \cdots Q_{r-1}$, then $Q_i = AQ_i'$ and $Q_i' = A' \cap Q_i$ for $i = 0, 1, \ldots, r - 1$.

Proof: We have $Aq = Q_0 Q_1 \cdots Q_{r-1}$, the product of distinct prime ideals of inertial degree f. Each prime ideal Q' of A' is the restriction to A' of a prime ideal of A dividing Aq. Thus $Q' = Q_i \cap A'$ for some i, $0 \le i \le r - 1$.

The inertial degree of Q' is 1 (in the extension $K'|\mathbb{Q}$), because

$$A'/Q' = \mathbb{Z}[\eta_0, \eta_1, \ldots, \eta_{r-1}]/Q' = \mathbb{F}_q,$$

since by (O) every period η_j is congruent modulo $Q' = Q_i \cap A'$ to an integer.

Since each ideal Q_i is unramified (in the extension $K|\mathbb{Q}$), the same is true for each ideal $Q' = Q_i \cap A$ (in $K'|\mathbb{Q}$). Noting that $[K' : \mathbb{Q}] = r$, by the fundamental relation, there are precisely r prime ideals in A' dividing $A'q$. By the preceding considerations they must be the ideals $Q_i' = Q_i \cap A'$, which are therefore distinct. Finally, Q_i divides AQ_i' and since there are exactly r prime ideals in A dividing Aq then each one of the ideals AQ' must be prime, since they cannot be further decomposed. Hence $Q_i = AQ_i'$ for $i = 0, 1, \ldots, r - 1$.   ∎

We shall also require the following proposition:

Q.   *With the same notations as before, let Q be a prime ideal of A dividing Aq, let $\alpha = \alpha_0' + \alpha_1'\zeta, \ldots, \alpha_{f-1}'\zeta^{f-1} \in A$, with $\alpha_j' \in A'$. If $t \ge 1$ then Q^t divides $A\alpha$ if and only if $Q^t|A\alpha_j'$ for $j = 0, 1, \ldots, f - 1$.*

Proof: It is clear that if $Q^t|A\alpha_j'$ for $j = 0, 1, \ldots, r - 1$, then $Q^t|A\alpha$. Now we prove the converse.

Let $Q = Q_0$, Q_1, $\ldots$, Q_{r-1} be the distinct prime ideals of A dividing Aq where $fr = p - 1$; let $I = Q_1 \cdots Q_{r-1}$, so $I \not\subseteq Q$ and there exists $\beta \in I$, $\beta \notin Q$. Let $\beta' = N_{K|K'}(\beta) \in A'$, so $\beta' \in I \cap A'$. Since Q is the only prime ideal of A containing $Q \cap A'$ (as proved in (**P**)) and σ^r is a generator of the Galois group of $K|K'$, then $\sigma^r(Q)$ is a prime ideal of A dividing $Q \cap A'$; therefore $\sigma^{ir}(Q) = Q$ for every $i = 0, 1, \ldots, f - 1$. It follows that $\sigma^{ir}(\beta) \notin \sigma^{ir}(Q) = Q$ for $i = 0, 1, \ldots, f - 1$, so $\beta' = N_{K|K'}(\beta) \notin Q \cap A'$.

Since Q^t divides $A\alpha$ then $Aq^t = Q^t I^t$ divides $A\alpha \cdot I^t$, hence also Aq^t divides $A\alpha\beta'^t$ because I divides $A\beta'$. But

$$\alpha\beta'^t = (\alpha_0'\beta'^t) + (\alpha_1'\beta'^t)\zeta + \cdots + (\alpha_{f-1}'\beta'^t)\zeta^{f-1}$$

with $\alpha_j'\beta'^t \in A'$. Then

$$\frac{\alpha\beta'^t}{q^t} = \frac{\alpha_0'\beta'^t}{q^t} + \frac{\alpha_1'\beta'^t}{q^t}\zeta + \cdots + \frac{\alpha_{f-1}'\beta'^t}{q^t}\zeta^{f-1} \in A.$$

The above expression is unique as a linear combination with coefficients in K', and since, by (**B**), $\{1, \zeta, \ldots, \zeta^{f-1}\}$ is a basis of the free A'-module A, then the coefficients $\alpha_j'\beta'^t/q^t$ must be in A', that is, Aq^t divides $A\alpha_j'\beta'^t$.

But $Q^t | Q^t I^t = A q^t$ and Q does not divide β', so Q^t divides $A \alpha'_j$ for $j = 0, 1, \ldots, f - 1$.   ∎

The next result of Kummer has the special feature that it establishes that the product of an appropriate set of conjugates of Q is a principal ideal, even better, the principal ideal generated by the pth power of the Lagrange resolvent $\langle \zeta, \rho \rangle_\tau$. Moreover, products of appropriate sets of conjugates of Q are principal ideals generated by $\psi_d(\zeta)$ (as defined in Section 20.3).

Stickelberger generalized Kummer's result by considering resolvents associated to natural numbers m, not assumed to be prime.

For each natural number $j \geq 0$, let g_j be the unique integer such that $1 \leq g_j \leq p - 1$ and $g_j \equiv g^j \pmod{p}$. If $j < 0$, let g_j be the unique integer such that $1 \leq g_j \leq p - 1$ and $g_j g^{-j} \equiv 1 \pmod{p}$. Since p does not divide g, the last condition may be written as $g_j \equiv g^j \pmod{p}$. Clearly, if $i, j \in \mathbb{Z}$ then $g_{i+j} \equiv g_i g_j \pmod{p}$.

Henceforth we assume that $f = 1$, that is, $q \equiv 1 \pmod{p}$; let $q - 1 = 2kp$, with $k \geq 1$. For simplicity, we write $\pi = (p - 1)/2$ (this should not cause any confusion with the number $\pi = 3.14 \ldots$).

R.   *Let Q be the ideal of A generated by q and $\zeta - h^{2k}$. Then:*

(1)
$$A \langle \zeta, \rho \rangle^p = \prod_{i=0}^{p-2} (\sigma^i(Q))^{g_{\pi - i}}.$$

(2) *If $1 \leq d \leq p - 2$ and*

$$I_d = \{ i \mid 0 \leq i \leq p - 2 \text{ such that } g_{\pi-i} + g_{\pi-i+\mathrm{ind}_g(d)} > p \}$$

then

$$A \psi_d(\zeta) = \prod_{i \in I_d} \sigma^i(Q).$$

Proof: We have seen in (**J**) that $\langle \zeta, \rho \rangle^p = q \psi_1(\zeta) \cdots \psi_{p-2}(\zeta)$. By (20.15) $Aq = \prod_{i=0}^{p-2} \sigma^i(Q)$, so each $\sigma^i(Q)$ divides Aq.

Conversely, if a prime ideal of A divides $\langle \zeta, \rho \rangle^p \in A$ then it divides the ideal generated by $\langle \zeta, \rho \rangle^p \cdot \langle \zeta^{-1}, \rho \rangle^p = q^p$ (see (**G**)) hence it divides Aq, so it must be one of the ideals $\sigma^i(Q)$.

Thus

$$A \langle \zeta, \rho \rangle^p = \prod_{i=0}^{p-2} [\sigma^i(Q)]^{r_i},$$

where the exponents $r_i \geq 1$ have to be determined.

We shall prove that $r_i = g_{\pi - i}$. Since $Aq = \prod_{i=0}^{p-2} \sigma^i(Q)$ and $\langle \zeta, \rho \rangle^p = q \psi_1(\zeta) \cdots \psi_{p-2}(\zeta)$, it is equivalent to proving that

$$A \prod_{d=1}^{p-2} \psi_d(\zeta) = \prod_{i=0}^{p-2} [\sigma^i(Q)]^{g_{\pi - i} - 1}.$$

Let i be such that $0 \le i \le p-2$. Since $h^{2k} \equiv \zeta \pmod{Q}$, then

$$h^{2k(p-g_{\pi-i})} \equiv \zeta^{p-g_{\pi-i}} = \zeta^{-g_{\pi-i}} = \zeta^{g^\pi g^{\pi-i}} = \zeta^{g^{-i}} \pmod{Q}.$$

Applying the automorphism σ^i we have

$$h^{2k(p-g_{\pi-i})} \equiv \sigma^i(\zeta^{g^{-i}}) = \zeta \pmod{\sigma^i(Q)}.$$

For any integer d, $1 \le d \le p-2$, we have defined in (20.10):

$$\psi_d(\zeta) = \sum_{t=1}^{q-2} \zeta^{\mathrm{ind}_h(t)-(d+1)\,\mathrm{ind}_h(t+1)} = \sum_{t=1}^{q-2} \zeta^{\mathrm{ind}_h(t)+(q-1-(d+1))\,\mathrm{ind}_h(t+1)},$$

since p divides $q-1$. So if $0 \le i \le p-2$ then

$$\psi_d(\zeta) \equiv \sum_{t=1}^{q-2} h^{2k(p-g_{\pi-i})\{\mathrm{ind}_h(t)+[q-1-(d+1)]\,\mathrm{ind}_h(t+1)\}} \pmod{\sigma^i(Q)},$$

Let $\ell = \ell_{d,i} = 2kg_{\mathrm{ind}_g(d)-i}$ and $m = m_i = 2k(p-g_{\pi-i})$, so $0 < \ell < 2kp = q-1$, $0 < m < 2kp = q-1$. Moreover, $\ell + m \ne q-1$, otherwise $q-1 = 2kp + 2k(g_{\mathrm{ind}_g(d)-i} - g_{\pi-i})$, hence $g_{\mathrm{ind}_g(d)-i} = g_{\pi-i}$, thus $\mathrm{ind}_g(d) = \pi$ and $d \equiv g^\pi \equiv -1 \pmod{p}$, contrary to the hypothesis. We may therefore consider the integer $J_{\ell,m}$, which was defined in (20.13):

$$J_{\ell,m} = \sum_{t=1}^{q-2} h^{m\,\mathrm{ind}_h(t)+[2(q-1)-\ell-m]\,\mathrm{ind}_h(t+1)}$$

$$\equiv \sum_{t=1}^{q-2} h^{2k(p-g_{\pi-i})\{\mathrm{ind}_h(t)+[(q-1)-(d+1)]\,\mathrm{ind}_h(t+1)\}} \pmod{q},$$

because

$$2(q-1) - \ell - m \equiv -\ell - m \equiv -2kg_{\mathrm{ind}_g(d)-i} - 2kp + 2kg_{\pi-i}$$

$$\equiv 2k(g^{\pi-i+\mathrm{ind}_d(g)} + g^{\pi-i}) \equiv 2kg_{\pi-i}(d+1)$$

$$\equiv 2k(p-g_{\pi-i})[(q-1)-(d+1)] \pmod{q-1}.$$

Therefore, $\psi_d(\zeta) \equiv J_{\ell_{d,i},m_i} \pmod{\sigma^i(Q)}$ for $d = 1, \ldots, p-2$. By (M), $J_{\ell_{d,i},m_i} \equiv 0 \pmod{q}$ if and only if $\ell_{d,i} + m_i < q-1$. So $\psi_d(\zeta) \equiv 0 \pmod{\sigma^i(Q)}$ if and only if $\ell_{d,i} + m_i < q-1$. From (K), the exact power of $\sigma^i(Q)$ dividing $\psi_d(\zeta)\psi_d(\zeta^{-1}) = q$ is $\sigma^i(Q)$. So $\left[\sigma^i(Q)\right]^2$ does not divide $A\psi_d(\zeta)$ and therefore

$$A\psi_d(\zeta) = \prod_{i=0}^{p-2} \{\sigma^i(Q) \mid \ell_{d,i} + m_i < q-1\}.$$

For each index i, $0 \le i \le p-2$, we show that there are exactly $g_{\pi-i}-1$ values of d such that $\ell_{d,i} + m_i < q-1$; this implies that

$$A\left(\prod_{d=1}^{p-2} \psi_d(\zeta)\right) = \prod_{i=0}^{p-2} [\sigma^i(Q)]^{g_{\pi-i-1}}.$$

In fact,

$$\ell_{d,i} + m_i = 2kg_{\text{ind}_g(d)-i} + 2kp - 2kg_{\pi-i} = (q-1) - 2k[g_{\pi-i} - g_{\text{ind}_g(d)-i}].$$

For each i the mapping $d \mapsto g_{\text{ind}_g(d)-i}$ is a permutation of the set $\{1, 2, \ldots, p-1\}$, and $p-1$ has image $g_{\pi-i}$. So there exist exactly $g_{\pi-i}-1$ values of d such that $g_{\text{ind}_g(d)-i} < g_{\pi-i}$, or equivalently $\ell_{d,i} + m_i < q-1$. This proves the first assertion.

To conclude the proof of the second assertion, we note that $\ell_{d,i} + m_i < q-1$, or equivalently $g_{\text{ind}_g(d)-i} < g_{\pi-i}$ holds if and only if $g_{\pi+\text{ind}_g(d)-i} + g_{\pi-i} > p$. Indeed,

$$g_{\pi+\text{ind}_g(d)-i} \equiv g^{\pi+\text{ind}_g(d)-i} \equiv -g^{\text{ind}_g(d)-i} \equiv -g_{\text{ind}_g(d)-i} \pmod{p},$$

so $g_{\pi+\text{ind}_g(d)-i} = p - g_{\text{ind}_g(d)-i}$. Therefore $g_{\pi+\text{ind}_g(d)-i} + g_{\pi-i} = p - g_{\text{ind}_g(d)-i} + g_{\pi-i} > p$ if and only if $g_{\text{ind}_g(d)-i} < g_{\pi-i}$. So

$$A\psi_d(\zeta) = \prod_{i \in I_d} \sigma^i(Q). \qquad \blacksquare$$

By applying σ^j, $j = 0, 1, \ldots, p-2$, we deduce from (**R**) that

$$A\langle \zeta^{g^j}, \rho \rangle^p = \prod_{i=0}^{p-1} (\sigma^{i+j}(Q))^{g_{\pi-i}}. \qquad (20.19)$$

We also remark:

S. Let i, i' be such that $0 \le i, i' \le p-2$, and $i' \equiv i + \pi \pmod{p-1}$. For each d, $1 \le d \le p-2$, we have $i \in I_d$ if and only if $i' \notin I_d$. In particular, I_d has $(p-1)/2$ elements.

Proof: Let us note that

$$p - g_{\pi-i} \equiv -g^{\pi-i} \equiv g^{-i} \equiv g^{\pi-i'} \equiv g_{\pi-i'} \pmod{p}$$

and similarly

$$p - g_{\pi-i+\text{ind}_g(d)} \equiv g_{\pi-i'+\text{ind}_g(d)} \pmod{p}.$$

Therefore, if $i \in I_d$, that is, if

$$p < g_{\pi-i} + g_{\pi-i+\text{ind}_g(d)} < 2p$$

then

$$0 < (p - g_{\pi-i}) + (p - g_{\pi-i+\text{ind}_g(d)}) < p$$

so

$$0 < g_{\pi-i'} + g_{\pi-i'+\mathrm{ind}_g(d)} < p$$

hence $i' \notin I_d$. And conversely.   ∎

The result (**R**) may be rephrased as follows:

T.   *If $1 \le d \le p-2$ then*

$$A\psi_d(\zeta^{-1}) = \prod_{r=1}^{d} \prod_{rp/(d+1)<s<rp/d} \sigma^{-\mathrm{ind}_g(s)}(Q).$$

Proof: By (**K**) and (**R**) we have

$$A\psi_d(\zeta)\psi_d(\zeta^{-1}) = Aq = \prod_{i \in I_d} \sigma^i(Q) \prod_{i \notin I_d} \sigma^i(Q)$$

hence it suffices to show that

$$\prod_{i \notin I_d} \sigma^i(Q) = \prod_{r=1}^{d} \prod_{rp/(d+1)<s<rp/d} \sigma^{-\mathrm{ind}_g(s)}(Q).$$

For every i, $0 \le i \le p-2$, let s be such that $1 \le s \le p-1$ and $\mathrm{ind}_g(s) = p-1-i$, so $\sigma^{-\mathrm{ind}_g(s)} = \sigma^i$. Hence it suffices to show that $i \notin I_d$ if and only if there exists r, $1 \le r \le d$, such that $rp/(d+1) < s < rp/d$.

We have $g_{\pi-i} \equiv g^{\pi-i} \equiv -g^{-i} \equiv -g^{\mathrm{ind}_g(s)} \equiv -s \pmod{p}$ so $g_{\pi-i} = p-s$.

Similarly, $g_{\pi-i+\mathrm{ind}_g(d)} \equiv -ds \pmod{p}$ and if r is such that $(r-1)p < ds < rp$ then $g_{\pi-i+\mathrm{ind}_g(d)} = rp - ds$.

Thus $g_{\pi-i} + g_{\pi-i+\mathrm{ind}_g(d)} = (r+1)p - (d+1)s$. Therefore $i \notin I_d$ exactly when $0 < (r+1)p - (d+1)s < p$ that is, $rp/(d+1) < s$. We also note that $s < rp/d$ because $g_{\pi-i+\mathrm{ind}_g(d)} > 0$ and $1 \le r \le d$. Thus

$$\prod_{r=1}^{d} \prod_{rp/(d+1)<s<rp/d} \sigma^{-\mathrm{ind}_g(s)}(Q) = \prod_{i \notin I_d} \sigma^i(Q).$$ ∎

20.5 Generation of the Class Group of the Cyclotomic Field

In this section we shall prove a theorem of Kummer on the class group of the cyclotomic field. Even though a much stronger theorem will be proved, we include this result as an illustration of the achievements of Kummer, at a much earlier date. In this connection, see Chapter 24, Theorem 2.

U.   *The group of ideal classes of the cyclotomic field $K = \mathbb{Q}(\zeta)$ is generated by the classes of the prime ideals of inertial degree 1.*

Proof: Let J be a nonzero ideal of A, so it is a product of prime ideals. If we show that each prime ideal Q of inertial degree $f > 1$ is in the same class as a product of prime ideals each having inertial degree less than f, then by induction on the inertial degree, we conclude that J is in the same class a product of prime ideals of inertial degree 1.

So let $Q = Q_0$ be a prime ideal with norm $N(Q) = q^f$, and let r be defined by $fr = p - 1$.

Let $\eta_0, \eta_1, \ldots, \eta_{r-1}$ be the r periods of f terms (with respect to the primitive root g). Let $K' = \mathbb{Q}(\eta_0, \eta_1, \ldots, \eta_{r-1})$, $A' = \mathbb{Z}[\eta_0, \eta_1, \ldots, \eta_{r-1}]$. We consider a polynomial similar to $G(X)$, defined in the proof of (**B**).

For every $k = 0, 1, \ldots, r - 1$, let

$$G_k(X) = \prod_{l=0}^{f-1} (X - \zeta^{g^{ir+k}}) \tag{20.20}$$

be the polynomial whose roots are the f summands of the period η_k. Its coefficients are invariant under σ^r, hence they belong to $A' = K' \cap A$.

Since A' is a free Abelian group with the basis $\{\eta_0, \eta_1, \ldots, \eta_{r-1}\}$ (by (**B**)), each coefficient of $G_k(X)$ may be written, in a unique way, as a linear combination of the periods, with coefficients in $\mathbb{Z}$.

Let

$$G(X) = G_0(X) = X^f + \alpha_1 X^{f-1} + \alpha_2 X^{f-2} + \cdots + \alpha_f,$$

so we may write $\alpha_i = \sum_{j=0}^{r-1} b_{ij}\eta_j$, with $b_{ij} \in \mathbb{Z}$. Since

$$G_k(X) = \prod_{i=0}^{f-1} (X - \zeta^{g^{ir+k}}) = \prod_{i=0}^{f-1} (X - \sigma^k(\zeta^{g^{ir}}))$$

the coefficients of $G_k(X)$ are obtained from those of $G(X)$ by applying σ^k, that is, they are

$$\sigma^k(\alpha_i) = \sum_{j=0}^{r-1} b_{ij}\sigma^k(\eta_j) = \sum_{j=0}^{r-1} b_{ij}\eta_{j+k}.$$

Let us note also that $\alpha_f = (-1)^f$. Indeed,

$$\alpha_f = (-1)^f \prod_{i=0}^{f-1} \zeta^{g^{ir}} = (-1)^f \zeta^s,$$

where

$$s = \sum_{i=0}^{f-1} g^{ir} = \frac{g^{rf} - 1}{g^r - 1} \equiv 0 \pmod{p}.$$

As was shown in (**O**), for every $j = 0, 1, \ldots, r - 1$ there exists $u_j \in \mathbb{Z}$ such that $\eta_j \equiv u_j \pmod{Q}$. Applying the automorphism σ^k, $k = 1, \ldots, r - 1$, and letting $Q_k = \sigma^k(Q)$, we have $\eta_{j+k} \equiv u_j \pmod{Q_k}$, that is, $\eta_j \equiv u_{j-k} \pmod{Q_k}$ with the convention that $u_{j+r} = u_j$ for any integer j.

For $k = 0, 1, \ldots, r - 1$ let

$$H_k(X) = X^f + \sum_{i=1}^{f-1} \left(\sum_{j=0}^{r-1} b_{ij} u_{j-k} \right) X^{f-i} + (-1)^f + q,$$

so $H_k(X)$ has coefficients in $\mathbb{Z}$. For simplicity, we write $H(X) = H_0(X)$. Then

$$H_k(\zeta) = \zeta^f + \sum_{i=1}^{f-1} \left(\sum_{j=0}^{r-1} b_{ij} u_{j-k} \right) \zeta^{f-i} + (-1)^f + q, \qquad (20.21)$$

is an element of A.

Since $G(\zeta) = 0$ then

$$H_k(\zeta) = H_k(\zeta) - G(\zeta) = \sum_{i=1}^{f-1} \left(\sum_{j=0}^{r-1} b_{ij} (u_{j-k} - \eta_j) \right) \zeta^{f-i} + q. \qquad (20.22)$$

Let us note that (20.22) is an expression of $H_k(\zeta) \in A$ as a linear combination of $1, \zeta, \ldots, \zeta^{f-1}$ with coefficients in the ring A'.

We shall prove successively:

(1) Q divides $A \cdot H(\zeta)$.

(2) Q^2 does not divide $A \cdot H(\zeta)$.

(3) $Q_1, \ldots, Q_{r-1}$ (the prime ideals conjugate to Q) do not divide $A \cdot H(\zeta)$.

(4) If a prime ideal Q' of A divides $A \cdot H(\zeta)$, and $Q' \neq Q$, then Q' has inertial degree smaller than f.

Proof: (1)   Since Q divides Aq and $\eta_j \equiv u_j \pmod{Q}$ then Q divides each coefficient in the expression (20.22) with $k = 0$. It follows that Q divides the ideal $A \cdot H(\zeta)$.

(2)   Since Q divides $A \cdot H(\zeta)$ but Q^2 does not divide Aq, it follows from (**Q**) and the expression (20.22) that Q^2 does not divide $A \cdot H(\zeta)$.

(3)   Let us assume, for example, that Q_{r-1} divides $A \cdot H(\zeta)$. Then $Q \cdot Q_{r-1}$ divides $A \cdot H(\zeta)$. But for every $k = 1, 2, \ldots, r - 2$, Q_k divides Aq and $u_{j-k} \equiv \eta_j \pmod{Q_k}$. Hence Q_k divides $A \cdot H(\zeta)$. Altogether, $Aq = Q \cdot Q_1 \cdots Q_{r-1}$ divides

$$A \cdot H(\zeta) \cdots H_{r-2}(\zeta)$$

(we have excluded the factor $H_{r-1}(\zeta)$). The latter product is of the form

$$\zeta^{f(r-1)} + c_1 \zeta^{f(r-1)-1} + \cdots + c_{f(r-1)}$$

with coefficients $c_i \in \mathbb{Z}$, this follows from (20.21). Since $f(r-1) < p-1$ and $\{1, \zeta, \ldots, \zeta^{p-2}\}$ is a basis of the Abelian group A, it follows that q divides each coefficient. In particular, q divides 1, an absurdity.

(4)   Let Q' be a prime ideal of A, $Q' \neq Q$, of inertial degree f', and dividing $A \cdot H(\zeta)$. Then Q' does not divide Aq and f' divides p.

If $f' > f$, we note that (20.21) with $k = 0$, is an expression of $H(\zeta) \in A$ as a linear combination of $1, \zeta, \ldots, \zeta^f, \ldots, \zeta^{f'-1}$ with coefficients in $\mathbb{Z}$ (hence, a fortiori, in the ring of integers of the field generated by the periods of f' terms). We may apply (Q), so Q' divides each coefficient in (20.21), in particular, Q' divides 1, an absurdity.

If $f' = f$ it follows from (Q) and the expression (20.22) for $H(\zeta)$ that Q' divides each coefficient, hence Q' divides Aq, contrary to the hypothesis.

Therefore $f' < f$, as we had to prove.

We conclude the proof as follows. We have $A \cdot H(\zeta) = Q \cdot J$, where J is a product of prime ideals different from Q, and therefore of inertial degree less than f. After multiplication with the conjugates of the ideal J, we have

$$Q \cdot N(J) = A \cdot H(\zeta) \prod_{i \neq 0} \sigma^i(J).$$

So Q and $\prod_{i \neq 0} \sigma^i(J)$ are in the same ideal class, and the latter ideal is a product of prime ideals with inertial degree less than f.   ∎

A much stronger theorem will be indicated in Chapter 24, Theorem 2, and the following comments. The proof requires analytical methods or class field theory, which were nonexistent in Kummer's time.

We deduce the following corollary:

V.   *For every ideal J of A and integer d, $1 \leq d \leq p-2$, the product*

$$\prod_{i \in I_d} \sigma^i(J)$$

is a principal ideal.

Proof: We have seen that J is in the same ideal class as a product of prime ideals Q_j, each of inertial degree 1: $J = A\alpha \times \prod_{j=1}^s Q_j$.

Let $N(Q_j) = q_j$ so $q_j \equiv 1 \pmod{p}$ and we may write $q_j = 2k_j p + 1$. Let Q_j^* be the ideal generated by q_j and $h_j^{2k_j} - \zeta$, where h_j is a primitive root modulo q_j.

So Q_j is a conjugate of Q_j^*, say $Q_j^* = \sigma^{\ell(j)}(Q_j)$, for some integer $\ell(j)$, $0 \leq \ell(j) \leq p-2$.

In (R) we have shown that

$$\prod_{i \in I_d} \sigma^i(Q_j^*) = \sigma^{\ell(j)}\left(\prod_{i \in I_d} \sigma^i(Q_j)\right)$$

is a principal ideal of A. Hence

$$\prod_{i \in I_d} \sigma^i(J) = A \left(\prod_{i \in I_d} \sigma^i(\alpha) \right) \times \prod_{j=1}^{s} \left(\prod_{i \in I_d} \sigma^i(Q_j) \right)$$

is also a principal ideal of A.   ∎

This result was generalized by Stickelberger.

E X E R C I S E S

1. Use the method indicated in the text to obtain a formula involving radicals for the roots of the equation $X^3 + aX^2 + bX + c = 0$.

   Hint:

   (a) Put $X = Y - a/3$ and obtain an equation of the form $Y^3 + uY + v = 0$.

   (b) Show that the roots y_1, y_2, y_3 are given by

$$y_1 = \alpha + \beta, \qquad y_2 = \omega^2 \alpha + \omega \beta, \qquad y_3 = \omega \alpha + \omega^2 \beta,$$

    where

$$\alpha = \sqrt[3]{-\frac{v}{2} + \sqrt{\frac{u^3}{27} + \frac{v^2}{4}}}, \qquad \beta = \sqrt[3]{-\frac{v}{2} - \sqrt{\frac{u^3}{27} + \frac{v^2}{4}}},$$

$$\omega = \frac{-1 + \sqrt{-3}}{2}, \qquad \omega^2 = \frac{-1 - \sqrt{-3}}{2}.$$

2. Use the method indicated in the text to obtain a formula involving radicals for the roots of the equation $X^4 + aX^3 + bX^2 + cX + d = 0$.

   Hint:

   (a) Put $X = Y - a/4$ and obtain an equation of the form $Y^4 + sY^2 + uY + v = 0$.

   (b) Show that the roots y_1, y_2, y_3, y_4 are given by

$$y_1 = \tfrac{1}{2}\left[\sqrt{-\alpha_1} + \sqrt{-\alpha_2} + \sqrt{-\alpha_3} \right],$$

$$y_2 = \tfrac{1}{2}\left[\sqrt{-\alpha_1} - \sqrt{-\alpha_2} - \sqrt{-\alpha_3} \right],$$

$$y_3 = \tfrac{1}{2}\left[\sqrt{-\alpha_1} + \sqrt{-\alpha_2} - \sqrt{-\alpha_3} \right],$$

$$y_4 = \tfrac{1}{2}\left[\sqrt{-\alpha_1} - \sqrt{-\alpha_2} + \sqrt{-\alpha_3} \right],$$

    where $\alpha_1, \alpha_2, \alpha_3$ are the roots of the equation $X^3 - 2sX^2 + (s^2 - 4v)X + u^2 = 0$.

3. Find the expression involving radicals for the primitive fifth root of 1.

Hint: Use Gaussian periods to solve the equation $X^4 + X^3 + X^2 + X + 1 = 0$. Obtain

$$\zeta = \frac{1}{2}\left[\frac{-1 + \sqrt{5}}{2} + \sqrt{-\frac{5 + \sqrt{5}}{2}} \right].$$

4. Find the expression involving radicals for the primitive seventh root of 1.

Hint: Use Gaussian periods to solve the equation $X^6 + X^5 + X^4 + X^3 + X^2 + X + 1 = 0$ and obtain $\zeta = \frac{1}{3}[\alpha + \beta + \gamma]$ where

$$\alpha = \frac{-1 + \sqrt{-7}}{2},$$

$$\beta = \sqrt[3]{\frac{13 + 3\sqrt{-3}}{2}} - \alpha(1 - 3\sqrt{-3}),$$

$$\gamma = \sqrt[3]{\frac{13 - 3\sqrt{-3}}{2}} - \alpha(1 + 3\sqrt{-3}).$$

5. Let $q = 5$, $h = 2$. Compute the polynomial whose roots are the two periods with two terms defined by h and a primitive fifth root of 1. Obtain an expression of $\sqrt{5}$ in terms of sines.

6. Let $q = 7$, $h = 3$. Compute the polynomial whose roots are the three periods with two terms, respectively, the two periods with three terms, defined by h and a primitive seventh root of 1. Obtain an expression of $\sqrt{7}$ in terms of sines.

7. Let $q = 7$, $h = 3$, and let μ_0, μ_1, μ_2 be the three periods with two terms associated to the primitive seventh root of 1.
Compute μ_1, μ_2 as polynomials in μ_0 with rational coefficients. Verify the statement **(C)** by explicit calculation.

8. Let $q = 11$, $h = 2$, $p = 5$. With the notations of the text, compute $\langle \zeta, \rho \rangle$ and then $\langle \zeta, \rho \rangle^5$.

9. Let $q = 13$, $h = 2$, $p = 3$. With the notations of the text, compute $\langle \zeta, \rho \rangle$ and then $\langle \zeta, \rho \rangle^3$. Verify by explicit calculation that $\langle \zeta^2, \rho \rangle \langle \zeta^{-2}, \rho \rangle = 13$ (in agreement with **(G)**).

10. Let $q = 11$, $h = 2$, $p = 5$. With the notations of the text, verify by explicit calculation that $|\langle \zeta^n, \rho \rangle| = \sqrt{11}$ for $n = 1, 2, 3, 4$.

11. Let $q = 13$, $h = 2$, $p = 3$. Compute the Jacobi cyclotomic function $\psi_{n,m}(\zeta)$ for $nm(n + m)$ not a multiple of 3.

12. Let $q = 7$, $h = 3$. Compute $J_{\ell,m}$ for $0 < \ell$, $m < 6$, $\ell + m \neq 6$, and verify explicitly the congruences indicated in (**M**).

13. Let $p = 11$, $g = 2$, $q = 3$, so q has order 5 modulo 11. Let η_0, η_1 be two periods with five terms defined by ζ (a eleventh root of 1) and by g (see the text).

Compute the period polynomial $F_{\eta_0}(X)$ and find $u_0, u_1 \in \mathbb{Z}$ such that $F_{\eta_0}(X) \equiv (X - u_0)(X - u_1) \pmod{3}$.

14. Repeat the preceding exercise with $p = 11$, $g = 2$, $q = 23$.

15. Let $p = 11$, $g = 2$, $q = 5$, σ defined by $\sigma(\zeta) = \zeta^2$, ζ a primitive 11th root of 1. For each $d = 1, 2, \ldots, q$ compute the set I_d and verify by numerical calculation that

$$A\psi_d(\zeta) = \prod_{i \in I_d} \sigma^i(Q),$$

where $\Phi_{11}(X) \equiv F_0(X)\, F_1(X) \pmod{5}$ and $Q = A \times 5 + AF_0(\zeta)$.

Part Four

21

Characters and Gaussian Sums

The Jacobi symbol is of great importance in the study of quadratic fields. For cyclotomic extensions or more general Abelian extensions, the role of the Jacobi symbol is played by characters.

We begin by studying the characters of finite Abelian extensions, then we consider the modular characters. In the last section, we introduce the Gaussian sums associated to modular characters.

21.1 Characters of Finite Abelian Groups

Let G be an Abelian group (with operations written multiplicatively) having n elements. Let $\mathbb{C}^{\cdot}$ be the multiplicative group of nonzero complex numbers.

Definition 1. Every homomorphism $\chi: G \to \mathbb{C}^{\cdot}$ is called a *character* of G (with complex values).

Thus $\chi(a) \neq 0$ for every $a \in G$ and $\chi(ab) = \chi(a) \cdot \chi(b)$ for $a, b \in G$. In particular, if e is the unit element of G then $\chi(e) = 1$.

If χ is any character of G, we also have the *complex conjugate character* $\overline{\chi}$, defined by $\overline{\chi}(a) = \overline{\chi(a)}$. Clearly, $\overline{\chi}$ is also a character and $\overline{\overline{\chi}} = \chi$.

If χ is a character, since $a^n = e$ for every $a \in G$, we deduce that $(\chi(a))^n = \chi(a^n) = \chi(e) = 1$; hence $\chi(G)$ is a subgroup of the multiplicative cyclic group of nth roots of unity, so $\chi(G)$ is a multiplicative cyclic group.

The set $\widehat{G}$ of characters has a multiplication defined as follows: $(\chi \cdot \chi')(a) = \chi(a) \cdot \chi'(a)$ for every $a \in G$.

$\widehat{G}$ is a multiplicative group with unit element χ_0, defined by $\chi_0(a) = 1$ for every $a \in G$, and the inverse of χ is given by $\chi^{-1}(a) = \chi(a)^{-1} = \overline{\chi(a)} = \overline{\chi}(a)$ for every $a \in G$.

We want to compare character groups of different groups. If φ is a group-homomorphism, $\varphi: G' \to G$, where G', G are finite Abelian groups, then we define a group-homomorphism $\widehat{\varphi}: \widehat{G} \to \widehat{G'}$ as follows: $\widehat{\varphi}(\chi) = \chi \circ \varphi$. It is immediate that $\widehat{\varphi}(\chi)$ is a character of G' and $\widehat{\varphi}$ is a group-homomorphism. In particular, if G' is a subgroup of G and φ is the inclusion mapping, then $\widehat{\varphi}(\chi)$ is just the restriction of the character χ to the subgroup G'.

The kernel of the homomorphism $\widehat{\varphi}$ consists of all characters $\chi \in \widehat{G}$ such that $\chi(\varphi(a')) = 1$ for every $a' \in G'$. If G' is a subgroup of G and φ is the inclusion map, then the kernel of $\widehat{\varphi}$ consists of all characters $\chi \in \widehat{G}$ such that if $aG' = bG'$ then $\chi(a) = \chi(b)$.

We may also describe the kernel of $\widehat{\varphi}$ in the following way. Let $\psi \colon G \to G/G'$ be the canonical homomorphism onto the quotient group. Then $\widehat{\psi} \colon \widehat{G/G'} \to \widehat{G}$ is the lifting homomorphism, namely if $\widetilde{\chi} \in \widehat{G/G'}$ then $\widehat{\varphi}(\widetilde{\chi}) = \chi = \widetilde{\chi} \circ \psi$, $\chi(a) = \widetilde{\chi}(aG')$ for every $a \in G$. The mapping $\widehat{\psi}$ is one-to-one, because if $\widehat{\psi}(\widetilde{\chi}) = \chi_0$ (unit character of G) then $1 = \chi_0(a) = \widetilde{\chi}(aG')$ for every coset. The characters of G/G' correspond by $\widehat{\psi}$ to the characters of G, whose restriction to G' is the unit character of G', that is, $\widehat{\psi}(\widehat{G/G'}) = \ker(\widehat{\varphi})$. Hence we have an isomorphism $\widehat{G}/\widehat{\psi}(\widehat{G/G'}) \cong \widehat{\varphi}(\widehat{G}) \subseteq \widehat{G'}$. We shall see soon that this isomorphism is onto $\widehat{G'}$.

We shall describe explicitly how to obtain the characters of a finite Abelian group. First we consider the following particular case:

A.   *Let G be a cyclic group of order n, with generator a, and let ζ be a primitive nth root of unity. Then the characters of G are $\chi_0, \ldots, \chi_{n-1}$, where $\chi_r(a^s) = \zeta^{rs}$ for every $s = 1, \ldots, n$, $r = 0, 1, \ldots, n-1$. Moreover, $\widehat{G} \cong G$.*

Proof: Each of the above mappings χ_r is a character of G. For $r \neq r'$ we have $\chi_r \neq \chi_{r'}$ since ζ is a primitive nth root of unity.

On the other hand, if χ is any character of G then $[\chi(a)]^n = \chi(a^n) = \chi(e) = 1$, so $\chi(a) = \zeta^r$ where $0 \leq r \leq n-1$. Then $\chi(a^s) = [\chi(a)]^s = \zeta^{rs} = \chi_r(a^s)$ for $s = 1, \ldots, n$, showing that $\chi = \chi_r$.

The mapping $\theta_a \colon G \to \widehat{G}$ defined by $\theta_a(a^s) = \chi_s$ for $s = 1, \ldots, n$ is clearly an isomorphism between G and $\widehat{G}$.   ∎

It is worthwhile noting that for each generator a of G there is one such isomorphism θ_a between G and $\widehat{G}$.

If G is any finite Abelian group, then G is isomorphic to the Cartesian product of finitely many cyclic groups. In this situation we have:

B.   *If $\theta : G \to \prod_{i=1}^r G_i$ is a group-isomorphism then it induces a group-isomorphism*

$$\widehat{\theta} : \widehat{G} \to \prod_{i=1}^r \widehat{G_i}.$$

Proof: Let $\nu_i : G_i \to G$ be defined by $\nu_i(x_i) = \theta^{-1}(1, \ldots, x_i, \ldots, 1)$, so ν_i is an isomorphism from G_i to a subgroup of G.

If $\chi \in \widehat{G}$ let $\chi_i = \chi \circ \nu_i$ for $i = 1, \ldots, r$, so $\chi_i \in \widehat{G}_i$. We define

$$\widehat{\theta} : \widehat{G} \to \prod_{i=1}^{r} \widehat{G}_i$$

by letting $\widehat{\theta}(\chi) = (\chi_1, \ldots, \chi_r)$. It is obvious that $\widehat{\theta}$ is a group-homomorphism. It is one-to-one, for if $\chi_i = \chi \circ \nu_i$ is the unit character of G_i for $i = 1, \ldots, r$ then

$$\chi(x) = \chi \left(\prod_{i=1}^{r} \nu_i \, \mathrm{pr}_i(\theta(x)) \right) = \prod_{i=1}^{r} (\chi \circ \nu_i)(\mathrm{pr}_i(\theta(x))) = 1$$

for every $x \in G$.

Finally, given any characters $\chi^{(i)} \in \widehat{G}_i$ for $i = 1, \ldots, r$ let $\chi : G \to \mathbb{C}^{\cdot}$ be defined by $\chi(x) = \prod_{i=1}^{r} \chi^{(i)}(\mathrm{pr}_i(\theta(x)))$ for every $x \in G$. Then $\chi \in \widehat{G}$ and

$$\chi \circ \nu_i(x_i) = \prod_{j=1}^{r} \chi^{(j)} \, \mathrm{pr}_j(\theta(\nu_i(x_i)))$$

$$= \prod_{j=1}^{r} \chi^{(j)} \, \mathrm{pr}_j(1, \ldots, x_i, \ldots, 1) = \chi^{(i)}(x_i)$$

for every $x_i \in G_i$. So $\widehat{\theta}(\chi) = (\chi^{(1)}, \ldots, \chi^{(r)})$ proving that $\widehat{\theta}$ is a group-isomorphism.   ∎

As a corollary we have:

C. $\widehat{G} \cong G$, hence $\#(\widehat{G}) = \#(G)$.

Proof: Writing $G \cong \prod_{i=1}^{r} G_i$ where each group G_i is cyclic, we deduce from (**A**) and (**B**) that

$$\widehat{G} \cong \prod_{i=1}^{r} \widehat{G}_i \cong \prod_{i=1}^{r} G_i \cong G.$$ ∎

As an application of these results we may explicitly determine the group of characters of the multiplicative group $(\mathbb{Z}/m)^{\cdot}$ of prime residue classes modulo $m > 1$.

If $m = m_1 \cdots m_r$ is decomposition into pairwise relatively prime integers (for example, $m = \prod_{i=1}^{r} p_i^{e_i}$ the prime decomposition of the integer m) let

$$\theta : (\mathbb{Z}/m)^{\cdot} \to \prod_{i=1}^{r} (\mathbb{Z}/m_i)^{\cdot}$$

be the natural group-isomorphism given by

$$\theta(x \ (\mathrm{mod} \ m)) = (x \ (\mathrm{mod} \ m_1), \ldots, x \ (\mathrm{mod} \ m_r)).$$

Then $\widehat{\theta} = (\chi_1, \ldots, \chi_r)$, where $\chi_i(x \pmod{m_i}) = \chi(x_i \pmod{m})$, x_i being defined by the congruences

$$\begin{cases} x_i \equiv x \pmod{m_i}, \\ x_i \equiv 1 \pmod{m_j} \quad \text{for } j \neq i. \end{cases}$$

We need only to describe explicitly the character groups of $(\mathbb{Z}/p^e)^{\cdot}$, where p is any prime number.

If $p \neq 2$, it was shown in Chapter 3, (**J**), that $(\mathbb{Z}/p^e)^{\cdot}$ is a cyclic group. $(\mathbb{Z}/p^e)^{\cdot} = B \times C$, where B is the multiplicative cyclic group of order $p - 1$ with a generator $b \pmod{p^e}$, and C is the multiplicative cyclic group of order p^{e-1} with generator $(1 + p) \pmod{p^e}$. Thus for every integer a we have

$$a \equiv b^{\alpha'}(1 + p)^{\alpha''} \pmod{p^e} \qquad \text{with} \ \ 0 \leq \alpha' < p - 1, \ \ 0 \leq \alpha'' < p^{e-1}.$$

For every $(p - 1)$th root of unity ζ_{p-1} and for every (p^{e-1})th root of unity $\zeta_{p^{e-1}}$ we have a character χ of $(\mathbb{Z}/p^e)^{\cdot}$, which is defined by $\chi(a \pmod{p^e}) = \zeta_{p-1}^{\alpha'} \cdot \zeta_{p^{e-1}}^{\alpha''}$.

Similarly, if $p = 2$ and $e \geq 2$, then by Chapter 3, (**K**), $(\mathbb{Z}/2^e)^{\cdot} \cong \{1, -1\} \times C$, where C is the multiplicative cyclic group of order 2^{e-2} generated by $5 \pmod{2^e}$. Then for every integer a we have

$$a \equiv (-1)^{\alpha'} 5^{\alpha''} \pmod{2^e} \qquad \text{with} \ \ \alpha' = 0, \ \ \text{or} \ 1 \ \ \text{and} \ \ 0 \leq \alpha'' < 2^{e-2}.$$

For every (2^{e-2})th root of unity $\zeta_{2^{e-2}}$ we have the characters χ, χ' of $(\mathbb{Z}/2^e)^{\cdot}$ which are defined by

$$\chi(a \pmod{2^e}) = (-1)^{\alpha'} \zeta_{2^{e-2}}^{\alpha''},$$

$$\chi'(a \pmod{2^e}) = \zeta_{2^{e-2}}^{\alpha''}.$$

In the determination of the characters of a group, sometimes we proceed inductively by extending to the whole group characters of a subgroup. In this respect let us note:

D.   *If G' is a subgroup of order m of the group G of order n then every character χ' of G' admits n/m extensions to characters of G.*

Proof: Let $\varphi : G' \to G$ be the inclusion mapping and let $\psi \colon G \to G/G'$ be the canonical homomorphism. We have seen that $\widehat{\varphi} \colon \widehat{G} \to \widehat{G'}$ induces an isomorphism from $\widehat{G}/\widehat{\psi(G/G')}$ into $\widehat{G'}$. But $\#(\widehat{G}) = \#(G)$, $\#(\widehat{G'}) = \#(G')$, $\#(\widehat{\psi(G/G')}) \leq \#(\widehat{G/G'}) = \#(G/G') = \#(G)/\#(G')$, hence $\widehat{G}/\widehat{\psi(G/G')} \cong \widehat{G'}$ so $\widehat{\psi(\widehat{G})} = \widehat{G'}$. This means that every character of G' is the restriction of a character of G. Since the kernel of $\widehat{\psi}$ is $\widehat{\psi(G/G')}$ which has order n/m then every character of G' admits n/m extensions to characters of G. ∎

We summarize the above results as follows:
If

$$1 \longrightarrow G' \xrightarrow{\varphi} G \xrightarrow{\psi} G/G' \longrightarrow 1$$

is an exact sequence of finite Abelian groups, then the sequence

$$1 \longrightarrow \widehat{G/G'} \xrightarrow{\widehat{\psi}} \widehat{G} \xrightarrow{\widehat{\varphi}} \widehat{G'} \longrightarrow 1$$

is also exact.

One of the most important facts about characters is the *separating property*, which states that there are enough characters to distinguish the elements of the group:

E.   *If $a, a' \in G$, $a \neq a'$, then there exists a character $\chi \in \widehat{G}$ such that $\chi(a) \neq \chi(a')$.*

Proof: It is clearly equivalent to show that if $a \in G$, $a \neq e$ then there exists $\chi \in \widehat{G}$ such that $\chi(a) \neq 1$.

Thus, we need to prove that

$$G' = \{a' \in G \mid \chi(a') = 1 \text{ for all characters } \chi \in \widehat{G}\}$$

is equal to $\{e\}$. At any rate, G' is a subgroup of G.

For every $\chi \in \widehat{G}$ let $\widetilde{\chi}: G/G' \to \mathbb{C}^{\cdot}$ be defined by $\widetilde{\chi}(aG') = \chi(a)$. We note that $\widetilde{\chi}$ is well defined, because if $aG' = a'G'$ then $a'^{-1}a \in G'$ hence $\chi(a'^{-1}a) = 1$ so $\chi(a') = \chi(a)$. Also, it is clear that $\widetilde{\chi}$ is a character of the group G/G'. This defines a mapping $\theta: \widehat{G} \to \widehat{G/G'}$, $\theta(\chi) = \widetilde{\chi}$. Clearly, θ is a group-homomorphism. Finally θ is one-to-one, because if $\widetilde{\chi}(aG') = 1$ for every coset aG' then $\chi(a) = 1$ for every element $a \in G$, so χ is the unit character.

Therefore $\#(G) = \#(\widehat{G}) \leq \#(\widehat{G/G'}) = \#(G/G') \leq \#(G)$, so $\#(G/G') = \#(G)$ and therefore G' is the trivial subgroup.   ■

With the separating property we may show:

F.   *There exists a natural isomorphism $\iota : G \to \widehat{\widehat{G}}$.*

Proof: For every $a \in G$ let $\iota_a: \widehat{G} \to \mathbb{C}^{\cdot}$ be defined by $\iota_a(\chi) = \chi(a)$ for every character χ of G. Then ι_a is a group-homomorphism, so $\iota_a \in \widehat{\widehat{G}}$.

Let $\iota: G \to \widehat{\widehat{G}}$ be the mapping defined by $\iota(a) = \iota_a$. Clearly, ι is a homomorphism. If $\iota(a)$ is the unit character of $\widehat{G}$, then $\iota_a(\chi) = 1$ for every $\chi \in \widehat{G}$, that is, $\chi(a) = 1$ for every $\chi \in \widehat{G}$. By (**E**), we must have $a = e$. Thus ι is one-to-one. Since $\#(\widehat{\widehat{G}}) = \#(G)$ then ι is onto $\widehat{\widehat{G}}$, so ι is an isomorphism.   ■

A useful result is the characterization of the kth powers of elements of G by means of characters:

G.   *The element a is a kth power in G (where $k \geq 1$) if and only if $\chi(a) = 1$ for every character of order dividing k.*

Proof: If $a = b^k$ with $b \in G$, if $\chi^k = \chi_0$ (unit character) then

$$\chi(a) = \chi(b^k) = [\chi(b)]^k = \chi^k(b) = \chi_0(b) = 1.$$

Conversely, let G^k denote the group consisting of kth powers of elements of G. By **(E)**, it is enough to show that if λ is any character of G/G^k then $\lambda(aG^k) = 1$, so $a \in G^k$.

If $\varphi: G \to G/G^k$ is the canonical homomorphism then $\lambda \circ \varphi \in \widehat{G}$ and this character has order dividing k, because

$$(\lambda \circ \varphi)^k(x) = [(\lambda \circ \varphi)(x)]^k = [\lambda(xG^k)]^k = \lambda(x^k G^k) = 1$$

for every $x \in G$. By hypothesis, $(\lambda \circ \varphi)(a) = 1$ hence $\lambda(aG^k) = 1$.   ∎

If the order of a character χ is equal to 2, we call it a *quadratic character*. The values of χ are equal to 1 or -1. An element of G is a square if and only if $\chi(a) = 1$ for every quadratic character of G. The quadratic characters of G may be identified with the characters of G/G^2.

For example, if $G = (\mathbb{Z}/p)^{\cdot}$, the multiplicative group of nonzero residue classes modulo the prime number $p \neq 2$, let $\chi(\overline{a}) = (a/p)$ where $a \in \mathbb{Z}$, and (a/p) denotes the Legendre symbol. Then χ is a quadratic character. Moreover, it is the only nontrivial quadratic character of $(\mathbb{Z}/p)^{\cdot}$. Indeed, if $\overline{w} \in (\mathbb{Z}/p)^{\cdot}$ is a generator of this cyclic group, if $\overline{a} = \overline{w}^\alpha$ with $0 \leq \alpha \leq p-1$, then $\chi(\overline{a}) = \chi(\overline{w})^\alpha$. Since χ is not the unit character χ_0 then $\chi(\overline{w}) \neq 1$. From $\chi^2 = \chi_0$ it follows that $[\chi(\overline{w})]^2 = 1$, so $\chi(\overline{w}) = -1$. Thus $\chi(\overline{a}) = 1$ when α is even, $\chi(\overline{a}) = -1$ when α is odd, and therefore $\chi(\overline{a}) = (a/p)$ (Legendre symbol).

The following is worthwhile noting:

H.   *The set $\{\chi_1, \ldots, \chi_s\}$ is a system of generators of $\widehat{G}$ if and only if $\chi_i(a) = 1$, for all $i = 1, \ldots, s$, implies $a = e$.*

Proof: If $\{\chi_1, \ldots, \chi_s\}$ generates $\widehat{G}$ and $\chi_i(a) = 1$, for all $i = 1, \ldots, s$, if $\chi \in \widehat{G}$ we may write

$$\chi = \prod_{i=1}^{s} \chi_i^{t_i} \qquad \text{hence} \qquad \chi(a) = \prod_{i=1}^{s} \chi_i(a)^{t_i} = 1.$$

Since this holds for every $\chi \in \widehat{G}$, by **(E)** we have $a = e$.

Conversely, let H be the subgroup of $\widehat{G}$ generated by the set $\{\chi_1, \ldots, \chi_s\}$ and consider the quotient group $\widehat{G}/H$. Let $\alpha \in \widehat{\widehat{G}/H}$ and let $\varphi: \widehat{G} \to \widehat{G}/H$ be the canonical group-homomorphism. Thus $\alpha \circ \varphi: \widehat{G} \to \mathbb{C}^{\cdot}$ and actually $\alpha \circ \varphi \in \widehat{\widehat{G}}$. By **(F)**, there exists an element $a \in G$ such that $\alpha \circ \varphi = \iota_a$.

Since $\varphi(\chi_i) = \chi_i H = H$, unit element of $\widehat{G}/H$, then $\chi_i(a) = \iota_a(\chi_i) = \alpha(\varphi(\chi_i)) = 1$ for every $i = 1, \ldots, s$. By the hypothesis we have $a = e$, hence $\alpha \circ \varphi(\chi) = 1$ for every character $\chi \in \widehat{G}$. Since φ maps $\widehat{G}$ onto $\widehat{G}/H$ then α is the unit character of $\widehat{G}/H$. We have shown that $\widehat{\widehat{G}/H}$ has only one element, so $\#(\widehat{G}/H) = 1$ and $H = \widehat{G}$.   ∎

The characters of G are elements of the set V of a complex-valued function defined on G. V is clearly a vector space of dimension n over $\mathbb{C}$. We consider on V an *inner product* defined as follows:

$$\langle f, g \rangle = \frac{1}{n} \sum_{a \in G} f(a)\overline{g(a)} \qquad \text{for} \quad f, g \in V.$$

Clearly,

$$\langle f, g + g' \rangle = \langle f, g \rangle + \langle f, g' \rangle,$$
$$\langle f + f', g \rangle = \langle f, g \rangle + \langle f', g \rangle,$$
$$\langle \alpha f, g \rangle = \alpha \langle f, g \rangle,$$
$$\langle f, \alpha g \rangle = \overline{\alpha} \langle f, g \rangle,$$
$$\langle g, f \rangle = \overline{\langle f, g \rangle}.$$

for $f, f', g, g' \in V$, $\alpha \in \mathbb{C}$. Hence $\langle f, f \rangle$ is a positive real number, for any $f \in V$, $f \neq 0$.

We say that $f, g \in V$ are *orthogonal functions* when $\langle f, g \rangle = 0$. The *length* of the function f is defined to be $\|f\| = \sqrt{\langle f, f \rangle}$.

Let $\widehat{G} = \{\chi_0, \ldots, \chi_{n-1}\}$ where χ_0 is the unit element of $\widehat{G}$.

I.   *We have:*

(1) $$\|\chi_i\| = 1, \quad \langle \chi_i, \chi_j \rangle = 0 \qquad \text{for} \quad i \neq j.$$

(2) $$\left\langle \sum_{i=0}^{n-1} \alpha_i \chi_i, \chi_j \right\rangle = \alpha_j$$

for every $j = 0, 1, \ldots, n - 1$, *and* $\alpha_i \in \mathbb{C}$.

(3) $\{\chi_0, \ldots, \chi_{n-1}\}$ *is a basis of* V.

Proof: (1)   Let $\alpha = \langle \chi_i, \chi_0 \rangle = (1/n) \sum_{a \in G} \chi_i(a)$. If $\chi_i = \chi_0$ then $\alpha = (1/n) \sum_{a \in G} \chi_0(a) = 1$. If $\chi_i \neq \chi_0$ let $b \in G$ be such that $\chi_i(b) \neq 1$ then

$$\alpha \chi_i(b) = \frac{1}{n} \sum_{a \in G} \chi_i(a)\chi_i(b) = \frac{1}{n} \sum_{a \in G} \chi_i(ab) = \alpha$$

and therefore $\alpha = 0$. Applying the relation to $\chi_i \overline{\chi}_j \in \widehat{G}$ we have

$$\langle \chi_i, \chi_j \rangle = \frac{1}{n} \sum_{a \in G} \chi_i(a)\overline{\chi_j}(a) = \langle \chi_i \chi_j^{-1}, \chi_0 \rangle.$$

If $\chi_i \chi_j^{-1} = \chi_0$ then $\langle \chi_i, \chi_i \rangle = 1$ so $\|\chi_i\| = 1$, however, if $\chi_i \chi_j^{-1} \neq \chi_0$ then $\chi_i \neq \chi_j$ and $\langle \chi_i, \chi_j \rangle = 0$.

(2)   By linearity, it follows at once that

$$\left\langle \sum_{i=0}^{n-1} \alpha_i \chi_i, \chi_j \right\rangle = \sum_{i=0}^{n-1} \alpha_i \langle \chi_i, \chi_j \rangle = \alpha_j$$

for $j = 0, 1, \ldots, n-1$.

(3)   If $\sum_{i=0}^{n-1} \alpha_i \chi_i = 0$ then $\alpha_j = \langle \sum_{i=0}^{n-1} \alpha_i \chi_i, \chi_j \rangle = 0$ for every $j = 0, 1, \ldots, n-1$. So $\chi_0, \ldots, \chi_{n-1}$ are linearly independent over $\mathbb{C}$. Since V has dimension n over $\mathbb{C}$ then $\chi_0, \ldots, \chi_{n-1}$ is a basis of V. ■

As a corollary we write down explicitly the *orthogonality relations* between characters:

J.   *We have:*

(1) $$\sum_{a \in G} \chi_i(a) = \begin{cases} n, & \text{when } i = 0, \\ 0, & \text{when } i \neq 0. \end{cases}$$

(2) $$\sum_{a \in G} \chi_i(a) \overline{\chi}_j(a) = \begin{cases} n, & \text{when } i = j, \\ 0, & \text{when } i \neq j. \end{cases}$$

(3) $$\sum_{i=0}^{n-1} \chi_i(a) = \begin{cases} n, & \text{when } a = e, \\ 0, & \text{when } a \neq e. \end{cases}$$

(4) $$\sum_{i=0}^{n-1} \chi_i(a) \overline{\chi}_i(b) = \begin{cases} n, & \text{when } a = b, \\ 0, & \text{when } a \neq b. \end{cases}$$

Proof: Parts (1) and (2) have already been established, (3) and (4) are the same as (1), (2), respectively, if we replace G by $\widehat{G}$ and note that $\widehat{\widehat{G}} \cong G$. ■

The orthogonality relations are a tool for solving systems of linear equations.

K.   *Let* $G = \{a_0, a_1, \ldots, a_{n-1}\}$ *and* $\widehat{G} = \{\chi_0, \ldots, \chi_{n-1}\}$, *where* $a_0 = e$ *is the unit element of* G *and* χ_0 *is the unit character of* G. *The system of* n *linear equations*

$$\sum_{j=0}^{n-1} \chi_i(a_j) X_j = \beta_i, \qquad i = 0, 1, \ldots, n-1$$

(where $\beta_i \in \mathbb{C}$ *for every index* $i = 0, 1, \ldots, n-1$*) has unique solution*

$$x_j = \frac{1}{n} \sum_{i=0}^{n-1} \overline{\chi}_i(a_j) \beta_i \qquad \text{for} \quad j = 0, 1, \ldots, n-1.$$

Proof: Let $A = (\chi_i(a_j))_{i,j}$ denote the matrix of coefficients, and let $\overline{A}'$ denote the transpose of the conjugate of A. Then the orthogonality formulas may be translated by stating that $A \cdot \overline{A}' = nI$ where I is the identity $n \times n$ matrix. Hence $|\det(A)|^2 = n^n$ so $\det(A) \neq 0$. Thus the above system has unique solution.

The inverse of the matrix $(1/\sqrt{n})A$ is equal to $(1/\sqrt{n})\overline{A}'$ (by the above relation), hence the solution $(x_1, \ldots, x_n)$ of the system of equations satisfies

$$\sum_{j=0}^{n-1} \frac{1}{\sqrt{n}} \chi_i(a_j)x_j = \frac{1}{\sqrt{n}} \beta_i \qquad \text{for} \quad i = 0, 1, \ldots, n-1.$$

Multiplying by the inverse matrix

$$x_j = \sum_{i=0}^{n-1} \frac{1}{\sqrt{n}} \overline{\chi}_i(a_j) \frac{1}{\sqrt{n}} \beta_i = \frac{1}{n} \sum_{i=0}^{n-1} \overline{\chi}_i(a_j)\beta_i$$

for $j = 0, 1, \ldots, n-1$. ∎

The last result is used in the proof of Dirichlet's theorem to separate the primes in the various prime residue classes.

On the space V of complex-valued functions defined on the group G, we shall consider the *shifting operators* S_a, defined for every element $a \in G$ as follows: $S_a(f)$ is the function of G such that $S_a(f)(b) = f(ab)$ for every $b \in G$. S_a is a linear transformation of V into itself.

With every element $f \in V$ and the shifting operators S_a we may build a new operator S_f of V, in the following way:

$$S_f = \sum_{a \in G} f(a)S_a.$$

For every character χ of G we have $S_a(\chi)(b) = \chi(ab) = \chi(a)\chi(b)$ so $S_a(\chi) = \chi(a) \cdot \chi$ and this shows that each character χ is an eigenvector of the operator S_a, having eigenvalue $\chi(a)$. Hence

$$[S_f(\chi)](b) = \sum_{a \in G} f(a)[S_a(\chi)(b)] = \left[\sum_{a \in G} f(a)\chi(a) \cdot \chi \right](b)$$

so each character χ is again an eigenvector of S_f having eigenvalue $\sum_{a \in G} f(a)\chi(a)$.

Now we show the following relation:

L.   *If $f \in V$ then*

$$\prod_{\chi \in \widehat{G}} \left(\sum_{a \in G} f(a)\chi(a) \right) = \det(f(ba^{-1})).$$

Proof: Let $S = \sum_{a \in G} f(a)S_a$. By **(H)**, $\{\chi_0, \chi_1, \ldots, \chi_{n-1}\}$ is a basis of V which consists of eigenvectors of S and the sums $\sum_{a \in G} f(a)\chi_i(a)$ are the corresponding eigenvalues. So

$$\prod_{i=0}^{n-1}\left(\sum_{a\in G} f(a)\chi_i(a)\right)$$

is equal to the determinant of the operator S. Now we compute this determinant by considering the standard basis $(h_a)_{a\in G}$ of V, where

$$h_a(b) = \begin{cases} 1, & \text{when } b = a, \\ 0, & \text{when } b \neq a. \end{cases}$$

Since

$$[S_f(h_b)](c) = \left[\sum_{a\in G} f(a)S_a(h_b)\right](c) = \sum_{a\in G} f(a)h_b(ac)$$

$$= f(bc^{-1}) = \left[\sum_{a\in G} f(ba^{-1})h_a\right](c)$$

for $a, b, c \in G$, we have $S_f(h_b) = \sum_{a\in G} f(ba^{-1})h_a$. Thus the matrix of S with respect to the standard basis $(h_a)_{a\in G}$ has entry $f(ba^{-1})$ at row a and column b.

Hence

$$\det(f(ba^{-1})) = \prod_{i=0}^{n-1}\left(\sum_{a\in G} f(a)\chi_i(a)\right). \qquad \blacksquare$$

21.2 Modular Characters

Definition 2.   Let $m > 1$ be an integer. A mapping $\chi\colon \mathbb{Z} \to \mathbb{C}$ is called *modular character (with modulus m)* when it satisfies the following conditions:

(1) $\chi(a) = 0$ if and only if $\gcd(a, m) > 1$;

(2) if $a \equiv b \pmod{m}$ then $\chi(a) = \chi(b)$; and

(3) $\chi(ab) = \chi(a)\chi(b)$.

The *support* of χ is $\{a \in \mathbb{Z} \mid \gcd(a, m) = 1\}$. Clearly, $\chi(a) = 1$ for every $a \in \mathbb{Z}$ such that $a \equiv 1 \pmod{m}$.

Among characters modulo m we distinguish the *trivial character* χ_0 modulo m, which is defined as follows:

$$\begin{cases} \chi_0(a) = 1, & \text{when } \gcd(a, m) = 1, \\ \chi_0(a) = 0, & \text{when } \gcd(a, m) > 1. \end{cases}$$

M.   *For every $m > 1$ there is a natural one-to-one correspondence between the characters of the multiplicative group $P(m) = (\mathbb{Z}/m)^{\cdot}$ and the modular characters with modulus m.*

Proof: Let χ be a character modulo m. We define $\widetilde{\chi} \colon P(m) \to \mathbb{C}$ by letting $\widetilde{\chi}(\overline{a}) = \chi(a)$ for any $\overline{a} \in P(m)$; note that $\widetilde{\chi}$ is well defined and it is obvious that $\widetilde{\chi}$ is a character of $P(m)$. Clearly, the mapping $\chi \mapsto \widetilde{\chi}$ is injective.

Given any character ρ of $P(m)$, let $\chi \colon \mathbb{Z} \to \mathbb{C}$ be defined as follows:

$$\begin{cases} \chi(a) = 0, & \text{when } \gcd(a, m) \neq 1, \\ \chi(a) = \rho(\overline{a}), & \text{when } \gcd(a, m) = 1. \end{cases}$$

It is again immediate to check that χ is a character modulo m and that $\widetilde{\chi} = \rho$, concluding the proof. ∎

If χ is a character modulo m we consider the set M_{χ} of integers $m' \geq 1$ with the following property: if $\gcd(a, m) = \gcd(b, m) = 1$ and $a \equiv b \pmod{m'}$ then $\chi(a) = \chi(b)$. Each element of M_{χ} is called a *defining modulus* of χ.

For example, $m \in M_{\chi}$. If $m_1 \in M_{\chi}$ and m_1 divides m_2 then $m_2 \in M_{\chi}$.

The smallest positive integer belonging to M_{χ} is called the *conductor* of χ and denoted by f_{χ}.

If $\chi_0(a) = 1$ when $\gcd(a, m) = 1$, $\chi_0(a) = 0$ when $\gcd(a, m) \neq 1$, then $f_{\chi_0} = 1$.

N.   *M_{χ} consists of the positive multiples of the conductor of χ.*

Proof: It is enough to show that if $m_1, m_2 \in M_{\chi}$ then $d = \gcd(m_1, m_2)$ also belongs to M_{χ}. This implies indeed that every integer $m' \in M_{\chi}$ must be a multiple of the conductor f_{χ}, because $\gcd(m', f_{\chi}) \in M_{\chi}$.

Let $\gcd(a, m) = \gcd(b, m) = 1$ and $a \equiv b \pmod{d}$. Let m' be the product of all prime numbers p which divide m but do not divide m_2; then $d = \gcd(m'm_1, m_2)$. Hence there exists an integer x such that

$$\begin{cases} x \equiv a \pmod{m'm_1}, \\ x \equiv b \pmod{m_2}. \end{cases}$$

Moreover, $\gcd(x, m) = 1$ because if a prime number p divides x and m it cannot divide m_2 (since $\gcd(b, m) = 1$); then p divides m' hence also $m'm_1$ and a, impossible because $\gcd(a, m) = 1$.

From $m'm_1 \in M_{\chi}$ it follows that $\chi(x) = \chi(a)$. From $m_2 \in M_{\chi}$ it follows that $\chi(x) = \chi(b)$ hence $\chi(a) = \chi(b)$, showing that d is also a defining modulus for χ. ∎

O.   *The conductor of a modular character cannot be equal to 2.*

Proof: Assume that the character χ, with modulus m, has conductor 2. By (**N**), m is even, hence $\chi(a) = 0$ if a is even, while $\chi(a) = 1$ if a is odd. Thus, if a, b are any odd integers (so $a \equiv b \pmod{1}$) then $\chi(a) = \chi(b)$. Thus $1 \in M_{\chi}$, which is a contradiction. ∎

P.   *If χ is a character modulo m with conductor f, there exists a unique character ψ modulo f such that if $\gcd(a, m) = 1$ then $\psi(a) = \chi(a)$.*

Proof: Since f divides m, then if $\gcd(a, m) = 1$ then $\gcd(a, f) = 1$. In order to define ψ we show that if $\gcd(a, f) = 1$ there exists a' relatively prime to m such that $a' \equiv a \pmod{f}$. Indeed, let m_0 be the product of all primes p dividing m but not dividing f. Then there exists an integer a' such that

$$\begin{cases} a' \equiv a \pmod{f}, \\ a' \equiv 1 \pmod{m_0}. \end{cases}$$

It follows that a' is relatively prime to m.

If $\gcd(a'', m) = 1$ and $a'' \equiv a \pmod{f}$ then $\chi(a') = \chi(a'')$ since $f \in M_\chi$.

We define $\psi(a) = \chi(a')$ when $\gcd(a, f) = 1$ and where a' is defined above; $\psi(a) = 0$ when $\gcd(a, f) \neq 1$. Thus ψ is well defined.

It is easy to verify that $\psi(ab) = \psi(a)\psi(b)$ and if $a \equiv b \pmod{f}$ then $\psi(a) = \psi(b)$. So ψ is a character modulo f and if $\gcd(a, m) = 1$ then $\psi(a) = \chi(a)$.

The conductor of ψ is equal to f. Indeed, if d is a defining modulus for ψ dividing f, we show that it is also a defining modulus for χ, hence $d = f$. Let $\gcd(a, m) = \gcd(b, m) = 1$, $a \equiv b \pmod{d}$, so a, b are relatively prime to f, so $\chi(a) = \psi(a) = \psi(b) = \chi(b)$.

It remains to show that if ψ' is a character modulo f such that $\gcd(a', m) = 1$ implies $\psi'(a) = \chi(a')$ then $\psi' = \psi$. This is immediate, because if $\gcd(a, f) = 1$ and $\gcd(a', m) = 1$, $a' \equiv a \pmod{f}$ then $\psi'(a) = \psi'(a') = \chi(a') = \psi(a)$.   ∎

A character ψ modulo f, with conductor $f_\psi = f$ is called a *primitive character*.

With this definition we may rephrase (**P**) as follows:

Given any character χ modulo m, there exists a unique primitive character with the same conductor and which coincides with χ on the support of χ.

We characterize primitive characters:

Q.   *χ is a primitive character with conductor m if and only if for every divisor d of m, $1 < d < m$, there exists an integer a such that $\gcd(a, m) = 1$, $a \equiv 1 \pmod{d}$, and $\chi(a) \neq 1$.*

Proof: We suppose first there exists a character ψ modulo d, such that $1 < d < m$, d divides m if $\gcd(a, m) = 1$ then $\psi(a) = \chi(a)$. In particular, if $a \equiv 1 \pmod{d}$ then $\chi(a) = \psi(a) = \psi(1) = 1$.

Conversely, let us assume that there exists a divisor d of m, $1 < d < m$, such that if $\gcd(a, m) = 1$, $a \equiv 1 \pmod{d}$ then $\chi(a) = 1$. We shall define a character ψ modulo d, such that if $\gcd(b, m) = 1$ then $\chi(b) = \psi(b)$.

In the proof of (**P**) we have seen that if b is an integer relatively prime to d then there exists an integer b', relatively prime to m and such that $b' \equiv b \pmod{d}$. In particular, if $\gcd(b,m) = 1$ we just take $b' = b$. We define $\psi(b) = \chi(b')$. The mapping ψ is well defined, because if b', b'' are relatively prime to m and $b' \equiv b \pmod{d}$, $b'' \equiv b \pmod{d}$ there exists $a \in \mathbb{Z}$, such that $ab' \equiv b'' \pmod{m}$. Thus $a \equiv 1 \pmod{d}$, $\gcd(a,m) = 1$. By hypothesis, $\chi(a) = 1$ hence $\chi(b'') = \chi(a)\chi(b') = \chi(b')$.

From this, it is immediate to check that ψ is a character modulo d and if $\gcd(b,m) = 1$ then $\chi(b) = \psi(b)$. Since $d \neq m$, then χ is not a primitive character modulo m. ∎

In order to describe the characters modulo m we prove the following result:

R.   *Let $m = m_1 \cdots m_r$, where $m_1, \ldots, m_r$ are pairwise relatively prime integers. If χ is a character modulo m, then it may be written uniquely in the form $\chi = \chi_1 \cdots \chi_r$, where χ_i is a character modulo m_i. Moreover, for the conductors, we have the relation $f_\chi = f_{\chi_1} \cdots f_{\chi_r}$. If χ is a primitive character then each character χ_i is also primitive.*

Proof: If a is such that $\gcd(a, m_i) = 1$, there exists an a_i, $\gcd(a_i, m) = 1$, such that
$$\begin{cases} a_i \equiv a \pmod{m_i}, \\ a_i \equiv 1 \pmod{m_j}, \quad \text{for } j \neq i. \end{cases}$$

We define $\chi_i(a) = \chi(a_i)$ when $\gcd(a, m_i) = 1$ and $\chi_i(a) = 0$ otherwise. The mapping χ_i is well defined, because if a_i' satisfies
$$\begin{cases} a_i' \equiv a \pmod{m_i}, \\ a_i' \equiv 1 \pmod{m_j}, \quad \text{for } j \neq i, \end{cases}$$
then $a_i' \equiv a_i \pmod{m}$ so $\chi(a_i) = \chi(a_i')$.

It is easy to verify that χ_i is a character modulo m_i. Moreover, $\chi = \chi_1 \cdots \chi_r$. Indeed, given a, $\gcd(a, m) = 1$ we have $a \equiv a_1 \cdots a_r \pmod{m}$ where each a_i has been determined by the above congruences. Thus $\chi(a) = \chi(a_1) \cdots \chi(a_r) = \chi_1(a) \cdots \chi_r(a)$, proving that $\chi = \chi_1 \cdots \chi_r$.

This multiplicative representation of χ is unique. In fact, let us assume that $\chi = \chi_1' \cdots \chi_r'$ where each χ_i' is a character modulo m_i. Then for every a, $\gcd(a, m_i) = 1$, and a_i defined as above, we have $\chi_i'(a) = \chi_i'(a_i) = \chi_1'(a_i) \cdots \chi_i'(a_i) \cdots \chi_r'(a_i) = \chi(a_i) = \chi_i(a)$.

Finally we prove the assertion about the conductors. First we note that if d_i is any defining modulus for χ_i $(i = 1, \ldots, r)$ then $d = d_1 \cdots d_r$ is a defining modulus for χ: if a, b are such that $\gcd(a, m) = \gcd(b, m) = 1$ and $a \equiv b \pmod{d}$ then also $\gcd(a, m) = 1$, $\gcd(b, m) = 1$, $a \equiv b \pmod{d_i}$, so $\chi_i(a) = \chi_i(b)$ for $i = 1, \ldots, r$; hence $\chi(a) = \chi(b)$.

Conversely, let d be a defining modulus for χ dividing m, let $d_i = \gcd(d, m_i)$. Then $d = d_1 \cdots d_r$ and each d_i is a defining modulus for χ_i.

Indeed, let $\gcd(a, m) = \gcd(b, m) = 1$ and $a \equiv b \pmod{d_i}$; let a_i, b_i with $\gcd(a_i, m) = \gcd(b_i, m) = 1$ be defined as already indicated. Then $a_i \equiv b_i \pmod{d_i}$, $a_i \equiv b_i \pmod{m_j}$, so $a_i \equiv b_i \pmod{d_j}$ for $j \neq i$ and therefore $a_i \equiv b_i \pmod{d}$. Hence $\chi(a_i) = \chi(b_i)$ and by definition $\chi_i(a) = \chi_i(b)$.

Altogether we have shown that every defining modulus of χ is the product of defining moduli of the characters χ_i, and conversely. In view of (N) the conductor of χ is the product of the conductors of χ_i $(i = 1, \ldots, r)$.

The last assertion is now immediate. ∎

To describe the characters with conductor $f = p_1^{e_1} \cdots p_r^{e_r}$ we need only describe those characters whose conductor is a power of a prime. By means of (M) they correspond to the characters of the multiplicative group $P(p^e)$ (for some prime p and integer $e \geq 1$), and these have already been explicitly indicated in Section 21.1.

If the character χ modulo m corresponds to a quadratic character of the group $P(m)$, we say that χ is a *quadratic modular character*.

S.   *The conductor of a quadratic modular character is either m, $4m$, or $8m$, where m is an odd number without square factor.*

Proof: First we consider the case when the conductor of χ is a power p^e of a prime number p, $e \geq 1$.

Let $p \neq 2$. If $\gcd(a, p^e) = 1$ then we may write $a \equiv b^{\alpha'}(1 + p)^{\alpha''}$ $\pmod{p^e}$ and if ζ_{p-1}, $\zeta_{p^{e-1}}$ are roots of unity of orders $p - 1$, p^{e-1}, respectively, then $\chi(a) = \zeta_{p-1}^{\alpha'}\zeta_{p^{e-1}}^{\alpha''}$. χ is a quadratic modular character when for every α', α'' we have

$$1 = \chi^2(a) = \zeta_{p-1}^{2\alpha'}\zeta_{p^{e-1}}^{2\alpha''},$$

that is, $2\alpha' \equiv 0 \pmod{p-1}$ and $2\alpha'' \equiv 0 \pmod{p^{e-1}}$. This is possible if and only if $e = 1$, so the conductor of χ is equal to p.

If $p = 2$, $e \neq 1$, since by (O) we have no modular character with conductor 2. Since $e \geq 2$, if $\gcd(a, 2^e) = 1$, if $a \equiv (-1)^{\alpha'}5^{\alpha''} \pmod{2^e}$ and $\zeta_{2^{e-2}}$ is a root of unity of order 2^{e-2}, then

$$\chi(a) = (-1)^{\alpha'}\zeta_{2^{e-2}}^{\alpha''} \qquad \text{or} \qquad \chi(a) = \zeta_{2^{e-2}}^{\alpha''}.$$

We have $1 = \chi^2(a)$ for every a exactly when $2\alpha'' \equiv 0 \pmod{2^{e-2}}$ for every α''. This is possible exactly when $e = 2$ or 3, so the conductor of χ is equal to 4 or to 8.

Now we investigate the general case. By (R), if $f_\chi = \prod_{i=1}^{r} p_i^{e_i}$ then $\chi = \chi_1 \cdots \chi_r$ where χ_i has conductor $p_i^{e_i}$. From the definition of χ_i we see that $\chi_i^2(a) = [\chi_i(a)]^2 = [\chi(a_i)]^2 = \chi^2(a_i) = 1$. So either χ_i is the principal character modulo f_{χ_i} or it is a quadratic modular character. By the preceding discussion, it follows that the conductor of χ is either equal to a product of distinct odd primes, or it is 4 or 8 times a product of distinct odd primes. ∎

We note explicitly:

T.   *Let p be an odd prime. The only quadratic character modulo p is χ given by the Legendre symbol: $\chi(a) = (a/p)$ for every $a \in \mathbb{Z}$; this character is primitive with conductor p.*

Proof: It is clear that the Legendre symbol defines a primitive character with conductor p. Conversely, let g be a primitive root modulo p, $\eta = \cos(2\pi/(p-1)) + i\sin(2\pi/(p-1))$. The character χ is completely determined by its value $\chi(g) = \eta^e$ ($0 < e < p - 1$). If χ is quadratic then $1 = \chi^2(g) = \eta^{2e}$, so $e = (p - 1)/2$.

On the other hand, $(g/p) = -1 = \eta^{(p-1)/2} = \chi(g)$, therefore if p does not divide a then $a \equiv g^c \pmod{p}$, for some c, $0 \le c \le p - 2$, and

$$\left(\frac{a}{p}\right) = \left(\frac{g}{p}\right)^c = (-1)^c = \eta^{((p-1)/2)\times c} = \chi(g^c) = \chi(a),$$

proving the statement.   ∎

A character χ modulo m is said to be *even* if $\chi(-a) = \chi(a)$ for every $a \in \mathbb{Z}$ with $\gcd(a, m) = 1$. The character χ is said to be *odd* when $\chi(-a) = -\chi(a)$ for every $a \in \mathbb{Z}$, $\gcd(a, m) = 1$.

Since $\chi(-a) = \chi(-1)\chi(a)$ then χ is even if and only if $\chi(-1) = 1$, while χ is odd if and only if $\chi(-1) = -1$. Since $\chi(-1) = \pm 1$, then every character is either even or odd.

If χ, χ' are both even or both odd, then $\chi\chi'$ is even. On the other hand, if χ is even, χ' is odd (or vice versa), then $\chi\chi'$ is an odd character.

If p is an odd prime, the quadratic character defined by the Legendre symbol $\chi(a) = (a/p)$ is even if and only if $p \equiv 1 \pmod{4}$, because $(-1/p) = 1$ exactly in this case.

To conclude the section we follow Hecke and introduce characters associated to groups of classes of ideals in number fields.

We recall (see Chapter 8, Section 2) that if J is a nonzero integral ideal of K, then the group $\mathcal{C}_{J,+} = \mathcal{F}_J/\mathcal{P}r_{J,+}$ of classes of ideals modulo J is a finite Abelian group. Here

$$\mathcal{F}_J = \left\{ \frac{I}{I'} \;\middle|\; I, I' \text{ integral ideals, } I' \ne 0, \ \gcd(I, J) = \gcd(I', J) = A \right\}$$

and

$$\mathcal{P}r_{J,+} = \{Ax \mid x \in K, \ x \ne 0, \ x \equiv 1 \pmod{J} \text{ and } x \text{ is totally positive}\}.$$

If $I \in \mathcal{F}_J$ let $[I]$ denote its image in $\mathcal{C}_{J,+}$. If $\tilde{\chi}$ is any character of $\mathcal{C}_{J,+}$, let χ be defined as follows:

$$\chi(I) = \begin{cases} \tilde{\chi}([I]), & \text{when } I \in \mathcal{F}_J, \\ 0, & \text{when } I \notin \mathcal{F}_J. \end{cases}$$

χ is called a *Hecke character of K with modulus J*.

We see that if $I, I' \in \mathcal{F}_J$, and $I = I' \cdot Ax$ (with $Ax \in \mathcal{P}r_{J,+}$) then $\chi(I) = \chi(I')$. Also, if $I, I' \in \mathcal{F}_J$ then $\chi(II') = \chi(I)\chi(I')$.

Hecke characters are a generalization of modular characters. Indeed, if $K = \mathbb{Q}$, $J = \mathbb{Z}m$, then $\mathcal{C}_{J,+}$ is naturally isomorphic to $P(m)$.

The concepts of conductor of a character and primitive Hecke character are defined like the ones of modular characters.

21.3 Gaussian Sums

In Chapter 4, we encountered Gaussian sums associated to the quadratic characters given by the Legendre symbol. We now introduce Gaussian sums associated to arbitrary modular characters.

Let $m \geq 2$, let χ be a character modulo m, and let $\zeta = \cos(2\pi/m) + i\sin(2\pi/m)$.

Definition 3.   The expressions

$$\tau_k(\chi) = \sum_{\substack{\overline{a} \in P(m) \\ 1 \leq a < m}} \chi(a)\zeta^{ak} \qquad \text{for} \quad k = 0, 1, \ldots, m-1, \qquad (21.1)$$

are called the *Gaussian sums belonging to the character* χ (and the primitive mth root of unity ζ).

For $k = 0$ we have

$$\tau_0(\chi) = \sum_{\substack{\overline{a} \in P(m) \\ 1 \leq a < m}} \chi(a) = \begin{cases} \varphi(m), & \text{when } \chi = \chi_0, \\ 0, & \text{when } \chi \neq \chi_0, \end{cases} \qquad (21.2)$$

as follows from (**J**).

The *principal Gaussian sum* is

$$\tau_1(\chi) = \sum_{\substack{\overline{a} \in P(m) \\ 1 \leq a < m}} \chi(a)\zeta^a. \qquad (21.3)$$

In Chapter 4, Section 2, we considered the Gaussian sums associated to the Legendre symbol, namely $\sum_{a=1}^{p-1}(a/p)\zeta^{ak}$, for which we used the notation $\tau(k)$.

The different Gaussian sums of χ are related as follows:

U.   *Let χ be a character modulo m and $1 \leq k < m$. If $\gcd(k, m) = 1$ then*

$$\tau_k(\chi) = \frac{1}{\chi(k)} \tau_1(\chi).$$

If χ is primitive and $\gcd(k, m) \neq 1$ then $\tau_k(\chi) = 0$.

Proof: Let $\gcd(k, m) = 1$, then the multiplication by $\overline{k}$ establishes a one-to-one correspondence from $P(m)$ onto $P(m)$, so

$$\chi(k)\tau_k(\chi) = \sum_{\substack{\overline{a}\in P(m) \\ 1\le a<m}} \chi(k)\chi(a)\zeta^{ak} = \sum_{\substack{\overline{b}\in P(m) \\ 1\le b<m}} \chi(b)\zeta^b = \tau_1(\chi).$$

Now let $d = \gcd(k, m) > 1$, so $m = dm'$, $1 < m' < m$. Since χ is a primitive character modulo m, by (**Q**) there exists an integer b, such that $\gcd(b, m) = 1$, $b \equiv 1 \pmod{m'}$ and $\dot{\chi}(b) \neq 1$; therefore $\zeta^{kb} = \zeta^k$. The multiplication with $\overline{b}$ establishes a one-to-one correspondence from $P(m)$ onto $P(m)$ and we have

$$\tau_k(\chi) = \sum_{\substack{\overline{a}\in P(m) \\ 1\le a<m}} \chi(a)\zeta^{ak} = \sum_{\substack{\overline{a}\in P(m) \\ 1\le a<m}} \chi(b)\chi(a)\zeta^{bak}$$

$$= \chi(b) \sum_{\substack{\overline{a}\in P(m) \\ 1\le a<m}} \chi(a)\zeta^{ak} = \chi(b)\tau_k(\chi).$$

Since $\chi(b) \neq 1$ then $\tau_k(\chi) = 0$.   ∎

Reducing to the case of characters modulo powers of primes makes the evaluation of the Gaussian sum of a character. Let χ be a character modulo m, let $m = m_1 \cdots m_r$, where $m_1, \ldots, m_r$ are pairwise relatively prime positive integers. Following (**R**), we may write $\chi = \chi_1 \cdots \chi_r$ where χ_i is a character modulo m_i. Let

$$m_i' = \frac{m}{m_i} \qquad \text{and} \qquad \zeta_i = \zeta^{m_i'} = \cos\frac{2\pi}{m_i} + i\sin\frac{2\pi}{m_i},$$

for $i = 1, \ldots, r$.

With these notations:

V.

$$\tau_1(\chi) = \prod_{i=1}^{r} \chi_i(m_i') \prod_{i=1}^{r} \tau_1(\chi_i).$$

Proof: For every $\overline{a} \in P(m)$ we may write $a \equiv a_1 \cdots a_r \pmod{m}$ where $\gcd(a_i, m_i) = 1$, $a \equiv a_i \pmod{m_i}$, $a_i \equiv 1 \pmod{m_i'}$ and $\chi(a) = \chi_1(a_1)\cdots\chi_r(a_r)$ (see (**R**)). Since $\gcd(m_i', m_i) = 1$ there exists an integer b_i such that $b_i m_i' \equiv 1 \pmod{m_i}$ for $i = 1, \ldots, r$. We also have $b_i m_i' \equiv 0 \pmod{m_j}$ for every $j \neq i$. Then $a \equiv \sum_{i=1}^{r} b_i m_i' a_i \pmod{m}$ hence

$$\tau_1(\chi) = \sum_{\substack{\overline{a}\in P(m) \\ 1\le a<m}} \chi(a)\zeta^a = \prod_{i=1}^{r} \left[\sum_{\overline{a}\in P(m_i)} \chi_i(a_i)\zeta_i^{b_i a_i} \right]$$

as we see by multiplying out the last expression and noting that

$$\zeta_1^{b_1 a_1} \cdot \zeta_2^{b_2 a_2} \cdots \zeta_r^{b_r a_r} = \zeta^{b_1 m_1' a_1 + b_2 m_2' a_2 + \cdots + b_r m_r' a_r} = \zeta^a.$$

Since

$$\sum_{\substack{\overline{a}_i \in P(m_i) \\ 1 \le a_i < m_i}} \chi_i(a_i)\zeta_i^{b_i a_i} = \frac{1}{\chi_i(b_i)} \sum_{\substack{\overline{a}_i \in P(m_i) \\ 1 \le a_i < m_i}} \chi_i(b_i a_i)\zeta_i^{b_i a_i} = \chi_i(m_i') \cdot \tau_1(\chi_i)$$

we conclude that

$$\tau_1(\chi) = \prod_{i=1}^r \chi_i(m_i') \cdot \prod_{i=1}^r \tau_1(\chi_i). \qquad \blacksquare$$

W.   *If χ is a primitive character with conductor f then $|\tau_1(\chi)|^2 = f$.*

Proof: Let $f = p_1^{e_1} \cdots p_r^{e_r}$, where $p_1, \ldots, p_r$ are distinct primes and $e_i > 0$. Let $\chi = \chi_1 \cdots \chi_r$ be the decomposition of χ into primitive characters with conductors $p_i^{e_i}$.

It is enough to show that $|\tau_1(\chi_i)|^2 = p_i^{e_i}$ for $i = 1, \ldots, r$.

In fact, letting $f_i' = f/p_i^{e_i}$, this implies that

$$|\tau_1(\chi)|^2 = \overline{\tau_1(\chi)} \cdot \tau_1(\chi)$$

$$= \left[\prod_{i=1}^r \overline{\chi_i(f_i')} \prod_{i=1}^r \overline{\tau_1(\chi_i)} \right] \left[\prod_{i=1}^r \chi_i(f_i') \prod_{i=1}^r \tau_1(\chi_i) \right]$$

$$= \prod_{i=1}^r \overline{\tau_1(\chi_i)} \cdot \tau_1(\chi_i) = \prod_{i=1}^r p_i^{e_i} = f.$$

So, we may assume that the conductor of χ is $f = p^e$ and we have

$$\overline{\tau_1(\chi)} \cdot \tau_1(\chi) = \left[\sum_{\overline{a} \in P(p^e)} \overline{\chi(a)}\zeta^{-a} \right] \cdot \left[\sum_{\overline{b} \in P(p^e)} \chi(b)\zeta^b \right]$$

$$= \sum_{\overline{a}} \sum_{\overline{b}} \chi(a)^{-1}\chi(b)\zeta^{b-a}.$$

For any $\overline{a}, \overline{b} \in P(p^e)$ let t be such that $b \equiv at \pmod{p^e}$, so $\chi(b) = \chi(a)\chi(t)$ and

$$\overline{\tau_1(\chi)} \cdot \tau_1(\chi) = \sum_{\substack{\overline{a} \in P(p^e) \\ 1 \le a \le p^e - 1}} \sum_{\substack{\overline{t} \in P(p^e) \\ 1 \le t \le p^e - 1}} \chi(t)\zeta^{a(t-1)}$$

$$= \sum_{t=1}^{p^e} \chi(t) \left[\sum_{\overline{a} \in P(p^e)} \zeta^{a(t-1)} \right]$$

$$= \sum_{t=1}^{p^e} \chi(t) \left[\sum_{a=1}^{p^e} \zeta^{a(t-1)} \right] - \sum_{t=1}^{p^e} \chi(t) \left[\sum_{c=1}^{p^{e-1}} \zeta^{pc(t-1)} \right].$$

But

$$\sum_{a=1}^{p^e} \zeta^{a(t-1)} = \begin{cases} p^e, & \text{when } t = 1, \\ 0, & \text{when } t \neq 1, \end{cases}$$

and

$$\sum_{c=1}^{p^{e-1}} \zeta^{pc(t-1)} = \begin{cases} p^{e-1}, & \text{when } t \equiv 1 \pmod{p^{e-1}}, \\ 0, & \text{when } t \not\equiv 1 \pmod{p^{e-1}}. \end{cases}$$

Hence

$$\overline{\tau_1(\chi)} \cdot \tau_1(\chi) = p^e - p^{e-1} \left(\sum_{\substack{t=1 \\ t \equiv 1 \,(\text{mod } p^{e-1})}}^{p^e} \chi(t) \right).$$

To evaluate the last sum we recall that χ is a primitive character modulo p^e and we refer to Chapter 3, (J) and (K). If $p \neq 2$, every integer t, prime to p, $1 \leq t < p^e$, satisfies a congruence $t \equiv b^{\alpha'}(1+p)^{\alpha''} \pmod{p^e}$ where $0 < \alpha' \leq p-1$, $0 \leq \alpha'' < p^{e-1}$, and $\chi(t) = \zeta_{p-1}^{\alpha'}\zeta_{p^{e-1}}^{\alpha''}$ where ζ_{p-1}, $\zeta_{p^{e-1}}$ are primitive roots of unity of orders $p-1$, p^{e-1} (since χ is a primitive character). We have $t \equiv 1 \pmod{p^{e-1}}$ if and only if $\alpha' = 0$, $\alpha'' = kp^{e-2}$, $0 \leq k \leq p-1$. In this case

$$\chi(t) = \zeta_{p^{e-1}}^{kp^{e-2}} = \eta^k \qquad \text{where} \quad \eta = \zeta_{p^{e-1}}^{p^{e-2}}$$

is a primitive pth root of unity. Hence

$$\sum_{\substack{t=1 \\ t \equiv 1 \,(\text{mod } p^{e-1})}}^{p^e} \chi(t) = \sum_{k=0}^{p-1} \eta^k = 0.$$

If $p = 2$, $e \geq 3$, every integer t, prime to p, $1 \leq t \leq p^e$, satisfies a congruence $t \equiv (-1)^{\alpha'}5^{\alpha''} \pmod{2^e}$ with $\alpha' = 0$ or 1, $0 \leq \alpha'' < 2^{e-2}$, and for the characters χ with conductor 2^e we have

$$\chi(t) = (-1)^{\alpha'}\zeta_{2^{e-2}}^{\alpha''} \qquad \text{or} \qquad \chi(t) = \zeta_{2^{e-2}}^{\alpha''}.$$

We have $t \equiv 1 \pmod{2^{e-1}}$ if and only if $\alpha' = 0$ and $\alpha'' = 0$, 2^{e-3}. Thus

$$\sum_{\substack{t=1 \\ t \equiv 1 \,(\text{mod } 2^{e-1})}}^{2^e} \chi(t) = 1 + \zeta_{2^{e-2}}^{2^{e-3}} = 0$$

since $\zeta_{2^{e-2}}^{2^{e-3}}$ is not equal to 1, hence it is equal to -1. Finally, for the case where χ has conductor 4 we see immediately that

$$\sum_{\substack{t=1 \\ t \equiv 1 \,(\text{mod } 2)}}^{4} \chi(t) = \chi(1) + \chi(3) = 0,$$

since $\chi(3) = -1$.

Concluding, we have shown that $|\tau_1(\chi)|^2 = p^e$ as required. ∎

If χ is a quadratic modular character then $\chi(a) = \pm 1$ when a is relatively prime to the conductor of χ. Thus

$$\overline{\tau_1(\chi)} = \sum_{\bar{a} \in P(f)} \chi(a)\zeta^{-a} = \sum_{\bar{a} \in P(f)} \chi(-a)\zeta^a = \chi(-1)\tau_1(\chi)$$

and so

$$[\tau_1(\chi)]^2 = \chi(-1)\overline{\tau_1(\chi)}\tau_1(\chi) = \chi(-1)f.$$

Therefore

$$\tau_1(\chi) = \pm\sqrt{f} \quad \text{or} \quad \tau_1(\chi) = \pm i\sqrt{f}.$$

It is quite subtle to decide what is the sign of $\tau_1(\chi)$. We shall deal with this question in Chapter 26, (E).

EXERCISES

1. Determine explicitly the characters modulo $4, 6, 7, 8$. Indicate the orders of the characters.

2. Determine explicitly the group of characters of the multiplicative group of invertible residue classes modulo n, where $n = 7, 8, 10, 12$. In each case determine the structure of the group of characters.

3. Let $m = 2^e \prod_{i=1}^r p_i^{e_i}$, where each p_i is an odd prime, $e \geq 0$, $e_i > 0$. According to Chapter 3, (K), for every odd integer n and each $e > 2$ there exists $b > 0$ such that

$$n \equiv (-1)^{(n-1)/2}5^b \pmod{2^e}.$$

For $e = 0, 1$ let $\chi_{2^e}(n) = 1$, and for $e = 2$ let $\chi_{2^e}(n) = (-1)^{(n-1)/2}$ for every odd integer n. For $e > 2$ let

$$\chi_{2^e}(n) = e^{2\pi i b/2^{e-2}}$$

(where b was defined above), for every odd integer n. If $f \geq 1$ and p is an odd prime, let

$$\chi_{p^f}(n) = e^{2\pi \, \mathrm{ind}_g(n)/\varphi(p^f)}$$

(where g is a primitive root modulo p and $\mathrm{ind}_g(n)$ denotes the index of n with respect to g) for every n not a multiple of p. Show:

    (a) χ_{2^e} is a character modulo 2^e (for all $e \geq 0$) and χ_{p^f} is a character modulo p^f (for $f \geq 1$, p odd prime).

(b) With the above notations if χ is a character modulo m, there exist a, a', a_i $(1 \leq i \leq r)$ with

$$a = 0, 1, \qquad 0 \leq a' < 2^{e-2}, \qquad 0 \leq a_i < \varphi(p_i^{e_i}),$$

such that:
If $e = 0, 1$ then

$$\chi(n) = \prod_{i=1}^{r} [\chi_{p_i'}(n)]^{a_i}.$$

If $e = 2$ then

$$\chi(n) = (-1)^{a \cdot (n-1)/2} \prod_{i=1}^{r} [\chi_{p_i'}(n)]^{a_i}.$$

If $e > 2$ then

$$\chi(n) = (-1)^{a \cdot (n-1)/2} [\chi_{2'}(n)]^{a'} \cdot \prod_{i=1}^{r} [\chi_{p_i'}(n)]^{a_i}.$$

4. Let G be an Abelian group of order $m > 1$. Let $G = \{a_1, \ldots, a_m\}$ and $\widehat{G} = \{\chi_0, \chi_1, \ldots, \chi_{m-1}\}$. Show that the matrix M with entries $(1/m)\chi_i(a_j)$ is unitary.

5. Let χ be a modular character for the moduli m_1, m_2. Show that χ is also a character modulo $\gcd(m_1, m_2)$.

6. Let m be odd. Prove that there is no real primitive character modulo $2m$.

7. Describe explicitly the primitive characters modulo 16 as well as those modulo 25 and modulo 360.

8. Determine explicitly all the primitive quadratic characters.

9. Let χ be a primitive quadratic character with even conductor m. Show that $\chi(a + m/2) = -\chi(a)$ for all a, with $\gcd(a, m) = 1$.

10. Show that the number of primitive characters with conductor $m > 1$ is equal to

$$\sum_{d|m} \mu(d)\varphi\left(\frac{m}{d}\right),$$

where μ denotes the Möbius function.

11. Let χ be a character modulo m, which is not the principal character. Show that if $0 \leq k < h$ then

$$\left| \sum_{n=k}^{h} \chi(n) \right| \leq \frac{\varphi(m)}{2}.$$

In particular, for the Legendre symbol modulo $p > 2$:

$$\left| \sum_{n=k}^{h} \left(\frac{n}{p} \right) \right| \leq \frac{p-1}{2}.$$

12. Let $m \geq 2$, $a \geq 1$ with $\gcd(a, m) = 1$. Let $b \geq 1$. Show that

$$\sum_{\chi} \frac{\chi(b)}{\chi(a)} = \begin{cases} \varphi(m) & \text{if } b \equiv a \pmod{m}, \\ 0 & \text{if } b \not\equiv a \pmod{m}, \end{cases}$$

sum over all characters modulo m.

13. Let χ be a modular character with conductor f. Show that if $1 \leq a < b$ then

$$\left| \sum_{k=a+1}^{b} \frac{\chi(k)}{k} \right| < \frac{2}{a+1} \sqrt{f} \log f.$$

14. Let $n > 1$ and let ζ be a primitive nth root of 1. Show that

$$\sum_{r=1}^{n-1} r\zeta^r = \frac{n}{\zeta - 1}.$$

15. Write explicitly the principal Gaussian sums $\tau_1(\chi)$ associated to each character modulo 7 and compute $|\tau_1(\chi)|^2$ as well as $\tau_1(\chi)$.

16. Same as the preceding exercise for characters modulo 9.

17. For $m > 1$ let ζ be a primitive mth root of 1. For each $n \in \mathbb{Z}$ let

$$C_m(n) = \sum_{\substack{a=1 \\ \gcd(a,m)=1}}^{m-1} \zeta^{an}.$$

Show:

(a) If $m_1, m_2 > 1$ and $\gcd(m_1, m_2) = 1$ then

$$C_{m_1}(n) \cdot C_{m_2}(n) = C_{m_1 m_2}(n)$$

for every $n \in \mathbb{Z}$.

(b) If p is a prime, $e \geq 1$ then

$$C_{p^e}(n) = \begin{cases} \varphi(p^e) & \text{if } p^e \mid n, \\ -p^{e-1} & \text{if } p^e \nmid n, \ p^{e-1} \mid n, \\ 0 & \text{if } p^{e-1} \nmid n. \end{cases}$$

(c) $C_m(1) = \mu(m)$, where μ denotes the Möbius function.

18. Let χ be a real primitive character with odd conductor m. Then

$$\tau(\chi) = \begin{cases} \pm\sqrt{m} & \text{if } m \equiv 1 \pmod 4, \\ \pm\sqrt{-m} & \text{if } m \equiv 3 \pmod 4. \end{cases}$$

19. Let k, n be integers, with $n \geq 1$. Define the *quadratic Gauss sum* by

$$G(k, n) = \sum_{r=1}^{n} \zeta^{kr^2},$$

where ζ is a primitive nth root of 1.
  Show:

(a) If $\gcd(m, n) = 1$ then

$$G(k, mn) = G(km, n) \cdot G(kn, m).$$

(b) Let p be an odd prime, $p \nmid k$, let $a \geq 2$. Then

$$G(k, p^a) = p\, G(k, p^{a-2})$$

and

$$G(k, p^a) = \begin{cases} p^{a/2} & \text{if } a \text{ is even}, \\ p^{(a-1)/2} G(k, p) & \text{if } a \text{ is odd}. \end{cases}$$

22

Zeta-Functions and L-Series

Many deep results about algebraic numbers require Riemann and Dedekind zeta-functions, as well as Dirichlet and Hecke L-series. In this chapter we gather some basic results to be used later.

22.1 The Riemann Zeta-Function

One of the seminal ideas for application of analytical methods occurred in Euler's proof for the existence of infinitely many primes.

A. *There exist infinitely many primes.*

Proof: Suppose that $p_1, p_2, \ldots, p_r$ are all the primes. For each $i = 1, \ldots, r$ we have

$$\sum_{k=0}^{\infty} \frac{1}{p_i^k} = \frac{1}{1 - 1/p_i}.$$

Multiplying these r equalities, we obtain

$$\prod_{i=1}^{r} \left(\sum_{k=0}^{\infty} \frac{1}{p_i^k} \right) = \prod_{i=1}^{r} \frac{1}{1 - 1/p_i}.$$

The right-hand side is a rational number. If $p_1, \ldots, p_r$ are assumed to be all the primes, by the fundamental theorem of unique factorization of integers as products of primes, the left-hand side is the sum (in some order) of all fractions $1/n$ for $n = 1, 2, 3 \ldots$. This sum of positive numbers is independent of the order and is infinite (as we know the harmonic series $\sum_{n=1}^{\infty} 1/n$ is divergent). This is a contradiction. ∎

For each $x > 0$ let $\pi(x) = \#\{p \text{ prime} \mid p \leq x\}$.

The fundamental *Prime Number Theorem* was observed experimentally by Gauss and later proved independently by de la Vallée Poussin and by Hadamard in 1896. There are elementary proofs of the prime number theorem by Selberg and also by Erdős. See, for example, Trost [28, Chapter VII], Gelfond and Linnik [5, Chapter 3], and for further references Hardy and Wright [6, Chapter XXII], and Ribenboim [25, pp. 429–430].

The theorem states

$$\pi(x) \sim \frac{x}{\log x}, \qquad \text{that is,} \qquad \lim_{x \to \infty} \frac{\pi(x)}{x/\log x} = 1.$$

The proof of the theorem may be found in most books on analytic number theory; see, for example, [1, Chapter 13], [2, Chapter 2].

Euler considered the *zeta-series* $\sum_{n=1}^{\infty} 1/n^s$ (s a positive real number) which is convergent for $s > 1$ and divergent for $0 < s \leq 1$. Euler related the series with an infinite product involving all the prime numbers. Riemann considered a far-reaching generalization, by letting s be a complex number and he showed that for every $\delta > 0$ the series is uniformly convergent for the half-plane $\{s \in \mathbb{C} \mid \mathfrak{Re}(s) \geq 1 + \delta\}$. Thus it defines a function which is holomorphic when $\mathfrak{Re}(s) > 1$. This function may be analytically extended to the whole complex plane to a meromorphic function, which is called *Riemann's zeta-function*; its value at s is denoted by $\zeta(s)$. The zeta-function satisfies a functional equation, and it has a unique pole of order 1 at $s = 1$ with residue 1. Besides zeros at negative integers, the exact location of the other zeros—which is the object of the so-called *Riemann's hypothesis*—is of the utmost importance in the theory of the distribution of prime numbers.

These matters are beyond our aims and are treated in the books on Analytic Number Theory.

Here we shall content ourselves to consider the series $\sum_{n=1}^{\infty} 1/n^s$, and similar series, when s is real and positive. Among the results of the rich theory of these series, we indicate only the properties which will be used in the sequel.

Besides the zeta-series we shall also consider later the L-series associated to characters. They are particular examples of Dirichlet's series, which we define now.

A series

$$\sum_{n=1}^{\infty} \frac{a_n}{n^s},$$

where $s > 0$ and each a_n is a complex number, is called a *Dirichlet series*. If $a_n = 1$ for each $n \geq 1$, we obtain the *zeta-series* $\sum_{n=1}^{\infty} 1/n^s$.

For a given Dirichlet series, it is important to determine the domain of convergence, as well as zeros, poles (if any), their order, residue, etc.

The first result concerns the domain of convergence.

B.   *Let $S(m) = a_1 + \cdots + a_m$ for every $m \geq 1$. If there exists $s_0 > 0$ and a real number $\alpha > 0$ such that*

$$\left| \frac{S(m)}{m^{s_0}} \right| < \alpha$$

for every $m \geq 1$ then for every $\delta > 0$ the series $\sum_{n=1}^{\infty} a_n/n^s$ converges uniformly in the interval $[s_0 + \delta, \infty)$ and defines a continuous function of s in (s_0, ∞).

Proof: Let $s \geq s_0 + \delta$. We have

$$\sum_{n=m}^{m+h} \frac{a_n}{n^s} = \sum_{n=m}^{m+h} \frac{S(n) - S(n-1)}{n^s}$$

$$= \frac{S(m+h)}{(m+h)^s} - \frac{S(m-1)}{m^s} + \sum_{n=m}^{m+h-1} S(n) \left[\frac{1}{n^s} - \frac{1}{(n+1)^s} \right].$$

Taking absolute values:

$$\left| \sum_{n=m}^{m+h} \frac{a_n}{n^s} \right| \leq \left| \frac{S(m+h)}{(m+h)^s} \right| + \left| \frac{S(m-1)}{m^s} \right| + \sum_{n=m}^{m+h-1} |S(n)| \left[\frac{1}{n^s} - \frac{1}{(n+1)^s} \right]$$

$$\leq \frac{(m+h)^{s_0}\alpha}{(m+h)^s} + \frac{(m-1)^{s_0}\alpha}{m^s} + \alpha s \sum_{n=m}^{m+h-1} n^{s_0} \int_n^{n+1} \frac{dx}{x^{s+1}}$$

$$\leq \frac{\alpha}{m^{s-s_0}} + \frac{\alpha}{m^{s-s_0}} + \alpha s \int_m^{\infty} \frac{dx}{x^{s-s_0+1}}.$$

Now

$$s \int_m^{\infty} \frac{dx}{x^{s-s_0+1}} = \frac{1}{s-s_0} \cdot \frac{s}{m^{s-s_0}} \leq \frac{s}{s-s_0} \cdot \frac{1}{m^s}.$$

We note that the function f given by

$$f(s) = \frac{s}{s - s_0} \qquad \text{for} \quad s \geq s_0 + \delta > s_0,$$

is a decreasing function, since $df/ds = -s_0/(s - s_0)^2 < 0$; hence $f(s) \leq f(s_0 + \delta) = (s_0 + \delta)/\delta$.
    Thus

$$\left| \sum_{n=m}^{m+h} \frac{a_n}{n^s} \right| \leq \frac{2\alpha}{m^{\delta}} + \frac{\alpha}{m^{\delta}} \cdot \frac{s_0 + \delta}{\delta}.$$

Since the right-hand side is independent of s and tends to 0 when m tends to infinity, the given series is uniformly convergent on $[s_0 + \delta, \infty)$. By a general theorem of Analysis, $\sum_{n=1}^{\infty} a_n/n^s$ defines a continuous function of s on (s_0, ∞).   ∎

We apply this result to the zeta-series.

C.   *For every $\delta > 0$ the zeta-series converges uniformly on the interval $[1 + \delta, \infty)$ and defines a continuous function on the interval $(1, \infty)$:*

$$\zeta(s) = \sum_{n=1}^{\infty} \frac{1}{n^s} \qquad \text{for} \quad 1 < s. \qquad (22.1)$$

This function is the Riemann zeta-function (we reiterate that we restricted s to be a real number).

Proof: This is a simple corollary of (**B**), noting that $S(m) = m$, so we may take $s_0 = 1$. ∎

Clearly if $0 < s < 1$, the series $\sum_{n=1}^{\infty} 1/n^s$ is divergent, since

$$\sum_{n=1}^{\infty} \frac{1}{n^s} > \sum_{n=1}^{\infty} \frac{1}{n} = \infty.$$

D.   *The difference $\zeta(s) - 1/(s - 1)$ remains bounded when s tends to 1, from the right. This is written:*

$$\zeta(s) \approx \frac{1}{s - 1} \qquad \text{for} \quad s \to 1 + 0. \qquad (22.2)$$

In particular,

$$\lim_{s \to 1+0} (s - 1)\zeta(s) = 1. \qquad (22.3)$$

Proof: We have the inequalities (where $s > 1$):

$$\int_n^{n+1} \frac{du}{u^s} \leq \frac{1}{n^s} \leq \int_{n-1}^{n} \frac{du}{u^s}$$

for every $n \geq 2$; hence

$$\int_1^{\infty} \frac{du}{u^s} < \sum_{n=1}^{\infty} \frac{1}{n^s} < 1 + \int_1^{\infty} \frac{du}{u^s}.$$

From

$$\zeta(s) = \sum_{n=1}^{\infty} \frac{1}{n^s}, \qquad \int_1^{\infty} \frac{du}{u^s} = \frac{1}{s - 1},$$

we obtain $\zeta(s) - 1/(s - 1) < 1$ for $s > 1$. Hence $\lim_{s \to 1+0}(s - 1)\zeta(s) = 1$. ∎

Another basic result about Dirichlet series is the following:

E.   *Assume that*

$$\lim_{m \to \infty} \frac{S(m)}{m} = c, \qquad \text{where} \quad S(m) = a_1 + \cdots + a_m.$$

Then the Dirichlet series is convergent for $s > 1$ and moreover

$$\lim_{s \to 1+0} (s-1) \sum_{n=1}^{\infty} \frac{a_n}{n^s} = c. \tag{22.4}$$

Proof: The first assertion is just (**B**), with $s_0 = 1$.

Since

$$\lim_{m \to \infty} \frac{S(m)}{m} = c,$$

then we may write $S(m) = cm + \nu(m)m$ where $\lim_{m \to \infty} \nu(m) = 0$. In view of (**D**) it is enough to show that

$$\lim_{s \to 1+0} (s-1) \sum_{n=1}^{\infty} \frac{a_n}{n^s} = c \lim_{s \to 1+0} (s-1) \sum_{n=1}^{\infty} \frac{1}{n^s} = c.$$

Thus, we evaluate

$$\sum_{n=1}^{\infty} \frac{a_n}{n^s} - c \sum_{n=1}^{\infty} \frac{1}{n^s}$$

$$= \sum_{n=1}^{\infty} \frac{S(n) - S(n-1)}{n^s} - c \sum_{n=1}^{\infty} \frac{n - (n-1)}{n^s}$$

$$= \sum_{n=1}^{\infty} S(n) \left[\frac{1}{n^s} - \frac{1}{(n+1)^s} \right] - \sum_{n=1}^{\infty} cn \left[\frac{1}{n^s} - \frac{1}{(n+1)^s} \right]$$

$$= \sum_{n=1}^{\infty} [S(n) - cn] \left[\frac{1}{n^s} - \frac{1}{(n+1)^s} \right]$$

$$= \sum_{n=1}^{\infty} \nu(n)n \left[\frac{1}{n^s} - \frac{1}{(n+1)^s} \right]$$

$$= s \sum_{n=1}^{\infty} \nu(n)n \int_n^{n+1} \frac{dx}{x^{s+1}} \leq s \sum_{n=1}^{\infty} \nu(n) \int_n^{n+1} \frac{dx}{x^s}.$$

Taking absolute values:

$$\left| \sum_{n=1}^{\infty} \frac{a_n}{n^s} - c \sum_{n=1}^{\infty} \frac{1}{n^s} \right| \leq s \sum_{n=1}^{\infty} |\nu(n)| \int_n^{n+1} \frac{dx}{x^s}.$$

Given $\delta > 0$ and s such that $1 < s < 1 + \delta$ we have

$$\left| (s-1) \sum_{n=1}^{\infty} \frac{a_n}{n^s} - (s-1)c \sum_{n=1}^{\infty} \frac{1}{n^s} \right| \leq s(s-1) \sum_{n=1}^{\infty} |\nu(n)| \int_n^{n+1} \frac{dx}{x^s}.$$

Since $\lim_{n \to \infty} \nu(n) = 0$ there exists $\beta > 0$ such that $|\nu(n)| < \beta$ for every integer $n \geq 1$. For the given $\delta > 0$ let $N > 0$ be an integer such that if $n \geq N$ then $|\nu(n)| < \delta$.

Then

$$s(s-1) \sum_{n=1}^{\infty} |\nu(n)| \int_n^{n+1} \frac{dx}{x^s}$$

$$\leq s(s-1)\beta \sum_{n=1}^{N-1} \int_n^{n+1} \frac{dx}{x} + s(s-1)\delta \sum_{n=N}^{\infty} \int_n^{n+1} \frac{dx}{x^s}$$

$$= s(s-1)\beta \log N + \frac{s\delta}{N^{s-1}} < (1+\delta)\delta\beta \log N + \frac{(1+\delta)\delta}{N^{\delta}}.$$

since the function $f(s) = s/N^{s-1}$ is increasing. Thus

$$\lim_{s \to 1+0} (s-1) \sum_{n=1}^{\infty} \frac{a_n}{n^s} = \lim_{s \to 1+0} (s-1)c \sum_{n=1}^{\infty} \frac{1}{n^s} = c,$$

as we wanted to prove. ∎

Now we give Euler's product representation for certain series.

Let $\mathcal{P}$ be a set of prime numbers, let $\mathcal{N}$ be the set of integers which are products of the primes $p \in \mathcal{P}$. To exclude the trivial case, we assume that $\mathcal{P}$ is infinite. Let $f : \mathcal{N} \to \mathbb{C}$ be a function such that $f(1) = 1$ and $f(nn') = f(n)f(n')$ for all $n, n' \in \mathcal{N}$.

Lemma 1.   *Assume that the series $\sum_{n \in \mathcal{N}} f(n)$ is absolutely convergent. Then the product $\prod_{p \in \mathcal{P}} 1/(1 - f(p))$ is absolutely convergent and*

$$\sum_{n \in \mathcal{N}} f(n) = \prod_{p \in \mathcal{P}} \frac{1}{1 - f(p)}. \tag{22.5}$$

Proof: From

$$\sum_{p \in \mathcal{P}} |f(p)| \leq \sum_{n \in \mathcal{N}} |f(n)|,$$

the series $\sum_{p \in \mathcal{P}} f(p)$ is absolutely convergent. Hence the product

$$\prod_{p \in \mathcal{P}} (1 - f(p))$$

is absolutely convergent with nonzero limit; so

$$\prod_{p \in \mathcal{P}} \frac{1}{1 - f(p)}$$

is absolutely convergent.

For $m \geq 1$, let $\mathcal{P}(m) = \{p \in \mathcal{P} \mid p \leq m\}$, let $\mathcal{N}_m$ be the set of natural numbers which are products of primes in $\mathcal{P}_m$. For each $p \in \mathcal{P}$:

$$\frac{1}{1 - f(p)} = 1 + f(p) + f(p^2) + \cdots$$

because $f(p^k) = f(p)^k$ for $k \geq 1$. Then

$$\prod_{p \in \mathcal{P}_m} \frac{1}{1 - f(p)} = \prod_{p \in \mathcal{P}_m} (1 + f(p) + f(p^2) + \cdots) = \sum_{n \in \mathcal{N}_m} f(n),$$

as follows from the unique prime factorization of integers and the multiplicative property of f. Then, by the absolute convergence

$$\prod_{p \in \mathcal{P}} \frac{1}{1 - f(p)} = \lim_{m \to \infty} \prod_{p \in \mathcal{P}_m} \frac{1}{1 - f(p)}$$

$$= \lim_{m \to \infty} \sum_{n \in \mathcal{N}_m} f(n) = \sum_{n \in \mathcal{N}} f(n). \qquad \blacksquare$$

We apply this lemma with $\mathcal{P}$, $\mathcal{N}$ as above.

F.   *For $s > 1$, the infinite product*

$$\prod_{p \in \mathcal{P}} \frac{1}{1 - 1/p^s}$$

is absolutely convergent and

$$\sum_{n \in \mathcal{N}} \frac{1}{n^s} = \prod_{p \in \mathcal{P}} \frac{1}{1 - 1/p^s}. \qquad (22.6)$$

Proof: In the above lemma, let $f(n) = 1/n^s$ for every $n \in \mathcal{N}$. Since $\sum_{n=1}^{\infty} 1/n^s$ is absolutely convergent, then so is

$$\sum_{p \in \mathcal{P}} \frac{1}{p^s}$$

for $s > 1$. The result follows from the lemma. $\qquad \blacksquare$

The most important case of the above result is when $\mathcal{P}$ is the set of all prime numbers. In this situation we obtain *Euler's product representation* of the zeta-function

$$\zeta(s) = \sum_{n=1}^{\infty} \frac{1}{n^s} = \prod_{p} \frac{1}{1 - 1/p^s} \qquad \text{for} \quad s > 1. \qquad (22.7)$$

22.2 *L*-Series

Associated with the modular characters, we define other important Dirichlet series. Let χ be a character modulo $m > 1$. The Dirichlet series $\sum_{n=1}^{\infty} \chi(n)/n^s$ is called the *L-series of χ*.

We recall that $\chi(n) = 0$ if and only if m, n are not relatively prime; moreover, if $\chi(n) \neq 0$ then $|\chi(n)| = 1$.

Combining the previous results, we have:

G.   *Let χ be a character modulo $m > 1$. The L-series associated with χ converges absolutely for every $s > 1$. For every $\delta > 0$, the L-series of χ converges uniformly on the interval $[1+\delta, \infty)$. Hence it defines a continuous function $L(s|\chi)$ of s on $(1, \infty)$:*

$$L(s|\chi) = \sum_{n=1}^{\infty} \frac{\chi(n)}{n^s} \qquad for \quad 1 < s. \qquad (22.8)$$

Moreover, $L(s|\chi)$ admits the multiplicative representation

$$L(s|\chi) = \prod_{p} \frac{1}{1 - \chi(p)/p^s} \qquad for \quad 1 < s. \qquad (22.9)$$

In particular, for the trivial character χ_0 modulo m:

$$L(s|\chi_0) = \prod_{p|m} \left(1 - \frac{1}{p^s}\right) \zeta(s) \qquad for \quad 1 < s, \qquad (22.10)$$

and the series

$$\sum_{n=1}^{\infty} \frac{\chi_0(n)}{n^s}$$

is divergent when $0 < s \leq 1$.

Proof: Since $|\chi(n)|$ is equal to 0 or 1, by (**D**), for every $s > 1$ the series

$$\sum_{n=1}^{\infty} \frac{\chi(n)}{n^s}$$

is absolutely convergent and for every $\delta > 0$ it converges uniformly on $[1 + \delta, \infty)$, so it defines the continuous function $L(s|\chi)$ on the interval $(1, \infty)$.
    Since

$$\sum_{n=1}^{\infty} \frac{\chi(n)}{n^s}$$

is absolutely convergent (for $s > 1$) we may apply Lemma 1 with $f(n) = \chi(n)/n^s$, which is a complex-valued multiplicative function. We deduce that the infinite product

$$\prod_{p \in \mathcal{P}} \frac{1}{1 - \chi(p)/p^s}$$

(where $\mathcal{P}$ is the set of all prime numbers) is absolutely convergent and (22.9) holds.

If $\chi = \chi_0$ (the trivial character modulo m), then $\chi_0(p) = 0$ exactly when p divides m. Then

$$L(s\,|\,\chi_0) = \prod_p \frac{1}{1 - \chi_0(p)/p^s} = \prod_p \frac{1}{1 - 1/p^s}$$

$$\times \prod_{p|m} \left(1 - \frac{1}{p^s}\right) = \zeta(s) \prod_{p|m} \left(1 - \frac{1}{p^s}\right).$$

Finally, we note that if $0 < s \le 1$ the series

$$\sum_{n=1}^{\infty} \frac{\chi_0(n)}{n^s}$$

is divergent. Otherwise it would be convergent, hence absolutely convergent, because $\chi_0(n) = 1$ or 0. By Lemma 1, (22.9) holds with $0 < s \le 1$; hence by (22.10), $\zeta(s)$ would be convergent for $0 < s \le 1$, which is false. ∎

Actually, if $\chi \ne \chi_0$, the domain of convergence of $L(s\,|\,\chi)$ is $(0, \infty)$. This will follow from a general convergence test due to Abel:

H.   *Let $(a_n)_n$ be a sequence of complex numbers for which there exists $\varepsilon > 0$ such that*

$$\left| \sum_{n=1}^{\ell} a_n \right| < \varepsilon \qquad \text{for every} \quad \ell \ge 1.$$

Let $(f_n(s))_n$ be a monotonically decreasing sequence of functions of s, defined on an interval I of $\mathbb{R}$, with positive values, and converging uniformly to 0 on I. Then the series $\sum_{n=1}^{\infty} a_n f_n(s)$ converges uniformly on I.

Proof: Let $\alpha_\ell = \sum_{n=1}^{\ell} a_n$ for $\ell \ge 1$. Then $a_n = \alpha_n - \alpha_{n-1}$ and if $k < \ell$ we have

$$\left| \sum_{n=k}^{\ell} a_n f_n(s) \right| = \left| \sum_{n=k}^{\ell} (\alpha_n - \alpha_{n-1}) f_n(s) \right|$$

$$= \left| \sum_{n=k}^{\ell-1} \alpha_n (f_n(s) - f_{n+1}(s)) + \alpha_\ell f_\ell(s) - \alpha_{k-1} f_k(s) \right|$$

$$\le \varepsilon [f_k(s) - f_\ell(s) + f_\ell(s) + f_k(s)] = 2\varepsilon f_k(s).$$

Given $\delta > 0$, by hypothesis there exists j_0 such that if $j \ge j_0$ then $2\varepsilon f_j(s) \le \delta$ for every $s \in I$. Thus the series $\sum_{n=1}^{\infty} a_n f_n(s)$ converges uniformly on I. ∎

We apply this result to obtain:

I.   *Let χ be a character modulo m, $\chi \neq \chi_0$. Then for every $\delta > 0$:*

$$\sum_{n=1}^{\infty} \frac{\chi(n)}{n^s}$$

converges uniformly on $[\delta, \infty)$. Hence it defines a continuous function $L(s|\chi)$ on $(0, \infty)$.

Proof: Taking $a_n = \chi(n)$ we have $|\sum_{n=1}^{\ell} \chi(n)| \leq \varphi(m)$. Indeed, since χ is a character modulo m we have $\chi(j) = \chi(k)$ where $j \equiv k \pmod{m}$. Since $\chi \neq \chi_0$ we have $\sum_{n=1}^{m} \chi(n) = 0$ (by Chapter 21, (**J**)). If $\ell = qm + r$, $0 \leq r < m$, then

$$\left| \sum_{n=1}^{\ell} \chi(n) \right| = \left| \sum_{n=1}^{r} \chi(n) \right| \leq \varphi(m).$$

Next, the sequence of functions $f_n(s) = 1/n^s$ is monotonically decreasing and uniformly convergent to 0 on $[\delta, \infty)$. By (**H**) we conclude that $\sum_{n=1}^{\infty} \chi(n)/n^s$ converges uniformly on $[\delta, \infty)$ where $\delta > 0$. So it defines a continuous function $L(s|\chi)$ on $(0, \infty)$.   ∎

We shall use later the following result.
In the sequel it is agreed that if $|x| < 1$ then

$$\log \frac{1}{1-x} = \sum_{n=1}^{\infty} \frac{x^n}{n}.$$

Also, if an infinite product of functions $\prod f_n(s)$ is absolutely convergent (for $s > 1$) then

$$\log \prod_n f_n(s) = \sum_n \log f_n(s).$$

J.   *Let χ be a character modulo m. Then*

$$\log L(s|\chi) \approx \sum_p \frac{\chi(p)}{p^s} \qquad \text{for} \quad s \to 1 + 0. \tag{22.11}$$

Proof: By (**G**), we have

$$L(s|\chi) = \prod_p \frac{1}{1 - \chi(p)/p^s} \qquad \text{for} \quad s > 1. \tag{22.12}$$

Taking logarithms, and due to the absolute convergence, we have (for $s > 1$):

$$\log L(s\,|\,\chi) = \log \prod_p \frac{1}{1 - \chi(p)/p^s} = \sum_p \log \frac{1}{1 - \chi(p)/p^s}$$

$$= \sum_p \sum_{\nu=1}^{\infty} \frac{1}{\nu} \cdot \frac{\chi(p^\nu)}{p^{\nu s}} = \sum_p \frac{\chi(p)}{p^s} + \sum_p \sum_{\nu=2}^{\infty} \frac{1}{\nu} \cdot \frac{\chi(p^\nu)}{p^{\nu s}}.$$

We have the following bounds when $s \to 1 + 0$:

$$\left| \sum_p \sum_{\nu=2}^{\infty} \frac{1}{\nu} \cdot \frac{\chi(p^\nu)}{p^{\nu s}} \right| \leq \frac{1}{2} \sum_p \left(\sum_{\nu=2}^{\infty} \frac{1}{p^{\nu s}} \right)$$

$$\leq \frac{1}{2} \sum_p \frac{1/p^{2s}}{1 - 1/p^s} < \sum_p \frac{1}{p^{2s}} \leq \sum_p \frac{1}{p^2} < \zeta(2).$$

Thus

$$\left| \log L(s\,|\,\chi) - \sum_p \frac{\chi(p)}{p^s} \right| < \zeta(2). \qquad \blacksquare$$

To conclude this short section on *L*-series, we give an expression for *L*-series involving Gaussian sums.

Let χ be a character modulo m. Let $\zeta = \cos(2\pi/m) + i \sin(2\pi/m)$.

K.   *If $\chi \neq \chi_0$ then*

$$L(s\,|\,\chi) = \frac{1}{m} \sum_{k=0}^{m-1} \tau_k(\chi) \sum_{n=1}^{\infty} \frac{\zeta^{-nk}}{n^s} \qquad \text{for} \quad s > 0. \qquad (22.13)$$

Proof: By definition, since χ is a character modulo m, we have

$$L(s\,|\,\chi) = \sum_{n=1}^{\infty} \frac{\chi(n)}{n^s} = \sum_{\substack{a \in P(m) \\ 1 \leq a < m}} \chi(a) \left[\sum_{n \equiv a \ (\mathrm{mod}\ m)} \frac{1}{n^s} \right] \qquad \text{for} \quad s > 1.$$

$$(22.14)$$

We may also write

$$\sum_{n \equiv a \ (\mathrm{mod}\ m)} \frac{1}{n^s} = \sum_{n=1}^{\infty} \frac{c_n}{n^s}, \qquad (22.15)$$

where $c_n = 1$ when $n \equiv a$ (mod m), $c_n = 0$ otherwise (the coefficients c_n depend on the class of a modulo m). It is possible to express c_n in terms of the mth roots of unity.

If $\zeta = \cos(2\pi/m) + i\sin(2\pi/m)$ is a fixed primitive mth root of unity, we know that

$$\sum_{k=0}^{m-1} \zeta^{rk} = \begin{cases} m, & \text{when } m \text{ divides } r, \\ 0, & \text{when } m \text{ does not divide } r. \end{cases}$$

Rewriting this sum, we have

$$\sum_{k=0}^{m-1} \zeta^{(a-n)k} = \begin{cases} m, & \text{when } n \equiv a \pmod{m}, \\ 0, & \text{when } n \not\equiv a \pmod{m}. \end{cases}$$

Therefore

$$c_n = \frac{1}{m} \sum_{k=0}^{m-1} \zeta^{(a-n)k}. \tag{22.16}$$

From (22.14), (22.15), and (22.16) we deduce

$$L(s|\chi) = \frac{1}{m} \sum_{\substack{a \in P(m) \\ 1 \le a < m}} \chi(a) \sum_{n=1}^{\infty} \sum_{k=0}^{m-1} \frac{\zeta^{(a-n)k}}{n^s}$$

$$= \frac{1}{m} \sum_{k=0}^{m-1} \left\{ \sum_{\substack{a \in P(m) \\ 1 \le a < m}} \chi(a)\zeta^{ak} \right\} \sum_{n=1}^{\infty} \frac{\zeta^{-nk}}{n^s} \quad \text{for } s > 1. \tag{22.17}$$

Using the definition of the Gaussian sums given in Chapter 21, (21.1), we have

$$L(s|\chi) = \frac{1}{m} \sum_{k=0}^{m-1} \tau_k(\chi) \sum_{n=1}^{\infty} \frac{\zeta^{-nk}}{n^s} \quad \text{for } s > 1. \tag{22.18}$$

By Chapter 21, (21.1), we have $\tau_0(\chi) = 0$ because $\chi \neq \chi_0$. Taking into account the definition of Gaussian sums, the statement follows at once. ∎

In particular, since $\chi \neq \chi_0$, the L-series $L(s|\chi)$ is defined and continuous on the interval $(0, \infty)$. Hence

$$L(1|\chi) = \frac{1}{m} \sum_{k=1}^{m-1} \tau_k(\chi) \log \frac{1}{1 - \zeta^{-k}}. \tag{22.19}$$

E X E R C I S E S

1. Show that the series $\sum_p 1/p$ (sum for all primes) is divergent.

2. Show that $\zeta(s) \neq 0$ when $s > 1$.

3. Let

$$\sum_{n=1}^{\infty} \frac{a_n}{n^s}$$

(s real, $s > 0$, and each a_n a complex number) be a Dirichlet series. Let s_0 be the abscissa of convergence (so the series converges for $s > s_0$ and diverges for $s < s_0$). Show that

$$s_0 = \limsup_{k} \frac{\log \left| \sum_{n=1}^{k} a_n \right|}{\log k}.$$

4. Let s_0 (respectively, s_0^*) be the abscissa of convergence of the Dirichlet series

$$\sum_{n=1}^{\infty} \frac{a_n}{n^s}$$

(as in the preceding exercise) (respectively, of

$$\sum_{n=1}^{\infty} \frac{|a_n|}{n^s} \text{).}$$

Show that $s_0^* \leq s_0 + 1$ (this holds also when s_0 or s_0^* is infinity).

5. Determine the abscissa of convergence of the following Dirichlet series:

(a)
$$\sum_{n=1}^{\infty} \frac{(-1)^{n-1}}{n^s},$$

(b)
$$\sum_{n=1}^{\infty} \frac{a^n}{n^s} \qquad \text{where} \quad |a| < 1,$$

(c)
$$\sum_{n=1}^{\infty} \frac{a^n}{n^s} \qquad \text{where} \quad |a| > 1,$$

(d)
$$\sum_{n=1}^{\infty} \frac{1}{(\log n)^2 \, n^s}.$$

6. Let $(F_n)_{n \geq 0}$ be the sequence of Fibonacci numbers (see Chapter 1, Exercise 17). Let $\mathcal{F}$ be the set of all natural numbers which are products of Fibonacci numbers. Define $a_n = 1$ when $n \in \mathcal{F}$ and $a_n = 0$ when $n \notin \mathcal{F}$.
Determine the abscissa of convergence of the Dirichlet series

$$\sum_{n=1}^{\infty} \frac{a_n}{n^s}.$$

Hint: Use the expression $F_n = (\alpha^n - \beta^n)/(\alpha - \beta)$ where α, β are the roots of $X^2 - X - 1$.

7. Let $f: \mathbb{N} \to \mathbb{C}$ be a function such that the series

$$\sum_{n=1}^{\infty} \frac{f(n)}{n^s}$$

converges absolutely for $s > s_0$ (where $s_0 > 0$). Show that

$$\sum_{n=1}^{\infty} \frac{\mu(n)f(n)}{n^s}$$

also converges absolutely for $s > s_0$ and that

$$\left[\sum_{n=1}^{\infty} \frac{f(n)}{n^s}\right] \cdot \left[\sum_{n=1}^{\infty} \frac{\mu(n)f(n)}{n^s}\right] = 1.$$

8. Show that if $s > 1$ then

(a) $\zeta(s) = s \int_1^{\infty} \frac{[x]}{x^{s+1}} dx = \frac{1}{s-1} - s \int_1^{\infty} \frac{x - [x]}{x^{s+1}} dx.$

(b) $\sum_p \frac{1}{p^s} = s \int_1^{\infty} \frac{\pi(x)}{x^{s+1}} dx.$

9. Show that if χ_0 is the trivial character modulo m then

$$\lim_{s \to 1+0} (s - 1) \cdot L(s|\chi_0) = \frac{\varphi(m)}{m}.$$

Hint: Use the preceding exercise.

10. Assume that $(a_n)_{n \geq 1}$ is a sequence of complex numbers and that $\alpha > 0$ is a real number such that

$$\left| \sum_{i=1}^{n} a_i \right| < \alpha \qquad \text{for every} \quad n \geq 1.$$

Let $(f_n(s))_{n \geq 1}$ be a sequence of functions of s, with real values, such that there is an interval $[s_0, s_1]$ on which the sequence $(f_n(s))_n$ converges uniformly and monotonically to 0. Then the series $\sum_{n=1}^{\infty} a_n f_n(s)$ converges uniformly on $[s_0, s_1]$.

11. Let χ be a character modulo m. Show that

$$L(s|\chi) = \sum_{n=1}^{\infty} \frac{\chi(n) \log n}{n^s}$$

converges uniformly in every interval $[1 + \delta, \infty)$ where $\delta > 0$ and converges absolutely for $s > 1$. $L(s|\chi)$ has a continuous derivative for $s > 1$, which is equal to

$$L'(s|\chi) = -\sum_{n=1}^{\infty} \frac{\chi(n) \log n}{n^s} \qquad \text{for} \quad s > 1.$$

Moreover, if $\chi \neq \chi_0$ then the above result holds for $[\delta, \infty)$, $\delta > 0$, and $s > 0$.

Hint: For $\chi \neq \chi_0$ use the preceding exercise, considering the series obtained from $L(s|\chi)$ by termwise differentiation.

12. Let χ be a character modulo m. Let $L'(s|\chi)$ denote the derivative of $L(s|\chi)$ for $s > 1$. Show that

$$\frac{L'(s|\chi)}{L(s|\chi)} = -\sum_{n=1}^{\infty} \frac{\chi(n)\Lambda(n)}{n^s} \qquad \text{for} \quad s > 1,$$

where Λ denotes the von Mangoldt function (see Chapter 3, Exercise 51).

Hint: Use Exercise 11 and Chapter 3, Exercise 51.

13. Let χ be a character modulo m. Show that

$$\frac{L'(s|\chi)}{L(s|\chi)} + \sum_{p} \frac{\chi(p) \log p}{p^s}$$

is bounded for $s \geq 1$.

Hint: Use Exercise 12.

14. Show that

$$\sum_{\substack{m=1 \\ \gcd(m,n)=1}}^{\infty} \sum_{n=1}^{\infty} \frac{1}{m^2 n^2} = \frac{[\zeta(2)]^2}{\zeta(4)}.$$

15. Compute the sum

$$\sum_{\substack{m=1 \\ \gcd(m,n)=1}}^{\infty} \sum_{n=1}^{\infty} \frac{1}{m^s n^t},$$

where $s > 1$, $t > 1$.

16. If χ is not the trivial character modulo m, show that

$$\left| \sum_{n=1}^{\infty} \frac{\chi(n) \log n}{n^s} \right| < \varphi(m) \qquad \text{for} \quad s \geq 1.$$

Hint: Use Exercise 5 of Chapter 21 and estimate

$$\left| \sum_{n=3}^{h} \frac{\chi(n) \, \log n}{n^s} \right|.$$

17. Let χ_0 be the trivial character modulo m. Show that

$$\lim_{s \to 1+0} \frac{L'(s|\chi_0)}{L(s|\chi_0)} = -\infty.$$

Hint: Use Exercises 11 and 1.

18. If χ is a character modulo m, $\chi \neq \chi_0$, show that

$$\frac{L'(s|\chi)}{L(s|\chi)} \qquad \text{is bounded for} \quad s \geq 1.$$

Hint: Use Exercises 11 and 2 to bound $L'(s|\chi)$; to bound $1/L(s|\chi)$ use Exercise 7 and the fact that $L(1|\chi) \neq 0$ for $\chi \neq \chi_0$.

19. Let $m \geq 1$ and $a \geq 1$ be integers, and $\gcd(a, m) = 1$; let Λ be the von Mangoldt arithmetic function (see Chapter 3, Exercise 51). Show that if $s > 1$ then

$$-\frac{1}{\varphi(m)} \sum_{\chi} \frac{1}{\chi(a)} \cdot \frac{L'(s|\chi)}{L(s|\chi)} = \sum_{n \equiv a \ (\mathrm{mod}\ m)} \frac{\Lambda(n)}{n^s}.$$

Hint: Use Exercise 13 and an orthogonality relation for characters.

20. Let $(a_n)_{n \geq 1}$ be a sequence of complex numbers, for each $x \geq 1$ let $S(x) = \sum_{n \leq x} a_n$. Assume that there exists $\delta > 0$ such that

$$\left| \frac{S(x)}{x^\delta} \right|$$

remains bounded for all x sufficiently large.
    Show that if $s > \delta$ then

$$\sum_{n=1}^{\infty} \frac{a_n}{n^s} = s \int_1^{\infty} \frac{S(t)}{t^{s+1}} \, dt.$$

21. Let $(a_n)_{n \geq 1}$ be a sequence of complex numbers, and let $f(x)$ be a differentiable complex-valued function defined for $x \geq 1$. Let $S(x) = \sum_{n \leq x} a_n$.
    Show that

$$\sum_{n \leq x} a_n f(n) = S(x) f(x) - \int_1^{x} S(t) f'(t) \, dt.$$

22. In this exercise consider a Dirichlet series with complex argument namely

$$\sum_{n=1}^{\infty} \frac{a_n}{n^s},$$

where a_n and $s = \sigma + it$ are complex numbers (σ, t real numbers). Extend to these series the results of the text which were proved for s real number.

23. Show that the Riemann zeta-series

$$\sum_{n=1}^{\infty} \frac{1}{n^s} \qquad \text{with} \quad s = \sigma + it, \ \sigma, t \text{ real},$$

is an oscillating series for every point $s = it, \ t \neq 0$.

24. Show that if $n \geq 2$ then

$$\zeta(2)\zeta(2n-2) + \zeta(4)\zeta(2n-4) + \cdots + \zeta(2n-2)\zeta(2) = (n + \tfrac{1}{2})\zeta(n).$$

23

The Dedekind Zeta-Function

In this chapter we introduce the important Dedekind zeta-function of an algebraic number field and obtain an asymptotic expression for the class number.

23.1 Asymptotic Expression for the Class Number

Let K be an algebraic number field. Our purpose is to obtain an asymptotic expression for the class number h of K. It will involve other invariants of K as well as norms of integral ideals.

By Chapter 8, **(G)**, for every $m \geq 1$:

$$\nu(m) = \#\{J \text{ integral ideal of } K \mid N(J) = m\}.$$

is finite.

For every real number $t > 0$, let $\sigma(t)$ denote the number of integral ideals J of K such that $N(J) \leq t$.

Thus

$$\sigma(t) = \sum_{m=1}^{[t]} \nu(m).$$

Let $C_1, C_2, \ldots, C_h$ be the h classes of ideals of K. We recall (Chapter 9, statement equivalent to **(C)**) that every nonzero ideal is equivalent to some nonzero integral ideal $J_i \in C_i$ such that $N(J_i) < \sqrt{|\delta|}$, where δ is the discriminant of K. The ideals of the class C_i are of the form $Ax \cdot J_i$, where $x \in K$, $x \neq 0$, and A is the ring of integers of K.

Since $N(Ax \cdot J_i) = |N_{K|\mathbb{Q}}(x)| \cdot N(J_i)$ then the norms of the ideals of the class C_i are obtained from $N(J_i)$ by multiplying with the absolute values of the norms of the elements $x \in K$, $x \neq 0$.

For every real number $t > 0$ let $\sigma(t; C_i)$ denote the number of integral ideals I of the class C_i such that $N(I) \leq t$. Since $N(J_i) \leq \sqrt{|\delta|}$ for every

$i = 1, \ldots, h$, it is plausible that $\sigma(t; C_i)/t$ has a limit, which is independent of the class C_i. This is indeed true, as we shall prove soon.

Let n be the degree of the algebraic number field K, let r_1 be the number of real fields conjugate to K, and $2r_2$ the number of complex (nonreal) fields conjugate to K (they appear in pairs of complex conjugate fields). So $0 \le r_1$, $0 \le r_2$, $n = r_1 + 2r_2$.

We denote by $K^{(1)}, \ldots, K^{(r_1)}$ the real fields conjugate to K, and by $K^{(r_1+1)}, \ldots, K^{(n)}$ the nonreal fields conjugate to K; these are numbered in such a way that

$$\overline{K^{(r_1+j)}} = K^{(r_1+r_2+j)} \qquad \text{for} \quad j = 1, \ldots, r_2.$$

We also put

$$\ell_1 = \cdots = \ell_{r_1} = 1, \qquad \ell_{r_1+1} = \cdots = \ell_{r_1+r_2} = 2.$$

Let U denote the group of units of A, and W the subgroup of roots of unity, $w = \#(W)$. Let $\{u_1, \ldots, u_r\}$ be a fundamental system of units of infinite order, where $r = r_1 + r_2 - 1$ (see Chapter 10, Theorem 1). Finally, let $R = |\det(\ell_i \log |u_j^{(i)}|)_{i,j}|$ be the regulator of K.

If $a_1, \ldots, a_n$ is any basis of the $\mathbb{Q}$-vector space K, and if $\xi_1, \ldots, \xi_n$ are real numbers, let

$$x^{(j)} = \sum_{k=1}^{n} \xi_k a_k^{(j)} \qquad \text{for} \quad j = 1, \ldots, n.$$

The elements $x^{(j)} \in \mathbb{C}$ belong to the $\mathbb{R}$-vector space generated by $\{a_1^{(j)}, \ldots, a_n^{(j)}\}$. If $\xi_1, \ldots, \xi_n \in \mathbb{Q}$ then $x^{(j)} \in K^{(j)}$ and $x^{(1)}, \ldots, x^{(n)}$ constitute a set of conjugate elements.

Since the regulator R is different from 0, for each n-tuple $(\xi_1, \ldots, \xi_n)$ of real numbers such that $x^{(1)} \ne 0, \ldots, x^{(n)} \ne 0$, there exist uniquely defined real numbers $\alpha_1, \ldots, \alpha_r$ such that

$$\sum_{k=1}^{r} \alpha_k \log |u_k^{(j)}| = \log \left| \frac{x^{(j)}}{|x^{(1)} \cdots x^{(n)}|^{1/n}} \right|. \qquad (23.1)$$

$\alpha_1, \ldots, \alpha_r$ are called the *exponents of* $x^{(1)}, \ldots, x^{(n)}$ with respect to the fundamental system of units $\{u_1, \ldots, u_r\}$. If $\xi_1, \ldots, \xi_r \in \mathbb{Q}$ and $x = x^{(1)} \in K$ we simply say that $\alpha_1, \ldots, \alpha_r$ are exponents of x.

If $x \in K^{\cdot}$ has exponents $\alpha_1, \ldots, \alpha_r$ and $x' \in K^{\cdot}$ has exponents $\alpha_1', \ldots, \alpha_r'$ then the exponents of xx' are obviously $\alpha_1 + \alpha_1', \ldots, \alpha_r + \alpha_r'$.

If v is a unit of K, we may write $v = \zeta u_1^{m_1} u_2^{m_2} \cdots u_r^{m_r}$ where ζ is a root of unity and each m_i is an integer. Since $|N_{K|\mathbb{Q}}(v)| = 1$ then the exponents of v are $m_1, \ldots, m_r$. Moreover, a unit v has exponents all equal to 0 if and only if $v = \zeta$ is a root of unity.

A. *For every class of ideals C, we have*

$$\lim_{t \to \infty} \frac{\sigma(t; C)}{t} = \frac{2^{r_1 + r_2} \pi^{r_2} R}{w \sqrt{|\delta|}}. \tag{23.2}$$

Proof: Let C^{-1} be the inverse of the class C in the group $\mathcal{C}$ of classes of ideals, and let J be an integral ideal in C^{-1} (which clearly exists).

For every integer $t \geq 1$, there is a one-to-one correspondence between the following sets:

$$\mathcal{E}_t = \{ I \in C \mid I \text{ is integral and } N(I) \leq t \}$$

and

$$\mathcal{E}_t' = \{ Ax \mid 0 \neq Ax \subseteq J, \ |N_{K|\mathbb{Q}}(x)| \leq N(J) \cdot t \}.$$

In fact, given $I \in C$, an integral ideal such that $N(I) \leq t$, then $I \cdot J \in C \cdot C^{-1}$, which is the class of principal fractional ideals. But $I \cdot J$ is a nonzero integral ideal, so $I \cdot J = Ax$ where $x \in I \cdot J \subseteq J$ and

$$|N_{K|\mathbb{Q}}(x)| = N(Ax) = N(I) \cdot N(J) \leq t \cdot N(J).$$

The correspondence is one-to-one because I, I' are distinct, $I \cdot J = Ax$, $I' \cdot J = Ax'$ and, therefore, $Ax \neq Ax'$.

Conversely, given $Ax \neq 0$, $Ax \subseteq J$ such that $|N_{K|\mathbb{Q}}(x)| \leq N(J) \cdot t$, let $I = J^{-1} \cdot Ax$; so $I \in (C^{-1})^{-1} = C$, I is integral and $N(I) \leq t$.

Since $\sigma(t; C) = \#(\mathcal{E}_t)$ it is enough to count the number of elements in $\mathcal{E}_t'$. For this purpose, let $\{u_1, \ldots, u_r\}$ be a fundamental system of units of infinite order of A. We associate with $\mathcal{E}_t'$ the set $\mathcal{E}_t'' = \{ x \in J \mid 0 < |N_{K|\mathbb{Q}}(x)| \leq N(J) \cdot t$ and the exponents $\alpha_1, \ldots, \alpha_r$ of x with respect to $\{u_1, \ldots, u_r\}$ satisfy $0 \leq \alpha_k < 1$ for $k = 1, \ldots, r\}$.

We show that $\#(\mathcal{E}_t'') = w \cdot \#(\mathcal{E}_t')$, where w is the number of roots of unity in K. Indeed, if $x, y \in \mathcal{E}_t''$ and $Ax = Ay$ then $x = vy$, where v is a unit. Considering the exponents α_k of x, m_k of v, and β_k of y, we have $\alpha_k = m_k + \beta_k$ with $0 \leq \alpha_k < 1$, $0 \leq \beta_k < 1$, and $m_k \in \mathbb{Z}$. This implies that each m_k is equal to 0 hence v is a root of unity. On the other hand, if $y \in \mathcal{E}_t''$ and v is a root of unity, then $x = vy \in \mathcal{E}_t''$. This shows that $\#(\mathcal{E}_t'') = w \cdot \#(\mathcal{E}_t')$ and we need to evaluate $\#(\mathcal{E}_t'')$.

We shall construct a closed and bounded domain D_t contained in the real n-dimensional space $\mathbb{R}^n$, in such a way that the points of $\mathcal{E}_t''$ are in one-to-one correspondence with the points of D_t, distinct from the origin, and having integral coordinates.

By Chapter 6, (**K**), J is a free Abelian group of rank n. Let $\{a_1, \ldots, a_n\}$ be a basis. For every n-tuple $(\xi_1, \ldots, \xi_n) \in \mathbb{R}^n$, let

$$x^{(j)} = \sum_{k=1}^{n} \xi_k a_k^{(j)} \in \mathbb{C}, \qquad j = 1, \ldots, n.$$

We define E_t to be the set of all n-tuples $(\xi_1, \ldots, \xi_n) \in \mathbb{R}^n$ such that:

(1) $$0 < \left| \prod x^{(j)} \right| \leq N(J) \cdot t; \qquad (23.3)$$

and

(2) the exponents $\alpha_1, \ldots, \alpha_r$ of $x^{(1)}, \ldots, x^{(n)}$ satisfy $0 \leq \alpha_k < 1$ for $k = 1, \ldots, r$.

E_t is a bounded set. Indeed, since $\{a_1, \ldots, a_n\}$ is linearly independent over $\mathbb{Q}$, $\operatorname{discr}_{K|\mathbb{Q}}(a_1, \ldots, a_n) = [\det(a_k^{(j)})]^2 \neq 0$. Then the linear transformation $\theta: (\xi_1, \ldots, \xi_n) \to (x^{(1)}, \ldots, x^{(n)})$ is invertible. From the definition of the exponents and conditions (1) and (2) it follows that

$$|x^{(j)}| = |x^{(1)} \cdots x^{(n)}|^{1/n} \exp \left(\sum_{k=1}^{r} \alpha_k \log |u_k^{(j)}| \right)$$

$$\leq |x^{(1)} \cdots x^{(n)}|^{1/n} \exp(rM) \leq [N(J) \cdot t]^{1/n} \exp(rM),$$

where $M = \max\{\log |u_k^{(j)}| \mid j = 1, \ldots, n; \; k = 1, \ldots, r\}$.

The image of E_t under θ is a bounded set, hence considering θ^{-1} we deduce that E_t is bounded.

In order to obtain a closed set D_t we consider the set E_t' of all points $(\xi_1, \ldots, \xi_n) \in \mathbb{R}^n$ such that for some i, $1 \leq i \leq n$, we have

$$x^{(i)} = \sum_{k=1}^{n} \xi_k a_k^{(i)} = 0$$

and moreover

$$|x^{(j)}| \leq [N(J) \cdot t]^{1/n} \cdot e^{rM} \qquad \text{for} \quad j = 1, \ldots, n.$$

Let $D_t = E_t \cup E_t'$. Then D_t is a bounded and closed subset of $\mathbb{R}^n$.

The elements of $\mathcal{E}_t''$ are in one-to-one correspondence with the points of D_t having integral coordinates distinct from $(0, \ldots, 0)$. In fact, $x \in \mathcal{E}_t''$ implies that $x = \sum_{k=1}^{n} m_k a_k$ with $m_1, \ldots, m_n \in \mathbb{Z}$, not all equal to 0. The n-tuple $(m_1, \ldots, m_n)$ belongs to $E_t \subseteq D_t$. The correspondence $x \mapsto (m_1, \ldots, m_n)$ is one-to-one, because $\{a_1, \ldots, a_n\}$ is a basis of the free Abelian group J. Every point of D_t with integral coordinates and different from the origin may be obtained from some element $x \in \mathcal{E}_t''$ in this way; we need only to note that if $(m_1, \ldots, m_n) \in \mathcal{E}_t'$ then $x = 0$ and $(m_1, \ldots, m_n) = (0, \ldots, 0)$.

Therefore $1 + \#(\mathcal{E}_t'')$ is equal to the number of points with integral coordinates in D_t.

It is our aim to compute $\lim_{t \to \infty} \sigma(t; C)/t$.

We have

$$\lim_{t \to \infty} \frac{w \cdot \sigma(t; C)}{t} = \lim_{t \to \infty} [1 + w \cdot \sigma(t; C)] \frac{1}{t}$$

$$= \lim_{t \to \infty} [1 + \#(\mathcal{E}_t'')] \cdot \frac{1}{t} = \lim_{t \to \infty} \#(D_t) \cdot \frac{1}{t}.$$

We show that $\lim_{t \to \infty} \#(D_t)/t$ is equal to the volume of the closed and bounded set D_1. For every $t > 0$ let $\theta : \mathbb{R}^n \to \mathbb{R}^n$ be the linear transformation defined by $\theta(\xi_1, \ldots, \xi_n) = (\eta_1, \ldots, \eta_n)$ with $\eta_k = (1/\sqrt[n]{t})\xi_k$ for $k = 1, \ldots, n$. Then $\theta(D_t) = D_1$ as one verifies at once. Moreover,

$$\theta(m_1, \ldots, m_n) = \left(\frac{m_1}{\sqrt[n]{t}}, \ldots, \frac{m_1}{\sqrt[n]{t}} \right),$$

each hypercube H of side 1 and center $(m_1, \ldots, m_n)$ has image $\theta(H)$ equal to the hypercube of side $1/\sqrt[n]{t}$ and center $\theta(m_1, \ldots, m_n)$, hence of volume $1/t$. Therefore $1 + \#(\mathcal{E}_t'')$ is the number of hypercubes of the type described with center in the domain D_1 each having volume $1/t$. Thus $(1/t)[1 + \#(\mathcal{E}_t'')]$ is an approximation for the volume of the closed domain D_1. Therefore

$$\text{vol}(D_1) = \lim_{t \to \infty} \frac{1}{t} \#(D_t). \tag{23.4}$$

Our task now is the computation of the volume of D_1. This will be performed by effecting successive changes of variables.

Let $\zeta_1, \ldots, \zeta_n$ be new variables defined by

$$\begin{cases} \sum_{k=1}^{n} \xi_k a_k^{(j)} = \zeta_j & \text{for } j = 1, \ldots, r_1, \\ \sum_{k=1}^{n} \xi_k a_k^{(j)} = \zeta_j + i\zeta_{j+r_2} & \text{for } j = r_1 + 1, \ldots, r_1 + r_2. \end{cases}$$

So

$$\zeta_j = \sum_{k=1}^{n} \xi_k a_k^{(j)} \qquad \text{for } j = 1, \ldots, r_1,$$

$$\zeta_j = \sum_{k=1}^{n} \xi_k \left[\frac{a_k^{(j)} + a_k^{(j+r_2)}}{2} \right] \qquad \text{for } j = r_1 + 1, \ldots, r_1 + r_2,$$

$$\zeta_{j+r_2} = \sum_{k=1}^{n} \xi_k \left[\frac{a_k^{(j)} - a_k^{(j+r_2)}}{2i} \right] \qquad \text{for } j = r_1 + 1, \ldots, r_1 + r_2.$$

The absolute value of the Jacobian of the change of variables is

$$\left| \frac{\partial(\zeta_1, \ldots, \zeta_n)}{\partial(\xi_1, \ldots, \xi_n)} \right| = 2^{-r_2} |\det(a_k^{(j)})| = 2^{-r_2} \cdot N(J) \cdot \sqrt{|\delta|}$$

by Chapter 8, (**E**). Hence

$$\text{vol}(D_1) = \int_{D_1} d\xi_1 \cdots d\xi_n = \int_{D_1'} \left| \frac{\partial(\xi_1, \ldots, \xi_n)}{\partial(\zeta_1, \ldots, \zeta_n)} \right| d\zeta_1 \cdots d\zeta_n$$

$$= \frac{2^{r_2}}{N(J) \cdot \sqrt{|\delta|}} \int_{D_1'} d\zeta_1 \cdots d\zeta_n, \tag{23.5}$$

where D_1' denotes the domain obtained from D_1 by the change of variables.

A new change to polar coordinates

$$\begin{cases} \zeta_j = \rho_j & \text{for } j = 1, \ldots, r_1, \\ \zeta_j = \rho_j \cos \varphi_j & \text{for } j = r_1 + 1, \ldots, r_1 + r_2, \\ \zeta_{r_2+j} = \rho_j \sin \varphi_j & \text{for } j = r_1 + 1, \ldots, r_1 + r_2, \end{cases}$$

with $0 \le \varphi_j < 2\pi$, $0 < \rho_j$, has Jacobian

$$\frac{\partial(\zeta_1, \ldots, \zeta_n)}{\partial(\rho_1, \ldots, \rho_{r_1+r_2}, \varphi_{r_1+1}, \ldots, \varphi_{r_1+r_2})}$$

equal to the determinant of the matrix

$$\begin{pmatrix} I_{r_1} & & & & 0 & & & \\ \hline & \cos \varphi_{r_1+1} & 0 & 0 \cdots -\rho_{r_1+1} \sin \varphi_{r_1+1} & 0 & 0 \\ & 0 & \cos \varphi_{r_1+2} & 0 \cdots & 0 & -\rho_{r_1+2} \sin \varphi_{r_1+2} & 0 \\ 0 & \cdots\cdots\cdots\cdots\cdots\cdots\cdots\cdots\cdots\cdots\cdots\cdots\cdots\cdots & \\ & \sin \varphi_{r_1+1} & 0 & 0 \cdots \rho_{r_1+1} \cos \varphi_{r_1+1} & 0 & 0 \\ & 0 & \sin \varphi_{r_1+2} & 0 \cdots & 0 & \rho_{r_1+2} \cos \varphi_{r_1+2} & 0 \\ & \cdots\cdots\cdots\cdots\cdots\cdots\cdots\cdots\cdots\cdots\cdots\cdots\cdots\cdots & \end{pmatrix},$$

where I_{r_1} is the identity matrix of r_1 rows and r_1 columns. A simple computation shows that the absolute value of the Jacobian is equal to $\rho_{r_1+1}\rho_{r_1+2} \cdots \rho_{r_1+r_2}$.

In the new variables, D_1' becomes the set D_1'' of points $(\rho_1, \ldots, \rho_{r_1+r_2}, \varphi_{r_1+1}, \ldots, \varphi_{r_1+r_2})$ such that $0 < \rho_j$, $0 \le \varphi_j < 2\pi$, for $j = r_1 + 1, \ldots, r_1 + r_2$, and

$$\begin{cases} 0 < \displaystyle\prod_{j=1}^{r_1+r_2} |\rho_j|^{\ell_j} \le N(J), \\ \log |\rho_j| = \dfrac{1}{n} \log \displaystyle\prod_{k=1}^{r_1+r_2} |\rho_k|^{\ell_k} + \sum_{k=1}^{r} \alpha_k \log |u_k^{(j)}|, \end{cases} \tag{23.6}$$

with $0 \le \alpha_k < 1$.

Thus

$$\text{vol}(D_1)$$

$$= \frac{2^{r_2}}{N(J) \cdot \sqrt{|\delta|}} \int_{D_1''} \left| \frac{\partial(\zeta_1, \ldots, \zeta_n)}{\partial(\rho_1, \ldots, \varphi_{r_1+r_2})} \right| d\rho_1 \cdots d\rho_{r_1+r_2} d\varphi_{r_1+1} \cdots d\varphi_{r_1+r_2}$$

$$= \frac{2^{r_2}}{N(J) \cdot \sqrt{|\delta|}} \int_{D_1''} \rho_{r_1+1} \rho_{r_1+2} \cdots \rho_{r_1+r_2} d\rho_1 \cdots d\rho_{r_1+r_2} d\varphi_{r_1+1} \cdots d\varphi_{r_1+r_2}$$

$$= \frac{2^{r_2} \pi^{r_2}}{N(J) \cdot \sqrt{|\delta|}} \int_{D_1''} \rho_{r_1+1} \cdots \rho_{r_1+r_2} d\rho_1 \cdots d\rho_{r_1+r_2}. \tag{23.7}$$

We consider a part of the domain of integration, namely D_1''', where $\rho_1 > 0, \ldots, \rho_{r_1} > 0$. Then

$$\int_{D_1''} \rho_{r_1+1} \cdots \rho_{r_1+r_2} d\rho_1 \cdots d\rho_{r_1+r_2} = 2^{r_1} \int_{D_1'''} \rho_{r_1+1} \cdots \rho_{r_1+r_2} d\rho_1 \cdots d\rho_{r_1+r_2}. \tag{23.8}$$

Let new variables be defined by $\tau_j = \rho_j^{\ell_j}$ for $j = 1, \ldots, r_1 + r_2$, where $\ell_1 = \cdots = \ell_{r_1} = 1$ and $\ell_{r_1+1} = \cdots = \ell_{r_1+r_2} = 2$. The absolute value of the Jacobian of this change of variables is given by

$$\left| \frac{\partial(\rho_1, \ldots, \rho_{r_1+r_2}, \varphi_{r_1+1}, \ldots, \varphi_{r_1+r_2})}{\partial(\tau_1, \ldots, \tau_{r_1+r_2}, \varphi_{r_1+1}, \ldots, \varphi_{r_1+r_2})} \right| = 2^{-r_2} \rho_{r_1+1}^{-1} \cdots \rho_{r_1+r_2}^{-1}.$$

In the new variables, D_1''' becomes the set $D_1^{(iv)}$ of points

$$(\tau_1, \ldots, \tau_{r_1+r_2}, \varphi_{r_1+1}, \ldots, \varphi_{r_1+r_2})$$

such that $\tau_j > 0$ for $j = 1, \ldots, r_1 + r_2$, $0 \le \varphi_j < 2\pi$ for $j = r_1 + 1, \ldots, r_1 + r_2$, and $0 < \tau_1 \cdots \tau_{r_1+r_2} \le N(J)$:

$$\log(\tau_j) = \frac{\ell_j}{n} \log(\tau_1 \cdots \tau_{r_1+r_2}) + \ell_j \sum_{k=1}^{r} \alpha_k \log |u_k^{(j)}| \tag{23.9}$$

with $0 \le \alpha_k < 1$.

Thus

$$\text{vol}(D_1) = \frac{2^{r_1+r_2} \pi^{r_2}}{N(J) \cdot \sqrt{|\delta|}} \int_{D_1^{(iv)}} d\tau_1 \cdots d\tau_{r_1+r_2}. \tag{23.10}$$

Let a new set of variables be $\alpha_1, \ldots, \alpha_r$, $\omega = \tau_1 \tau_2 \cdots \tau_{r_1+r_2}$, $\varphi_{r_1+1}, \ldots, \varphi_{r_1+r_2}$. To determine the Jacobian of this change of variables, we note that

$$\log(\tau_j) = \frac{\ell_j}{n} \log(\omega) + \ell_j \sum_{k=1}^{r} \alpha_k \log |u_k^{(j)}|$$

hence

$$\frac{1}{\tau_j} \cdot \frac{\partial \tau_j}{\partial \alpha_k} = \ell_j \log |u_k^{(j)}| \quad \text{and} \quad \frac{1}{\tau_j} \cdot \frac{\partial \tau_j}{\partial \omega} = \frac{\ell_j}{n} \cdot \frac{1}{\omega}.$$

Therefore

$$\frac{\partial(\tau_1, \ldots, \tau_{r_1+r_2}, \varphi_{r_1+1}, \ldots, \varphi_{r_1+r_2})}{\partial(\alpha_1, \ldots, \alpha_r, \omega, \varphi_{r_1+1}, \ldots, \varphi_{r_1+r_2})} = \frac{\tau_1 \cdots \tau_{r_1+r_2}}{\omega}$$

$$\times \det \left(\begin{array}{c|c} \begin{array}{cccc} \ell_1 \log |u_1^{(1)}| & \cdots & \ell_1 \log |u_r^{(1)}| & \ell_1/n \\ \ell_2 \log |u_1^{(2)}| & \cdots & \ell_2 \log |u_r^{(2)}| & \ell_2/n \\ \cdots\cdots\cdots\cdots & \cdots\cdots\cdots\cdots & \cdots\cdots\cdots & \cdots \\ \ell_{r_1+r_2} \log |u_1^{(r_1+r_2)}| & \cdots & \ell_{r_1+r_2} \log |u_r^{(r_1+r_2)}| & \ell_{r_1+r_2}/n \end{array} & 0 \\ \hline 0 & I_{r_2} \end{array} \right) ,$$

where I_{r_2} is the identity matrix with r_2 rows and r_2 columns.

But

$$1 = |N_{K|\mathbb{Q}}(u_k)| = \prod_{j=1}^{r_1+r_2} |u_k^{(j)}|^{\ell_j}$$

hence $\sum_{j=1}^{r_1+r_2} \ell_j \log |u_k^{(j)}| = 0$. Similarly $\sum_{j=1}^{r_1+r_2} \ell_j = n$, hence the absolute value of the determinant

$$\det \left(\begin{array}{ccc} \ell_1 \log |u_1^{(1)}| & \cdots & \ell_1 \log |u_r^{(1)}| \\ \cdots\cdots\cdots\cdots & & \cdots\cdots\cdots\cdots \\ \cdots\cdots\cdots\cdots & & \cdots\cdots\cdots\cdots \\ \ell_r \log |u_1^{(r)}| & \cdots & \ell_r \log |u_r^{(r)}| \end{array} \right)$$

is equal to the regulator R. Hence

$$\text{vol}(D_1) = \frac{2^{r_1+r_2} \pi^{r_2} R}{N(J) \cdot \sqrt{|\delta|}} \int_0^{N(J)} d\omega \int_0^1 \cdots \int_0^1 d\alpha_1 \cdots d\alpha_r$$

$$= \frac{2^{r_1+r_2} \pi^{r_2} R}{\sqrt{|\delta|}}.$$

Concluding, we have

$$\lim_{t\to\infty} \frac{\sigma(t;C)}{t} = \frac{2^{r_1+r_2} \pi^{r_2} R}{w \sqrt{|\delta|}}. \qquad \blacksquare$$

We recall that $\sigma(t)$ denotes the number of integral ideals I of K such that $N(I) \le t$. We obtain at once

B.
$$\lim_{t\to\infty} \frac{\sigma(t)}{t} = h \cdot \frac{2^{r_1+r_2} \pi^{r_2} R}{w \sqrt{|\delta|}}.$$

Proof:
$$\sigma(t) = \sum_{i=1}^{h} \sigma(t;C_i)$$

hence

$$\lim_{t\to\infty} \frac{\sigma(t)}{t} = \sum_{i=1}^{h} \lim_{t\to\infty} \frac{\sigma(t;C_i)}{t}$$

$$= \sum_{i=1}^{h} \frac{2^{r_1+r_2}\pi^{r_2}R}{w\sqrt{|\delta|}} = h \cdot \frac{2^{r_1+r_2}\pi^{r_2}R}{w\sqrt{|\delta|}}. \qquad \blacksquare$$

If we want to use the above formula to compute the class number, we need to know r_1, r_2, w, δ, R as well as the value of the limit on the left-hand side. In many cases there is no great obstacle in determining r_1, r_2, w, and δ. It is more difficult to find the regulator, since it requires the knowledge of a fundamental system of units. The computation of $\lim_{t\to\infty} \sigma(t)/t$ is even more awkward. We shall therefore express this limit in another way, more appropriate for calculations.

The limit $\lim_{t\to\infty} \sigma(t)/t$ expresses the distribution of integral ideals in K with respect to the norm. We recall that $\nu(m)$ denoted the number of integral ideals of K with norm equal to the integer $m \geq 1$. Thus

$$\sigma(t) = \sum_{m=1}^{[t]} \nu(m) \quad \text{and} \quad \frac{\sigma(t)}{t} \leq \sum_{m=1}^{[t]} \frac{\nu(m)}{m}$$

hence

$$\lim_{t\to\infty} \frac{\sigma(t)}{t} \leq \sum_{m=1}^{\infty} \frac{\nu(m)}{m}.$$

However, the series $\sum_{m=1}^{\infty} \nu(m)/m$ is usually divergent. We get convergent series by considering large denominators; for example, when $s > 0$ is sufficiently large then $\sum_{m=1}^{\infty} \nu(m)/m^s$ will be convergent. The idea is to study the function of s defined by such a series and then investigate its behavior when s tends to 1. This leads us to the consideration of the Dedekind zeta-series of the field K.

23.2 The Dedekind Zeta-Series

The *Dedekind zeta-series* of the algebraic number field K is

$$\sum_{m=1}^{\infty} \frac{\nu(m)}{m^s},$$

where s is a positive real number and for every integer $m \geq 1$, $\nu(m)$ denotes the number of integral ideals J of K with norm $N(J) = m$.

Thus, the Dedekind zeta-series is a Dirichlet series. If $K = \mathbb{Q}$, the Dedekind zeta-series is the usual Riemann zeta-series (with s restricted to be real, as is sufficient for our purposes). We may apply the results of Chapter 22 to determine the domain of convergence of the Dedekind zeta-series and to obtain the Euler product representation.

C. *For every $\delta > 0$ the Dedekind zeta-series converges uniformly on the interval $[1 + \delta, \infty)$ and defines a continuous function of s in $(1, \infty)$, called* the Dedekind zeta-function of K and denoted by $\zeta_K(s)$. Moreover,

$$\lim_{s \to 1+0} (s - 1)\zeta_K(s) = h \cdot \frac{2^{r_1+r_2}\pi^{r_2} R}{w\sqrt{|\delta|}} \neq 0. \qquad (23.11)$$

Proof: With the notations of Chapter 22, (B), we have

$$S(m) = \sum_{n=1}^{m} \nu(n) = \sigma(m).$$

By (B) we have

$$\lim_{m \to \infty} \frac{\sigma(m)}{m} = h \cdot \frac{2^{r_1+r_2}\pi^{r_2} R}{w\sqrt{|\delta|}},$$

hence the result follows immediately from Chapter 22, (B) and (E).   ∎

We shall be able to express the Dedekind zeta-series as an infinite product, involving the norms of the prime ideals in K and valid for $s > 1$.

More generally, let $\mathcal{J}$ be a set of nonzero ideals in K. For every $m \geq 1$, let $\mathcal{J}_m = \{J \in \mathcal{J} \mid N(J) \leq m\}$.

For every $k \geq 1$, let $\nu_{\mathcal{J}}(k) = \#\{J \in \mathcal{J} \mid N(J) = k\}$. Thus

$$\#(\mathcal{J}_m) = \sum_{k=1}^{m} \nu_{\mathcal{J}}(k).$$

We define

$$S(m) = S_m(\mathcal{J}) = \sum_{k=1}^{m} \frac{\nu_{\mathcal{J}}(k)}{k^s} = \sum_{J \in \mathcal{J}_m} \frac{1}{N(J)^s} \qquad \text{for} \quad m \geq 1, \quad s > 0.$$

The sequence of positive real numbers $(S_m)_{m \geq 1}$ has a limit (which may be infinite). We define

$$\sum_{J \in \mathcal{J}} \frac{1}{N(J)^s} = \lim_{m \to \infty} S_m(\mathcal{J}).$$

Let

$$T_m = T_m(\mathcal{J}) = \prod_{k=1}^{m} \left(\frac{1}{1 - 1/k^s} \right)^{\nu_{\mathcal{J}}(k)} = \prod_{J \in \mathcal{J}_m} \left(\frac{1}{1 - 1/N(J)^s} \right)$$

for $m \geq 1$, $s > 0$.

The sequence of real numbers $(T_m)_{m \geq 1}$, each satisfying $1 < T_m$, has a limit (which may be infinite). We define

$$\prod_{J \in \mathcal{J}} \frac{1}{1 - 1/N(J)^s} = \lim_{m \to \infty} T_m(\mathcal{J}).$$

Now let $\mathcal{P}$ be a set of nonzero prime ideals of K, let $\mathcal{J}$ be the set of all integral ideals which are products of ideals in $\mathcal{P}$.

Concerning the convergence of

$$\sum_{J \in \mathcal{J}} \frac{1}{N(J)^s} \quad \text{and of the product} \quad \prod_{P \in \mathcal{P}} \frac{1}{1 - 1/N(P)^s},$$

we have:

D. *Let $\mathcal{P}$ be a set of nonzero prime ideals in K. Then the product*

$$\prod_{P \in \mathcal{P}} \frac{1}{1 - 1/N(P)^s}$$

is absolutely convergent for $s > 1$ and

$$\prod_{P \in \mathcal{P}} \frac{1}{1 - 1/N(P)^s} = \sum_{J \in \mathcal{J}} \frac{1}{N(J)^s} \quad \text{for} \quad s > 1.$$

Proof: Let $\nu(k)$ denote, as before, the number of integral ideals in K with norm equal to k.

We have

$$\sum_{P \in \mathcal{P}} \frac{1}{N(P)^s} = \lim_{m \to \infty} \sum_{P \in \mathcal{P}_m} \frac{1}{N(P)^s} = \lim_{m \to \infty} \sum_{k=1}^{m} \frac{\nu_{\mathcal{P}}(k)}{k^s}$$

$$\leq \lim_{m \to \infty} S_m(\mathcal{J}) = \sum_{k=1}^{\infty} \frac{\nu_{\mathcal{J}}(k)}{k^s} \leq \sum_{k=1}^{\infty} \frac{\nu(k)}{k^s}$$

and by (**C**) the last series is convergent, hence absolutely convergent, when $s > 1$.

Hence

$$\prod_{P \in \mathcal{P}} \frac{1}{1 - 1/N(P)^s} \neq 0$$

and so

$$\prod_{P \in \mathcal{P}} \frac{1}{1 - 1/N(P)^s}$$

is absolutely convergent.

For each $P \in \mathcal{P}$, we have

$$\frac{1}{1 - 1/N(P)^s} = 1 + \frac{1}{N(P)^s} + \frac{1}{N(P^2)^s} + \cdots$$

because $N(P^{ks}) = \left[N(P^k) \right]^s$ for all $k \geq 1$.

Then, for $m \geq 1$:

$$T_m(\mathcal{P}) = \prod_{P \in \mathcal{P}_m} \frac{1}{1 - 1/N(P)^s} = \prod_{P \in \mathcal{P}_m} \left(\sum_{k=0}^{\infty} \frac{1}{N(P^k)^s} \right)$$

$$= \sum \left\{ \frac{1}{N(J)^s} \,\bigg|\, J \in \mathcal{J}(\mathcal{P}_m) \right\},$$

where $\mathcal{J}(\mathcal{P}_m)$ denotes the set of ideals which are products of ideals in $\mathcal{P}_m$. This is a consequence of Dedekind's theorem (Chapter 7, Theorem 2). Hence

$$\prod_{P \in \mathcal{P}} \frac{1}{1 - 1/N(P)^s} = \lim_{m \to \infty} T_m(\mathcal{P}) = \lim_{m \to \infty} \sum \left\{ \frac{1}{N(J)^s} \,\middle|\, J \in \mathcal{J}(\mathcal{P}_m) \right\}$$

$$= \sum_{J \in \mathcal{J}} \frac{1}{N(J)^s},$$

as follows from the absolute convergence of the series.   ∎

Taking $\mathcal{P}$ equal to the set of all prime ideals in K, we obtain

$$\zeta_K(s) = \prod_P \frac{1}{1 - 1/N(P)^s} \qquad \text{for} \quad s > 1. \tag{23.12}$$

In particular, taking $K = \mathbb{Q}$ we have the multiplicative representation of Riemann's zeta-function, already given in Chapter 22:

$$\zeta(s) = \prod_p \frac{1}{1 - 1/p^s} \qquad \text{for} \quad s > 1. \tag{23.13}$$

Combining (**C**) and (**D**) we obtain:

Theorem 1.

$$h = \frac{w\sqrt{|\delta|}}{2^{r_1 + r_2} \pi^{r_2} R} \cdot \lim_{s \to 1+0} (s - 1) \prod_P \frac{1}{1 - 1/N(P)^s}, \tag{23.14}$$

where the product is extended over all nonzero prime ideals of K.

In the above expression we have reduced the computation of the left-hand side to the determination of the norms of the prime ideals in K. For practical purposes, Theorem 1 is not yet satisfactory since it contains an infinite product. Later, we shall obtain more explicit formulas for the class number in the special cases of quadratic and cyclotomic fields.

We now give some estimates which will be useful. Let $L|K$ be an extension of number fields, $[L : K] = n$. Let $\mathcal{P}$ be a set of nonzero prime ideals in L. For each $f = 1, \ldots, n$, let $\mathcal{P}_f$ be the set of all $P \in \mathcal{P}$ with inertial degree f in $L|K$.

E.   *For $f \geq 1$ and $s > 1$, we have*

$$1 \leq \prod_{P \in \mathcal{P}_f} \frac{1}{1 - 1/N(P)^s} \leq [\zeta_K(fs)]^n$$

and if $f \geq 2$ then

$$1 \leq \prod_{P \in \mathcal{P}_f} \frac{1}{1 - 1/N(P)^s} \leq [\zeta_K(f)]^n$$

for all $s > 1$.

Proof: Let $\underline{\mathcal{P}}$ denote the set of all nonzero prime ideals in K. We have, for $f \geq 1$, $s > 1$:

$$1 \leq \prod_{P \in \mathcal{P}_f} \frac{1}{1 - 1/N(P)^s} \leq \prod_{P \in \underline{\mathcal{P}}} \left(\frac{1}{1 - 1/N(P)^{fs}} \right)^n = [\zeta_K(fs)]^n$$

noting that there exist at most n prime ideals with a given norm.

If $f \geq 2$, since $\zeta_K(f) > \zeta_K(fs)$ when $s > 1$, then

$$1 \leq \prod_{P \in \mathcal{P}_f} \frac{1}{1 - 1/N(P)^s} < [\zeta_K(f)]^n$$

for all $s > 1$. ∎

Let $\mathcal{P}$ be a set of nonzero prime ideals in K, and let $\mathcal{J}$ be the set of all integral ideals in K which are products of prime ideals in $\mathcal{P}$.

With these hypotheses, we have:

F.

$$\log \sum_{J \in \mathcal{J}} \frac{1}{N(J)^s} \approx \sum_{P \in \mathcal{P}_1} \frac{1}{N(P)^s} \qquad for \quad s \to 1 + 0. \qquad (23.15)$$

Proof: By **(D)**, we have for $s > 1$:

$$\sum_{J \in \mathcal{J}} \frac{1}{N(J)^s} = \prod_{P \in \mathcal{P}} \frac{1}{1 - 1/N(P)^s} = \prod_{f=1}^{n} \prod_{P \in \mathcal{P}_f} \frac{1}{1 - 1/N(P)^s}.$$

Taking logarithms, we obtain

$$\log \sum_{J \in \mathcal{J}} \frac{1}{N(J)^s} = \sum_{f=1}^{n} \log \prod_{P \in \mathcal{P}_f} \frac{1}{1 - 1/N(P)^s}$$

$$\approx \log \prod_{P \in \mathcal{P}_1} \frac{1}{1 - 1/N(P)^s} \qquad for \quad s \to 1 + 0,$$

as follows from **(E)**. Next

$$\log \prod_{P \in \mathcal{P}_1} \frac{1}{1 - 1/N(P)^s} = \sum_{P \in \mathcal{P}_1} \log \frac{1}{1 - 1/N(P)^s}$$

$$= \sum_{P \in \mathcal{P}_1} \sum_{\nu=1}^{\infty} \frac{1}{\nu} \cdot \frac{1}{N(P)^{\nu s}} = \sum_{\nu=1}^{\infty} \frac{1}{\nu} \sum_{P \in \mathcal{P}_1} \frac{1}{N(P)^{\nu s}}$$

$$= \sum_{P \in \mathcal{P}_1} \frac{1}{N(P)^s} + \sum_{\nu=2}^{\infty} \sum_{P \in \mathcal{P}_1} \frac{1}{\nu} \cdot \frac{1}{N(P)^{\nu s}}.$$

Now we have the following bounds when $s \to 1 + 0$:

$$\sum_{\nu=2}^{\infty} \sum_{P \in \mathcal{P}_1} \frac{1}{\nu} \cdot \frac{1}{N(P)^{\nu s}} \leq \sum_{P \in \mathcal{P}_1} \frac{1}{2} \left(\sum_{\nu=2}^{\infty} \frac{1}{N(P)^{\nu s}} \right)$$

$$= \sum_{P \in \mathcal{P}_1} \frac{1}{2} \cdot \frac{1/N(P)^{2s}}{1 - 1/N(P)^s} \leq \sum_{P \in \mathcal{P}_1} \frac{1}{N(P)^{2s}} \leq n \sum_{p} \frac{1}{p^{2s}} \leq n\zeta(2s)$$

(noting that $N(P) = p \geq 2$, $s > 1$, and that there exist at most n prime ideals with given norm). So

$$\log \sum_{J \in \mathcal{J}} \frac{1}{N(J)^s} \approx \sum_{P \in \mathcal{P}_1} \frac{1}{N(P)^s} \qquad \text{for} \quad s \to 1 + 0. \qquad \blacksquare$$

Taking $\mathcal{P}$ to be the set of all nonzero prime ideals in K we have the special case

$$\log \zeta_K(s) \approx \sum_{P \in \mathcal{P}_1} \frac{1}{N(P)^s} \qquad \text{for} \quad s \to 1 + 0. \qquad (23.16)$$

In particular,

$$\log \zeta(s) \approx \sum_{p} \frac{1}{p^s} \qquad \text{for} \quad s \to 1 + 0 \qquad (23.17)$$

(see also Chapter 22, (22.11)).

23.3 Hecke L-Series

For later use we introduce the Hecke L-series, associated to Hecke characters.

Just like the Dedekind zeta-function of a number field K generalizes the Riemann zeta-function, Hecke L-series are an extension to number fields of the Dirichlet L-series associated to modular characters.

The theory of Hecke L-series is important, for example, in relation to class field theory, but that goes beyond the level of this book.

We recall from Chapter 8, Section 2, that if J is a nonzero integral ideal of the number field K, we may consider the associated group of classes of ideals $\mathcal{C}_{J,+} = \mathcal{F}_J/\mathcal{P}r_{J,+}$ where $\mathcal{F}_J$ is the multiplicative group of nonzero fractional ideals of K, which are relatively prime to J (in the sense already explained) and $\mathcal{P}r_{J,+}$ is the subgroup of those principal ideals Ax, where $x \equiv 1 \pmod{J}$ and x is totally positive. The number of elements of $\mathcal{C}_{J,+}$ is denoted by $h_{J,+}$.

In Chapter 21, Section 2, we considered the Hecke characters with modulus J, which are generalizations of the Dirichlet modular characters.

So Hecke was led to introduce L-series associated with Hecke characters (which are generalizations of the Dirichlet L-series), defined as follows:

$$L(s|\chi) = \sum_I \frac{\chi(I)}{N(I)^s} \qquad (23.18)$$

(sum extended over all nonzero integral ideals of K) where χ is a Hecke character of the number field K, associated to the nonzero integral ideal J, and s is a real number.

The following results may be proved in the same way as for the case of Dirichlet L-series:

(1) For every Hecke character χ, for every $\delta > 0$, the L-series $L(s|\chi)$ converges uniformly and absolutely on the interval $[1 + \delta, \infty)$, hence it defines a continuous function for $s > 1$.

(2) If χ is any Hecke character different from the trivial character χ_0, for every $\delta > 0$, the series $L(s|\chi)$ converges uniformly on the interval $[\delta, \infty)$, hence it defines a continuous function for $s > 0$.

(3) For $s > 1$ there is the Euler product representation

$$\sum_I \frac{\chi(I)}{N(I)^s} = \prod_P \frac{1}{1 - \chi(P)/N(P)^s} \qquad (23.19)$$

(the product is extended over the set of prime ideals P not dividing the ideal J).

(4) For the trivial Hecke character χ_0:

$$L(s|\chi_0) = \prod_{P|J} \left(1 - \frac{1}{N(P)^s}\right) \zeta_K(s) \qquad \text{for} \quad s > 1, \qquad (23.20)$$

where $\zeta_K(s)$ denotes Dedekind's zeta-function.

(5) For any Hecke character χ:

$$\log L(s|\chi) \approx \sum_P \frac{\chi(P)}{N(P)^s} \qquad \text{as} \quad s \to 1 + 0. \qquad (23.21)$$

EXERCISES

1. Calculate $\lim_{t\to\infty} \sigma(t)/t$ for the following fields:
   (a) $\mathbb{Q}$;
   (b) $\mathbb{Q}(\sqrt{-1})$;
   (c) $\mathbb{Q}(\sqrt{-3})$;
   (d) $\mathbb{Q}(\sqrt{-2})$;

(e) $\mathbb{Q}(\sqrt{2})$;

(f) $\mathbb{Q}(\sqrt{5})$.

2. Calculate $\lim_{t\to\infty} \sigma(t)/t$ for the following fields:

(a) $\mathbb{Q}(\zeta_5)$, where ζ_5 is a primitive fifth root of 1;

(b) $\mathbb{Q}(\zeta_5 + \zeta_5^{-1})$.

3. With the notations in the text, show that if $m, n \geq 1$ and $\gcd(m, n) = 1$, then $\nu(mn) = \nu(m)\nu(n)$.

4. Calculate the first terms (for $1 \leq n \leq 11$) of the Dedekind zeta-series of the following fields:

(a) $\mathbb{Q}$;

(b) $\mathbb{Q}(\sqrt{-1})$;

(c) $\mathbb{Q}(\sqrt{-3})$;

(d) $\mathbb{Q}(\sqrt{-2})$;

(e) $\mathbb{Q}(\sqrt{2})$;

(f) $\mathbb{Q}(\sqrt{5})$;

(g) $\mathbb{Q}(\zeta_5)$;

(h) $\mathbb{Q}(\zeta_5 + \zeta_5^{-1})$.

5. Let K be a field of algebraic numbers. Show that if $s > 1$ then

$$\frac{\zeta_K(s)}{\zeta(s)} = \sum_{n=1}^{\infty} \frac{c_n}{n^s},$$

where

$$c_n = \sum_{d|n} \mu(d)\nu\left(\frac{n}{d}\right)$$

and $\nu(m)$ denotes the number of integral ideals I of K such that $N(I) = m$.

6. Let K be an algebraic number field, and let I be an integral ideal of K, $m \geq 1$. Let $T(m, I)$ denote the number of pairwise nonassociated elements x of I such that $|N_{K|\mathbb{Q}}(x)| \leq m$.

Show that

$$\lim_{m\to\infty} \frac{T(m, I)}{m} = \frac{2^{r_1+r_2}\pi^{r_2} R}{w\sqrt{|\delta|N(I)}}.$$

7. Let $K = \mathbb{Q}(\sqrt{d})$ where d is a square-free nonzero integer. With the notations of the text, show that

$$\nu(n) = \sum_{k|\delta} \left(\frac{\delta}{k}\right),$$

where δ denotes the discriminant of K.

8. Let $d \neq 0$ be square-free, $K = \mathbb{Q}(\sqrt{d})$. Show that for every $x > 0$:

$$\left| \sum_{n \leq x} \left(\frac{\delta}{n}\right) \right| \leq |\delta|.$$

24

Primes in Arithmetic Progressions

In this chapter we shall prove Dirichlet's theorem on primes in arithmetic progressions, which we already stated and used in Chapter 4.

Theorem 1.   *Let a, m be integers such that $1 \leq a \leq m$, $\gcd(a, m) = 1$. Then the arithmetic progression*

$$\{a, \ a + m, \ a + 2m, \ \ldots, \ a + km, \ \ldots\}$$

contains infinitely many prime numbers.

The hypothesis that a, m be relatively prime is necessary, because if d is the greatest common divisor of a, m, and $d > 1$, then there exists at most one prime in the progression, namely, when $d = a$ is prime.

If $m = 1$ then the above progression consists of all integers $n \geq 1$, and the theorem reduces to the fact that there exist infinitely many prime numbers.

24.1 Proof of Dirichlet's Theorem

We use the Dedekind zeta-series to prove first a special case.

A.   *There exist infinitely many prime numbers p such that $p \equiv 1$ (mod m). We have*

$$\varphi(m) \sum_{p \equiv 1 \ (\text{mod } m)} \frac{1}{p^s} \approx \log \frac{1}{s - 1} \qquad \text{for} \quad s \to 1 + 0.$$

Proof: Let ζ be a primitive mth root of unity, and let $K = \mathbb{Q}(\zeta)$, so $[K : \mathbb{Q}] = \varphi(m)$. In Chapter 16, (**D**), we have seen that p is ramified in $K|\mathbb{Q}$ if and only if p divides m, and p is totally decomposed in $K|\mathbb{Q}$, that is, there are $\varphi(m)$ ideals P such that $N(P) = p$ if and only if $p \equiv 1$ (mod m). Hence

$$\varphi(m) \sum_{p \equiv 1 \ (\text{mod } m)} \frac{1}{p^s} = \sum_{P \in \mathcal{P}_1} \frac{1}{N(P)^s},$$

for $f = 1, \ldots, n$, $\mathcal{P}_f$ denotes the set of prime ideals P in K having inertial degree f over $\mathbb{Q}$.

By Chapter 23, (23.16):

$$\sum_{P \in \mathcal{P}_1} \frac{1}{N(P)^s} \approx \log \zeta_K(s) \qquad \text{for} \quad s \to 1 + 0. \tag{24.1}$$

By Chapter 23, (23.11), we have

$$\lim_{s \to 1+0} (s-1)\zeta_K(s) = c \neq 0.$$

Thus, given α, such that $0 < \alpha < c$, there exists $\delta > 0$ such that if $1 < s < 1 + \delta$ then $c - \alpha < (s-1)\zeta_K(s) < c + \alpha$. In particular, for $1 < s < 1 + \delta$, $(s-1)\zeta_K(s)$ remains bounded and also bounded away from zero. Taking logarithms, we have

$$\log \zeta_K(s) \approx \log \frac{1}{s-1} \qquad \text{for} \quad s \to 1 + 0. \tag{24.2}$$

Hence

$$\varphi(m) \sum_{p \equiv 1 \ (\text{mod } m)} \frac{1}{p^s} \approx \log \frac{1}{s-1} \qquad \text{for} \quad s \to 1 + 0.$$

This implies that the series on the left-hand side is divergent, so there exist infinitely many primes p such that $p \equiv 1 \ (\text{mod } m)$. ∎

To deal with arithmetic progressions $\{k + rm \mid r = 0, 1, \ldots\}$, $1 \leq k \leq m$, $\gcd(k, m) = 1$, we shall use characters modulo m to distinguish between the primes of the different arithmetic progressions. Therefore we consider the L-series of these characters.

If χ_0 is the trivial character modulo m, it follows from Chapter 22, (**D**) and (**G**), that

$$\log L(s|\chi_0) \approx \log \zeta(s) \approx \sum_p \frac{1}{p^s} \approx \log \frac{1}{s-1} \tag{24.3}$$

for $s \to 1 + 0$.

Proof of Dirichlet's theorem: To show that there exist infinitely many primes in the arithmetic progression $\{a + km \mid k = 1, 2, \ldots\}$ where $\gcd(a, m) = 1$, we consider the group $P(m)$ (of invertible residue classes modulo m), a system of representatives $\{a_1, \ldots, a_{\varphi(m)}\}$, the characters of $P(m)$, and the corresponding modular characters $\chi_0, \chi_1, \ldots, \chi_{\varphi(m)-1}$ modulo m.

We have, for $i = 0, 1, \ldots, \varphi(m) - 1$:

$$\sum_p \frac{\chi_i(p)}{p^s} = \sum_{j=1}^{\varphi(m)} \chi_i(a_j) \left[\sum_{p \equiv a_j \ (\text{mod } m)} \frac{1}{p^s} \right]. \tag{24.4}$$

By Chapter 21, (**K**), this system of equations has solution

$$\sum_{p\equiv a_j \ (\mathrm{mod}\ m)} \frac{1}{p^s} = \frac{1}{\varphi(m)} \sum_{i=0}^{\varphi(m)-1} \overline{\chi_i(a_j)} \left[\sum_p \frac{\chi_i(p)}{p^s} \right]$$

for $j = 1, \ldots, \varphi(m)$.

By Chapter 22, (**I**), we have, for $j = 1, \ldots, \varphi(m)$:

$$\sum_{p\equiv a_j \ (\mathrm{mod}\ m)} \frac{1}{p^s} \approx \frac{1}{\varphi(m)} \sum_{i=0}^{\varphi(m)-1} \overline{\chi_i(a_j)} \log L(s|\chi_i) \qquad (24.5)$$

for $s \to 1 + 0$. In order to show that there exist infinitely many prime numbers p such that $p \equiv a_j$ (mod m), we prove that the right-hand side of (24.5) is unbounded for $s \to 1 + 0$.

For $i = 0$ we have

$$\log L(s|\chi_0) \approx \log \frac{1}{s-1} \qquad (24.6)$$

hence the term corresponding to the principal character χ_0 is unbounded when $s \to 1 + 0$.

It will be enough to show that if $\chi_i \neq \chi_0$ then $\log L(s|\chi_i)$ is bounded when $s \to 1 + 0$.

In Chapter 23, (**H**), we have seen that if $\chi_i \neq \chi_0$ then $L(s|\chi_i)$ is a continuous function on $(0, \infty)$; thus $\lim_{s\to 1} \log L(s|\chi_i) = \log L(1|\chi_i)$.

So we have to prove the following crucial fact:

If χ_i is a modular character modulo m, $\chi_i \neq \chi_0$, then $L(1|\chi_i) \neq 0$.

Taking $a_j \equiv 1$ (mod m) in (24.5) we have

$$\sum_{p\equiv 1 \ (\mathrm{mod}\ m)} \frac{1}{p^s} \approx \frac{1}{\varphi(m)} \sum_{i=0}^{\varphi(m)-1} \log L(s|\chi_i)$$

and by (**A**):

$$\log \frac{1}{s-1} \approx \sum_{i=0}^{\varphi(m)-1} \log L(s|\chi_i) \qquad \text{for} \quad s \to 1 + 0.$$

But

$$\log \frac{1}{s-1} \approx \log \zeta(s) \approx \log L(s|\chi_0) \qquad \text{for} \quad s \to 1 + 0$$

as recalled in (24.3). Hence $H(s) = \sum_{\chi_i\neq\chi_0} \log L(s|\chi_i)$ remains bounded for $s \to 1 + 0$. Therefore

$$\prod_{\chi_i\neq\chi_0} L(1|\chi_i) = \lim_{s\to 1} \prod_{\chi_i\neq\chi_0} \log L(s|\chi_i) = \lim_{s\to 1} e^{H(s)} \neq 0$$

hence necessarily $L(1|\chi_i) \neq 0$ for $\chi_i \neq \chi_0$.   ∎

We shall indicate in Chapter 26, (**F**), another proof that $L(1|\chi_i) \neq 0$ for $\chi_i \neq \chi_0$.

We now give a more precise quantitative version of Dirichlet's theorem. We have seen in (24.3) that

$$\sum_p \frac{1}{p^s} \approx \log \frac{1}{s-1} \qquad \text{for} \quad s \to 1+0.$$

Hence

$$\lim_{s \to 1+0} \left(\sum_p \frac{1}{p^s} \right) \Big/ \left(\log \frac{1}{s-1} \right) = 1.$$

If S is a set of prime numbers such that the limit

$$\lim_{s \to 1+0} \left(\sum_{p \in S} \frac{1}{p^s} \right) \Big/ \left(\log \frac{1}{s-1} \right)$$

exists and is equal to d, we say that d is the (Dirichlet) *density* of S. Thus $0 \leq d \leq 1$ and the density of the set of all prime numbers is equal to 1.

B.   *If a, m are integers, $1 \leq a \leq m$ and $\gcd(a,m) = 1$, then the set S_a of prime numbers p such that $p \equiv a$ (mod m) has density equal to $1/\varphi(m)$.*

Proof: From (24.5), (24.6) we have

$$\sum_{p \in S_a} \frac{1}{p^s} \approx \frac{1}{\varphi(m)} \left[\log \frac{1}{s-1} + \sum_{\chi \neq \chi_0} \chi(a) \log L(s|\chi) \right] \qquad \text{for} \quad s \to 1+0.$$

Finally, since $L(1|\chi) \neq 0$ then

$$\sum_{p \in S_a} \frac{1}{p^s} \approx \frac{1}{\varphi(m)} \log \frac{1}{s-1}$$

and the density of S_a is equal to $1/\varphi(m)$.   ∎

It is worthwhile to stress that this density is independent of the particular arithmetic progression.

The theorem of Dirichlet on primes in arithmetic progressions may be generalized by using the L-series associated to Hecke characters. Explicitly, let K be a number field, and J a nonzero fractional ideal. Let $C_{J,+} = \mathcal{F}_J/\mathcal{P}r_{J,+}$ and $C_J = \mathcal{F}_J/\mathcal{P}r_J$, as introduced in Chapter 8, Section 2, and recalled in Chapter 21, Section 2.

For each Hecke character associated to $C_{J,+}$ (or to C_J), we consider the corresponding L-series $L(s|\chi)$. For $\chi \neq \chi_0$, it was stated in Chapter 23, Section 2, that $L(s|\chi)$ converges for $s > 1$.

With a proof similar to the one for modular characters, the following holds:

$L(1|\chi) \neq 0$ *for every* $\chi \neq \chi_0$.

Again with a proof analogous to the above proof of Dirichlet's theorem we obtain:

Theorem 2.   *Each class in* $\mathcal{C}_{J,+}$ *contains infinitely many prime ideals.*

A fortiori, each class in $\mathcal{C}_J$ contains infinitely many prime ideals.

If $K = \mathbb{Q}$, $J = \mathbb{Z}m$ (with $m > 1$) the classes in $\mathcal{C}_J$ (or $\mathcal{C}_{J,+}$) correspond to the residue classes a modulo m, where $\gcd(a, m) = 1$. So Theorem 2 becomes, in this case, Dirichlet's theorem on primes in arithmetic progression.

Moreover, for each class $[I] \in \mathcal{C}_J$ (respectively, class $[I]_+ \in \mathcal{C}_{J,+}$) we have

$$\lim_{s \to 1+0} \frac{\sum_{P \in [I]} 1/N(P)^s}{\log(1/(s-1))} = \frac{1}{h_J}, \tag{24.7}$$

where h_J is the number of elements of $\mathcal{C}_J$, and

$$\lim_{s \to 1+0} \frac{\sum_{P \in [I]_+} 1/N(P)^s}{\log(1/(s-1))} = \frac{1}{h_{J,+}}, \tag{24.8}$$

where $h_{J,+}$ is the number of elements of $\mathcal{C}_{J,+}$.

We note that the limits are independent of the class in consideration and they are called the *Dirichlet density of the set of prime ideals* in $[I]$ (respectively, $[I]_+$).

It is worth noting that Theorem 2 is a substantial strengthening of the result of Kummer of Chapter 20, (**T**).

Since the set of prime ideals of inertial degree greater than 1 has density 0 (see Chapter 23, (**E**)), it follows that each ideal class contains, in fact, infinitely many prime ideals of inertial degree 1.

In the next chapter, we shall prove a related generalization of Dirichlet's theorem on primes in arithmetic progressions.

Now we give an unexpected application of Theorem 2, which is due to Carlitz.

C.   *Let K be a number field. Then the following properties are equivalent:*
  (1) *The class number of K is $h = 2$.*
  (2) *The ring A of integers of K is not a unique factorization domain and if*

$$a = p_1 p_2 \cdots p_r = p_1' p_2' \cdots p_{r'}',$$

  where $a, p_i, p_j' \in A$ with p_i, p_j' indecomposable elements, not necessarily distinct, then $r = r'$.

Proof: (1) $\to$ (2) Let $h = 2$.
  (1°) If q is an indecomposable element, then Aq is a prime ideal or the product of two (not necessarily distinct) prime ideals.

Indeed, if Q, Q' are nonprincipal ideals, then Q^2, QQ' are principal ideals, because from $[Q] \neq [A]$ then $[Q^2] \neq [Q]$ so $[Q]^2 = [A]$. Also from $[Q]$, $[Q'] \neq [A]$ then $[Q] = [Q']$ hence $[QQ'] = [Q]^2 = [A]$.

Let $Aq = Q_1 Q_2 \cdots Q_s$, where each Q_j is a prime ideal. Let Q_1, Q_2, $\ldots$, Q_r be principal ideals and let Q_{r+1}, $\ldots$, Q_s be nonprincipal ideals. We show that $s - r$ is even. Otherwise, $s - r$ is odd and $Q_{r+1}Q_{r+2}, \ldots$, $Q_{s-2}Q_{s-1}$ are principal ideals. Therefore, Q_s would also be principal, which is a contradiction.

Let $Q_i = Aq_i$ for $i = 1, \ldots, r$ and $Q_{r+1}Q_{r+2} = Aq_{r+1}, \ldots, Q_{s-1}Q_s$ $= Aq_{s-1}$ thus $Aq = Aq_1 \cdots Aq_r Aq_{r+1} Aq_{r+3} \cdots Aq_{s-1}$. Since q is an indecomposable element, either $r = s = 1$, or $r = 0$, $s = 2$.

(2°)   Let $a = p_1 \cdots p_r q_1 \cdots q_s$ where p_i, q_j are indecomposable elements, and $Ap_i = P_i$ is a prime ideal while Aq_j is not a prime ideal for $i = 1, \ldots, r$, $j = 1, \ldots, s$. By (1°), $Aq_j = Q_j \overline{Q}_j$ where Q_j, $\overline{Q}_j$ are prime ideals. Thus

$$Aa = P_1 \cdots P_r Q_1 \overline{Q}_1 \cdots Q_s \overline{Q}_s.$$

If $a = p'_1 \cdots p'_{r'} q'_1 \cdots q'_{s'}$ where $Ap'_i = P'_i$, $Aq'_j = Q'_j \overline{Q}'_j$ with P_i, Q'_j, $\overline{Q}'_j$ prime ideals, then

$$Aa = P'_1 \cdots P'_{r'} Q'_1 \overline{Q}'_1 \cdots Q'_{s'} \overline{Q}'_{s'}.$$

By Dedekind's theorem of unique factorization into prime ideals, $r + 2s = r' + 2s'$ and the set of principal prime ideals dividing Aa is $\{P_1, \ldots, P_r\}$ but also $\{P'_1, \ldots, P'_{r'}\}$, so $r = r'$. From $r + 2s = r' + 2s'$ then $s = s'$, proving that (2) holds if $h = 2$.

(2) → (1)   Let $h > 2$. Then either there exists an ideal class $[J]$ with order $m > 2$ or all nonprincipal ideal classes have order 2 and there exist at least two ideal classes of order 2.

Case 1:

By Theorem 2 there exists a prime ideal P such that $[P] = [J]$. Let $[J']$ be the inverse of $[J]$, and let P' be a prime ideal such that $[P'] = [J']$. Then $P^m = Ap$, $P'^m = Ap'$, $PP' = Aa$, so $Aa^m = App'$.

Now we observe that a, p, p' are indecomposable elements. Indeed, if $Abc = Ap = P^m$, by the unique factorization, $Ab = P^k$, $Ac = P^\ell$ with $k + \ell = m$. So $k = m$, $\ell = 0$ (or vice versa) because m is the order of $[P]$ in the class group.

Thus p is an indecomposable element. Similarly, p' is an indecomposable element. Also if $Aa = Abc$ then from $Aa = PP'$, it follows that $Ab = P$, $Ac = P'$ (or vice versa). This is a contradiction, because P is not a principal ideal.

Since $m > 2$ and $a^m = pp'$ we conclude that (2) does not hold.

Case 2:

Let $[J]$, $[J']$ be distinct classes having order 2. Then $[JJ'] \neq [A]$, otherwise $[JJ'] = [A]$, so $[J'] = [J]^{-1} = [J]$, which is not true. Thus $[JJ']$ has order 2.

Let P, P', P'' be prime ideals such that $[P] = [J]$, $[P'] = [J']$, $[P''] = [JJ']$. Then $[PP'P''] = [A]$. Let $P^2 = Ap$, $P'^2 = Ap'$, $P''^2 = Ap''$. As proved before (Case 1) p, p', p'' are indecomposable elements.

Let $PP'P'' = Aq$ so q is an indecomposable element, because if $Aq = Aab$ then $PP'P'' = Aab$, so either Aa or Ab is equal to one of the ideals P, P', P''—which is not possible, because they are not principal ideals.

Then $Aq^2 = P^2 P'^2 P''^2 = App'p''$, so $q^2 = upp'p''$, where u is a unit. This contradicts condition (2).   ■

24.2 Special Cases

We digress from the main line of development of the exposition, in order to discuss interesting proofs of special cases of Dirichlet's theorem. For example, we have the following easy proof:

D.   *The arithmetic progressions $\{4k + 3 \mid k = 0, 1, 2, \ldots\}$ and $\{6k + 5 \mid k = 0, 1, 2, \ldots\}$ each contain infinitely many primes.*

Proof: Assume that p_1, p_2, $\ldots$, p_n are primes of the form $4k+3$ with $k \geq 0$. Then $N = 4p_1p_2 \cdots p_n + 3 > 1$, so there exists a prime p dividing N and such that $p \not\equiv 1 \pmod 4$ —because $N \not\equiv 1 \pmod 4$; so $p \equiv 3 \pmod 4$ and clearly $p \neq p_1, p_2, \ldots, p_n$. This is enough to prove the statement for the progression $\{4k + 3 \mid k = 0, 1, 2, \ldots\}$.

For the progression $\{6k + 5 \mid k = 0, 1, 2, \ldots\}$ we proceed in a similar way, considering this time $N = 6p_1p_2 \cdots p_n + 5$.   ■

Using simple properties of quadratic residues, it is also easy to show that there exist infinitely many primes in the arithmetic progressions $\{mk + a \mid k = 0, 1, 2, \ldots\}$ in each of the following cases: $(m, a) = (4, 1)$, $(6, 1)$, $(8, 1)$, $(8, 3)$, $(8, 5)$, $(8, 7)$ (these last four cases include the progressions with difference 4), $(12, 5)$, $(12, 7)$, $(12, 11)$ (these include the progressions with difference 6).

We also have:

E.   *For every $r \geq 3$ there exist infinitely many primes in the arithmetic progression $\{2^r k + 1 \mid k = 0, 1, 2, \ldots\}$.*

Proof: Assume that p_1, p_2, $\ldots$, p_n are primes in the given arithmetic progression. Let $N = (2p_1 \cdots p_n)^{2^{r-1}} + 1$ and let p be a prime dividing N. So $(2p_1 \cdots p_n)^{2^{r-1}} \equiv -1 \pmod p$.

Hence the order of $2p_1 \cdots p_n \pmod p$ is equal to 2^r and therefore 2^r divides $p - 1$, that is, $p \equiv 1 \pmod{2^r}$.

Clearly $p \neq p_1, \ldots, p_n$, which suffices to conclude the proof.   ■

Now we consider the arithmetic progressions $\{mk + 1 \mid k = 0, 1, 2, \ldots\}$. The proof involves cyclotomic polynomials. We need more properties of cyclotomic polynomials than those indicated in Chapter 2, Section 8.

Let

$$\Phi_m(X) = \prod_{\gcd(j,m)=1} (X - \zeta^j), \qquad (24.9)$$

where ζ is a primitive mth root of 1. Since ζ is an algebraic integer, then $\Phi_m \in \mathbb{Z}[X]$ and it is also monic, and has degree $\varphi(m)$.

Let $m = p_1^{e_1} \cdots p_r^{e_r}$ with $e_i \geq 1$ for $i = 1, \ldots, r$, $p_1 < p_2 < \cdots < p_r$, each p_i being a prime. We define the polynomials P_j for $j \geq 0$. Let $P_0(X) = X^m - 1$. If $1 \leq j \leq r$, let

$$P_j(X) = \prod_{i_1 < i_2 < \cdots < i_j} (X^{m/p_{i_1} p_{i_2} \cdots p_{i_j}} - 1). \qquad (24.10)$$

If $r < j$, let $P_j(X) = 1$. When necessary, we shall use the notation $P_j^{(m)} = P_j$.

F.   *We have*

$$\Phi_m = \frac{P_0 P_2 P_4 \cdots}{P_1 P_3 P_5 \cdots}.$$

Proof: Let ζ be a primitive root of unity of order d dividing m. Then ζ is a root of

$$X^{m/p_{i_1} p_{i_2} \cdots p_{i_j}} - 1 \quad \text{if and only if } d \text{ divides } \frac{m}{p_{i_1} p_{i_2} \cdots p_{i_j}},$$

that is, each $p_{i_1}, p_{i_2}, \ldots, p_{i_j}$ divides m/d. If $d = m$ then $j = 0$ and also clearly $X - \zeta$ divides $P_0 = X^m - 1$. If $d < m$, let s be the number of distinct prime factors of m/d, so $s \geq 1$. Then the exact power of $X - \zeta$ dividing $P_j(X)$ is equal to $(X - \zeta)^{\binom{s}{j}}$ (we note that $\binom{s}{j} = 0$ if $s < j$). Hence the exact power of $X - \zeta$ dividing $(P_0 P_2 P_4 \cdots)/(P_1 P_3 P_5 \cdots)$ is $(X - \zeta)^e$ with

$$e = 1 - \binom{s}{1} + \binom{s}{2} - \binom{s}{3} + \cdots = (1 - 1)^s = 0.$$

This shows the identity.   ■

G.   (1) *If $p \mid m$ then $\Phi_{pm}(X) = \Phi_m(X^p)$.*

(2) *If p does not divide m and $e \geq 1$ then*

$$\Phi_{p^e m}(X) = \frac{\Phi_m(X^{p^e})}{\Phi_m(X^{p^{e-1}})}.$$

Proof: (1)   Let ε be a primitive mpth root of 1, so ε^p is a primitive mth root of 1, and ε^m is a primitive pth root of 1. Then

$$\Phi_m(X^p) = \prod_{\gcd(k,m)=1} (X^p - \varepsilon^{pk}).$$

But

$$X^p - \varepsilon^{pk} = \prod_{j=1}^{p-1}(X - \varepsilon^k \varepsilon^{mj}),$$

so

$$\Phi_m(X^p) = \prod_{\gcd(k,m)=1} \prod_{j=1}^{p-1}(X - \varepsilon^{k+mj}).$$

Note that $\gcd(k + mj, pm) = 1$: let q be a prime which divides pm and $k + mj$. If $q = p$ divides m then $q|k$ hence q divides $\gcd(k, m) = 1$, which is impossible. So $q \neq p$, hence $q|m$ so $q|k$, which is again impossible.

We note that the number of factors in the above product expression of $\Phi_m(X^p)$ is $\varphi(m)p$; if $m = p^e m'$, where p does not divide m', then

$$\varphi(m)p = \varphi(p^e)\varphi(m')p = \varphi(p^{e+1})\varphi(m')$$
$$= \varphi(p^{e+1}m') = \varphi(pm).$$

Thus ε^{k+mj} for all k, j, runs through the set of all primitive pmth roots of 1. Thus

$$\Phi_{pm}(X) = \Phi_m(X^p).$$

(2)   First we show that if p does not divide m then

$$\Phi_{pm}(X) = \frac{\Phi_m(X^p)}{\Phi_m(X)}.$$

For $j \geq 1$, we have

$$P_j^{(mp)}(X) = \prod_{p_j \neq p} (X^{pm/p_{i_1} \cdots p_{i_j}} - 1) \prod_{i_1 < \cdots < i_{j-1}} (X^{m/p_{i_1} \cdots p_{i_{j-1}}} - 1)$$

$$= P_j^{(m)}(X^p)P_{j-1}^{(m)}(X).$$

If $j = 0$ then $P_0^{(mp)}(X) = X^{mp} - 1 = P_0^{(m)}(X^p)$. Then

$$\Phi_{mp}(X) = \frac{P_0^{(mp)}(X)P_2^{(mp)}(X)P_4^{(mp)}(X)\cdots}{P_1^{(mp)}(X)P_3^{(mp)}(X)P_5^{(mp)}(X)\cdots}$$

$$= \frac{P_0^{(m)}(X^p)P_2^{(m)}(X^p)P_1^{(m)}(X)P_4^{(m)}(X^p)P_3^{(m)}(X)\cdots}{P_1^{(m)}(X^p)P_0^{(m)}(X)P_3^{(m)}(X^p)P_2^{(m)}(X)P_5^{(m)}(X^p)P_4^{(m)}(X)\cdots}$$

$$= \frac{\Phi_m(X^p)}{\Phi_m(X)}.$$

Assume already shown that

$$\Phi_{p^{e-1}m}(X) = \frac{\Phi_m(X^{p^{e-1}})}{\Phi_m(X^{p^{e-2}})},$$

where $e \geq 2$. Then

$$\Phi_{p^e m}(X) = \Phi_{p^{e-1}m}(X^p) = \frac{\Phi_m(X^{p^e})}{\Phi_m(X^{p^{e-1}})}. \qquad \blacksquare$$

H.   *If $m > 1$ then $\Phi_m(0) = 1$.*

Proof: Let s be the number of distinct prime factors of m. Then $P_j(m)(0) = (-1)^{\binom{s}{j}}$, so, by **(F)**, $\Phi_m(0) = (-1)^e$, where $e = 1 - \binom{s}{1} + \binom{s}{2} - \cdots = 0$. Thus $\Phi_m(0) = 1$. $\qquad \blacksquare$

I.   *If $m > 1$ then*

$$\Phi_m(1) = \begin{cases} p, & \text{when } m \text{ is a power of } p, \\ 1, & \text{when } m \text{ has two distinct prime factors.} \end{cases}$$

Proof: If $m = p^e$ then

$$\Phi_{p^e}(X) = X^{p^{e-1}(p-1)} + X^{p^{e-1}(p-2)} + \cdots + X^{p^{e-1}} + 1,$$

so $\Phi_{p^e}(1) = p$.

We proceed by induction on the number of distinct prime factors of m. Let $m = p^e m'$, $e \geq 1$, and p does not divide $m' > 1$. So $\Phi_{m'}(1) = 1$ or q where q is a prime distinct from p. By **(G)**, we have

$$\Phi_m(X) = \frac{\Phi_{m'}(X^{p^e})}{\Phi_{m'}(X^{p^{e-1}})} \qquad \text{and} \qquad \Phi_m(1) = \frac{\Phi_{m'}(1)}{\Phi_{m'}(1)} = 1. \qquad \blacksquare$$

J.   *For integers $m > 1$ and $a > 1$ we have $|\Phi_m(a)| > 1$.*

Proof:

$$|\Phi_m(a)| = \prod_{\gcd(j,m)=1} |a - \zeta^j|.$$

But

$$|a - \zeta^j| \geq ||a| - |\zeta^j|| = |a - 1| = a - 1 \geq 1.$$

We sharpen the estimate. In fact, for $j \neq 0$, $|a - \zeta^j| > 1$, otherwise, $|a - \zeta^j| = a - 1 = 1$, so $a = 2$. Let $\zeta^j = x + iy$ with $i = \sqrt{-1}$, so $x^2 + y^2 = 1$. Then if $|2 - (x + iy)| = 1$ we have $(2 - x)^2 + y^2 = 1$ so $4 - 4x + x^2 + y^2 = 1$ and therefore $x = 1$, $y = 0$, thus $\zeta^j = 1$, which is a contradiction. $\qquad \blacksquare$

For each $m \geq 1$ we consider the polynomial $\Psi_m(X, Y)$ obtained by homogenizing $\Phi_m(X)$. Explicitly

$$\Psi_m(X, Y) = Y^{\varphi(m)} \Phi_m \left(\frac{X}{Y} \right).$$

Thus

$$\Psi_m(X, Y) = \prod_{\gcd(j,m)=1} (X - \zeta^j Y). \tag{24.11}$$

From the corresponding formulas for the cyclotomic polynomials, we obtain at once:

$$X^m - Y^m = \prod_{d|m} \Psi_d(X, Y). \tag{24.12}$$

Let $a > b \geq 1$ be relatively prime integers. If p is a prime divisor of $a^n - b^n$ with $1 \leq n$, but p does not divide $a^m - b^m$ for all $m = 1, \ldots, n-1$, we say that p is a *primitive factor* of $a^n - b^n$. Then p does not divide a nor b. If $bb' \equiv 1 \pmod{p}$, then $(ab')^n \equiv 1 \pmod{p}$, but $(ab')^m \not\equiv 1 \pmod{p}$ for each m, $1 \leq m < n$; and conversely, if this happens, then p is a primitive factor of $a^n - b^n$.

Lemma 1.   *Let $a > b \geq 1$, with $\gcd(a, b) = 1$. Let $n \geq 2$. Then the following statements are equivalent:*

(1) *p is a primitive factor of $a^n - b^n$.*

(2) *p divides $a^n - b^n$ but if $1 \leq d < n$ and d divides n, then p does not divide $a^d - b^d$.*

(3) *$p | \Psi_n(a, b)$, but p does not divide $\Psi_m(a, b)$ for all m, $1 \leq m < n$.*

(4) *p divides $\Psi_n(a, b)$ and if $d|n$, $1 \leq d < n$, then p does not divide $\Psi_d(a, b)$.*

Proof: The equivalence of (1) and (3), as well as the equivalence of (2) and (4), both follow from (24.12), by replacing X, Y by a, b, respectively. Clearly (1) implies (2). On the other hand, it was already noted that (1) holds exactly when n is the order of $ab' \pmod{p}$, where $bb' \equiv 1 \pmod{p}$. Thus if (2) holds then so also does (1).   ∎

The next result is due to Legendre:

K.   *Let $a > b \geq 1$, with $\gcd(a, b) = 1$ and let $n \geq 2$. Then the following sets coincide:*

$E_1 = \{p \text{ prime } | \ p \text{ is a primitive factor of } a^n - b^n\}$,
$E_2 = \{p \text{ prime } | \ p \text{ divides } \Psi_n(a, b) \text{ and } p \equiv 1 \pmod{n}\}$,
$E_3 = \{p \text{ prime } | \ p \text{ divides } \Psi_n(a, b) \text{ and } p \text{ does not divide } n\}$.

Proof: Let $p \in E_1$, so if $bb' \equiv 1 \pmod{n}$ then $ab' \pmod{p}$ has order n, thus n divides $p - 1$, so $p \equiv 1 \pmod{n}$.

By Lemma 1, p divides $\Psi_n(a,b)$. This shows that $E_1 \subseteq E_2$. It is trivial that $E_2 \subseteq E_3$.

Let p divide $\Psi_n(a,b)$, so p divides $a^n - b^n$. If $d < n$, d divides n, and p divides $a^d - b^d$, then from

$$a^n - b^n = \Psi_n(a,b)(a^d - b^d) \prod_{\substack{e \mid d, \, e \mid n \\ e \neq n}} \Psi_e(a,b),$$

it follows that p divides $(a^n - b^n)/(a^d - b^d)$.

Now $n = dm$; if $a^d = r$, $b^d = s$ then $r \equiv s \pmod{p}$. Finally, p divides

$$\frac{r^m - s^m}{r - s} = r^{m-1} + r^{m-2}s + \cdots + rs^{m-2} + s^{m-1} \equiv mr^{s-1} \pmod{p}.$$

Thus p divides m, since p does not divide r. Then p divides $n = dm$, which is absurd, proving that $E_3 \subseteq E_1$.   ∎

Now we give the proof of a special case of Dirichlet's theorem (which does not require analytical methods):

L.   *If $m \geq 2$ the arithmetical progression $\{km + 1 \mid k = 0, 1, 2, \ldots\}$ contains infinitely many primes.*

Proof: Let p_1, p_2, $\ldots$, p_r be primes such that $p_i \equiv 1 \pmod{m}$ for $i = 1, 2, \ldots, r$. Let $N = mp_1p_2 \cdots p_r$. By (**J**), $|\Phi_m(N)| > 1$. Let p be a prime dividing $\Phi_m(N)$. Since $\Phi_m(N) \equiv \Phi_m(0) = 1 \pmod{N}$, then p does not divide N. Hence p does not divide m. By (**K**), $p \equiv 1 \pmod{m}$ and also $p \neq p_1, \ldots, p_r$. This suffices to conclude that there exist infinitely many primes in the arithmetic progression $\{km + 1 \mid k = 1, 2, \ldots\}$.   ∎

In order to prove that the arithmetic progression $\{km - 1 \mid k = 1, 2, \ldots\}$ also contains infinitely many primes, we first need some more properties of polynomials. For each $m \geq 1$ we define the polynomials $R_m(X, Y)$ and $S_m(X, Y)$ by the relation

$$(X + iY)^m = R_m(X, Y) + iS_m(X, Y). \tag{24.13}$$

So R_m, S_m have integral coefficients. In particular, $R_1(X, Y) = X$, $S_1(X, Y) = Y$.

Taking conjugates

$$(X - iY)^m = R_m(X, Y) - iS_m(X, Y)$$

and multiplying

$$(X^2 + Y^2)^m = R_m^2 + S_m^2.$$

Also

$$S_m(X, Y) = \frac{1}{2i}[(X + iY)^m - (X - iY)^m].$$

Let

$$\Omega_m(X,Y) = \Psi_m(X+iY, X-iY)$$

and

$$Q_j^{(m)}(X,Y) = \prod_{p_{i_1}<\cdots<p_{i_j}} (X^{m/p_{i_1}\cdots p_{i_j}} - Y^{m/p_{i_1}\cdots p_{i_j}}),$$

where $j \geq 0$ and $p_{i_1},\ldots,p_{i_j}$ are among the r distinct prime factors of m. So $Q_j^{(m)}(X,Y) \in \mathbb{Z}[X,Y]$ and

$$Q_j^{(m)}(X,Y) = Y^{e_j} \prod \left(\left(\frac{X}{Y}\right)^{m/p_{i_1}\cdots p_{i_j}} - 1\right) = Y^{e_j} P_j^{(m)}\left(\frac{X}{Y}\right)$$

with

$$e_j = \sum_{p_{i_1}<\cdots<p_{i_j}} \frac{m}{p_{i_1}\cdots p_{i_j}}.$$

M. *We have*

$$\Psi_m(X,Y) = \frac{Q_0^{(m)}(X,Y)Q_2^{(m)}(X,Y)Q_4^{(m)}(X,Y)\cdots}{Q_1^{(m)}(X,Y)Q_3^{(m)}(X,Y)Q_5^{(m)}(X,Y)\cdots}.$$

Proof: By (**F**), and the above expression for $Q_j^{(m)}(X,Y)$, the right-hand side is equal to

$$Y^e \frac{P_0^{(m)}(X/Y)P_2^{(m)}(X/Y)P_4^{(m)}(X/Y)\cdots}{P_1^{(m)}(X/Y)P_3^{(m)}(X/Y)P_5^{(m)}(X/Y)\cdots} = Y^e\Phi_m\left(\frac{X}{Y}\right),$$

where $e = e_0 - e_1 + e_2 - e_3 + \cdots = \varphi(m)$. This is equal to

$$Y^{\varphi(m)}\Phi_m\left(\frac{X}{Y}\right) = \Psi_m(X,Y). \qquad \blacksquare$$

We note that:

N. $\Omega_1(X,Y) = 2iY$ *and* $\Omega_m(X,Y) \in \mathbb{Z}[X,Y]$ *for* $m \geq 2$.

Proof: Clearly

$$\Omega_1(X,Y) = (X+iY) - (X-iY) = 2iY.$$

Let $m \geq 2$, then

$$Q_j^{(m)}(X+iY, X-iY) = \prod_{p_{i_1}<\cdots<p_{i_j}} ((X+iY)^{m/p_{i_1}\cdots p_{i_j}} - (X-iY)^{m/p_{i_1}\cdots p_{i_j}}).$$

Its conjugate is

$$Q_j^{(m)}(X-iY, X+iY) = (-1)^{\binom{r}{j}} Q_j^{(m)}(X+iY, X-iY),$$

where r is the number of distinct prime factors of $m \geq 2$; note that $r \geq 1$.

Hence

$$\Psi_m(X - iY, X + iY) = \Psi_m(X + iY, X - iY)$$

because

$$1 - \binom{r}{1} + \binom{r}{2} - \binom{r}{3} + \cdots = 0,$$

since $r \geq 1$.

Therefore if

$$\Omega_m(X, Y) = A(X, Y) + iB(X, Y),$$

with $A(X, Y), B(X, Y) \in \mathbb{Z}[X, Y]$, its conjugate is

$$\overline{\Omega}_m(X, Y) = A(X, Y) - iB(X, Y).$$

Since $\overline{\Omega}_m(X, Y) = \Omega_m(X, Y)$ then $B(X, Y) = 0$, so $\Omega_m(X, Y) \in \mathbb{Z}[X, Y]$. ∎

It follows that $\Omega_m(X, Y)$ divides $S_m(X, Y)$, when $m \geq 2$.

Indeed, $\Omega_m(X, Y) = \Psi_m(X + iY, X - iY)$ divides

$$(X + iY)^m - (X - iY)^m = 2iS_m(X, Y),$$

and since $\Omega_m(X, Y)$, $S_m(X, Y) \in \mathbb{Z}[X, Y]$ then $\Omega_m(X, Y)$ divides $S_m(X, Y)$.

We also note that from $(X + iY)^m = Y^m(X/Y + i)^m$ it follows that

$$R_m(X, Y) = Y^m R_m\left(\frac{X}{Y}, 1\right), \qquad S_m(X, Y) = Y^m S_m\left(\frac{X}{Y}, 1\right).$$

O. $\Omega_m(X, Y) = Y^{\varphi(m)}\Omega_m(X/Y, 1).$

Proof: For convenience, let $f_j = m/p_{i_1} \cdots p_{i_j}$ where $p_{i_1} < \cdots < p_{i_j}$ (prime factors of m). We have

$$Q_j^{(m)}(X + iY, X - iY) = \prod_{p_{i_1} < \cdots < p_{i_j}} 2iS_{f_j}(X, Y)$$

$$= Y^{\sum f_j} \prod 2iS_{f_j}\left(\frac{X}{Y}, 1\right) = Y^{\sum f_j} Q_j^{(m)}\left(\frac{X}{Y} + i, \frac{X}{Y} - i\right).$$

Therefore

$$\Omega_m(X, Y) = Y^e \frac{Q_0^{(m)}(X/Y + i, X/Y - i)Q_2^{(m)}(X/Y + i, X/Y - i)\cdots}{Q_1^{(m)}(X/Y + i, X/Y - i)Q_3^{(m)}(X/Y + i, X/Y - i)\cdots}$$

$$= Y^e\Omega_m\left(\frac{X}{Y}, 1\right),$$

with

$$e = m - \sum_{p_1}\frac{m}{p_1} + \sum_{p_1 < p_2}\frac{m}{p_1p_2} - \sum_{p_1 < p_2 < p_3}\frac{m}{p_1p_2p_3} + \cdots = \varphi(m). \quad ∎$$

P.   *Let $m = dm'$. Then:*

(1) *$S_{m'}$ divides S_m.*

(2) *Ω_m divides S_m.*

(3)
$$S_m = S_{m'}\left[\binom{d}{1}R_{m'}^{d-1} - \binom{d}{3}R_{m'}^{d-3}S_{m'}^2 \pm \cdots\right]$$

(4) *If $n_1, n_2 \geq 1$ then $S_{n_1+n_2} = R_{n_1}S_{n_2} + R_{n_2}S_{n_1}$.*

Proof: (1) $X^{m'} - Y^{m'}$ divides $X^m - Y^m$, hence
$$S_{m'}(X,Y) = \frac{1}{2i}[(X+iY)^{m'} - (X-iY)^{m'}]$$

divides
$$S_m(X,Y) = \frac{1}{2i}[(X+iY)^m - (X-iY)^m].$$

(2)
$$X^m - Y^m = \Psi_m(X,Y)(X^{m'} - Y^{m'})\prod_{\substack{e \neq m, e|m \\ e \nmid m'}} \Psi_e(X,Y).$$

So replacing X, Y, respectively, by $X+iY$, $X-iY$, then $\Omega_m(X,Y)$ divides
$$\frac{S_m(X,Y)}{S_{m'}(X,Y)}.$$

(3)
$$(X+iY)^{dm'} = (R_{m'} + iS_{m'})^d$$
$$= \left[R_{m'}^d - \binom{d}{2}R_{m'}^{d-2}S_{m'}^2 + \binom{d}{4}R_{m'}^{d-4}S_{m'}^4 - \cdots\right]$$
$$+ i\left[\binom{d}{1}R_{m'}^{d-1} - \binom{d}{3}R_{m'}^{d-3}S_{m'}^3 + \cdots\right].$$

Thus
$$S_m = S_{m'}\left[\binom{d}{1}R_{m'}^{d-1} - \binom{d}{3}R_{m'}^{d-3}S_{m'}^2 \pm \cdots\right].$$

(4)
$$(X+iY)^{n_1+n_2} = (R_{n_1} + iS_{n_1})(R_{n_2} + iS_{n_2})$$
$$= (R_{n_1}R_{n_2} - S_{n_1}S_{n_2}) + i(R_{n_1}S_{n_2} + S_{n_1}R_{n_2}),$$

hence
$$S_{n_1+n_2} = R_{n_1}S_{n_2} + S_{n_1}R_{n_2}.\qquad\blacksquare$$

Q. *Let $m \geq 2$.*

(1)
$$\Omega_m(1,0) = \begin{cases} 1, & \text{if } m \text{ is not a prime-power,} \\ p, & \text{if } m \text{ is a power of } p. \end{cases}$$

(2) *If $c \geq 1$ is any integer, if $x, y \geq 1$, $\gcd(x,y) = 1$, and x/y is sufficiently large, then $\Omega_m(x,y) > c$.*

(3) *There exist integers $a, b > 0$ with $\gcd(a,b) = 1$, such that $\Omega_m(a,b) < 0$.*

Proof: (1)

$$\Omega_m(1,0) = \Psi_m(1,1) = \prod_{\gcd(j,m)=1} (1 - \zeta^j)$$

$$= \Phi_m(1) = \begin{cases} 1, & \text{when } m \text{ is not a prime-power,} \\ p, & \text{when } m \text{ is a power of } p, \end{cases}$$

(see (**I**)).

(2) Now

$$\Omega_m(X,Y) = \Psi_m(X + iY, X - iY)$$

$$= \prod_{\gcd(j,m)=1} [(X + iY) - \zeta^j(X - iY)]$$

$$= \prod_{\gcd(j,m)=1} [X(1 - \zeta^j) + i(1 + \zeta^j)Y]$$

$$= Y^{\varphi(m)} \prod_{\gcd(j,m)=1} (1 - \zeta^j) \cdot \prod_{\gcd(j,m)=1} \left(\frac{X}{Y} - i\frac{\zeta^j + 1}{\zeta^j - 1} \right)$$

$$= Y^{\varphi(m)} \Phi_m(1) \prod_{\gcd(j,m)=1} \left(\frac{X}{Y} - i\frac{\zeta^j + 1}{\zeta^j - 1} \right),$$

since $\prod(1 - \zeta^j) = \Phi_m(1) > 0$.

Let $\alpha_j = i(\zeta^j + 1)/(\zeta^j - 1)$. We note that α_j is real:

$$\overline{\alpha}_j = -i\frac{\zeta^{-j} + 1}{\zeta^{-j} - 1} = -i\frac{1 + \zeta^j}{1 - \zeta^j} = i\frac{\zeta^j + 1}{\zeta^j - 1} = \alpha_j.$$

Let $x, y \geq 1$, $\gcd(x,y) = 1$ be such that

$$\frac{x}{y} > \max\{\alpha_j \mid \gcd(j,m) = 1\} + c.$$

By (**I**), $\Phi_m(1) \geq 1$, so

$$\Omega_m(x,y) \geq \prod_{\gcd(j,m)=1} \left(\frac{x}{y} - \alpha_j \right) > c.$$

(3)   Let $W_m(X) = \Omega_m(X, 1) \in \mathbb{Z}[X]$. $W_m(X)$ has degree $\varphi(m)$ and the roots $\alpha_j = i(\zeta^j + 1)/(\zeta^j - 1)$. If $j \neq k$ then $\alpha_j \neq \alpha_k$: if $\alpha_j = \alpha_k$ then

$$\frac{\zeta^j + 1}{\zeta^j - 1} = \frac{\zeta^k + 1}{\zeta^k - 1} \qquad \text{so} \qquad \zeta^{j+k} - \zeta^j + \zeta^k - 1 = \zeta^{j+k} + \zeta^j - \zeta^k - 1.$$

Thus $\zeta^j = \zeta^k$, a contradiction.

The derivative $W'_m(X)$ has degree $\varphi(m) - 1$, so it cannot vanish at all α_j; if $W'_m(\alpha_{j_0}) \neq 0$ then there exists a/b such that $W_m(a/b) = \Omega_m(a/b, 1) < 0$. Hence $\Omega_m(a, b) = b^{\varphi(m)}\Omega_m(a/b, 1) < 0$. ∎

R.   Let $x, y \in \mathbb{Z}$, $\gcd(x, y) = 1$, and $m \geq 2$. If $p \nmid m$, $p|\Omega_m(x, y)$, and $p \equiv -1 \pmod{4}$ then $p \equiv -1 \pmod{m}$.

Proof: Since $p|\Omega_m(x, y)$ then $p|S_m(x, y)$. We show that $p \nmid S_{m'}(x, y)$ for any m', where $m = dm'$, $d \neq 1$.

Otherwise, let $m' = m/d < m$ and assume that $p|S_{m'}(x, y)$. By (**P**):

$$\frac{S_m(x, y)}{S_{m'}(x, y)}$$

$$= \binom{d}{1}[R_{m'}(x, y)]^{d-1} - \binom{d}{3}[R_{m'}(x, y)]^{d-3}[S_{m'}(x, y)]^2 \pm \cdots$$

$$\equiv d[R_{m'}(x, y)]^{d-1} \pmod{p}.$$

By (**P**):

$$p|\Omega_m(x, y), \quad \Omega_m(x, y)\Big|\frac{S_m(x, y)}{S_{m'}(x, y)}, \qquad \text{so} \quad p|d[R_{m'}(x, y)]^{d-1}.$$

If $p|d$ then $p|m$, which is contrary to the hypothesis. So $p|R_{m'}(x, y)$. Hence

$$p|[R_{m'}(x, y)]^2 + [S_{m'}(x, y)]^2 = (x^2 + y^2)^{m'},$$

so p divides $x^2 + y^2$. If $p|x$ then $p|y$, again contrary to the hypothesis that $\gcd(x, y) = 1$. So $p \nmid x$.

Let $x' \in \mathbb{Z}$ be such that $xx' \equiv 1 \pmod{p}$, so $1 + (yx')^2 \equiv 0 \pmod{p}$ hence $(-1/p) = 1$ and therefore $p \equiv 1 \pmod{4}$, which is a contradiction.

Now we compute $S_{p+1}(X, Y)$; noting that $R_1(x, y) = x$, $S_1(x, y) = y$, and using (**P**), we have

$$S_{p+1}(x, y)$$

$$= y\left[\binom{p+1}{1}x^p - \binom{p+1}{3}x^{p-2}y^2 + \cdots + (-1)^{(p-1)/2}\binom{p+1}{p}xy^{p-1}\right].$$

If $k = 2, 3, \ldots, p - 1$ then

$$\binom{p+1}{k} = \frac{(p+1)p(p-1)\cdots(p-k+2)}{1 \cdot 2 \cdots k} \equiv 0 \pmod{p}.$$

Also $(p - 1)/2$ is odd, since $p \equiv -1 \pmod{4}$.

Hence

$$S_{p+1}(x, y) \equiv (p + 1)(x^p y - x y^p)$$
$$= (p + 1)xy(x^{p-1} - y^{p-1}) \equiv 0 \pmod{p}$$

using the little Fermat theorem.

Let $m' = \gcd(p + 1, m)$. We shall show that $m' = m$, i.e. m divides $p + 1$, so $p \equiv -1 \pmod{m}$.

Let $m = m'd$, $d \geq 1$. Let r, s be integers such that $m' = r(p + 1) - sm$. We may choose $r, s \geq 1$. Indeed, let $t > 0$ be an integer such that $tm > -r$, $t(p + 1) > -s$, then we still have

$$(r + tm)(p + 1) - (s + t(p + 1))m = r(p + 1) - sm = m'.$$

Since $r(p + 1) = m' + sm$, then by (**P**):

$$S_{r(p+1)}(x, y) = S_{m'+sm}(x, y)$$
$$= R_{m'}(x, y)S_{sm}(x, y) + R_{sm}(x, y)S_{m'}(x, y).$$

Since $S_m(x, y) | S_{sm}(x, y)$ then $p | S_{sm}(x, y)$. Similarly, $p | S_{r(p+1)}(x, y)$. Thus $p | R_{sm}(x, y)S_{m'}(x, y)$. But

$$[R_{sm}(x, y)]^2 + [S_{sm}(x, y)]^2 = (x^2 + y^2)^{sm}.$$

If $p | R_{sm}(x, y)$ then $p | x^2 + y^2$ and we conclude as before that $p \equiv 1 \pmod{4}$, which is contrary to the hypothesis. So $p \nmid R_{sm}(x, y)$, hence $p | S_{m'}(x, y)$. As was shown above, this is only possible if $d = 1$, i.e., $m | p + 1$ and $p \equiv -1 \pmod{m}$. ∎

Now we are ready to show:

S.   *Let $m \geq 2$. There exist infinitely many primes p such that $p \equiv -1 \pmod{m}$.*

Proof: By (**Q**), there exist integers $a, b > 0$, $\gcd(a, b) = 1$ such that $\Omega_m(a, b) < 0$.

Let $c = -\Omega_m(a, b) > 0$ and let $T(X) = (1/c)\Omega_m(cbX + a, b)$. Since $\Omega_m(X, Y) \in \mathbb{Z}[X, Y]$, the coefficient of X^j in $T(X)$ for $j \geq 1$ belongs to $\mathbb{Z}$. It follows that $T(X) \in \mathbb{Z}[X]$. We note that for every sufficiently large x we have $T(x) = (1/c)\Omega_m(cbx + a, b) > 1$.

Now suppose that p_1, ..., p_n are primes such that $p_i \equiv -1 \pmod{m}$. Let $t \geq 1$ be such that if $N = 4tmp_1 \cdots p_n$ then $T(N) > 1$. We have $T(N) \equiv T(0) = -1 \pmod{N}$ since $T(X) \in \mathbb{Z}[X]$.

But $T(N) \equiv -1 \pmod{4}$, $T(N) > 1$, so there exists a prime p such that $p | T(N)$, $p \equiv -1 \pmod{4}$. Then $p \nmid N$. So $p \neq p_1, p_2, \ldots, p_n$, $p \nmid m$. Noting that $\gcd(cbN + a, b) = 1$ and $p | cT(N) = \Omega_m(cbN + a, b)$, $p \nmid m$, $p \equiv -1 \pmod{4}$, then, by (**R**), $p \equiv -1 \pmod{m}$. This is enough to establish the proposition. ∎

The proofs of (**L**) and (**S**) do not use analytical methods, but just properties of cyclotomic and related polynomials. A proof using properties of

Lucas sequences, of the existence of infinitely many primes of the form $kq - 1$ (where q is a given odd prime) may be found in [25, Chapter 4, Section IV].

There are also elementary proofs of Erdős and of Selberg of the theorem of Dirichlet, in its full generality, along the lines of their proof of the prime number theorem. See the book of Gelfond and Linnik [5, Chapter 3], and, for further references Ribenboim [25, Chapter 4, Section IV].

E X E R C I S E S

1. Write explicitly, following the model of the text, an analytic proof of Dirichlet's theorem, for the arithmetic progressions with modules $4, 5$, and 6.

2. Following the proof of (**E**) given in the text, write explicitly the proof in the case of the arithmetic progression of integers n such that $n \equiv 1$ (mod 8) and $n \equiv 1$ (mod 16).

3. Following the model of (**L**) given in the text, write explicitly the proof that the arithmetic progressions $\{5k + 1 \mid k = 0, 1, \ldots\}$, $\{7k + 1 \mid k = 0, 1, \ldots\}$, and $\{12k + 1 \mid k = 0, 1, \ldots\}$ contain infinitely many primes.

4. Following the model of (**S**) given in the text, write the proofs that the arithmetic progressions $\{8k - 1 \mid k = 0, 1, \ldots\}$, $\{16k - 1 \mid k = 0, 1, \ldots\}$, $\{5k - 1 \mid k = 0, 1, \ldots\}$, and $\{7k - 1 \mid k = 0, 1, \ldots\}$ contain infinitely many primes.

5. Complete the details of the following proof of Dirichlet's theorem:
   (i) By Exercises 11, 9, 10 of Chapter 22 deduce that

$$\lim_{s \to 1+0} \left\{ \sum_{n \equiv a \ (\mathrm{mod}\ m)} \frac{\Lambda(n)}{n^s} \right\} = \infty.$$

(ii) Show that if $h \geq 2$, the sum

$$\sum_{h=2}^{\infty} \sum_{p^h \equiv a \ (\mathrm{mod}\ m)} \frac{\log p}{p^s}$$

remains bounded for $s \to 1 + 0$.
   (iii) Conclude that

$$\sum_{p \equiv a \ (\mathrm{mod}\ m)} \frac{\log p}{p^s}$$

is a sum of infinitely many terms.

6. Let p be an odd prime. Show:
   (a) $p!$ and $(p-1)! - 1$ are relatively prime integers.
   (b) If $n > 0$ and $n \equiv (p-1)! - 1 \pmod{p!}$ then every integer m, such that $m \neq n$ and $n - p + 2 \leq m \leq n + p$, is not prime.

   Hint: Use Wilson's theorem (Chapter 3, Exercise 18).

7. Let $r > 0$ be any integer. Show that there exist infinitely many prime numbers p such that if q is any prime number, $q \neq p$, then $|p - q| > r$.

   Hint: Use the preceding exercise and Dirichlet's theorem on primes in arithmetic progressions.

8. Let $m \geq 1$, $1 \leq a < m$, with $\gcd(a, m) = 1$ and am is even. Show that there exists an infinite set of pairwise relatively prime integers $\{k_1, k_2, \ldots\}$ such that each $p_i = k_i m + a$ is a prime number.

9. Do not use Dirichlet's theorem to prove:
   (a) If $a, d \geq 1$, $\gcd(a, d) = 1$, and $n \geq 2$ there exist infinitely many $k \geq 1$ such that $\gcd(a + kd, n) = 1$.
   (b) There is an infinite sequence $k_1 < k_2 < \cdots$ such that if $i \neq j$ then $\gcd(a + k_i d, a + k_j d) = 1$.

10. Let $a, d \geq 1$, $\gcd(a, d) = 1$. Do not use Dirichlet's theorem to show:
   (a) There exists a geometric progression $\{br^n \mid n \geq 0\}$ (with $b \geq 1$, $r \geq 2$) which is contained in the arithmetic progression $A = \{a + kd \mid k \geq 1\}$.
   (b) A contains an infinite subset whose elements have the same set of prime factors.

11. Do not use Dirichlet's theorem to show: Let $a, d \geq 1$, $\gcd(a, d) = 1$, and let S be an infinite subset of $\{a + kd \mid k = 0, 1, \ldots\}$; then for every $n \geq 1$ there exists $m \in S$ which is the product of at least n distinct factors in S.

12. Suppose that for all pairs of positive integers (a, d), with $\gcd(a, d) = 1$ there exists one prime in the arithmetic progression $\{a + kd \mid k = 0, 1, \ldots\}$. Deduce Dirichlet's theorem on primes in arithmetic progressions.

13. Let $f \in \mathbb{Z}[X]$ with degree $n \geq 1$ and assume that $f(p)$ is a prime-power for each prime p. Show that $f(X) = X^n$.

   Hint: Observe that if p, q are distinct primes, $m \geq 1$ and $f(p) = q^m$ then q^{m+1} divides $f(p + kq^{m+1}) - f(p)$ for all $k = 1, 2, \ldots$.

25

The Frobenius Automorphism and the Splitting of Prime Ideals

25.1 The Frobenius Automorphism

Let K be a number field, and let $L|K$ be a Galois extension of degree n. We denote by A_L, (respectively, A_K), the ring of algebraic integers of L (respectively, K).

Let $\mathcal{U}(L|K)$ be the set of all prime ideals P in K which are unramified in $L|K$.

If $\widetilde{P}$ is any prime ideal in L, $P = \widetilde{P} \cap A_K$, let $\overline{K} = \overline{K}_P = A_K/P$, $\overline{L} = \overline{L}_{\widetilde{P}} = A_L/\widetilde{P}$. The fields $\overline{K}$, $\overline{L}$ have $\#(\overline{K}) = N(P)$, $\#(\overline{L}) = N(\widetilde{P})$ elements. The inertial degree of $\widetilde{P}$ in $L|K$ is $f = f_{\widetilde{P}}(L|K) = [\overline{L}_{\widetilde{P}} : \overline{K}_P]$, so $N(\widetilde{P}) = N(P)^f$.

If $P \in \mathcal{U}(L|K)$, by Chapter 11, Theorem 1, $n = fg$, where g is the number of distinct prime ideals $\widetilde{P}$ in L which divide $A_L P$, or equivalently, $\widetilde{P} \cap A_K = P$.

In this situation, we say that P has *splitting type* (f, g) in $L|K$.

If $f = 1$, $g = n$, then P is totally decomposed (or also, splits completely) in $L|K$. If $f = n$, $g = 1$, then $A_L P$ is a prime ideal in L and P is inert in $L|K$.

Let $\mathcal{L} = G(L|K)$ and $\overline{\mathcal{L}} = \overline{\mathcal{L}}_{\widetilde{P}} = G(\overline{\mathcal{L}}_{\widetilde{P}}|\overline{K}_P)$. So $\overline{\mathcal{L}}$ is a cyclic group of order f, generated by the automorphism φ defined by

$$\varphi(\overline{x}) = \overline{x}^{N(P)} \tag{25.1}$$

for every $\overline{x} \in \overline{L} = A_L/\widetilde{P}$. Let $\mathcal{Z} = \mathcal{Z}_{\widetilde{P}}(L|K)$ be the decomposition group of $\widetilde{P}$, so $\mathcal{Z} = \{\sigma \in \mathcal{L} \mid \sigma(\widetilde{P}) = \widetilde{P}\}$. Each $\sigma \in \mathcal{L}$ is such that $\sigma(A_L) = A_L$. Hence, if $\sigma \in \mathcal{Z}$ it induces $\overline{\sigma} : \overline{L} \to \overline{L}$ given by $\overline{\sigma}(\overline{x}) = \overline{\sigma(x)}$ (where $\overline{y} = y + \widetilde{P} \in \overline{L}$, for every $y \in A_L$). Then $\overline{\sigma} \in \overline{\mathcal{L}}$. The mapping $\sigma \mapsto \overline{\sigma}$ is a surjective group homomorphism and since P is unramified, it is in fact an isomorphism (see Chapter 14, (D)).

Let $\left(\dfrac{L|K}{\widetilde{P}}\right)$ denote the unique element of $\mathcal{Z}$ which corresponds to φ by the above isomorphism. Thus

$$\left(\frac{L|K}{\widetilde{P}}\right)(x) \equiv x^{N(P)} \pmod{\widetilde{P}} \tag{25.2}$$

for every $x \in A_L$.

Definition 1. $\left(\dfrac{L|K}{\widetilde{P}}\right)$ is called the *Frobenius automorphism* associated to $\widetilde{P}$ in $L|K$.

We shall indicate in this section several properties of the Frobenius automorphism.

A. *With the above hypotheses and notations:*

(1) $\left(\dfrac{L|K}{\widetilde{P}}\right)$ *has an order* $f = f_{\widetilde{P}}(L|K)$.

(2) $\left(\dfrac{L|K}{\widetilde{P}}\right)$ *is the identity automorphism if and only if* P *is totally decomposed in* $L|K$.

(3) $\mathcal{L}$ *is cyclic and* $\left(\dfrac{L|K}{\widetilde{P}}\right)$ *is a generator of* $\mathcal{L}$ *if and only if* P *is inert in* $L|K$.

Proof: (1) The order of $\left(\dfrac{L|K}{\widetilde{P}}\right)$ is equal to the order of φ, which is f, since φ is a generator of the cyclic Galois group $\overline{\mathcal{L}}$ of order f.

(2) By (1), $\left(\dfrac{L|K}{\widetilde{P}}\right)$ is the identity automorphism if and only if $f = 1$, so $g = n$, that is, P is totally decomposed in $L|K$.

(3) If $\mathcal{L}$ is a cyclic group of order n and $\left(\dfrac{L|K}{\widetilde{P}}\right)$ is a generator of $\mathcal{L}$, by (1), $f = n$, so $g = 1$ and P is inert. Conversely, if $g = 1$, then $f = n$, so $\mathcal{L} = \mathcal{Z}$ is a cyclic group generated by $\left(\dfrac{L|K}{\widetilde{P}}\right)$. ∎

Let

$$\left(\frac{L|K}{P}\right) = \left\{\left(\frac{L|K}{\widetilde{P}}\right) \Bigm| \widetilde{P} \text{ is a prime ideal in } L \text{ dividing } A_L P\right\}.$$

We have:

B. $\left(\dfrac{L|K}{P}\right)$ *is a conjugacy class of automorphisms in* $\mathcal{L}$.

Proof: Let $\widetilde{P} = \widetilde{P}_1, \widetilde{P}_2, \ldots, \widetilde{P}_g$ be the distinct prime ideals in L which divide $A_L P$. By Chapter 11, (E), for every $i = 1, \ldots, g$, there exists $\sigma_i \in \mathcal{L}$

(with $\sigma_1 = \varepsilon$ the identity automorphism) such that $\sigma_i(\widetilde{P}) = \widetilde{P}_i$. Then

$$\left(\frac{L|K}{\widetilde{P}_i}\right) = \sigma_i\left(\frac{L|K}{\widetilde{P}}\right)\sigma_i^{-1}.$$

Indeed, let $\tau = \sigma_i\left(\dfrac{L|K}{\widetilde{P}}\right)\sigma_i^{-1}$ so $\tau \in \mathcal{Z}_{\widetilde{P}_i}(L|K)$. Since $\overline{K}$ is a finite field and $[\overline{L}_{\widetilde{P}} : \overline{K}] = [\overline{L}_{\widetilde{P}_i} : \overline{K}] = f$, then $\overline{L}_{\widetilde{P}_i} = \overline{L}_{\widetilde{P}}$ for every $i = 1, 2, \ldots, g$. The image of τ is $\overline{\tau} \in \overline{\mathcal{L}}$, given by

$$\overline{\tau}(\overline{x}) = \overline{\sigma}_i(\varphi(\overline{\sigma}_i^{-1}(\overline{x}))) = \overline{\sigma}_i\left[(\overline{\sigma}_i^{-1}(\overline{x}))^{N(P)}\right] = \overline{x}^{N(P)},$$

thus $\overline{\tau} = \varphi$, hence $\tau = \left(\dfrac{L|K}{\widetilde{P}_i}\right)$. ∎

Definition 2.   The mapping from the set of nonzero prime ideals P of K to the set of conjugacy classes of $G(L|K)$, defined by $P \mapsto \left(\dfrac{L|K}{P}\right)$, is called the *Frobenius symbol* of $L|K$.

If $L|K$ is an Abelian extension, then $\left(\dfrac{L|K}{\widetilde{P}}\right) = \left(\dfrac{L|K}{\widetilde{P}_i}\right)$ for all $i =$ $1, \ldots, g$, so $\left(\dfrac{L|K}{P}\right)$ consists of only one automorphism and we write, more simply, $\left(\dfrac{L|K}{P}\right) = \left(\dfrac{L|K}{\widetilde{P}}\right)$.

Before the next result we fix the notation. Let K be a number field, and let $K \subset L \subset L'$, with $[L : K] = n$, $[L' : K] = n'$. Assume that $L|K$, $L'|K$ are Galois extensions, $\mathcal{L} = G(L|K)$, $\mathcal{L}' = G(L'|K)$. If P' is a prime ideal in L', let $\widetilde{P} = P' \cap L$, $P = P' \cap K$. We assume that P is unramified in $L'|K$, hence also in $L|K$. Let $\rho : \mathcal{L}' \to \mathcal{L}$ be the surjective group-homomorphism of restriction.

For each subset $\mathcal{H}'$ of $\mathcal{L}'$, let $\rho(\mathcal{H}') = \{\rho(\sigma') \mid \sigma' \in \mathcal{H}'\}$.

C.   *With the above hypothesis and notations*

$$\left(\frac{L|K}{P}\right) = \rho\left(\frac{L'|K}{P}\right).$$

Proof: We show that

$$\rho\left(\frac{L'|K}{P'}\right) = \left(\frac{L'|K}{\widetilde{P}}\right).$$

Indeed, by definition,

$$\left(\frac{L'|K}{P'}\right)(x) \equiv x^{N(P)} \pmod{P'}$$

for every $x \in A_{L'}$, hence for every $x \in A_L$:

$$\left[\rho\left(\frac{L'|K}{P'}\right)\right](x) \equiv x^{N(P)} \pmod{\widetilde{P}},$$

so

$$\rho\left(\frac{L'|K}{P'}\right) = \left(\frac{L|K}{\widetilde{P}}\right).$$

It follows from (**B**) that

$$\rho\left(\frac{L'|K}{P}\right) \subseteq \left(\frac{L|K}{P}\right).$$

But every automorphism in $\mathcal{L}$ belonging to $\left(\dfrac{L|K}{\widetilde{P}}\right)$ is the restriction of a conjugate in $\mathcal{L}'$ to $\left(\dfrac{L'|K}{P'}\right)$. This concludes the proof. ∎

For the next proposition, let $L|K$, $L'|K$ be Galois extensions of number fields, where we no longer assume that $L \subset L'$. Let $\mathcal{L} = G(L|K)$, $\mathcal{L}' = G(L'|K)$. Let $M = LL'$, so $M|K$ is a Galois extension, with Galois group denoted by $\mathcal{M}$. Let $\rho : \mathcal{M} \to \mathcal{L}$, $\rho' : \mathcal{M} \to \mathcal{L}'$ be the canonical restriction mappings, and let $\widetilde{\rho} : \mathcal{M} \to \mathcal{L} \times \mathcal{L}'$ be the homomorphism

$$\sigma \mapsto (\rho(\sigma), \rho'(\sigma)) = (\rho(\sigma), \varepsilon) \cdot (\varepsilon, \rho'(\sigma)),$$

where ε is the identity automorphism.

Identifying $\mathcal{L}$ with the subgroup $\mathcal{L} \times \{\varepsilon\}$ of $\mathcal{L} \times \mathcal{L}'$ and $\mathcal{L}'$ with $\{\varepsilon\} \times \mathcal{L}'$, then $\widetilde{\rho}$ has the image contained in $\mathcal{L}\mathcal{L}'$.

D.   *With these hypotheses and notations:*

$$\widetilde{\rho}\left(\frac{M|K}{P}\right) \subseteq \left(\frac{L|K}{P}\right)\left(\frac{L'|K}{P}\right).$$

Proof: Let Q be a prime ideal in M such that Q divides $A_M P$. Let $R = Q \cap L$, $R' = Q \cap L'$. Let $\sigma = \left(\dfrac{M|K}{Q}\right)$. By (**C**), $\rho(\sigma) = \left(\dfrac{L|K}{R}\right)$, $\rho'(\sigma) = \left(\dfrac{L'|K}{R'}\right)$. So

$$\widetilde{\rho}(\sigma) = (\rho(\sigma), \varepsilon)(\varepsilon, \rho'(\sigma)) = \left(\frac{L|K}{R}\right)\left(\frac{L'|K}{R'}\right).$$

This shows that

$$\widetilde{\rho}\left(\frac{M|K}{P}\right) \subseteq \left(\frac{L|K}{P}\right)\left(\frac{L'|K}{P}\right).$$ ∎

Now let $L|K$ be a Galois extension of number fields, and $M|K$ an arbitrary extension of number fields, so $LM|M$ is a Galois extension. Let

$\mathcal{L} = G(L|K)$, $\mathcal{M} = G(LM|M)$, so the restriction mapping ρ identifies $\mathcal{M}$ with the subgroup $G(L|L \cap M)$ of $\mathcal{L}$. Let P be a prime ideal in K, and Q a prime ideal in M such that $Q \cap K = P$. For any subset $\mathcal{H}$ of $\mathcal{L}$ and $f \geq 1$, let $\mathcal{H}^f = \{\sigma^f \mid \sigma \in \mathcal{H}\}$.

E.   *With the above notations, we have:*

If $P \in \mathcal{U}(L|K)$ then $Q \in \mathcal{U}(LM|M)$ and

$$\rho\left(\frac{LM|M}{Q}\right) \subseteq \left(\frac{L|K}{P}\right)^f,$$

where f is the inertial degree of Q in $M|K$.

Proof: Let $\widetilde{Q}$ be a prime ideal in LM such that $\widetilde{Q} \cap M = Q$; let $\widetilde{P} = \widetilde{Q} \cap L$, so $\widetilde{P} \cap K = P$. By hypothesis, $\widetilde{P}$ is unramified over P, so the inertial group $\mathcal{T}_{\widetilde{P}}(L|K)$ is trivial, hence also $\mathcal{T}_{\widetilde{P}}(L|L \cap M)$ is trivial. If $\sigma \in \mathcal{T}_{\widetilde{Q}}(LM|M)$ by Chapter 14, remark after (**D**), $\sigma(x) \equiv x \pmod{\widetilde{Q}}$ for every $x \in A_M$. So $(\rho(\sigma))(x) \equiv x \pmod{\widetilde{P}}$ for every $x \in A_L$. Since $(\rho(\sigma))(\widetilde{P}) = \widetilde{P}$ by the above remark, $\rho(\sigma) \in \mathcal{T}_{\widetilde{P}}(L|L \cap M)$ so $\rho(\sigma)$ is the identity automorphism.

Since ρ is an isomorphism, then σ is the identity automorphism, showing that $\mathcal{T}_{\widetilde{Q}}(LM|M) = \{\varepsilon\}$ and therefore Q is unramified in $LM|M$.

Let $\sigma = \left(\dfrac{LM|M}{\widetilde{Q}}\right)$, so $\sigma(x) \equiv x^{N(Q)} \pmod{\widetilde{Q}}$ for every $x \in A_{LM}$.

From $N(Q) = N(P)^f$ then $(\rho(\sigma))(x) \equiv x^{N(P)^f} \pmod{\widetilde{P}}$, for every $x \in A_L$. Thus $\rho(\sigma) = \left(\dfrac{L|K}{\widetilde{P}}\right)^f$. This shows that

$$\rho\left(\frac{LM|M}{Q}\right) \subseteq \left(\frac{L|K}{P}\right)^f. \qquad \blacksquare$$

Now let $L|K$ be an extension of number fields. Let $L = L_1, L_2, \ldots, L_m$ be the conjugates of L over K and let $M = L_1 L_2 \cdots L_m$, so $M|K$ is a Galois extension of number fields; M is the smallest Galois extension of K containing L. Let Q be a prime ideal in A_M, $P_i = Q \cap L_i$, $P = Q \cap K$.

Assume that $P = \mathcal{U}(L|K)$. Then by applying conjugation, $P \in \mathcal{U}(L_i|K)$ for $i = 1, \ldots, m$. By Chapter 13, (**U**), $P \in \mathcal{U}(M|K)$.

F.   *With the above hypotheses and notations: If $\sigma = \left(\dfrac{M|K}{Q}\right)$ then the smallest power of σ in $G(M|L)$ is σ^f, where f is the inertial degree of $P = Q \cap L$ in $L|K$.*

Proof: The decomposition group $\mathcal{Z}_Q(M|K)$ is cyclic with generator σ and order $f_Q(M|K)$. The decomposition group $\mathcal{Z}_Q(M|L) = G(M|L) \cap$

$\mathcal{Z}_Q(M|K)$ is cyclic with order $f_Q(M|L)$. Then

$$f_P(L|K) = \frac{f_Q(M|K)}{f_Q(M|L)} = (\mathcal{Z}_Q(M|K) : \mathcal{Z}_Q(M|L))$$

is the smallest integer k such that $\sigma^k \in \mathcal{Z}_Q(M|L)$, that is, $\sigma^k \in G(M|L)$. ∎

25.2 Density Results on the Decomposition of Prime Ideals

In Chapter 24, before (**B**), we introduced the density of a set of prime numbers. We shall extend this notion and define the density of a set of prime ideals in the field K.

We begin with a preliminary result.

G.   *Let $\mathcal{P}$ be a set of nonzero prime ideals in the number field K. Then the limit*

$$\lim_{s \to 1+0} \frac{\sum_{P \in \mathcal{P}} 1/N(P)^s}{\log(1/(s-1))} \tag{25.3}$$

exists if and only if the limit

$$\lim_{s \to 1+0} \frac{\log \prod_{P \in \mathcal{P}} 1/(1 - 1/N(P)^s)}{\log \zeta_K(s)} \tag{25.4}$$

exists. In this case, these limits are equal.

Proof: Let $\mathcal{J}$ be the set of ideals in K which are products of ideals $P \in \mathcal{P}$. By Chapter 23, (**D**) and (**F**), we have

$$\log \prod_{P \in \mathcal{P}} \frac{1}{1 - 1/N(P)^s} = \log \sum_{J \in \mathcal{J}} \frac{1}{N(J)^s}$$

$$\approx \sum_{P \in \mathcal{P}} \frac{1}{N(P)^s} \qquad \text{for} \quad s \to 1+0.$$

By Chapter 24, (24.2), we have

$$\log \zeta_K(s) \approx \log \frac{1}{s-1} \qquad \text{for} \quad s \to 1+0.$$

From these facts, the result follows at once. ∎

Definition 3.   A set $\mathcal{P}$ of nonzero prime ideals in K has *Dirichlet density* α if the limit

$$\lim_{s \to 1+0} \frac{\sum_{P \in \mathcal{P}} 1/N(P)^s}{\log(1/(s-1))}$$

exists and is equal to α. By (**G**), it is equivalent to saying that the limit

$$\lim_{s \to 1+0} \frac{\log \prod_{P \in \mathcal{P}} 1/(1 - 1/N(P)^s)}{\log \zeta_K(s)}$$

exists and is equal to α.

We use the notation $\mathrm{Dd}(\mathcal{P}) = \alpha$ to express the above fact.

We note that if the set $\mathcal{P}$ has Dirichlet density α, then $0 \le \alpha \le 1$. Indeed,

$$1 \le \prod_{P \in \mathcal{P}} \frac{1}{1 - 1/N(P)^s} \le \zeta_K(s).$$

Also, if $\mathrm{Dd}(\mathcal{P}) = \alpha \ne 0$, then the set $\mathcal{P}$ is infinite.

We recall that it was proved in Chapter 24, (**B**), that the set of prime numbers in the arithmetic progression $\{a + km \mid k = 0, 1, 2, \ldots\}$ (where $1 \le a < m$, $\gcd(a, m) = 1$), has Dirichlet density $1/\varphi(m)$.

We state explicitly the following fact, which is trivial to verify. Let $\mathcal{S}_1$, $\mathcal{S}_2$, $\ldots$, $\mathcal{S}_k$ be pairwise disjoint sets of nonzero prime ideals in K. Assume that each set $\mathcal{S}_i$ has Dirichlet density. Then $\mathcal{S}_1 \cup \cdots \cup \mathcal{S}_k$ has Dirichlet density, namely

$$\mathrm{Dd}(\mathcal{S}_1 \cup \cdots \cup \mathcal{S}_k) = \sum_{i=1}^{k} \mathrm{Dd}(\mathcal{S}_i).$$

Let $L|K$ be an extension of degree n of number fields. For $f = 1, \ldots, n$ let

$$\mathcal{P}_f = \{P \text{ nonzero prime ideal in } L \mid P \text{ has inertial degree } f \text{ in } L|K\}.$$

Then the set $\mathcal{P}$ of all nonzero prime ideals in L is the union of the pairwise disjoint sets $\mathcal{P}_f$ for $f = 1, \ldots, n$.

We have:

H.   *For $f \ge 2$: $\mathrm{Dd}(\mathcal{P}_f) = 0$; for $f = 1$: $\mathrm{Dd}(\mathcal{P}_1) = 1$.*

Proof: By Chapter 23, (**E**), if $f \ge 2$, $s > 1$:

$$1 \le \prod_{P \in \mathcal{P}_f} \frac{1}{1 - 1/N(P)^s} \le [\zeta_K(f)]^n.$$

Since $\lim_{s \to 1+0} \zeta_K(s) = \infty$, it follows that

$$\lim_{s \to 1+0} \frac{\log \prod_{P \in \mathcal{P}_f} 1/(1 - 1/N(P)^s)}{\log \zeta_K(s)} = 0,$$

i.e., $\mathrm{Dd}(\mathcal{P}_f) = 0$.

By a previous remark, since $\mathcal{P}$ has Dirichlet density equal to 1, then $\mathrm{Dd}(\mathcal{P}_1) = 1$.   ∎

In particular, there exist infinitely many prime ideals P in L with inertial degree 1.

Let $S(L|K)$ be the set of nonzero prime ideals in K which are totally decomposed in $L|K$. Thus $S(L|K) \subseteq U(L|K)$ and if $\underline{P} \in S(L|K)$ and $A_L\underline{P} = P_1 \cdots P_n$ then each P_i has inertial degree 1 over K, so $P_i \in \mathcal{P}_1$. First we note:

I.   *The sets $S(\cdot|\cdot)$ have the following properties:*

   (1) *If $K \subset L \subset M$ are number fields, then $S(M|K) \subseteq S(L|K)$.*

   (2) *If $L_1|K$, $L_2|K$ are Galois extensions of number fields, then*

$$S(L_1 L_2|K) = S(L_1|K) \cap S(L_2|K).$$

Proof: (1)   Let $\underline{P} \in U(M|K) \subseteq U(L|K)$, and let $A_L\underline{P} = P_1 \cdots P_g$. For each P_i let $A_M P_i = Q_{i1} \cdots Q_{ig_i}$ where the ideals Q_{ij} ($i = 1, \ldots, g$; $j = 1, \ldots, g_i$) are all distinct. The number of prime ideals Q_{ij} dividing $A_M\underline{P}$ is $g' = \sum_{i=1}^{g} g_i$ where each $g_i \leq [M : L]$, $g \leq [L : K]$. If $g' = [M : K]$ then necessarily $g = [L : K]$, thus $\underline{P} \in S(L|K)$.
   (2)   By (1), we have the inclusion

$$S(LL'|K) \subseteq S(L|K) \cap S(L'|K).$$

Conversely, let

$$\underline{P} \in S(L|K) \cap S(L'|K),$$

so the decomposition groups of $\underline{P}$ in $L|K$ and in $L'|K$ are trivial. If σ is in the decomposition group of $\underline{P}$ in $LL'|K$, then its restrictions to $G(L|K)$ and to $G(L'|K)$ are in the decomposition groups of $\underline{P}$ in $L|K$ (respectively, $L'|K$); so they are equal to the identity automorphisms. Therefore σ is the identity and this implies that $\underline{P}$ is totally decomposed in $LL'|K$, proving the reverse inclusion.   ∎

Now we show:

J.   *Let $L|K$ be an extension of number fields of degree n. Then*

$$\mathrm{Dd}(S(L|K)) = \frac{1}{n}.$$

Proof: For every $\underline{P} \in S(L|K)$ we have $A_L\underline{P} = P_1 \cdots P_n$ with $P_i \in \mathcal{P}_1$, hence $N(P_i) = N(\underline{P})$ where $\underline{P} = N_{L|K}(P_i)$. Thus

$$\prod_{P \in \mathcal{P}_1} \frac{1}{1 - 1/N(P)^s} = \prod_{\underline{P} \in S(L|K)} \left(\frac{1}{1 - 1/N(\underline{P})^s} \right)^n.$$

It follows that

$$\lim_{s\to 1+0} \frac{\log \displaystyle\prod_{\underline{P}\in \mathcal{S}(L|K)} \frac{1}{1-1/N(\underline{P})^s}}{\log \zeta_K(s)} = \lim_{s\to 1+0} \frac{\frac{1}{n}\log \displaystyle\prod_{P\in \mathcal{P}_1} \frac{1}{1-1/N(P)^s}}{\log \zeta_K(s)}$$

$$= \lim_{s\to 1+0} \frac{\frac{1}{n}\displaystyle\sum_{P\in \mathcal{P}_1} \frac{1}{N(P)^s}}{\log(1/(s-1))} = \frac{1}{n},$$

where we need (**H**) and Chapter 23, (**D**), (**F**), as well as the fact that $\log \zeta_K(s) \approx \log(1/(s-1))$ for $s \to 1+0$. ∎

We obtain the following corollary, which takes into account Theorem 2 of Chapter 11. We keep the notations of the theorem. Let K be a number field, $L = K(t)$, where t is algebraic over K. Let $F \in A_K[X]$ be the minimal polynomial of t over K. Let n be the degree of F. Assume that either:

(a) A_K is a principal ideal domain; or

(b) $A_L = A_K[t]$ holds.

K.   *With the above hypotheses and notations, the set of prime ideals $\underline{P}$ in K such that F modulo $\underline{P}$ is the product of distinct linear factors, has density $1/n$.*

Proof: According to Theorem 2 of Chapter 11 with the possible exception, in case (a), of the finitely many prime ideals $\underline{P}$ which divide $A_K\alpha$ (where α was defined in Chapter 11), the set of prime ideals under consideration is equal to $\mathcal{S}(L|K)$. By (**J**), it has density $1/n$, where $n = [L : K]$ is the degree of F. ∎

The above result is applicable when $K = \mathbb{Q}$ and tells us that if $F \in \mathbb{Z}[X]$ is an irreducible polynomial of degree n, the set of prime numbers p such that F modulo p is a product of distinct linear factors has Dirichlet density $1/n$, so it is infinite.

The result which follows tells us that a Galois extension of a number field is uniquely determined by the knowledge of the prime ideals which are totally decomposed.

Precisely:

L.   *Let $L|K$, $L'|K$ be Galois extensions of number fields. Then the following conditions are equivalent:*

(1) $L = L'$;

(2) $\mathcal{S}(L|K) = \mathcal{S}(L'|K)$; and

(3) $\mathrm{Dd}(\mathcal{S}(L|K) \setminus \mathcal{S}(L'|K)) = \mathrm{Dd}(\mathcal{S}(L'|K) \setminus \mathcal{S}(L|K)) = 0$.

Proof: It suffices to show that condition (3) implies (1). Let $L'' = LL'$ so $L''|K$ is a Galois extension. Let $[L : K] = n$, $[L' : K] = n'$, and $[L'' : K] =$

n''. By (**I**), $\mathcal{S}(L''|K) = \mathcal{S}(L|K) \cap \mathcal{S}(L'|K)$, hence $\mathcal{S}(L|K) \setminus \mathcal{S}(L''|K) = \mathcal{S}(L|K) \setminus \mathcal{S}(L'|K)$ so $\mathrm{Dd}(\mathcal{S}(L|K) \setminus \mathcal{S}(L''|K)) = 0$.

From (**I**):

$$\mathcal{S}(L|K) = \mathcal{S}(L''|K) \cup (\mathcal{S}(L|K) \setminus \mathcal{S}(L''|K)),$$

so by (**J**), $1/n = 1/n''$, hence $n = n''$. Similarly $n' = n''$, thus $L = L'$. ∎

Thus, if $L \neq L'$ there exist infinitely many prime ideals $\underline{P}$ in K which are totally decomposed in one, but not in the other, of the fields L, L'. This result will be extended later in Section 25.4.

25.3 The Theorem of Chebotarev

Let $L|K$ be a Galois extension of number fields. For each conjugacy class C of $G(L|K)$, let

$$\mathcal{A}_C = \left\{ P \text{ prime ideal of } K \ \Bigg| \ \left(\frac{L|K}{P} \right) = C \right\}.$$

In this section, we shall prove the following theorem of Chebotarev:

Theorem 1.   *With the above hypothesis and notations:*

$$\mathrm{Dd}(\mathcal{A}_C) = \frac{\#(C)}{[L : K]}.$$

It is appropriate to recall that if $C = \{\varepsilon\}$ then $\mathcal{A}_{\{\varepsilon\}} = \mathcal{S}(L|K)$. In this case, the theorem becomes the statement (**J**). If $K = \mathbb{Q}$, $L = \mathbb{Q}(\zeta_m)$ where ζ_m is a primitive root of 1 of order m, then the theorem is a rephrasement of Dirichlet's theorem on primes in arithmetic progressions.

We shall prove Chebotarev's theorem by considering in succession several special cases.

It is convenient to begin with some lemmas about cyclic groups.

If $\mathcal{G}$ is a finite group, $h \geq 1$, let $\nu(\mathcal{G}, h) = \#\{\sigma \in \mathcal{G} \mid h \text{ divides the order of } \sigma\}$.

Lemma 1.   *Let $\mathcal{G}$ be a cyclic group of order $g = p_1^{d_1} \cdots p_r^{d_r}$ with $p_1, \ldots, p_r$ distinct primes, $d_i \geq 1$ for all $i = 1, \ldots, r$. Let $h = p_1^{e_1} \cdots p_r^{e_r}$ with $1 \leq e_i \leq d_i$ for all $i = 1, \ldots, r$. Then*

$$\nu(\mathcal{G}, h) = \prod_{i=1}^{r} (p_i^{d_i} - p_i^{e_i - 1}).$$

Proof: First we assume that $r = 1$, that is, $\mathcal{G}$ has order $g = p^d$, $h = p^e$, $1 \leq e \leq d$. Let σ be a generator of $\mathcal{G}$. The order of σ^s for $0 \leq s \leq p^d$ does not divide p^e if and only if $\sigma^{sp^{e-1}} = \varepsilon$; equivalently, p^d divides sp^{e-1}. This happens for $s = p^{d-e+1}, 2p^{d-e+1}, \ldots, p^{e-1}p^{d-e+1}$. So there are exactly p^{e-1} elements with order not divisible by p^e, hence $p^d - p^{e-1}$ elements with order divisible by p^e. This proves the lemma when $r = 1$. Now we consider the general case.

For $i = 1, \ldots, r$, let $g_i = g/p_i^{d_i}$. Since these integers are relatively prime there exist integers n_i such that $1 = \sum_{i=1}^r n_i g_i$; we note that p_i does not divide n_i for each $i = 1, \ldots, r$. Let $\sigma_i = \sigma^{n_i g_i}$ where σ is a generator of $\mathcal{G}$; then σ_i has order $p_i^{d_i}$. Let $\mathcal{G}_i$ be the cyclic group generated by σ_i. The group $\mathcal{G}$ is isomorphic to $\mathcal{G}_1 \times \cdots \times \mathcal{G}_r$ by the mapping $\sigma^s \mapsto (\sigma_1^{s_1}, \ldots, \sigma_r^{s_r})$, where $s \equiv s_i \pmod{p_i^{d_i}}$, $0 \le s_i < p_i^{d_i}$.

Indeed, the mapping is a group-homomorphism and a bijection, as follows from the Chinese remainder theorem.

We note that if $0 \le s < g$ the order of σ^s is divisible by h if and only if for every $i = 1, \ldots, r$ the order of $\sigma_i^{s_i}$ is divisible by $p_i^{e_i}$. This follows easily, because the order of σ^s is the product of the orders of $\sigma_i^{s_i}$ for $i = 1, \ldots, r$. By the first part of the proof

$$\nu(\mathcal{G}, h) = \prod_{i=1}^r (p_i^{d_i} - p_i^{e_i-1}). \qquad \blacksquare$$

Lemma 2.   Let $h \ge 1$. For every $\varepsilon > 0$ there exists a cyclic group $\mathcal{G}$ such that

$$1 - \varepsilon < \frac{\nu(\mathcal{G}, h)}{\#(\mathcal{G})} \le 1.$$

Proof: Let $h = p_1^{e_1} \cdots p_r^{e_r}$ where $p_1, \ldots, p_r$ are distinct primes, each $e_i \ge 1$. Given $\varepsilon > 0$, let $d_1, \ldots, d_r$ be sufficiently large. If $\mathcal{G}$ is a cyclic group of order $p_1^{d_1} \cdots p_r^{d_r}$ then, by Lemma 1, we have

$$\frac{\nu(\mathcal{G}, h)}{\#(\mathcal{G})} = \prod_{i=1}^r (1 - p_i^{e_i-1-d_i}) > 1 - \varepsilon. \qquad \blacksquare$$

We remark that in order to prove the theorem, it suffices to show that for every conjugacy class C of $G(L|K)$:

$$\mathrm{Dd}(\mathcal{A}_C) \ge \#(C)/[L : K].$$

Indeed, this implies that

$$1 = \sum_C \mathrm{Dd}(\mathcal{A}_C) \ge \sum_C \frac{\#(C)}{[L : K]} = \frac{1}{[L : K]} \sum_C \#(C)$$

$$= \frac{\#G(L|K)}{[L : K]} = 1$$

hence necessarily $\mathrm{Dd}(\mathcal{A}_C) = \#(C)/[L : K]$ for every conjugacy class C of $G(L|K)$.

Now we prove the theorem by establishing in succession several special cases.

Special Case 1: $K = \mathbb{Q}$, $L = \mathbb{Q}(\zeta)$, where $\zeta = \zeta_m$ is a primitive root of 1 of order $m > 1$.

For each r, $1 \leq r < m$, with $\gcd(r, m) = 1$, let $\sigma_r \in G(L|\mathbb{Q})$ be uniquely defined by $\sigma_r(\zeta) = \zeta^r$. We recall from Chapter 16, (**D**), that a prime p is unramified in $L|\mathbb{Q}$ if and only if p does not divide m.

We show: if p is unramified in $L|\mathbb{Q}$ then $\left(\dfrac{L|\mathbb{Q}}{p}\right) = \{\sigma_r\}$ if and only if $p \equiv r \pmod{m}$.

Indeed, let P be a prime ideal in L such that $P \cap \mathbb{Q} = \mathbb{Z}p$. If $p \equiv r \pmod{m}$ then $\zeta^p = \zeta^r$, hence $\sigma_r(\zeta) = \zeta^p$. By Chapter 5, (**X**), the ring of algebraic integers of L is $A_L = \mathbb{Z}[\zeta]$, so for every

$$x = \sum_{i=0}^{\varphi(m)-1} a_i \zeta^i \in A_L,$$

$$\sigma_r(x) = \sum_{i=0}^{\varphi(m)-1} a_i \zeta^{ip} \equiv \left(\sum_{i=0}^{\varphi(m)-1} a_i \zeta^i \right)^p \equiv x^p \pmod{P}.$$

Then $\left(\dfrac{L|\mathbb{Q}}{P}\right) = \{\sigma_r\}$.

Conversely, if $\left(\dfrac{L|\mathbb{Q}}{P}\right) = \{\sigma_r\}$, since $p \nmid m$ there exists r', $1 \leq r' < m$, with $\gcd(r', m) = 1$ such that $p \equiv r' \pmod{m}$. By the above proof $\left(\dfrac{L|\mathbb{Q}}{P}\right) = \{\sigma_{r'}\}$ hence $r' \equiv r \pmod{m}$.

Let

$$\mathcal{A}_r = \mathcal{A}_{\{\sigma_r\}} = \left\{ p \text{ prime} \ \middle| \ \left(\frac{L|\mathbb{Q}}{P}\right) = \{\sigma_r\} \right\}.$$

It follows that

$$\sum_{p \in \mathcal{A}_r} \frac{1}{p^s} = \sum_{p \equiv r \pmod{m}} \frac{1}{p^s}$$

and by Chapter 24, (**B**):

$$\mathrm{Dd}(\mathcal{A}_r) = \lim_{s \to 1+0} \frac{\sum_{P \in \mathcal{A}_r} 1/p^s}{\log(1/(s-1))} = \lim_{s \to 1+0} \frac{\sum_{p \equiv r \pmod{m}} 1/p^s}{\log(1/(s-1))} = \frac{1}{\varphi(m)}$$

as it was required to prove.

Special Case 2:   Let $m > 1$, and let $\zeta = \zeta_m$ be a primitive root of 1 of order m. Let $L = K(\zeta)$ and denote by $A = A_K$ the ring of integers of K.

So $L|K$ is a Galois extension with

$$G(L|K) \cong G(\mathbb{Q}(\zeta)|\mathbb{Q}(\zeta) \cap K) \subseteq G(\mathbb{Q}(\zeta)|\mathbb{Q}).$$

By restriction, for every $\sigma \in G(L|K)$ there exists r, $1 \leq r < m$, with $\gcd(r, m) = 1$, such that σ is identified with $\sigma_r \in G(\mathbb{Q}(\zeta)|\mathbb{Q})$.

Let

$$\mathcal{A}_r = \left\{ P \text{ prime ideal in } K \,\middle|\, \left(\frac{L|K}{P}\right) = \{\sigma_r\} \right\}.$$

We wish to show that $\mathrm{Dd}(\mathcal{A}_r) = 1/[L : K]$. The proof will be subdivided into steps.

(1°)   We need to evaluate

$$\sum_{P \in \mathcal{A}_r} \frac{1}{N(P)^s}.$$

For each $f = 1, 2, \ldots, [K : \mathbb{Q}]$ let $\mathcal{P}_f$ be the set of all prime ideals P in K with inertial degree f over $\mathbb{Q}$.

It follows from Chapter 23, (E), that

$$\sum_{P \in \mathcal{A}_r} \frac{1}{N(P)^s} \approx \sum_{P \in \mathcal{A}_r \cap \mathcal{P}_1} \frac{1}{N(P)^s} \qquad \text{as} \quad s \to 1 + 0.$$

We show that $P \in \mathcal{A}_r \cap \mathcal{P}_1$ if and only if $N(P) \equiv r \pmod{m}$ and P has inertial degree 1 over $\mathbb{Q}$.

Indeed, if $P \in \mathcal{A}_r \cap \mathcal{P}_1$ then, on the one hand, $\left(\frac{L|K}{P}\right)(\zeta) \equiv \zeta^p \pmod{P}$ where $p = N(P)$; on the other hand, $\left(\frac{L|K}{P}\right)(\zeta) = \zeta^r$, so $\zeta^p \equiv \zeta^r$ $\pmod{P}$. If $p \not\equiv r \pmod{m}$, then P divides $A(1 - \zeta^{p-r}) = A(1 - \zeta)$ which in turn divides Am (by Chapter 24, (I)); but this is not possible since P is unramified in $L|K$. So $N(P) \equiv r \pmod{m}$.

Conversely, if $N(P) = p \equiv r \pmod{m}$, then $\left(\frac{L|K}{P}\right)(\zeta) \equiv \zeta^p = \zeta^r$ $\pmod{P}$; if $\left(\frac{L|K}{P}\right) = \sigma_{r'}$, then $\left(\frac{L|K}{P}\right)(\zeta) = \zeta^{r'}$ hence P divides $A(\zeta^r - \zeta^{r'})$, which by the above argument implies that $r' \equiv r \pmod{m}$, so $\left(\frac{L|K}{P}\right)(\zeta) = \zeta^r$ and, as before, we conclude that $\left(\frac{L|K}{P}\right) = \sigma_r$.

From this we deduce that

$$\sum_{P \in \mathcal{A}_r \cap \mathcal{P}_1} \frac{1}{N(P)^s} = \sum_{\substack{N(P) \equiv r \pmod{m} \\ P \in \mathcal{P}_1}} \frac{1}{N(P)^s} \approx \sum_{N(P) \equiv r \pmod{m}} \frac{1}{N(P)^s},$$

with the preceding argument.

(2°)   We shall evaluate

$$\sum_{N(P) \equiv r \pmod{m}} \frac{1}{N(P)^s}.$$

in an indirect way. For this purpose, as in Chapter 8, let $\mathcal{F}_m$ denote the multiplicative group of nonzero fractional ideals of K which are relatively prime to Am; let $\mathcal{P}r_{m,+}$ be the subgroup of $\mathcal{F}_m$ consisting of the principal ideals Ax such that x is totally positive and $x \equiv 1 \pmod{Am}$. The quotient group $\mathcal{C}_m$ is finite with order denoted by $h^* = h_{m,+}$. The elements of $\mathcal{C}_m$ are equivalence classes of ideals belonging to $\mathcal{F}_m$; namely I, J are equivalent if there exists $Ax \in \mathcal{P}r_{m,+}$, such that $I = Ax \cdot J$.

We observe that in every class there is some integral ideal. Indeed, given $I = J/a$ with $\gcd(Aa, Am) = \gcd(J, Am) = A$, then $a^{2N(Am)}$ is totally positive and $a^{2N(Am)} \equiv 1 \pmod{Am}$, since $N(Am) = \#(A/Am)$.

Let $I' = Aa^{2N(Am)-1}J$, so I' is an integral ideal and $I' = a^{2N(Am)}I$.

This remark allows us to consider a system of representatives $\{I_1, I_2, \ldots, I_{h^*}\}$ of $\mathcal{C}_m$ consisting of integral ideals.

(3°)   For each r, $1 \leq r < m$, $\gcd(r, m) = 1$, let

$$\mathcal{I}_r = \{I \in \mathcal{F}_m \mid N(I) \equiv r \pmod{m}\}.$$

Let $M = \{r \mid \mathcal{I}_r \neq \varnothing\}$. We remark that $M \neq \varnothing$. Indeed, let q be a prime number not dividing m, let Q be a prime ideal in K dividing Aq. So $N(Q)$ is a power of q, hence there exists r, as above, such that $N(Q) \equiv r \pmod{m}$, so $\mathcal{I}_r \neq \varnothing$ and therefore $M \neq \varnothing$. Let $\mu \geq 1$ be the number of elements of M.

(4°)   We show that if $\mathcal{I}_r \neq \varnothing$ then $\mathcal{I}_r$ is the union of equivalence classes of $\mathcal{F}_m$.

Indeed, let $I \in \mathcal{I}_r$ and let $J \in \mathcal{F}_m$ belong to the same class as I. So there exists $Ax \in \mathcal{P}r_{m,+}$ such that $J = Ax \cdot I$. Then $x \equiv 1 \pmod{Am}$. For every conjugation τ of K, we have $\tau(x) \equiv 1 \pmod{\tau(A)m}$. Hence, as easily seen, $N_{K|\mathbb{Q}}(x) \equiv 1 \pmod{m}$. Since x is totally positive, then $N(Ax) = |N_{K|\mathbb{Q}}(x)| = N_{K|\mathbb{Q}}(x) \equiv 1 \pmod{m}$. Hence $N(J) \equiv N(I) \pmod{m}$ and $J \in \mathcal{I}_r$.

Let $\nu_r = \#(\mathcal{I}_r)$. We show that if $\nu_r \neq 0$, $\nu_{r'} \neq 0$ then $\nu_r = \nu_{r'}$.

Indeed, let $I \in \mathcal{I}_r$, $I' \in \mathcal{I}_{r'}$, and let $J \in \mathcal{F}_m$ be such that $I' = JI$. Then I, I_1 belong to the same equivalence class in $\mathcal{F}_m$ if and only if $I' = JI$ and $I'_1 = JI_1$ belong to the same equivalence class. This shows that $\nu_r = \nu_{r'}$.

Let $\nu = \nu_r$ for any $r \in M$.

(5°)   Let $\widetilde{\chi}_0$ (trivial character), $\widetilde{\chi}_1$, $\ldots$, $\widetilde{\chi}_{h^*-1}$ be the characters of the Abelian group $\mathcal{C}_m$. Let χ_0 (trivial character), χ_1, $\ldots$, χ_{h^*-1} be the associated Hecke characters (see Chapter 21, Section 2).

For the purpose of evaluating the sums

$$\sum_{N(P) \equiv r \pmod{m}} \frac{1}{N(P)^s}$$

for $1 \le r < m$, $\gcd(r, m) = 1$, we shall consider the following sums for each $i = 0, 1, \ldots, h^* - 1$:

$$\sum_P \frac{\chi_i(P)}{N(P)^s} = \sum_{j=1}^{h^*} \chi_i(I_j) \left[\sum_{P \equiv I_j} \frac{1}{N(P)^s} \right], \qquad (25.5)$$

where $\{I_1, I_2, \ldots, I_{h^*}\}$ is a system of representatives of classes of $\mathcal{F}_m$, each I_j is an integral ideal, and $P \equiv I_j$ means that P is in the class of I_j

By Chapter 21, (**J**), for each $j = 1, \ldots, h^*$:

$$\sum_{P \equiv I_j} \frac{1}{N(P)^s} = \frac{1}{h^*} \sum_{i=0}^{h^*-1} \overline{\chi_i(I_j)} \left[\sum_P \frac{\chi_i(P)}{N(P)^s} \right]. \qquad (25.6)$$

If $r \in M$ then

$$\sum_{N(P) \equiv r \ (\text{mod } m)} \frac{1}{N(P)^s} = \sum_{I_j \in \mathcal{I}_r} \sum_{P \equiv I_j} \frac{1}{N(P)^s}$$

$$= \frac{1}{h^*} \sum_{I_j \in \mathcal{I}_r} \sum_{i=0}^{h^*-1} \overline{\chi_i(I_j)} \left[\sum_P \frac{\chi_i(P)}{N(P)^s} \right]$$

(due to the absolute convergence, for $s > 1$, the order of summation is irrelevant).

From Chapter 23, (23.21), if $\chi_i \neq \chi_0$ the sum

$$\sum_P \frac{\chi_i(P)}{N(P)^s}$$

remains bounded when $s \to 1 + 0$, while for χ_0 we have

$$\frac{1}{h^*} \sum_{I_j \in \mathcal{I}_r} \overline{\chi_0(I_j)} \left[\sum_P \frac{\chi_0(P)}{N(P)^s} \right] \approx \frac{1}{h^*} \sum_{I_j \in \mathcal{I}_r} \log L(s|\chi_0) = \frac{\nu}{h^*} \log L(s|\chi_0)$$

$$\approx \frac{\nu}{h^*} \log \frac{1}{s-1}$$

by Chapter 24, (24.3), when $s \to 1 + 0$.

Hence

$$\sum_{N(P) \equiv r \ (\text{mod } m)} \frac{1}{N(P)^s} \approx \frac{\nu}{h^*} \log \frac{1}{s-1} \qquad \text{for} \quad s \to 1 + 0,$$

for every $r \in M$. Thus

$$\mathrm{Dd}(\mathcal{A}_r) = \mathrm{Dd}(\{P \mid N(P) \equiv r \ (\text{mod } m)\}) = \frac{\nu}{h^*}.$$

Adding up for all $r \in M$, $1 = \mu\nu/h^*$.

But

$$\mu \le \#G(\mathbb{Q}(\zeta) \mid \mathbb{Q}(\zeta) \cap K) = [L : K].$$

So $\mathrm{Dd}(\mathcal{A}_r) = \nu/h^* \ge 1/[L : K]$ and by a previous remark, this suffices to prove the theorem in the Special Case 2.

Special Case 3:   We assume that $K \subseteq L \subseteq K(\zeta) = M$, where ζ is a root of 1.

Let $\rho : G(M|K) \to G(L|K)$ be the restriction mapping. If $\sigma \in G(L|K)$ then $\rho^{-1}(\sigma) = \{\tilde{\sigma}_1, \dots, \tilde{\sigma}_m\}$ where $m = [M : L]$. Let

$$\mathcal{A}_{\tilde{\sigma}_i} = \left\{ P \text{ prime ideal in } K \,\middle|\, \left(\frac{M|K}{P}\right) = \tilde{\sigma}_i \right\}$$

for $i = 1, \dots, m$. Then the sets $\mathcal{A}_{\tilde{\sigma}_i}$ are pairwise disjoint and $\mathcal{A}_\sigma = \bigcup_{i=1}^m \mathcal{A}_{\tilde{\sigma}_i}$. Therefore

$$\sum_{P \in \mathcal{A}_\sigma} \frac{1}{N(P)^s} = \sum_{i=1}^m \sum_{P \in \mathcal{A}_{\tilde{\sigma}_i}} \frac{1}{N(P)^s}.$$

By the Special Case 2, we conclude that

$$\lim_{s \to 1+0} \frac{\sum_{P \in \mathcal{A}_\sigma} 1/N(P)^s}{\log(1/(s-1))} = \lim_{s \to 1+0} \frac{\sum_{i=1}^m \sum_{P \in \mathcal{A}_{\tilde{\sigma}_i}} 1/N(P)^s}{\log(1/(s-1))}$$

$$= \sum_{i=1}^m \frac{1}{[M : K]} = \frac{[M : L]}{[M : K]} = \frac{1}{[L : K]}.$$

Special Case 4:   We assume that $L|K$ is a cyclic extension.

(1°)   By Chapter 11, Section 3, for each cyclic group $\mathcal{G}$ there exists a cyclic extension $M|K$ such that $M \cap L = K$, $G(M|K) = \mathcal{G}$, and $K \subseteq M \subseteq K(\zeta)$ where ζ is a root of 1. Let $\mathcal{M}$ be the set of such extensions $M|K$.

Before Lemma 1, we introduced the following notation: $\nu(G(M|K), h)$ denotes the number of $\sigma \in G(M|K)$ whose order is divisible by the integer $h \ge 1$.

If $M \in \mathcal{M}$, $\sigma \in G(L|K)$, and $\tau \in G(M|K)$, let

$$\mathcal{A}_{\sigma,\tau} = \left\{ P \text{ prime ideal in } K \,\middle|\, \left(\frac{L|K}{P}\right) = \{\sigma\}, \left(\frac{M|K}{P}\right) = \{\tau\} \right\}.$$

Let $o(\sigma)$ (respectively, $o(\tau)$), denote the order of σ in $G(L|K)$ (respectively, of τ in $G(M|K)$).

(2°)   We shall show that the Special Case 4 is implied by the following statements:

(a) For every $\varepsilon > 0$ and $h \ge 1$ dividing $[L : K]$, there exists $M \in \mathcal{M}$ such that

$$1 - \varepsilon < \frac{\nu(G(M|K), h)}{[M : K]} \le 1.$$

(b) If $o(\sigma)$ divides $o(\tau)$, then

$$\mathrm{Dd}(\mathcal{A}_{\sigma,\tau}) \geq \frac{1}{[L:K][M:K]}.$$

We show that (a) and (b) imply the Special Case 4. Let $\sigma \in G(L|K)$. Given $\varepsilon > 0$ let $M \in \mathcal{M}$ be such that

$$1 - \varepsilon < \frac{\nu(G(M|K), o(\sigma))}{[M:K]}.$$

For every $\tau \in G(M|K)$ we consider the set $\mathcal{A}_{\sigma,\tau}$. These sets are pairwise disjoint, so

$$\mathrm{Dd}(\mathcal{A}_{\sigma}) = \sum_{\tau \in G(M|K)} \mathrm{Dd}(\mathcal{A}_{\sigma,\tau}) \geq \sum_{o(\sigma)|o(\tau)} \mathrm{Dd}(\mathcal{A}_{\sigma,\tau})$$

$$\geq \sum_{o(\sigma)|o(\tau)} \frac{1}{[L:K][M:K]} = \frac{\nu(G(M|K), o(\sigma))}{[L:K][M:K]}$$

$$> (1 - \varepsilon)\frac{1}{[L:K]}.$$

Since this inequality holds for every $\varepsilon > 0$, then $\mathrm{Dd}(\mathcal{A}_{\sigma}) \geq 1/[L:K]$ for every $\sigma \in G(L|K)$. By a previous remark, this suffices to conclude that

$$\mathrm{Dd}(\mathcal{A}_{\sigma}) = \frac{1}{[L:K]} \qquad \text{for every} \quad \sigma \in G(L|K).$$

(3°)   Proof of (a).
Let $\varepsilon > 0$ and $h \geq 1$, where h divides $[L:K]$. By Lemma 2 there exists a cyclic group $\mathcal{G}$ such that

$$1 - \varepsilon \leq \frac{\nu(\mathcal{G}, h)}{\#(\mathcal{G})} \leq 1.$$

By Chapter 11, Section 3, there exists $M \in \mathcal{M}$ with $G(M|K) = \mathcal{G}$. Hence

$$1 - \varepsilon < \frac{\nu(G(M|K), h)}{[M:K]} \leq 1.$$

(4°)   Proof of (b).
Let $M \in \mathcal{M}$, $\sigma \in G(L|K)$, and $\tau \in G(M|K)$, such that the order $o(\sigma)$ divides the order $o(\tau)$. From $M \cap L = K$ then $G(LM|K) \cong G(L|K) \times G(M|K)$, the isomorphism associates to $\rho \in G(LM|K)$ the pair (ρ_L, ρ_M) of restrictions of ρ to L and to M. Let $\widetilde{\sigma}$ (respectively, $\widetilde{\tau}$) be the canonical extension of σ (respectively, τ) to $G(LM|M)$ (respectively, to $G(LM|L)$).

Consider the cyclic subgroup of $G(L|K) \times G(M|K)$ generated by (σ, τ). By the isomorphism, it corresponds to the cyclic subgroup U generated by $\widetilde{\sigma}\widetilde{\tau}$ since $(\widetilde{\sigma}\widetilde{\tau})_L = \sigma$, $(\widetilde{\sigma}\widetilde{\tau})_M = \tau$. Let H be the subfield of LM fixed by U.

Let

$$\mathcal{A}'_{\sigma,\tau} = \left\{ P \text{ prime ideal in } K \,\middle|\, \left(\frac{LM|K}{P}\right) \in U \text{ and } \left(\frac{M|K}{P}\right) = \tau \right\},$$

let

$$\mathcal{A}''_{\sigma,\tau} = \left\{ P \text{ prime ideal in } K \,\middle|\, \left(\frac{H|K}{P}\right) = \varepsilon \text{ and } \left(\frac{M|K}{P}\right) = \tau \right\},$$

where ε is the identity automorphism, and let

$$\mathcal{A}'''_{\sigma,\tau} = \left\{ P \text{ prime ideal in } K \,\middle|\, P \text{ splits completely in } H \text{ and } \left(\frac{M|K}{P}\right) = \tau \right\}.$$

We have

$$\mathcal{A}_{\sigma,\tau} \supseteq \mathcal{A}'_{\sigma,\tau} = \mathcal{A}''_{\sigma,\tau} = \mathcal{A}'''_{\sigma,\tau}.$$

Indeed, if $P \in \mathcal{A}'_{\sigma,\tau}$ let

$$\left(\frac{LM|K}{P}\right) = (\widetilde{\sigma\tau})^k \qquad \text{so} \qquad \left(\frac{M|K}{P}\right) = (\widetilde{\sigma\tau})^k_M = \tau^k.$$

But $\left(\dfrac{M|K}{P}\right) = \tau$, thus $k \equiv 1 \pmod{o(\tau)}$, hence $k \equiv 1 \pmod{o(\sigma)}$ and

so $\left(\dfrac{L|K}{P}\right) = (\widetilde{\sigma\tau})^k_L = \sigma^k = \sigma$, showing that $P \in \mathcal{A}_{\sigma,\tau}$. The equality $\mathcal{A}'_{\sigma,\tau} = \mathcal{A}''_{\sigma,\tau}$ follows from (**C**) and $\mathcal{A}''_{\sigma,\tau} = \mathcal{A}'''_{\sigma,\tau}$ follows from (**A**).

Now we observe that if P splits completely in $H|K$ there exist $[H : K]$ prime ideals Q in H such that $Q \cap K = P$ and moreover

$$N(Q) = N(N_{H|K}(Q)) = N(P).$$

At this point we note that $\tau \in G(M|M \cap H)$. Indeed, if $x \in M \cap H$ then $\widetilde{\sigma\tau}(x) = x$; but $\widetilde{\sigma}(x) = x$ so $\tau(x) = \widetilde{\tau}(x) = \widetilde{\sigma\tau}(x) = x$. Since $G(HM|M) \cong G(M|M \cap H)$ there exists $\tau' \in G(HM|M)$ such that $\tau'_M = \tau$. Let

$$\mathcal{B} = \left\{ Q \text{ prime ideal in } H \,\middle|\, \left(\frac{MH|H}{Q}\right) = \tau' \right\}.$$

Let

$$Q_1 = \{ Q \text{ prime ideal in } H \mid f_Q(H|K) = 1 \}.$$

Then

$$\sum_{P \in \mathcal{A}_{\sigma,\tau}} \frac{1}{N(P)^s} \geq \sum_{P \in \mathcal{A}'''_{\sigma,\tau}} \frac{1}{N(P)^s}.$$

Since $H \subseteq HM \subseteq H(\zeta)$, by the Special Case 3:

$$\lim_{s \to 1+0} \frac{\sum_{Q \in \mathcal{B}} 1/N(P)^s}{\log(1/(s-1))} = \frac{1}{[MH : H]}$$

hence

$$\lim_{s \to 1+0} \frac{\sum_{P \in \mathcal{A}_{\sigma,r}} 1/N(P)^s}{\log(1/(s-1))} \geq \frac{1}{[H:K]} \cdot \frac{1}{[MH:H]} = \frac{1}{[MH:K]}$$

$$\geq \frac{1}{[LM:K]} = \frac{1}{[L:K]} \cdot \frac{1}{[M:K]}.$$

Proof of the General Case: Let $L|K$ be a Galois extension with Galois group $\mathcal{G}$, let C be a conjugacy class of $\mathcal{G}$, and let $\mathcal{A}_C$ be as indicated. Let $\sigma \in C$, $\mathcal{H}$ the cyclic subgroup of $\mathcal{G}$ generated by σ, and M the subfield of L fixed by σ, so $L|M$ is a Galois extension with Galois group $\mathcal{H}$. Let

$$\mathcal{B}_\sigma = \left\{ Q \text{ prime ideal in } M \,\middle|\, \left(\frac{L|M}{Q}\right) = \sigma \right\}.$$

For each $r = 1, 2 \ldots$, let

$$\mathcal{B}_{\sigma,r} = \{ Q \in \mathcal{B}_\sigma \mid f_Q(M|K) = r \};$$

so $\mathcal{B}_{\sigma,r} = \varnothing$ when $r > (\mathcal{G} : \mathcal{H})$, the sets $\mathcal{B}_{\sigma,r}$ are pairwise disjoint and

$$\mathcal{B}_\sigma = \bigcup_{r \geq 1} \mathcal{B}_{\sigma,r}.$$

(1°)   If $Q \in \mathcal{B}_\sigma$ there exists a unique prime ideal $\widetilde{P}$ in L such that $\widetilde{P} \cap M = Q$ and the decomposition group of $\widetilde{P}$ in $L|M$ is $\mathcal{Z}_{\widetilde{P}}(L|M) = \mathcal{H}$.

Indeed, since $\left(\dfrac{L|M}{Q}\right) = \sigma$ has order equal to $\#(\mathcal{H})$ and, by **(A)**, its order is $f_Q(L|M)$, from $\#(\mathcal{H}) = f_Q(L|M)$ it follows by the fundamental relation that there exists only one prime ideal $\widetilde{P}$ in L with $\widetilde{P} \cap M = Q$. Thus the decomposition group of $\widetilde{P}$ in $L|M$ is the Galois group $\mathcal{H}$.

(2°)   We observe that if $Q \in \mathcal{B}_\sigma$ and $P = Q \cap K$ then $P \in \mathcal{A}_C$.

Indeed, let $\widetilde{P}$ be the prime ideal of L dividing $A_L Q$, so $\widetilde{P} \cap K = P$ and $\left(\dfrac{L|M}{\widetilde{P}}\right) = \left(\dfrac{L|M}{Q}\right) = \sigma$ so $\left(\dfrac{L|K}{\widetilde{P}}\right) = \sigma$ hence $\left(\dfrac{L|K}{P}\right) = C$.

We define $\psi : \mathcal{B}_{\sigma,1} \to \mathcal{A}_C$ by letting $\psi(Q) = P$, where $P = Q \cap K$.

(3°)   *The mapping ψ is surjective.*

Indeed, given $P \in \mathcal{A}_C$, let $\widetilde{P}$ be a prime ideal of L such that $\widetilde{P} \cap K = P$ and $\left(\dfrac{L|K}{\widetilde{P}}\right) = \sigma$; the existence of $\widetilde{P}$, as above, follows at once from **(B)**.

To show this, we let $\widetilde{P}'$ be such that $\widetilde{P}' \cap K = P$, $\left(\dfrac{L|K}{\widetilde{P}'}\right) = \sigma' \in C$.

There exists $\tau \in \mathcal{G}$ such that $\sigma = \tau \sigma' \tau^{-1}$. Let $\widetilde{P} = \tau(\widetilde{P}')$, so $\widetilde{P} \cap K = P$ and

$$\left(\frac{L|K}{\widetilde{P}}\right) = \tau \left(\frac{L|K}{\widetilde{P}'}\right) \tau^{-1} = \tau \sigma' \tau^{-1} = \sigma.$$

Let $Q = \widetilde{P} \cap M$ so $Q \cap K = P$, $\left(\dfrac{L|M}{Q}\right) = \sigma$ thus $Q \in \mathcal{B}_\sigma$.

We still need to show that $Q \in \mathcal{B}_{\sigma,1}$, that is, $f_Q(M|K) = 1$ or, equivalently, $f_{\widetilde{P}}(L|M) = f_{\widetilde{P}}(L|K)$. We have $\left(\dfrac{L|K}{\widetilde{P}}\right) = \sigma$ then, by (**A**),

$f_{\widetilde{P}}(L|K) = \#(\mathcal{H})$; on the other hand, also $\left(\dfrac{L|M}{\widetilde{P}}\right) = \left(\dfrac{L|M}{Q}\right) = \sigma$

so $f_{\widetilde{P}}(L|M) = \#(\mathcal{H})$.

($4°$)   *For every $P \in \mathcal{A}_C$ the number of elements in $\psi(P)$ is $(\mathcal{C}(\sigma) : \mathcal{H})$ where $\mathcal{C}(\sigma)$ is the centralizer of σ in $\mathcal{G}$*

For this purpose we define a mapping $\theta : \mathcal{C}(\sigma) \to \mathcal{B}_{\sigma,1}$. First we note that if $\tau \in \mathcal{C}(\sigma)$ then $\tau(M) = M$ because $\sigma(\tau(x)) = \tau(\sigma(x)) = \tau(x)$ where $x \in M$.

If $\tau \in \mathcal{C}(\sigma)$ let $\theta(\tau) = \tau(Q)$, which is a prime ideal in M. Then

$$\left(\frac{L|M}{\tau(Q)}\right) = \tau\left(\frac{L|M}{Q}\right)\tau^{-1} = \tau\sigma\tau^{-1} = \sigma,$$

so $\tau(Q) \in \mathcal{B}_\sigma$. Also by applying $\tau \in \mathcal{G}$ to Q induces an A_K/P-isomorphism of residue fields $A_M/Q \cong A_M/\tau(Q)$ so $f_Q(M|K) = f_{\tau(Q)}(M|K)$, hence $\tau(Q) \in \mathcal{B}_{\sigma,1}$.

The mapping θ is surjective: if $Q' \in \mathcal{B}_{\sigma,1}$ let $\widetilde{P}'$ be the unique prime ideal of L such that $\widetilde{P}' \cap M = Q'$. Let $\tau \in \mathcal{G}$ be such that $\widetilde{P}' = \tau(\widetilde{P})$. Since

$$\left(\frac{L|K}{\widetilde{P}'}\right) = \left(\frac{L|M}{Q'}\right) = \sigma \quad \text{and} \quad \left(\frac{L|K}{P'}\right) = \tau\left(\frac{L|K}{\widetilde{P}}\right)\tau^{-1} = \tau\sigma\tau^{-1},$$

then $\tau \in \mathcal{C}(\sigma)$ and so $Q' \in \theta(\tau)$.

If $\tau, \tau' \in \mathcal{Z}(\sigma)$ and $\tau^{-1}\tau' \in \mathcal{H}$ then $\tau^{-1}\tau'(Q) = Q$ so $\tau(Q) = \tau'(Q)$, that is, $\theta(\tau) = \theta(\tau')$.

Conversely, if $\tau, \tau' \in \mathcal{C}(\sigma)$ with $\theta(\tau) = \theta(\tau')$, then $\tau^{-1}\tau' \in \mathcal{H}$. Indeed, $\tau(Q) = \tau'(Q)$. If $\widetilde{P}$ is the unique prime ideal of L with $\widetilde{P} \cap M = Q$ then

$$\tau(\widetilde{P}) \cap M = \tau(Q) = \tau'(Q) = \tau'(\widetilde{P}) \cap M,$$

so by ($1°$), $\tau(\widetilde{P}) = \tau'(\widetilde{P})$, hence $\tau^{-1}\tau' \in \mathcal{Z}_{\widetilde{P}}(L|K)$ (decomposition group of $\widetilde{P}$ in $L|K$). By ($1°$):

$$\mathcal{H} = \mathcal{Z}_{\widetilde{P}}(L|M) = \mathcal{H} \cap \mathcal{Z}_{\widetilde{P}}(L|K) \subseteq \mathcal{Z}_{\widetilde{P}}(L|K).$$

But

$$\#\mathcal{Z}_{\widetilde{P}}(L|K) = f_{\widetilde{P}}(L|K) = f_{\widetilde{P}}(L|M) = \#\mathcal{Z}_{\widetilde{P}}(L|M) = \#(\mathcal{H})$$

because $f_Q(M|K) = 1$. So $\tau^{-1}\tau' \in \mathcal{Z}_{\widetilde{P}}(L|K) = \mathcal{H}$. This proves ($4°$).

($5°$)   We finish the proof. Since

$$\{Q \in \mathcal{B}_\sigma \mid f_Q(M|\mathbb{Q}) = 1\} \subseteq \mathcal{B}_{\sigma,1}$$

and $N(Q) = N(P)$ for $Q \in \mathcal{B}_{\sigma,1}$, $P = Q \cap K$, from $(4°)$ and Chapter 23, $(\mathbf{E})$:

$$\sum_{Q \in \mathcal{B}_\sigma} \frac{1}{N(Q)^s} \approx \sum_{Q \in \mathcal{B}_{\sigma,1}} \frac{1}{N(Q)^s} = \sum_{P \in \mathcal{A}_C} \sum_{Q \in \psi^{-1}(P)} \frac{1}{N(Q)^s}$$

$$= (\mathcal{C}(\sigma) : \mathcal{H}) \sum_{P \in \mathcal{A}_C} \frac{1}{N(P)^s}.$$

By the Special Case 4, since $\mathcal{H} = G(L|M)$ is a cyclic group, then $\mathrm{Dd}(\mathcal{B}_\sigma) = 1/\#(\mathcal{H})$. Hence

$$\mathrm{Dd}(\mathcal{A}_C) = \frac{1}{\#(\mathcal{H})} \cdot \frac{1}{(\mathcal{C}(\sigma) : \mathcal{H})} = \frac{1}{\#\mathcal{C}(\sigma)} = \frac{(\mathcal{G} : \mathcal{C}(\sigma))}{\#\mathcal{G}} = \frac{\#(C)}{[L : K]}.$$

This concludes the proof of the theorem.   ∎

25.4 Bauerian Extensions of Fields

Let $L|K$ be an extension of number fields, and let $T(L|K)$ denote the set of prime ideals $P \in \mathcal{U}(L|K)$ such that there exists a prime ideal $\tilde{P}$ in L, dividing $A_L P$ and with inertial degree equal to 1. If $L|K$ is a Galois extension, then $T(L|K) = \mathcal{S}(L|K)$.

In this section we focus on the sets $T(L|K)$ and examine to what extent they determine the extension $L|K$.

Definition 4.   A number field extension $L|K$ is *Bauerian* when the following condition is satisfied: if $M|K$ is any number field extension and $\mathrm{Dd}(T(M|K) \setminus T(L|K)) = 0$ then there exists a field L', $K \subset L' \subset M$, such that L, L' are K-isomorphic.

M.   *If $L|K$ is a Galois extension of number fields, then $L|K$ is Bauerian.*

Proof: Let $M|K$ be an extension of number fields and assume that

$$\mathrm{Dd}(T(M|K) \setminus T(L|K)) = 0.$$

Let M' be the smallest Galois extension of K containing M.

Let $\sigma \in G(L|L \cap M)$, $\varepsilon \in G(M'|K)$ (the identity automorphism), so $\sigma_{L \cap M} = \varepsilon_{L \cap M}$ (restrictions to $L \cap M$). By Chapter 2, Section 7, there exists a unique $\sigma' \in G(LM'|K)$ such that

$$\sigma'_L = \sigma, \qquad \sigma'_{M'} = \varepsilon.$$

Let C' be the conjugacy class of σ' in $G(LM'|K)$ and let C be the conjugacy class of σ in $G(L|K)$. So C is the set of restrictions to L of the elements in C'.

Let

$$\mathcal{A} = \left\{ P \in \mathcal{U}(LM'|K) \,\middle|\, \left(\frac{LM'|K}{P} \right) = C' \right\}.$$

By Chebotarev's theorem, $\mathrm{Dd}(\mathcal{A}) > 0$.

Let

$$\mathcal{B} = \left\{ P \in \mathcal{U}(L|K) \cap \mathcal{U}(M'|K) \,\middle|\, P \in \mathcal{T}(M|K) \text{ and } \left(\frac{L|K}{P} \right) = C \right\}.$$

Then $\mathcal{A} \subseteq \mathcal{B}$. Indeed, if $P \in Q$, by (C), $\left(\dfrac{L|K}{P} \right) = C$, $\left(\dfrac{M'|K}{P} \right) = \varepsilon$, so $P \in \mathcal{T}(M'|K) \subseteq \mathcal{T}(M|K)$. Thus $\mathrm{Dd}(\mathcal{B}) > 0$.

Let

$$\mathcal{C} = \left\{ P \in \mathcal{U}(L|K) \,\middle|\, \left(\frac{L|K}{P} \right) = C, \; P \in \mathcal{T}(L|K) \right\}.$$

By hypothesis, $\mathrm{Dd}(\mathcal{C}) = \mathrm{Dd}(\mathcal{B}) > 0$.

But if $P \in \mathcal{C}$ then necessarily $\left(\dfrac{L|K}{P} \right) = \varepsilon$, so from $\sigma \in C$ it follows that $\sigma = \varepsilon$. This shows that $L = L \cap M$, so $L \subseteq M$, and proves that $L|K$ is a Bauerian extension.   ∎

This result extends that which was proved in (**L**).

EXERCISES

1. Determine explicitly the Frobenius symbol of $\mathbb{Q}(\zeta)|\mathbb{Q}$ in the following cases:

   (a) ζ is a primitive fifth root of 1.

   (b) ζ is a primitive seventh root of 1.

   (c) ζ is a primitive eighth root of 1.

2. Determine explicitly the Frobenius symbol for the following extensions:

   (a) $\mathbb{Q}(\sqrt{3})|\mathbb{Q}$;

   (b) $\mathbb{Q}(\sqrt{3}, \sqrt{2})|\mathbb{Q}(\sqrt{3})$;

   (c) $\mathbb{Q}(\sqrt{3}, \sqrt{2})|\mathbb{Q}$.

3. Compute the Dirichlet density of the following sets of prime ideals in $\mathbb{Q}(\sqrt{2}, \sqrt{3})$:

   (a) the set of prime ideals P with norm $N(P) \equiv 1 \pmod 4$;

(b) same, with $N(P) \equiv 3 \pmod 4$.

4. Let $f(X) = X^4 - 2X^3 + 2$. Determine an infinite set of primes p such that $f(X)$ is congruent modulo p to the product of four distinct linear factors.

5. Find infinitely many primes p which are totally decomposed in $\mathbb{Q}(\sqrt{3})$ but not in $\mathbb{Q}(\sqrt{5})$, as well as an infinite set of primes p which are totally decomposed in $\mathbb{Q}(\sqrt{5})$ but not in $\mathbb{Q}(\sqrt{3})$. Are there infinitely many primes which are totally decomposed in both fields $\mathbb{Q}(\sqrt{3})$ and $\mathbb{Q}(\sqrt{5})$?

6. Let $L|K$ be an Abelian extension of degree n of number fields, with Galois group G. Let (f, g) be a splitting type of $L|K$, and let ν_f denote the number of elements of order f in the group G. Let $\mathcal{S}_f$ denote the set of prime ideals in K of splitting type (f, g) in $L|K$. Show:

(a) $\mathcal{S}_f \neq \varnothing$ if and only if G has an element of order f;

(b) $\mathcal{S}_f$ has Dirichlet density equal to ν_f/n.

7. Let $L|K$ be a Galois extension of number fields, and let $M|K$ be an extension of number fields. Let $\mathcal{C}$ be a conjugacy class in $G(L|K)$ and let $\mathcal{A} = \{\underline{P} \in \mathcal{U}(LM|K) \mid \underline{P} \in \mathcal{T}(M|K),\ \left(\dfrac{L|K}{\underline{P}}\right) = \mathcal{C}\}$. Show that $\mathrm{Dd}(\mathcal{A}) > 0$ if and only if $\mathcal{C} \cap G(L|L \cap M) \neq \varnothing$. In particular, the above condition is equivalent to $\mathcal{C} \subseteq G(L|L \cap M)$.

8. Let $L|K$, $M|K$ be arbitrary number field extensions. Let L' be the smallest Galois extension of K containing L. Let $H = G(L'|L)$ and let $H = H_1, \ldots, H_r$ be the conjugates in $G(L'|K)$ of the subgroup H.

Show that $\mathrm{Dd}(\mathcal{T}(M|K) \setminus \mathcal{T}(L|K)) = 0$ if and only if

$$G(L'|L' \cap M) \subseteq \bigcup_{i=1}^{r} H_i.$$

9. Let $L|K$ be an extension of number fields, and let L' be the smallest Galois extension of K containing L. Let $H = G(L'|L)$ and let $H = H_1, H_2, \ldots, H_r$ be the subgroups of $G(L'|K)$ which are conjugates to H.

Show that the following conditions are equivalent:

(a) $L|K$ is Bauerian.

(b) If H' is a subgroup of $G(L'|K)$ and $H' \subseteq \bigcup_{i=1}^{r} H_i$, there exists i, $1 \leq i \leq r$, such that $H' \subseteq H_i$.

10. Let $F(X) = 2X^5 - 32X + 1$, let a be a root of $F(X)$, and let $L = \mathbb{Q}(a)$.

Show:
- (a) If L' is the smallest Galois extension of $\mathbb{Q}$ containing L, then $G(L'|\mathbb{Q}) \cong S_5$, the symmetric group on the set $\{1, 2, 3, 4, 5\}$.
- (b) $G(L'|L)$ consists of all permutations fixing 1.
- (c) The conjugates of H in $G(L'|\mathbb{Q}) = S_5$ are H_i $(i = 1, \ldots, 5)$, where H_i consists of the permutations fixing i.
- (d) Let $H' = \{\text{identity}, (12), (45), (123), (132)\}$. Show that $H' \subseteq \bigcup_{i=1}^{5} H_i$ but H' is not contained in any H_i $(i = 1, \ldots, 5)$. Conclude that $L|K$ is not a Bauerian extension.

26

Class Numbers of Quadratic Fields

Let $K = \mathbb{Q}(\sqrt{d})$ where d is a square-free nonzero integer. We recall that in Chapter 23, (**C**), we showed that

$$\lim_{s \to 1+0} (s - 1)\zeta_K(s) = h \times \frac{2^{r_1 + r_2} \pi^{r_2} R}{w\sqrt{|\delta|}}, \qquad (26.1)$$

where

$$
\begin{aligned}
r_1 &= \text{number of real conjugates of } K, \\
2r_2 &= \text{number of nonreal conjugates of } K, \\
w &= \text{number of roots of unity in } K, \\
\delta &= \text{discriminant of } K, \\
R &= \text{regulator of } K.
\end{aligned}
$$

Our aim is to obtain a formula for the class number of the quadratic fields. In this study we distinguish two cases.

If $d > 0$ then K is a subfield of $\mathbb{R}$, so $r_1 = 2$, $r_2 = 0$, $w = 2$, $R = \log u$, where u is a fundamental unit, $u > 1$. Hence

$$\lim_{s \to 1+0} (s - 1)\zeta_K(s) = \frac{h \cdot 2 \log u}{\sqrt{\delta}} \qquad \text{when} \quad d > 0. \qquad (26.2)$$

If $d < 0$ then K is a quadratic imaginary field, so $r_1 = 0$, $r_2 = 1$, every unit is a root of unity, so $R = 1$. Hence

$$\lim_{s \to 1+0} (s - 1)\zeta_K(s) = \frac{h \cdot 2\pi}{w\sqrt{|\delta|}} \qquad \text{when} \quad d < 0. \qquad (26.3)$$

Moreover, we have seen that if $d = -1$ then $w = 4$, if $d = -3$ then $w = 6$, and if $d \neq -1, -3$, $d < 0$ then $w = 2$. Also $\delta = d$, when $d \equiv 1 \pmod 4$ or $\delta = 4d$ when $d \equiv 2$ or $3 \pmod 4$.

In order to compute h explicitly we need to know the limit on the left-hand side and a fundamental unit when $d > 0$.

For this purpose, we shall express the limit

$$\lim_{s \to 1+0} (s - 1)\zeta_K(s)$$

in terms of a certain character attached to the field K.

26.1 The Quadratic Character Attached to the Quadratic Field

The definition of the character χ involves the Jacobi symbol (see Chapter 4). Let $d = (-1)^\varepsilon |d|$, so ε is even when $d > 0$ and ε is odd otherwise. We define

$$\chi(a) = \begin{cases} 0, & \text{when } \gcd(a, \delta) \neq 1, \\ (-1)^{\alpha\varepsilon} \cdot (-1)^{\frac{d^2-1}{8}\ell} \cdot (d/a'), & \text{when } \gcd(a, \delta) = 1 \text{ and} \\ & a = (-1)^\alpha |a|, \ |a| = 2^\ell a', \\ & \ell \geq 0, \ a' \text{ odd}. \end{cases}$$

We may express $\chi(a)$ in a different way, using the Jacobi reciprocity law:

A.   Let $\gcd(a, \delta) = 1$.

If $d \equiv 1 \pmod 4$ then $\chi(a) = (a/|d|)$.

If $d \equiv 3 \pmod 4$ then $\chi(a) = (-1)^{(a-1)/2} \cdot (a/|d|)$.

If $d = 2d'$, d' odd, then $\chi(a) = (-1)^{(a^2-1)/8} \cdot (-1)^{\frac{a-1}{2} \cdot \frac{d'-1}{2}} \cdot (a/|d'|)$.

Proof: Let $d \equiv 1 \pmod 4$ then $\delta = d$ and by the Jacobi reciprocity law

$$\chi(a) = (-1)^{\alpha\varepsilon} \cdot (-1)^{\frac{d^2-1}{8}\ell} \cdot \left(\frac{d}{a'}\right)$$

$$= (-1)^{\alpha\varepsilon} \cdot (-1)^{\frac{d^2-1}{8}\ell} \cdot (-1)^{\frac{a'-1}{2} \cdot \frac{d-1}{2}} \cdot \left(\frac{a'}{|d|}\right)$$

$$= (-1)^{\alpha\varepsilon} \cdot (-1)^{\frac{d^2-1}{8}\ell} \left(\frac{a'}{|d|}\right)$$

$$= (-1)^{\alpha\varepsilon} \cdot \left(\frac{2}{|d|}\right)^\ell \left(\frac{a'}{|d|}\right) = (-1)^{\alpha\varepsilon} \left(\frac{|a|}{|d|}\right).$$

However, in this case, $(-1/|d|) = (-1)^\varepsilon$. In fact,

$$\left(\frac{-1}{|d|}\right) = (-1)^{(|d|-1)/2} = (-1)^{(d-1)/2 + (|d|-1)/2}.$$

If $d > 0$ we have $(d - 1)/2 + (|d| - 1)/2 = d - 1$ even, if $d < 0$ we have $(d - 1)/2 + (|d| - 1)/2 = -1$ odd, so in both cases $(d - 1)/2 + (|d| - 1)/2 \equiv \varepsilon \pmod 2$ and

$$(-1)^{\alpha\varepsilon} \left(\frac{|a|}{|d|}\right) = \left(\frac{-1}{|d|}\right)^\alpha \left(\frac{|a|}{|d|}\right) = \left(\frac{a}{|d|}\right).$$

Let $d \equiv 3 \pmod 4$ then $\delta = 4d$ and a is odd, so

$$\chi(a) = (-1)^{\alpha\varepsilon}\left(\frac{d}{|a|}\right) = (-1)^{\alpha\varepsilon}(-1)^{\frac{|a|-1}{2}\cdot\frac{d-1}{2}}\left(\frac{|a|}{|d|}\right)$$

$$= (-1)^{\alpha\varepsilon}(-1)^{(|a|-1)/2}\left(\frac{|a|}{|d|}\right).$$

However, in this case

$$\left(\frac{-1}{|d|}\right) = (-1)^{(|d|-1)/2} = -(-1)^{(d-1)/2+(|d|-1)/2}.$$

As we have shown above $(d-1)/2 + (|d|-1)/2 \equiv \varepsilon \pmod 2$ and

$$(-1)^{\alpha\varepsilon}(-1)^{(|a|-1)/2}\left(\frac{|a|}{|d|}\right) = (-1)^{\alpha+(|a|-1)/2}\left(\frac{-1}{|d|}\right)^{\alpha}\left(\frac{|a|}{|d|}\right)$$

$$= (-1)^{\alpha+(|a|-1)/2}\left(\frac{a}{|d|}\right).$$

But a is odd hence $\alpha + (|a|-1)/2 \equiv (a-1)/2 \pmod 2$, therefore if $d \equiv 3 \pmod 4$ we have $\chi(a) = (-1)^{(a-1)/2}(a/|d|)$.

Now let $d \equiv 2 \pmod 4$ then $d = 2d'$, where d' is odd (since d has no square factor), $\delta = 4d$, a is odd, so

$$\chi(a) = (-1)^{\alpha\varepsilon}\left(\frac{d}{|a|}\right) = (-1)^{\alpha\varepsilon}\left(\frac{2}{|a|}\right)\left(\frac{d'}{|a|}\right)$$

$$= (-1)^{\alpha\varepsilon}(-1)^{(a^2-1)/8}(-1)^{\frac{|a|-1}{2}\cdot\frac{d'-1}{2}}\left(\frac{|a|}{|d'|}\right).$$

Since d' is odd, we deduce as before that $(d'-1)/2 + (|d'|-1)/2 \equiv \varepsilon \pmod 2$ hence $(-1)^{\varepsilon} = (-1)^{(d'-1)/2}(-1/|d'|)$ and since a is odd then $\alpha + (|a|-1)/2 \equiv (a-1)/2 \pmod 2$, therefore

$$\chi(a) = (-1)^{\frac{d'-1}{2}\alpha}\cdot(-1)^{\frac{|a|-1}{2}\cdot\frac{d'-1}{2}}\cdot(-1)^{(a^2-1)/8}\cdot\left(\frac{-1}{|d'|}\right)^{\alpha}\left(\frac{|a|}{|d'|}\right)$$

$$= (-1)^{\frac{a-1}{2}\cdot\frac{d'-1}{2}}\cdot(-1)^{(a^2-1)/8}\cdot\left(\frac{a}{|d'|}\right). \qquad \blacksquare$$

B. χ *is a primitive quadratic character with conductor* $|\delta|$. *If* $d > 0$ *then* χ *is even, if* $d < 0$ *then* χ *is odd.*

Proof: By definition, $\chi(a) = 0$ if and only if $\gcd(a, \delta) \neq 1$.

If a, b are relatively prime to δ then $\chi(ab) = \chi(a)\chi(b)$. Indeed, let $a = (-1)^{\alpha}2^{\ell}a'$, $b = (-1)^{\beta}2^{m}b'$, with $\ell \geq 0$, $m \geq 0$, a', b' odd and positive. Then $\gcd(ab, \delta) = 1$, $ab = (-1)^{\alpha+\beta}2^{\ell+m}a'b'$ so

$$\chi(ab) = (-1)^{(\alpha+\beta)\varepsilon} \cdot (-1)^{\frac{d^2-1}{8}(\ell+m)} \cdot \left(\frac{d}{a'b'}\right)$$

$$= (-1)^{\alpha\varepsilon} \cdot (-1)^{(d^2-1)/8} \cdot \left(\frac{d}{a'}\right)(-1)^{\beta\varepsilon} \cdot (-1)^{\frac{d^2-1}{8}m} \cdot \left(\frac{d}{b'}\right)$$

$$= \chi(a)\chi(b).$$

Now, if $a \equiv b \pmod{|\delta|}$ and $d \equiv 1 \pmod 4$ then

$$\chi(a) = \left(\frac{a}{|d|}\right) = \left(\frac{b}{|d|}\right) = \chi(b).$$

If $d \equiv 3 \pmod 4$ then $\delta = 4d$ so $a \equiv b \pmod{|d|}$ and $a \equiv b \pmod 4$, thus $(a-1)/2 \equiv (b-1)/2 \pmod 2$ and

$$\chi(a) = (-1)^{(a-1)/2}\left(\frac{a}{|d|}\right) = (-1)^{(b-1)/2}\left(\frac{b}{|d|}\right) = \chi(b).$$

Finally, if $d \equiv 2 \pmod 4$ then $\delta = 4d = 8d'$ and again we have $a \equiv b \pmod{|d'|}$, $a \equiv b \pmod 8$ so $(a^2-1)/8 \equiv (b^2-1)/8 \pmod 2$ and $\chi(a) = \chi(b)$.

Thus we have shown that χ is a character modulo $|\delta|$. It is obviously a quadratic character. To show that χ is primitive with conductor $|\delta|$, we use the criterion of Chapter 21, (**Q**). It suffices to show that if ℓ is an integer dividing $|\delta|$, $1 < \ell < |\delta|$, there exists an integer a such that $\gcd(a, \delta) = 1$, $a \equiv 1 \pmod \ell$, and $\chi(a) = -1$.

Now, since $\ell < |\delta|$ there exists a prime number p such that ℓ divides $|\delta|/p = m$.

If $p \neq 2$ let b be a quadratic nonresidue modulo p and let a satisfy the congruences

$$\begin{cases} a \equiv b \pmod p, \\ a \equiv 1 \pmod{2m}. \end{cases}$$

Then $\gcd(a, \delta) = 1$, $a \equiv 1 \pmod \ell$ and $\chi(a)$ may be computed by the above formulas. If $d \equiv 1 \pmod 4$ then

$$\chi(a) = \left(\frac{a}{|\delta|}\right) = \left(\frac{a}{m}\right)\left(\frac{a}{p}\right) = \left(\frac{a}{p}\right) = \left(\frac{b}{p}\right) = -1.$$

If $d \equiv 3 \pmod 4$ then $\delta = 4d$ so $|d| = p \cdot (m/4)$ hence 4 divides m and

$$\chi(a) = (-1)^{(a-1)/2}\left(\frac{a}{|d|}\right) = (-1)^{(a-1)/2}\left(\frac{a}{p}\right)\left(\frac{a}{m/4}\right) = \left(\frac{a}{p}\right) = \left(\frac{b}{p}\right) = -1$$

because $a \equiv 1 \pmod 4$ and $a \equiv 1 \pmod{m/4}$.

If $d \equiv 2 \pmod 4$ then $d = 2d'$, d' odd, $\delta = 4d$, $|d| = p \cdot (m/4)$, 4 divides m and

$$\chi(a) = (-1)^{\frac{a^2-1}{8} + \frac{a-1}{2} \cdot \frac{d'-1}{2}} \cdot \left(\frac{a}{|d'|}\right)$$

$$= (-1)^{\frac{a^2-1}{8} + \frac{a-1}{2} \cdot \frac{d'-1}{2}} \cdot \left(\frac{a}{p}\right)\left(\frac{a}{m/2}\right) = \left(\frac{a}{p}\right) = \left(\frac{b}{p}\right) = -1$$

because $a \equiv 1 \pmod 8$ and $a \equiv 1 \pmod{m/2}$.

Now we assume that $p = 2$ so $d \not\equiv 1 \pmod 4$ and $\delta = 4d$.

If $d \equiv 3 \pmod 4$ let a satisfy the congruences

$$\begin{cases} a \equiv 3 \pmod 4, \\ a \equiv 1 \pmod{2|d|}. \end{cases}$$

Then $\gcd(a, \delta) = 1$, $a \equiv 1 \pmod \ell$, and

$$\chi(a) = (-1)^{(a-1)/2}\left(\frac{a}{|d|}\right) = (-1)^{(a-1)/2} = -1.$$

If $d \equiv 2 \pmod 4$ then $d = 2d'$, d' odd, $\delta = 4d$. Let a satisfy the congruences

$$\begin{cases} a \equiv 5 \pmod 8, \\ a \equiv 1 \pmod{2|d|}. \end{cases}$$

Then $\gcd(a, \delta) = 1$, $a \equiv 1 \pmod \ell$, and

$$\chi(a) = (-1)^{\frac{a^2-1}{8} + \frac{a-1}{2} \cdot \frac{d'-1}{2}}\left(\frac{a}{|d'|}\right) = (-1)^{(a^2-1)/8} = -1$$

since $a \equiv 1 \pmod 4$, and $a \equiv 1 \pmod{|d'|}$.

The last assertion follows at once from the definition:

$\chi(-1) = (-1)^\varepsilon$ hence if $d > 0$ then $\chi(-1) = 1$, if $d < 0$ then $\chi(-1) = -1$. ∎

We recall that the character $\chi = \chi_d$ depends on the quadratic field $K = \mathbb{Q}(\sqrt{d})$. It is important to relate the characters belonging to different quadratic fields. Let us write

$$d = (-1)^\varepsilon 2^{\varepsilon'} p_1 \cdots p_s p_{s+1} \cdots p_r,$$

where ε, ε' are equal to 0 or 1, $0 \le s \le r$, each p_i is an odd prime, $p_i \equiv 3 \pmod 4$ for $i = 1, \ldots, s$, and $p_i \equiv 1 \pmod 4$ for $i = s+1, \ldots, r$.

If $\varepsilon' = 0$ then the conductor $|\delta|$ of χ_d is

$$|\delta| = \begin{cases} (-1)^\varepsilon d, & \text{when } s + \varepsilon \text{ is even,} \\ (-1)^\varepsilon 4d, & \text{when } s + \varepsilon \text{ is odd.} \end{cases}$$

Indeed, $d \equiv (-1)^{s+\varepsilon} \pmod 4$ so $\delta = d$ when $s + \varepsilon$ is even and $\delta = 4d$ when $s + \varepsilon$ is odd.

If $\varepsilon' = 1$ then $\delta = 4d$ and the conductor of χ_d is $|\delta| = (-1)^\varepsilon 4d = 8(p_1 \cdots p_r)$.

With these notations, we obtain the explicit decomposition of χ_d into a product of characters belonging to quadratic fields and pairwise relatively prime conductors (see Chapter 21, (**R**)):

C.   *If $\varepsilon' = 0$ then*

$$\chi_d = (\chi_{-1})^{\varepsilon+s}\chi_{-p_1} \cdots \chi_{-p_s}\chi_{p_{s+1}} \cdots \chi_{p_r}.$$

If $\varepsilon' = 1$ then $\chi_d = \chi_{(-1)^{\varepsilon+s}\cdot 2}\chi_{-p_1} \cdots \chi_{-p_s}\chi_{p_{s+1}} \cdots \chi_{p_r}.$

Proof: Let $\varepsilon' = 0$ and assume that $\gcd(a, \delta) = 1$. We have

$$\chi_{-1}(a) = (-1)^\alpha \left(\frac{-1}{a'}\right) = (-1)^\alpha (-1)^{(a'-1)/2}.$$

$$\chi_{-p_i}(a) = (-1)^\alpha (-1)^{\ell(p_i^2-1)/8} \left(\frac{-p_i}{a'}\right) = (-1)^\alpha (-1)^{\ell(p_i^2-1)/8} \left(\frac{-1}{a'}\right)\left(\frac{p_i}{a'}\right)$$

for $i = 1, \ldots, s$, and

$$\chi_{p_i}(a) = (-1)^{\ell(p_i^2-1)/8}\left(\frac{p_i}{a'}\right)$$

for $i = s + 1, \ldots, r$.

Hence

$$((\chi_{-1})^{\varepsilon+s}\chi_{-p_1} \cdots \chi_{p_r})(a) = \left[(-1)^\alpha\left(\frac{-1}{a'}\right)\right]^{\varepsilon+s}(-1)^{\alpha s}$$

$$\times (-1)^{\sum_{i=1}^r ((p_i^2-1)/8)\cdot\ell}\left(\frac{-1}{a'}\right)\left(\frac{|d|}{a'}\right) \quad (26.4)$$

If $\varepsilon + s$ is even then (26.4) becomes

$$(-1)^{\alpha\varepsilon}(-1)^{\ell(d^2-1)/8}\left(\frac{d}{a'}\right) = \chi_d(a),$$

noting that ε, s have the same parity and

$$\sum_{i=1}^r \frac{p_i^2 - 1}{8} \equiv \frac{d^2 - 1}{8} \pmod 2$$

(see Chapter 4, (**R**)).

Similarly, if $\varepsilon + s$ is odd, ε, s have different parity, and (26.4) is equal to

$$(-1)^\alpha\left(\frac{-1}{a'}\right)(-1)^\alpha(-1)^{\alpha\varepsilon}(-1)^{\ell(d^2-1)/8}\left(\frac{-1}{a'}\right)\left(\frac{d}{a'}\right)$$

$$= (-1)^{\alpha\varepsilon}(-1)^{\ell(d^2-1)/8}\left(\frac{d}{a'}\right) = \chi_d(a).$$

Now we take $\varepsilon' = 1$ and $\gcd(a, \delta) = 1$, hence a is odd, $a = (-1)^{\alpha}|a|$ and $|a| = a'$. We have

$$\chi_{(-1)^{\varepsilon+\varkappa}\cdot 2}(a) = (-1)^{\alpha(\varepsilon+s)}\left(\frac{-1}{a'}\right)^{s+\varepsilon}\left(\frac{2}{a'}\right),$$

$$\chi_{-p_i}(a) = (-1)^{\alpha}\left(\frac{-p_i}{a'}\right) = (-1)^{\alpha}\left(\frac{-1}{a'}\right)\left(\frac{p_i}{a'}\right)$$

for $i = 1, \ldots, s$ and

$$\chi_{p_i}(a) = \left(\frac{p_i}{a'}\right)$$

for $i = s + 1, \ldots, r$. Hence

$$(\chi_{(-1)^{\varepsilon+\varkappa}\cdot 2}\chi_{-p_1}\cdots\chi_{p_r})(a) = (-1)^{\alpha(\varepsilon+s)+\alpha s}\left(\frac{2}{a'}\right)\left(\frac{-1}{a'}\right)^{2s+\varepsilon}$$

$$\times \left(\frac{p_1}{a'}\right)\cdots\left(\frac{p_r}{a'}\right) = (-1)^{\alpha\varepsilon}\left(\frac{-1}{a'}\right)^{\varepsilon}\left(\frac{|d|}{a'}\right) = (-1)^{\alpha\varepsilon}\left(\frac{d}{a'}\right) = \chi_d(a). \blacksquare$$

In the next section we express the Dedekind zeta-function in terms of the *L*-series of the characters.

26.2 The *L*-Series and the Gaussian Sum of the Quadratic Character

We return to the limit

$$\lim_{s\to 1+0}(s - 1)\zeta_K(s)$$

for a quadratic field K and express it in terms of the *L*-series of the quadratic character of K. Then we compute this expression by determining the value of the corresponding Gaussian sum.

D.   Let χ be the (*nontrivial*) character of the quadratic field $K = \mathbb{Q}(\sqrt{d})$. Then, for $s > 1$:

$$\zeta_K(s) = \zeta(s)L(s|\chi) = \zeta(s)\prod_p \frac{1}{1 - \chi(p)/p^s} \qquad (26.5)$$

and

$$\lim_{s\to 1+0}(s - 1)\zeta_K(s) = L(1|\chi). \qquad (26.6)$$

Proof: We have

$$\zeta_K(s) = \prod_P \frac{1}{1 - 1/N(P)^s} = \prod_p\prod_{P|p} \frac{1}{1 - 1/N(P)^s}$$

(product extended over all prime ideals P of the ring of integers A of K). As we have seen in Chapter 11, Section 2, for every prime number p, one of the following three cases occurs:

(1) $Ap = P \cdot P'$, where P, P' are distinct prime ideals of A; then $N(P) = N(P') = p$.

(2) $Ap = P$; then $N(P) = p^2$.

(3) $Ap = P^2$; then $N(P) = p$.

Cases (1), (2), and (3) occur, respectively, when $(d/p) = 1$, $(d/p) = -1$, and $(d/p) = 0$. In terms of the character χ, this means that $\chi(p) = 1$, $\chi(p) = -1$, and $\chi(p) = 0$. So, we have in all cases

$$\prod_{P|p} \frac{1}{1 - 1/N(P)^s} = \frac{1}{1 - 1/p^s} \cdot \frac{1}{1 - \chi(p)/p^s}$$

for every prime number p. Hence

$$\zeta_K(s) = \prod_p \frac{1}{1 - 1/p^s} \prod_p \frac{1}{1 - \chi(p)/p^s} = \zeta(s) \cdot L(s|\chi) \qquad \text{for} \quad s > 1.$$

Since χ is different from the trivial character, by Chapter 22, (I), the L-series $L(s|\chi)$ is defined and continuous on $(0, \infty)$. From this, we deduce

$$\lim_{s \to 1+0} (s - 1)\zeta_K(s) = L(1|\chi)$$

because

$$\lim_{s \to 1+0} (s - 1)\zeta(s) = 1. \qquad \blacksquare$$

From (26.2) and (26.3) it follows that: If $d > 0$ then

$$h = \frac{\sqrt{\delta}}{2 \log u} L(1|\chi) \tag{26.7}$$

and if $d < 0$ then

$$h = \frac{w\sqrt{|\delta|}}{2\pi} L(1|\chi). \tag{26.8}$$

We shall express

$$L(1|\chi) = \sum_{n=1}^{\infty} \frac{\chi(n)}{n}$$

in a form not involving an infinite sum. This is done exactly as in the case of a cyclotomic field, and we obtain the following expression (see Chapter 22, (22.19)):

$$L(1|\chi) = \frac{1}{|\delta|} \sum_{\substack{k=1 \\ \gcd(k,\delta)=1}}^{|\delta|-1} \tau_k(\chi) \log \frac{1}{1 - \zeta^{-k}}, \tag{26.9}$$

where ζ is a primitive root of unity of order $|\delta|$ and

$$\tau_k(\chi) = \sum_{\substack{a=1 \\ \gcd(a,\delta)=1}}^{|\delta|-1} \chi(a)\zeta^{ak} \qquad \text{for} \quad k = 1, \ldots, |\delta| - 1, \qquad (26.10)$$

is the kth Gaussian sum of the character χ.

By Chapter 21, (**U**), each Gaussian sum is expressible in terms of the principal Gaussian sum

$$\tau_1(\chi) = \sum_{\substack{a=1 \\ \gcd(a,\delta)=1}}^{|\delta|-1} \chi(a)\zeta^{a},$$

namely

$$\tau_k(\chi) = \frac{1}{\chi(k)}\tau_1(\chi) \qquad \text{when} \quad \gcd(k, |\delta|) = 1.$$

Following Chapter 27, (**D**), and noting that the values of χ are ± 1, we have:

If $d > 0$ then χ is an even character, hence

$$L(1|\chi) = -\frac{\tau_1(\chi)}{\delta} \sum_{\substack{k=1 \\ \gcd(k,\delta)=1}}^{\delta-1} \chi(k) \log \sin \frac{k\pi}{\delta}. \qquad (26.11)$$

If $d < 0$ then χ is an odd character, so

$$L(1|\chi) = \frac{\pi i \tau_1(\chi)}{\delta^2} \sum_{\substack{k=1 \\ \gcd(k,\delta)=1}}^{|\delta|-1} \chi(k) \cdot k. \qquad (26.12)$$

It remains now to find the value of the Gaussian sum $\tau_1(\chi)$.

E.
$$\tau_1(\chi) = \begin{cases} \sqrt{\delta}, & \text{when } d > 0, \\ i\sqrt{|\delta|}, & \text{when } d < 0. \end{cases}$$

Proof: By (**B**) we know that $d > 0$ exactly when χ is even. By the remark following Chapter 21, (**W**), we have $[\tau_1(\chi)]^2 = \chi(-1)|\delta|$. Hence, if $d > 0$ then $\tau_1(\chi) = \pm\sqrt{\delta}$ and if $d < 0$ then $\tau_1(\chi) = \pm i\sqrt{|\delta|}$.

We still need to determine the sign of the Gaussian sum. This is the most important point of the proof and we shall follow Kronecker's method.

Case 1: Let us assume that $|\delta| = p$ is a prime number.

Thus $d \equiv 1 \pmod 4$ and $\delta = d$ (so $p \neq 2$). Let $\eta = \zeta^{(p+1)/2}$ so $\eta^2 = \zeta$. In a first step we determine the value of

$$\rho = \prod_{a=1}^{t} (\eta^a - \eta^{-a}) \qquad \text{where} \quad t = \frac{p-1}{2}.$$

We have

$$\eta^a - \eta^{-a} = \eta^a(1 - \eta^{-2a}) = \eta^a(1 - \zeta^{-a})$$

and also

$$\eta^a - \eta^{-a} = \eta^{-a}(\eta^{2a} - 1) = \eta^{-a}(\zeta^a - 1);$$

hence

$$\rho^2 = \left[\prod_{a=1}^{t}(\eta^a - \eta^{-a})\right]^2$$

$$= \eta^{1+2+\cdots+t} \cdot \eta^{-(1+2+\cdots+t)} \cdot (-1)^t \prod_{a=1}^{t}(1 - \zeta^{-a})(1 - \zeta^a)$$

$$= (-1)^t \prod_{a=1}^{p-1}(1 - \zeta^a) = (-1)^t \Phi_p(1) = (-1)^{(p-1)/2}p,$$

where Φ_p denotes the pth cyclotomic polynomial. But

$$\eta^a = \cos\left(\frac{2\pi a}{p} \cdot \frac{p+1}{2}\right) + i\sin\left(\frac{2\pi a}{p} \cdot \frac{p+1}{2}\right),$$

$$\eta^{-a} = \cos\left(\frac{2\pi a}{p} \cdot \frac{p+1}{2}\right) - i\sin\left(\frac{2\pi a}{p} \cdot \frac{p+1}{2}\right),$$

hence

$$\eta^a - \eta^{-a} = 2i\sin\left(\frac{2\pi a}{p} \cdot \frac{p+1}{2}\right) = 2i\sin\left(\frac{\pi a}{p} + \pi a\right).$$

Thus the signs of the numbers

$$2\sin\left(\frac{\pi a}{p} + \pi a\right), \qquad a = 1, 2, \ldots, t,$$

alternatively, the first sign (for $a = 1$) being negative. Hence among these numbers there are k negative if t is even, $t = 2k$, or t is odd, $t = 2k - 1$.

Consequently

$$\rho = \prod_{a=1}^{t}(\eta^a - \eta^{-a}) = i^t(-1)^k \cdot c,$$

where c is a real positive number. Moreover, from $|\rho^2| = p$, it follows that $c = \sqrt{p}$.

Hence $\rho = \sqrt{p}$ when $t = 2k$ is even, that is, $p \equiv 1 \pmod 4$ or still $d > 0$, while $\rho = -i\sqrt{p}$ when $t = 2k - 1$ is odd, that is, $p \equiv 3 \pmod 4$ or still $d < 0$. So

$$\rho = \begin{cases} \sqrt{p} = \sqrt{\delta}, & \text{when } d > 0, \\ -i\sqrt{p} = -i\sqrt{|\delta|}, & \text{when } d < 0. \end{cases}$$

Hence we need to show that

$$\tau_1(\chi) = (-1)^t \rho = \begin{cases} \sqrt{\delta}, & \text{when } d > 0, \\ i\sqrt{|\delta|}, & \text{when } d < 0. \end{cases}$$

It will be enough to find an ideal J in A (ring of integers of $\mathbb{Q}(\zeta)$) such that

$$\begin{cases} \tau_1(\chi) \equiv (-1)^t \rho \pmod{J}, \\ \rho \not\equiv -\rho \pmod{J}. \end{cases} \qquad (26.13)$$

Indeed, from $\tau_1(\chi)^2 = \rho^2$ we know that $\tau_1(\chi) = \pm(-1)^t \rho$. If $\tau_1(\chi) \equiv (-1)^t \rho \pmod{J}$ and $\tau_1(\chi) = -(-1)^t \rho$ we would deduce that $(-1)^t \rho \equiv -(-1)^t \rho \pmod{J}$, that is, $\rho \equiv -\rho \pmod{J}$, a contradiction.

We take $J = A\xi^{t+1}$, where $\xi = 1 - \zeta \in A$. Then

$$\rho \equiv (-1)^t t! \xi^t \pmod{J}. \qquad (26.14)$$

In fact, for $a = 1, \ldots, t$, we have

$$\eta^a - \eta^{-a} = -\eta^{-a}(1 - \zeta^a) = -\eta^{-a}[1 - (1 - \xi)^a]$$
$$\equiv -\eta^{-a} a\xi \equiv -a\xi \pmod{A\xi^2}$$

because $1 - \eta^{-a} = 1 - \zeta^{-a(p+1)/2}$ is associated with $1 - \zeta = \xi$, so $\eta^{-a} \equiv 1 \pmod{A\xi}$. Hence $\eta^a - \eta^{-a} = (-a + b_a \xi)\xi$ where $b_a \in A$ for $a = 1, \ldots, t$. Multiplying out we have

$$\rho = \prod_{a=1}^{t} (\eta^a - \eta^{-a}) \equiv (-1)^t t! \xi^t \pmod{A\xi^{t+1}}$$

which shows (26.14).

Now, if $\rho \equiv -\rho \pmod{J}$ then $2\rho \equiv 0 \pmod{A\xi^{t+1}}$. From $p \neq 2$ and $Ap = A\xi^{p-1}$ (by Chapter 5, Section 5) it follows that $A\xi$ does not divide 2, hence $\rho \equiv 0 \pmod{A\xi^{t+1}}$ and therefore $t! \equiv 0 \pmod{A\xi}$. Again, since $t < p$ we know that $A\xi$ does not divide a, for $a = 1, \ldots, t$. So, we have a contradiction.

It remains to show that $\tau_1(\chi) \equiv t! \xi^t \pmod{J}$, hence from (26.14) we conclude that $\tau_1(\chi) \equiv (-1)^t \rho \pmod{J}$.

By Euler's criterion for the Legendre symbol, $(a/p) \equiv a^t \pmod{p}$. From $Ap = A\xi^{p-1} \subseteq A\xi^{t+1} = J$ for $p \neq 2$, we have

$$\tau_1(\chi) = \sum_{a=1}^{p-1} \left(\frac{a}{p}\right) \zeta^a \equiv \sum_{a=1}^{p-1} a^t (1 - \xi)^a = \sum_{a=1}^{p-1} a^t \sum_{k=0}^{a} (-1)^k \binom{a}{k} \xi^k$$

$$= \sum_{k=0}^{t} \left[\sum_{a=1}^{p-1} a^t \binom{a}{k} \right] (-1)^k \xi^k \pmod{A\xi^{t+1}},$$

recalling that $\binom{a}{k} = 0$ when $a < k$.

We want to find an appropriate expression for the binomial coefficients $\binom{a}{k}$ modulo p. From Wilson's theorem (see Chapter 3, Exercise 18) we have

$$-1 \equiv (p-1)! = \prod_{a=1}^{t} a(p-a) \equiv (-1)^t (t!)^2 \pmod{p}$$

hence $1/t! \equiv -(-1)^t t! \pmod{p}$. Therefore

$$\binom{a}{k} = \frac{a(a-1)\cdots(a-k+1)}{k!} = \frac{(k+1)\cdots t \cdot a(a-1)\cdots(a-k+1)}{t!}.$$

Writing $c_k = (k+1)(k+2)\cdots t$ for $k \leq t$ (then $c_t = 1$) we have

$$\binom{a}{k} \equiv -(-1)^t t! \, c_k a(a-1)\cdots(a-k+1) \pmod{p}.$$

Hence

$$\tau_1(\chi) \equiv -(-1)^t t! \sum_{k=0}^{t} \sum_{a=1}^{p-1} \left[a^t a(a-1)\cdots(a-k+1) \right]$$

$$\times c_k (-1)^k \xi^k \pmod{A\xi^{t+1}}. \tag{26.15}$$

But

$$\sum_{a=1}^{p-1} a^r \equiv 0 \pmod{p}, \qquad \text{when} \quad r \not\equiv 0 \pmod{p-1},$$

and

$$\sum_{a=1}^{p-1} a^r \equiv -1 \pmod{p}, \qquad \text{when} \quad r \equiv 0 \pmod{p-1}.$$

Hence in the above sum for $k = 0, 1, \ldots, t-1$ we have

$$\sum_{a=1}^{p-1} a^{t+1}(a-1)\cdots(a-k+1) \equiv 0 \pmod{p}$$

while

$$\sum_{a=1}^{p-1} a^{t+1}(a-1)\cdots(a-t+1) \equiv \sum_{a=1}^{p-1} a^{p-1} \equiv -1 \pmod{p}.$$

Replacing this in (26.15), we deduce that

$$\tau_1(\chi) \equiv -(-1)^t t! \, c_t (-1)^t \xi^t = -t! \xi^t \pmod{J}.$$

Altogether, we have shown the proposition in the case where $|\delta|$ is a prime number.

Case 2:   Let us assume that $|\delta| = 4$.

Then necessarily $d = -1$, $\delta = -4$.

The character χ with conductor 4 is given by (**A**) explicitly as

$$\begin{cases} \chi(a) = (-1)^{(a-1)/2}, & \text{when } a \text{ is odd,} \\ \chi(a) = 0, & \text{when } a \text{ is even.} \end{cases}$$

Taking the primitive fourth root of unity $\zeta = i$ then

$$\tau_1(\chi) = \chi(1)\zeta + \chi(3)\zeta^3 = i + (-1)(-i) = 2i.$$

Case 3:   Let us assume now that $|\delta| = 8$.

Then $d = \pm 2$.

If $d = 2$ then the character χ with conductor 8 is given in (**A**):

$$\begin{cases} \chi(a) = (-1)^{(a^2-1)/8}, & \text{for } a \text{ odd,} \\ \chi(a) = 0, & \text{for } a \text{ even.} \end{cases}$$

A primitive eighth root of unity is $\zeta = (\sqrt{2}/2)(1 + i)$. Then

$$\tau_1(\chi) = \chi(1)\zeta + \chi(3)\zeta^3 + \chi(5)\zeta^5 + \chi(7)\zeta^7$$

$$= \frac{\sqrt{2}}{2}(1 + i) - \frac{\sqrt{2}}{2}(-1 + i) - \frac{\sqrt{2}}{2}(-1 - i) + \frac{\sqrt{2}}{2}(1 - i) = 2\sqrt{2}.$$

If $d = -2$ then the character χ with conductor 8 is given in (**A**):

$$\begin{cases} \chi(a) = (-1)^{(a^2-1)/8-(a-1)/2}, & \text{for } a \text{ odd,} \\ \chi(a) = 0, & \text{for } a \text{ even.} \end{cases}$$

Then a similar computation gives

$$\tau_1(\chi) = \frac{\sqrt{2}}{2}(1 + i) + \frac{\sqrt{2}}{2}(-1 + i) - \frac{\sqrt{2}}{2}(-1 - i) - \frac{\sqrt{2}}{2}(1 - i)$$

$$= 2\sqrt{2}\,i.$$

Case 4:   General Case.

Let $d = (-1)^\varepsilon 2^{\varepsilon'} p_1 \cdots p_s p_{s+1} \cdots p_r$ where ε, ε' are equal to 0 or 1, $0 \le s \le r$, each p_i is an odd prime, $p_i \equiv 3 \pmod 4$ for $i = 1, \ldots, s$, $p_i \equiv 1 \pmod 4$ for $i = s + 1, \ldots, r$.

First let $\varepsilon' = 0$. By (**C**):

$$\chi = (\chi_{-1})^{\varepsilon+s}\chi_{-p_1} \cdots \chi_{-p_s}\chi_{p_{s+1}} \cdots \chi_{p_r}.$$

From Chapter 21, (**V**), we have

$$\tau_1(\chi) = \tau_1(\chi_{-1})^{\varepsilon+s}\tau_1(\chi_{-p_1}) \cdots \tau_1(\chi_{p_r})$$

$$\times \left[\chi_{-1}\left(\frac{|\delta|}{4}\right)\right]^{\varepsilon+s} \cdot \chi_{-p_1}\left(\frac{|\delta|}{p_1}\right) \cdots \chi_{-p_s}\left(\frac{|\delta|}{p_s}\right)$$

$$\times \chi_{p_{s+1}}\left(\frac{|\delta|}{p_{s+1}}\right) \cdots \chi_{p_r}\left(\frac{|\delta|}{p_r}\right).$$

If $\varepsilon + s$ is even, $|\delta| = (-1)^\varepsilon d = p_1 \cdots p_r$, and

$$\chi_{-p_1}\left(\frac{(-1)^\varepsilon d}{p_1}\right) = \chi_{-p_1}(p_2 \cdots p_r)$$

$$= \left(\frac{-1}{p_2}\right) \cdots \left(\frac{-1}{p_r}\right)\left(\frac{p_1}{p_2}\right) \cdots \left(\frac{p_1}{p_r}\right),$$

$$\chi_{-p_2}\left(\frac{(-1)^\varepsilon d}{p_2}\right) = \chi_{-p_2}(p_1 p_3 \cdots p_r)$$

$$= \left(\frac{-1}{p_1}\right)\left(\frac{-1}{p_3}\right) \cdots \left(\frac{-1}{p_r}\right)\left(\frac{p_2}{p_1}\right) \cdots \left(\frac{p_2}{p_r}\right),$$

. . . .

$$\chi_{-p_s}\left(\frac{(-1)^\varepsilon d}{p_s}\right) = \chi_{-p_s}(p_1 \cdots p_{s-1} p_{s+1} \cdots p_r)$$

$$= \left(\frac{-1}{p_1}\right) \cdots \left(\frac{-1}{p_{s-1}}\right)\left(\frac{-1}{p_{s+1}}\right) \cdots \left(\frac{-1}{p_r}\right)\left(\frac{p_s}{p_1}\right) \cdots \left(\frac{p_s}{p_r}\right),$$

$$\chi_{p_{s+1}}\left(\frac{(-1)^\varepsilon d}{p_{s+1}}\right) = \chi_{p_{s+1}}(p_1 \cdots p_s p_{s+2} \cdots p_r)$$

$$= \left(\frac{p_{s+1}}{p_1}\right) \cdots \left(\frac{p_{s+1}}{p_s}\right)\left(\frac{p_{s+1}}{p_{s+2}}\right) \cdots \left(\frac{p_{s+1}}{p_r}\right),$$

. . . .

$$\chi_{p_r}\left(\frac{(-1)^\varepsilon d}{p_r}\right) = \chi_{p_r}(p_1 \cdots p_{r-1}) = \left(\frac{p_r}{p_1}\right) \cdots \left(\frac{p_r}{p_{r-1}}\right).$$

By Jacobi's reciprocity law (Chapter 4, (S)):

$$\left(\frac{p_i}{p_j}\right)\left(\frac{p_j}{p_i}\right) = (-1)^{\frac{p_i-1}{2} \cdot \frac{p_j-1}{2}}$$

so if i or j is greater than s then $(p_i/p_j)(p_j/p_i) = 1$, while if i, $j \le s$ then $(p_i/p_j)(p_j/p_i) = -1$. From $(-1/p_j) = 1$ for $j \ge s + 1$ and $(-1/p_j) = -1$ for $j \le s$, we have

$$\tau_1(\chi) = \tau_1(\chi_{-p_1}) \cdots \tau_1(\chi_{p_r})\left(\frac{-1}{p_1}\right)^{s-1} \cdots \left(\frac{-1}{p_r}\right)^{s-1} (-1)^{1+2+\cdots+(s-1)}$$

$$= \tau_1(\chi_{-p_1}) \cdots \tau_1(\chi_{p_r}) \cdot (-1)^{s(s-1)+s(s-1)/2}.$$

By the previous cases of the proof, we have

$$\tau_1(\chi) = i^s \sqrt{p_1 \cdots p_r}(-1)^{3s(s-1)/2} = i^{3s(s-1)+s}\sqrt{|\delta|}.$$

If $d > 0$ then $\varepsilon = 0$ hence s is even, so $3s^2 - 2s$ is a multiple of 4, hence $\tau_1(\chi) = \sqrt{\delta}$.

If $d < 0$ then $\varepsilon = 1$ hence s is odd, so $3s^2 - 2s \equiv 1 \pmod{4}$ hence $\tau_1(\chi) = i\sqrt{|\delta|}$.

If $\varepsilon + s$ is odd then $|\delta| = (-1)^\varepsilon 4d = 4p_1 \cdots p_r$.
Now

$$\chi_{-1}\left(\frac{|\delta|}{4}\right) = \chi_{-1}(p_1 \cdots p_r) = \left(\frac{-1}{p_1}\right) \cdots \left(\frac{-1}{p_r}\right).$$

A similar computation gives

$$\tau_1(\chi) = \tau_1(\chi_{-1})\tau_1(\chi_{-p_1}) \cdots \tau_1(\chi_{p_r})\left(\frac{-1}{p_1}\right)^s \cdots \left(\frac{-1}{p_r}\right)^s (-1)^{s(s-1)/2}$$

$$= \tau_1(\chi_{-1})\tau_1(\chi_{-p_1}) \cdots \tau_1(\chi_{p_r}) \cdot (-1)^{s^2 + s(s-1)/2}.$$

By the previous cases of the proof, we have

$$\tau_1(\chi) = 2i^{s+1}\sqrt{p_1 \cdots p_r}(-1)^{(3s-1)s/2} = i^{s+1+s(3s-1)}\sqrt{|\delta|}.$$

If $d > 0$ then $\varepsilon = 0$ hence s is odd and $s + 1 + s(3s - 1) \equiv 0 \pmod 4$, thus $\tau_1(\chi) = \sqrt{\delta}$.
If $d < 0$ then $\varepsilon = 1$ hence s is even and $s + 1 + s(3s - 1) \equiv 1 \pmod 4$, thus $\tau_1(\chi) = i\sqrt{|\delta|}$.
Now we assume that $\varepsilon' = 1$, hence $|\delta| = 8(p_1 \cdots p_r)$. By (**C**):

$$\chi_d = \chi_{(-1)^{\varepsilon+\varkappa} \cdot 2}\chi_{-p_1} \cdots \chi_{-p_\varkappa}\chi_{p_{s+1}} \cdots \chi_{p_r}$$

and from Chapter 21, (**V**), we have

$$\tau_1(\chi) = \tau_1(\chi_{(-1)^{\varepsilon+\varkappa} \cdot 2})\tau_1(\chi_{-p_1}) \cdots \tau_1(\chi_{p_r})$$
$$\times \chi_{(-1)^{\varepsilon+\varkappa} \cdot 2}\left(\frac{|\delta|}{8}\right) \cdot \chi_{-p_1}\left(\frac{|\delta|}{p_1}\right) \cdots \chi_{-p_\varkappa}\left(\frac{|\delta|}{p_s}\right)$$
$$\times \chi_{p_{s+1}}\left(\frac{|\delta|}{p_{s+1}}\right) \cdots \chi_{p_r}\left(\frac{|\delta|}{p_r}\right).$$

Now

$$\chi_{(-1)^{\varepsilon+\varkappa} \cdot 2}(p_1 \cdots p_r) = \left(\frac{(-1)^{\varepsilon+s} \cdot 2}{p_1}\right) \cdots \left(\frac{(-1)^{\varepsilon+s} \cdot 2}{p_r}\right)$$

$$= \left[\left(\frac{-1}{p_1}\right) \cdots \left(\frac{-1}{p_r}\right)\right]^{\varepsilon+s}\left(\frac{2}{p_1}\right) \cdots \left(\frac{2}{p_r}\right) = (-1)^{s(\varepsilon+s)} \cdot (-1)^{\sum_{i=1}^r (p_i^2 - 1)/8},$$

$$\chi_{-p_1}(8p_2 \cdots p_r) = (-1)^{3(p_1^2-1)/8}\left(\frac{-1}{p_2}\right) \cdots \left(\frac{-1}{p_r}\right)\left(\frac{p_1}{p_2}\right) \cdots \left(\frac{p_1}{p_r}\right),$$

. . . .

$$\chi_{-p_s}(8p_1 \cdots p_{s-1}p_{s+1} \cdots p_r) = (-1)^{3(p_s^2-1)/8}\left(\frac{-1}{p_1}\right) \cdots \left(\frac{-1}{p_{s-1}}\right)\left(\frac{-1}{p_{s+1}}\right)$$

$$\cdots \left(\frac{-1}{p_r}\right)\left(\frac{p_s}{p_1}\right) \cdots \left(\frac{p_s}{p_r}\right),$$

$$\chi_{p_{s+1}}(8p_1 \cdots p_s p_{s+2} \cdots p_r) = (-1)^{3(p_{s+1}^2-1)/8}\left(\frac{p_{s+1}}{p_1}\right) \cdots \left(\frac{p_{s+1}}{p_r}\right),$$

. . . .

$$\chi_{p_r}(8p_1 \cdots p_{r-1}) = (-1)^{3(p_r^2-1)/8}\left(\frac{p_r}{p_1}\right) \cdots \left(\frac{p_r}{p_{r-1}}\right).$$

By Jacobi's reciprocity law and a similar computation

$$\tau_1(\chi) = \tau_1(\chi_{(-1)^{\epsilon+\kappa}\cdot 2})\tau_1(\chi_{-p_1}) \cdots \tau_1(\chi_{p_r})$$

$$\times (-1)^{s(\epsilon+s)}(-1)^{\sum_{i=1}^r (p_i^2-1)/8}(-1)^{3\sum_{i=1}^r (p_i^2-1)/8}(-1)^{s(s-1)+s(s-1)/2}$$

$$= (-1)^{s(\epsilon+s)+3s(s-1)/2}\tau_1(\chi_{(-1)^{\epsilon+\kappa}\cdot 2})\tau_1(\chi_{-p_1}) \cdots \tau_1(\chi_{p_r}).$$

By the previous case of the proof

$$\tau_1(\chi) = i^s\sqrt{p_1 \cdots p_r}\,\tau_1(\chi_{(-1)^{\epsilon+\kappa}\cdot 2})(-1)^{s(\epsilon+s)+3s(s-1)/2}.$$

If $\epsilon + s$ is even then $\tau_1(\chi_2) = 2\sqrt{2}$, so

$$\tau_1(\chi) = \sqrt{8p_1 \cdots p_r}\,i^{s+3s(s-1)}.$$

If $d > 0$ then $\epsilon = 0$, s is even so $s + 3s(s - 1) \equiv 0 \pmod 4$, hence

$$\tau_1(\chi) = \sqrt{\delta}.$$

If $d < 0$ then $\epsilon = 1$, s is odd so $s + 3s(s - 1) \equiv 1 \pmod 4$, hence

$$\tau_1(\chi) = i\sqrt{|\delta|}.$$

Now, if $\epsilon + s$ is odd then $\tau_1(\chi_{-2}) = 2\sqrt{2}\,i$, so

$$\tau_1(\chi) = \sqrt{8p_1 \cdots p_r}\,i^{1+s}(-1)^{s+3s(s-1)/2} = \sqrt{|\delta|}\,i^{1+3s+3s(s-1)}.$$

If $d > 0$ then $\epsilon = 0$, s is odd, so $1 + 3s + 3s(s - 1) \equiv 0 \pmod 4$, hence

$$\tau_1(\chi) = \sqrt{\delta}.$$

If $d < 0$ then $\epsilon = 1$, s is even so $1 + 3s + 3s(s - 1) \equiv 1 \pmod 4$, hence

$$\tau_1(\chi) = i\sqrt{|\delta|}. \qquad \blacksquare$$

In view of (**E**) and (26.11), (26.12) we have:
If $d > 0$ then

$$L(1|\chi) = -\frac{1}{\sqrt{\delta}} \sum_{\substack{k=1 \\ \gcd(k,\delta)=1}}^{\delta-1} \chi(k) \log\left(\sin\frac{k\pi}{\delta}\right) \qquad (26.16)$$

and if $d < 0$ then

$$L(1\,|\,\chi) = -\frac{\pi}{\sqrt{|\delta|}^3} \sum_{\substack{k=1 \\ \gcd(k,\delta)=1}}^{|\delta|-1} \chi(k)k. \qquad (26.17)$$

26.3 The Class Number Formula and the Distribution of Quadratic Residues

We obtain from (26.7), (26.8) and (26.16), (26.17) the *class number formula* of Dirichlet:

Theorem 1.   *If $d > 0$ then*

$$h = -\frac{1}{\log u} \sum_{1 \le k < \delta/2} \chi(k) \log \left(\sin \frac{k\pi}{\delta} \right) \qquad (26.18)$$

and if $d < 0$ then

$$h = -\frac{w}{2|\delta|} \sum_{k=1}^{|\delta|-1} \chi(k)k. \qquad (26.19)$$

Proof: As stated, the above formulas follow at once from (26.7), (26.8) and (26.16), (26.17) noting that if $d > 0$ then χ is an even character.   ∎

The formula for h when $d > 0$ involves a combination of logarithms of sines and it is not at all obvious a priori that it gives rise to the natural number h. In practice, besides the knowledge of the fundamental unit u of $\mathbb{Q}(\sqrt{d})$ (see Chapter 10, Section 2), one would need to compute explicitly the logarithms of the sines. This is awkward, so we still need to indicate a method which leads to the effective computation of the class number.

Things are easier when $d < 0$ and therefore we shall first deal with this case.

For example, using the values of χ given in (**A**), we obtain the following results:

If $d = -1$ we have $h = -\frac{1}{2}(\chi(1) + \chi(3) \cdot 3) = 1$.
If $d = -2$ we have $h = -\frac{1}{8}(\chi(1) + \chi(3) \cdot 3 + \chi(5) \cdot 5 + \chi(7) \cdot 7) = -\frac{1}{8}(1 + 3 - 5 - 7) = 1$.
If $d = -3$ we have $h = -(\chi(1) + \chi(2) \cdot 2) = 1$.
If $d = -5$ we have $h = -\frac{1}{20}(\chi(1)+\chi(3)\cdot3)+\chi(7)\cdot7+\chi(9)\cdot9+\chi(11)\cdot11+\chi(13)\cdot13+\chi(17)\cdot17+\chi(19)\cdot19) = -\frac{1}{20}(1+3+7+9-11-13-17-19) = 2$.

In general, the computation of h is reduced to that of Jacobi symbols, which is effective by means of the Jacobi reciprocity law.

By rewriting the formula for h, when $d < 0$, we shall be able to deduce an interesting consequence, involving the distribution of quadratic residues, and leading to a very straightforward method for computing the class number.

F.   *If $d \neq -1, -3$, $d < 0$, then*

$$h = \frac{1}{2 - \chi(2)} \sum_{1 \leq k < |\delta|/2} \chi(k). \tag{26.20}$$

Proof: From the hypothesis we have $w = 2$, hence

$$|\delta|h = - \sum_{k=1}^{|\delta|-1} \chi(k)k.$$

If $d \equiv 2 \pmod 4$ or $d \equiv 3 \pmod 4$ then $\delta = 4d$. Letting $t = |\delta|/2 = 2|d|$ we may write

$$|\delta|h = - \sum_{1 \leq k < t} \chi(k)k - \sum_{1 \leq k < t} \chi(k + t) \cdot (k + t). \tag{26.21}$$

But $\chi(k + t) = -\chi(k)$ when $\gcd(k, \delta) = 1$, as we verify now using the expression (**A**) of the character.
   If $d \equiv 3 \pmod 4$ then

$$\chi(k + 2|d|) = (-1)^{(k+2|d|-1)/2} \left(\frac{k + 2|d|}{|d|} \right) = -(-1)^{(k-1)/2} \left(\frac{k}{|d|} \right) = -\chi(k)$$

because $|d|$ is odd.
   If $d \equiv 2 \pmod 4$ and $d = 2d'$ (d' odd) then

$$\chi(k + 2|d|) = (-1)^{((k+2|d|)^2 - 1)/8}(-1)^{\frac{k+2|d|-1}{2} \cdot \frac{d'-1}{2}} \left(\frac{k + 2|d|}{|d'|} \right)$$

$$= -(-1)^{(k^2-1)/8}(-1)^{\frac{k-1}{2} \cdot \frac{d'-1}{2}} \left(\frac{k}{|d'|} \right) = -\chi(k)$$

because

$$\frac{(k + 2|d|)^2 - 1}{8} + \frac{k + 2|d| - 1}{2} \cdot \frac{d' - 1}{2} - \frac{k^2 - 1}{8} - \frac{k - 1}{2} \cdot \frac{d' - 1}{2}$$

$$= k|\,d'| + 2|d'|^2 + |d'|(|d'| - 1) \equiv 1 \pmod 2$$

since k is odd.
   Returning to (26.21) we may write

$$|\delta|h = - \sum_{1 \leq k < t} \chi(k)k + \sum_{1 \leq k < t} \chi(k) \cdot (k + t) = t \left(\sum_{1 \leq k < t} \chi(k)k \right)$$

and therefore

$$h = \frac{1}{2} \sum_{1 \leq k < |\delta|/2} \chi(k)$$

as we intended to show (because $\chi(2) = 0$ in this case).

Now let $d \equiv 1 \pmod 4$ then $\delta = d$ and we have similarly

$$|\delta|h = - \sum_{1 \le k < t} \chi(k)k - \sum_{1 \le k < t} \chi(|\delta| - k) \cdot (|\delta| - k)$$

$$= -2 \sum_{1 \le k < t} \chi(k)k + |\delta| \sum_{1 \le k < t} \chi(k) \qquad (26.22)$$

because χ is an odd character. On the other hand, since $|\delta|$ is odd, we also have

$$|\delta|h = - \sum_{k \text{ even}} \chi(k)k - \sum_{k \text{ even}} \chi(|\delta| - k) \cdot (|\delta| - k)$$

$$= - \sum_{1 \le j < t} \chi(2j) \cdot 2j + \sum_{1 \le j < t} \chi(2j) \cdot (|\delta| - 2j)$$

$$= -4 \sum_{1 \le j < t} \chi(2j) \cdot j + |\delta| \sum_{1 \le j < t} \chi(2j)$$

and multiplying by $\chi(2)$ we have

$$|\delta|h\,\chi(2) = -4 \sum_{1 \le j < t} \chi(j)j + |\delta| \sum_{1 \le j < t} \chi(j). \qquad (26.23)$$

From (26.22) and (26.23) we deduce that

$$|\delta|h(2 - \chi(2)) = |\delta| \sum_{1 \le k < t} \chi(k)$$

hence

$$h = \frac{1}{2 - \chi(2)} \sum_{1 \le k < |\delta|/2} \chi(k). \qquad \blacksquare$$

Now we discuss the distribution of quadratic residues. Let

$$Q^+ = \#\{k \mid 1 \le k < |\delta|,\ \chi(k) = 1\},$$
$$Q^- = \#\{k \mid 1 \le k < |\delta|,\ \chi(k) = -1\}.$$

Since $\sum_{k=1}^{|\delta|} \chi(k) = 0$ then $Q^+ = Q^-$.

Now, let

$$Q_1^+ = \#\left\{ k \,\middle|\, 1 \le k < \frac{|\delta|}{2},\ \chi(k) = 1 \right\},$$

$$Q_1^- = \#\left\{ k \,\middle|\, 1 \le k < \frac{|\delta|}{2},\ \chi(k) = -1 \right\},$$

$$Q_2^+ = \#\left\{ k \,\middle|\, \frac{|\delta|}{2} < k < |\delta|,\ \chi(k) = 1 \right\},$$

$$Q_2^- = \#\left\{ k \,\middle|\, \frac{|\delta|}{2} < k < |\delta|,\ \chi(k) = -1 \right\}.$$

So $Q^+ = Q_1^+ + Q_2^+$, and $Q^- = Q_1^- + Q_2^-$. Since χ is an odd character then $Q_1^+ = Q_2^-$ and $Q_1^- = Q_2^+$. We shall compare the numbers Q_1^+, Q_1^- in the following special case:

Theorem 2.   *Let $p \neq 3$ be a prime number, $p \equiv 3 \pmod 4$.*
   If $p \equiv 7 \pmod 8$ then $h = Q_1^+ - Q_1^-$.
   If $p \equiv 3 \pmod 8$ then $h = \frac{1}{3}[Q_1^+ - Q_1^-]$.

Proof: Let $K = \mathbb{Q}(\sqrt{-p})$. From $-p \equiv 1 \pmod 4$ we have $\delta = -p$. By (**F**):

$$h = \frac{1}{2 - \chi(2)} \sum_{1 \leq k < p//2} \chi(k).$$

But

$$\chi(2) = \left(\frac{2}{p}\right) = \begin{cases} 1, & \text{when } p \equiv 7 \pmod 8, \\ -1, & \text{when } p \equiv 3 \pmod 8, \end{cases}$$

and

$$\sum_{1 \leq k < p/2} \chi(k) = Q_1^+ - Q_1^-.$$

Hence

$$h = \begin{cases} Q_1^+ - Q_1^-, & \text{when } p \equiv 7 \pmod 8, \\ \frac{1}{3}(Q_1^+ - Q_1^-), & \text{when } p \equiv 3 \pmod 8. \end{cases} \qquad \blacksquare$$

The above result tells us that in the first half of the interval $(0, p)$, for $p \neq 3$, $p \equiv 3 \pmod 4$, there are more quadratic residues than nonresidues modulo p. The excess is either h or $3h$ when $p \neq 3$, $p \equiv 3 \pmod 4$. Despite this simple-minded statement, no elementary proof for this result is known as yet.

For example, for the field $\mathbb{Q}(\sqrt{-23})$ the class number is $h = 3$, since 1, 2, 3, 4, 6, 8, and 9 are the quadratic residues modulo 23 and 5, 7, 10, and 11 are the nonresidues in the interval $(0, \frac{23}{2})$.

Now we turn our attention to the real quadratic fields and proceed to indicate a practical method to compute the class number.

G.   *Let $d > 0$ and let*

$$\eta = \frac{\prod_k \sin(k\pi/\delta)}{\prod_j \sin(j\pi/\delta)},$$

where k, j are relatively prime to δ, $1 \leq k, j < \delta/2$, $\chi(k) = -1$, and $\chi(j) = 1$. Then η is a unit of $K = \mathbb{Q}(\sqrt{d})$, $\eta > 1$, and $\eta = u^h$.

Proof: This result is an almost immediate consequence of formula (26.18). Indeed, we obtain from it

$$u^h = \frac{\prod_k \sin(k\pi/\delta)}{\prod_j \sin(j\pi/\delta)} = \eta,$$

where j, k belong to the sets indicated in the statement. Therefore η is a unit and $u > 1$, $h \geq 1$ imply that $\eta > 1$. ∎

The fact that η is an algebraic number is already nontrivial. No elementary proof is known of the fact that $\eta > 1$.

Concerning the distribution of values of the character χ, we make the following preliminary remark. As before, $Q^+ = Q^-$ and since χ is now an even character then $Q_1^+ = Q_2^+$ and $Q_1^- = Q_2^-$. Hence $2Q_1^+ = Q_1^+ + Q_2^+ = Q^+ = Q^- = Q_1^- + Q_2^- = 2Q_1^-$. So, no result similar to Theorem 2 holds when $d > 0$. However, we may still arrive at a qualitative remark concerning the class number and the distribution of quadratic residues.

For this purpose, let $K = \mathbb{Q}(\sqrt{p})$, where p is a prime number, $p \equiv 1 \pmod 4$. Then $\delta = p$ and the character χ belonging to K coincides with the Legendre symbol for p. Since $\eta > 1$ then

$$\prod_k \sin \frac{k\pi}{p} > \prod_j \sin \frac{j\pi}{p}, \qquad (26.24)$$

where $1 \leq k, j < p/2$, $(k/p) = -1$, $(j/p) = 1$. But the sine function is monotonic increasing in the interval $(0, \pi/2)$. Thus in order to have the relation (26.24) the quadratic residues modulo p must appear mostly near 0, while the quadratic nonresidues are more likely to be closer to $p/2$. This behavior is more accentuated the larger η is, that is, the larger the class number h of $\mathbb{Q}(\sqrt{p})$ is.

Now we shall express the unit η rationally in terms of the root of unity $\zeta = \cos(2\pi/\delta) + i\sin(2\pi/\delta)$. Let $\xi = \cos(\pi/\delta) + i\sin(\pi/\delta)$, so ξ is a primitive root of unity of order 2δ and $\xi^2 = \zeta$.

H. *If δ is odd then*

$$\eta = (-\zeta^{(\delta+1)/2})^s \frac{\prod_k (1 - \zeta^k)}{\prod_j (1 - \zeta^j)}.$$

If δ is even then

$$\eta = \zeta^{s/2} \frac{\prod_k (1 - \zeta^k)}{\prod_j (1 - \zeta^j)},$$

where $s = \sum_{1 \leq a < \delta/2} \chi(a)a$, s is even when δ is even, $1 \leq j, k < \delta/2$, $\chi(j) = 1$, and $\chi(k) = -1$.

Proof: Since $Q_1^+ = Q_1^-$ we may write

$$\eta = \frac{\prod_k \sin(k\pi/\delta)}{\prod_j \sin(j\pi/\delta)} = \frac{\prod_k (2i\sin(k\pi/\delta))}{\prod_j (2i\sin(j\pi/\delta))}.$$

But

$$2i \sin \frac{k\pi}{\delta} = \xi^k - \xi^{-k} = \xi^{-k}(\xi^{2k} - 1) = -\xi^{-k}(1 - \zeta^k).$$

Hence

$$\eta = \frac{\prod_k \xi^{-k}(1 - \zeta^k)}{\prod_j \xi^{-j}(1 - \zeta^j)} = \xi^s \frac{\prod_k (1 - \zeta^k)}{\prod_j (1 - \zeta^j)}$$

with

$$s = \sum_{1 \le a < \delta/2} \chi(a)a, \qquad 1 \le j, k < \delta/2, \quad \chi(j) = 1, \quad \chi(k) = -1.$$

If δ is odd then $\xi = -\zeta^{(\delta+1)/2}$ because $\xi^\delta = -1$, hence $-\zeta^{(\delta+1)/2} = \xi^{\delta+\delta+1} = \xi$.

If δ is even, we show that s is also even. Indeed, if $1 \le a < \delta/2$ then either $\chi(a) = 0$ or $\chi(a)a \equiv a \pmod 2$. Moreover, $\chi(a) = 0$ if and only if $\gcd(a, \delta) \ne 1$ and this is equivalent to $\chi(\delta/2 - a) = 0$. Hence

$$s = \sum_{1 \le a < \delta/2} \chi(a)a = \sum_{1 \le a < \delta/4} \chi(a)a + \sum_{1 \le a < \delta/4} \chi\left(\frac{\delta}{2} - a\right)\left(\frac{\delta}{2} - a\right)$$

$$\equiv \sum_{\substack{\chi(a) \ne 0 \\ 1 \le a < \delta/4}} a + \sum_{\substack{\chi(a) \ne 0 \\ 1 \le a < \delta/4}} \left(\frac{\delta}{2} - a\right) \pmod 2,$$

But δ is a multiple of 4 therefore

$$s \equiv 2 \sum_{\substack{\chi(a) \ne 0 \\ 1 \le a < \delta/2}} a \equiv 0 \pmod 2.$$

Summarizing, we have shown that if δ is odd then

$$\eta = (-\zeta^{(\delta+1)/2})^s \frac{\prod_k (1 - \zeta^k)}{\prod_j (1 - \zeta^j)}$$

while if δ is even then

$$\eta = \zeta^{s/2} \frac{\prod_k (1 - \zeta^k)}{\prod_j (1 - \zeta^j)}. \qquad \blacksquare$$

These formulas are very appropriate for the practical computation of the class number. We observe that we may group by pairs the factors of the numerator and denominator of η. For the pair of exponents k, j (relatively prime to δ) we let g be such that $jg \equiv k \pmod \delta$.
Then

$$\frac{1 - \zeta^k}{1 - \zeta^j} = \frac{1 - \zeta^{jg}}{1 - \zeta^j} = 1 + \zeta^j + \zeta^{2j} + \cdots + \zeta^{(g-1)j}.$$

It is possible to express the above sum in terms of the powers of $\zeta + \zeta^{-1}$ and in terms of the Gaussian sum of χ, whose value has been computed in (E). Comparing the expression of η with the fundamental unit, we finally find the value of h.

Numerical Examples: (1°) If $d = 2$ then $\delta = 8$, $\chi(1) = 1$, $\chi(3) = -1$, and $s = 1 - 3 = -2$:

$$\eta = \zeta^{-1} \frac{1 - \zeta^3}{1 - \zeta} = \zeta^{-1}(1 + \zeta + \zeta^2) = 1 + (\zeta + \zeta^{-1}).$$

But

$$2\sqrt{2} = \tau_1(\chi) = \zeta - \zeta^3 - \zeta^5 + \zeta^7 = 2(\zeta + \zeta^{-1})$$

hence $\zeta + \zeta^{-1} = \sqrt{2}$, so $\eta = 1 + \sqrt{2}$. The fundamental unit of $\mathbb{Q}(\sqrt{2})$ is $u = 1 + \sqrt{2}$, hence $u^h = \eta$ implies that $h = 1$.

(2°) If $d = 3$ then $\delta = 12$, $\chi(1) = 1$, $\chi(5) = -1$, and $s = 1 - 5 = -4$:

$$\eta = \zeta^{-2} \frac{1 - \zeta^5}{1 - \zeta} = \zeta^{-2}(1 + \zeta + \zeta^2 + \zeta^3 + \zeta^4)$$
$$= \zeta^{-2} + \zeta^{-1} + 1 + \zeta + \zeta^2 = 1 + (\zeta + \zeta^{-1}) + (\zeta^2 + \zeta^{-2})$$
$$= -1 + (\zeta + \zeta^{-1}) + (\zeta + \zeta^{-1})^2.$$

But

$$2\sqrt{3} = \tau_1(\chi) = \zeta - \zeta^5 - \zeta^7 + \zeta^{11} = 2(\zeta + \zeta^{-1})$$

since $\zeta^6 = -1$. Hence $\eta = -1 + \sqrt{3} + 3 = 2 + \sqrt{3}$. The fundamental unit is $u = 2 + \sqrt{3}$, hence $h = 1$.

(3°) If $d = 5$ then $\delta = 5$, $\chi(1) = 1$, $\chi(2) = -1$, and $s = 1 - 2 = -1$:

$$\eta = (-\zeta^{(5+1)/2})^{-1} \frac{1 - \zeta^2}{1 - \zeta} = -\zeta^2(1 + \zeta) = -\zeta^2 - \zeta^3 = 1 + (\zeta + \zeta^{-1})$$

because ζ satisfies the relation $\zeta^4 + \zeta^3 + \zeta^2 + \zeta + 1 = 0$.

But

$$\sqrt{5} = \tau_1(\chi) = \zeta - \zeta^2 - \zeta^3 + \zeta^4 = 1 + 2(\zeta + \zeta^{-1})$$

hence $\zeta + \zeta^{-1} = (-1 + \sqrt{5})/2$ and $\eta = (1 + \sqrt{5})/2$. The fundamental unit is $u = (1 + \sqrt{5})/2$ hence $h = 1$.

It should be observed that for larger values of $d > 0$ the above procedure becomes difficult to perform, since it is awkward to express the quotients $(1 - \zeta^k)/(1 - \zeta^j)$ and the Gaussian sum in terms of $\zeta + \zeta^{-1}$. There are more efficient methods described in the literature but we shall not consider this matter here.

Even without explicit computation of the class number, we obtain the following useful property:

I.   *The fields* $\mathbb{Q}(\sqrt{-1})$, $\mathbb{Q}(\sqrt{-2})$, $\mathbb{Q}(\sqrt{-p})$ *where* $p \equiv 3 \pmod 4$, $\mathbb{Q}(\sqrt{2})$, $\mathbb{Q}(\sqrt{p})$ *where* $p \equiv 1 \pmod 4$ *have odd class number. For the above real quadratic fields the norm of the fundamental unit* u *is* -1.

Proof: The assertion has already been established for the fields $\mathbb{Q}(\sqrt{-1})$, $\mathbb{Q}(\sqrt{-2})$, and $\mathbb{Q}(\sqrt{2})$. For $\mathbb{Q}(\sqrt{-p})$ where $p \equiv 3 \pmod 4$ we make use of

Theorem 2. Since $Q_1^+ + Q_1^- = (p-1)/2$ is odd then $Q_1^+ - Q_1^-$ is also odd. Therefore h is odd.

For $K = \mathbb{Q}(\sqrt{p})$ where $p \equiv 1 \pmod 4$, we prove that $N_{K|\mathbb{Q}}(\eta) = -1$. This implies that $N_{K|\mathbb{Q}}(u)^h = -1$ hence $N_{K|\mathbb{Q}}(u) = -1$ and h is odd.

As before, let $\xi = \cos(\pi/p) + i\sin(\pi/p)$ be a primitive $(2p)$th root of unity, $\zeta = \xi^2$, hence

$$\eta = \frac{\prod_k (2i \sin(k\pi/\delta))}{\prod_j (2i \sin(j\pi/\delta))} = \frac{\prod_k (\xi^k - \xi^{-k})}{\prod_j (\xi^j - \xi^{-j})},$$

where $1 \le j,\ k < p/2$, $(j/p) = 1$, $(k/p) = -1$. Thus we may rewrite

$$\eta = \prod_{1 \le a < p/2} (\xi^a - \xi^{-a})^{-\left(\frac{a}{p}\right)}.$$

Since $\xi^p = -1$ then $\xi^{p-a} - \xi^{-(p-a)} = \xi^a - \xi^{-a}$; noting that a is even exactly when $p - a$ is odd then we may write

$$\eta = \prod_{\substack{1 \le a < p \\ a \text{ odd}}} (\xi^a - \xi^{-a})^{-\left(\frac{a}{p}\right)}.$$

The unit $\eta = u^h$ belongs to $\mathbb{Q}(\sqrt{p})$. Since

$$\sqrt{p} = \tau_1(\chi) = \sum_{k=1}^{p-1} \chi(k)\zeta^k \in \mathbb{Q}(\zeta) = \mathbb{Q}(\xi)$$

then $\mathbb{Q}(\sqrt{p})$ is a subfield of $\mathbb{Q}(\zeta)$.

The Galois group of $\mathbb{Q}(\zeta)$ over $\mathbb{Q}$ is

$$G = \{\sigma_{\bar{a}} \mid \gcd(a, p) = 1\},$$

where $\sigma_{\bar{a}}(\zeta) = \zeta^a$. The subgroup

$$H = \left\{ \sigma_{\bar{a}} \,\middle|\, \gcd(a, p) = 1 \text{ and } \left(\frac{a}{p}\right) = 1 \right\}$$

has order $(p-1)/2$ and leaves $\sqrt{p} = \tau_1(\chi)$ invariant:

$$\sigma_{\bar{a}}(\tau_1(\chi)) = \sum_{k=1}^{p-1} \chi(k)\zeta^{ak} = \sum_{k=1}^{p-1} \chi(ak)\zeta^{ak} = \tau_1(\chi).$$

Hence $\mathbb{Q}(\sqrt{p})$ is the field of invariants of H and since $\eta \in \mathbb{Q}(\sqrt{p})$, its only conjugate (different from η) is $\eta' = \sigma_{\bar{b}}(\eta)$, for any b such that $\gcd(b, p) = 1$ and $(b/p) = -1$. We may choose b odd (for if b is even then $b + p$ is odd).

Let us show that $\sigma_{\bar{b}}(\xi) = -\eta^{-1}$; this will imply that the norm of η is $\eta(-\eta^{-1}) = -1$. Since b is odd, then

$$\sigma_{\bar{b}}(\xi) = \sigma_{\bar{b}}(-\zeta^{(p+1)/2}) = -\zeta^{((p+1)/2)b} = (-\zeta^{(p+1)/2})^b = \xi^b.$$

Hence

$$\sigma_{\bar{b}}(\eta) = \prod_{\substack{1 \le a < p \\ a \text{ odd}}} (\xi^{ba} - \xi^{-ba})^{-\left(\frac{a}{p}\right)}.$$

For every odd a, $1 \le a < p$, there exists a', odd, $1 \le a' < p$, such that

$$ba \equiv a' \pmod{2p} \qquad \text{or} \qquad ba \equiv -a' \pmod{2p}. \qquad (26.25)$$

Let r be the number of integers a, $1 \le a < p$, for which $ba \equiv -a'$ (mod $2p$).

If $ba \equiv a'$ (mod $2p$) then $\xi^{ba} - \xi^{-ba} = \xi^{a'} - \xi^{-a'}$.

If $ba \equiv -a'$ (mod $2p$) then $\xi^{ba} - \xi^{-ba} = \xi^{-a'} - \xi^{a'} = (-1)(\xi^{a'} - \xi^{-a'})$.

Multiplying the congruences (26.25) for all odd a. $1 \le a' < p$, we have

$$b^{(p-1)/2}(1 \cdot 3 \cdots (p-2)) \equiv (-1)^r (1 \cdot 3 \cdots (p-2)) \pmod{2p}$$

hence $b^{(p-1)/2} \equiv (-1)^r$ (mod $2p$). But $-1 = (b/p) \equiv b^{(p-1)/2}$ (mod p) hence $(-1)^r = -1$ showing that r is odd. Taking into account that $(a/p) = -(ba/p) = -(a'/p)$ then

$$\sigma_{\bar{b}}(\eta) = (-1)^r \prod_{\substack{1 \le a' < p/2 \\ a' \text{ odd}}} (\xi^{a'} - \xi^{-a'})^{\left(\frac{a'}{p}\right)} = -\eta^{-1}.$$

This concludes the proof. ∎

The fact that the fundamental unit of $\mathbb{Q}(\sqrt{p})$ (when $p \equiv 1$ (mod 4)) has norm -1 had been proved, in an elementary manner, in Chapter 10, (**H**).

We indicate now a proof of the preceding result, which does not involve analytical considerations. To begin, a few easy remarks about ideals.

Let K be an algebraic number field, and let A be the ring of integers of K and J a nonzero fractional ideal. For every prime number p let $m(p) \ge 0$ be the largest integer such that $Ap^{m(p)}$ divides J. Let $m = \prod p^{m(p)}$, so $J = Am \cdot J_0$. The ideal J_0, which is completely determined by J, is called the *primitive part* of J. If $J = Ax$ is a principal ideal then J_0 is also a principal ideal, generated by $x_0 = x/m$.

If $K|\mathbb{Q}$ is a Galois extension and $\sigma(J) = J$ for every $\sigma \in G(K|\mathbb{Q})$ then we also have $\sigma(J_0) = J_0$. It follows from Chapter 11, Section 2, that the only prime ideals P of A which divide J_0 are ramified. Indeed, if P is unramified and divides J_0 then all its conjugates divide J_0, hence Ap divides J_0 (where $P \cap \mathbb{Z} = \mathbb{Z}p$), contrary to the definition of J_0.

If $K = \mathbb{Q}(\sqrt{-p})$ or $\mathbb{Q}(\sqrt{p})$ the only ramified prime is p and $Ap = P^2$ where $P = A\sqrt{-p}$ or $P = A\sqrt{p}$. In both cases P is a principal ideal. So if J is a nonzero fractional ideal, invariant by conjugation, then its primitive part J_0 is a principal ideal.

New proof of (**I**):   We need only to consider the fields $K = \mathbb{Q}(\sqrt{p})$ for $p \equiv 1 \pmod 4$ and $K = \mathbb{Q}(\sqrt{-p})$ for $p \equiv 3 \pmod 4$. In order to prove that h is odd, we shall show that if J is any nonzero fractional ideal of A and J^2 is principal, then J is already principal. This implies that in the group $\mathcal{C}$ of classes of ideals every class of ideals has odd order, so $\mathcal{C}$ itself has odd order h.

We may assume that J is an integral ideal. Let J' be its conjugate ideal. From $J \cdot J' = Am$, where $m = N_{K|\mathbb{Q}}(J)$ it follows that the ideal J' belongs to the class of ideals inverse of the class of J. Since J^2 is a principal ideal, we deduce that J and J' are in the same class, that is, $J = Ac \cdot J'$. Then $N_{K|\mathbb{Q}}(c) = \pm 1$. Writing $c = c_1 + c_2\sqrt{d}$ (where $d = \pm p$) we have $N_{K|\mathbb{Q}}(c) = c_1{}^2 - c_2{}^2 d$. If $d < 0$ then $N_{K|\mathbb{Q}}(c) = 1$. If $d > 0$ and $N_{K|\mathbb{Q}}(c) = -1$ we may replace c by uc which has norm $N_{K|\mathbb{Q}}(uc) = (-1)(-1) = 1$, because it was shown that $N_{K|\mathbb{Q}}(u) = -1$ (see Chapter 10, (**H**)). So in any case we may assume that $N_{K|\mathbb{Q}}(c) = 1$ and we may write $c = (1 + c)/(1 + c')$ (when $c \neq -1$) or $c = \sqrt{d}/(-\sqrt{d})$, (when $c = -1$), so $c = x/x'$, where $x \in A$. Therefore $J = Ac \cdot J'$ implies that $\dfrac{J}{Ax} = \dfrac{J'}{Ax'}$ and the ideal $\dfrac{J}{Ax}$ is invariant by conjugation. By the remark at the beginning, the primitive part of $\dfrac{J}{Ax}$ is also invariant by conjugation, hence it is a principal ideal. This shows that J is principal and concludes the proof.   ∎

EXERCISES

1. Write explicitly the decomposition of χ_d, as in (**C**), for $d = -30, 30, 105$, and -33.

2. Use the formula for h, given in Theorem 1, to determine the class number when $d = -7, -11, -163, -10$, and -105.

3. Same as the preceding exercise, for $d = 2, 3, 5, 11, 163, 6, 10$, and 105.

4. Let $d = 5, 10, 11$, and 30. Compute explicitly the unit η (in (**G**)), a fundamental unit, and derive the value of the class number of $\mathbb{Q}(\sqrt{d})$.

5. Use the formula indicated in (**H**) to calculate explicitly the unit η, where $d = 5, 10, 11$, and 105.

6. Let p be a prime. Show that the equation $X^2 - pY^2 = -1$ has a solution in integers if and only if $p \equiv 1 \pmod 4$.

7. Let χ be the quadratic character with conductor 4. By explicit calculation show that the associated principal Gaussian sum is $\tau_1(\chi) = 2i$.

8. Let χ', χ'' be the following characters:

$$\chi'(a) = (-1)^{(a^2-1)/8},$$

$$\chi''(a) = (-1)^{(a^2-1)/8+(a-1)/2},$$

for all odd a. By explicit calculation show that the associated principal Gaussian sums are $\tau_1(\chi') = 2\sqrt{2}$ and $\tau_1(\chi'') = 2\sqrt{2}\,i$.

9. Show that in each quadratic field $\mathbb{Q}(\sqrt{d})$ there exist infinitely many prime ideals whose norm is the square of a prime.

Hint: Apply **(D)**.

27

Class Number of Cyclotomic Fields

In this chapter we shall derive formulas for the class number of cyclotomic fields generated by pth roots of unity, where p is a prime. They involve L-series and Gaussian sums associated to characters.

27.1 The Class Number Formula

Let $m > 2$ and let $\zeta = \zeta_m = \cos(2\pi/m) + i\sin(2\pi/m)$ be a primitive root of unity of order m. Let $K = K_m = \mathbb{Q}(\zeta_m)$. Our aim in this section is to give a formula for the class number $h_m = h(K_m)$ of K (sometimes we simply denote it by h). From Chapter 23, (C), it suffices to determine $\lim_{s \to 1+0}(s - 1)\zeta_K(s)$, where $\zeta_K(s)$ denotes the Dedekind zeta-function of K. So we begin by determining this function in terms of the L-series of the modular character χ modulo m.

Let $A = \mathbb{Z}[\zeta]$ be the ring of integers of K. Let

$$J(s) = \prod_{P|m} \frac{1}{1 - 1/N(P)^s} \cdot \prod_{p|m}\left(1 - \frac{1}{p^s}\right), \qquad (27.1)$$

where the first product is over all prime ideals P of A dividing Am and the second product is over all prime numbers p dividing m.

A. *For $s > 1$ we have:*

(1) $$\zeta_K(s) = \prod_{\chi} L(s|\chi) \prod_{P|m} \frac{1}{1 - 1/N(P)^s}, \qquad (27.2)$$

(2) $$\zeta_K(s) = \prod_{\chi \neq \chi_0} L(s|\chi) \cdot J(s)\zeta(s), \qquad (27.3)$$

where $\prod_{\chi}$ (respectively, $\prod_{\chi \neq \chi_0}$), indicate the product over all characters χ modulo m, respectively, $\chi \neq \chi_0$.

Proof: (1)   By Chapter 23, (23.12), we have

$$\zeta_K(s) = \prod_{P \in \mathcal{P}} \frac{1}{1 - 1/N(P)^s} \qquad \text{for} \quad s > 1, \tag{27.4}$$

where $\mathcal{P}$ denotes the set of nonzero prime ideals of $A = \mathbb{Z}[\zeta]$.

We recall that p is ramified in $K|\mathbb{Q}$ exactly when p divides m. On the other hand, if p does not divide m then $Ap = P_1 \cdots P_{g(p)}$, where $N(P_i) = p^{f(p)}$ for every $i = 1, \ldots, g(p)$, $f(p)g(p) = \varphi(m)$, and $f(p)$ is the order of $\bar{p}$ (class of p modulo m) in the multiplicative group $P(m)$ of prime residue classes modulo m (Chapter 16, Section 2).

So

$$\zeta_K(s) = F(s) \cdot \prod_{p \nmid m} \prod_{P|p} \frac{1}{1 - 1/N(P)^s} \qquad \text{for} \quad s > 1, \tag{27.5}$$

where

$$F(s) = \prod_{p|m} \prod_{P|p} \frac{1}{1 - 1/N(P)^s}. \tag{27.6}$$

Now we evaluate the other products appearing in (27.5). If p does not divide m and P divides p then

$$1 - \frac{1}{N(P)^s} = 1 - \left(\frac{1}{p^s}\right)^{f(p)} = \prod_{k=0}^{f(p)-1} \left(1 - \frac{\xi_p^k}{p^s}\right),$$

where ξ_p is a primitive root of unity of order $f(p)$. Hence if p does not divide m then

$$\prod_{P|p} \left(1 - \frac{1}{N(P)^s}\right)^{-1} = \prod_{k=0}^{f(p)-1} \left(1 - \frac{\xi_p^k}{p^s}\right)^{-g(p)}. \tag{27.7}$$

We shall relate these products with the characters modulo m. For this purpose, we prove that, for every $k = 0, 1, \ldots, f(p) - 1$, there exist precisely $g(p)$ characters χ modulo m such that $\chi(p) = \xi_p^k$.

Indeed, let G be the subgroup of $P(m)$ generated by $\bar{p} = p \pmod{m}$; it has order $f(p)$. Given ξ_p^k, where $0 \leq k \leq f(p) - 1$, there exists a unique character $\bar{\chi}'$ of G such that $\bar{\chi}'(\bar{p}) = \xi_p^k$. By Chapter 21, (**D**), $\bar{\chi}'$ has precisely $\varphi(m)/f(p) = g(p)$ extensions to characters $\bar{\chi}$ of $P(m)$. Finally, from Chapter 22, (**G**), our assertion follows immediately for the modular characters.

Returning to (27.7), we may write it as follows:

$$\prod_{k=0}^{f(p)-1} \left(1 - \frac{\xi_p^k}{p^s}\right)^{-g(p)} = \prod_{\chi} \left(1 - \frac{\chi(p)}{p^s}\right)^{-1}, \tag{27.8}$$

where the last product is extended over all the $\varphi(m)$ characters modulo m.

Hence

$$\zeta_K(s) = F(s) \prod_{p \nmid m} \prod_{\chi} \frac{1}{1 - \chi(p)/p^s} = F(s) \prod_{\chi} \prod_{p \nmid m} \frac{1}{1 - \chi(p)/p^s}$$

$$= F(s) \prod_{\chi} L(s|\chi) \qquad \text{for} \quad s > 1,$$

noting that if p divides m then $\chi(p) = 0$ and using the multiplicative expressions for the L-series (Chapter 22, (G)).

(2)   We have seen in Chapter 22, (G), for the trivial character modulo m:

$$L(s|\chi_0) = \zeta(s) \prod_{p|m} \left(1 - \frac{1}{p^s}\right) \qquad \text{for} \quad s > 1.$$

Substituting this in (27.2) we have at once the above expression (27.3). ∎

Now we obtain an expression for the class number, involving the L-series and the invariants of the field K:

B.

$$h = \frac{w\sqrt{|\delta|}}{2^{r_1+r_2}\pi^{r_2}R} J(1) \prod_{\chi \neq \chi_0} L(1|\chi). \tag{27.9}$$

Proof: From (27.3) we have

$$\lim_{s \to 1+0} (s-1)\zeta_K(s) = J(1) \lim_{s \to 1+0} (s-1)\zeta(s) \prod_{\chi \neq \chi_0} L(s|\chi).$$

By Chapter 23, (C), and by Chapter 22, (D) and (I), we have

$$h = \frac{w\sqrt{|\delta|}}{2^{r_1+r_2}\pi^{r_2}R} J(1) \prod_{\chi \neq \chi_0} L(1|\chi). \qquad ∎$$

We note incidentally that (B) provides another proof that $L(1|\chi) \neq 0$ for every character $\chi \neq \chi_0$ (see Chapter 24, Section 1, proof of Dirichlet's theorem).

It also follows from $J(1) > 0$ that $\prod_{\chi \neq \chi_0} L(1|\chi)$ is a positive real number.

In Chapter 22, (J), we obtained the following expression of $L(s|\chi)$ in terms of the principal Gaussian sums $\tau_k(\chi)$:

$$L(s|\chi) = \frac{1}{m} \sum_{k=0}^{m-1} \tau_k(\chi) \sum_{n=1}^{\infty} \frac{\zeta^{-nk}}{n^s} \qquad \text{for} \quad s > 1. \tag{27.10}$$

Combining this with the previous relations we have the following *class number formula for cyclotomic fields*:

Theorem 1.

$$h = \frac{w\sqrt{|\delta|}}{2^{r_1+r_2}\pi^{r_2}R} J(1) \frac{1}{m^{\varphi(m)-1}} \prod_{\chi\neq\chi_0} \left\{ \sum_{k=1}^{m-1} \tau_k(\chi) \log \frac{1}{1-\zeta^{-k}} \right\}. \quad (27.11)$$

Proof: In view of (**B**) it is enough to evaluate $L(1|\chi)$ when $\chi \neq \chi_0$. By (27.10):

$$L(1|\chi) = \frac{1}{m}\sum_{k=0}^{m-1}\tau_k(\chi)\sum_{n=1}^{\infty}\frac{\zeta^{-nk}}{n} = \frac{1}{m}\sum_{k=1}^{m-1}\tau_k(\chi)\log\frac{1}{1-\zeta^{-k}}, \quad (27.12)$$

noting that $\tau_0(\chi) = 0$ for $\chi \neq \chi_0$ and

$$\sum_{n=1}^{\infty}\frac{\zeta^{-nk}}{n} = \log\frac{1}{1-\zeta^{-k}}.$$

Indeed, for every $\ell \geq 1$ we have $|\sum_{n=1}^{\ell}\zeta^{-nk}| \leq m$ since ζ is an mth root of unity. By Chapter 22, (**B**), it follows that the series $\sum_{n=1}^{\infty}\zeta^{-nk}/n^s$ converges uniformly on every interval $[\delta,\infty)$, where $\delta > 0$. In particular, $\sum_{n=1}^{\infty}\zeta^{-nk}/n$ converges and by the definition of the logarithmic function

$$\sum_{n=1}^{\infty}\frac{\zeta^{-nk}}{n} = \log\frac{1}{1-\zeta^{-k}}.$$

From (**B**) we conclude that (27.11) holds.   ∎

The expression (27.11) involves no infinite product, and therefore it is more appropriate than (27.10) for explicit computations. However, it is of a somewhat awkward nature, since h is a natural number and it is expressed in terms of complex numbers and logarithms.

C.   *If ζ is a primitive pth root of 1, the class number of $K = \mathbb{Q}(\zeta)$ is*

$$h = \frac{p^{p/2}}{2^{(p-3)/2}\pi^{(p-1)/2}R} \prod_{\chi\neq\chi_0} L(1|\chi). \quad (27.13)$$

Proof: In this case, we have $\varphi(p) = p-1$, $r_1 = 0$, $r_2 = (p-1)/2$, $w = 2p$ (see Chapter 10, (**F**)), and $\delta = (-1)^{(p-1)/2}p^{p-2}$ (see Chapter 6, (**R**)); moreover, $P = A(1-\zeta)$ is the only prime ideal of $A = \mathbb{Z}[\zeta]$ dividing p, and $Ap = P^{p-1}$ so $N(P) = p$; hence

$$J(1) = \frac{1}{1-1/p}\left(1-\frac{1}{p}\right) = 1.$$

Therefore, by (**B**) we deduce (27.13).   ∎

27.2 The Two Factors of the Class Number

We recall from Chapter 21 that modular characters are either even or odd.

Let χ be a primitive character modulo m and let $\tau_1(\chi)$ be the associated principal Gaussian sum (see Chapter 21, Section 3).

D.   *If χ is even then*

$$L(1|\chi) = \frac{\tau_1(\chi)}{m} \sum_{\substack{k=1 \\ k \in P(m)}}^{m-1} \overline{\chi}(k) \log \frac{1}{|1 - \zeta^k|}$$

$$= \frac{\tau_1(\chi)}{m} \sum_{\substack{k=1 \\ k \in P(m)}}^{m-1} \overline{\chi}(k) \log \left(\sin \frac{k\pi}{m} \right). \tag{27.14}$$

If χ is odd then

$$L(1|\chi) = \frac{\pi i \tau_1(\chi)}{m^2} \sum_{\substack{k=1 \\ k \in P(m)}}^{m-1} \overline{\chi}(k)k. \tag{27.15}$$

Proof: Since χ is a primitive character modulo m, it follows from Chapter 22, (22.19), and Chapter 21, (**U**), that

$$L(1|\chi) = \frac{\tau_1(\chi)}{m} \sum_{\substack{k=1 \\ k \in P(m)}}^{m-1} \frac{1}{\chi(k)} \log \frac{1}{1 - \zeta^{-k}}.$$

We have $\zeta = \cos(2\pi/m) + i \sin(2\pi/m)$ hence

$$1 - \zeta^{-k} = \zeta^{-k/2}(\zeta^{k/2} - \zeta^{-k/2}) = \left(\cos \frac{k\pi}{m} - i \sin \frac{k\pi}{m} \right) 2i \sin \frac{k\pi}{m}$$

$$= \left[\cos \left(\frac{\pi}{2} - \frac{k\pi}{m} \right) + i \sin \left(\frac{\pi}{2} - \frac{k\pi}{m} \right) \right] \cdot 2 \sin \frac{k\pi}{m},$$

where $-\pi/2 < \pi/2 - k\pi/m < \pi/2$ since $1 \leq k \leq m - 1$. The above expression shows that $|1 - \zeta^{-k}| = 2 \sin(k\pi/m)$ and taking logarithms, we obtain from the polar form of $1 - \zeta^{-k}$:

$$\log(1 - \zeta^{-k}) = i \left(\frac{\pi}{2} - \frac{k\pi}{m} \right) + \log |1 - \zeta^{-k}|.$$

For the complex conjugate $1 - \zeta^k$ we have

$$\log(1 - \zeta^k) = -i \left(\frac{\pi}{2} - \frac{k\pi}{m} \right) + \log |1 - \zeta^{-k}|.$$

If χ is an even character then $\chi(k) = \chi(-k)$ so

$$S = \sum_{\substack{k=1 \\ k \in P(m)}}^{m-1} \frac{1}{\chi(k)} \log \frac{1}{1 - \zeta^{-k}} = \sum_{\substack{k=1 \\ k \in P(m)}}^{m-1} \frac{1}{\chi(k)} \log \frac{1}{1 - \zeta^k}$$

hence

$$2S = \sum_{\substack{k=1 \\ k \in P(m)}}^{m-1} \frac{1}{\chi(k)} \left[\log \frac{1}{1-\zeta^{-k}} + \log \frac{1}{1-\zeta^k} \right]$$

$$= 2 \sum_{\substack{k=1 \\ k \in P(m)}}^{m-1} \frac{1}{\chi(k)} \log \frac{1}{|1-\zeta^k|} = -2 \sum_{\substack{k=1 \\ k \in P(m)}}^{m-1} \overline{\chi}(k) \log \left(2 \sin \frac{k\pi}{m} \right).$$

But since $\chi \neq \chi_0$ then

$$\left(\sum_{\substack{k=1 \\ k \in P(m)}}^{m-1} \chi(k) \right) \log 2 = 0$$

hence

$$S = - \sum_{\substack{k=1 \\ k \in P(m)}}^{m-1} \overline{\chi}(k) \log \left(\sin \frac{k\pi}{m} \right)$$

and therefore

$$L(1|\chi) = \frac{\tau_1(\chi)}{m} \sum_{\substack{k=1 \\ k \in P(m)}}^{m-1} \frac{1}{\chi(k)} \log \frac{1}{|1-\zeta^k|}$$

$$= -\frac{\tau_1(\chi)}{m} \sum_{\substack{k=1 \\ k \in P(m)}}^{m-1} \overline{\chi}(k) \log \left(\sin \frac{k\pi}{m} \right).$$

If now χ is odd then $\chi(k) = -\chi(-k)$ so

$$S = \sum_{\substack{k=1 \\ k \in P(m)}}^{m-1} \frac{1}{\chi(k)} \log \frac{1}{1-\zeta^{-k}} = - \sum_{\substack{k=1 \\ k \in P(m)}}^{m-1} \frac{1}{\chi(k)} \log \frac{1}{1-\zeta^k}$$

hence

$$2S = \sum_{\substack{k=1 \\ k \in P(m)}}^{m-1} \frac{1}{\chi(k)} \left[\log \frac{1}{1-\zeta^{-k}} - \log \frac{1}{1-\zeta^k} \right]$$

$$= -2 \sum_{\substack{k=1 \\ k \in P(m)}}^{m-1} \frac{1}{\chi(k)} i \left(\frac{\pi}{2} - \frac{k\pi}{m} \right)$$

$$= -i\pi \sum_{k=1}^{m-1} \overline{\chi}(k) + \frac{2i\pi}{m} \sum_{k=1}^{m-1} \overline{\chi}(k)k.$$

But χ is not the trivial character modulo m, hence $\sum_{k=1}^{m-1} \overline{\chi}(k) = 0$ and so

$$S = \frac{i\pi}{m} \sum_{k=1}^{m-1} \overline{\chi}(k)k$$

showing that

$$L(1|\chi) = \frac{\tau_1(\chi)i\pi}{m^2} \sum_{k=1}^{m-1} \overline{\chi}(k)k. \qquad \blacksquare$$

Taking $m = p$, an odd prime, we have:

Theorem 2.   *If ζ is a primitive pth root of unity, the class number h of $\mathbb{Q}(\zeta)$ may be written in the form $h = h^- h^+$, where*

$$h^- = \frac{1}{(2p)^{(p-3)/2}} |G(\eta)G(\eta^3) \cdots G(\eta^{p-2})| \qquad (27.16)$$

and

$$h^+ = \frac{2^{(p-3)/2}}{R} \prod_{k=1}^{(p-3)/2} \left| \sum_{j=0}^{(p-3)/2} \eta^{2kj} \log |1 - \zeta^{r^j}| \right|. \qquad (27.17)$$

Here r is a primitive root modulo p, η is a primitive $(p-1)$th root of unity, and $G(X) = \sum_{j=0}^{p-2} r_j X^j$, $r_j \in \mathbb{Z}$, being such that $1 \leq r_j < p$, $r_j \equiv r^j$ (mod p).

Proof: Let r be a primitive root modulo p, so $r^{(p-1)/2} \equiv -1$ (mod p). Since $P(p)$ is cyclic of order $p - 1$ then the group $\widehat{P(p)}$ of characters is also cyclic of order $p - 1$. Let η be a primitive $(p - 1)$th root of unity and let χ be the unique character modulo p defined by $\chi(r) = \eta^{-1}$; it is the character modulo p which corresponds to a generator of $\widehat{P(p)}$. Thus $\chi, \chi^2, \ldots, \chi^{p-2}, \chi^{p-1} = \chi_0$ are the characters modulo p and for every integer s we have

$$\chi^s(-1) = \chi^s(r^{(p-1)/2}) = \eta^{-((p-1)/2)s} = (-1)^s.$$

So χ^s is an even character if and only if s is even.
    By (**D**), we have, for $k = 1, 2, \ldots, (p - 3)/2$:

$$L(1|\chi^{2k}) = \frac{\tau_1(\chi^{2k})}{p} \sum_{j=0}^{p-2} \overline{\chi^{2k}}(r^j) \log \frac{1}{|1 - \zeta^{r^j}|}.$$

By Chapter 21, (**W**), $|\tau_1(\chi^{2k})| = \sqrt{p}$ hence

$$|L(1|\chi^{2k})| = \frac{1}{\sqrt{p}} \left| \sum_{j=0}^{p-2} \eta^{2kj} \log |1 - \zeta^{r^j}| \right|.$$

But we note that $r^t \equiv -1 \pmod{p}$ where $t = (p-1)/2$. So $\zeta^{r^{t+j}} = \zeta^{-r^j}$ for $j = 0, 1, \ldots, t-1$ (since $r^j(r^t + 1) \equiv 0 \pmod{p}$). Similarly, η is a $(p-1)$th root of unity, so $\eta^{2k(t+j)} = \eta^{2kj}$. Hence we reach the expression

$$|L(1|\chi^{2k})| = \frac{2}{\sqrt{p}} \left| \sum_{j=0}^{(p-3)/2} \eta^{2kj} \log|1 - \zeta^{r^j}| \right|. \tag{27.18}$$

For odd characters χ^{2k-1}, $k = 1, 2, \ldots, (p-1)/2$, we have

$$\sum_{a=1}^{p-1} \overline{\chi^{2k-1}}(a) a = \sum_{j=0}^{p-2} [\chi^{2k-1}(r^j)]^{-1} r_j = \sum_{j=0}^{p-2} \eta^{(2k-1)j} r_j = G(\eta^{2k-1}).$$

By (27.15) and Chapter 21, (**W**), we deduce that

$$|L(1|\chi^{2k-1})| = \frac{\pi\sqrt{p}}{p^2} |G(\eta^{2k-1})|. \tag{27.19}$$

Replacing these values in (27.13), after taking absolute values, we have

$$
\begin{aligned}
h &= \frac{p^{p/2}}{2^{(p-3)/2}\pi^{(p-1)/2}R} \prod_{k=1}^{(p-1)/2} |L(1|\chi^{2k-1})| \cdot \prod_{k=1}^{(p-3)/2} |L(1|\chi^{2k})| \\
&= \frac{p^{(p+3)/4}}{2^{(p-3)/2}\pi^{(p-1)/2}} \prod_{k=1}^{(p-1)/2} |L(1|\chi^{2k-1})| \cdot \frac{p^{(p-3)/4}}{R} \prod_{k=1}^{(p-3)/2} |L(1|\chi^{2k})| \\
&= \frac{1}{(2p)^{(p-3)/2}} \prod_{k=1}^{(p-1)/2} |G(\eta^{2k-1})| \cdot \frac{2^{(p-3)/2}}{R} \prod_{k=1}^{(p-3)/2} \left| \sum_{j=0}^{(p-3)/2} \eta^{2kj} \log|1 - \zeta^{r^j}| \right|.
\end{aligned}
$$
∎

We may rewrite the expression for h^- in the following form:

$$h^- = \gamma(p) \prod_{\chi \text{ odd}} L(1|\chi), \tag{27.20}$$

where

$$\gamma(p) = 2p \left(\frac{p}{4\pi^2} \right)^{(p-1)/4} = \frac{p^{(p+3)/4}}{2^{(p-3)/2}\pi^{(p-1)/2}}. \tag{27.21}$$

We shall prove that h^-, h^+ are integers and also that h^+ is the class number of $K^+ = K \cap \mathbb{R}$, which is the maximal real subfield of K. So, we first study the field K^+.

E. (1) $K^+ = \mathbb{Q}(\lambda)$, where $\lambda = \zeta + \zeta^{-1}$ and $[K^+ : \mathbb{Q}] = (p-1)/2 = t$.

  (2) The ring of integers A^+ of K^+ has basis $\{\lambda_0 = 1, \lambda_1 = \lambda, \ldots, \lambda_{t-1}\}$ where $\lambda_j = \zeta^j + \zeta^{-j}$ for $j = 1, 2, \ldots, t-1$. It has also the basis $\{1, \lambda, \lambda^2, \ldots, \lambda^{t-1}\}$.

  (3) The discriminant of K^+ is $\delta_{K^+} = p^{t-1}$.

(4) *The regulator of K^+ is $R^+ = R/2^{t-1}$ where R is the regulator of K.*

Proof: (1)   From $\lambda = \zeta + \zeta^{-1}$ we deduce that $\zeta^2 - \lambda\zeta + 1 = 0$. Since $\mathbb{Q}(\lambda) \subseteq K^+ = K \cap \mathbb{R} \subset K$ and $\zeta \notin \mathbb{R}$ then $[K : \mathbb{Q}(\lambda)] = 2$ and, necessarily, $\mathbb{Q}(\lambda) = K^+$. Therefore $[K^+ : \mathbb{Q}] = (p-1)/2$.

(2)   Each element λ_j is an algebraic integer and therefore $\lambda_j \in A^+$. Assume that $\sum_{j=0}^{t-1} a_j \lambda_j = 0$ with $a_j \in \mathbb{Z}$, that is

$$a_0 + \sum_{j=1}^{t-1} a_j \zeta^j + \sum_{j=1}^{t-1} a_j \zeta^{p-j} = 0;$$

noting that $\zeta^{p-1} = -(1 + \zeta + \cdots + \zeta^{p-2})$ then

$$(a_0 - a_1) + (a_2 - a_1)\zeta^2 + \cdots + (a_{t-1} - a_1)\zeta^{t-1} - a_1\zeta^t$$
$$- a_1\zeta^{t+1} + (a_{t-1} - a_1)\zeta^{t+2} + \cdots + (a_2 - a_1)\zeta^{p-2} = 0$$

and therefore $a_0 = a_1 = a_2 = \cdots = a_{t-1}$ and $a_1 = 0$, showing the linear independence over $\mathbb{Q}$ of the elements $1, \lambda, \ldots, \lambda_{t-1}$.

Now we show that these elements generate the Abelian group A^+. It is convenient to consider the following integral basis:

$$\{\zeta^{-t+1}, \zeta^{-t+2}, \ldots, \zeta^{-1}, 1, \zeta, \ldots, \zeta^{t-1}, \zeta^t\}$$

of K over $\mathbb{Q}$ (which is obtained from the integral basis $\{1, \zeta, \zeta^2, \ldots, \zeta^{p-2}\}$ by multiplication with the unit ζ^{-t+1}). If $x \in A^+ \subseteq A$ we may write

$$x = a_{-t+1}\zeta^{-t+1} + a_{-t+2}\zeta^{-t+2} + \cdots + a_{-1}\zeta^{-1} + a_0 + a_1\zeta + \cdots + a_t\zeta^t$$

with coefficients $a_j \in \mathbb{Z}$. The complex conjugate is

$$\overline{x} = a_{-t+1}\zeta^{t-1} + a_{-t+2}\zeta^{t-2} + \cdots + a_{-1}\zeta + a_0 + a_1\zeta^{-1} + \cdots + a_t\zeta^{-t}.$$

Since x is real then $x - \overline{x} = 0$ hence

$$0 = (a_1 - a_{-1})(\zeta - \zeta^{-1}) + (a_2 - a_{-2})(\zeta^2 - \zeta^{-2})$$
$$+ \cdots + (a_{t-1} - a_{-t+1})(\zeta^{t-1} - \zeta^{-t+1}) + a_t(\zeta^t - \zeta^{-t}).$$

We shall obtain a linear relation for $1, \lambda_1, \lambda_2, \ldots, \lambda_{t-1}$ with coefficients in $\mathbb{Q}$—these must therefore be equal to 0. For this purpose, we note that

$$\frac{\zeta^2 - \zeta^{-2}}{\zeta - \zeta^{-1}} = \zeta + \zeta^{-1} = \lambda_1,$$

$$\frac{\zeta^3 - \zeta^{-3}}{\zeta - \zeta^{-1}} = (\zeta^2 + \zeta^{-2}) + 1 = \lambda_2 + 1,$$

$$\frac{\zeta^4 - \zeta^{-4}}{\zeta - \zeta^{-1}} = (\zeta^3 + \zeta^{-3}) + (\zeta + \zeta^{-1}) = \lambda_3 + \lambda_1,$$

$$\cdots \cdots \cdots \cdots \cdots \cdots \cdots \cdots \cdots \cdots \cdots$$

$$\frac{\zeta^{t-1} - \zeta^{-(t-1)}}{\zeta - \zeta^{-1}} = (\zeta^{t-2} + \zeta^{-(t-2)}) + (\zeta^{t-4} + \zeta^{-(t-4)}) + \cdots$$

$$= \lambda_{t-2} + \lambda_{t-4} + \cdots,$$

$$\frac{\zeta^t - \zeta^{-t}}{\zeta - \zeta^{-1}} = (\zeta^{t-1} + \zeta^{-(t-1)}) + (\zeta^{t-3} + \zeta^{-(t-3)}) + \cdots$$

$$= \lambda_{t-1} + \lambda_{t-3} + \cdots.$$

In the above sum of linearly independent elements $1, \lambda_1, \lambda_2, \ldots, \lambda_{t-1}$, we have

coefficient of λ_{t-1} is $a_t = 0$,

coefficient of λ_{t-2} is $a_{t-1} - a_{-t+1} = 0$,

coefficient of λ_{t-3} is $a_t + (a_{t-2} - a_{-t+2}) = 0$,

coefficient of λ_{t-4} is $(a_{t-1} - a_{-t+1}) + (a_{t-3} - a_{-t+3}) = 0$,

coefficient of λ_{t-5} is $a_t + (a_{t-2} - a_{-t+2}) + (a_{t-4} - a_{-t+4}) = 0$,

and so on. From this, we deduce that $a_t = 0$, $a_{t-1} = a_{-t+1}$, $a_{t-2} = a_{-t+2}$, ... Therefore

$$x = a_0 + a_1(\zeta + \zeta^{-1}) + a_2(\zeta^2 + \zeta^{-2}) + \cdots + a_{t-1}(\zeta^{t-1} + \zeta^{-t+1})$$

$$= a_0 + a_1\lambda_1 + a_2\lambda_2 + \cdots + + a_{t-1}\lambda_{t-1}$$

with $a_j \in \mathbb{Z}$.

To show that $\{1, \lambda, \lambda^2, \ldots, \lambda^{t-1}\}$ is also an integral basis of K^+, we note:

$$\lambda^2 = (\zeta + \zeta^{-1})^2 = (\zeta^2 + \zeta^{-2}) + 2 = \lambda_2 + 2,$$

$$\lambda^3 = (\zeta + \zeta^{-1})^3 = (\zeta^3 + \zeta^{-3}) + 3(\zeta + \zeta^{-1}) = \lambda_3 + 3\lambda_1,$$

$$\lambda^4 = (\zeta + \zeta^{-1})^4 = (\zeta^4 + \zeta^{-4}) + 4(\zeta^2 + \zeta^{-2}) + 6 = \lambda_4 + 4\lambda_2 + 6,$$

$$\cdots \cdots \cdots \cdots \cdots \cdots \cdots \cdots \cdots \cdots \cdots \cdots \cdots$$

Thus $1, \lambda_1, \lambda_2, \lambda_3, \ldots, \lambda_{t-1}$, may be expressed as linear combinations with integral coefficients in terms of the powers of λ, and conversely.

(3)   To compute the discriminant δ_{K^+} we use the known value $\delta_K = (-1)^{(p-1)/2} p^{p-2}$ as well as the differents Δ_K, Δ_{K^+}.

Since $A = \mathbb{Z}[\zeta]$ and since $\Phi(X) = X^{p-1} + X^{p-2} + \cdots + X + 1$ is the minimal polynomial of ζ then $\Delta_K = A\Phi'(\zeta)$.

The minimal polynomial Ψ of λ over $\mathbb{Q}$ is such that $\Phi(X) = X^t\Psi(X + X^{-1})$. Indeed, $X^t\Psi(X+X^{-1})$ vanishes on ζ, it is monic of degree $2t = p-1$ so it coincides with Φ.

The different of K^+ is $\Delta_{K^+} = A^+\Psi'(\lambda)$ hence

$$\Delta_K = A\Phi'(\zeta) = A\Psi'(\lambda) \cdot A(1 - \zeta^{-2}) = \Delta_{K^+} \cdot A(1 - \zeta^{-2}).$$

But

$$1 - \zeta^{-2} = -\zeta^{-2}(1 - \zeta^2) = -\zeta^{-2}(1 + \zeta)(1 - \zeta)$$

and $1 + \zeta$ is a unit. On the other hand, $Ap = A(1 - \zeta)^{p-1}$ and $|N_{K|\mathbb{Q}}(1 - \zeta)| = p$.

Thus taking the norms of the differents we obtain the discriminants

$$\mathbb{Z}\delta_K = N_{K|\mathbb{Q}}(\Delta_K) = N_{K|\mathbb{Q}}(\Delta_{K^+}) \cdot N_{K|\mathbb{Q}}(A(1 - \zeta^{-2}))$$
$$= [N_{K^+|\mathbb{Q}}(\Delta_{K^+})]^2 \cdot N_{K|\mathbb{Q}}(A(1 - \zeta)) = \mathbb{Z}\delta_{K^+}^2 \cdot \mathbb{Z}p$$

and so $\mathbb{Z}\delta_{K^+}^2 = \mathbb{Z}p^{p-3}$, hence $|\delta_{K^+}| = p^{(p-3)/2}$.

Moreover, since K^+ is contained in $\mathbb{R}$ then δ_{K^+} is the square of a determinant with real entries, so δ_{K^+} is positive and therefore $\delta_{K^+} = p^{(p-3)/2}$.

(4)   The field K has a fundamental system of $t - 1$ units of infinite order $\{u_1, \ldots, u_{t-1}\}$. By Chapter 10, (F), we may write each unit in the form $u_i = \zeta^{k_i} v_i$, where v_i is a real unit, $0 \leq k_i \leq p - 1$. It follows that $\{v_1, \ldots, v_{t-1}\}$ is a fundamental system of units of infinite order of K^+; they are obviously independent and if v is a unit of K^+ then it is also a unit of K, hence we may write it in the form

$$v = \zeta^b \prod_{i=1}^{t-1} u_i^{a_i} = \zeta^c \prod_{i=1}^{t-1} v_i^{a_i} \quad \text{with} \quad a_i \in \mathbb{Z}, \quad 0 \leq b, c \leq p - 1.$$

We are able to compare regulators R and R^+ of K, K^+, respectively. Since K is not a real field, we have $\ell_1 = \ell_2 = \cdots = \ell_{t-1} = 2$ (see the notations in Chapter 10, Definition 5); on the other hand, K^+ is real, hence for K^+ the corresponding exponents ℓ_i^+ are equal to 1. Since $|u_j^{(i)}| = |v_j^{(i)}|$ for all conjugates of u_j, v_j, we have

$$R = |\det(2 \log |u_j^{(i)}|)| = 2^{t-1}|\det(\log |v_j^{(i)}|)| = 2^{t-1}R^+. \quad \blacksquare$$

Now we indicate the decomposition of the prime numbers $q \neq p$ in the extension $K^+|\mathbb{Q}$.

F.   *Let q be a prime number different from p, and let f be the order of $\bar{q} = q \pmod{p}$ in the multiplicative group $P(p)$. Then $A^+q = Q_1^+ \cdots Q_{g^+}^+$ where Q_1^+, ..., $Q_{g^+}^+$ are distinct prime ideals of A^+, $g^+ = (p - 1)/2f^+$ and $f^+ = f$ when f is odd, while $f^+ = f/2$ when f is even.*

Proof: We have $Aq = Q_1 \cdots Q_g$, where $Q_1, \ldots, Q_g$ are distinct prime ideals of A and f is the order of q modulo p in $P(p)$ (see Chapter 11, (**O**)).

From $2 = [K : K^+] = f_{Q_1}(K|K^+) \cdot g_{Q_1}(K|K^+)$, it follows that $2 = (f/f^+) \cdot (g/g^+)$ (by Chapter 11, Theorem 1) and either $f = f^+$, $g = 2g^+$ or $f = 2f^+$, $g = g^+$. If f is odd then $f = f^+$, $g = 2g^+$. If f is even we shall prove that $f^+ = f/2$.

Since f is the order of q modulo p in $P(p)$ and $f = 2k$ then $q^k \equiv -1 \pmod{p}$. The decomposition group $\mathcal{Z}$ of Q_1 consists of all automorphisms σ of K such that $\sigma(Q_1) = Q_1$; since q is unramified, the inertial group of Q_1 is trivial. By Chapter 14, Theorem 1, $\mathcal{Z} \cong G(\overline{K}|\mathbb{F}_q)$, where $\overline{K} = A/Q_1$; hence $\mathcal{Z}$ is a cyclic group with generator σ defined by $\sigma(\zeta) = \zeta^q$. It follows that $\tau = \sigma^k \in \mathcal{Z}$; but $\tau(\zeta) = \zeta^{q^k} = \zeta^{-1} = \overline{\zeta}$ so τ is the complex conjugation. Therefore Q_1 is invariant by complex conjugation. But $G(K|K^+) = \{\varepsilon, \tau\}$ so Q_1 is the only prime ideal of K lying over Q_1^+. This shows that $g = g^+$ and therefore $f = 2f^+$. ∎

Now we determine the Dedekind zeta-series of the field K^+.

G.
$$\zeta_{K^+}(s) = \zeta(s) \cdot \prod_{\substack{\chi \neq \chi_0 \\ \chi \text{ even}}} L(s|\chi) \tag{27.22}$$

and

$$\lim_{s \to 1+0} (s - 1)\zeta_{K^+}(s) = \prod_{\substack{\chi \neq \chi_0 \\ \chi \text{ even}}} L(1|\chi) \tag{27.23}$$

(*products extended over the nontrivial even characters modulo p*).

Proof: By Chapter 23, (23.12):

$$\zeta_{K^+}(s) = \frac{1}{1 - 1/p^s} \prod_{q \neq p} \prod_{Q^+|q} \frac{1}{1 - 1/N(Q^+)^s}$$

because $A^+ p = (P^+)^t$ and $N(P^+) = p$.

Now

$$\frac{1}{1 - 1/N(Q^+)^s} = 1 - \left(\frac{1}{q^s}\right)^{f^+(q)} = \prod_{k=0}^{f^+(q)-1} \left(1 - \frac{\xi_q^k}{q^s}\right),$$

where $f^+(q)$ denotes the inertial degree of Q^+ over q, $g^+(q)$ is the number of prime ideals of A^+ dividing q, and ξ_q is a primitive root of unity of order $f^+(q)$. Let $f(q)$ be the inertial degree of Q over q; thus $f(q)$ is the order of q modulo p in the group $P(p)$. Let $g(q)$ be the number of prime ideals Q of A dividing q.

If $f(q)$ is even, we have seen in (**F**) that $f(q) = 2f^+(q)$, $g(q) = g^+(q)$. We choose a primitive root of unity ξ_q' of order $f(q)$ and let $\xi_q = \xi_q'^2$, so ξ_q has order $f^+(q)$. Let G be the subgroup of $P(p)$ generated by $\overline{q}$, so it has order $f(q)$ and $(P(p) : G) = (p - 1)/f(q) = g(q) = g^+(q)$.

For every integer $k = 0, 1, 2, , \ldots, f^+(q) - 1$, there exists exactly one character $\widetilde{\chi}_{2k}$ of G such that $\widetilde{\chi}_{2k}(\overline{q}) = \xi_q'^{2k}$ (because $(\xi_q'^{2k})^{f(q)} = 1$) and these characters are distinct. Each such character $\widetilde{\chi}_{2k}$ admits $g^+(q)$ extensions to characters $\widetilde{\chi}$ of $P(p)$ (by Chapter 21, (D)). The corresponding modular characters χ are even, because

$$\chi(-1) = \widetilde{\chi}(-\overline{1}) = \widetilde{\chi}_{2k}(\overline{q}^{f^+(q)}) = (\xi_q')^{2kf^+(q)} = 1.$$

Different characters $\widetilde{\chi}$ of $P(p)$ give different even characters modulo p. But we have $g^+(q)f^+(q) = (p-1)/2$ such characters $\widetilde{\chi}$, so we obtain all even characters χ modulo p.

Thus if $f(q)$ is even then

$$\prod_{Q^+|q} \left(1 - \frac{1}{N(Q^+)^s}\right)^{-1}$$

$$= \prod_{Q^+|q} \left[1 - \left(\frac{1}{q^s}\right)^{f^+(q)}\right]^{-1} = \prod_{Q^+|q} \prod_{k=0}^{f^+(q)-1} \left(1 - \frac{\xi_q^k}{q^s}\right)^{-1}$$

$$= \prod_{Q^+|q} \prod_{k=0}^{f^+(q)-1} \left(1 - \frac{\xi_q'^{2k}}{q^s}\right)^{-1} = \prod_{Q^+|q} \prod_{k=0}^{f^+(q)-1} \left(1 - \frac{\widetilde{\chi}_{2k}(\overline{q})}{q^s}\right)^{-1}$$

$$= \prod_{k=0}^{f^+(q)-1} \left(1 - \frac{\widetilde{\chi}_{2k}(\overline{q})}{q^s}\right)^{-g^+(q)} = \prod_{\chi \text{ even}} \left(1 - \frac{\chi(q)}{q^s}\right)^{-1}.$$

Now let $f(q)$ be odd, so by (F) we have $f(q) = f^+(q)$, and $g(q) = 2g^+(q)$. Again let ξ_q be a primitive root of unity of order $f(q)$. Since $f(q)$ is odd then $(-\overline{q})^{f(q)} = -\overline{1}$, so $-\overline{q}$ has order $2f(q)$ in $P(p)$.

Let G be the subgroup of $P(p)$ generated by $-\overline{q}$. For every integer $k = 0, 1, \ldots, f^+(q) - 1$ there exists exactly one character $\widetilde{\chi}_k$ of G such that $\widetilde{\chi}(-\overline{q}) = \xi_q^k$. Each character of G admits $g^+(q)$ extensions to characters $\widetilde{\chi}$ of $P(p)$. The corresponding modular characters χ are even, because

$$\chi(-1) = \widetilde{\chi}(-\overline{1}) = \chi_k((-\overline{q}^{f(q)}) = \xi_q^{kf(q)} = 1.$$

So in this way we obtain all the $(p-1)/2$ even characters modulo p.

If $f(q)$ is odd then

$$\prod_{Q^+|q}\left(1-\frac{1}{N(Q^+)^s}\right)^{-1}=\prod_{Q^+|q}\left[1-\left(\frac{1}{q^s}\right)^{f^+(q)}\right]^{-1}$$

$$=\prod_{Q^+|q}\prod_{k=0}^{f^+(q)-1}\left(1-\frac{\xi_q^k}{q^s}\right)^{-1}=\prod_{Q^+|q}\prod_{k=0}^{f^+(q)-1}\left(1-\frac{\tilde{\chi}_k(-\bar{q})}{q^s}\right)^{-1}$$

$$=\prod_{k=0}^{f^+(q)-1}\left(1-\frac{\tilde{\chi}_k(-\bar{q})}{q^s}\right)^{-g^+(q)}=\prod_{\chi\ \text{even}}\left(1-\frac{\chi(-q)}{q^s}\right)^{-1}$$

$$=\prod_{\chi\ \text{even}}\left(1-\frac{\chi(q)}{q^s}\right)^{-1}.$$

Putting together these expressions, we have

$$\zeta_{K^+}(s)=\frac{1}{1-1/p^s}\prod_{q\neq p}\prod_{\chi\ \text{even}}\left(1-\frac{\chi(q)}{q^s}\right)^{-1}$$

$$=\frac{1}{1-1/p^s}\prod_{\chi\ \text{even}}\prod_{q\neq p}\left(1-\frac{\chi(q)}{q^s}\right)^{-1}$$

$$=\prod_{q}\frac{1}{1-1/q^s}\prod_{\substack{\chi\neq\chi_0\\ \chi\ \text{even}}}\left(1-\frac{\chi(q)}{q^s}\right)^{-1}$$

$$=\zeta(s)\cdot\prod_{\substack{\chi\neq\chi_0\\ \chi\ \text{even}}}L(s\,|\,\chi).$$

Finally, recalling that $\lim_{s\to1+0}(s-1)\zeta(s)=1$, we have

$$\lim_{s\to1+0}(s-1)\zeta_K^+(s)=\prod_{\substack{\chi\neq\chi_0\\ \chi\ \text{even}}}L(1\,|\,\chi). \qquad\blacksquare$$

Now we give a first interpretation for the factor h^+ of the class number:

Theorem 3.   *The factor h^+ of the class number of $K=\mathbb{Q}(\zeta)$ is equal to the class number of $K^+=\mathbb{Q}(\zeta+\zeta^{-1})$; in particular, h^+ is a positive integer.*

Proof: In Chapter 23, (C), we have shown that

$$\lim_{s\to1+0}(s-1)\zeta_K^+(s)=h_{K^+}\frac{2^tR^+}{w^+\sqrt{|\delta^+|}},$$

where h_{K^+} is the class number of K^+, R^+ is the regulator of K^+, δ^+ is its discriminant, w^+ is the number of roots of unity in K^+, and $t=(p-1)/2$, noting that all t conjugates of K^+ are real.

By (**G**), we have

$$\lim_{s\to 1+0}(s-1)\zeta_K^+(s) = \prod_{\substack{\chi\neq\chi_0\\ \chi\ \text{even}}} L(1\,|\,\chi).$$

Thus by the above

$$h_{K^+} = \frac{w^+\sqrt{|\delta^+|}}{2^t R^+} \prod_{\substack{\chi\neq\chi_0\\ \chi\ \text{even}}} L(1\,|\,\chi).$$

By (**E**) we have $\delta^+ = p^{t-1}$, $R^+ = R/2^{t-1}$ where R is the regulator of K; moreover, $w^+ = 2$ since K^+ is a real field.

Thus

$$h_{K^+} = \frac{p^{(t-1)/2}}{R} \prod_{\substack{\chi\neq\chi_0\\ \chi\ \text{even}}} L(1\,|\,\chi).$$

By (27.18) and (27.17):

$$h_{K^+} = \frac{p^{(t-1)/2}}{R}\,\frac{2^{t-1}}{p^{(t-1)/2}} \prod_{k=1}^{t-1}\left|\sum_{j=0}^{t-1}\eta^{2kj}\log|1-\zeta^{r^j}|\right| = h^+. \qquad\blacksquare$$

We intend to give another interpretation of the factor h^+ by means of a certain subgroup of the group of units.

In Chapter 10, Section 3, we encountered the following: $t-1 = (p-3)/2$ positive real units of A (called the circular units):

$$v_k = \sqrt{\frac{1-\zeta^k}{1-\zeta}\cdot\frac{1-\zeta^{-k}}{1-\zeta^{-1}}}, \qquad k=2,\dots,t.$$

Let V be the subgroup of U generated by these units. By Dirichlet's theorem, U has a system of $t-1$ fundamental units; each unit of infinite order of A is the product of a root of unity with a positive real unit (see Chapter 10, (**F**)). Let U^* denote the subgroup of U consisting of the positive real units; so U^* has a fundamental system of units and V is a subgroup of U^*.

Theorem 4. $h^+ = (U^* : V)$. *The above units $v_2, \dots, v_t$ are independent; they constitute a fundamental system of units of K if and only if $h^+ = 1$.*

Proof: We consider the numbers $\alpha_j = \log|1-\zeta^{r^j}|$, $j=0,1,\dots,t-1$. Since $1-\zeta^{r^{t+j}} = 1-\zeta^{-r^j}$ and $|\zeta^{r^j}| = 1$ then

$$\log|1-\zeta^{r^{t+j}}| = \log\left|\frac{\zeta^{r^j}-1}{\zeta^{r^j}}\right| = \log|1-\zeta^{r^j}|,$$

that is, $\alpha_{t+j} = \alpha_j$ for $j=0,1,\dots,t-1$. More generally, for every $k\in\mathbb{Z}$ we define $\alpha_k = \alpha_j$ when $k\equiv j\pmod{t}$, $0\le j\le t-1$.

We abbreviate (27.17), writing it as follows:

$$h^+ = \frac{2^{t-1}}{R}\,|\alpha|,\qquad\qquad (27.24)$$

where

$$\alpha = \prod_{k=1}^{t-1}\left(\sum_{j=0}^{t-1}\eta^{2kj}\alpha_j\right).\qquad\qquad (27.25)$$

We want to compute $|\alpha|$. Let G be the multiplicative group of order t generated by η^2. For every $k = 0, 1, \ldots, t-1$ there is a character χ_k of G, defined by $\chi_k(\eta^2) = \eta^{2k}$; then $\hat{G} = \{\chi_0, \chi_1, \ldots, \chi_{t-1}\}$. Let $f : G \to \mathbb{C}$ be the function defined by $f(\eta^{2j}) = \alpha_j$ for $j = 0, 1, \ldots, t-1$. Thus

$$\alpha = \prod_{k=1}^{t-1}\left(\sum_{j=0}^{t-1}\chi_k(\eta^{2j})f(\eta^{2j})\right)$$

and hence

$$(\alpha_0 + \alpha_1 + \cdots + \alpha_{t-1})\alpha = \prod_{k=0}^{t-1}\left(\sum_{j=0}^{t-1}\chi_k(\eta^{2j})f(\eta^{2j})\right) = \det(M),$$

where M is the $t \times t$ matrix with $f(\eta^{2j-2i}) = \alpha_{j-i}$ at row i and column j, $0 \le i, j \le t-1$. The matrix

$$M' = \begin{pmatrix} \alpha_0 & \alpha_1 & \cdots & \alpha_{t-1} \\ \alpha_1 & \alpha_2 & \cdots & \alpha_0 \\ \alpha_2 & \alpha_3 & \cdots & \alpha_1 \\ \cdots\cdots\cdots\cdots\cdots \\ \alpha_{t-1} & \alpha_0 & \cdots & \alpha_{t-2} \end{pmatrix}$$

with entry α_{i+j} at row i and column j, is obtained from M by interchanging rows; thus $|\det(M)| = |\det(M')|$. We observe that if we add all the rows of M' to its last row, this one becomes equal to the row of elements $\alpha_0 + \alpha_1 + \cdots + \alpha_{t-1}$; hence $\det(M') = (\alpha_0 + \alpha_1 + \cdots + \alpha_{t-1}) \cdot \det(M'')$ where M'' is obtained from M' tby replacing the last row by a row of elements equal to 1. If we subtract the first column from all other columns of M'' then

$$|\alpha| = \frac{|\det(M')|}{|\alpha_0 + \alpha_1 + \cdots + \alpha_{t-1}|} = |\det(M''')|,$$

where

$$M''' = \begin{pmatrix} \alpha_1 - \alpha_0 & \alpha_2 - \alpha_0 & \cdots & \alpha_{t-1} - \alpha_0 \\ \alpha_2 - \alpha_1 & \alpha_3 - \alpha_1 & \cdots & \alpha_0 - \alpha_1 \\ \cdots\cdots\cdots\cdots\cdots\cdots\cdots\cdots\cdots\cdots\cdots \\ \alpha_{t-1} - \alpha_{t-2} & \alpha_0 - \alpha_{t-2} & \cdots & \alpha_{t-3} - \alpha_{t-2} \end{pmatrix}.$$

So M''' is a $(t-1) \times (t-1)$ matrix with entry $\alpha_{i+j} - \alpha_i$ at row i and column j $(i = 0, 1, \ldots, t-2;\; j = 1, 2, \ldots, t-1)$.

Let σ_k, $k = 0, 1, \ldots, p-1$, be the automorphisms of $\mathbb{Q}(\zeta)$, numbered in such a way that $\sigma_k(\zeta) = \zeta^{r^k}$. Thus $\sigma_0 = \varepsilon$ and $\sigma_{t+k}(\zeta) = \zeta^{r^{t+k}} = \zeta^{-r^k}$. Let r_j be the only integer such that $r_j \neq 0$, $-t < r_j \leq t$, and $r^j \equiv r_j$ (mod p); let $r'_j = |r_j|$. For $j = 1, 2, \ldots, t-1$ and $i = 0, 1, \ldots, t-2$ we have

$$\left| \frac{1 - \zeta^{r^{i+j}}}{1 - \zeta^{r^i}} \right| = \sqrt{\frac{1 - \zeta^{r^{i+j}}}{1 - \zeta^{r^i}} \cdot \frac{1 - \zeta^{-r^{i+j}}}{1 - \zeta^{-r^i}}} = \sqrt{\sigma_i \left(\frac{1 - \zeta^{r^j}}{1 - \zeta} \cdot \frac{1 - \zeta^{-r^j}}{1 - \zeta^{-1}} \right)}$$

$$= \left| \sigma_i \sqrt{\frac{1 - \zeta^{r_j}}{1 - \zeta} \cdot \frac{1 - \zeta^{-r_j}}{1 - \zeta^{-1}}} \right| = |\sigma_i(v_{r_j})| = |\sigma_i(v_{r'_j})|$$

and, since $j \neq 0$, we have $r'_j \neq 1$, so $2 \leq r'_j \leq t$. Therefore

$$\alpha_{i+j} - \alpha_i = \log \left| \frac{1 - \zeta^{r^{i+j}}}{1 - \zeta^{r^i}} \right| = \log |\sigma_i(v_{r_{j'}})|$$

and so $|\alpha|$ is equal to the absolute value of the determinant of the matrix

$$M^{(\mathrm{iv})} = \begin{pmatrix} \log|\sigma_0(v_2)| & \log|\sigma_0(v_3)| & \cdots & \log|\sigma_0(v_t)| \\ \log|\sigma_1(v_2)| & \log|\sigma_1(v_3)| & \cdots & \log|\sigma_1(v_t)| \\ \cdots\cdots\cdots\cdots\cdots\cdots\cdots\cdots\cdots\cdots\cdots\cdots \\ \log|\sigma_{t-2}(v_2)| & \log|\sigma_{t-2}(v_3)| & \cdots & \log|\sigma_{t-2}(v_t)| \end{pmatrix}.$$

Let $\{u_1, \ldots, u_{t-1}\}$ be a fundamental system of positive real units of K (see Chapter 10, Section 3). Then we may write

$$v_j = \prod_{k=1}^{t-1} u_k{}^{c_{kj}} \quad \text{(for } j = 2, \ldots, t), \quad \text{with } c_{kj} \in \mathbb{Z}.$$

Then

$$\log|\sigma_i(v_j)| = \sum_{k=1}^{t-1} c_{kj} \log|\sigma_i(u_k)|$$

for $i = 0, \ldots, t-2$ and $k = 2, \ldots, t$.

Therefore

$$|\alpha| = |\det(\log|\sigma_i(v_j)|)| = |\det(c_{kj})| \cdot |\det(\log|\sigma_i(u_k)|)|.$$

But

$$|\det(\log|\sigma_i(u_k)|) \substack{i=0,1,\ldots,t-2 \\ k=1,2,\ldots,t-1}| = \frac{R}{2^{t-1}}$$

by the definition of the regulator.

So

$$h^+ = \frac{2^{t-1}}{R} |\alpha| = |\det(c_{kj})|,$$

which proves anew that h^+ is a positive integer.

It remains to show that $|\det(c_{kj})| = (U^* : V)$. This follows from considerations of Linear Algebra. By Chapter 10, Section 3, there exists a fundamental system of positive real units $\{u'_1, , \ldots, u'_{t-1}\}$ of U^* and integers $m_1, \ldots, m_{t-1}$ such that $\{u'^{m_1}_1, , \ldots, u'^{m_{t-1}}_{t-1}\}$ is an independent system of units belonging to V and generating V.

Then

$$U^*/V \cong \prod_{i=1}^{t-1} \mathbb{Z}/\mathbb{Z}m_i \quad \text{and} \quad (U^* : V) = |m_1 \cdots m_{t-1}|.$$

But the units u'_i are expressible in terms of the units

$$u_1, \ldots, u_{t-1}$$

in the form

$$u'_i = \prod_{k=1}^{t-1} u_k^{a_{ki}} \quad \text{with} \quad a_{ki} \in \mathbb{Z}, \quad i = 1, \ldots, t-1,$$

and, conversely, we also have

$$u_k = \prod_{i=1}^{t-1} u_i'^{a'_{ki}} \quad \text{with} \quad a'_{ki} \in \mathbb{Z}.$$

As in Chapter 10, (**H**), we show that $|\det(a_{ki})| = |\det(a'_{ki})| = 1$.

Similarly, the units $u'^{m_1}_1, , \ldots, u'^{m_{t-1}}_{t-1}$ of V are expressible in terms of the units $v_2, \ldots, v_t$ and conversely; so we have

$$u'^{m_i}_i = \prod_{k=2}^{t} v_k^{b_{ki}} \quad \text{with} \quad b_{ki} \in \mathbb{Z}, \quad i = 1, \ldots, t-1,$$

and

$$v_k = \prod_{i=1}^{t-1} (u'_i)^{b'_{ik} m_i} \quad \text{with} \quad b'_{ik} \in \mathbb{Z}, \quad k = 2, \ldots, t.$$

Again $|\det(b_{ki})| = |\det(b'_{ik})| = 1$.

Expressing the units v_k in terms of the units u_i, by expressing successively the units v_k in terms of $u'^{m_i}_i$, these in terms of u'_i and finally these units in terms of u_i, gives rise to the following relation between determinants:

$$h^+ = |\det(c_{kj})| = |\det(a_{ki})| \cdot \left| \prod_{i=1}^{t-1} m_i \right| \cdot |\det(b'_{ik})| = \left| \prod_{i=1}^{t-1} m_i \right| = (U^* : V).$$

Since the index $(U^* : V)$ is finite, it follows that the maximal number of independent units in V and U^* are both equal to $t - 1$. Hence the system of generators $\{v_2, \ldots, v_t\}$ of V is a system of independent units. The last assertion is now obvious. ∎

Now we shall study the factor h^- of the class number of $\mathbb{Q}(\zeta)$.

Theorem 5. h^- *is a positive integer.*

Proof: Let $\gamma = G(\eta)G(\eta^3) \cdots G(\eta^{p-2})$, where

$$G(X) = \sum_{j=0}^{p-2} r_j X^j, \qquad r_j \in \mathbb{Z}, \quad 1 \le r_j < p, \quad r_j \equiv r^j \pmod{p},$$

r is a primitive root modulo p, and η is a primitive $(p-1)$th root of unity. Thus $\bar{\eta} = \eta^{-1} = \eta^{p-2}$ and $\bar{\gamma} = G(\eta^{p-2})G(\eta^{p-4}) \cdots G(\eta) = \gamma$, so $\gamma \in \mathbb{R}$.
From (27.16) we have

$$|\gamma| = (2p)^{t-1}h^- = (2p)^{t-1}\frac{h}{h^+} \in \mathbb{Q}$$

by Theorem 2. Thus $\gamma \in \mathbb{Q}$. On the other hand, $\gamma \in \mathbb{Z}[\eta]$, so $\gamma \in \mathbb{Z}[\eta] \cap \mathbb{Q} = \mathbb{Z}$. Moreover, $|\gamma|$ is independent of the choice of r and of η.

To show that h^- is a positive integer, we shall prove that 2^{t-1} and p^{t-1} divide γ.

We shall compute the power of 2 which divides

$$\gamma(1 - \eta)(1 - \eta^3) \cdots (1 - \eta^{p-2}).$$

From

$$r_{t+j} + r_j \equiv r^{t+j} + r^j = r^j(r^t + 1) \equiv 0 \pmod{p}$$

we deduce that $r_{t+j} + r_j = p$, so r_{t+j} and r_j have different parity.
If k is odd then

$$G(\eta^k) = \sum_{j=0}^{p-2} r_j \eta^{kj} = \sum_{j=0}^{t-1}[r_j \eta^{kj} + r_{t+j}\eta^{k(t+j)}]$$

$$= \sum_{j=0}^{t-1}(r_j - r_{t+j})\eta^{kj} \equiv \sum_{j=0}^{t-1} \eta^{kj} \pmod{2}$$

hence

$$G(\eta^k)(1 - \eta^k) \equiv \sum_{j=0}^{t-1}\eta^{kj} - \sum_{j=0}^{t-1}\eta^{k(j+1)} = 1 - \eta^{tk}$$

$$= 1 - (-1)^k \equiv 0 \pmod{2}$$

because k is odd. Therefore 2^t divides

$$\gamma(1-\eta)(1-\eta^3)\cdots(1-\eta^{p-2}) = \prod_{\substack{k=1 \\ k \text{ odd}}}^{p-2} G(\eta^k)(1-\eta^k).$$

But $(1-\eta)(1-\eta^3)\cdots(1-\eta^{p-2}) = 2$. Indeed, η is a primitive $(p-1)$th root of unity, while η^2 is a primitive tth root of unity.

Thus

$$p-1 = \prod_{k=1}^{p-2}(1-\eta^k), \qquad \frac{p-1}{2} = \prod_{k=1}^{t-1}(1-\eta^{2k}),$$

hence $2 = (1-\eta)(1-\eta^3)\cdots(1-\eta^{p-2})$. This shows that 2^{t-1} divides γ.

Now we shall prove that p^{t-1} divides γ. For this purpose, we study the prime decomposition of p in the field $\mathbb{Q}(\eta)$.

Let B be the ring of integers of $\mathbb{Q}(\eta)$. By Chapter 16, (**B**), p is unramified in $\mathbb{Q}(\eta)$ and since $p \equiv 1 \pmod{p-1}$ then Bp is the product of $\varphi(p-1)$ distinct prime ideals of norm p.

For any choice of the primitive root r modulo p and of the $(p-1)$th root of unity η, we shall indicate a numbering of the set of prime ideals P of B dividing Bp. We prove that given P there exists a unique integer k, such that $1 \leq k \leq p-2$ and P divides $B(1-r\eta^k)$; moreover, $\gcd(k, p-1) = 1$. Indeed, p divides

$$1 - r^{p-1} = \prod_{k=1}^{p-1}(1-r\eta^k)$$

hence P divides some ideal $B(1-r\eta^k)$, with $1 \leq k \leq p-1$. Let $d = \gcd(k, p-1)$. From $r\eta^k \equiv 1 \pmod{P}$, raising to the power $(p-1)/d$ we have

$$r^{(p-1)/d} = r^{(p-1)/d} \cdot \eta^{(k/d)(p-1)} \equiv 1 \pmod{P} \quad \text{so also } r^{(p-1)/d} \equiv 1 \pmod{p}.$$

This is only possible when $d = 1$, because r is a primitive root modulo p.

To show the uniqueness of k, assume that $1 \leq j < k \leq p-2$ and $1 - r\eta^j$, $1 - r\eta^k \in P$. Then $r\eta^j(1-\eta^{k-j}) \in P$ and since η is a unit of B and $r \notin P$ (otherwise P would contain r^{p-1} and $r^{p-1} - 1$ hence $1 \in P$), then $1 - \eta^{k-j} \in P$, so $\eta^k - \eta^j \in P$. In the quotient field B/P we have $\overline{\eta}^k = \overline{\eta}^j$. But

$$X^{p-1} - 1 = \prod_{i=0}^{p-2}(X - \eta^i)$$

hence in $(B/P)[X]$ we have

$$X^{p-1} - \overline{1} = \prod_{i=0}^{p-2}(X - \overline{\eta}^i),$$

so $X^{p-1} - \overline{1}$ would have a double root $\overline{\eta}^k = \overline{\eta}^j$; however, the derivative of this polynomial is $\overline{(p-1)X^{p-2}}$ and it has no common root with $X^{p-1} - \overline{1}$ because B/P has characteristic p.

We have established a mapping from the set of prime ideals of B dividing Bp into the set of integers k such that $\gcd(k, p-1) = 1$, $1 \le k \le p-2$; both sets have $\varphi(p-1)$ elements. Explicitly, the image of P is k when $1 - r\eta^k \in P$.

If $1 \le k, j \le p-2$, $\gcd(k, p-1) = \gcd(j, p-1) = 1$ then η^k, η^j are primitive $(p-1)$th roots of unity, and there exists an automorphism σ of $\mathbb{Q}(\eta)$ such that $\sigma(\eta^k) = \eta^j$. If P contains $1 - r\eta^k$ then $\sigma(P)$ contains $1 - r\eta^j$.

This shows that the mapping $P \mapsto k$ which we considered above has image equal to the set of all integers k such that $1 \le k \le p-2$, $\gcd(k, p-1) = 1$. Hence the mapping is also one-to-one.

Thus, we have a numbering of the prime ideals dividing Bp, by the condition that $1 - r\eta^k \in P_k$ for $1 \le k \le p-2$, $\gcd(k, p-1) = 1$.

If we choose the primitive root r such that $r^{p-1} \not\equiv 1 \pmod{p^2}$ (see Chapter 19, Lemma 1) then $B(1 - r\eta^k) = P_k I_k$ where Bp is relatively prime to I_k. In fact, if $j \ne k$, $\gcd(j, p-1) = 1$, then P_j does not divide I_k, otherwise $1 - r\eta^k \in P_j$, contrary to the numbering. Taking norms, then p^2 divides

$$N(B(1 - r\eta^k)) = \left| \prod_{\gcd(j, p-1)=1} (1 - r\eta^j) \right|$$

hence p^2 divides

$$\prod_{j=0}^{p-2} (1 - r\eta^j) = 1 - r^{p-1},$$

contrary to the choice of r.

To prove that p^{t-1} divides γ in B, we show that p divides $G(\eta^k)(1 - r\eta^k)$ in B for $k = 1, \ldots, p-1$.

Indeed,

$$G(\eta^k)(1 - r\eta^k) = \sum_{j=0}^{p-2} r_j \eta^{kj}(1 - r\eta^k) \equiv \sum_{j=0}^{p-2} (r\eta^k)^j (1 - r\eta^k)$$

$$= 1 - (r\eta^k)^{p-1} = 1 - r^{p-1} \equiv 0 \pmod{p}.$$

Hence Bp^t divides

$$\prod_{k \text{ odd}} BG(\eta^k) \cdot \prod_{k \text{ odd}} B(1 - r\eta^k) = B\gamma \cdot \prod_{k=1}^{p-1} B(1 - r\eta^k).$$

But p does not divide $1 - r\eta^k$ when $\gcd(k, p-1) > 1$, otherwise P would divide $B(1 - r\eta^k)$ and this has been shown to be impossible. Therefore

p^{t-1} divides

$$B\gamma \prod_{\gcd(k,p-1)=1} B(1 - r\eta^k) = B\gamma \left(\prod_{\gcd(k,p-1)=1} P_k \right) \left(\prod_{\gcd(k,p-1)=1} I_k \right)$$

$$= B\gamma \cdot Bp \cdot I,$$

where Bp and I are relatively prime ideals. Thus p^{t-1} divides γ in B, concluding the proof.   ∎

Definition 1. h^- is called the *relative class number* of K (or *the first factor* of the class number of K). h^+ is called the *real class number* of K (or *the second factor* of the class number of K).

EXERCISES

1. Write explicitly the class number h_m, using the formula (27.11) for $m = 3, 4, 5, 7, 9$, and 15.

2. Write explicitly the expression $L(1|\chi)$ for $m = 3, 5$, and 15 and all characters $\chi \neq \chi_0$. Compute numerical approximations to $L(1|\chi)$.

3. Use the formulas of Theorem 2 to compute h^-, h^+ for $p = 3, 5, 7$, and 11.

4. Let $m > 1$, let ζ be a primitive mth root of 1, and let $K = \mathbb{Q}(\zeta)$, $K^+ = K \cap \mathbb{R}$. Let U, W, R, U^+, W^+, and R^+ be as in the text. Let $\nu = (U/W: U^+/W^+)$ be the index of the subgroup U^+/W^+ in U/W. Show

(a) $$(U: U^+) = \begin{cases} m\nu, & \text{when } m \text{ is odd,} \\ (m/2)\nu, & \text{when } m \text{ is even.} \end{cases}$$

(b) $$2^{\varphi(m)/2-1} R^+ = \nu R,$$

(c) Let

$$N(U) = \{N_{K|K^+}(\varepsilon) \mid \varepsilon \text{ unit in } K\}.$$

Then $\nu = (N(U): (U^+)^2)$, so $\nu = 1$ or 2.

(d) Let $m = p^e$ where $p \neq 2$, $e \geq 1$. Then $\nu = 1$ and every fundamental system of units of K^+ is a fundamental system of units of K.

28

Miscellaneous Results About the Class Number of Quadratic Fields

In this chapter, our purpose is to present a sample of results about the class number of quadratic fields. Due to their nature, several proofs have to be omitted. The reader is encouraged to study the original papers listed in the Bibliography.

There are many aspects in the study of the class number of quadratic fields, as well as numerous applications. So what we present is not to be construed as an attempt to treat the topic fully, but rather to let the reader have a glimpse of possible directions for further study.

28.1 Divisibility Properties

The first result, which is rather general, is due to Gut (1929); we follow his proof of 1973.

A. *Let K be an algebraic number field and q any prime number. Then there exist infinitely many extensions $L|K$ of degree q such that q divides the class number of L.*

Proof:

Case 1: $q \neq 2$.

Assume that $L_1, \ldots, L_s$ (with $s \geq 0$) are extensions of degree q of K and with class number a multiple of q. We shall determine an extension $L_{s+1}|K$ with the same properties and $L_{s+1} \neq L_i$ for all $i = 1, \ldots, s$.

Let δ be the discriminant of K and δ_i the discriminant of L_i for $i = 1, \ldots, s$. Let S_q be the set of all primes p such that $p \equiv 1 \pmod{q}$ and p does not divide $\delta\delta_1 \cdots \delta_s$. By Dirichlet's theorem, S_q is an infinite set.

Let $p_1, p_2 \in S_q$, $p_1 \neq p_2$. For $i = 1, 2$ let ζ_i be a primitive root of 1 of order p_i. So $\mathbb{Q}(\zeta_i)|\mathbb{Q}$ is a Galois extension of degree $p_i - 1$. Let g_i, $1 \leq g_i \leq p_i - 1$, be a primitive root modulo p_i. Thus a generator of the Galois group G_i of $\mathbb{Q}(\zeta_i)|\mathbb{Q}$ is $\overline{\sigma}_i$, defined by $\overline{\sigma}_i(\zeta_i) = \zeta_i^{g_i}$. Since p_i is totally ramified in $\mathbb{Q}(\zeta_i)|\mathbb{Q}$ and unramified in the other extension $\mathbb{Q}(\zeta_j)|\mathbb{Q}$ (with

$j \neq i$), then $\mathbb{Q}(\zeta_1) \cap \mathbb{Q}(\zeta_2) = \mathbb{Q}$. Hence the Galois extension $\mathbb{Q}(\zeta_1, \zeta_2)|\mathbb{Q}$ has Galois group $G \cong G_1 \times G_2$. So there exist $\sigma_1, \sigma_2 \in G$ such that $\sigma_1(\zeta_1) = \zeta_1^{g_1}$, $\sigma_1(\zeta_2) = \zeta_2$ and $\sigma_2(\zeta_1) = \zeta_1$, $\sigma_2(\zeta_2) = \zeta_2^{g_2}$.

Clearly, σ_i has order $p_i - 1$ for $i = 1, 2$. Let $f_i = (p_i - 1)/q$ and

$$\eta_i = \zeta_i + \sigma_i^q(\zeta_i) + \sigma_i^{2q}(\zeta_i) + \cdots + \sigma_i^{(f_i-1)q}(\zeta_i)$$

for $i = 1, 2$.

The element η_i is a period of f_i terms of the cyclotomic field $\mathbb{Q}(\zeta_i)$. Clearly, $\sigma_1^q(\eta_1) = \eta_1$, $\sigma_2(\eta_1) = \eta_1$ and $\sigma_1(\eta_2) = \eta_2$, $\sigma_2^q(\eta_2) = \eta_2$.

The elements η_1, $\sigma_1(\eta_1)$, $\sigma_1^2(\eta_1)$, $\ldots$, $\sigma_1^{q-1}(\eta_1)$ are distinct and so are the elements η_2, $\sigma_2(\eta_2)$, $\sigma_2^2(\eta_2)$, $\ldots$, $\sigma_2^{q-1}(\eta_2)$. We have $[\mathbb{Q}(\zeta_1, \zeta_2) : \mathbb{Q}(\zeta_1)] = p_2 - 1$ and $[\mathbb{Q}(\zeta_1, \zeta_2) : \mathbb{Q}(\zeta_2)] = p_1 - 1$.

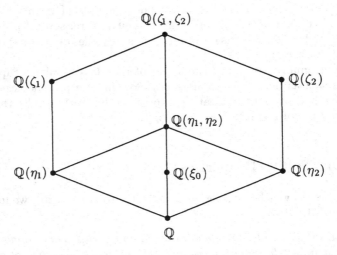

$\mathbb{Q}(\eta_i)$ is the field fixed by the subgroup generated by σ_i^q and σ_j for $i, j = 1, 2$, $i \neq j$, so $[\mathbb{Q}(\zeta_1, \zeta_2) : \mathbb{Q}(\eta_i)] = f_i(p_j - 1)$ and $[\mathbb{Q}(\eta_i) : \mathbb{Q}] = q$. Since the elements η_i and $\sigma_i^\ell(\eta_i)$ for $\ell = 1, \ldots, q - 1$ are conjugate and the extensions $\mathbb{Q}(\eta_j)|\mathbb{Q}$ are Galois extensions, then $\sigma_i^\ell(\eta_i) \in \mathbb{Q}(\eta_i)$.

It follows that $\{\eta_i, \sigma_i(\eta_i), \ldots, \sigma_i^{q-1}(\eta_i)\}$ is a basis for $\mathbb{Q}(\eta_i)|\mathbb{Q}$ and actually an integral basis.

From $\sum_{\ell=0}^{p_i-1} \zeta_i^\ell = 0$ we deduce that

$$\sum_{m=0}^{q-1} \sigma_i^m(\eta_i) = -1, \qquad i = 1, 2.$$

For every $m \in \mathbb{Z}$ let

$$\xi_m = \sum_{\ell=0}^{q-1} \sigma_1^{m+\ell}(\eta_1)\sigma_2^\ell(\eta_2) \in \mathbb{Q}(\eta_1, \eta_2);$$

then ξ_m is an algebraic integer, $\xi_m = \xi_{m+q}$, and

$$\sum_{m=0}^{q-1} \xi_m = \sum_{m=0}^{q-1} \sum_{\ell=0}^{q-1} \sigma_1^{m+\ell}(\eta_1)\sigma_2^\ell(\eta_2)$$

$$= \sum_{\ell=0}^{q-1} \sigma_2^\ell(\eta_2)\sigma_1^\ell \left\{ \sum_{m=0}^{q-1} \sigma_1^m(\eta_1) \right\}$$

$$= -\sum_{\ell=0}^{q-1} \sigma_2^\ell(\eta_2) = 1.$$

Also, for $n = 0, 1, \ldots, q - 1$:

$$\sigma_1^n(\xi_m) = \xi_{m+n} \qquad \text{and} \qquad \sigma_2^n(\xi_m) = \xi_{m-n},$$

since $\sigma_1\sigma_2(\xi_m) = \sigma_2\sigma_1(\xi_m) = \xi_m$. It follows that ξ_0, ξ_1, $\ldots$, ξ_{q-1} are all distinct. Indeed, if $\xi_h = \xi_k$, $0 \le h < k \le q - 1$, let $n = k - h$; we have $1 \le n < q$ and $\sigma_1^n(\xi_h) = \xi_k = \xi_h$, hence $\sigma_1^{rn} = \xi_{h+rn} = \xi_h$; but the set of positive residues modulo q of $\{n, 2n, \ldots, (q-1)n, qn\}$ is $\{0, 1, \ldots, q-1\}$, hence for every ℓ, $0 \le \ell \le q - 1$ there is an integer r, $1 \le r \le q$, such that if $0 \le \ell \le q - 1$, then $\xi_\ell = \xi_{h+rn} = \xi_h$.

Hence

$$\sum_{m=0}^{q-1} \xi_m = q\xi_0 = -1$$

and $\xi_0 = -1/q$, which is not possible since ξ_0 is an algebraic integer.

The elements ξ_0, ξ_1, $\ldots$, ξ_{q-1} are conjugates and all distinct, since the Galois group is Abelian. $\mathbb{Q}(\xi_0) = \mathbb{Q}(\xi_1) = \cdots = \mathbb{Q}(\xi_{q-1})$ and the field must have degree at least equal to q. On the other hand, the elementary symmetric functions on ξ_0, ξ_1, $\ldots$, ξ_{q-1} are invariant under the action of the elements σ_1, σ_2, which generate the Galois group of $\mathbb{Q}(\zeta_1, \zeta_2)|\mathbb{Q}$. So they are rational numbers and therefore ξ_0, ξ_1, $\ldots$, ξ_{q-1} are the roots of a polynomial of degree q over $\mathbb{Q}$, showing that $\mathbb{Q}(\xi_0)|\mathbb{Q}$ has degree q.

Since p_i is the only prime ramified in $\mathbb{Q}(\zeta_i)|\mathbb{Q}$, then p_1, p_2 are the only primes ramified in $\mathbb{Q}(\zeta_1, \zeta_2)|\mathbb{Q}$ (see Chapter 13, (U)).

From $\mathbb{Q}(\xi_0) \ne \mathbb{Q}$, by Minkowski's theorem (see Chapter 9, (D)), and what we just proved, either p_1 or p_2 is ramified in $\mathbb{Q}(\xi_0)|\mathbb{Q}$.

Actually, by symmetry, both primes are ramified. Since p_1, p_2 are unramified in $K|\mathbb{Q}$ (because they do not divide the discriminant of K) then $\xi_0 \notin K$ so $K \cap \mathbb{Q}(\xi_0) = \mathbb{Q}$, because $[\mathbb{Q}(\xi_0) : \mathbb{Q}] = q$ is a prime number. Hence $K(\xi_0)|K$ is a Galois extension of degree q.

Similarly, since p_i is totally ramified in $\mathbb{Q}(\zeta_i)|\mathbb{Q}$, hence also in $\mathbb{Q}(\eta_i)|\mathbb{Q}$, then $K \cap \mathbb{Q}(\eta_i) = \mathbb{Q}$ and $K(\eta_i)|K$ is an extension of degree q. Next we see that $K(\eta_1) \ne K(\eta_2)$, otherwise since p_2 is unramified in $K|\mathbb{Q}$ and in

$\mathbb{Q}(\eta_1)|\mathbb{Q}$, by Chapter 13, (U), p_2 is unramified in $K(\eta_1)|\mathbb{Q}$ and a fortiori, it is unramified in $\mathbb{Q}(\eta_1)|\mathbb{Q}$, which is a contradiction.

From $K(\eta_1) \neq K(\eta_2)$ it follows that $K(\eta_1, \eta_2)|K$ has degree q^2 and so $[K(\eta_1, \eta_2) : K(\xi_0)] = q$.

We let $L_{s+1} = K(\xi_0)$ and we shall prove that q divides the class number h of $K(\xi_0)$. According to class field theory (see Chapter 15, Theorem 2), h is the degree over $K(\xi_0)$ of the Hilbert absolute class field H of $K(\xi_0)$: $h = [H : K(\xi_0)]$.

Moreover, every unramified Abelian extension of $K(\xi_0)$ is contained in H. Hence it suffices to prove that $K(\eta_1, \eta_2)|K(\xi_0)$ is an unramified Abelian extension.

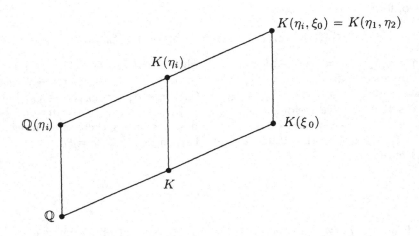

for $i = 1, 2$.

We have $K(\eta_i, \xi_0) = K(\eta_1, \eta_2)$ since

$$K(\xi_0) \neq K(\eta_i, \xi_0) \subseteq K(\eta_1, \eta_2).$$

Hence the Galois groups of $\mathbb{Q}(\eta_i)|\mathbb{Q}$, $K(\eta_i)|K$, and $K(\eta_1, \eta_2)|K(\xi_0)$ are canonically isomorphic; in particular this last extension is Abelian.

To show that it is unramified, by Chapter 13, Theorem 2, it is equivalent to proving that the relative different D is equal to the unit ideal. By Chapter 13, (T), D is the ideal generated by the relative differents of the algebraic integers of $K(\eta_1, \eta_2)$. In particular, D contains the relative differents of η_1, η_2; these are, respectively, equal to

$$\prod_{\ell=1}^{q-1}(\eta_1 - \sigma_1^\ell(\eta_1)), \qquad \prod_{\ell=1}^{q-1}(\eta_2 - \sigma_2^\ell(\eta_2)).$$

In view of the isomorphism of the Galois groups, these are also the relative differents of η_1, η_2 in $\mathbb{Q}(\eta_1)|\mathbb{Q}$ (respectively, $\mathbb{Q}(\eta_2)|\mathbb{Q}$). Then D contains the norm in $\mathbb{Q}(\eta_i)|\mathbb{Q}$ of the different of η_i, that is, the absolute discriminant of $\mathbb{Q}(\eta_i)$. Since p_i is the only prime ramified in $\mathbb{Q}(\eta_i)$ then D contains a power of p_i, for $i = 1, 2$; so D must be the unit ideal.

We have therefore shown that $K(\xi_0) \subset K(\eta_1, \eta_2) \subseteq H$, hence q divides h.

We conclude noting that p_1, p_2 divide the discriminant of $\mathbb{Q}(\xi_0)$, hence also the discriminant of $K(\xi_0) = L_{s+1}$. On the other hand, p_1, p_2 do not divide the discriminant of K, L_1, ..., L_s by hypothesis, so L_{s+1} is not equal to any one of the fields K, L_1, ..., L_s.

Case 2: $q = 2$.

Let L_1, ..., L_s be quadratic extensions of K (with $s \geq 0$) and having even class numbers. We shall determine another quadratic extension $L_{s+1}|K$ having even class number.

Let δ, δ_1, ..., δ_s be, respectively, the discriminants of K, L_1, ..., L_s. Let p_1, p_2 be distinct primes not dividing $\delta\delta_1 \cdots \delta_s$ and such that $p_1 \equiv 1 \pmod 4$, $p_2 \equiv 1 \pmod 4$.

For $i = 1, 2$ let

$$\eta_i = \frac{-1 + \sqrt{p_i}}{2} \qquad \text{and} \qquad \xi_0 = \frac{-1 + \sqrt{p_1 p_2}}{2}.$$

Then the discriminants of $\mathbb{Q}(\sqrt{p_1})$, $\mathbb{Q}(\sqrt{p_2})$, and $\mathbb{Q}(\sqrt{p_1 p_2})$ are, respectively, p_1, p_2, and $p_1 p_2$ and so the primes p_i ($i = 1, 2$) are ramified in both extensions $\mathbb{Q}(\sqrt{p_i})$, $\mathbb{Q}(\sqrt{p_1 p_2})$, and

$$\mathbb{Q}(\sqrt{p_1 p_2}) \cap K = \mathbb{Q}.$$

So $K(\sqrt{p_i})|K$ and $K(\sqrt{p_1 p_2})|K$ are quadratic extensions.

Moreover, $K(\sqrt{p_1}) \neq K(\sqrt{p_2})$. Otherwise, since p_2 is unramified in $\mathbb{Q}(\sqrt{p_1})$ and in K, by Chapter 13, (U), it would be unramified in

$$\mathbb{Q}(\sqrt{p_1})K = K(\sqrt{p_1}) = K(\sqrt{p_2}),$$

thus also in $\mathbb{Q}(\sqrt{p_2})$, which is not true.

We conclude that

$$\left[K(\sqrt{p_1}, \sqrt{p_2}) : K\right] = 4, \qquad \text{so} \qquad \left[K(\sqrt{p_1}, \sqrt{p_2}) : K(\sqrt{p_1 p_2})\right] = 2.$$

It remains to show that the relative different D of the extension

$$K(\sqrt{p_1}, \sqrt{p_2})|K(\sqrt{p_1 p_2}))$$

is the unit ideal. This will imply that $K(\sqrt{p_1}, \sqrt{p_2}) \subseteq H$ (the Hilbert class field of $K(\sqrt{p_1 p_2})$), therefore 2 divides the class number of $K(\sqrt{p_1 p_2})$.

Now D contains the relative different of $\eta_i = (-1 + \sqrt{p_i})/2$ for $i = 1, 2$ which are algebraic integers of the field $K(\sqrt{p_1}, \sqrt{p_2})$. This relative different is

$$\frac{-1 + \sqrt{p_i}}{2} - \frac{-1 - \sqrt{p_i}}{2} = \sqrt{p_i}, \qquad \text{for} \quad i = 1, 2.$$

Thus D contains their norms p_1, p_2 and so D is the unit ideal.

We take $L_{s+1} = K(\sqrt{p_1 p_2})$. The primes p_1, p_2 are ramified in L_{s+1}, but in none of the fields K, L_1, ..., L_s, so L_{s+1} is different from K, L_1, ..., L_s. ∎

In particular, we have shown that there exist infinitely many quadratic number fields with even class number. As a matter of fact, Gauss had obtained a much more precise theorem, which is derived as a by-product of the theory of genera of binary quadratic forms (see Bibliography to Chapter 28 at the end of the book, where we quote as references, Gauss, Flath, and Ribenboim).

Let $d \neq 0, 1$ be a square-free integer, and let s be the number of prime factors of d. Let

$$h_d^* = \begin{cases} h_d, & \text{if } d < 0, \\ 2h_d, & \text{if } d > 0, \end{cases}$$

where h_d is the class number of $\mathbb{Q}(\sqrt{d})$. Then Gauss proved:

B.   *With the above notations:*
   (1) *If $d \equiv 1$ or $2 \pmod 4$ then 2^{s-1} divides h_d^*.*
   (2) *If $d \equiv 3 \pmod 4$ then 2^s divides h_d^*.*

This is a very interesting result, the proof coming after the development of the beautiful theory of genera of binary quadratic forms.

Since there are infinitely many primes $p \equiv 1 \pmod 4$, it follows from the above result of Gauss, for every $k \geq 1$ and $m = 1$ (respectively, $m = 2$, $m = 3$) there exist infinitely many positive (and also negative) square-free integers d such that $d \equiv m \pmod 4$ and 2^k divides the class number h_d of $\mathbb{Q}(\sqrt{d})$.

Actually, h_d may be divisible by a higher power of 2 than the one guaranteed by (**B**).

Many interesting papers were devoted to the study of this more precise divisibility property.

In 1922, Nagell proved that for every $n > 1$ there exist infinitely many real (and also imaginary) quadratic fields having class number divisible by n. There have been many papers giving new proofs or variants of this result, requiring also specified behavior of a given finite set of primes of $\mathbb{Q}$ in each of the quadratic extensions with a class number multiple of n. For $n = 3$ see, for example, Humbert (for imaginary quadratic fields), Gut; in 1974, Hartung constructed explicitly infinitely many imaginary quadratic fields with a class number divisible by 3. See also Ankeny and Chowla, and Kuroda.

Here we shall follow Yamamoto's paper (1970) who succeeds in giving a general proof of the theorem, both in the imaginary and real cases, with supplementary conditions on finite sets of primes.

Let $n > 1$ be a natural number, let d be a square-free integer ($d \neq 0, 1$), and let $K = \mathbb{Q}(\sqrt{d})$. We denote by σ the conjugation automorphism of

$K : \sigma(a + b\sqrt{d}) = a - b\sqrt{d}$ for any $a, b \in \mathbb{Q}$. Let A be the ring of integers of K.

If K is a real field, let ε be a fundamental unit, while $\varepsilon = 1$ if K is imaginary.

We denote by $\mathcal{C}\ell_K$ the class-group of K and by h its class number.

Lemma 1.   *Assume that the equation*

$$X^2 - Y^2 d = 4Z^n \qquad (28.1)$$

has the solution in integers (x, y, z) with $\gcd(x, z) = 1$. Then:
  (1) $\alpha = (x + y\sqrt{d})/2$ *is an algebraic integer of* $K = \mathbb{Q}(\sqrt{d})$.
  (2) *The principal ideal (α) is the nth power of an (integral) ideal I:* $(\alpha) = I^n$.
  (3) $I + \sigma(I) = (1)$.

Proof: Since $x^2 - y^2 d = 4z^n \equiv 0 \pmod{4}$ then $x \equiv y \pmod{2}$, because d is square-free. So α is an algebraic integer.

We have $\alpha + \sigma(\alpha) = x$ and $\alpha \cdot \sigma(\alpha) = (x^2 - y^2 d)/4 = z^n$.

But the ideals (α), $(\sigma(\alpha))$ are relatively prime, since $(\alpha) + (\sigma(\alpha))$ contains x and also z^n, hence it contains $\gcd(x, z^n) = 1$.

Writing the decomposition of (z) into prime ideals of K it follows that $(\alpha) = I^n$ for some ideal I, which is necessarily an integral ideal. Finally, since $\gcd((\alpha), (\sigma(\alpha))) = (1)$ then $\gcd(I, \sigma(I)) = (1)$, that is, $I + \sigma(I) = (1)$.   ∎

Let p be a prime number such that $p|n$; let ℓ be a prime such that

$$\begin{cases} \ell \equiv 1 \pmod{p}, & \text{when } p \neq 2, \\ \ell \equiv 1 \pmod{4}, & \text{when } p = 2, \end{cases}$$

(the existence of such a prime ℓ follows from a special case of Dirichlet's theorem on primes in arithmetic progressions). In particular, $\ell \neq 2$. Also: -1 is a pth power modulo ℓ. Indeed, if $p = 2$ then $(-1/\ell) = 1$ (Legendre symbol) because $\ell \equiv 1 \pmod{4}$; so -1 is a square modulo ℓ. If $p \neq 2$ since $p|\ell - 1$ and $\mathbb{F}_\ell$ is a cyclic group of order $\ell - 1$ there exists a subgroup of order $(\ell - 1)/p$, say, generated by ξ; since $2p|\ell - 1$ then $(\xi^{(\ell-1)/2p})^p = -\bar{1}$ (in $\mathbb{F}_\ell$), that is, $-1 \equiv a^p \pmod{\ell}$ where $a \pmod{\ell}$ is equal to $\xi^{(\ell-1)/2p}$.

Lemma 2.   *Let p, ℓ be chosen as above. Assume that (x, y, z) is a solution of (28.1) such that:*
  (1) $\gcd(x, z) = 1$;
  (2) $\ell|z$; *and*
  (3) x *is not a pth power modulo ℓ.*

Let $\alpha = (x + y\sqrt{d})/2$. Then $(\ell) = L \cdot \sigma(L)$ where L, $\sigma(L)$ are distinct prime ideals of K and $(\ell) = L \cdot \sigma(L)$ divides $(\alpha) \cdot (\sigma(\alpha))$.

624 28. Miscellaneous Results About the Class Number of Quadratic Fields

Proof: We show that (ℓ) is decomposed in the quadratic extension $K|\mathbb{Q}$. Indeed, $\ell \nmid d$, otherwise ℓ would also divide x, which contradicts (28.1). From $x^2 - y^2 d = 4z^n$ and (2) we have $x^2 - y^2 d \equiv 0 \pmod{\ell}$. But $\ell \nmid x$ so $\ell \nmid y$, thus d is a square modulo ℓ, that is, $(d/\ell) = 1$. According to the general theory of quadratic fields the prime ℓ is decomposed, i.e., $(\ell) = L \cdot \sigma(L)$ where L, $\sigma(L)$ are distinct prime ideals of K.

From Lemma 1 and (2), $(\ell) = L \cdot \sigma(L)$ divides $(z)^n = (\alpha) \cdot (\sigma(\alpha))$. ∎

We may assume that $L|\sigma(\alpha)$ and therefore $L \nmid (\alpha)$.

Lemma 3. *With the hypotheses of Lemma 2, if ε is a pth power modulo L then (α) is not the pth power of a principal ideal of K.*

Proof: We keep the same notations. If $\sigma(\alpha) = (x - y\sqrt{d})/2 \in L$ then $x \equiv y\sqrt{d} \pmod{L}$, so $\alpha = (x + y\sqrt{d})/2 \equiv x \pmod{L}$. But the residue class field A/L is equal to $\mathbb{F}_\ell$ (since (ℓ) is decomposed in K), so it follows from (3) and the above congruence that α is not a pth power modulo L.

If $(\alpha) = (\beta)^p$, where $\beta \in K$, then β is also an integer of K, so we may write $\alpha = \pm \varepsilon^k \beta^p$ where k is some integer. By hypothesis ε is a pth power modulo L. Also -1 is a pth power modulo ℓ, hence also modulo L. Thus α would be a pth power modulo L, which is a contradiction. ∎

Let $n = p_1^{e_1} \cdots p_s^{e_s}$, where $p_1, \ldots, p_s$ are distinct primes, $e_1 \geq 1, \ldots, e_s \geq 1$ and $s \geq 1$.

For every $i = 1, \ldots, s$ let ℓ_i be a prime chosen as before, namely:

$$\begin{cases} \ell_i \equiv 1 \pmod{p_i} & \text{if } p_i \neq 2, \\ \ell_i \equiv 1 \pmod{4} & \text{if } p_i = 2. \end{cases}$$

Assume that (x, y, z) is a solution in integers of (28.1) such that:

(1) $\gcd(x, z) = 1$;

(2) $\ell_i | z$ for $i = 1, \ldots, s$; and

(3) x is not a (p_i)th power modulo ℓ_i for $i = 1, \ldots, s$.

Let $\alpha = (x + y\sqrt{d})/2$, so by Lemma 1, $(\alpha) = I^n$ where I is an integral ideal. By Lemma 2, $(\ell_i) = L_i \cdot \sigma(L_i)$ where L_i, $\sigma(L_i)$ are distinct prime ideals of K, and $L_i|(\sigma(\alpha))$, $L_i \nmid (\alpha)$.

Lemma 4. *If ε is a (p_i)th power modulo L_i for $i = 1, \ldots, s$, then the ideal class $[I]$ has order n in the group $\mathcal{C}\ell_K$.*

Proof: Since $I^n = (\alpha)$ then $[I]^n$ is the unit element in the group $\mathcal{C}\ell_K$. Now let m be the order of $[I]$, so $m|n$. If $m < n$ there exists a prime p_i (with $1 \leq i \leq s$) such that $p_i m|n$, so $[I]^{n/p_i} = 1$, that is, $I^{n/p_i} = (\beta)$ for some $\beta \in K$, $\beta \neq 0$. Hence $(\alpha) = I^n = (\beta)^{p_i}$. Since ε is a (p_i)th power modulo L_i by Lemma 3, (α) is not the (p_i)th power of a principal ideal of K. A contradiction. ∎

Lemma 5.   *If q is any odd prime there exists an integer x such that $x^2 - 4$ is not a square modulo q.*

Proof: If $q \equiv 1 \pmod 4$ we write $q - 1 = 4k$, so $(k/q) = (-1/q) = 1$.

If $q \equiv -1 \pmod 4$ we write $q + 1 = 4k$, and again $(k/q) = (1/q) = 1$. Hence $k \equiv h^2 \pmod q$ for some h, $h \not\equiv 0 \pmod q$.

Just by counting, we note that there exists y, $y \not\equiv 0 \pmod q$ such that $y^2 + 1$ (respectively, $y^2 - 1$) is not a square modulo q.

Let x be such that $hx \equiv y \pmod q$. So

$$k(x^2 - 4) = kx^2 - q \pm 1 \equiv kx^2 \pm 1 \equiv (hx)^2 \pm 1 \equiv y^2 \pm 1 \pmod q,$$

hence $k(x^2 - 4)$, and also $x^2 - 4$, is not a square modulo q. ∎

We shall now prove the first result, which concerns imaginary quadratic fields.

C.   *Let $n > 1$, and let S_1, S_2, S_3 be pairwise disjoint finite sets of primes. Then there exist infinitely many imaginary quadratic fields K such that:*

(1) *$\mathcal{C}\ell_K$ has an element of order n, so n divides the class number of K.*

(2) *If $q \in S_1$ then q decomposes totally in $K|\mathbb{Q}$.*

(3) *If $q \in S_2$ then q is inert in $K|\mathbb{Q}$.*

(4) *If $q \in S_3$ then q is ramified in $K|\mathbb{Q}$.*

Proof: We may assume without loss of generality that $S_3 \not\subseteq \{2,3\}$; indeed, if $S_3 \subseteq \{2,3\}$, it suffices to replace it by $S_3' = S_3 \cup \{q\}$, where q is a prime, $q \notin S_1 \cup S_2 \cup S_3$, and $q > 3$.

Let $S = \{p_1, \ldots, p_s\}$ be the set of primes dividing n.

Let k be the product of all primes $q \in S_1$.

By the Chinese remainder theorem there exists an integer ℓ such that

$$\begin{cases} \ell \equiv 1 \pmod q & \text{for every } q \in (S \cup S_2) \setminus \{2\}, \\ \ell \equiv 1 \pmod{q^2} & \text{for every } q \in S_3 \cup \{2\}. \end{cases}$$

If

$$t = 4 \prod_{\substack{q \mid n \\ q \neq 2}} q \times \prod_{\substack{q \in S_2 \\ q \neq 2}} q \times \prod_{\substack{q \in S_3 \\ q \neq 2}} q^2$$

then $\ell \equiv 1 \pmod t$. By Dirichlet's theorem on primes in arithmetic progressions, there is a prime number $\ell \notin S_1 \cup S_2 \cup S_3$ satisfying the above congruences.

Let w be a primitive root modulo ℓ.

We observe that w is not a pth power modulo ℓ, for every $p \in S$. Indeed, let $p|n$, $p \neq 2$, then $p|\ell - 1$. If w is a pth power modulo ℓ, there exists r such that $w \equiv (w^r)^p \pmod{\ell}$, so $w^{rp-1} \equiv 1 \pmod{\ell}$, therefore $p|\ell - 1|rp - 1$, which is impossible. If $2|n$, then $4|\ell - 1$. If w is a square modulo ℓ, then $w \equiv (w^r)^2 \pmod{\ell}$ so $4|\ell - 1|2r - 1$, again impossible.

For every $q \in S_2 \setminus \{2\}$, by Lemma 5 there exists an integer y_q such that $y_q^2 - 4$ is not a square modulo q.

Since $\ell \notin S_1 \cup S_2 \cup S_3$ by the Chinese remainder theorem there exists an integer x such that:

$$\begin{cases} x \equiv w \pmod{\ell}, \\ x \equiv 1 \pmod{q} & \text{for every } q \in S_1, \\ x \equiv y_q \pmod{q} & \text{for every } q \in S_2 \setminus \{2\}, \\ x \equiv 1 \pmod{2} & \text{if } 2 \in S_2, \\ x \equiv q + 2 \pmod{q^2} & \text{if } q \in S_3. \end{cases}$$

In particular, x is not a pth power modulo ℓ for every p dividing n. Also $x^2 - 4$ is not a square modulo q for every $q \in S_2 \setminus \{2\}$. Moreover, $x \equiv 1 \pmod{2}$ if $2 \in S_2$, $x \equiv 0 \pmod{4}$ if $2 \in S_3$.

By the Chinese remainder theorem there exists an integer z such that

$$\begin{cases} z \equiv 1 \pmod{q} & \text{for every prime } q \text{ dividing } x \\ & \text{and such that } q \notin S_1 \cup S_2 \cup S_3 \cup \{\ell\}, \\ z \equiv 0 \pmod{q\ell} & \text{for every } q \in S_1, \\ z \equiv 1 \pmod{q^2} & \text{for every } q \in S_2 \cup S_3. \end{cases}$$

Moreover, z is defined up to a multiple of the moduli of the above congruences. So it is possible to choose such an element z satisfying also the inequality

$$4z^n > x^2.$$

Let $K = \mathbb{Q}(\sqrt{x^2 - 4z^n})$, so this is an imaginary quadratic field.

Let $x^2 - 4z^n = y^2 d$ where d is a square-free integer, so $d < 0$ and $K = \mathbb{Q}(\sqrt{d})$. Therefore the integers x, y, z are a solution of (28.1).

By the requirement $\gcd(x, z) = 1$, so the hypotheses of Lemma 2 are satisfied. Let $\alpha = (x + y\sqrt{d})/2$, thus $(\alpha) = I^n$ by Lemma 1.

Since $\varepsilon = 1$ by Lemma 4, $[I]$ has order n in the class group $\mathcal{C}\ell_K$, so n divides the class number of K.

Now we prove the other assertions.

If $q \in S_1$, $q \neq 2$ then $y^2 d = x^2 - 4z^n \equiv x^2 \pmod{q}$ so $(d/q) = 1$ and thus q decomposes in the extension $K|\mathbb{Q}$.

If $q \in S_2$, $q \neq 2$ then $y^2 d = x^2 - 4z^n \equiv x^2 - 4 \pmod{q}$. Since $x^2 - 4$ is not a square modulo q then $(d/q) = -1$, that is, q is inert in $K|\mathbb{Q}$.

If $q \in S_3$, $q \neq 2$ then $y^2 d = x^2 - 4z^n \equiv (q + 2)^2 - 4 \equiv 4q \pmod{q^2}$. Since $q \neq 2$ then $q|d$, thus q is ramified in the extension $K|\mathbb{Q}$.

If $2 \in S_1$ then x is odd, z is even, and $y^2 d = x^2 - 4z^n \equiv x^2 \equiv 1$ (mod 8). Since y has to be odd, then $y^2 \equiv 1$ (mod 8) so $d \equiv 1$ (mod 8). This means that 2 is decomposed in $K|\mathbb{Q}$.

If $2 \in S_2$ then x is odd, $z \equiv 1$ (mod 4), $y^2 d = x^2 - 4z^n \equiv 1 - 4 \equiv 5$ (mod 8), so $d \equiv 5$ (mod 8) hence 2 is inert in $K|\mathbb{Q}$.

If $2 \in S_3$ then $x \equiv 0$ (mod 4), $z \equiv 1$ (mod 4), $y^2 d = x^2 - 4z^n \equiv -4$ (mod 16), so $4|y^2 d$ and since d is square-free then y is even, therefore $4|y^2$. Writing $y^2 = 4u^2$ we have $u^2 d \equiv -1$ (mod 4). Thus u is odd, $u^2 \equiv 1$ (mod 4) hence $d \equiv 3$ (mod 4) and therefore 2 is ramified in $K|\mathbb{Q}$ (because the discriminant is equal to $4d$).

We have therefore proved the existence of a quadratic field K belonging to the set $\mathcal{K}(S_1, S_2, S_3)$ of fields satisfying the conditions of the statement.

We shall prove that the nonempty set $\mathcal{K}(S_1, S_2, S_3)$ is infinite.

Assume on the contrary that this set is finite, namely, equal to $\{K_1, \ldots, K_\nu\}$ where $\nu \geq 1$.

Let $\bar{S}_3$ be the set of all prime numbers which are ramified in some of the extensions $K_i|\mathbb{Q}$ ($1 \leq i \leq \nu$). Then $\bar{S}_3$ is a finite set, $S_3 \subseteq \bar{S}_3$, $S_1 \cap \bar{S}_3 = \varnothing$, and $S_2 \cap \bar{S}_3 = \varnothing$.

Let q' be a prime, $q' \notin S_1 \cup S_2 \cup \bar{S}_3 \cup \{2, 3\}$, and let $S_3' = \bar{S}_3 \cup \{q'\}$. Then $S_1 \cap S_3' = \varnothing$, $S_2 \cap S_3' = \varnothing$ so there exists some field $K' \in \mathcal{K}(S_1, S_2, S_3')$. It follows that $K' \in \mathcal{K}(S_1, S_2, S_3)$. Since q' is ramified in K' then $q' \in \bar{S}_3$, which is a contradiction.

This proves that the set $\mathcal{K}(S_1, S_2, S_3)$ is infinite. ∎

To handle the case of real quadratic fields we need some more preliminary lemmas.

As before, let $n = p_1^{e_1} \cdots p_s^{e_s}$ ($s \geq 1$) and for every $i = 1, \ldots, s$, let ℓ_i, ℓ_i' be primes such that $\ell_i \neq \ell_i'$ and

$$\begin{cases} \ell_i \equiv 1 \ (\text{mod } p_i) & \text{when } p_i \neq 2, \\ \ell_i \equiv 1 \ (\text{mod } 4) & \text{when } p_i = 2, \end{cases}$$

and

$$\begin{cases} \ell_i' \equiv 1 \ (\text{mod } p_i) & \text{when } p_i \neq 2, \\ \ell_i' \equiv 1 \ (\text{mod } 4) & \text{when } p_i = 2. \end{cases}$$

In particular, $\ell_i \neq 2$, $\ell_i' \neq 2$.

Lemma 6.   *Let $x, x', z, z' \in \mathbb{Z}$ be integers satisfying the equation*

$$X^2 - 4Z^n = X'^2 - 4Z'^n, \tag{28.2}$$

and such that:

(1) $\gcd(x, z) = 1$ *and* $\gcd(x', z') = 1$.

(2) $\ell_i | z$, $\ell_i' | z'$ *for every* $i = 1, \ldots, s$.

(3) *x is not a (p_i)th power modulo ℓ_i and x' is not a (p_i)th power modulo ℓ_i' for $i = 1, \ldots, s$.*

(4) $(x + x')/2$ is a (p_i)th power modulo ℓ_i for $i = 1, \ldots, s$.

Let $K = \mathbb{Q}(\sqrt{x^2 - 4z^n})$. Then:

If $x^2 - 4z^n < 0$ the ideal class group $C\ell_K$ has a subgroup isomorphic to $\mathbb{Z}/n \times \mathbb{Z}/n$.

If $x^2 - 4z^n > 0$ then $C\ell_K$ has a subgroup isomorphic to $\mathbb{Z}/n$.

Proof: Let $x^2 - 4z^n = x'^2 - 4z'^n = y^2 d$ where d, y are integers and d is square-free. So $x^2 - y^2 d = 4z^n$ and $x'^2 - y^2 d = 4z'^n$. Let

$$\alpha = \frac{x + y\sqrt{d}}{2}, \qquad \alpha' = \frac{x' + y\sqrt{d}}{2}.$$

By Lemma 1, $(\alpha) = I^n$ and $(\alpha') = I'^n$ where I, I' are integral ideals of $K = \mathbb{Q}(\sqrt{d})$. Since $\ell_i | z$ then $x^2 - y^2 d \equiv 0 \pmod{\ell_i}$. But $\ell_i \nmid x$ hence $\ell_i \nmid y$, $\ell_i \nmid d$ and therefore d is a square modulo ℓ_i, that is, $(d/\ell_i) = 1$. So ℓ_i decomposes in the extension $K | \mathbb{Q}$. Similarly, ℓ'_i decomposes in $K | \mathbb{Q}$.

By Lemma 2, $(\ell_i) = L_i \cdot \sigma(L_i)$ and $(\ell'_i) = L'_i \cdot \sigma(L'_i)$ where L_i, $\sigma(L_i)$, L'_i, and $\sigma(L'_i)$ are distinct ideals of K and we may assume that $L_i | (\sigma(\alpha))$, $L_i \nmid (\alpha)$, $L'_i | (\sigma(\alpha'))$, and $L'_i \nmid (\alpha')$.

Let $R_i = \{\beta \in A \mid \beta \text{ is a } (p_i)\text{th power modulo } L_i\}$ and $R'_i = \{\beta' \in A \mid \beta' \text{ is a } (p_i)\text{th power modulo } L'_i\}$.

We note that $(x - y\sqrt{d})/2 = \sigma(\alpha) \in L_i$ so $x \equiv y\sqrt{d} \pmod{L_i}$, hence $\alpha = (x + y\sqrt{d})/2 \equiv x \pmod{L_i}$. By (3), $\alpha \notin R_i$ for every $i = 1, \ldots, s$. Similarly, $\alpha' \notin R'_i$ for $i = 1, \ldots, s$. Also $\alpha' = (x' + y\sqrt{d})/2 \equiv (x' + x)/2 \pmod{L_i}$ and from (4) it follows that $\alpha' \in R_i$.

If $x^2 - 4z^n < 0$ then K is an imaginary field, $\varepsilon = 1$. By Lemma 4, the ideal classes $[I]$, $[I']$ have order n in the group $C\ell_K$.

Now we show that the subgroup of $C\ell_K$ generated by $[I]$ and $[I']$ is isomorphic to $\mathbb{Z}/n \times \mathbb{Z}/n$. If $m > 0$, $m' > 0$ are such that $[I]^m [I']^{m'} = 1$ then there exists $\beta \in K$, $\beta \neq 0$, such that $I^m I'^{m'} = (\beta)$. So $I^{mn} I'^{m'n} = (\alpha)^m (\alpha')^{m'} = (\beta)^n$, hence $\alpha^m \alpha'^{m'} = \pm\beta^n$ since the only units are 1, -1. We show now that $n | m$, $n | m'$, which implies that the subgroup generated by $[I]$, $[I']$ is isomorphic to the direct product of the subgroups generated by $[I]$ and $[I']$—and the latter group is isomorphic to $\mathbb{Z}/n \times \mathbb{Z}/n$.

Let v_{p_i} denote the p_i-adic valuation and let $v_{p_i}(\gcd(m, m')) = d_i$, $i = 1, \ldots, s$. We are required to prove that $e_i \leq d_i$ for every $i = 1, \ldots, s$. Suppose, on the contrary, that there exists an index i such that $d_i < e_i$. Let $m = p_i^{d_i} m_0$, $m' = p_i^{d_i} m'_0$, $n = p_i^{d_i} n_0$ so $p_i | n_0$ and p_i does not divide both m_0, m'_0. We have

$$(\alpha^{m_0} \alpha'^{m'_0})^{p_i^{d_i}} = \pm(\beta^{n_0})^{p_i^{d_i}}.$$

Since 1, -1 are the only roots of unity in K then

$$\alpha^{m_0} \alpha'^{m'_0} = \pm\beta^{n_0}.$$

But $\alpha' \in R_i$ so $\alpha'^{m_0'} \in R_i$. We show also that $\pm\beta^{n_0} \in R_i$. Indeed, if $p_i = 2$, since $\ell_i \equiv 1 \pmod 4$ then $(-1/\ell_i) = 1$, that is, -1 is a square modulo ℓ_i hence it is a square modulo L_i. If $p_i \neq 2$ then $-1 = (-1)^{p_i}$. On the other hand, $p_i | n_0$ thus $\beta^{n_0} \in R_i$. From this, it follows that $\alpha^{m_0} = \pm\beta^{n_0}(\alpha'^{m_0'})^{-1} \in R_i$. If w_i is a primitive root modulo ℓ_i, then $\alpha \equiv w_i^s \pmod{L_i}$ for some integer s, $1 \leq s \leq \ell_i - 1$; since $\alpha \notin R_i$ then $p_i \nmid s$. Also $\alpha^{m_0} \equiv w_i^{rp_i} \pmod{L_i}$ for some integer r. Therefore $sm_0 \equiv rp_i \pmod{\ell_i - 1}$; thus $p_i | \ell_i - 1 | sm_0 - rp_i$ and since $p_i \nmid s$ then $p_i | m_0$.

Similarly, $\pm\beta^{n_0} \in R_i'$ and $\alpha^{m_0} \in R_i'$ therefore $\alpha'^{m_0'} \in R_i'$ hence $p_i | m_0'$. This is a contradiction and we have proved the theorem for the case of imaginary quadratic fields.

Let $x^2 - 4z^n > 0$, let m be the order of $[I]$, and let m' be the order of $[I']$ in $\mathcal{C}\ell_K$. Since $I^n = (\alpha)$, $I'^n = (\alpha')$ then $m|n$, $m'|n$.

If ε is the fundamental unit of K, let $J = \{i \mid 1 \leq i \leq s, \ \varepsilon \text{ is a } (p_i)\text{th}$ power modulo $L_i\}$. By Lemma 3, if $i \in J$ then (α) is not the (p_i)th power of a principal ideal of K. Hence $p_i^{e_i}|m$, otherwise $p_i|(n/m)$ so $m|(n/p_i)$ and $(\alpha) = ((I)^{n/p_i})^{p_i}$ where $(I)^{n/p_i}$ is a principal ideal, a contradiction. This shows that

$$\prod_{i \in J} p_i^{e_i} \Big| m.$$

Also, if $i \notin J$ then $p_i^{e_i}|m'$. Otherwise $p_i m'|n$ hence $(\alpha') = I'^n = (\beta')^{p_i}$ for some $\beta' \in K$, $\beta' \neq 0$. Thus $\alpha' = \pm\varepsilon^k \beta'^{p_i}$ with some integer k. But $\alpha' \in R_i$, $\beta'^{p_i} \in R_i$ hence $\pm\varepsilon^k \in R_i$. Since $i \notin J$ then $\varepsilon \notin R_i$ hence by a previous argument $p_i|k$, and $\alpha' = \pm\varepsilon^k \beta'^{p_i} \in R_i'$, which is a contradiction.

Now let

$$m = m_0 \prod_{i \in J} p_i^{e_i} \qquad \text{and} \qquad m' = m_0' \prod_{i \notin J} p_i^{e_i}.$$

Then $[I^{m_0}]$ has order $\prod_{i \in J} p_i^{e_i}$ and $[I'^{m_0'}]$ has order $\prod_{i \notin J} p_i^{e_i}$. So $[I^{m_0} I'^{m_0'}]$ has order $n = \prod_{i=1}^s p_i^{e_i}$.

This proves that $\mathcal{C}\ell_K$ contains a subgroup isomorphic to $\mathbb{Z}/n$. ∎

From the above proof, it also follows that if $d > 0$, if the fundamental unit ε is a (p_i)th power modulo L_i (and modulo L_i') for every $i = 1, \ldots, s$ then $[I]$ has order n ($[I']$ has order n). And according to the proof for the case where $d < 0$, it follows now that the subgroup of $\mathcal{C}\ell_K$ generated by $[I]$ and $[I']$ is isomorphic to their direct product, hence to $\mathbb{Z}/n \times \mathbb{Z}/n$.

Lemma 7.   *Given any prime p, there exist infinitely many primes ℓ such that:*

(1)
$$\begin{cases} \ell \equiv 1 \pmod p & \text{when } p \neq 2, \\ \ell \equiv 1 \pmod 4 & \text{when } p = 2. \end{cases}$$

(2) *2 and -1 are pth powers modulo ℓ.*

(3) 3 *is not a pth power modulo* ℓ.

Proof: Let ζ_p be a primitive pth root of 1, and let $M = \mathbb{Q}(\zeta_p)$ when $p \neq 2$, or $M = \mathbb{Q}(\sqrt{-1})$ when $p = 2$. Let $\overline{M} = M(\sqrt[p]{2}, \sqrt[p]{3})$. So $\overline{M}|\mathbb{Q}$ is a Galois extension of degree $p^2(p-1)$ when $p \neq 2$, or of degree 8 when $p = 2$.

For every prime number ℓ, unramified in the extension $\overline{M}|\mathbb{Q}$, let $\left(\dfrac{\overline{M}|\mathbb{Q}}{\ell}\right)$ denote the Artin symbol, which is a conjugacy class of the Galois group of $\overline{M}|\mathbb{Q}$, say the one containing the element τ. According to class field theory, the decomposition group of ℓ in $\overline{M}|\mathbb{Q}$ is, up to isomorphism, the subgroup generated by τ (see Chapter 25, Section 1).

By Chebotarev's density theorem (see Chapter 25), if τ is the generator of the Galois group $G(\overline{M}|M(\sqrt[p]{2}))$ there exist infinitely many (unramified) primes ℓ in $\mathbb{Q}$ such that $\left(\dfrac{\overline{M}|\mathbb{Q}}{\ell}\right)$ is the conjugacy class of τ (in $G(\overline{M}|\mathbb{Q})$).

Hence the decomposition field of ℓ is $M(\sqrt[p]{2})$. We have $\ell \neq p$, $\ell \neq 2$, $\ell \neq 3$, since ℓ is unramified.

It results that each such prime ℓ is totally decomposed in the extension $M(\sqrt[p]{2})$ since the decomposition field of ℓ in $M(\sqrt[p]{2})|\mathbb{Q}$ is $M(\sqrt[p]{2})$. On the other hand, if L is any prime ideal of M such that L divides ℓ, then the decomposition field of L in $M(\sqrt[p]{3})|M$ is $M(\sqrt[p]{2}) \cap M(\sqrt[p]{3}) = M$. This means that L is inert in the extension $M(\sqrt[p]{3})|M$.

From the theory of cyclotomic extensions (Chapter 11, (O)), we deduce that for every ℓ (as above) we have $\ell \equiv 1 \pmod{p}$, when $p \neq 2$, or $\ell \equiv 1 \pmod 4$ when $p = 2$. Indeed, in the extension $M|\mathbb{Q}$ we have $(\ell) = L_1 \cdots L_g$ where $gf = p - 1$ and f is the order of ℓ modulo p. Since ℓ is totally decomposed in $M(\sqrt[p]{2})|\mathbb{Q}$ and a fortiori totally decomposed in $M|\mathbb{Q}$, then $g = p - 1$, $f = 1$ so $\ell \equiv 1 \pmod{p}$ when $p \neq 2$. Now, if $p = 2$ then ℓ is totally decomposed in $\mathbb{Q}(\sqrt{-1}) = M$, that is, $(-1/\ell) = 1$, so $\ell \equiv 1 \pmod 4$.

Now we recall from Chapter 16, Section 6, and some facts from the theory of relative binomial extension $M(\sqrt[p]{a})|M$, where M contains a primitive pth root of 1, or $\sqrt{-1}$ when $p = 2$.

If L is a prime ideal of M, L does not divide (p) nor (a), then:

(1) L decomposes totally in $M(\sqrt[p]{a})|M$ if and only if the congruence $X^p \equiv a \pmod{L}$ has a solution in M.

(2) L is inert in $M(\sqrt[p]{a})|M$ if and only if the above congruence has no solution in M.

Taking $a = 2$, since each prime L, which divides (ℓ), decomposes totally in $M(\sqrt[p]{a})|M$ then 2 is a pth power modulo L of an element in M. But L and (ℓ) define the same residue class field, so 2 is a pth power modulo ℓ of an integer in $\mathbb{Z}$.

On the other hand, taking $a = 3$, if L divides (ℓ) then L is inert in $M(\sqrt[p]{3})|M$, hence 3 is not a pth power modulo L of any element of M, and a fortiori, it is not a pth power modulo ℓ of any integer in $\mathbb{Z}$.

Finally, if $p \neq 2$ then $-1 = (-1)^p$ and if $p = 2$ then -1 is a square modulo ℓ, because $\ell \equiv 1 \pmod{4}$. ∎

Now we prove the main result concerning real quadratic fields:

D.   *Let $n > 1$ and let S_1, S_2, S_3 be pairwise disjoint finite sets of primes. Then there exist infinitely many real (respectively, imaginary) quadratic fields K such that:*

(1) *$\mathcal{C}\ell_K$ has a subgroup isomorphic to $\mathbb{Z}/n$ (respectively, to $\mathbb{Z}/n \times \mathbb{Z}/n$).*

(2) *If $q \in S_1$ then q decomposes totally in $K|\mathbb{Q}$.*

(3) *If $q \in S_2$ then q is inert in $K|\mathbb{Q}$.*

(4) *If $q \in S_3$ then q is ramified in $K|\mathbb{Q}$.*

Proof: We shall prove the existence of a field K with the properties indicated. From this, it follows, as in the proof of (**C**), that there exist infinitely many such fields.

We may assume, without loss of generality, that $S_3 \not\subseteq \{2,3\}$, as in the proof of (**C**).

Let S be the set of prime factors p_1, ..., p_s of n. By Lemma 7 there exist distinct primes ℓ_1, ℓ_1', ..., ℓ_s, ℓ_s', not belonging to $S_1 \cup S_2 \cup S_3 \cup \{2,3\}$ and such that, for every $i = 1, \ldots, s$, ℓ_i and ℓ_i' satisfy the conditions of the lemma.

For every prime p let $e_p = v_p(n)$ be the exponent of the exact power of p dividing n.

Let

$$f_q = \begin{cases} e_q + 2 & \text{when } q \in S_2 \setminus \{2\}, \\ e_q + 1 & \text{when } q \in S_3 \setminus \{2\}. \end{cases}$$

For every q let a_q be an integer, not a multiple of q, satisfying the following conditions:

If $q \in S_2 \setminus \{2\}$ then $2(nq^{-e_q})a_q$ is not a quadratic residue modulo q.

If $q \in S_3 \setminus \{2\}$ then $a_q \not\equiv 0 \pmod{q}$.

If $2 \in S_1$ then $a_2(n2^{-e_2}) \equiv 1 \pmod{8}$.

If $2 \in S_2$ then $a_2(n2^{-e_2}) \equiv 5 \pmod{8}$.

If $2 \in S_3$ then $a_2(n2^{-e_2}) \equiv 3 \pmod{4}$.

By the Chinese remainder theorem there exists an integer a such that

$$\begin{cases} \ell_i \nmid a,\ \ell_i' \mid a & \text{for } i = 1, \ldots, s, \\ a \equiv 1 \pmod{q} & \text{for every } q \in S_1 \setminus \{2\}, \\ a \equiv a_q q^{f_q} \pmod{q^{f_q+1}} & \text{for every } q \in S_2 \cup S_3 \setminus \{2\}, \\ a \equiv a_2 2^{e_2+5} \pmod{2^{e_2+8}} & \text{if } 2 \in S_1 \cup S_2 \cup S_3, \\ a \equiv 0 \pmod{2} & \text{if } 2 \notin S_1 \cup S_2 \cup S_3. \end{cases}$$

Similarly, there exists an integer b such that

$$\begin{cases} \ell_i \mid b,\ \ell_i' \nmid b & \text{for } i = 1, \ldots, s, \\ b \equiv a \pmod{q} & \text{for every } q \in S_1 \setminus \{2\}, \\ b \equiv 0 \pmod{q^{f_q+1}} & \text{for every } q \in S_2 \cup S_3 \setminus \{2\}, \\ b \equiv 0 \pmod{2^{e_2+8}} & \text{if } 2 \in S_1 \cup S_2 \cup S_3, \\ b \equiv 0 \pmod{2} & \text{if } 2 \notin S_1 \cup S_2 \cup S_3. \end{cases}$$

Now we show that there exists an integer t satisfying the system of congruences:

$$\begin{cases} t \equiv a \pmod{\ell_i} & \text{for } i = 1, \ldots, s, \\ t \equiv b \pmod{\ell_i'} & \text{for } i = 1, \ldots, s, \\ t \equiv a \pmod{q} & \text{for } q \in S_1 \setminus \{2\}, \\ t \equiv 1 \pmod{q} & \text{for } q \in S_2 \cup S_3 \setminus \{2\}, \\ t \equiv 1 \pmod{8}, \\ t \equiv 1 \pmod{a^n - b^n}, \\ t \equiv 1 + a \pmod{2a^n - \frac{1}{2}(a-b)^n}, \\ t \equiv 1 + b \pmod{2b^n - \frac{1}{2}(b-a)^n}. \end{cases}$$

To prove the existence of t we need to consider the pairs of moduli in the above system, which are not relatively prime. We note first that all primes $\ell_1, \ell_1', \ldots, \ell_s, \ell_s'$ and of $S_1 \cup S_2 \cup S_3 \cup \{2\}$ are distinct. Also $\ell_i \nmid a^n - b^n$, as seen at once. Again, $\ell_i \nmid 2a^n - \frac{1}{2}(a-b)^n$, because $\ell_i \mid b$ hence if $\ell_i \mid 2a^n - \frac{1}{2}(a-b)^n$ then $\ell_i \mid 2a^n - \frac{1}{2}a^n = 3a^n/2$; but $\ell_i \neq 3$ hence $\ell_i \mid a^n$ so $\ell_i \mid a$, contrary to the hypothesis.

Similarly, $\ell_i \nmid 2b^n - \frac{1}{2}(b-a)^n$, otherwise from $\ell_i \mid b$ it follows that $\ell_i \mid a$, a contradiction.

In the same way we show that

$$\ell_i' \nmid a^n - b^n, \qquad \ell_i' \nmid 2a^n - \frac{1}{2}(a-b)^n, \qquad \ell_i' \nmid 2b^n - \frac{1}{2}(b-a)^n.$$

If $q \in S_1 \setminus \{2\}$ and $q \mid a^n - b^n$ then we have $a \equiv 1 \pmod{q}$ by hypothesis.

If $q \in S_1 \setminus \{2\}$ then $q \nmid 2a^n - \frac{1}{2}(a-b)^n$, otherwise from $q \mid a - b$ it would follow that $q \mid a$, a contradiction to $a \equiv 1 \pmod{q}$.

If $q \in S_1 \setminus \{2\}$ then $q \nmid 2b^n - \frac{1}{2}(b-a)^n$, which is seen in the same way.

If $q \in S_2 \cup S_3 \cup \{2\}$, noting that $a \equiv 0 \pmod{q}$ and $b \equiv 0 \pmod{q}$, if $q \mid 2a^n - \frac{1}{2}(a-b)^n$ then $1 \equiv 1 + a \pmod{q}$ and if $q \mid 2b^n - \frac{1}{2}(b-a)^n$ then $1 \equiv 1 + b \pmod{q}$.

Finally, 2 divides both $2a^n - \frac{1}{2}(a-b)^n$ and $2b^n - \frac{1}{2}(b-a)^n$ but $1 + a \equiv 1$ (mod 2) and $1 + b \equiv 1$ (mod 2), whether $2 \in S_1 \cup S_2 \cup S_3$ or $2 \notin S_1 \cup S_2 \cup S_3$.

Altogether we have seen that if two moduli are not relatively prime the system of the corresponding two congruences has a solution. By the Chinese remainder theorem there exists an integer and therefore infinitely many integers, t satisfying all the congruences indicated.

We define the numbers:

$$\begin{cases} x = 2t^n + \frac{1}{2}[(t-a)^n - (t-b)^n], \\ x' = 2t^n - \frac{1}{2}[(t-a)^n - (t-b)^n], \\ z = t(t-a), \\ z' = t(t-b). \end{cases}$$

Note that since a, b are even then $t - a$, $t - b$ have the same parity and therefore x, x', as well as z, z', are integers.

Then

$$x^2 - x'^2 = 4 \times t^n[(t-a)^n - (t-b)^n] = 4z^n - 4z'^n,$$

that is, $x^2 - 4z^n = x'^2 - 4z'^n$.

We shall prove now that x, x', z, z' satisfy the four conditions in Lemma 6.

Condition (1). First we see that $2 \nmid z = t(t-a)$. Indeed, $2 \nmid t$, a is even, so $2 \nmid t - a$. Now, if q is a prime dividing x and z then $q \neq 2$ and either $q|t$ or $q|t - a$. If $q|t$ and $q|x$ then

$$0 \equiv 2t^n + \frac{1}{2}[(t-a)^n - (t-b)^n] \equiv \frac{(-1)^n}{2}(a^n - b^n) \pmod{q},$$

so $q|a^n - b^n$ and $q|t$, impossible. On the other hand, if $q|t - a$, $q|x$ then

$$0 \equiv 2t^n + \frac{1}{2}[(t-a)^n - (t-b)^n] \equiv 2a^n - \frac{1}{2}(a-b)^n \pmod{q},$$

that is, $q|t - a$ and $q|2a^n - \frac{1}{2}(a-b)^n$ which is impossible. This shows that $\gcd(x, z) = 1$.

Similarly, $\gcd(x', z') = 1$.

Condition (2). Since $t \equiv a \pmod{\ell_i}$ and $t \equiv b \pmod{\ell_i'}$ then $\ell_i|z$ and $\ell_i'|z'$ for $i = 1, \ldots, s$.

Condition (3). We have $x \equiv 2a^n - \frac{1}{2}(a-b)^n \equiv \frac{3}{2}a^n \pmod{\ell_i}$, and similarly, $x' \equiv \frac{3}{2}b^n \pmod{\ell_i'}$. From this, it follows that $x + x' \equiv 4t^n \equiv 4a^n \pmod{\ell_i}$.

From this, we deduce that x is not a (p_i)th power modulo ℓ_i. Otherwise, $\frac{3}{2}a^n$ is a (p_i)th power modulo ℓ_i. But, by Lemma 7, 2 is a (p_i)th power modulo ℓ_i.

Since $p_i|n$ then a^n is also a (p_i)th power modulo ℓ_i, from which it follows that so is 3, and this contradicts Lemma 7. Similarly, x' is not a (p_i)th power modulo ℓ_i'.

Condition (4). From $(x + x')/2 \equiv 2a^n$ (mod ℓ_i) it follows by the same lemma and the fact that $p_i | n$ that $(x + x')/2$ is a (p_i)th power modulo ℓ_i for every $i = 1, \ldots, s$.

It is possible to choose t such that $x^2 - 4z^n < 0$, or such that $x^2 - 4z^n > 0$. Indeed, $x^2 - 4z^n = 2n(a + b)t^{2n-1} + $ (terms of lower degree in t), so for large values of $|t|$, $x^2 - 4z^n$ and $2n(a + b)t^{2n-1}$ have the same sign.

Let

$$K = \mathbb{Q}(\sqrt{x^2 - 4z^n}).$$

We write $x^2 - 4z^n = y^2 d$, where d is a square-free integer, so $K = \mathbb{Q}(\sqrt{d})$. By Lemma 6, if $x^2 - 4z^n < 0$ the ideal class group $\mathcal{C}\ell_K$ contains a subgroup isomorphic to $\mathbb{Z}/n \times \mathbb{Z}/n$, while if $x^2 - 4z^n > 0$ it contains a subgroup isomorphic to $\mathbb{Z}/n$. This proves the first assertion.

Now let $q \in S_1 \setminus \{2\}$. Then $q | t - a$ hence $q | z$ so $x^2 \equiv y^2 d$ (mod q). But $\gcd(x, z) = 1$, so $q \nmid x$ hence $q \nmid y$ and d is a square modulo q, that is, $(d/q) = 1$. Therefore q is totally decomposed in the extension $K | \mathbb{Q}$.

Let $q \in S_2 \cup S_3 \setminus \{2\}$. Then

$$t \equiv 1 \ (\text{mod } q), \qquad a \equiv a_q q^{f_q} \ (\text{mod } q^{f_q+1}), \qquad b \equiv 0 \ (\text{mod } q^{f_q+1}).$$

We have

$$x^2 - 4z^n = \{2t^n + \tfrac{1}{2}[(t - a)^n - (t - b)^n]\}^2 - 4t^n(t - a)^n. \qquad (28.3)$$

Since

$$t - a \equiv t - a_q q^{f_q} \ (\text{mod } q^{f_q+1}),$$
$$t - b \equiv t \ (\text{mod } q^{f_q+1}),$$

and q^{e_q} divides n, then we deduce, noting that $2f_q \geq f_q + e_q + 1$:

$$(t - a)^n \equiv (t - a_q q^{f_q})^n \equiv t^n - nt^{n-1}a_q q^{f_q} \ (\text{mod } q^{e_q + f_q + 1})$$

and also

$$(t - b)^n \equiv t^n \ (\text{mod } q^{e_q + f_q + 1}).$$

Thus

$$\begin{aligned}
x^2 - 4z^n &= 4t^{2n} + 2t^n[(t - a)^n - (t - b)^n] \\
&\quad + \tfrac{1}{4}[(t - a)^n - (t - b)^n]^2 - 4t^n(t - a)^n \\
&\equiv 4t^{2n} + 2t^n[-nt^{n-1}a_q q^{f_q}] - 4t^{2n} + 4t^{2n-1}na_q q^{f_q} \\
&\equiv 2t^{2n-1}na_q q^{f_q} \equiv 2na_q q^{f_q} \ (\text{mod } q^{e_q + f_q + 1}).
\end{aligned}$$

Hence

$$\frac{x^2 - 4z^n}{q^{e_q + f_q}} \equiv 2(nq^{-e_q})a_q \ (\text{mod } q)$$

so we conclude that q does not divide $(x^2 - 4z^n)/q^{e_q + f_q}$.

If $q \in S_2$ then $f_q = e_q + 2$ so $q^{e_q + f_q}$ is a square, therefore it divides y^2, that is

$$\frac{x^2 - 4z^n}{q^{e_q + f_q}} = \frac{y^2}{q^{e_q + f_q}} \times d.$$

Since $2(nq^{-e_q})a_q$ is not a quadratic residue modulo q then d is not a square modulo q. This means that q is inert in the extension $K|\mathbb{Q}$.

If $q \in S_3$ then $a_q \not\equiv 0 \pmod{q}$ so q does not divide $(x^2 - 4z^n)/q^{e_q + f_q}$. But $f_q = e_q + 1$ so $e_q + f_q$ is odd, hence

$$\frac{x^2 - 4z^n}{q^{e_q + f_q}} = \frac{y^2}{q^{e_q + f_q - 1}} \times \frac{d}{q}.$$

We have proved above that the left hand side is not divisible by q, so we conclude that $q|d$ and therefore q is ramified in $K|\mathbb{Q}$.

It remains to study the behavior of the prime 2, if $2 \in S_1 \cup S_2 \cup S_3$.

We need to establish the following congruence:

$$x^2 - 4z^n \equiv na_2 2^{e_2 + 6} \pmod{2^{2e_2 + 9}}. \tag{28.4}$$

We have

$$t \equiv 1 \pmod 8, \quad a \equiv a_2 2^{e_2 + 5} \pmod{2^{e_2 + 8}} \quad \text{and} \quad b \equiv 0 \pmod{2^{2e_2 + 8}}.$$

We proceed in the same manner. From

$$t - a \equiv t - a_2 2^{e_2 + 5} \pmod{2^{e_2 + 8}},$$
$$t - b \equiv t \pmod{2^{2e_2 + 8}},$$

and 2^{e_2} divides n, we deduce

$$(t - a)^n \equiv (t - a_2 2^{e_2 + 5})^n \equiv t^n - nt^{n-1} a_2 2^{e_2 + 5} \pmod{2^{2e_2 + 8}},$$
$$(t - b)^n \equiv t \pmod{2^{2e_2 + 8}}.$$

Thus

$$x^2 - 4z^n = 4t^n + 2t^n[-nt^{n-1} a_2 2^{e_2 + 5}] - 4t^{2n} + 4t^{2n-1} na_2 2^{e_2 + 5}$$
$$\equiv 2t^{n-1} na_2 2^{e_2 + 5} \equiv na_2 2^{e_2 + 6} \pmod{2^{2e_2 + 8}}.$$

Now if $2 \in S_1$ then

$$\frac{x^2 - 4z^n}{2^{2e_2 + 6}} \equiv (n2^{-e_2})a_2 \equiv 1 \pmod 8$$

so it is a square modulo 8. Since

$$\frac{x^2 - 4z^n}{2^{2e_2 + 6}} = \frac{y^2}{2^{2e_2 + 6}} \times d$$

then d is a square modulo 8; since d is square-free then $d \equiv 1 \pmod 8$ therefore 2 decomposes totally in $K|\mathbb{Q}$.

If $2 \in S_2$ then

$$\frac{y^2}{2^{2e_2 + 6}} \times d = \frac{x^2 - 4z^n}{2^{2e_2 + 6}} \equiv (n2^{-e_2})a_2 \equiv 5 \pmod 8,$$

so it is not a square modulo 8. As above d is not a square modulo 8, hence $d \equiv 5 \pmod 8$ and 2 is inert in $K|\mathbb{Q}$.

Finally, if $2 \in S_3$ then

$$\frac{y^2}{2^{2e_2+6}} \times d = \frac{x^2 - 4z^n}{2^{2e_2+6}} \equiv (n2^{-e_2})a_2 \equiv 3 \pmod 4.$$

Therefore $d \equiv 3 \pmod 4$ and the discriminant of $K = \mathbb{Q}(\sqrt{d})$ is $4d$; therefore 2 divides the discriminant and it is ramified in $K|\mathbb{Q}$.

This concludes the proof of the theorem.   ■

In particular, we have also shown that for every $n > 1$ there exist infinitely many real quadratic extensions $K|\mathbb{Q}$ with class number a multiple of n.

As an extra information (beyond quadratic extensions), we quote that Honda has completely determined all the pure cubic fields $\mathbb{Q}(\sqrt[3]{n})$ (where n is a cube-free natural number, $n \neq 1$) with the class number divisible by 3.

28.2 Quadratic Fields with Class Number 1

In his extensive and brilliant research on binary quadratic forms, published in *Disquisitiones Arithmeticae* (when he was 19 years old) Gauss was led to concepts, which may be translated into class groups and class numbers of quadratic fields. The topic is of the greatest importance and beauty, but is not treated in the present book.

Gauss found the following nine imaginary quadratic fields $\mathbb{Q}(\sqrt{d})$ with class number 1, namely when $-d = 3$, 4, 7, 8, 11, 19, 43, 67, and 163.

Gauss conjectured that no other $\mathbb{Q}(\sqrt{d})$, $d < 0$, (d square-free) exists with class number 1.

This was first proved by Heegner, but his proof, which presented obscurities, was later recognized to be correct. Baker gave other proofs, as an application of his theory of linear forms of logarithms, as did Deuring and Stark.

Gauss also conjectured that for each integer $h > 1$, there exist only finitely many imaginary quadratic fields $\mathbb{Q}(\sqrt{d})$, $d < 0$ (d square-free), with class number h.

The work of Goldfeld and of Gross and Zagier led to the proof of the conjecture.

The lists of imaginary quadratic fields with low class numbers like 2, 3, and 4 are now completely known. For example, this is the complete list of imaginary quadratic fields $\mathbb{Q}(\sqrt{d})$ having class number 2: $-d = 5$, 6, 10, 13, 15, 22, 35, 37, 51, 58, 91, 115, 133, 187, 235, 267, 403, and 467.

Concerning real quadratic fields, Gauss conjectured that there exist infinitely many real quadratic fields with class number 1.

This is a difficult open problem. It should be noted that it is not yet known if there exist infinitely many number fields (of arbitrary degree) having class number 1. Numerical calculations support Gauss' conjecture.

29

Miscellaneous Results About the Class Number of Cyclotomic Fields

Class numbers of cyclotomic fields have been the subject of considerable investigations. It is not our intention to present systematically the results obtained, but just to sample a few of them. By following our presentation the reader will consolidate his understanding of the subject treated in this book.

Once again, the reader is encouraged to study the papers listed in the Bibliography, thus learning the proofs of the numerous results which will be described below.

29.1 Miscellanea About the Relative Class Number of $\mathbb{Q}(\zeta_p)$

This section concerns the relative class number h_p^- of the cyclotomic field $K_p = \mathbb{Q}(\zeta_p)$ where

$$\zeta_p = \cos \frac{2\pi}{p} + i \sin \frac{2\pi}{p}$$

and p is an odd prime. Sometimes, for simplicity, we write only h^-, K, ζ.

We shall indicate successively new formulas for h^-, involving certain determinants, then give estimates for h^- and values at 1 of the L-series. Next, we determine the cyclotomic fields K_p for which $h_p^- = 1$, etc.

We recall from Chapter 27, (27.20) and (27.21), that

$$h_p^- = \gamma(p) \prod_{\chi \in S^-} L(1|\chi), \qquad (29.1)$$

where

$$\gamma(p) = 2p \left(\frac{p}{4\pi^2} \right)^{(p-1)/4} = \frac{p^{(p+3)/4}}{2^{(p-3)/2} \pi^{(p-1)/2}} \qquad (29.2)$$

and S^- is the set of odd characters modulo p.

29.1.1 Determinantal Formulas for the Relative Class Number

Carlitz expressed the relative class number h_p^- in terms of the determinant of Maillet's matrix. The computation will lead to circulant matrices, which we discuss first, for the convenience of the reader.

Let K be a field of characteristic 0. The *circulant* of the n-tuple $(a_0, a_1, \ldots, a_{n-1})$, where each $a_i \in K$, is defined to be the determinant of the matrix

$$
C = \begin{pmatrix}
a_0 & a_1 & \cdots & a_{n-1} \\
a_{n-1} & a_0 & \cdots & a_{n-2} \\
\cdots\cdots\cdots\cdots\cdots \\
a_1 & a_2 & \cdots & a_0
\end{pmatrix}.
$$

We denote it by $\mathrm{Circ}(a_0, a_1, \ldots, a_{n-1})$.

Let $F(X) = a_0 + a_1 X + \cdots + a_{n-1} X^{n-1}$, and let ξ be a primitive root of unity of order n (in the algebraic closure K^* of K).

Lemma 1.   *With the above notations:*

$$
\mathrm{Circ}(a_0, a_1, \ldots, a_{n-1}) = \prod_{j=0}^{n-1} F(\xi^j) = R(F(X), X^n - 1).
$$

Proof: Let $G = \{1, \sigma, \ldots, \sigma^{n-1}\}$ be the cyclic group of order n and A the $n \times n$ matrix

$$
A = \begin{pmatrix}
0 & 1 & 0 & \cdots & 0 \\
0 & 0 & 1 & \cdots & 0 \\
\cdots\cdots\cdots\cdots\cdots \\
0 & 0 & 0 & \cdots & 1 \\
1 & 0 & 0 & \cdots & 0
\end{pmatrix}.
$$

So $A^n = I$ (unit $n \times n$ matrix), but $A^m \neq I$ for $1 \leq m < n$.

The characteristic polynomial of A is $X^n - 1$; it has distinct roots, so A may be diagonalized (in K^*).

Thus there exists an $n \times n$ invertible matrix U (with entries in K^*) such that

$$
UAU^{-1} = \mathrm{diag}(1, \xi, \ldots, \xi^{n-1}).
$$

Since $C = a_0 I + a_1 A + \cdots + a_{n-1} A^{n-1}$ then

$$
UCU^{-1} = \mathrm{diag}(F(1), F(\xi), \ldots, F(\xi^{n-1})),
$$

and we have

$$
\det(C) = \det(UCU^{-1}) = \prod_{j=0}^{n-1} F(\xi^j) = R(F(X), X^n - 1),
$$

by Chapter 2, Section 12.   ∎

Now we introduce Maillet's matrix.

If p is an odd prime, and if $p \nmid r$, let $R(r)$ denote the least positive residue of r modulo p, thus $1 \leq R(r) \leq p - 1$ and $r = [r/p]p + R(r)$. It is easily seen that $R(-r) + R(r) = p$. For every r, $1 \leq r \leq p - 1$, let r' be the unique integer such that $1 \leq r' \leq p - 1$ and $rr' \equiv 1 \pmod{p}$. Then $(p - r)' = p - r'$.

The *Maillet matrix* for p is by definition

$$M_p = (R(rs'))_{r,s=1,\ldots,(p-1)/2}.$$

Its determinant will be denoted by

$$D_p = \det(M_p).$$

Lemma 2.   (1) $D_p = (-p)^{(p-3)/2} \det\left(\left[\dfrac{rs'}{p}\right]\right)_{r,s=2,\ldots,(p-1)/2}$

(2) $D_p = (-p)^{(p-3)/2} \det\left(\left[\dfrac{rs'}{p}\right] - \left[\dfrac{(r-1)s'}{p}\right]\right)_{r,s=2,\ldots,(p-1)/2}$

and the entries of this matrix are either 0 or 1.

(3) $D_p = \pm(-p)^{(p-3)/2} \det\left(\left[\dfrac{rs}{p}\right] - \left[\dfrac{(r-1)s}{p}\right]\right)_{r,s=3,\ldots,(p-1)/2}$

and the entries of this matrix are either 0 or 1.

Proof: (1) Replace the row r ($r \geq 2$) of the Maillet matrix by the row r less r times the row 1. So the new row has entries

$$R(rs') - rR(s') = R(rs') - rs' = -p\left[\dfrac{rs'}{p}\right].$$

Let M_p' be the matrix so obtained. Its first column has entries $R(1) = 1$ and $-p[r/p] = 0$ for $r \geq 2$:

$$M_p' = \begin{pmatrix} 1 & 2' & \cdots & ((p-1)/2)' \\ 0 & \cdots & & \cdots \\ \vdots & \vdots & -p[rs'/p] & \vdots \\ 0 & \cdots & & \cdots \end{pmatrix}.$$

Thus

$$D_p = (-p)^{(p-3)/2} \det(N_p) \quad \text{where} \quad N_p = \left(\left[\dfrac{rs'}{p}\right]\right)_{r,s=2,\ldots,(p-1)/2}.$$

(2) Replace the row $r \geq 3$ of N_p by the row r less the row $r - 1$. Since $[1s'/p] = 0$ we may write the entries of the row 2 as $[2s'/p] - [1s'/p]$. Let

$$N_p' = \left(\left[\dfrac{rs'}{p}\right] - \left[\dfrac{(r-1)s'}{p}\right]\right)_{r,s=2,\ldots,(p-1)/2}.$$

Then $\det(N_p) = \det(N_p')$.

Moreover, $rs' - (r-1)s' = s' < p$ so $rs'/p - (r-1)s'/p < 1$. If $[rs'/p] - [(r-1)s'/p] \geq 2$ then necessarily $rs'/p - (r-1)s'/p > 1$, a contradiction. So the entries of N_p' are either 0 or 1.

(3)   Let $S = \{s \mid 1 \leq s \leq (p-1)/2\}$, let $S_1 = \{s \in S \mid s' \in S\}$, and let $S_2 = \{s \in S \mid s' \notin S\}$. If $s \in S_2$ then $(p-s)' = p - s' \in S$ so $p - s \in S_1$.

Given the matrix

$$N_p = \left(\left[\frac{rs'}{p} \right] \right)_{r,s=2,\ldots,(p-1)/2}$$

we shall form a new matrix $N_p^{(1)}$ by permuting the columns of N_p as follows. Let t, $2 \leq t \leq (p-1)/2$. If $t = s'$ where $s \in S$, then $t \in S_1$ and the column t of $N_p^{(1)}$ has entries $[rs'/p] = [rt/p]$. If $t \neq s'$ for every $s \in S$ then $t' \notin S$ so $t \in S_2$ and $p - t \in S_1$ so $(p-t)' = p - t' \in S$ so $p - t = s'$ for $s \in S$. In this case, the column t of $N_p^{(1)}$ has entries $-[rs'/p]$. Now we note that

$$p\left[\frac{rs'}{p} \right] = rs' - R(rs') = r(p-t) - R(rp - rt)$$

$$= r(p-t) - R(-rt) = r(p-t) + R(rt) - p$$

$$= (r-1)p + R(rt) - rt = (r-1)p - p\left[\frac{rt}{p} \right],$$

hence

$$-\left[\frac{rs'}{p} \right] = 1 - r + \left[\frac{rt}{p} \right].$$

Let $N_p^{(2)}$ be the matrix obtained from $N_p^{(1)}$ replacing row r (for $r \geq 3$) by row r less row $(r-1)$.

The row 2 of $N_p^{(2)}$ has entry $[2t/p] = 0$ when $t \in S_1$ or $1 - 2 + [2t/p] = -1$ when $t \in S_2$. The row r of $N_p^{(2)}$ has entry $[rt/p] - [(r-1)t/p]$ when $t \in S_1$ or $-1 + [rt/p] - [(r-1)t/p]$ when $t \in S_2$.

In particular, since $2' = (p+1)/2$ then $2 \in S_2$ so the column 2 of $N_p^{(2)}$ has all entries equal to -1, since $[2r/p] - [2(r-1)/p] = 0$.

Let $N_p^{(3)}$ be the matrix obtained from $N_p^{(2)}$ by replacing column t (where $t \geq 3$ and $t \in S_2$) by column t less column 2.

Then

$$
N_p^{(3)} = \begin{pmatrix} & t \in S_1 & & & t \in S_2 & \\ & \downarrow & & & \downarrow & \\ -1 & 0 & \cdots & 0 & & \cdots \\ \cdots\cdots\cdots\cdots\cdots\cdots\cdots\cdots\cdots\cdots\cdots \\ -1 & \left[\dfrac{rt}{p}\right] - \left[\dfrac{(r-1)t}{p}\right] & \cdots & \left[\dfrac{rt}{p}\right] - \left[\dfrac{(r-1)t}{p}\right] & & \cdots \\ -1 & \cdots & \cdots & \cdots & & \cdots \end{pmatrix}.
$$

Hence

$$
D_p = (-p)^{(p-3)/2} \det(N_p) = \pm(-p)^{(p-3)/2} \det(N_p^{(4)}),
$$

where

$$
N_p^{(4)} = \left(\left[\frac{rt}{p}\right] - \left[\frac{(r-1)t}{p}\right] \right)_{r,s=3,\ldots,(p-1)/2}.
$$

As before, these entries are either 0 or 1. ∎

For every r, not a multiple of p, let $\{r\} = R(r) - p/2$, so $-p/2 < \{r\} < p/2$. If $r_1 \equiv r_2 \pmod{p}$ then $\{r_1\} = \{r_2\}$. Also $\{-r\} = -\{r\}$ as easily seen.

Let g be a primitive root modulo p. So $g^{(p-1)/2} \equiv -1 \pmod{p}$. If $0 \le i \le j \le p-1$ then the following statements are obviously equivalent:

(1) $j = i$ (respectively, $i + (p-1)/2$).

(2) $g^j \equiv g^i \pmod{p}$, (respectively, $g^j \equiv -g^i \pmod{p}$).

(3) $\{g^j\} = \{g^i\}$, (respectively, $\{g^j\} = -\{g^i\}$).

It follows that the set of absolute values

$$
\{|\{1\}|, \ |\{g\}|, \ |\{g^2\}|, \ \ldots, \ |\{g^{(p-3)/2}\}|\}
$$

coincides with the set of absolute values

$$
\left\{ |\{1\}|, \ |\{2\}|, \ \ldots, \ \left|\left\{\frac{p-1}{2}\right\}\right| \right\},
$$

and both sets have $(p-1)/2$ elements.

Let η be a primitive $(p-1)$st root of 1. Consider the following matrices:

$$
H = (\{rs'\})_{r,s=1,2,\ldots,(p-1)/2},
$$

$$
G = (\{g^{j-i}\})_{i,j=0,1,\ldots,(p-3)/2},
$$

and

$$
G' = (\{g^{j-i}\}\eta^{j-i})_{i,j=0,1,\ldots,(p-1)/2}.
$$

Lemma 3.   (1) $|\det(H)| = |\det(G)|$.

(2) $|\det(G)| = |\det(G')|$.

(3) *The matrix G' is a circulant and has determinant*

$$|\det(G')| = \prod_{j=0}^{(p-3)/2} \sum_{i=0}^{(p-3)/2} \{g^i\} \eta^{i(2j+1)} = |\det(H)|.$$

Proof: (1)   The first row of H is

$$\{1\}, \ \{2\}, \ \ldots, \ \left\{\frac{p-1}{2}\right\}$$

and the first row of G is

$$\{1\}, \ \{g\}, \ \ldots, \ \{g^{(p-3)/2}\}.$$

A permutation of the columns of G distinct from the first, followed if necessary by changing signs of some of these columns, gives a new matrix $G^{(1)}$ with the same first row as H, and $|\det(G)| = |\det(G^{(1)})|$.

The first column of H is

$$\{1\}, \ \{2\}, \ \ldots, \ \left\{\frac{p-1}{2}\right\}$$

while the first column of $G^{(1)}$ is, up to a permutation and possibly some changes of signs, equal to

$$\{1\}, \ \{g\}, \ \ldots, \ \{g^{(p-3)/2}\}.$$

A permutation of the rows of $G^{(1)}$, distinct from the first, followed if necessary by changing signs of some of these rows, gives a new matrix $G^{(2)}$ with the same first column as H and $|\det(G^{(2)})| = |\det(G^{(1)})| = |\det(G)|$.

Actually the matrices $G^{(2)}$ and H coincide, as we now verify. Consider an arbitrary entry $\{rs'\}$ of H with $r > 1$, $s > 1$. Let $\{s'\} = \varepsilon_1\{g^j\}$, $\{r\} = \varepsilon_2\{g^{-i}\}$, with ε_1, ε_2 equal to 1 or -1, $0 < i, j \le (p-3)/2$. Let $r \equiv g^k \pmod{p}$, $s' \equiv g^h \pmod{p}$, where $0 < k, h \le p-2$. Then $rs' \equiv g^{k+h} \pmod{p}$. Since $g^{-i} = -g^{(p-1)/2-i}$ then $\{g^k\} = \varepsilon_2\{g^{-i}\} = \varepsilon_2\{g^{(p-1)/2-i}\}$ so

$$k = \begin{cases} p-1-i & \text{when } \varepsilon_2 = +1, \\ \dfrac{p-1}{2} - i & \text{when } \varepsilon_2 = -1. \end{cases}$$

Similarly $\{g^h\} = \varepsilon_1\{g^j\}$ so

$$h = \begin{cases} j & \text{when } \varepsilon_1 = +1, \\ \dfrac{p-1}{2} + j & \text{when } \varepsilon_1 = -1. \end{cases}$$

Therefore

$$k + h = \begin{cases} p - 1 + j - i & \text{when } \varepsilon_1 = 1,\ \varepsilon_2 = 1, \\ \dfrac{p-1}{2} + j - i & \text{when } \varepsilon_1 = 1,\ \varepsilon_2 = -1, \\ \dfrac{3(p-1)}{2} + j - i & \text{when } \varepsilon_1 = -1,\ \varepsilon_2 = 1, \\ p - 1 + j - i & \text{when } \varepsilon_1 = -1,\ \varepsilon_2 = -1, \end{cases}$$

hence

$$\{rs'\} = \{g^{k+h}\} = \begin{cases} \{g^{j-i}\} & \text{when } \varepsilon_1 = 1,\ \varepsilon_2 = 1, \\ -\{g^{j-i}\} & \text{when } \varepsilon_1 = 1,\ \varepsilon_2 = -1, \\ -\{g^{j-i}\} & \text{when } \varepsilon_1 = -1,\ \varepsilon_2 = 1, \\ \{g^{j-i}\} & \text{when } \varepsilon_1 = -1,\ \varepsilon_2 = -1. \end{cases}$$

So in all cases $\{rs'\} = \varepsilon_1\varepsilon_2\{g^{j-i}\}$ proving that the matrices $G^{(2)}$ and H coincide. Therefore $|\det(H)| = |\det(G)|$.

(2)   Multiplying the column j of G by η^j for $j = 0, 1, \ldots, (p-3)/2$, we obtain a new matrix $G^{(1)}$ such that

$$\det(G^{(1)}) = \eta^{1+2+\cdots+(p-3)/2}\det(G).$$

Multiplying the row i of $G^{(1)}$ by η^{-i} for $i = 0, 1, \ldots, (p-3)/2$, we obtain the matrix G' and

$$\det(G') = \eta^{-(1+2+\cdots+(p-3)/2)} \cdot \det(G^{(1)}) = \det(G).$$

(3)   The entry in position i, j of G' shall be denoted by $c_{i,j}$. Then if $0 \le i \le (p-5)/2$, $0 \le j \le (p-5)/2$, we have

$$c_{i+1,j+1} = \{g^{(j+1)-(i+1)}\}\eta^{(j+1)-(i+1)} = c_{i,j}.$$

Moreover, $c_{i+1,0} = c_{i,(p-3)/2}$. Indeed,

$$\frac{p-3}{2} - i = \frac{p-1}{2} - (i+1), \quad g^{(p-1)/2} \equiv -1 \pmod{p}, \text{ and } \eta^{(p-1)/2} = -1.$$

Then

$$\{g^{(p-3)/2-i}\}\eta^{(p-3)/2-i} = -\{g^{-(i+1)}\} \cdot (-\eta^{-(i+1)}) = \{g^{-(i+1)}\}\eta^{-(i+1)}$$

$$= c_{i+1,0}.$$

This proves that G' is a circulant. Let

$$f(X) = \{1\} + \{g\}\eta X + \{g^2\}\eta^2 X^2 + \cdots + \{g^{(p-3)/2}\}\eta^{(p-3)/2}X^{(p-3)/2}.$$

By Lemma 1:

$$\det(G') = \pm \prod_{j=0}^{(p-3)/2} f(\eta^{2j})$$

noting that η^2 is a primitive root of 1 of order $(p-1)/2$.

Since

$$f(\eta^{2j}) = \sum_{i=0}^{(p-3)/2} \{g^i\}\eta^{i(2j+1)}$$

we conclude that

$$|\det(G')| = \prod_{j=0}^{(p-3)/2} \left(\sum_{i=0}^{(p-3)/2} \{g^i\}\eta^{i(2j+1)} \right) = |\det(H)|. \blacksquare$$

Carlitz and Olson gave the following relation between Maillet's determinant and the relative class number:

A.

$$h_p^- = \frac{1}{p^{(p-3)/2}} |D_p|.$$

Proof: Let X be an indeterminate and consider the matrix $N_p = (X + R(rs'))_{r,s=1,\ldots,(p-1)/2}$. Thus $N_p(0) = M_p$, the Maillet matrix. Since $((p-1)/2)' = p-2$ the last column of M_p has entries $R(r(p-2)) = p - 2r$ while its first column has entries $R(r) = r$. Let $M_p^{(1)}$ be the matrix obtained from M_p by replacing the last column, by the sum of the last column with the double of the first column. Then the last column of $M_p^{(1)}$ has entries all equal to p. Proceeding similarly with N_p we obtain a matrix $N_p^{(1)}$ with the last column having entries equal to $3X + p$. Let $M_p^{(2)}$ be the matrix obtained from $M_p^{(1)}$ by replacing each row r, $r > 1$, by the difference between row r and the first row. Then

$$M_p^{(2)} = \begin{pmatrix} & \cdots & p \\ & & 0 \\ M_p^{(3)} & & \vdots \\ & & 0 \end{pmatrix}$$

so

$$D_p = \det(M_p) = \det(M_p^{(1)}) = (-1)^{(p-1)/2} p \det(M_p^{(3)}),$$

where $M_p^{(3)}$ has entries $R(rs') - R(s')$ for $2 \leq r \leq (p-1)/2$, $1 \leq s \leq (p-3)/2$. Similarly, we reach a matrix

$$N_p^{(2)} = \begin{pmatrix} & \cdots & 3X + p \\ & & 0 \\ N_p^{(3)} & & \vdots \\ & & 0 \end{pmatrix}$$

with

$$E_p(X) = \det(N_p) = \det(N_p^{(1)}) = (-1)^{(p-1)/2}(3X + p) \det(N_p^{(3)}),$$

where $N_p^{(3)}$ has entries

$$[X + R(rs')] - [X + R(s')] = R(rs') - R(s').$$

So $N_p^{(3)} = M_p^{(3)}$ and $E_p(X) = ((3X + p)/p)D_p$.

Taking $X = -p/2$ we have $E_p(-p/2) = -\frac{1}{2}D_p$.

But the entries of $N_p(-p/2)$ are $-p/2 + R(rs') = \{rs'\}$, that is, $N_p(-p/2) = H$. According to Lemma 3:

$$|D_p| = 2|\det(H)| = 2 \prod_{j=0}^{(p-3)/2} \left(\sum_{i=0}^{(p-3)/2} \{g^i\}\eta^{i(2j+1)} \right).$$

By Chapter 27, Theorem 2:

$$h_p^- = \frac{1}{(2p)^{(p-3)/2}} \left| \prod_{j=0}^{(p-3)/2} G(\eta^{2j+1}) \right|,$$

where

$$G(X) = \sum_{i=0}^{p-2} g_i X^i \quad \text{and} \quad 1 \le g_i \le p-1, \qquad g_i \equiv g^i \pmod{p}.$$

Since $g_i = R(g^i)$ and $\eta^{(p-1)/2} = -1$ then, for every $\alpha = \eta^{2j+1}$:

$$G(\alpha) = \sum_{i=0}^{p-2} R(g^i)\alpha^i = \sum_{i=0}^{(p-3)/2} R(g^i)\alpha^i - \sum_{i=0}^{(p-3)/2} R(-g^i)\alpha^i$$

$$= 2\sum_{i=0}^{(p-3)/2} R(g^i)\alpha^i - p\sum_{i=0}^{(p-3)/2} \alpha^i = 2\sum_{i=0}^{(p-3)/2} \{g^i\}\alpha^i.$$

Therefore

$$h_p^- = \frac{1}{(2p)^{(p-3)/2}} \left| \prod_{j=0}^{(p-3)/2} G(\eta^{2j+1}) \right|$$

$$= \frac{2}{p^{(p-3)/2}} \prod_{j=0}^{(p-3)/2} \left(\sum_{i=0}^{(p-3)/2} \{g^i\}\eta^{(2j+1)i} \right) = \frac{1}{p^{(p-3)/2}} |D_p|. \qquad \blacksquare$$

In particular, we deduce that $D_p \neq 0$ for every odd prime p, a fact which is not obvious a priori. We also note that $p^{(p-1)/2}$ divides D_p if and only if p divides h_p^-, that is, p is not a regular prime.

Inkeri expressed h_p^- in terms of another determinant, with integral coefficients proving thereby, in an independent way, that h_p^- is an integer.

Inkeri's determinant is defined as follows. If g is a primitive root modulo p, and if $g_i = R(g^i)$ is the least positive residue of g^i, then p divides

$gg_i - g_{i+1}$; let $q_i = (gg_i - g_{i+1})/p$, $\mu = (p-1)/2$, and consider the matrix

$$Q = \begin{pmatrix} q_\mu & q_{\mu-1} & \cdots & q_1 & q_0 \\ q_{\mu+1} & q_\mu & \cdots & q_2 & q_1 \\ \cdots & \cdots & \cdots & \cdots & \cdots \\ q_{2\mu-2} & q_{2\mu-3} & \cdots & q_{\mu-1} & q_{\mu-2} \\ g_\mu & g_{\mu-1} & \cdots & g_1 & g_0 \\ 1 & 1 & \cdots & 1 & 1 \end{pmatrix}.$$

Inkeri showed:

B. $h_p^- = \det(Q)$, *in particular, h_p^- is an integer.*

From (**A**) and (**B**) follows the relation between the Maillet and Inkeri determinants, namely

$$|D_p| = p^{(p-3)/2} \det(Q).$$

29.1.2 Upper and Lower Bounds for the Relative Class Number

By estimating Maillet's determinant, Carlitz obtained an upper bound for h_p^-. The following lemma is due to Hadamard:

Lemma 4.   *Let $M = (a_{rs})_{r,s=1,\ldots,n}$ be a matrix with real entries. Let*

$$a_r = \sqrt{\sum_{s=1}^n a_{rs}^2} > 0 \qquad \text{for} \quad r = 1, \ldots, n.$$

Then

$$|\det(M)| \le a_1 a_2 \cdots a_n.$$

Proof: In $\mathbb{R}^n$ consider the paralellotope P having vertices $O = (0, \ldots, 0)$ and $P_r = (a_{r1}, \ldots, a_{rs})$ for $r = 1, \ldots, n$. Then $\overline{OP_r}$ has length a_r. Since $|\det(M)|$ is the volume of P, it is elementary that $|\det(M)| \le a_1 a_2 \cdots a_n$. ∎

C.
$$h_p^- \le \begin{cases} (m-1)! & \text{when } p = 4\mu + 1, \\ (m-1)!\sqrt{m} & \text{when } p = 4\mu + 3 \ge 7. \end{cases}$$

Proof: By (**A**):

$$h_p^- = \frac{1}{p^{(p-3)/2}} |D_p|.$$

By Lemma 2:

$$|D_p| = p^{(p-3)/2} \left| \det\left(\left[\frac{rs}{p} \right] - \left[\frac{(r-1)s}{p} \right] \right)_{r,s=3,\ldots,(p-1)/2} \right|$$

and the last matrix has entries equal to 0 or 1. For every $r = 3, \ldots, (p-1)/2$ we have

$$a_r^2 = \sum_{s=3}^{(p-1)/2} \left(\left[\frac{rs}{p} \right] - \left[\frac{(r-1)s}{p} \right] \right)^2 = \sum_{s=3}^{(p-1)/2} \left(\left[\frac{rs}{p} \right] - \left[\frac{(r-1)s}{p} \right] \right)$$

$$= \left[\frac{((p-1)/2)s}{p} \right] - \left[\frac{2s}{p} \right] = \left[\frac{(p-1)s/2}{p} \right].$$

If $s = 2t$ then $(p-1)2t/2 = (p-1)t = (t-1)p + p - t$.

If $s = 2t - 1$ then $(p-1)(2t-1)/2 = (p-1)t - (p-1)/2 = (t-1)p + (p+1)/2 - t$.

Hence

$$\left[\frac{(p-1)s/2}{p} \right] = t - 1 = \left[\frac{s-1}{2} \right]$$

in both cases.

By Lemma 4:

$$\left| \det \left(\left[\frac{rs}{p} \right] - \left[\frac{(r-1)s}{p} \right] \right) \right|^2 \leq \prod_{s=3}^{(p-1)/2} \left[\frac{s-1}{2} \right]$$

$$= \begin{cases} 1 \times 1 \times 2 \times 2 \times \cdots \times (m-1) \times (m-1) & \text{if } p = 4m + 1, \\ 1 \times 1 \times 2 \times 2 \times \cdots \times (m-1) \times (m-1) \times m & \text{if } p = 4m + 3 \geq 7, \end{cases}$$

that is,

$$h_p^- \leq \begin{cases} (m-1)! & \text{when } p = 4\mu + 1, \\ (m-1)! \sqrt{m} & \text{when } p = 4\mu + 3 \geq 7. \end{cases} \qquad \blacksquare$$

We give now another explicit upper bound for h_p^-, due to Lepistö. The proof given here is by Metsänkylä and we note that for $p > 19$ this bound is better than the bound of Carlitz.

D.

$$h_p^- < 2p \left(\frac{p}{24} \right)^{(p-1)/4}.$$

Proof: By Chapter 27, (**D**) and (27.20):

$$h_p^- = \frac{1}{(2p)^{(p-3)/2}} \left| \prod_{\chi \in S} \left(\sum_{k=1}^{p-1} \chi(k)k \right) \right|,$$

where S is the set of odd characters modulo p; we note that $\#(S) = (p-1)/2$. The orthogonality relation of characters gives (for $1 \leq k$, $m \leq$

$p - 1$):

$$\sum_{\chi \in S} \chi(k)\overline{\chi}(m) = \begin{cases} \dfrac{p-1}{2} & \text{when } m \equiv k \pmod{p}, \ p \nmid mk, \\ -\dfrac{p-1}{2} & \text{when } m \equiv -k \pmod{p}, \ p \nmid mk, \\ 0 & \text{otherwise.} \end{cases}$$

We now compute the following sum:

$$\sum_{\chi \in S} \left| \sum_{k=1}^{p-1} \chi(k)k \right|^2 = \sum_{\chi \in S} \left(\sum_{k=1}^{p-1} \chi(k)k \right) \left(\sum_{m=1}^{p-1} \overline{\chi}(m)m \right)$$

$$= \sum_{k=1}^{p-1} k \sum_{m=1}^{p-1} m \left(\sum_{\chi \in S} \chi(k)\overline{\chi}(m) \right) = \frac{p-1}{2} \left[\sum_{k=1}^{p-1} k^2 - \sum_{k=1}^{p-1} k(p-k) \right]$$

$$= \frac{p-1}{2} \left[2 \sum_{k=1}^{p-1} k^2 - p \sum_{k=1}^{p-1} k \right].$$

By Chapter 18, after (**K**), $\sum_{k=1}^{p-1} k^2 = (p-1)p(2p-1)/6$, hence the above sum is equal to

$$\frac{p-1}{2} \left[\frac{2(p-1)p(2p-1)}{6} - \frac{p(p-1)p}{2} \right] = \frac{p-1}{2} \times \frac{(p-2)(p-1)p}{6}.$$

The arithmetic-geometric mean inequality gives

$$\prod_{\chi \in S} \left| \sum_{k=1}^{p-1} \chi(k)k \right|^{4/(p-1)} \leq \frac{(p-2)(p-1)p}{6} < \frac{p^3}{6}.$$

Hence

$$h_p^- \leq \frac{1}{(2p)^{(p-3)/2}} \times \left(\frac{p^3}{6} \right)^{(p-1)/4} = 2p \left(\frac{p}{24} \right)^{(p-1)/4}. \qquad \blacksquare$$

Using Stirling's formula

$$n! \sim \sqrt{2\pi} e^{-n} n^{n+1/2} \tag{29.3}$$

it follows that for all sufficiently large p, the upper bound of Carlitz is worse than the one by Lepistö.

Indeed,

$$2p \left(\frac{p}{24} \right)^{(p-1)/4} = 2p \left(\frac{p}{p-5} \right)^{(p-1)/4} \left(\frac{p-5}{4} \right)^{(p-1)/4} \left(\frac{1}{6} \right)^{(p-1)/4}.$$

On the other hand, from Carlitz's upper bound, using Stirling's formula, we estimate

$$\left(\frac{p-5}{4}\right)! \sim \sqrt{2\pi}\,e^{-(p-5)/4}\left(\frac{p-5}{4}\right)^{(p-5)/4+1/2}$$

and the quotient is

$$\frac{2p\left(\dfrac{p}{p-5}\right)^{(p-1)/4}\left(\dfrac{p-5}{4}\right)^{(p-1)/4}\left(\dfrac{1}{6}\right)^{(p-1)/4}e^{(p-5)/4}}{\sqrt{2\pi}\left(\dfrac{p-5}{4}\right)^{(p-3)/4+1/2}}$$

$$= \left(1+\frac{1}{(p-5)/5}\right)^{(p-5)/5+5/4}p^2(p-5)^{-1/2}\left(\frac{e}{6}\right)^{(p-1)/4}(2\pi)^{-1/2}e^{-1},$$

and for p large this quantity is approximately

$$p^2(p-5)^{-1/2}e^{1/4}(2\pi)^{-1/2}\left(\frac{e}{6}\right)^{(p-1)/4}$$

and this tends to 0 as p increases, establishing the assertion.

Another method to obtain upper and lower bounds for h_p^- consists in estimating $L(1|\chi)$ and their products (for odd characters) and to use the formula (29.1):

$$h_p^- = \gamma(p)\prod_{\chi\in S^-}L(1|\chi),$$

where

$$\gamma(p) = \frac{p^{(p+3)/4}}{2^{(p-3)/2}\pi^{(p-1)/2}}$$

and S^- is the set of odd characters modulo p.

We note the following estimates due to Tatuzawa:

E. *There exists a constant $a > 0$ and for every $\varepsilon > 0$ there exists a constant $c(\varepsilon) > 0$ such that*

$$\frac{c(\varepsilon)}{p^\varepsilon} < \left|\prod_{\chi\in S^-}L(1|\chi)\right| < e^{a(\log\log p+1)}.$$

The constants in the above estimates were later computed explicitly by Lepistö.

Montgomery obtained the following simple (but not the best) effective bounds:

F. *If $p > 100$ then*

$$\frac{1}{40ep} < \prod_{\chi\neq\chi_0}L(1|\chi)$$

and

$$\prod_{\substack{\chi \text{ even} \\ \chi \neq \chi_0}} L(1|\chi) \leq \left(\frac{\pi^2}{6}\right)^{(p-3)/4} \exp\left(\tfrac{3}{10}\left(2 - (\log p)^2\right)\right).$$

Note that the above products of values of the L-series are nonnegative real numbers.

The proof of these bounds is beyond the scope of this book.

29.1.3 Cyclotomic Fields with Class Number 1

Montgomery, as well as Uchida, determined the cyclotomic fields K_p with class number 1:

Theorem 1. $h_p = 1$ *if and only if $p \leq 19$.*

Proof: By using the formula for h_p it is possible to show that if $p \leq 19$ then $h_p = 1$. (Actually, it should be noted that even for small values of p, the evaluation of h_p^+ is difficult, but we shall not discuss this here any further.)

We show the converse. We have

$$h_p^- = \frac{2p^{(p+3)/4}}{(2\pi)^{(p-1)/2}} \times \frac{\prod_{\chi \neq \chi_0} L(1|\chi)}{\prod_{\substack{\chi(1)=1 \\ \chi \neq \chi_0}} L(1|\chi)}.$$

Taking logarithms and applying (**F**) we obtain

$$\log h_p^- \geq \frac{p+3}{4} \log p - \frac{p-3}{2} \log 2 - \frac{p-1}{2} \log \pi - \log 40ep$$

$$- \frac{p-3}{4} \log\left(\frac{\pi^2}{6}\right) + \tfrac{3}{10}(\log p)^2 - \tfrac{3}{5}$$

$$= \frac{p}{4} \log p - \frac{p}{4} \log \frac{2\pi^4}{3} + \tfrac{3}{10}(\log p)^2$$

$$- \tfrac{1}{4} \log p + 2 \log \pi - \tfrac{9}{4} \log 2 - \tfrac{3}{4} \log 3 - \log 5 - \tfrac{8}{5}$$

$$> \frac{p}{4} \log \frac{p}{65} + \tfrac{3}{10}(\log p)^2 - \tfrac{1}{4} \log p - \tfrac{10}{3}.$$

for $p > 100$. The above function of p is increasing for $p > 100$ and its value for $p = 101$ is greater than $13 > \log(4 \times 10^5)$.

Thus $h_p^- > 4 \times 10^5$ for $p \geq 101$.

On the other hand, the values of h_p^- have already been calculated by Kummer for $p < 100$. It follows that for $p > 19$ then $h_p^- > 1$, hence $h_p > 1$, concluding the proof.   ∎

There is a similar result of Masley about cyclotomic fields $K_m = \mathbb{Q}(\zeta_m)$ where ζ_m is a primitive mth root of unity and $m > 2$. This will be considered in Section 29.3.

29.1.4 Growth of the Relative Class Number

Brauer proved that

$$\lim_{p \to \infty} h_p^- = \infty.$$

A natural question which arises is to ask whether the function h_p^- of p is ultimately monotonic.

Ankeny and Chowla proved:

G.   *There exists p_0 such that if $p_0 \leq p_1 < p_2$ (with p_1, p_2 prime numbers) then $h_{p_1}^- < h_{p_2}^-$.*

It should be noted that p_0 was not given explicitly.

Using the method of Montgomery, Metsänkylä showed:

H.   *If $p > 100$ then*

$$\frac{1}{30} \times \left(\frac{\pi^2}{6}\right)^{-(p-1)/4} < \frac{h_p^-}{\gamma(p)} < \left(\frac{\pi^2}{6}\right)^{(p-1)/4}.$$

This was used to obtain:

I.   *Let $0 < \varepsilon \leq 1$, let p, q be primes such that $p > (1 + \varepsilon)q$, and let $q \geq 67 \times 3^{1/\varepsilon}$. Then*

$$h_p^- > h_q^-.$$

Proof: Let $\delta > \varepsilon$ be such that $p = (1 + \delta)q$. Since $67 \times 3^{1/\varepsilon} > 200$, then $q > \max(200, 67 \times 3^{1/\delta})$.

We note that if a, b, c are any positive real numbers, the inequality $x > ab^{c/x}$ is satisfied for every sufficiently large real number x. We apply this remark and so

$$x > 4\pi^2 \left(\frac{\pi^2}{6}\right)^{1+2/\delta} \left(\frac{30\sqrt{6}}{\pi}\right)^{4/\delta x}$$

provided x is sufficiently large. A computation with logarithms tells us that it suffices that $x > \max(200, 67 \times 3^{1/\delta})$.

In particular,

$$q > 4\pi^2 \left(\frac{30\sqrt{6}}{\pi}\right)^{4/\delta q} \left(\frac{\pi^2}{6}\right)^{1+2/\delta}$$

so

$$q^{\delta q/4} > 30(4\pi^2)^{\delta q/4} \left(\frac{\pi^2}{6}\right)^{(\delta q + 2q - 2)/4}.$$

Since $p > q$ and $p - q = \delta q$ then

$$\frac{\gamma(p)}{\gamma(q)} = \frac{p^{(p+3)/4}}{2^{(p-3)/2}\pi^{(p-1)/2}} \times \frac{2^{(q-3)/2}\pi^{(q-1)/2}}{q^{(q+3)/4}} > \frac{q^{(p-q)/4}}{2^{(p-q)/2}\pi^{(p-q)/2}}$$

$$= \left(\frac{q}{4\pi^2}\right)^{(p-q)/4} > 30\left(\frac{\pi^2}{6}\right)^{(p+q-2)/4}.$$

By (**H**):

$$h_p^- > \frac{\gamma(p)}{30(\pi^2/6)^{(p-1)/4}} > \frac{\gamma(q)}{(\pi^2/6)^{(q-1)/4}} > h_q^-. \qquad \blacksquare$$

In particular, if $\varepsilon = 1$ and $p > 2q > 400$ then $h_p^- > h_q^-$.

Lepistö also concluded that if $p \geq q + 4$, $q \geq 2.4 \times 10^{66}$, or if $p \geq q + 64$, $q \geq 3$, then $h_p^- > h_q^-$. This suggested Metsänkylä and Lepistö to conjecture

If $p > q \geq 19$ then $h_p^- > h_q^-$.

Concerning the asymptotic behavior of h_p^-, as p increases Kummer conjectured

$$h_p^- \sim \gamma(p). \qquad (29.4)$$

However, Granville indicated in 1990 that Kummer's conjecture is not consistent with other conjectures in analytic number theory, for which there is more evidence. Granville also conjectured that

$$(\log\log p)^{-1/2+o(1)} < \frac{h_p^-}{\gamma(p)} < (\log\log p)^{1/2+o(1)}.$$

Ankeny and Chowla showed:

J.
$$\lim_{p\to\infty} \frac{\log(h_p^-/\gamma(p))}{\log p} = 0. \qquad (29.5)$$

Since $\lim_{p\to\infty} \log\gamma(p)/\log p = \infty$, it follows that $\lim_{p\to\infty} h_p^- = \infty$, as it had been proved earlier by Brauer.

The next result of Siegel uses the following estimate, which appears in Estermann's book:

K.   *If $\chi \neq \chi_0$ then*

$$|L(1\,|\,\chi)| \leq 2 + \log p.$$

For a lower bound, we quote:

L.   *There exists a constant c, $0 < c < 1$, such that for every nontrivial character χ modulo p:*

$$\frac{c}{(\log p)^5} < |L(1\,|\,\chi)|$$

and for all such χ, with possibly one exception

$$\frac{c}{\log p} < |L(1|\chi)|.$$

If χ' is the exceptional character, then χ' is a quadratic character.

Now we prove Siegel's theorem, which is a weaker form of Kummer's conjecture:

M.

$$\log h_p^- \sim \frac{p}{4} \log p.$$

Proof: From the formula for h_p^- we have

$$\log h_p^- = \frac{p+3}{4} \log p - \frac{p-3}{2} \log 2 - \frac{p-1}{2} \log \pi + \log \left(\prod_{\chi \in S^-} L(1|\chi) \right).$$

It is enough to show that

$$\lim_{p \to \infty} \frac{\log(\prod_{\chi \in S^-} L(1|\chi))}{p \log p} = 0.$$

This implies that

$$\lim_{p \to \infty} \frac{\log h_p^-}{(p/4) \log p} = 1.$$

It follows from Chapter 21, (**T**), that if $p \equiv 1 \pmod 4$ there is no odd quadratic character modulo p, while if $p \equiv 3 \pmod 4$ there is precisely one odd quadratic character χ modulo p. This is the quadratic character

$$\chi(k) = \left(\frac{k}{p} \right) \qquad \text{for every} \quad k = 1, 2, \ldots.$$

The formula of Dirichlet for the class number $H(-p)$ of the quadratic extension $\mathbb{Q}(\sqrt{-p})$ (Chapter 26, Theorem 1) gives

$$H(-p) = \frac{w\sqrt{p}}{2\pi} L(1|\chi),$$

where χ is the above quadratic character and w is the number of roots of unity in $\mathbb{Q}(\sqrt{-p})$. Thus $w \leq 6$ hence

$$L(1|\chi) = \frac{2\pi H(-p)}{w\sqrt{p}} > \frac{1}{\sqrt{p}}.$$

Putting together this information, we have

$$\frac{1}{\sqrt{p}} \left(\frac{c}{\log p} \right)^{2[(p-1)/4]} < \prod_{\chi \in S^-} L(1|\chi) < (2 \log p)^{(p-1)/2}$$

since there are exactly $(p-1)/2$ odd characters χ modulo p.

Taking logarithms, dividing by $p \log p$, and letting p tend to infinity, we conclude the proof of the proposition.   ∎

29.1.5 Some Divisibility Properties of the Relative Class Number

Now we shall indicate some divisibility properties of the relative class number. We begin with a lemma due to Kummer, which implies an expression of h_p^- as a product of norms.

Let p be an odd prime, and let η be a primitive root of 1 of order $p - 1$. Let m be any divisor of $p - 1$ and let $S_m = \{s \mid 1 \leq s \leq p - 1, \ \gcd(s, p-1) = m\}$. Then η^m is a primitive root of 1 of order $(p-1)/m$; let $L_m = \mathbb{Q}(\eta^m)$. If m divides s then $\eta^s \in L_m$. Moreover, η^s is conjugate to η^m in $L_m|\mathbb{Q}$ exactly when η^s is a primitive root of 1 of order $(p-1)/m$, that is, $\gcd(s/m, (p-1)/m) = 1$, or still, $\gcd(s, p-1) = m$, so $s \in S_m$. Let $S = \{s \text{ odd} \mid 1 \leq s \leq p - 2\}$.

Lemma 5.   Let $H \in \mathbb{Q}[X]$. Then

$$\prod_{m \in S} H(\eta^m) = \prod_{\substack{m \mid p-1 \\ m \text{ odd}}} N_{L_m|\mathbb{Q}}(H(\eta^m)).$$

Proof: Since S is the disjoint union of the subsets S_m, for m odd dividing $p - 1$, and since

$$\prod_{s \in S_m} H(\eta^s) = N_{L_m|\mathbb{Q}}(H(\eta^m))$$

then

$$\prod_{m \in S} H(\eta^m) = \prod_{\substack{m \mid p-1 \\ m \text{ odd}}} \prod_{s \in S_m} H(\eta^s) = \prod_{\substack{m \mid p-1 \\ m \text{ odd}}} N_{L_m|\mathbb{Q}}(H(\eta^m)). \quad ∎$$

The following result of Metsänkylä (1971) follows from the previous result of Iwasawa (1966) about ideal class groups:

N.   Let p be a prime and let $p - 1 = 2^c n$ where n is odd, $c \geq 1$. If h_p^- is even then $n \neq 1$ and h_p^- is a multiple of 2^d where d is the minimum of the orders of 2 modulo ℓ, for every prime factor ℓ of n. In particular, $d > 1$.

Proof: Let g be a primitive root modulo p, and let g_k $(0 \leq k \leq p - 2)$ be the least positive residue of g^k modulo p. Let η be a primitive root of 1 of order $p - 1$. By Chapter 27, Theorem 2:

$$h_p^- = \frac{1}{(2p)^{(p-3)/2}} |G(\eta)G(\eta^3) \cdots G(\eta^{p-2})|,$$

where

$$G(X) = \sum_{k=0}^{p-2} g_k X^k.$$

Let us note that

$$X^{p-1} - 1 = \left[\prod_{\substack{i=0 \\ i \text{ even}}}^{p-3} (X - \eta^i)\right] \left[\prod_{\substack{i=0 \\ i \text{ odd}}}^{p-2} (X - \eta^i)\right] = (X^{(p-1)/2} - 1) \prod_{\substack{i=1 \\ i \text{ odd}}}^{p-2} (X - \eta^i)$$

hence

$$\prod_{\substack{i=1 \\ i \text{ odd}}}^{p-2} (X - \eta^i) = \frac{X^{p-1} - 1}{X^{(p-1)/2} - 1} = X^{(p-1)/2} + 1$$

and therefore

$$\prod_{\substack{i=1 \\ i \text{ odd}}}^{p-2} (1 - \eta^i) = 2.$$

But

$$\prod_{\substack{i=1 \\ i \text{ odd}}}^{p-2} (\eta^{-i} - 1) = \left(\prod_{\substack{i=1 \\ i \text{ odd}}}^{p-2} \eta^i\right)^{-1} \prod_{\substack{i=1 \\ i \text{ odd}}}^{p-2} (1 - \eta^i)$$

and

$$\sum_{\substack{i=1 \\ i \text{ odd}}}^{p-2} i = \left(\frac{p-1}{2}\right)^2$$

so

$$\prod_{\substack{i=1 \\ i \text{ odd}}}^{p-2} \eta^i = \eta^{((p-1)/2)^2} = (-1)^{(p-1)/2} = \pm 1$$

and therefore

$$\prod_{\substack{i=1 \\ i \text{ odd}}}^{p-2} (\eta^{-i} - 1) = \pm 2.$$

Now, we compute (for $j = 1, 2, \ldots, (p-1)/2$):

$$(\eta^{1-2j} - 1) \left(\sum_{k=0}^{p-2} g_k \eta^{(2j-1)k}\right) = \sum_{k=0}^{p-2} g_k \eta^{(2j-1)(k-1)} - \sum_{k=0}^{p-2} g_k \eta^{(2j-1)k}$$

$$= \sum_{k=0}^{p-2} (g_{k+1} - g_k) \eta^{(2j-1)k} = 2 \sum_{k=0}^{(p-3)/2} (g_{k+1} - g_k) \eta^{(2j-1)k}$$

because if $k = (p-1)/2 + h$ then

$$(g_{k+1} - g_k)\eta^{(2j-1)k} = -(g_{h+1} - g_h)(-\eta^{(2j-1)h}) = (g_{h+1} - g_h)\eta^{(2j-1)h}.$$

Therefore

$$h_p^- = \frac{1}{(2p)^{(p-3)/2}} \left| \prod_{\substack{i=1 \\ i \text{ odd}}}^{p-2} (\eta^{-i} - 1) \right|^{-1} 2^{(p-1)/2} \left| \prod_{\substack{i=1 \\ i \text{ odd}}}^{p-2} H(\eta^i) \right|$$

$$= \frac{1}{p^{(p-3)/2}} \left| \prod_{\substack{i=1 \\ i \text{ odd}}}^{p-2} H(\eta^i) \right|,$$

where

$$H(X) = \sum_{k=0}^{(p-3)/2} (g_{k+1} - g_k) X^k \in \mathbb{Z}[X].$$

By Lemma 5 we may write

$$h_p^- = \frac{1}{p^{(p-3)/2}} \left| \prod_{\substack{m \mid p-1 \\ m \text{ odd}}} N_{L_m|\mathbb{Q}}(H(\eta^m)) \right| = \frac{1}{p^{(p-3)/2}} \left| \prod_{m \mid n} N_{L_m|\mathbb{Q}}(H(\eta^m)) \right|.$$

If h_p^- is even then there exists m dividing n such that $N_{L_m|\mathbb{Q}}(H(\eta^m)) \equiv 0$ (mod 2).

We must have $m \neq n$. Indeed, if $m = n$ then $L_n = \mathbb{Q}(\eta^n)$, where $\xi = \eta^n$ is a primitive root of 1 of order 2^c. Then 2 decomposes in the ring of integers of L_n as a power of the prime ideal $(1 - \xi)$ (see Chapter 11, (N)). Thus

$$H(\eta^n) = \sum_{k=0}^{(p-3)/2} (g_{k+1} - g_k)\eta^{kn} \equiv \sum_{k=0}^{(p-3)/2} (g_{k+1} - g_k)$$

$$= g_{(p-1)/2} - g_0 = (p-1) - 1 \equiv 1 \pmod{(1 - \xi)}$$

because $p - 1 = \prod_{j=1}^{p-2}(1 - \eta^j)$ and $-1 \equiv 1 \pmod{(1 - \xi)}$.

Hence $N_{L_n|\mathbb{Q}}H(\eta^n) \equiv 1 \pmod 2$, which is a contradiction.

Thus $m < n$ and so $n \neq 1$.

In the field $L_m = \mathbb{Q}(\eta^m)$ the ideal generated by 2 decomposes into the product of prime ideals $P_1, \ldots, P_r$, each with norm 2^f, where f is the order of 2 modulo n/m (see Chapter 11, Section 3). In particular, $d \leq f$. Since 2 divides the product

$$\prod_{\substack{i=1 \\ i \text{ odd}}}^{p-2} H(\eta^i)$$

then P_1 divides at least one factor, so $2^f = N_{L_m|\mathbb{Q}}(P_1)$ divides one norm $N_{L_m|\mathbb{Q}}(H(\eta^i))$ which in turn divides h_p^-. Thus 2^d divides h_p^-. ∎

Metsänkylä has also shown how a simple argument applied to the explicit expression of h_p^- leads to the following result:

O.   *Let p, q be odd primes. If $p = 2q + 1$ then q does not divide h_p^-.*

Vandiver gave the following interesting congruence for h_p^-:

P.   *For any integer $k \geq 1$:*

$$h_p^- \equiv (-1)^{(p-1)/2} 2^{(p-3)/2} p \prod_{\substack{s=1 \\ s \ \text{odd}}}^{p-2} B_{p^k s + 1} \pmod{p^k}.$$

Slavutskiĭ gave in 1969 a simple proof with p-adic methods.

29.2 Miscellanea About the Real Class Number of Cyclotomic Fields

Results about the factor h_p^+ of the class number of $K = \mathbb{Q}(\zeta_p)$ are essentially more difficult to obtain since this number is so closely related to the units. It is notorious that the determination of a fundamental system of units for K is a very delicate problem.

Until not very long ago, all computations of the h_p^+ have yielded the value 1. However, in 1965, Ankeny, Chowla, and Hasse found a very nice criterion for h_p^+ to be larger than 1. The proof involves class-field theory, but is otherwise elementary. We begin with a lemma of Davenport:

Lemma 6.   *Let ℓ, m be positive integers, m not a square. If the equation*

$$U^2 - (\ell^2 + 1)V^2 = \pm m$$

has solutions in integers u, v, then $m \geq 2\ell$.

Proof: If u, v are solutions then $v \neq 0$, since m is not a square. We may take a solution u, v with $u \geq 0$ and smallest positive v.

If $K = \mathbb{Q}(\sqrt{\ell^2 + 1})$ then $N_{K|\mathbb{Q}}(u - v\sqrt{\ell^2 + 1}) = \pm m$. But it is also true that $N_{K|\mathbb{Q}}(\ell + \sqrt{\ell^2 + 1}) = -1$. Multiplying these norms, we deduce that

$$N_{K|\mathbb{Q}}[(\ell u - (\ell^2 + 1)v) + (u - \ell v)\sqrt{\ell^2 + 1}] = \pm m,$$

that is,

$$[\ell u - (\ell^2 + 1)v]^2 - (\ell^2 + 1)(u - \ell v)^2 = \pm m.$$

Thus, we have obtained another solution. Since $v > 0$ was minimal, then either $u - \ell v \geq v$ so $u \geq (\ell + 1)v$, or $-u + \ell v \geq v$ so $(\ell - 1)v \geq u \geq 0$. Hence $\pm m = u^2 - (\ell^2 + 1)v^2 \geq 2\ell v^2 \geq 2\ell$ and $\pm m > 0$ or, respectively, $\pm m = u^2 - (\ell^2 + 1)v^2 \leq -2\ell v^2 \leq -2\ell$ and $\pm m < 0$. In both cases $m \geq 2\ell$. ∎

For every prime p let $H(p)$ denote the class-number of the quadratic field $\mathbb{Q}(\sqrt{p})$.

Lemma 7.   *Let q be a prime, $n > 1$, and assume that $p = (2qn)^2 + 1$ is also a prime. Then $H(p) > 1$.*

Proof: Let B denote the ring of integers of $K = \mathbb{Q}(\sqrt{p})$. Since $p \equiv 1$ (mod 4) the elements of B are of the form $(u + v\sqrt{p})/2$ where $u, v \in \mathbb{Z}$ and $u \equiv v$ (mod 2). Since $p \equiv 1$ (mod q) then $(p/q) = (1/q) = 1$. Hence $Bq = QQ'$ where Q, Q' are distinct prime ideals. Hence $NQ = q$.

If $H(p) = 1$ then B is a principal ideal domain, so Q is the principal ideal generated by an element $(u + v\sqrt{p})/2 \in B$. Hence

$$q = NQ = \left| N_{K|\mathbb{Q}}\left(\frac{u + v\sqrt{p}}{2}\right)\right| = \left|\frac{u^2 - pv^2}{4}\right|,$$

that is $(u^2 - pv^2)/4 = \pm q$. But $p = (2qn)^2 + 1$ so $u^2 - ((2qn)^2 + 1)v^2 = \pm 4q$. By the preceding lemma, $4q \geq 2(2qn) = 4qn$, with $n > 1$, which is impossible. This proves that $H(p) > 1$.   ∎

In the next result we shall use the Hilbert class field associated to any number field K (see Chapter 15, Section 2, (**A**), for the properties).

Lemma 8.   *Let K be an algebraic number field, and let $L|K$ be a finite extension. If no Abelian unramified extension of K is contained in L then $h(K)$ divides $h(L)$.*

Proof: Let $\overline{K}$, $\overline{L}$ denote, respectively, the Hilbert class fields of K, L. Since $(L \cap \overline{K})|K$ is an Abelian unramified extension then $L \cap \overline{K} = K$. Therefore $L\overline{K}|L$ is an Abelian unramified extension, so $L\overline{K} \subseteq \overline{L}$. We conclude that $h(K) = [\overline{K} : K] = [L\overline{K} : L]$ divides $[\overline{L} : L] = h(L)$.   ∎

Q.   *If $p \equiv 1$ (mod 4) then $H(p)$ divides h_p^+.*

Proof: The value of the principal Gauss sum is

$$\sum_{m \not\equiv 0 \;(\mathrm{mod}\; p)} \left(\frac{m}{p}\right)\zeta_p^m = \pm\sqrt{p}$$

(Chapter 21, (**W**)).

Moreover, since $p \equiv 1$ (mod 4) then $(-1/p) = 1$ so $(m/p) = (-m/p)$ for every m. Thus $\mathbb{Q}(\sqrt{p}) \subseteq \mathbb{Q}(\zeta_p + \zeta_p^{-1}) = K^+$.

The prime p is totally ramified in $K = \mathbb{Q}(\zeta_p)$ hence also in K^+, thus no Abelian unramified extension of $\mathbb{Q}(\sqrt{p})$ is contained in K^+. By Lemma 8, $H(p) = h(\mathbb{Q}(\sqrt{p}))$ divides the class number $h(K^+) = h_p^+$.

R.   *If q is a prime, if $n > 1$, and if $p = (2qn)^2 + 1$ is also prime, then $h_p^+ > 2$.*

Proof: By Lemma 7, $H(p) > 1$. Since $p \equiv 1 \pmod 4$ then $H(p)$ divides h_p^+ by **(Q)**. Thus $h_p^+ > 1$.

On the other hand, it follows from the theory of genera (see Borevich and Shafarevich [3, p. 354]) that if $p \equiv 1 \pmod 4$ then $H(p)$ is odd. Therefore $h_p^+ > 2$.   ∎

Taking, for example, $q = 2,\ 3,\ 5$ we see easily that for the following primes less than 10000 :

$$p = 257, 401, 577, 1297, 1601, 2917, 3137, 4357, 7057, 8101$$

we have $h_p^+ > 2$.

The work of Ankeny, Chowla, and Hasse was extended, but we shall not discuss this here (see Lang (1977), Takeuchi (1981)).

In connection with his study of Fermat's Last Theorem, Vandiver conjectured: *If $p > 2$ then p does not divide h_p^+.*

This statement has never been proved.

Since h_p^+ seems to be small compared with p, one might feel tempted to show that $h_p^+ < p$ thereby proving Vandiver's conjecture. However in 1985, assuming the generalized Riemann hypothesis, Cornell and Washington showed that for $p = 11\,290\,018\,777$, in fact, $h_p^+ > p$.

29.3 The Class Number of $\mathbb{Q}(\zeta_m)$, $m > 2$, and Miscellaneous Results

29.3.1 The Class Number Formula

Let $K_m = \mathbb{Q}(\zeta_m)$, where $m > 2$, $m \not\equiv 2 \pmod 4$, and ζ_m is a primitive root of 1 of order m. Let $K_m^+ = K_m \cap \mathbb{R}$, so $K_m | K_m^+$ has degree 2. Let $h_m = h(K_m)$ be the class number of K_m and let $h_m^+ = h(K_m^+)$ be the class number of K_m^+.

Just as for the case when $m = p$ is an odd prime, there are formulas for h_m, h_m^+ in terms of L-series of the characters modulo m and other invariants of the field.

Let

$$e(m) = \begin{cases} 1 & \text{if } m \text{ is a power of 2,} \\ 4 & \text{if } m \text{ is odd and divisible by at least} \\ & \quad 2 \text{ distinct primes,} \\ 2 & \text{otherwise.} \end{cases}$$

For each character χ modulo m, let $f(\chi)$ denote its conductor. Let $\zeta_{f(\chi)}$ be a primitive root of unity of order $f(\chi)$, and let R^+ denote the regulator of K_m^+. Then:

Theorem 2.   *We have* $h_m = h_m^- h_m^+$, *where*

$$h_m^- = m\, e(m) \prod_{\substack{\chi \text{ odd} \\ f(\chi)|m}} \frac{1}{2f(\chi)} \sum_{k=1}^{f(\chi)} (-\chi(k)k) \qquad (29.6)$$

and

$$h_m^+ = \frac{1}{R^+} \prod_{\substack{\chi \text{ even} \\ \chi \neq \chi_0 \\ f(\chi)|m}} \left(\frac{1}{2} \sum_{k=1}^{f(\chi)} -\chi(k) \log |1 - \zeta_{f(\chi)}^k| \right). \qquad (29.7)$$

The next statement follows from a classical result of Kronecker:

S.   *For every* $m > 2$, h_m^- *is a natural number.*

T. h_m^+ *is equal to the class number of* K_m^+.

In 1952 Hasse proved:

U. h_m^- *is the index of an appropriate subgroup of the ideal class group of* K_m^+.

29.3.2 Divisibility Properties

There are remarkable divisibility properties for the class numbers (and their factors) of K_{p^k}. The following result is due to Weber for $p = 2$, and to Westlund for $p > 2$:

V.   *If p is a prime and $k \geq 1$ then $h^-(K_{p^k})$ divides $h^-(K_{p^{k+1}})$, $h^+(K_{p^k})$ divides $h^+(K_{p^{k+1}})$, and $h(K_{p^k})$ divides $h(K_{p^{k+1}})$.*

Weber's famous theorem states:

Theorem 3.   (1) *The quotient $h^-(K_{2^{k+1}})/h^-(K_{2^k})$ is odd.*

    (2) *The quotient $h^+(K_{2^{k+1}})/h^+(K_{2^k})$ is odd.*

    (3) *The class number of K_{2^k} is odd for $k \geq 2$.*

A simpler proof of a more general statement may be found in Hasse [1952, pp. 26 and 101]. A surprisingly elementary proof of the first statement has been published by Lepistö [1966, 1970].

An interesting open problem to prove, or disprove, is the conjecture of H. Cohn:

$$h^+(K_{2^k}) = 1 \qquad \text{for all } k > 1.$$

The third statement of Theorem 3 is equivalent to the following one (see Weber [1899, p. 821] and Hasse [1952, p. 29]):

W.   *If a unit of the real cyclotomic field $K_{2^k}^+$ is totally positive (that is, all its conjugates are positive) then the unit is a square in the field.*

It was using this theorem that Weber provided a rigorous proof of the so-called Kronecker–Weber theorem: Weber [1899, p. 762] (see Chapter 15, Section 1).

In 1911 Furtwängler extended Weber's theorems as follows:

X.   *Let p be an odd prime, $k \geq 1$. The class number of K_{p^k} is a multiple of p if and only if the class number of K_p is a multiple of p.*

Y.   *If p does not divide the class number $h(K_{p^k})$ then every unit of K_{p^k} is the relative norm of a unit of $K_{p^{k+1}}$.*

Cornell and Rosen obtained many interesting results about the divisibility properties of h_m^+. For example:

If m is divisible by at least five distinct primes, then 2 divides h_m^+.

If p is an odd prime and m is divisible by four or more primes q, $q \equiv 1 \pmod{p}$, then p^2 divides h_m^+.

For every integer $a \geq 1$ there exist infinitely many m such that $h_m^+ > m^a$.

But there are, of course, many other known results of this kind in the literature, to which we will not allude here.

29.3.3 Fields with Class Number 1

Masley extended the result of Montgomery and Uchida and proved:

Theorem 4.   *There are exactly 29 distinct fields $K_m = \mathbb{Q}(\zeta_m) \neq \mathbb{Q}$ having class number 1, namely when*

$$m = 3 \text{ (or 6)}, 4, 5 \text{ (or 10)}, 7 \text{ (or 14)}, 8, 9 \text{ (or 18)}, 11 \text{ (or 22)}, 12, 13 \text{ (or 26)},$$
$$15 \text{ (or 30)}, 16, 17 \text{ (or 34)}, 19 \text{ (or 38)}, 20, 21 \text{ (or 42)}, 24, 25 \text{ (or 50)}, 27$$
$$\text{(or 54)}, 28, 32, 33 \text{ (or 66)}, 35 \text{ (or 70)}, 36, 40, 44, 45 \text{ (or 90)}, 48, 60, 84.$$

A Guide for Further Study

After reading the present book, which is an introduction to the theory of algebraic numbers, there are several possible interrelated topics for further study; the following books are recommended.

(1) For cyclotomic fields, local methods

Iwasawa, K., *Lectures on p-Adic L-Functions*, Princeton University Press, Princeton, RI, 1972.

Washington, L.C., *Introduction to Cyclotomic Fields*, Springer-Verlag, New York, 1997 (second edition).

Lang, S., *Cyclotomic Fields*, Springer-Verlag, New York, 1990 (second edition).

(2) For class field theory

Cassels, J.W.S., Fröhlich, A. (editors), *Algebraic Number Theory*, Academic Press, London, and Thompson, Washington, 1967.

Iyanaga, S. (editor), *The Theory of Numbers*, North-Holland, Amsterdam, and American Elsevier, New York, 1975.

Neukirch, J., *Algebraische Zahlentheorie*, Springer-Verlag, Berlin, 1991.

Neukirch, J., *Class Field Theory*, Springer-Verlag, Berlin, 1986.

(3) For analytic number theory

Apostol, T.M., *Introduction to Analytic Number Theory*, Springer-Verlag, New York, 1976.

Goldstein, L.J., *Analytic Number Theory*, Prentice Hall, Englewood Cliffs, NJ, 1971.

(4) For a wide variety of topics in number theory

Hardy, G.H., Wright, E.M., *An Introduction to the Theory of Numbers*, Clarendon Press, Oxford, 1979 (fifth edition).

Hua, L.K., *Introduction to Number Theory*, Springer-Verlag, New York, 1982.

(5) For complements on algebraic numbers

Hasse, H., *Number Theory*, Springer-Verlag, New York, 1980.

Weil, A., *Basic Number Theory*, Springer-Verlag, New York, 1967.

Narkiewicz, W., *Elementary and Analytic Theory of Algebraic Numbers*, Polish Scientific Publishers (PWN), Warsaw, and Springer-Verlag, Berlin, 1990.

(6) For the theory of binary quadratic forms

Gauss, C.F., *Disquisitiones Arithmeticae*, originally published in 1801. Reprinted in numerous editions.

Flath, D.F., *Introduction to Number Theory*, Wiley, New York, 1989.

Ribenboim, P., Gauss and the class number problems, in *My Numbers, My Friends*, Springer-Verlag, New York, 2000.

Bibliography

General Bibliography

[1] Apostol, T.M., *Introduction to Analytic Number Theory*, Springer-Verlag, New York, 1976.

[2] Ayoub, R., *An Introduction to the Analytic Theory of Numbers*, Amer. Math. Soc., Providence, RI, 1963.

[3] Borevich, Z.I., Shafarevich, I.R., *Number Theory*, Academic Press, New York, 1966.

[4] Bourbaki, N., *Algèbre*, Ch.V (Corps Commutatifs), Hermann, Paris, 1950.

[5] Gelfond, A.O., Linnik, Yu.V., *Elementary Methods in Analytic Number Theory* (English translation by A. Feinstein, revised and edited by L.J. Mordell), Rand McNally, Chicago, 1965.

[6] Hardy, G.H., Wright, E.M., *An Introduction to the Theory of Numbers*, Clarendon Press, Oxford, 1938 (fifth edition, 1979).

[7] Hasse, H., *Number Theory*, Akademie Verlag, Berlin, 1979.

[8] Hasse, H., *Number Theory*, Springer-Verlag, Berlin, 1980.

[9] Hasse, H., *Über die Klassenzahl Abelscher Zahlkörper*, Akademie Verlag, Berlin, 1952.

[10] Hasse, H., *Vorlesungen über Zahlentheorie*, Springer-Verlag, Berlin, 1964 (second edition).

[11] Hecke, E., *Vorlesungen über die Theorie der Algebraischen Zahlen*, Chelsea, New York, 1948 (reprint).

[12] Hilbert, D., *Gesammelte Abhandlungen* (*Zahlentheorie*), Chelsea, New York, 1968.

[13] Ireland, K., Rosen, M., *A Classical Introduction to Modern Number Theory*, Springer-Verlag, New York, 1990 (second edition).

[14] Iyanaga, S. (editor), *The Theory of Numbers*, North-Holland, Amsterdam, 1974, and American Elsevier, New York, 1974.

[15] Janusz, G.J., *Algebraic Number Fields*, Academic Press, New York, 1973.

[16] Lang, S., *Algebra*, Addison-Wesley, Reading, MA, 1965.

[17] Lang, S., *Algebraic Number Theory*, Springer-Verlag, New York, 1994 (second edition).

[18] LeVeque, W.J., *Topics in Number Theory*, Addison-Wesley, Reading, MA, 1956.

[19] McCarthy, P.J., *Algebraic Extensions of Fields*, Blaisdell, Waltham, MA, 1966.

[20] Nagell, T., *Introduction to Number Theory*, Wiley, New York, 1951.

[21] Narkiewicz, W., *Elementary and Analytic Theory of Algebraic Numbers*, Polish Scientific Publishers (PWN), Warsaw, and Springer-Verlag, Berlin, 1990 (second edition).

[22] Neukirch, J., *Klassenkörpertheorie*, Bibliographisches Institut, Mannheim, 1969.

[23] Ribenboim, P., *13 Lectures on Fermat's Last Theorem*, Springer-Verlag, New York, 1979.

[24] Ribenboim, P., The work of Kummer on Fermat's last theorem, in *Number Theory Related to Fermat's Last Theorem* (editor: N. Koblitz), pp. 1–20. Birkhäuser, Boston, 1982.

[25] Ribenboim, P., *The New Book of Prime Number Records*, Springer-Verlag, New York, 1996.

[26] Ribenboim, P., *The Theory of Classical Valuations*, Springer-Verlag, New York, 1999.

[27] Samuel, P., *Algebraic Theory of Numbers*, Houghton-Mifflin, Boston, MA, 1970.

[28] Trost, E., *Primzahlen*, Birkhäuser, Basel, 1953, 1968 (second edition).

[29] Weil, A., *Basic Number Theory*, Springer-Verlag, Berlin, 1967.

Bibliography to Chapter 28

1. Ankeny, N.C., Chowla, S., On the divisibility of the class number of quadratic fields, *Pacific J. Math.* **5** (1955), 321–324.

2. Baker, A., Imaginary quadratic fields with class number two, *Ann. of Math.* (2), **94** (1971), 139–157.

3. Baker, A., *Transcendental Number Theory*, Cambridge University Press, Cambridge, 1975.

4. Deuring, M., Imaginär-quadratische Zahlkörper mit der Klassenzahl 1, *Invent. Math.* **5** (1968), 169–179.

5. Flath, D.E., *Introduction to Number Theory*, Wiley, New York, 1989.

6. Gauss, C.F., *Disquisitiones Arithmeticae*, originally published in 1801. Reprinted in many editions.

7. Goldfeld, D., Gauss' class number problem for imaginary quadratic fields, *Bull. Amer. Math. Soc.* **134** (1985), 23–37.

8. Gross, B., Zagier, D., Heegner points and derivations of *L*-series, *Invent. Math.* **84** (1986), 225–320.

9. Gut, M., Die Zetafunktion, die Klassenzahl und die Kroneckersche Grenzformel eines beliebigen Kreiskörpers, *Comment. Math. Helv.* **1** (1929), 160–226.

10. Gut, M., Kubische Klassenkörper über quadratischen imaginären Grundkörpern, *Nieuw Arch. Wisk.* (2), **23** (1951), 185–189.

11. Gut, M., Erweiterungskörper von Primzahlgrad mit durch diese Primzahl teilbarer Klassenzahl, *Enseign. Math.* **19** (1973), 119–123.

12. Hartung, P., Explicit construction of a class of infinitely many imaginary quadratic fields whose class number is divisible by 3, *J. Number Theory* **6** (1974), 279–281.

13. Heegner, K., Diophantische Analysis und Modulfunktionen, *Math. Z.* **56** (1952), 227–253.

14. Honda, T., On real quadratic fields whose class numbers are multiples of 3, *J. Reine Angew. Math.* **233** (1968), 101–102.

15. Humbert, P., Sur les nombres de classes de certains corps quadratiques, *Comment. Math. Helv.* **12** (1940), 233–245.

16. Humbert, P., Note relative à l'article "Sur les nombres de classes de certains corps quadratiques", *Comment. Math. Helv.* **13** (1940), 67.

17. Kuroda, S., On the class-number of imaginary quadratic number fields, *Proc. Japan Acad. Sci.* **40** (1964), 365–367.

18. Nagell, T., Über die Klassenzahl imaginären quadratischer Zahlkörper, *Abh. Math. Sem. Univ. Hamburg* **1** (1922), 140–150.

19. Ribenboim, P., Gauss and the class number problem, *Symp. Gaussiana* **1** (1991), 13–63

20. Stark, H.M., A complete characterization of the complex quadratic fields of class-number one, *Michigan Math. J.* **14** (1967), 1–27.

21. Stark, H.M., On the "gap" in a theorem of Heegner, *J. Number Theory* **1** (1969), 16–27.

22. Yamamoto, Y., On unramified Galois extensions of quadratic number fields, *Osaka J. Math.* **7** (1970), 57–76.

Bibliography to Chapter 29

We have omitted several proofs, since they are long and technical. They may be found, together with more information, in the papers listed below. This is however not meant to be a complete bibliography on a topic which has been the object of such intensive research.

1. Ankeny, N.C., Chowla, S., The class number of the cyclotomic field, *Proc. Nat. Acad. Sci. U.S.A.* **35** (1949), 529–532.

2. Ankeny, N.C., Chowla, S., The class number of the cyclotomic field, *Canad. J. Math.* **3** (1951), 486–494.

3. Ankeny, N.C., Chowla, S., Hasse, H., On the class number of the maximal real subfield of a cyclotomic field, *J. Reine Angew. Math.* **217** (1965), 217–220.

4. Brauer, R., On the zeta functions of algebraic number fields, II, *Amer. J. Math.* **72** (1950), 739–746.

5. Carlitz, L., A generalization of Maillet's determinant and a bound for the first factor of the class number, *Proc. Amer. Math. Soc.* **12** (1961), 256–261.

6. Carlitz, L., Olson, F.R., Maillet's determinant, *Proc. Amer. Math. Soc.* **6** (1955), 265–269.

7. Cornell, G., Exponential growth of the ℓ-rank of the class group of the maximal real subfield of cyclotomic fields, *Bull. Amer. Math. Soc.* **8** (1983), 55–58.

8. Cornell, G., Rosen, M.L., The ℓ-rank of the real class group of cyclotomic fields, *Compositio Math.* **53** (1984), 133–141.

9. Cornell, G., Washington, L.C., Class numbers of cyclotomic fields, J. Number Theory **21** (1985), 260–273.

10. Estermann, T., On Dirichlet's L-functions, *J. London Math. Soc.* **23** (1948), 275–279.

11. Granville, A., On the size of the first factor of the class number of a cyclotomic field, *Invent. Math.* **100** (1990), 331–338.

12. Hasse, H., *Über die Klassenzahl abelscher Zahlkörper*, Akademie Verlag, Berlin, 1952.

13. Inkeri, K., Über die Klassenzahl des Kreiskörpers der ℓten Einheitswurzeln, *Ann. Acad. Sci. Fenn. Ser. A*, No. 199, 1955, 3–12.

14. Lang, S.D., Note on the class number of the maximal real subfield of a cyclotomic field, *J. Reine Angew. Math.* **290** (1977), 70–72.

15. Lepistö, T., On the product of the regulator and the class number of the cyclotomic field, *Ann. Univ. Turku. Ser. A*, No. 118, 1968, 5 pages.

16. Lepistö, T., On the class number of the cyclotomic field $k(\exp(2\pi i/p^h))$, *Ann. Univ. Turku. Ser. A*, No. 125, 1969, 11 pages.

17. Lepistö, T., On a cyclic determinant and the first factor of the class number of the cyclotomic field, *Ann. Univ. Turku. Ser. A*, No. 135, 1970, 3 pages.

18. Lepistö, T., On the growth of the first factor of the class number of the prime cyclotomic field, *Ann. Acad. Sci. Fenn. Ser. A I*, No. 577, 1974, 18 pages.

19. Masley, J.M., On the class number of cyclotomic fields, Thesis, Princeton University, 1972, 51 pages.

20. Masley, J.M., Montgomery, H.L., Cyclotomic fields with unique factorization, *J. Reine Angew. Math.* **216/7** (1976), 248–256.

21. Metsänkylä, T., Bemerkungen über den ersten Faktor der Klassenzahl des Kreiskörpers, *Ann. Univ. Turku. Ser. A I*, No. 105, 1967, 15 pages.

22. Metsänkylä, T., Über die Teilbarkeit des Relativklassenzahl des Kreiskörpers durch zwei, *Ann. Univ. Turku. Ser. A I*, No. 118, 1968, 8 pages.

23. Metsänkylä, T., Über die Teilbarkeit den ersten Faktors der Klassenzahl des Kreiskörpers, *Ann. Univ. Turku. Ser. A I*, No. 124, 1968, 6 pages.

24. Metsänkylä, T., On prime factors of the relative class numbers of cyclotomic fields, *Ann. Univ. Turku. Ser. A I*, No. 149, 1971, 8 pages.

25. Metsänkylä, T., On the growth of the first factor of the cyclotomic class number, *Ann. Univ. Turku. Ser. A I*, No. 155, 1972, 12 pages.

26. Metsänkylä, T., Class numbers and μ-invariants of cyclotomic fields, *Proc. Amer. Math. Soc.* **43** (1974), 299–300.

27. Newman, M., A table of the first factor for prime cyclotomic fields, *Math. Comp.* **24** (1970), 215–219.

28. Siegel, C.L., Zu zwei Bemerkungen Kummers, *Nachr. Akad. Wiss. Göttingen, Math. Phys. Kl.* **II** (1964), 51–57 (= *Gesammelte Abhandlungen,* Vol. III, Springer-Verlag, New York, 1966, pp. 436–442).

29. Slavutskiĭ, I.Sh., The simplest proof of Vandiver's theorem, *Acta Arith.* **15** (1969), 117–118.

30. Takeuchi, H., On the class number of the maximal real subfield of a cyclotomic field, *Canad. J. Math.* **33** (1981), 55–58.

31. Tatuzawa, T., On a theorem of Siegel, *Japan J. Math.* **21** (1951), 163–178.

32. Tatuzawa, T., On the product of $L(1|\chi)$, *Nagoya Math. J.* **5** (1953), 105–111.

33. Uchida, K., Class numbers of imaginary Abelian number fields, III *Tôhoku Math. J.* **23** (1971), 573–580.

34. Vandiver, H.S., On the first factor of the class number of a cyclotomic field, *Bull. Amer. Math. Soc.* **25** (1919), 158–161.

35. Westlund, J., On the class number of the cyclotomic field $\mathbb{Q}(e^{2\pi i/p^n})$, *Trans. Amer. Math. Soc.* **4** (1903), 201–212.

Index of Names

Subject Index

Universitext   *(continued)*